Data Handling in Science and Technology
Volume 29

# Fundamentals and Analytical Applications of Multiway Calibration

Data Handling in Science and Technology

Volume 29

# Fundamentals and Analytical Applications of Multiway Calibration


**Arsenio Muñoz de la Peña**

Department of Analytical Chemistry, University of Extremadura, Badajoz, Spain

**Héctor C. Goicoechea**

Laboratorio de Desarrollo Analítico y Quimiometría (LADAQ), Cátedra de Química Analítica I, Facultad de Bioquímica y Ciencias Biológicas, Universidad Nacional del Litoral, Ciudad Universitaria, Santa Fe, Argentina

**Graciela M. Escandar**

Instituto de Química Rosario (CONICET-UNR), Facultad de Ciencias Bioquímicas y Farmacéuticas, Universidad Nacional de Rosario, Rosario, Argentina

**Alejandro C. Olivieri**

Instituto de Química Rosario (CONICET-UNR), Facultad de Ciencias Bioquímicas y Farmacéuticas, Universidad Nacional de Rosario, Rosario, Argentina



ELSEVIER

AMSTERDAM • BOSTON • HEIDELBERG • LONDON • NEW YORK • OXFORD
PARIS • SAN DIEGO • SAN FRANCISCO • SINGAPORE • SYDNEY • TOKYO

Elsevier
Radarweg 29, PO Box 211, 1000 AE Amsterdam, Netherlands
The Boulevard, Langford Lane, Kidlington, Oxford OX5 1GB, UK
225 Wyman Street, Waltham, MA 02451, USA

First edition 2015



**Notices**
Knowledge and best practice in this field are constantly changing. As new research and experience broaden our understanding, changes in research methods, professional practices, or medical treatment may become necessary.

Practitioners and researchers must always rely on their own experience and knowledge in evaluating and using any information, methods, compounds, or experiments described herein. In using such information or methods they should be mindful of their own safety and the safety of others, including parties for whom they have a professional responsibility.

To the fullest extent of the law, neither the Publisher nor the authors, contributors, or editors, assume any liability for any injury and/or damage to persons or property as a matter of products liability, negligence or otherwise, or from any use or operation of any methods, products, instructions, or ideas contained in the material herein.

ISBN: 978-0-444-63527-3
ISSN: 0922-3487

# Contents

# Contributors

*Numbers in Parentheses indicate the pages on which the author's contributions begin.*

**Franco Allegrini** (541), Instituto de Química Rosario (CONICET-UNR), Facultad de Ciencias Bioquímicas y Farmacéuticas, Universidad Nacional de Rosario, Rosario, Argentina

**Fabio Augusto** (465), Institute of Chemistry, University of Campinas – UNICAMP, Campinas, São Paulo, Brazil

**Joan Ferré Baldrich** (7), Department of Analytical Chemistry and Organic Chemistry, Universitat Rovira i Virgili, Tarragona, Spain

**Santiago Bortolato** (541), Instituto de Química Rosario (CONICET-UNR), Facultad de Ciencias Bioquímicas y Farmacéuticas, Universidad Nacional de Rosario, Rosario, Argentina

**Andres D. Campiglia** (507), Department of Chemistry, University of Central Florida, Orlando, Florida, USA

**Luiz A.F. de Godoy** (465), Cristália Produtos Químicos Farmacêuticos Ltda, Itapira, São Paulo, Brazil

**Anna de Juan** (247), Department of Analytical Chemistry, Universitat de Barcelona, Barcelona, Spain

**Arsenio Muñoz de la Peña** (347, 365), Department of Analytical Chemistry, University of Extremadura, Badajoz, Spain

**Graciela M. Escandar** (347, 365), Instituto de Química Rosario (CONICET-UNR), Facultad de Ciencias Bioquímicas y Farmacéuticas, Universidad Nacional de Rosario, Rosario, Argentina

**Nicolaas (Klaas) M. Faber** (1), Chemometry Consultancy, Beek-Ubbergen, The Netherlands

**Héctor C. Goicoechea** (347, 365, 507), Laboratorio de Desarrollo Analítico y Quimiometría (LADAQ), Cátedra de Química Analítica I, Facultad de Bioquímica y Ciencias Biológicas, Universidad Nacional del Litoral, Ciudad Universitaria, Santa Fe, Argentina

**A. Herrero** (37), Department of Chemistry (Analytical Chemistry), University of Burgos, Burgos, Spain

**Chao Kang** (83, 167), State Key Laboratory of Chemo/Biosensing and Chemometrics, College of Chemistry and Chemical Engineering, Hunan University, Changsha, Hunan, China

**Yong Li** (83, 167), State Key Laboratory of Chemo/Biosensing and Chemometrics, College of Chemistry and Chemical Engineering, Hunan University, Changsha, Hunan, China

**Ricard Boqué Martí** (7), Department of Analytical Chemistry and Organic Chemistry, Universitat Rovira i Virgili, Tarragona, Spain

**Anthony F.T. Moore** (507), Department of Chemistry, University of Central Florida, Orlando, Florida, USA

**Alejandro C. Olivieri** (347, 365, 541), Instituto de Química Rosario (CONICET-UNR), Facultad de Ciencias Bioquímicas y Farmacéuticas, Universidad Nacional de Rosario, Rosario, Argentina

**M.C. Ortiz** (37), Department of Chemistry (Analytical Chemistry), University of Burgos, Burgos, Spain

**Hadi Parastar** (293), Department of Chemistry, Sharif University of Technology, Tehran, Iran

**Brendon A. Parsons** (427), Department of Chemistry, University of Washington, Seattle, Washington, USA

**Marcio P. Pedroso** (465), Department of Chemistry, Federal University of Lavras – UFLA, Lavras, Minas Gerais, Brazil

**Karisa M. Pierce** (427), Department of Chemistry and Biochemistry, Seattle Pacific University, Seattle, Washington, USA

**M. Pilar Callao** (399), Department of Analytical and Organic Chemistry, Rovira i Virgili University, Tarragona, Spain

**Ronei J. Poppi** (465), Institute of Chemistry, University of Campinas – UNICAMP, Campinas, São Paulo, Brazil

**C. Reguera** (37), Department of Chemistry (Analytical Chemistry), University of Burgos, Burgos, Spain

**S. Sanllorente** (37), Department of Chemistry (Analytical Chemistry), University of Burgos, Burgos, Spain

**L.A. Sarabia** (37), Department of Mathematics and Computation (Statistics and Operations Research), University of Burgos, Burgos, Spain

**M.S. Sánchez** (37), Department of Mathematics and Computation (Statistics and Operations Research), University of Burgos, Burgos, Spain

**M. Soledad Larrechi** (399), Department of Analytical and Organic Chemistry, Rovira i Virgili University, Tarragona, Spain

**Robert E. Synovec** (427), Department of Chemistry, University of Washington, Seattle, Washington, USA

**Romà Tauler** (247), Department of Environmental Chemistry, IDAEA-CSIC, Barcelona, Spain

**Walter B. Wilson** (507), National Institute of Standard and Technology, Gaithersburg, Maryland, USA

**Hai-Long Wu** (83, 167), State Key Laboratory of Chemo/Biosensing and Chemometrics, College of Chemistry and Chemical Engineering, Hunan University, Changsha, Hunan, China

**Ru-Qin Yu** (83, 167), State Key Laboratory of Chemo/Biosensing and Chemometrics, College of Chemistry and Chemical Engineering, Hunan University, Changsha, Hunan, China

# Preface

As a popular story goes, a representative of a shoe-making factory visits a town of the distant country Barefootland, with the purpose of opening up new markets, only to discover that people in the town do not wear any shoes. He then moves to other towns, finding the same hopeless reality. Greatly disappointed, he finally sent this message to the central factory: "don't send any shoes here, these people don't need them." After some time, another shoe-seller from a different country visits Barefootland, finds the same people wearing no shoes, but sends this exciting message to the headquarters: "please urgently send as many shoes as you can, these people need them!" This is just one version of the story, which includes sellers from many different countries. The moral is that great opportunities can be missed if one does not have the right attitude.

Precisely, the purpose of editing this book is to follow the example of the second shoe-seller and try to contribute to making multiway calibration popular. We strongly believe that if multiway calibration is not widespread enough among analytical chemists, it is not because they do not need it. We also think it is the duty of chemometricians and analytical chemists using multiway calibration to disseminate the multiway philosophy and its specific techniques to the analytical community. Opening up the analytical chemistry market would make a difference both ways. Analytical chemists will fully incorporate a set of powerful calibration tools into their protocols, and at the same time chemometricians will get the bonus of myriads of interesting real-world challenges.

Why is multiway calibration not as popular in analytical chemistry as we believe it should be? The late Henry A. Bent, distinguished chemistry professor and educator at the University of Pittsburgh, once compared learning chemistry with a foreign language. Learning a new language involves incorporating new words (and sometimes new symbols) to understand everyday concepts. Learning chemistry, on the other hand, implies using new words and symbols, but to represent *strange* concepts. In his own words, *chemistry is a foreign language twice over: strange terms for strange things* [1]. This applies to analytical chemists and multiway calibration: many unfamiliar concepts have to be grasped.

However, the rewards might be countless. By conducting multiway calibration, one may accomplish the dream of any analytical chemist: to be able to quantitate an analyte in a complex interfering sample only by calibrating with the pure analyte. This property has been named the "second-order advantage" because it was first observed in the domain of second-order (matrix) data, but it is shared by multiway data of *at least* second order [2]. It is amazing how profound changes take place in the properties of experimental data when their structure goes from scalars, vectors, and matrices to three- and multiway arrays. The sea change occurring when entering the multiway world has already been enjoyed by mathematicians and pure chemometricians. It is time for us chemists, and particularly analytical chemists, to join the banquet.

Multiway data analysis not only benefits analytical chemistry. Hidden factors within multiway data can be uncovered in a better way than in lower-dimensional data, thanks to the power of multiway algorithms, aiding data exploration, classification, and discrimination when it comes to the study of complex samples. Just some of the many other chemical activities which may equally improve from multiway data analysis are digital image processing, which is becoming increasingly popular in various fields, environmental data mining, metabolomics and related subdisciplines, where large amounts of data are being produced by modern instruments, and industrial chemical processes which depend on many different variables of economic and environmental importance.

We should bear in mind that the very first paper describing the second-order advantage was published in 1978 [3], nearly 40 years ago! Still, multiway calibration is striving to leave the scientific ghetto in which it has lived for so many years. Interestingly enough, the historic record of publications on multiway calibration shows that after emerging in the late 1970s, it practically disappeared from the literature and finally revived during the mid-1990s, with a nearly exponential growth of publications. This temporal evolution nicely fits the so-called Gartner cycle, which in computer science characterizes an initial overenthusiasm or "hype," followed by subsequent phases of disappointment and "enlightenment" that typically accompany the development of new technologies [4]. Was it so, or was it because few analytical chemists could foresee the power of the methods at the time they were first exposed?

The advantages of multiway calibration should be enough to convince analytical chemists that it is immensely more powerful than either univariate or first-order calibration. However, crowds are not knocking at the doors of multiway calibration and will not do so until the message is adequately spread. We believe that by editing multiauthor books such as the present one, we are doing our bit. In Spanish, the mother tongue of the editors, we would say that "we are adding our small grain of sand."

**The Editors**

## REFERENCES

[1] Bent HA. Uses (and abuses) of models in teaching chemistry. J Chem Educ 1984;61:774–7.

[2] Booksh KS, Kowalski BR. Theory of analytical chemistry. Anal Chem 1994;66:782A–791A.

[3] Ho CN, Christian GD, Davidson ER. Application of the method of rank annihilation to quantitative analyses of multicomponent fluorescence data from the video fluorometer. Anal Chem 1978;50:1108–13.

[4] Fenn J, Raskino M. Understanding Gartner's hype cycles. Harvard: Harvard Business Press; 2008.

# Views on Multiway Calibration: Its Past and Future

Nicolaas (Klaas) M. Faber
*Chemometry Consultancy, Beek-Ubbergen, The Netherlands*

The statistics-based methodology known as multiway calibration must appear like a fabled animal to the average practical analytical chemist. Fortunately, at its core this methodology amounts to estimating a calibration factor or sensitivity, essentially comparable to the slope of a straight-line fit, as detailed in the chapter on figures of merit. Let us therefore have a look at the natural progression of data complexity (Figure 1).

Univariate calibration, which is best exemplified by the familiar straight-line fit, commonly estimated using least-squares techniques, occupies a central place in the statistical part of an analytical chemist's curriculum. Univariate calibration works with scalars (plain single numbers) per sample and no index is required to mathematically characterize the data (it would always have the same value and therefore be noninformative). The analytical chemist should be comfortable here because the calibration methods and their results are as transparent as intuitive. I repeat: the calibration factor is the slope of the fitted line. How simple can it be to map signal space to concentration space?

Multivariate calibration, which digests vectors of data for each sample, already has a relatively high black-box character to the practical analytical chemist. What is the signal space? It is well known that if the data contain more than two elements, visualization is no longer straightforward, although

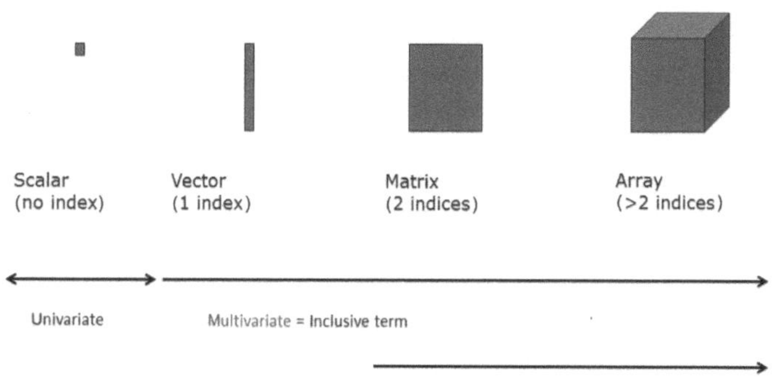


**FIGURE 1** Schematic view of the data measured for a single chemical sample.

Data Handling in Science and Technology, Vol. 29. http://dx.doi.org/10.1016/B978-0-444-63527-3.00014-X

technical escapes such as principal component analysis can be very useful. It is therefore not a surprise that the black-box character never fully disappeared since the introduction of calibration methods such as principal component regression and partial least squares (PLS) during the 1970s! Do these methods constitute black boxes to the seasoned chemometrician? Not that I know of and therefore this discrepancy constitutes a serious issue to pay attention to. There lie opportunities!

Multiway data require more than a single index. In the special case of two indices, the block of data is a mathematical matrix. Beyond that complexity, one refers to arrays. Here, I maintain the traditional distinction between multivariate and multiway calibration. Multiway data are also multivariate, but keeping the distinction is informative. It for example turns out that moving to multiway data leads to a real explosion of calibration methods. How is a practical analytical chemist going to find order in this obvious chaos? On his/her own? I conjecture: by learning from history. The problematic practical implementation of multivariate calibration methods led to fascinating opportunities for cross-fertilization between analytical chemists and chemometricians, and, from the increasing complexity of the data, I must infer that the opportunities for cross-fertilization must be even better for multiway calibration.

The analytical chemist is trained to know the mechanisms (e.g., chromatography, spectroscopy) leading to these highly complex data structures very well, whereas chemometricians are trained to even successfully work with rather "soft" data, about which little can be assumed. This appears to me as an important observation: chemometricians feel comfortable because they understand the methods to do the number crunching, while their understanding of the data might be at the same time "suboptimal" at best. Until now, however, for multiway calibration, even chemometricians need to construct their understanding about multiway calibration methods, when applied to analytical chemistry from what is scattered among "the literature." This situation leads to needless artificial differences in opinion among chemometricians. I can therefore think of two reasons why this volume is timely and relatable. First, the field has obviously matured so that analytical chemists *must* begin to recognize the full potential of the methodology applicable to their data. Second, this volume may contribute to a harmonization regarding some needless differing insights among chemometricians!

So far the admittedly long introduction to the real introduction! How did it all start in analytical chemistry? It is well known that the initial idea behind the most powerful model for multiway data, known as PARAFAC, was proposed in 1944 by the extremely influential psychologist Raymond Cattell [1]. The first numerical evaluation of multiway data was reported by the same researcher and his wife Alberta, a mathematician by training, in 1955 [2], shortly after the introduction of the first programmable computer, the ILLIAC I. It remained silent in analytical chemistry for almost 25 years. When asked by me about the origins (during a postdoc period 1994–1996), the late Bruce

Kowalski, a cofounder of the field of chemometrics but an analytical chemist by training, told the following story. Theoretical chemist Ernest Davidson had found, in collaboration with analytical chemists Chu-Ngi Ho and Gary Christian, a procedure for analyzing certain multiway data that enabled the determination of the analyte of interest in the presence of unsuspected interferents. Ernest Davidson plainly asked Bruce Kowalski: "Is this useful?" Bruce Kowalski immediately replied. "Yes, this is VERY useful!" That early work led to a series of three papers [3–5], and an improvement by Avraham Lorber [6]. Another landmark paper was published by Gene Sanchez and Bruce Kowalski in 1986 [7].[1]

What happened next? Of course, this development provoked a response by Svante Wold, organic chemist by training, leading chemometrician at the time and no doubt the father of the field of chemometrics, and therefore always in healthy competition with Bruce Kowalski's department. Paul Geladi explained to me during a stay at Wold's department in 1994 that they questioned the *validation* of the method developed in Seattle. The response came quickly [8,9] in the form of an addition to "their" favorite PLS for dealing with unsuspected interferents, now known as residual bilinearization. Interestingly, these two approaches are still among the main working horses to this very date!

The theoretical foundations were further disclosed by, among others, psychometrician Henk Kiers, Nicolas Sidiropoulos, who is highly active in Electrical and Computer Engineering, in collaboration with chemometrician Rasmus Bro and econometrician Age Smilde. Psychometrician Pieter Kroonenberg should not be left unmentioned here. And I very much appreciated the innovative work of statisticians Marie Linder and Rolf Sundberg on bilinear least squares. The algorithmic advances developed by physicist Pentti Paatero, who often collaborated with environmental scientist and chemometrician Phil Hopke, were also quite impressive and highly cited. (Of course, any omission here is a nonintended fault on my side.) And gradually, the number of applications started to boom, but notably so within chemometrics. One might say that it was the cross-fertilization between researchers from many seemingly unrelated fields that largely contributed to this topic becoming mainstream research throughout.

There is still one important article that I would like to pay attention to, written by mathematician Sue Leurgans and biochemist Robert Ross [10], which led one of the discussants, well-known psychometrician Jan de Leeuw, to exclaim in despair:

> *In a sense, the applicability of these models justifies the attention they have gotten so far. But from the mathematical and the statistical point of view, the available results are still very primitive. We have an elaborate notation*

1. Interestingly, close examination shows that Lorber, Sanchez, and Kowalski essentially reinvented the procedure developed by Cattell and Cattell in their little-known work.

*(actually, several elaborate notations), in many cases featuring specially defined operators, that make multiway arrays look like ordinary linear operators. As the efforts of Rice, Hitchcock and Oldenburger have shown, it is difficult to distill beautiful results out of this orgy of subscripts and ad hoc notation. Either there is not enough structure there, or we have not found the key yet.*

In my opinion, the pertinent methodology, as it is firmly founded in today's theory, has gradually moved quite far away from de Leeuw's sobering conclusion in 1992. A disclaimer is appropriate. Since I feel highly privileged and lucky to have worked with leading researchers on this topic, I should refer to this introduction as an opinion piece. Anyway, the reader always decides what to do with it!

To conclude, apart from promoting cross-fertilization between practical analytical chemists and chemometricians (or scientists with comparable skills), I take this opportunity to urge instrument developers to more closely look into this topic. That would constitute an immense leap ahead. As a first step, they should contemplate hiring chemometricians (or scientists with comparable skills), who show a keen awareness of the many barriers toward safely implementing this methodology in the instrument software; I mean full-time, not as mere consultants. As a major advantage, for example, better detection capabilities are in close reach, without additional experimentation, probably even less. Any chemometrician should feel motivated by the following visionary claim of Bruce Kowalski: "Chemometrics is only to have real impact when it is being built in commercial software." I can therefore be very short about the future: much is still left to be done as highly commendable efforts.

## REFERENCES

[1] Cattell RB. 'Parallel proportional profiles' and other principles for determining the choice of factors by rotation. Psychometrika 1944;9:267–83.

[2] Cattell RB, Cattell AKS. Factor rotation for proportional profiles: analytical solution and an example. Br J Math Stat Psychol 1955;8:83–92.

[3] Ho CN, Christian GD, Davidson ER. Application of the method of rank annihilation to quantitative analyses of multicomponent fluorescence data from the video fluorometer. Anal Chem 1978;50:1108–13.

[4] Ho CN, Christian GD, Davidson ER. Application of the method of rank annihilation to fluorescent multicomponent mixtures of polynuclear aromatic hydrocarbons. Anal Chem 1980;52:1071–9.

[5] Ho CN, Christian GD, Davidson ER. Simultaneous multicomponent rank annihilation and applications to multicomponent fluorescent data acquired by the video fluorometer. Anal Chem 1981;53:92–8.

[6] Lorber A. Quantifying chemical composition from two-dimensional data arrays. Anal Chim Acta 1984;164:293–7.

[7] Sanchez E, Kowalski BR. Generalized rank annihilation factor analysis. Anal Chem 1986;58:496–9.

[8] Öhman J, Geladi P, Wold S. Residual bilinearization. Part 1: theory and algorithms. J Chemometr 1990;4:79–90.

[9] Öhman J, Geladi P, Wold S. Residual bilinearization. Part 2: application to HPLC—diode array data and comparison with rank annihilation factor analysis. J Chemometr 1990;4:135–46.

[10] Leurgans S, Ross RT. Multilinear models: applications in spectroscopy. Statist Sci 1992;7:289–310.

Chapter 1

# Fundamentals of PARAFAC

**Ricard Boqué Martí[1] and Joan Ferré Baldrich**
*Department of Analytical Chemistry and Organic Chemistry, Universitat Rovira i Virgili, Tarragona, Spain*
*[1]Corresponding author: e-mail: ricard.boque@urv.cat*

**Chapter Outline**



## 1 INTRODUCTION

Parallel Factor Analysis (PARAFAC) is a decomposition method for multi-way data that goes back to 1970 when R. Harshman, following the work by Cattell [1], developed an alternating least-squares (ALS) method to uniquely decompose three-way datasets into triads or trilinear components [2]. Simultaneously, Carroll and Chang [3] developed a similar model, which they called Canonical Decomposition (CANDECOMP), and that was used for multidimensional scaling.

Since Appellof and Davidson [4] described the first application of PARAFAC to excitation–emission matrix (EEM) fluorescence spectra, the number of applications of PARAFAC in the analytical laboratory has increased steadily. Relevant developments of the theory and applications of PARAFAC in analytical chemistry can be found in monographs on multiway analysis [5,6], reviews [7,8], and general works about chemometrics [9–12]. Although in principle PARAFAC can be applied to any $N$-way array of data, we will focus in this chapter on three-way arrays made of responses of analytical

**Data Handling in Science and Technology, Vol. 29. http://dx.doi.org/10.1016/B978-0-444-63527-3.00001-1**

instruments, because they are, by now, the most common in the analytical laboratory. It must be noted, however, that PARAFAC is a general tool that is well described in multiway analysis books [13] and that can be applied to diverse types of data, such as sensory [14], environmental [15], and hyperspectral imaging [16] data.

## 2 NOTATION

Scalars are shown in nonbold italic characters (e.g., $y_i$, $A$), column vectors in bold lowercase characters (e.g., $\mathbf{x}$), and matrices in bold uppercase characters (e.g., $\mathbf{X}$). Row vectors are transposed column vectors. Italic boldface uppercase letters denote $N$th-order arrays ($N>2$), for example, $\boldsymbol{X}$. $\mathbf{I}$ denotes an identity matrix with the appropriate dimensions. Superscripts "T" and + denote the transposition of a matrix or vector (e.g., $\mathbf{X}^\mathrm{T}$ and $\mathbf{x}^\mathrm{T}$) and the Moore–Penrose pseudo-inverse of a matrix, respectively. $||\cdot||$ denotes the Frobenius norm, $\otimes$ and $|\otimes|$ denote the Kronecker product and the Khatri–Rao product (see Appendix). The running indexes are: $j=1$ to $J$ and $k=1$ to $K$ independent variables observed in $i=1$ to $I$ samples that contain $n=1$ to $N$ constituents.

## 3 SECOND-ORDER STRUCTURE AND MULTIWAY DATA

Second-order analytical techniques such as high-performance liquid chromatography with diode array detection (HPLC-DAD), gas chromatography with mass spectra detection (GC-MS), and EEM fluorescence provide a $J \times K$ array $\mathbf{X}$ of measured responses per sample, the so-called second-order data. A fundamental property of some types of second-order data is bilinearity. When second-order data are bilinear, the responses for pure compound $n$ can be written as the outer product (called a *bilinear component*) (see Appendix):

$$\mathbf{X} = a_n \mathbf{b}_n \mathbf{c}_n^\mathrm{T} + \mathbf{E} \tag{1}$$

where $\mathbf{b}_n$ represents a column profile, $\mathbf{c}_n^\mathrm{T}$ represents a row profile, $a_n$ is a scaling factor related to the concentration of the constituent, and $\mathbf{E}$ is the measurement noise. If the profiles are scaled in such a way that the product $\mathbf{b}_n\mathbf{c}_n^\mathrm{T}$ is the response of the pure constituent at unit concentration, then $a_n$ is the concentration of that constituent in the sample. Figure 1A shows the bilinear behavior of a pure constituent measured by HPLC-DAD; $\mathbf{b}_n$ is the elution profile of the compound, $\mathbf{c}_n$ its UV–vis spectrum, and $a_n$ is related to the concentration. Within the noise level, the spectrum is the same (except for a scale factor) at each elution time, and the elution profile is the same (except for a scale factor) at every wavelength. Likewise, Figure 1B shows a bilinear EEM fluorescence spectrum of a pure compound. In this case, $\mathbf{b}_n$ is the excitation profile of the fluorophore, $\mathbf{c}_n$ is its emission profile, and $a_n$ is the concentration-related

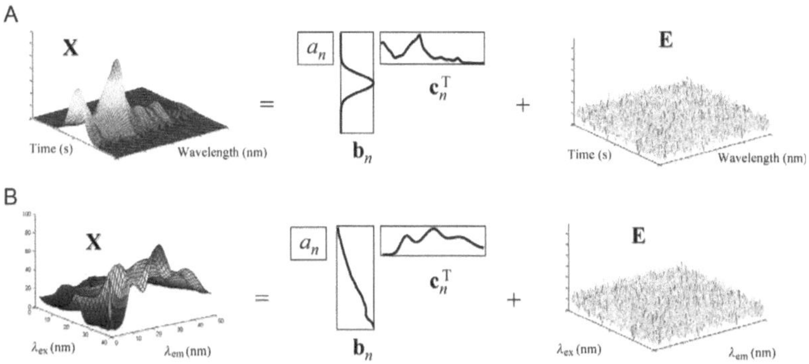


**FIGURE 1** (A) Bilinear chromatographic peak of a pure compound measured by HPLC-DAD, as the outer product of a column vector (elution profile) and a row vector (absorbance spectrum). (B) EEM fluorescence spectrum of a pure compound as the outer product of a column vector (excitation spectrum) and a row vector (emission spectrum).

scale. Again, if the emission spectrum does not depend on the excitation spectrum and the emission profiles are identical for all excitation wavelengths, then the data can be described as the outer product in Equation (1). In both examples, the difference between the underlying profiles and the measured data is attributed to random noise in matrix **E**. Note that not all two-dimensional techniques yield bilinear data. For example, two-dimensional nuclear magnetic resonance spectroscopy (2D-NMR) data and tandem (two-dimensional) mass spectrometry (MS/MS) data in their raw forms are not bilinear despite they generate matrices of data per sample.

Another fundamental property of some second-order data is rank additivity. For a mixture of $N$ constituents, if every constituent's response can be described by Equation (1) and there is no interaction between constituents (i.e., the presence of one constituent does not modify the response of the other constituent), then the contributions of each constituent in the two orders of the measurement can be added and the measured data can be described by

$$\mathbf{X} = \sum_{n=1}^{N} a_n \mathbf{b}_n \mathbf{c}_n^{\mathrm{T}} + \mathbf{E} \tag{2}$$

where $\mathbf{b}_n$, $\mathbf{c}_n$, and $a_n$ represent the pure profiles in both orders and the concentration-related scale of constituent $n$, respectively. Equation (2) describes, for example, overlapped peaks in a chromatographic separation and the EEM fluorescence of a mixture of $N$ constituents. Figure 2 represents Equation (2) for $N=2$. Since the matrix resulting from the outer product in Equation (1) has rank one, each new constituent increases by one the rank of **X** in Equation (2), so the rank of a noise-free **X** would indicate the number of constituents in the sample that give a response, here $N$. Then $N$ is called the true chemical rank (also called pseudo-rank) of the data.

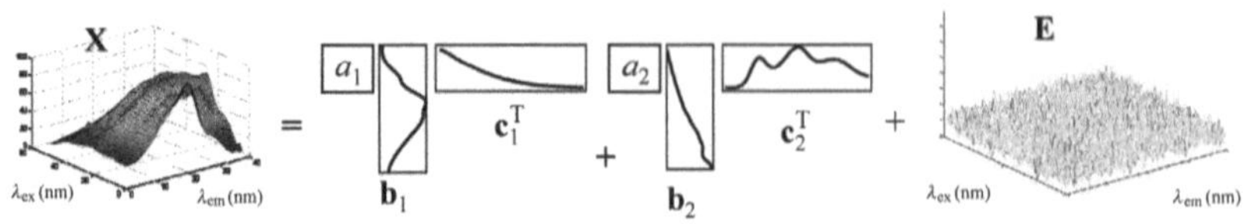


**FIGURE 2** Representation of bilinear EEM fluorescence spectrum of a mixture of two components.

Because of the noise in the experimental measurements, matrix **E** is of full mathematical rank and so is matrix **X**. An important part of the multiway algorithms is to find the underlying true chemical rank. Equation (2) can be written as

$$\mathbf{X} = \mathbf{B}\mathbf{C}^{\mathrm{T}} + \mathbf{E} \tag{3}$$

where $\mathbf{B}(J \times N)$ and $\mathbf{C}(K \times N)$ contain, in columns, the column and row profiles, respectively, and the scaling constant was absorbed in either **B** or **C**. Some methods, such as Iterative Target Transformation Factor Analysis or Heuristic Evolving Latent Projection try to decompose the single matrix **X** in their corresponding **B** and **C** profiles with chemical meaning. Others, such as Principal Component Analysis (PCA), provide a similar decomposition, although the profiles in **B** and **C** may not have chemical meaning (e.g., negative parts for spectra or elution profiles).

If measurements from different samples with common underlying contributions are available (e.g., measurements from mixtures of the same type, but different amount of constituents), all the matrices can be decomposed ensemble to find their common generating contributions. This is done by stacking the set of $I$ matrices as a three-way structure $\boldsymbol{X}$ $(I \times J \times K)$ and solving along the three dimensions (modes). Then, if the vectors describing the individual contributions in both dimensions are reproducible from sample to sample and identical for different samples (only varying in proportions), the concept of bilinearity extends to the third mode, that is trilinearity, and the three-way array can be decomposed into trilinear components as:

$$\boldsymbol{X} = \sum_{n=1}^{N} \mathbf{a}_n \otimes \mathbf{b}_n \otimes \mathbf{c}_n + \boldsymbol{E} \tag{4}$$

This data structure is represented in Figure 3 for a system of two constituents. The $i$th horizontal slice of the trilinear cube $\boldsymbol{X}$ is the $J \times K$ matrix of measured responses for the $i$th sample shown in Figure 2. An alternative form of Equation (4) also used in the literature is:

$$x_{ijk} = \sum_{n=1}^{N} a_{in} b_{jn} c_{kn} + e_{ijk} \tag{5}$$

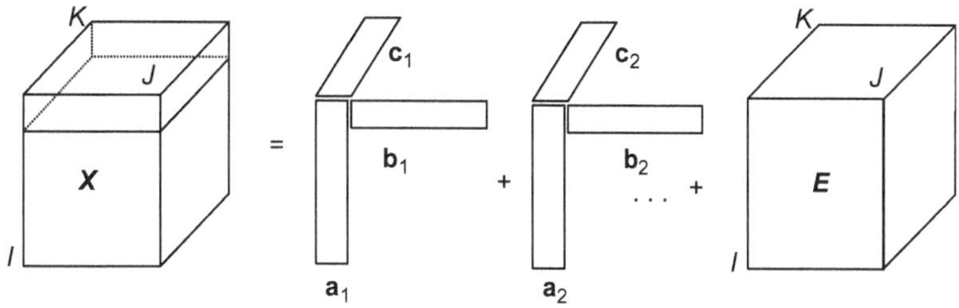


**FIGURE 3** Three-way array with trilinear structure. The horizontal slice represents a second-order data array obtained by measuring one sample.

where $a_{in}$, $b_{jn}$, and $c_{kn}$ are the $i$th, $j$th, and $k$th elements of $\mathbf{a}_n$, $\mathbf{b}_n$, and $\mathbf{c}_n$, and $e_{ijk}$ are the random errors.

## 4 THE PARAFAC MODEL

In the same way that a variety of methods can decompose **X** in Equation (3), there are also a variety of methods that perform the decomposition of multi-way arrays $\boldsymbol{X}$. Tucker [17] described several possibilities for decomposing a multiway array that are known as Tucker-1, Tucker-2, and Tucker-3 models. Of them, the Tucker-3 model is the most general, and the PARAFAC model can be seen as a particular case of Tucker-3.

The Tucker-3 model for a three-way array $\boldsymbol{X}(I \times J \times K)$ whose elements (measured values) are $x_{ijk}$ is defined as

$$x_{ijk} = \sum_{p=1}^{P} \sum_{q=1}^{Q} \sum_{r=1}^{R} a_{ip} b_{jq} c_{kr} g_{pqr} + e_{ijk} \tag{6}$$

where $e_{ijk}$ is the residual error term (the nonmodeled part of the measured value), $a_{ip}$, $b_{jq}$, and $c_{kr}$ are elements of loading matrices $\mathbf{A}(I \times P)$, $\mathbf{B}(J \times Q)$, and $\mathbf{C}(K \times R)$, respectively, and $g_{pqr}$ is an element of the core array $\boldsymbol{G}(P \times Q \times R)$ (Figure 4). The number of components $P$, $Q$, and $R$ can be different on each mode. The loading matrices are column-wise orthonormal. The elements of $\boldsymbol{G}$ represent the magnitude of the factors and the extent of their interaction.

The PARAFAC model is a particular case of the Tucker-3 model, with the same number of components, $N$, in each mode, and with the core array with size $(N \times N \times N)$ and superdiagonal, that is, with ones in the superdiagonal and zeros otherwise. Figure 5 represents the PARAFAC model of a three-way array. The decomposition of the data is made into triads or trilinear components. The contribution of each component to the three-way array consists of three loading vectors $\mathbf{a}_n$, $\mathbf{b}_n$, and $\mathbf{c}_n$ which are treated equally mathematically [18]. The loadings of all components are grouped in the three loading matrices: **A**, **B**, and **C** whose $n$th columns are $\mathbf{a}_n$ $\mathbf{b}_n$, and $\mathbf{c}_n$, respectively.

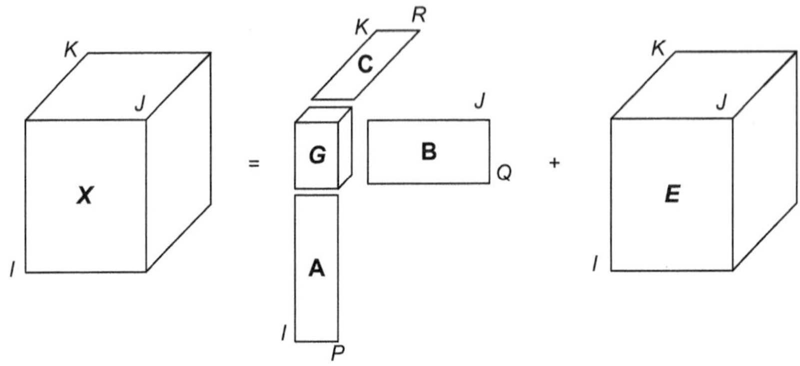


**FIGURE 4** Tucker-3 model.

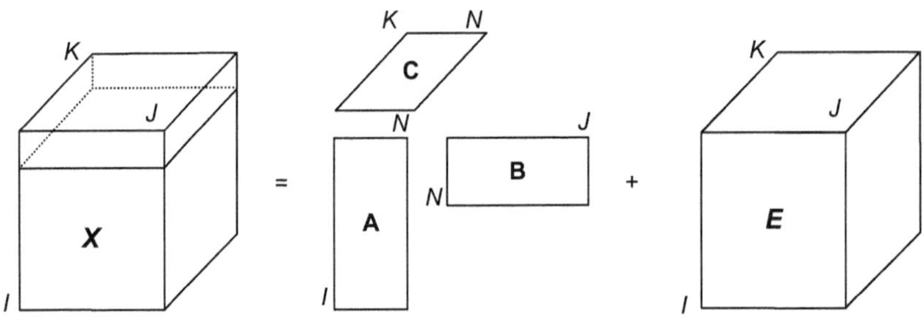


**FIGURE 5** PARAFAC model. Matrix **A** contains the loadings for mode $I$ (e.g., concentration ratios), matrix **B** contains the loadings for mode $J$ (e.g., pure excitation spectra), and matrix **C** contains the loadings for mode $K$ (e.g., pure emission spectra).

Under the ordering described in the previous section, where samples are horizontal slices, the **A**-mode represents differences between samples, and the loading vectors in the "sample" mode are sometimes referred to as *a*-scores or simply scores. The cube ***E*** contains the residuals.

The PARAFAC model can be written as a restricted version of a Tucker-3 model:

$$\mathbf{X} = \mathbf{AG}(\mathbf{C}|\otimes|\mathbf{B})^{\mathrm{T}} + \mathbf{E} \tag{7}$$

where the core array **G** ($N \times NN$), when refolded to ***G*** ($N \times N \times N$) is a three-way array with ones in the superdiagonal and zeros otherwise, and **X** is the unfolded (matricized) matrix ***X*** (Figure 6) in one of the modes. Actually, the PARAFAC model can also be expressed in any of the three following forms as:

$$\mathbf{X}_{\mathrm{a}} = \mathbf{A}(\mathbf{C}|\otimes|\mathbf{B})^{\mathrm{T}} + \mathbf{E}_{\mathrm{a}} \tag{8}$$

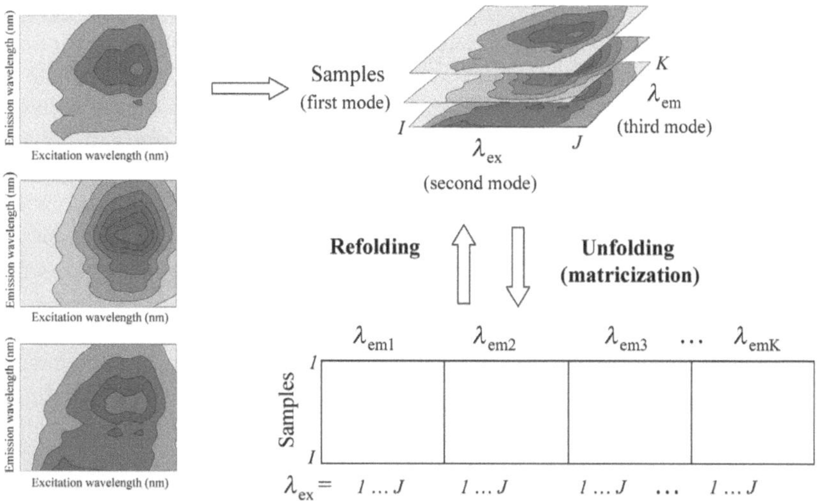


**FIGURE 6** Unfolding of a three-way array of fluorescence data.

$$\mathbf{X}_b = \mathbf{B}(\mathbf{A}|\otimes|\mathbf{C})^T + \mathbf{E}_b \quad (9)$$

$$\mathbf{X}_c = \mathbf{C}(\mathbf{B}|\otimes|\mathbf{A})^T + \mathbf{E}_c \quad (10)$$

where matrix $\mathbf{X}_a$ $(I \times JK)$ is $\mathbf{X}_a = [\mathbf{X}_1, \mathbf{X}_2, \ldots, \mathbf{X}_K]$, matrix $\mathbf{X}_b$ $(J \times IK)$ is $\mathbf{X}_b = [\mathbf{X}_1, \mathbf{X}_2, \ldots, \mathbf{X}_K]$, and matrix $\mathbf{X}_c$ $(K \times IJ)$ is $\mathbf{X}_c = [\mathbf{X}_1, \mathbf{X}_2, \ldots, \mathbf{X}_K]$, and $\mathbf{E}_a$ $(I \times JK)$, $\mathbf{E}_b$ $(J \times KI)$, and $\mathbf{E}_c$ $(K \times IJ)$ are the unfolded residual matrices.

An important characteristic of the PARAFAC model is the uniqueness of its solution (apart from trivial variations of scale and column order, see below) and the similarity of the PARAFAC solutions to the physical models underlying the chemical data (i.e., excitation–emission fluorescence profiles) [19]. Uniqueness means that it is not possible to rotate the optimal solution and obtain another set of estimated parameters without a loss of fit [12]. In methods such as PCA, the solution is not unique since the solution can be rotated without any loss of fit by taking an orthogonal transformation matrix $\mathbf{Q}$ $(R \times R)$, such that $\mathbf{Q}^T\mathbf{Q} = \mathbf{Q}\mathbf{Q}^T = \mathbf{I}$. Then the PCA solution can be written as $\mathbf{X} = \mathbf{T}\mathbf{P}^T = \mathbf{T}\mathbf{Q}\mathbf{Q}^T\mathbf{P}^T = (\mathbf{T}\mathbf{Q})(\mathbf{P}\mathbf{Q})^T$, and the amount of variance of $\mathbf{X}$ captured by the components does not change. In other words, $||\mathbf{X} - \mathbf{T}\mathbf{P}^T||^2 = ||\mathbf{X} - \mathbf{T}\mathbf{Q}\mathbf{Q}^T\mathbf{P}^T||^2$. This is known as rotational freedom and, together with the constraint that loadings $\mathbf{P}$ are orthogonal, makes PCA to not directly yield the pure underlying profiles that generated $\mathbf{X}$. The rotation of the PCA solution [20] or the use of other decomposition methods such as the Multivariate Curve Resolution Alternating Least-Squares (MCR-ALS) [21], which imposes restrictions (conditions) on the solutions, may be required to obtain more interpretable solutions. In PARAFAC, however, restrictions are not necessary to estimate the model and obtain the profiles. For trilinear data, the underlying profiles are recovered if the right number of components is used and the signal-to-noise ratio is appropriate [22] (although in practice, restrictions may be used during the iterations of the decomposition algorithm to help in reaching the optimal solution and control slight nonlinearities caused by measurement irreproducibility). For appropriate noise, i.e., random and not too severe, it also holds that the true underlying trilinear model will be the model with the best fit. Therefore, the true and estimated models must coincide when the right number of components is chosen. This allows the PARAFAC model to be the basis of a calibration method that can predict test samples even if new interferences are present. For a theoretical study of the sufficient condition under which least-squares decomposition methods achieve a unique resolution, see Kruskal [23]. The PARAFAC model is, however, indeterminate with respect to the order and the scaling of the loading vectors. The order of the components is not defined, and the profiles have to be identified after building the model (i.e., from reference spectra). The scale is neither defined, which means that the obtained profiles are not scaled and so a reference is needed if the PARAFAC decomposition is used to calculate a calibration model and predict test samples. It is possible, for example, to normalize the loading vectors **B** and **C** since they are only used for qualitative analysis and translate the scale to **A**. Another difference with respect to PCA is that PARAFAC is nonnested, i.e., the $N-1$ component model is not a subset of the $N$ component model so it cannot be calculated sequentially.

## 5 ALGORITHMS

The PARAFAC decomposition seeks to find matrices **A**, **B**, and **C** such that the residual error is minimized following a least-squares criterion:

$$\underset{\mathbf{A},\mathbf{B},\mathbf{C}}{\operatorname{argmin}} \|\mathbf{E}_{\mathrm{a}}\|^2 = \underset{\mathbf{A},\mathbf{B},\mathbf{C}}{\operatorname{argmin}} \left\|\mathbf{X}_{\mathrm{a}} - \mathbf{A}(\mathbf{B}|\otimes|\mathbf{C})^{\mathrm{T}}\right\|^2 \tag{11}$$

i.e., the trilinear model is found so that it minimizes the sum of squares of the residuals. Various algorithms calculate the PARAFAC decomposition [24,25]. Tomasi and Bro [26] grouped them into three classes:

**(1)** Alternating algorithms: ALS [2], Alternating Slice-wise Diagonalization [27], or Self-Weighted Alternating TriLinear Decomposition [28].
**(2)** Derivative-based algorithms: Positive Matrix Factorization for three-way arrays (PMF3) [29] or damped Gauss–Newton (dGN), also known as Levenberg–Marquardt [12].
**(3)** Direct (noniterative) algorithms: Generalized Rank Annihilation Method (GRAM) [30] or Direct TriLinear Decomposition (DTLD) [31].

A comparison of some of the algorithms for fitting the PARAFAC model mentioned above can be found in Faber *et al.* [24], Faber [32], and, for example, in Comas *et al.* [33]. The ALS algorithm is very versatile since it can handle higher-order arrays, constrained models, and weighted least-squares loss functions, so probably ALS is the most popular algorithm for PARAFAC decomposition. This algorithm is explained next, and the information about the other algorithms can be found in the references indicated above.

## 5.1 Alternating Least-Squares

The ALS algorithm calculates **A**, **B**, and **C** of Equation (7) iteratively to minimize the residual error. Let the unfolded matrices $\mathbf{X}_a$ $(I \times JK)$, $\mathbf{X}_b$ $(J \times IK)$, and $\mathbf{X}_c$ $(K \times IJ)$ be as defined in Equations (8)–(10). The ALS algorithm proceeds as follows:

1. Define starting values of **B** and **C**
2. $\mathbf{Z} = \mathbf{C} | \otimes | \mathbf{B}$
3. $\mathbf{A} = \mathbf{X}_a \, (\mathbf{Z}^{+})^{\mathrm{T}}$
4. $\mathbf{Z} = \mathbf{C} | \otimes | \mathbf{A}$
5. $\mathbf{B} = \mathbf{X}_b \, (\mathbf{Z}^{+})^{\mathrm{T}}$
6. $\mathbf{Z} = \mathbf{B} | \otimes | \mathbf{A}$
7. $\mathbf{C} = \mathbf{X}_c \, (\mathbf{Z}^{+})^{\mathrm{T}}$
8. Repeat steps 2–7 until convergence while $n_{\mathrm{iter}} < n_{\mathrm{iter,max}}$ where $n_{\mathrm{iter}}$ is the iteration number and $n_{\mathrm{iter,max}}$ is the maximum number of iterations allowed. Converge is reached when the relative decrease in the loss function or the relative change in the parameters is smaller than an arbitrarily small number.

In certain cases, the algorithm may converge to a local minimum instead of the global minimum. Although local minima are seldom a problem if the data are highly trilinear, a initialization with good starting values for **B** and **C** can significantly speed up the convergence to a global minimum. Commonly, starting values are obtained from the first column vectors of the singular value decomposition of the unfolded $\mathbf{X}_a$, $\mathbf{X}_b$, and $\mathbf{X}_c$ arrays, by the direct GRAM or DTLD methods or from spectral data of pure components that are known in advance. Alternatively, **B** and **C** can contain random values and the algorithm be started several times with new values each time. If the same solution is reached several times, then the probability that a local minimum has been found due to an unfortunate initial guess is small. Improved versions of the ALS algorithm have also been proposed [12]. Convergence speed is also a performance parameter to take into account, as the ALS algorithm may require thousands of iterations to converge, depending on the dimensions and type of data (number of components, similarity between profiles, ...). The optimization of the ALS convergence speed has been investigated by several authors and following different approaches, such as using the properties of the column-wise Khatri–Rao product, using compression techniques, or using line-search strategies [34].

## 5.2 Prediction

PARAFAC is an unsupervised method in the sense that concentration information is not used for calculating the decomposition. Optimization of the model is based solely on the minimization of the $x$-residuals. Since the PARAFAC scores represent relative proportions among the modeled samples, it is possible to go one step further and regress the PARAFAC scores for a set of calibration samples against the reference analyte concentrations in these samples and use this model to predict the analyte concentration in a new sample from its PARAFAC score. This is similar to multivariate Principal Component Regression in which PCA is first performed on the calibration samples and then the scores are related to the property values.

Of the several approaches that use PARAFAC as the basis for calibration (see Rinnan *et al.* [35] for a comparison of them), the two most used are described next. In the first approach, a PARAFAC model is derived from the calibration samples, and the obtained loadings are applied to the unknown sample data matrix to obtain the unknown sample scores and predictions. The steps followed are (see Figure 7):

1. Decompose the $I \times J \times K$ array of $I$ calibration samples with PARAFAC and obtain **A**, **B**, and **C**
2. Use the **B** and **C** profiles to identify the column of **A** corresponding to the $n$th analyte of interest, $\mathbf{a}_n$
3. Regress the $I$ elements of the column $\mathbf{a}_n$ versus known standard concentrations $\mathbf{y}_{n,\text{cal}}$ for analyte $n$: $\mathbf{a}_n = k\mathbf{y}_{n,\text{cal}}$. This regression graph is commonly known as pseudo-univariate representation of the model and plays a fundamental role in the calculation of some figures of merit (see below).

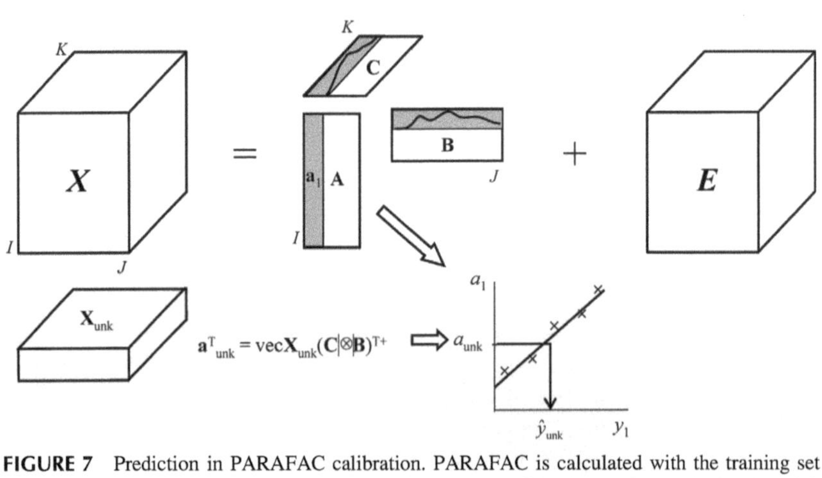


**FIGURE 7** Prediction in PARAFAC calibration. PARAFAC is calculated with the training set only, and the unknown sample is projected using the **B** and **C** loadings of the calculated model. The unknown sample score is then used to predict the concentration.

4. Calculate the scores (representing the relative concentrations) in the unknown as $\mathbf{a}_{\text{unk}} = \mathbf{Z}^{+} vec(\mathbf{X}_{\text{unk}})$ and select the one of interest: $a_{n,\text{unk}}$. Here, $\mathbf{Z} = (\mathbf{C}|\otimes|\mathbf{B})$ and $vec(\mathbf{X}_{\text{unk}})$ $(JK \times 1)$ is the vector obtained by unfolding the unknown sample data matrix $\mathbf{X}_{\text{unk}}$ column-wise into a column vector (see Appendix).
5. Convert relative to absolute concentration of $n$ in the unknown: $\hat{y}_{n,\text{unk}} = a_{n,\text{unk}}/k$.

In the second approach, the $I \times J \times K$ array of $I$ calibration samples is augmented with the sample to be predicted to form an array of dimensions $(I+1) \times J \times K$. PARAFAC simultaneously models both the calibration and the test sample. The steps followed are (see Figure 8):

1. Augment the $I \times J \times K$ array of $I$ training samples with the test sample to obtain an $(I+1) \times J \times K$ array and decompose it with PARAFAC to obtain **A**, **B**, and **C**
2. Use the **B** and **C** profiles to identify the column of **A** corresponding to the $n$th analyte of interest, $\mathbf{a}_n$. Note that the last element of $\mathbf{a}_n$ (element $I+1$) is the score of the unknown sample
3. Regress the elements 1 to $I$ of $\mathbf{a}_n$ against known standard concentrations $\mathbf{y}_{n,\text{cal}}$ for analyte $n$: $\mathbf{a}_n = k\mathbf{y}_{n,\text{cal}}$
4. Convert relative to absolute concentration of $n$ in the unknown using the $I$th+1 element of $\mathbf{a}_n$ (the one corresponding to the unknown sample) to obtain the prediction: $\hat{y}_{n,\text{unk}} = a_{n,\text{unk}}/k$

The first approach is like the one used in univariate and multivariate calibrations where the model is developed first using a calibration set and then

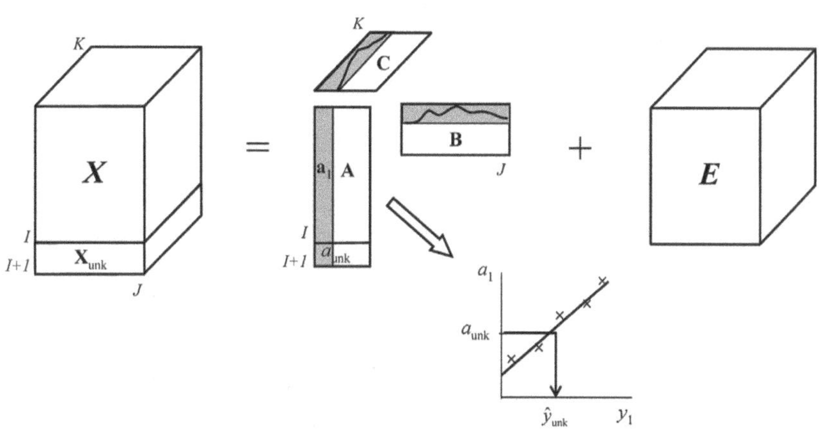


**FIGURE 8** Prediction in PARAFAC calibration. PARAFAC is calculated with the unknown sample joined with the training set. The **B** and **C** loadings now include also profiles only present in the unknown sample. The score of the unknown sample is calculated during the PARAFAC decomposition.

subsequently applied to all future samples. Since this approach always uses a previously calculated set of loadings **B** and **C**, it requires, like in first-order (multivariate) calibration, that all the constituents in the unknown sample be included in the training set and be spanned by the loading matrices. If the unknown sample includes constituents not present in the training data (and therefore not included in **B** and **C**), predictions may be severely biased. In the second approach, however, since the calibration and unknown sample are modeled together, the unknown sample may contain constituents not present in the calibration samples and still obtain correct predictions. This is possible because the loadings matrices **B** and **C** will contain additional columns corresponding to the new interferents present in the unknown sample. This important property (to estimate analyte concentrations in the presence of interferences not present in the calibration samples) is called the second-order advantage because it is obtained thanks to the rank additivity and bilinearity structure of the second-order data. The second-order advantage constitutes one of the main benefits of using second-order instrumental data in analytical chemistry. The shortcoming of the second approach is that for each new unknown sample, a new data cube must be created by replacing the previous unknown sample with the new sample. Since each factor in PARAFAC represents a unique constituent (e.g., analyte, background, interference) that contributes to the instrumental response in the data array, each new constituent in the unknown sample will increase the number of factors (i.e., the rank of the multiway array) by one as compared to the number used in a PARAFAC model for the calibration data alone (although it does not change the pure constituent profiles for the other constituents). Since the number of new constituents may be different for every unknown sample, the number of factors, $F$, must be chosen correctly for optimum performance of the model every time and performance measures of previous models may not be valid for the new model at hand. This makes the full process (calculation of a new PARAFAC model, optimization of the number of factors, identification of the constituent of interest by comparing the profiles in **B** and **C**, univariate regression and final prediction from scores) difficult to automate. The amount of profile overlap between constituents, as compared to the amount of noise in the data, may also severely influence the prediction.

## 6 PARAFAC MODELING AND VALIDATION

### 6.1 Data Preprocessing

Data preprocessing comprises a series of operations on the multiway data array pursuing two main objectives: (1) to remove constant contributions in the data (centering) and weight the signal contribution in the model (scaling) and (2) remove undesired effects that make the data deviate from trilinearity.

Although for PARAFAC calibration is not usually applied, it is convenient to center the data across the first mode (sample mode) for some exploratory or classification purposes or to remove undesired constant offsets. Centering the first mode is done by unfolding the $\boldsymbol{X}$ array to $\mathbf{X}_a$ $(I \times JK)$ (see Equation 8) and then column mean-centering $\mathbf{X}_a$, that is, by subtracting the column mean to every datum in the column. Obviously, the centering can be applied to any of the modes, depending on the organization of the data and the problem at hand, and centering can also be performed across more than one mode. If centering is to be performed across more than one mode, it is done by first centering one mode and then centering the outcome of this centering [36].

Scaling to remove differences of magnitude between variables or samples is also common in multiway data. Two common scaling methods are standardization and normalization. Standardization involves dividing each element of a matrix by its column standard deviation. Normalization is performed by scaling the data in the first (sample) mode to unit norm, i.e., dividing by the sum of the squared value of all variables for the sample. Normalization is performed to remove proportional variation, which is usually associated with the total amount of sample [37]. For three-way data, scaling must be done on the rows of a matrix and not across the columns like in centering (i.e., one has to scale whole matrices instead of columns) (Figure 9). For example, to remove differences in magnitude between the emission wavelengths of a set of EEMs, one can scale the three-way array within the emission mode. To do this, the three-way array has to be unfolded in such a way that emission wavelength (variable $j$) is in the rows and then the rows are scaled. This is because all the columns where variable $j$ occurs have to be scaled [37]. Finally, it must be reminded that not all combinations of centering and scaling are possible when working with three-way data [36].

Three-way data must follow the trilinear model in order to be effectively handled by PARAFAC, but not all bilinear data matrices when stacked in a three-way array have a trilinear structure at first. Instrumental artifacts or measurement inaccuracies among samples make data deviate from the required trilinearity. For instance, in EEM fluorescence, the emission region below excitation has zero intensity (or noise equivalent to zero) because the emitted energy is always lower than the excitation energy. Whenever these

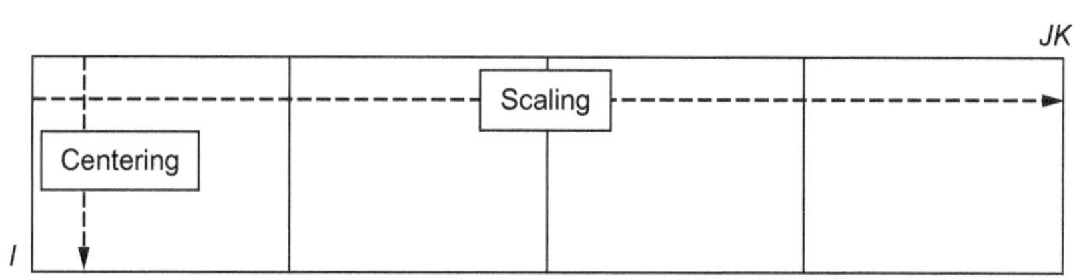


**FIGURE 9** A three-way unfolded array, showing the directions along which scaling and centering should be performed.

data are recorded, they should be treated as missing values and cannot be replaced with zeros to prevent the PARAFAC model from trying to fit them [38]. Other potential problems in EEM fluorescence are Rayleigh and Raman scattering. Scattering does not provide any information on the fluorescence of the sample and must be avoided. The Raman interference due to the solvent can often be removed by subtracting the solvent spectra from the sample spectra. Rayleigh scattering produces a diagonal line across the landscapes, at emission wavelengths equal to excitation wavelengths (first-order Rayleigh) and at emissions wavelengths around twice the excitation wavelength (second-order Rayleigh). Several strategies for solving this problem exist such as inserting missing values on the scattered area, plainly discarding the part of the matrix that includes the scatter, subtracting a blank spectra, or data point weighing [39]. In addition, fluorescence is affected by environmental factors such as solvent polarity, pH, and fluorescence quenching [40]. Hence, obtaining trilinearity in fluorescence may require precise work to keep the measurement conditions constant throughout the experiments.

Second-order data from hyphenated chromatography, such as HPLC-DAD or GC-MS, are another type of data commonly analyzed with PARAFAC that may suffer from lack of trilinearity. For these to be trilinear, the elution profiles of each constituent must coincide among different chromatographic runs (samples) except for a scaling constant related to the concentration. Although agreement in the spectral profile is achieved easily, complete agreement in retention times and shapes is difficult unless the instrument and the separation conditions are controlled tightly. So the main causes of deviations from trilinearity in chromatographic data are retention time shift, peak broadening, and peak skewing between chromatographic runs and baseline drifts and offsets. Sometimes small deviations from trilinearity pass unnoticed or are ignored if they introduce a slight structure in the estimated noise that does not apparently affect the conclusions of the study (estimated elution profiles, spectra, and concentrations). Often, however, deviations from trilinearity can be easily corrected by preprocessing by using alignment algorithms for chromatographic data [41,42] or are dealt with by using the PARAFAC2 model [43,44]. Baseline drifts and offsets can be removed using differentiation or centering methods and alternatively by adding additional components in the PARAFAC model. For stronger deviations, other PARAFAC-derived methods can be applied, such as the shifted factors PARAFAC model or the convoluted PARAFAC model [12].

## 6.2 Model Refinement and Constraints

The PARAFAC model building and validation steps may need recalculating, mainly for datasets that are difficult to fit. In such cases, PARAFAC may not find the right solution, and even repeated modeling starting from different initializations may produce different results. Such unstable or incorrect models may result from improper data preprocessing, lack of selectivity in any of the modes, or simply because there are many PARAFAC components to be recovered, including the constituents of interest and the interferences. Improvement of the models can be done in different ways: revising the preprocessing, starting the PARAFAC model with better initial loading estimates, adequately choosing the calibration samples to achieve as much selectivity as possible (the inclusion of pure standards, if available, may help), checking for outliers, or applying constraints during modeling [45]. A typical constraint for EEM fluorescence or chromatographic data is nonnegativity for concentration profiles and spectra. For chromatographic data, it can also be useful to apply a constraint to have only one peak (unimodality) in the elution profiles. Orthogonality is a constraint useful for explorative purposes, when the loadings do not need to be physically/chemically interpreted, and has several computational advantages, such as fast convergence to the final solution and no problems with degeneracies (i.e., loading vectors of the same mode having high correlations or factors nearly canceling out each other's contributions). Care must be taken when using constraints, because the results may be more visually appealing but it may happen that the recovered factors are not real. Constraints, then, need to be validated, and the best way to do it is to repeat the modeling several times changing the starting solutions, to finally choose the model with the minimum error.

## 6.3 Model Diagnostics and Interpretation of Results

Either if PARAFAC is used as an unsupervised method to explore a three-way array for relationships between samples and chemical profiles, or as a part of a calibration method to predict a property of interest in a sample from second-order instrumental responses, a PARAFAC model must always be validated.

The validity of the chemical interpretation of the PARAFAC decomposition relies, first of all, upon the right number of components being fitted. When fewer components/factors are used than there are independently varying chemical constituents responsible for the measured signal, the model may approximate the combined signal of chemically distinct compounds. Forcing the model to include too many components may generate two or more profiles that represent a split of a single constituent in two, often in combination with noise. There are different approaches to estimate the correct number of components in a PARAFAC model, although each may suggest a different optimal number. So it is sensible not to stick to a single validation criterion and to explore a variety of them and seek for common agreement. The main validation approaches that we will describe are residual analysis, loading visualization and comparison with external knowledge, the core consistency diagnostic (CORCONDIA), split-half analysis, and cross-validation (CV) strategies, the latter to be used mostly for calibration. The number of components of the PARAFAC model will be the same as the number of constituents in the samples provided that (1) the model is indeed trilinear, (2) there is no overfit, that is, the right number of components is chosen, and (3) the signal to noise is adequate.

### 6.3.1 Analysis of the Residuals

Analysis of residuals, matrix **E** in Equation (7), is performed similarly as in PCA models. For the correct number of PARAFAC components, significant systematic variation in the dataset is captured by the model, and the difference between the measured data, ***X***, and the reconstructed data is only due to random error. In this situation, residual plots for each sample (slices of the three-way array ***E***) show no consistent pattern as it can be seen in Figure 2. In practice for real datasets, systematic variation may be seen for at least some samples in the form of peaks (representing signals not captured by the model) or troughs (negative peaks). This may indicate slight lack of trilinearity or that more components could be extracted from the PARAFAC model. In the case of fluorescence EEMs, small peaks occurring along the diagonal due to incompletely removed scatter can be ignored since they do not respond to any constituent of interest in the sample. However, adjacent peaks and troughs in the residuals often indicate a problem whereby a peak is modeled using two or more poorly fitting components.

### 6.3.2 Analysis of Spectral Loadings

If the loadings (matrices **B** and **C**) of a PARAFAC model should have a chemical interpretation, it must be assessed whether they are physically reasonable with respect to the phenomenon being studied. For example, in EEM fluorescence spectra, the spectral loadings could be compared to reference library spectra or to spectra measured of pure standards. Additionally, if the EEMs correspond to noninteracting fluorophores, the emission spectrum should be shifted with respect to its excitation spectrum (Stokes shift), so excitation and emission spectrum for the same constituent should show a minimal overlap [40]. This happens because the energy of the emitted fluorescence is lower than the energy that excited the molecule. Other atypical fluorescence spectral results in water solutions include the appearance of multiple distinct emission peaks for the same species, the absence of excitation between consecutive absorption bands, and the presence of abrupt spectral changes over short wavelength distances [19]. For chromatographic data, the calculated elution profiles should be nonnegative and unimodal, while the spectral profiles should, again, resemble the spectra obtained when measuring pure standards.

### 6.3.3 Core Consistency

This is a commonly used diagnostic to determine the optimal number of components and evaluate the "appropriateness" of the PARAFAC model [46].

It assumes that the diagonal elements in the core array $\boldsymbol{G}$ associated to a PARAFAC solution for a given number of components will be much larger than the rest of solutions, whenever the number of components is correct. As shown in Equation (7), the PARAFAC model can be written as a restricted version of a Tucker-3 model, $\mathbf{X}=\mathbf{AG}(\mathbf{C}|\otimes|\mathbf{B})^{\mathrm{T}}$. Once the PARAFAC model has been fitted and matrices **A**, **B**, and **C** calculated, the core matrix **G** can be calculated via a least-squares procedure:

$$\underset{\mathbf{G}}{\operatorname{argmin}} \left\| \mathbf{X} - \mathbf{AG}(\mathbf{C}|\otimes|\mathbf{B})^{\mathrm{T}} \right\|^2 \tag{12}$$

A regression model can be built to solve the problem in such a way that:

$$\min_{\mathbf{G}} \left\| vec\boldsymbol{X} - (\mathbf{C}\otimes\mathbf{B}\otimes\mathbf{A})vec\mathbf{G} \right\|^2 \tag{13}$$

For a valid PARAFAC model, the core matrix **G** should ideally have values close to one in the superdiagonal and values close to zero in the off-superdiagonal. When this is not the case, the data are not properly modeled by a trilinear PARAFAC model, too many components have been used to fit the model or the data have outliers. A simple way to assess whether the model structure is reasonable is therefore to monitor the distribution of the superdiagonal and off-superdiagonal elements of **G**. The CORCONDIA, calculated as a single parameter, is defined as:

$$\text{Core consistency} = 100 \cdot \left( \frac{1 - \sum_{d=1}^{N}\sum_{e=1}^{N}\sum_{f=1}^{N}(g_{\text{def}} - t_{\text{def}})^2}{\sum_{d=1}^{N}\sum_{e=1}^{N}\sum_{f=1}^{N} t_{\text{def}}^2} \right) \tag{14}$$

where the summation involves all the values of core matrix **G**. The elements of $t_{\text{def}}$ are the theoretical values of the core array, that is, ones when $d=e=f$ and zeros otherwise. The core consistency takes values near to 100% for models with a low number of components and starts decreasing at the point when too many components are selected. The optimal number of components corresponds to the largest model still having a high core consistency.

Another way to visualize the consistency of the PARAFAC model is by means of the *core consistency plots* [18], which represent the distribution of the $N^3$ core elements for the different models. In these plots, if several off-superdiagonal elements of **G** are larger or of similar size as the superdiagonal elements, then the model chosen overfits the data and the model of choice should be one with fewer components.

### *6.3.4 Split-Half Analysis*

This is a powerful way to confirm that a PARAFAC model is appropriate. It consists on dividing the data into two halves in one of the modes (for PARAFAC calibration, typically the sample mode) and then building a PARAFAC model on both datasets. Due to the uniqueness of the PARAFAC model, if the correct number of components is used, the same loadings in the nonsplitted modes (for EEM data, the excitation and emission modes) on both data sets should be obtained [22]. The method is useful for relatively large datasets because the number of samples becomes a limiting condition on the number of components that can be identified. The method can be extended to multiple split-half tests, where various models are created and compared after dividing the dataset in half in different ways.

### 6.3.5 Cross-Validation

If the PARAFAC model is to be used for calibration and one knows the concentration of the constituent(s) of interest, then CV on the predicted concentrations can be used to find the optimal model. In the calibration step, the sample scores are obtained, which are the relative concentrations of the constituents in the mixtures. By running the leave-one-out CV procedure for the different validation samples, the prediction error of a PARAFAC calibration model can be calculated through the root mean squared error of cross-validation (RMSECV) as:

$$\mathrm{RMSECV} = \sqrt{\frac{\sum_{i=1}^{I_v}(y_i - \hat{y}_i)^2}{I_v}} \tag{15}$$

where $y_i$ and $\hat{y}_i$ are the actual and predicted concentrations, respectively, and $I_v$ is the number of training samples left out in the CV procedure. By plotting the RMSECV values versus the number of components of the PARAFAC models, the optimal model can be found when the plot reaches a minimum.

### 6.3.6 Departure from Trilinearity

Trilinearity may not be always achieved in practice. Some circumstances under which data may not be trilinear have already been discussed in Section 6.1, along with possible solutions. Here, it is only reminded that diagnostics of the outputs of PARAFAC must be checked in order to ensure that results are meaningful.

## 6.4 Figures of Merit

In chemical analysis, method validation is the provision of evidences that a method will perform as required for a certain analytical problem, i.e., that it is fit for purpose. After a PARAFAC model has been optimized (the number of factors selected, outliers removed, etc.), the so-called figures of merit accuracy, sensitivity, selectivity, limit of detection, limit of quantification (LOQ), and uncertainty or standard error of prediction are quantifiable measures of the final performance of the analytical method. The figures of merit that need to be characterized and their required values depend on customer requirements, prior experience of the method or the current guidelines in a particular sector of measurement, e.g., food analysis. When these figures of merit are used to prove that the procedure is suitable, they should be determined in a real matrix under routine conditions.

In this section, we focus on the figures of merit when the PARAFAC model is used for prediction. Accuracy is first described, since it is a general figure of merit and is similarly estimated for univariate, multivariate, or multiway calibration. For other figures of merit (sensitivity, selectivity, limit of detection, and LOQ), there are two alternative approaches to calculate them: one based on the concept of net analyte signal and one based on the use of the pseudo-univariate PARAFAC calibration plot, explained in Section 5.

### *6.4.1 Accuracy*

Accuracy is the most important figure of merit of any analytical method. It is defined as "the closeness of agreement between a test result and the accepted reference value." When applied to a set of test results, accuracy involves a combination of random components and a common systematic error or bias component [47]. So, conceptually, accuracy comprises precision and bias (lack of trueness). The degree to which the predicted value of a constituent in a sample corresponds to the true value may be checked against known (reference) concentrations, by analyzing reference materials or by spiking the samples with known concentrations of the constituent. A global estimation of the accuracy of a PARAFAC model can be obtained from the predictions of an external test set and reported as the root mean square error of prediction (RMSEP):

$$\mathrm{RMSEP} = \sqrt{\frac{\sum_{i=1}^{I_t} (y_i - \hat{y}_i)^2}{I_t}} \tag{16}$$

where $y_i$ and $\hat{y}_i$ are the actual and predicted concentrations, respectively, and $I_t$ is the total number of samples in the test set. RMSEP expresses the average error to be expected in the future predictions when the calibration model is applied to unknown samples. Accuracy can also be expressed as an average recovery, that is, the mean of the ratios predicted versus actual concentrations for the test set. An average recovery value close to one indicates an accurate method.

Bias represents the average difference between predicted and actual concentrations for the $I_t$ samples in the test set. It is calculated as:

$$\mathrm{Bias} = \frac{\sum_{i=1}^{I_t} (y_i - \hat{y}_i)}{I_t} \tag{17}$$

Note that this is an average bias for all test samples, which does not hold for each individual prediction.

Precision refers to the closeness of the agreement between the results of successive measurements of the same measurand carried out in certain conditions of measurement [47]. These conditions may be repeatability, intermediate, or reproducibility conditions. The precision of a method that uses PARAFAC calibration can be estimated from the standard error of performance (SEP), corrected for bias:

$$\mathrm{SEP} = \sqrt{\frac{\sum_{i=1}^{I_t}(y_i - \hat{y}_i - \mathrm{bias})^2}{I_t}} \tag{18}$$

Alternatively, the precision can be estimated from the residuals of the pseudo-univariate calibration plot (Section 5) as the standard deviation of the sample scores residuals for all calibration standards divided by the sensitivity, SEN (the slope of the calibration line) as:

$$\mathrm{Precision} = \frac{s_{\mathrm{res}}}{\mathrm{SEN}} = \mathrm{SEN}^{-1}\sqrt{\frac{\sum_{i=1}^{I}(a_i - \hat{a}_i)^2}{I-2}} \tag{19}$$

where $a_i$ are the sample scores (relative concentrations) for the given constituent estimated from the PARAFAC model, $\hat{a}_i$ are the estimated scores from the calibration line scores versus actual concentration, and $I$ is the number of calibration standards.

### 6.4.2 *Sensitivity*

Sensitivity is defined as the change in instrument response which corresponds to a change in the measured quantity (for example, an analyte concentration) [48]. In univariate calibration, the sensitivity for a given constituent is the slope of the analytical calibration curve which is usually constant at low concentrations where the calibration curve is a straight line. Analogously, for a PARAFAC calibration model, the sensitivity can be calculated as the slope of the pseudo-univariate calibration plot, built by regressing the sample scores (relative concentrations), versus the actual concentrations in the calibration samples. It is also possible to calculate a first-order sensitivity for the recovered first-order profiles, using Lorber's definition of sensitivity [49,50]:

$$\mathrm{SEN}_{\mathrm{B},n} = \|\mathbf{b}_n^*\| \tag{20}$$

where $\mathbf{b}_n^*$ is the net analyte contribution of constituent $n$ at unit concentration. A similar expression exists for the profile $\mathbf{c}$ of the same constituent $n$. Alternative definitions of sensitivity have also been proposed for multiway calibration [51], which have been condensed into a more general definition, conceived to take into account all possible second-order calibration situations [51,52]. As pointed out by the authors, the derivation of this sensitivity expression is not straightforward and cannot be extended in a simple way to higher-way calibration procedures.

Another approach to estimate the multiway sensitivity was developed by Olivieri and Faber [53] that can be extended to higher-order calibration situations. It is based on uncertainty propagation and is defined, for a constituent $n$, as the ratio of input-to-output noise:

$$\mathrm{SEN}_n = \frac{\sigma_x}{\sigma_y} \tag{21}$$

where $\sigma_x$ and $\sigma_y$ are the uncertainties in signal and concentration, respectively. From this initial approach, a general expression was derived including all sensitivity equations and covering from zeroth-order (univariate calibration) to higher-order calibration models [51]. For PARAFAC calibration, $\mathrm{SEN}_n$ can be calculated as:

$$\mathrm{SEN}_n = \left\{ \boldsymbol{\delta}_n^{\mathrm{T}} \left[ \mathbf{Z}_{\mathrm{exp}}^{\mathrm{T}} \left( \mathbf{I} - \mathbf{Z}_{\mathrm{unx}} \mathbf{Z}_{\mathrm{unx}}^{+} \right) \mathbf{Z}_{\mathrm{exp}} \right]^{-1} \boldsymbol{\delta}_n \right\}^{-1/2} \tag{22}$$

where $\boldsymbol{\delta}_n$ is a $(N \times 1)$ vector selecting the constituent of interest, that is, contains a one for the $n$th constituent and zeros for the rest of $(N-1)$ constituents and $\mathbf{Z}_{\mathrm{exp}}$ is a function of the loading vectors of the PARAFAC model; the subscript "exp" means expected constituents, that is, the constituents that have been modeled in the calibration step:

$$\mathbf{Z}_{\mathrm{exp}} = m_n \left( \mathbf{C}_{\mathrm{exp}} | \otimes | \mathbf{B}_{\mathrm{exp}} \right) \tag{23}$$

The term $m_n$ is the slope of the pseudo-univariate calibration graph (sample scores vs. concentrations in the calibration samples), and it is used as a scaling factor with respect to the $n$th constituent concentration when the profile vectors (**B** and **C** loadings) are normalized to unit length. Finally, the matrix $\mathbf{Z}_{\mathrm{unx}}$ contains the profiles for the unexpected (or potentially interfering) constituents, present in a given test sample:

$$\mathbf{Z}_{\mathrm{unx}} = [\mathbf{c}_1 \otimes \mathbf{I}_{\mathrm{b}} | \mathbf{I}_{\mathrm{c}} \otimes \mathbf{b}_1 | \mathbf{c}_2 \otimes \mathbf{I}_{\mathrm{b}} | \mathbf{I}_{\mathrm{c}} \otimes \mathbf{b}_2 | \ldots] \tag{24}$$

where the vectors $\mathbf{b}_1$, $\mathbf{b}_2$, $\mathbf{c}_1$, $\mathbf{c}_2$, ... are the profiles of the unexpected constituents in the various data modes, $\mathbf{I}_{\mathrm{b}}$ $(J \times J)$ and $\mathbf{I}_{\mathrm{c}}$ $(K \times K)$ are unit matrices, and the numbers 1, 2, ... indicate the index of the unexpected constituent.

### 6.4.3 Selectivity

According to the latest IUPAC recommendation, "selectivity refers to the extent to which the method can be used to determine particular analytes in mixtures or matrices without interferences from other compounds of similar behaviour" [54]. In other words, selectivity is the ability of a method to accurately and specifically determine the analyte of interest in the presence of other compounds in a sample matrix under given stated conditions.

A suitable numerical selectivity estimate should meet several requirements that have been described in Faber *et al.* [55]. A criterion that meets such requirements is the ratio between two constituent sensitivity values: the sensitivity in a real sample (with other constituents) and the sensitivity when all other sample constituents are absent:

$$\mathrm{SEL}_n = \frac{\mathrm{SEN}_n(\mathrm{sample})}{\mathrm{SEN}_n(\mathrm{pure})} \tag{25}$$

Accordingly, for PARAFAC selectivity has been defined as:

$$\mathrm{SEL}_n = \frac{\left\{ \boldsymbol{\delta}_n^{\mathrm{T}} \left[ \mathbf{Z}_{\mathrm{exp}}^{\mathrm{T}} \left( \mathbf{I} - \mathbf{Z}_{\mathrm{unx}} \mathbf{Z}_{\mathrm{unx}}^{+} \right) \mathbf{Z}_{\mathrm{exp}} \right]^{-1} \boldsymbol{\delta}_n \right\}^{-1/2}}{m_n} \tag{26}$$

For the first-order **B** profiles recovered by the PARAFAC model, the first-order Lorber selectivities can also be calculated for constituent $n$ as the net analyte profile divided by the recovered PARAFAC profile [49,56]:

$$\mathrm{SEL}_{\mathrm{B},n} = \|\mathbf{b}_n *\| / \|\mathbf{b}_n\| \tag{27}$$

A similar expression exists for the **C** profiles recovered by the PARAFAC model.

### 6.4.4 Limit of Detection

The limit of detection is an important figure of merit in any analytical method. According to ISO norm 11843-1:1997 [57], *the limit of detection or minimum detectable net concentration, is the true net concentration of the analyte in the material to be analyzed which will lead, with probability* 1-$\beta$, *to the correct conclusion that the concentration in the analyzed material is different from that in the blank material.*

The limit of detection can be estimated in PARAFAC models from the pseudo-univariate calibration plot, which takes into account the uncertainty of the model (through the calibration line) and considers $\alpha$ and $\beta$ probabilities of error, following also the IUPAC recommendations [58,59]:

$$\mathrm{LOD}_n = \delta(\alpha, \beta, \nu) \frac{s_{\mathrm{res}}}{\mathrm{SEN}} \sqrt{1 + \frac{1}{I} + \frac{\bar{y}^2}{\sum_{i=1}^{I} (y_i - \bar{y})^2}} \tag{28}$$

where $y_i$ is the concentration of the given analyte in the $i$th standard, $\bar{y}$ is the average concentration of the calibration standards, $I$ is the total number of calibration standards, $s_{\mathrm{res}}$ is the standard error or standard deviation of the sample scores residuals, and SEN is the sensitivity of the method, that is, the slope of the pseudo-univariate calibration line. $\delta(\alpha,\beta)$ is a statistical term that takes into account the risks of both false-positive ($\alpha$) and false-negative ($\beta$) decisions when stating the analyte is present or not in a given sample, for a given number $v$ of degrees of freedom [60]. When $v$ is large enough, $\delta(\alpha,\beta)$ can be expressed as $(t_{\alpha,v}+t_{\beta,v})$, from the corresponding $t$-Student distribution

Alternatively, and based on the recent definition of sensitivity by Olivieri [51], given by Equation (22), the LOD for a given analyte has been defined as:

$$\mathrm{LOD}_n = 3.3\sqrt{(1+h_0)\left(\frac{\sigma_x}{\mathrm{SEN}_n}\right)^2 + h_0\sigma_{\mathrm{ycal}}^2} \tag{29}$$

where $h_0$ is the leverage for the blank sample, and the factor 3.3 is equal to $(t_{\alpha,v}+t_{\beta,v})$ for $\alpha=\beta=0.05$ and a large value of $v$. The above equation is very similar to Equation (28), except for the calculation of the $\mathrm{SEN}_n$ term and the inclusion of a term taking into account the variance of the concentrations of the calibration samples (second term inside the square root).

Similarly, the $\mathrm{LOQ}_n$, which is defined as the concentration level for which the relative prediction error is 10%, can be easily calculated by replacing the value of 3.3 by the value of 10.

It has been argued that the reduction of PARAFAC calibration results to the univariate case, using the pseudo-univariate calibration plot of the sample scores versus actual constituent concentrations, for estimating detection capabilities results in just average, but not sample specific, values of $\mathrm{LOD}_n$ and $\mathrm{LOQ}_n$, corresponding to samples with low analyte concentrations but with varying levels of other constituents [51].

## 7 APPLICATION EXAMPLE OF PARAFAC

This section illustrates a simple application of PARAFAC for the simultaneous determination of pesticides carbendazim, fuberidazole, and thiabendazole in synthetic samples by EEM fluorescence. Experimental details can be found in Rodríguez-Cuesta *et al.* [61].

Twelve mixtures with different concentration of carbendazim (0–100 ng ml$^{-1}$), fuberidazole (0–0.7 ng ml$^{-1}$), and thiabendazole (0–40 ng ml$^{-1}$) were prepared in methanol. A standard of each pure pesticide was also prepared. In each sample, a methanol blank was subtracted to remove the interfering Raman effect of the solvent. Spectra were measured from 310 to 370 nm in the emission mode and from 260 to 306 nm in the excitation mode. The excitation signals below emission or Rayleigh scattering were not recorded. The data matrices had dimensions $50\times38$ (Figure 10).

Several PARAFAC models were tested using different combinations of the calibration standards, to test the ability of PARAFAC for decomposing the response mixture. Figure 11 shows the recovered excitation and emission profiles for the three pesticides for the optimal PARAFAC model calculated with three factors and nonnegativity constraints in all the modes. The results from the pseudo-calibration line fitted with the scores in the sample mode and the

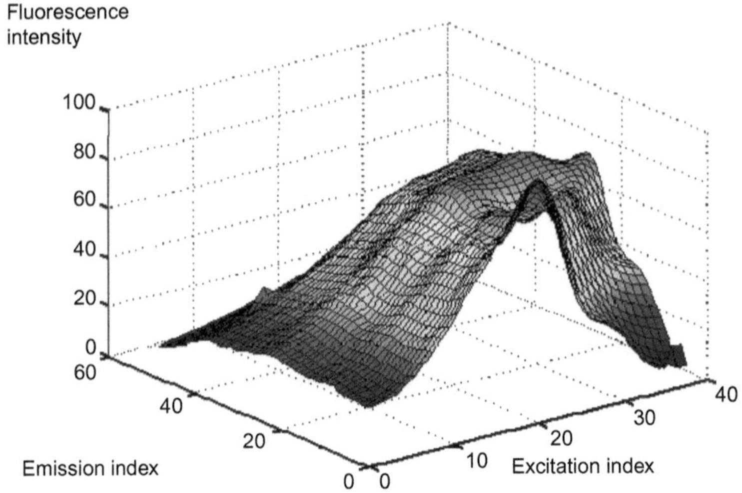


**FIGURE 10** EEM fluorescence profile of a mixture of carbendazim, fuberidazole, and thiabendazole.

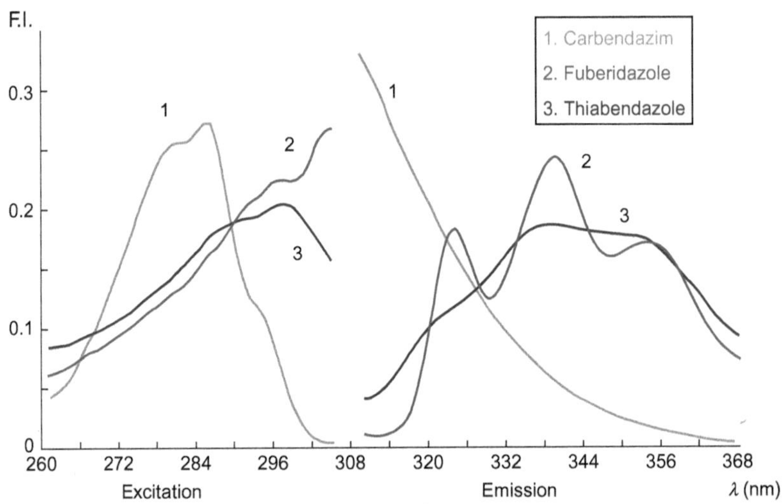


**FIGURE 11** Recovered excitation and emission spectra of carbendazim, fuberidazole, and thiabendazole from the optimal PARAFAC model, in fluorescence intensity (F.I.) units. *Reproduced with permission from Ref. [61].*

actual concentrations for the pesticides are in Table 1. The first thing to note is the similar excitation and emission profiles of fuberidazole and thiabendazole. This similarity is measured by the low first-order selectivity for fuberidazole (0.21 and 0.14 for the emission and excitation profiles, respectively)

**TABLE 1 Statistical Parameters and Figures of Merit from the Pseudo-Univariate Plot Between the *a*-Scores Calculated by the PARAFAC Model and the Actual Concentrations of the Three Pesticides**

| | Carbendazim | Thiabendazole | Fuberidazole |
|---|---|---|---|
| Intercept | $9.3 \times 10^{-3}$ | $-9.4 \times 10^{-3}$ | $-37.8 \times 10^{-3}$ |
| Sensitivity (slope, F.I. ml $ng^{-1}$) | $6.5 \times 10^{-3}$ | $16.4 \times 10^{-3}$ | $349 \times 10^{-3}$ |
| Standard error ($s_{res}$) | $22.3 \times 10^{-3}$ | $13.3 \times 10^{-3}$ | $8.3 \times 10^{-3}$ |
| Correlation coefficient ($r$) | 0.9969 | 0.999 | 0.9973 |
| Precision (%) | 3.4 | 2 | 3.4 |
| LOD (ng $ml^{-1}$) | 20 | 4.7 | 0.15 |

and for thiabendazole (0.20 and 0.12). The selectivity for carbendazim was higher (0.81 and 0.40) indicating that its spectra are the most different. Hence, if sensitivities were equal, the noise in the data would make it more difficult to correctly isolate the profiles of fuberidazole and thiabendazole, than those of carbendazim in mixtures of the three compounds. Fuberidazole was the most sensitive compound (highest slope of the pseudo-univariate curve), followed by thiabendazole and carbendazim, in agreement with the partial sensitivities calculated separately for each profile, as described by Faber *et al.* [56,62] and in Section 6.4: 15.4 and 7.6 for carbendazim, 19.2 and 11.2 for thiabendazole, and 356 and 241 for fuberidazole (F.I. ml $ng^{-1}$) (for the emission and excitation profiles, respectively). The correlation coefficients $r$ indicated good fit of the pseudo-univariate line. The precision for each pesticide was calculated in terms of concentration using Equation (19) as the standard deviation of the sample scores residuals for all standards divided by the sensitivity, SEN, i.e., the slope of the calibration line. The estimated precisions showed that the calibration results were in close agreement. The LOD for each pesticide was estimated from Equation (28), taking into account the uncertainty of the calibration line and $\alpha$ and $\beta$ probabilities of error. The LODs were in the order of magnitude of ng $ml^{-1}$. A set of test samples were predicted with this model using the loading matrices calculated from the calibration standards and interpolating in the corresponding regression line. The accuracy of the models was very good, with RMSEP values of 3.4% and 3.9% for thiabendazole and fuberidazole, respectively, although worse for carbendazim (5.6%), which increased the RMSEP to 5.6%. The second calibration strategy, commented in Section 5, was also studied, by including the test samples in the PARAFAC decomposition together with the calibration samples. The sample scores were recovered for each sample, although the calibration line was fitted only with the scores and concentrations of the calibration samples. The estimated excitation and emission spectral profiles were again correctly recovered, and the correlation coefficients with the reference spectra were very similar to those obtained with the model that did not include the test samples. This second model included more spectral variability, and although precision and limits of detection were of the same order for thiabendazole (1.9% and 4.4 ng $ml^{-1}$, respectively), they were slightly higher for carbendazim (3.7% and 22 ng $ml^{-1}$) and fuberidazole (6.4% and 0.29 ng $ml^{-1}$). The RMSEP values for the test samples were 3.7%, 4.8%, and 6.4% for carbendazim, thiabendazole, and fuberidazole, higher than those obtained by applying the previous calibration procedure, except for carbendazim.

## APPENDIX SOME USEFUL PRODUCTS AND OPERATORS IN THREE-WAY DATA ANALYSIS

The *outer product* of two vectors **u** and **v** is the rank one matrix given by:

$$\mathbf{u}\mathbf{v}^{\mathrm{T}} = \begin{bmatrix} u_1\mathbf{v}^{\mathrm{T}} \\ \vdots \\ u_I\mathbf{v}^{\mathrm{T}} \end{bmatrix}$$

Note that **u** and **v** may have different dimensions and that $\mathbf{u}\mathbf{v}^{\mathrm{T}} = (\mathbf{v}\mathbf{u}^{\mathrm{T}})^{\mathrm{T}}$

It is often convenient to transform a matrix to a vector that contains the elements of the matrix. The *vec-operator*, indicated by vec(·), transforms a matrix $\mathbf{A}$ $(m \times n)$ into a vector by stacking its columns one underneath the other. For example, if

$$\mathbf{A} = \begin{bmatrix} 1 & 3 & 5 \\ 2 & 4 & 6 \end{bmatrix}$$

Then

$$vec(\mathbf{A}) = \begin{bmatrix} 1 \\ 2 \\ 3 \\ 4 \\ 5 \\ 6 \end{bmatrix}$$

For two matrices $\mathbf{A} = [\mathbf{a}_1 \ \mathbf{a}_2 \ \ldots \ \mathbf{a}_n]$ $(m \times n)$ and $\mathbf{B} = [\mathbf{b}_1 \ \mathbf{b}_2 \ \ldots \ \mathbf{b}_q]$ $(p \times q)$, the *Kronecker product* of **A** and **B**, denoted by $\mathbf{A} \otimes \mathbf{B}$, is defined by the $mp \times nq$ matrix

$$\mathbf{A}\otimes\mathbf{B}=\begin{bmatrix} a_{11}\mathbf{B} & \cdots & a_{1n}\mathbf{B} \\ \vdots & \ddots & \vdots \\ a_{m1}\mathbf{B} & \cdots & a_{mn}\mathbf{B} \end{bmatrix}=\left[\mathbf{a}_1\otimes\mathbf{b}_1 \;\; \mathbf{a}_1\otimes\mathbf{b}_2 \;\; \cdots \;\; \mathbf{a}_n\otimes\mathbf{b}_{q-1} \;\; \mathbf{a}_n\otimes\mathbf{b}_q\right]$$

For two matrices $\mathbf{A}=[\mathbf{a}_1 \;\; \mathbf{a}_2 \;\; \ldots \;\; \mathbf{a}_n]$ and $\mathbf{B}=[\mathbf{b}_1 \;\; \mathbf{b}_2 \;\; \ldots \;\; \mathbf{b}_n]$ with the same number of columns, their *Khatri–Rao product*, denoted by $\mathbf{A}|\otimes|\mathbf{B}$, performs the following operation:

$$\mathbf{A}|\otimes|\mathbf{B}=[\mathbf{a}_1\otimes\mathbf{b}_1\,\mathbf{a}_2\otimes\mathbf{b}_2\cdots\mathbf{a}_n\otimes\mathbf{b}_n]=\left[vec\left(\mathbf{b}_1\mathbf{a}_1^{\mathrm{T}}\right)vec\left(\mathbf{b}_2\mathbf{a}_2^{\mathrm{T}}\right)\cdots vec\left(\mathbf{b}_n\mathbf{a}_n^{\mathrm{T}}\right)\right]$$

## REFERENCES


[1] Cattell RB. Parallel proportional profiles' and other principles for determining the choice of factors by rotation. Psychometrika 1944;9:267–83.

[2] Harshman RA. Foundations of the PARAFAC procedure: models and conditions for an 'Explanatory' multi-modal factor analysis. UCLA working papers in phonetics, vol. 16; 1970. p. 1–84.

[3] Carroll JD, Chang JJ. Analysis of individual differences in multidimensional scaling via an N-way generalization of Eckart-Young decomposition. Psychometrika 1970;35:283–319.

[4] Appellof CJ, Davidson ER. Strategies for analyzing data from video fluorometric monitoring of liquid-chromatographic effluents. Anal Chem 1981;53:2053–6.

[5] Smilde AK, Bro R, Geladi P. Multi-way analysis applications in the chemical sciences. 1st ed. Chichester: John Wiley & Sons; 2004.

[6] Olivieri AC, Escandar GM. Practical three-way calibration. Amsterdam: Elsevier; 2014.

[7] Bro R. Review on multiway analysis in chemistry—2000–2005. Crit Rev Anal Chem 2006;36:279–93.

[8] Amigo JM, Skov T, Bro R. ChroMATHography: solving chromatographic issues with mathematical models and intuitive graphics. Chem Rev 2010;110:4582–605.

[9] Massart DL, Vandeginste BGM, Buydens LMC, de Jong S, Lewi PJ, Smeyers-Verbeke J. Handbook of chemometrics and qualimetrics, part B. Amsterdam: Elsevier; 1998.

[10] Gemperline P, editor. Practical guide to chemometrics. 2nd ed. Boca Raton: CRC Press; 2006.

[11] Varmuza K, Filzmoser P. Introduction to multivariate statistical analysis in chemometrics. Boca Raton: CRC Press; 2009.

[12] Tomasi G, Bro R. Multilinear models: iterative methods. In: Brown S, Tauler R, Walczak B, editors. Comprehensive chemometrics: chemical and biochemical data analysis. Amsterdam: Elsevier; 2009. Vol. 2, p. 411–51.

[13] Kroonenberg PM. Applied multi-way data analysis. Hoboken, NJ: Wiley; 2008.

[14] Naes T, Risvik E. Multivariate analysis of data in sensory science. Amsterdam: Elsevier; 1996.

[15] Peré-Trepat E, Ginebreda A, Tauler R. Comparison of different multiway methods for the analysis of geographical metal distributions in fish, sediments and river waters in Catalonia. Chemom Intell Lab Syst 2007;88:69–83.

[16] Van Benthem MH, Lane TW, Davis RW, Lane PD, Keenan MR. PARAFAC modeling of three-way hyperspectral images: endogenous fluorophores as health biomarkers in aquatic species. Chemom Intell Lab Syst 2011;106:115–24.

[17] Tucker LR. Some mathematical notes on three-mode factor analysis. Psychometrika 1966;31:279–311.

[18] Bro R. Multi-way analysis in the food industry: models, algorithms, and applications. Ph.D. dissertation, Amsterdam, The Netherlands: University of Amsterdam; 1998.

[19] Murphy KR, Stedmon CA, Graeber D, Bro R. Fluorescence spectroscopy and multi-way techniques. PARAFAC. Anal Methods 2013;5:6557–66.

[20] Malinowski ER. Factor analysis in chemistry. 2nd ed. New York: John Wiley & Sons; 1991.

[21] de Juan A, Rutan SC, Tauler R. Two-way data analysis: multivariate curve resolution iterative resolution methods. In: Brown S, Tauler R, Walczak B, editors. Comprehensive chemometrics: chemical and biochemical data analysis. Amsterdam: Elsevier; 2009, p. 325–44.

[22] Bro R. PARAFAC. Tutorial and applications. Chemom Intell Lab Syst 1997;38:149–71.

[23] Kruskal JB. Three-way arrays: rank and uniqueness of trilinear decompositions, with application to arithmetic complexity and statistics. Linear Algebra Appl 1977;18:95–138.

[24] Faber NM, Bro R, Hopke PK. Recent developments in CANDECOMP/PARAFAC algorithms: a critical review. Chemom Intell Lab Syst 2003;65:119–37.

[25] Paulick C, Wright MN, Verleger R, Keller K. Decomposition of 3-way arrays: a comparison of different PARAFAC algorithms. Chemom Intell Lab Syst 2014;137:97–109.

[26] Tomasi G, Bro R. A comparison of algorithms for fitting the PARAFAC model. Comput Stat Data Anal 2006;50:1700–34.

[27] Jiang JH, Wu HL, Li Y, Yu RQ. Three-way data resolution by alternating slice-wise diagonalization (ASD) method. J Chemometr 2000;14:15–36.

[28] Chen ZP, Wu HL, Jiang JH, Li Y, Yu RQ. A novel trilinear decomposition algorithm for second-order linear calibration. Chemom Intell Lab Syst 2000;52:75–86.

[29] Paatero P. A weighted non-negative least squares algorithm for three-way 'PARAFAC' factor analysis. Chemom Intell Lab Syst 1997;38:223–42.

[30] Sanchez E, Kowalski BR. Generalized rank annihilation factor-analysis. Anal Chem 1986;58:496–9.

[31] Sanchez E, Kowalski BR. Tensorial resolution: a direct trilinear decomposition. J Chemometr 1990;4:29–45.

[32] Faber NM. Towards a rehabilitation of the generalized rank annihilation method (GRAM). Anal Bioanal Chem 2002;372:683–7.

[33] Comas E, Gimeno RA, Ferré J, Marcé RM, Borrull F, Rius FX. Quantification from highly drifted and overlapped chromatographic peaks using second-order calibration methods. J Chromatogr A 2004;1035:195–202.

[34] Tomasi G. Practical and computational aspects in chemometric data analysis. Ph.D. thesis, Frederiksberg, Denmark: The Royal and Agricultural University; 2006.

[35] Rinnan A, Riu J, Bro R. Multi-way prediction in the presence of uncalibrated interferents. J Chemometr 2007;21:76–86.

[36] Bro R, Smilde AK. Centering and scaling in component analysis. J Chemometr 2003;17:16–33.

[37] Guimet F. Olive oil characterization using excitation-emission fluorescence spectroscopy and three-way methods of analysis. Ph.D. thesis, Tarragona: Universitat Rovira i Virgili; 2005.

[38] Bro R, Sidiropoulos ND, Smilde AK. Maximum likelihood fitting using ordinary least squares algorithms. J Chemometr 2002;16:387–400.

[39] JiJi RD, Booksh KS. Mitigation of Rayleigh and Raman spectral interferences in multiway calibration of excitation-emission matrix fluorescence spectra. Anal Chem 2000;72:718–25.

[40] Lakowicz JR. Principles of fluorescence spectroscopy. 2nd ed. New York: Kluwer Academic Publishers/Plenum Press; 1999.

[41] Bylund D, Danielsson R, Malmquist G, Markides KE. Chromatographic alignment by warping and dynamic programming as a pre-processing tool for PARAFAC modelling of liquid chromatography-mass spectrometry data. J Chromatogr A 2002;961:237–44.
[42] Prazen BJ, Synovec RE, Kowalski BR. Standardization of second-order chromatographic/spectroscopic data for optimum chemical analysis. Anal Chem 1998;70:218–25.
[43] Bro R, Andersson CA, Kiers HAL. PARAFAC2—part II. Modelling chromatographic data with retention time shifts. J Chemometr 1999;13:295–309.
[44] Kiers HAL, Ten Berge JMF, Bro R. PARAFAC2—part I. A direct fitting algorithm for the PARAFAC2 model. J Chemometr 1999;13:275–94.
[45] Andersen CM, Bro R. Practical aspects of PARAFAC modeling of fluorescence excitation-emission data. J Chemometr 2003;17:200–15.
[46] Bro R, Kiers HAL. A new efficient method for determining the number of components in PARAFAC models. J Chemometr 2003;17:274–86.
[47] ISO Guide 3534-1:1993 Statistics—vocabulary and symbols. Part 1: probability and general statistical terms.
[48] Magnusson B, Ornemark U, editors. Eurachem guide: the fitness for purpose of analytical methods—a laboratory guide to method validation and related topics. 2nd ed.; 2014 Available from www.eurachem.org.
[49] Lorber A. Error propagation and figures of merit for quantification by solving matrix equations. Anal Chem 1986;58:1167–72.
[50] Ferré J, Faber NM. Net analyte signal calculation for multivariate calibration. Chemom Intell Lab Syst 2003;69:123–36.
[51] Olivieri AC. Analytical figures of merit: from univariate to multiway calibration. Chem Rev 2014;114:5358–78.
[52] Olivieri AC, Faber NM. A closed-form expression for computing the sensitivity in second-order bilinear calibration. J Chemometr 2005;19:583–92.
[53] Olivieri AC, Faber K. New developments for the sensitivity estimation in four-way calibration with the quadrilinear parallel factor model. Anal Chem 2012;84:186–93.
[54] Vessman J, Stefan RI, Van Staden JF, Danzer K, Lindner W, Burns DT, et al. Selectivity in analytical chemistry—(IUPAC Recommendations 2001). Pure Appl Chem 2001;73:1381–6.
[55] Faber NM, Ferré J, Boqué R, Kalivas JH. Quantifying selectivity in spectrophotometric multicomponent analysis. Trends Anal Chem 2003;22:352–61.
[56] Faber NM, Ferré J, Boqué R, Kalivas JH. Second-order bilinear calibration: the effects of vectorising the data matrices of the calibration set. Chemom Intell Lab Syst 2002;63:107–16.
[57] ISO Guide 11843-1:1997 Capability of detection—part 1: terms and definitions.
[58] Currie LA. Nomenclature in evaluation of analytical methods including detection and quantification capabilities (IUPAC Recommendations 1995). Pure Appl Chem 1995;67:1699–723.
[59] Ortiz MC, Sarabia LA, García I, Giménez D, Meléndez E. Capability of detection and three-way data. Anal Chim Acta 2006;559:124–36.
[60] Ortiz MC, Sarabia LA, Sánchez MS. Tutorial on evaluation of type I and type II errors in chemical analyses: from the analytical detection to authentication of products and process control. Anal Chim Acta 2010;674:123–42.
[61] Rodríguez-Cuesta MJ, Boqué R, Rius FX, Picón Zamora D, Martínez Galera M, Garrido Frenich A. Determination of carbendazim, fuberidazole and thiabendazole by three-dimensional excitation–emission matrix fluorescence and parallel factor analysis. Anal Chim Acta 2003;491:47–56.
[62] Faber K, Lorber A, Kowalski BR. Analytical figures of merit for tensorial calibration. J Chemometr 1997;11:419–61.

Chapter 2

# Usefulness of PARAFAC for the Quantification, Identification, and Description of Analytical Data

M.C. Ortiz*[,1], L.A. Sarabia[†], M.S. Sánchez[†], A. Herrero*, S. Sanllorente* and C. Reguera*

*Department of Chemistry (Analytical Chemistry), University of Burgos, Burgos, Spain

[†]Department of Mathematics and Computation (Statistics and Operations Research), University of Burgos, Burgos, Spain

[1]Corresponding author: e-mail: mcortiz@ubu.es

## Chapter Outline



## 1 THEORETICAL ELEMENTS

Although this chapter describes through specific cases some relevant aspects for the application of methods for analyzing three-way data, a reference to

Data Handling in Science and Technology, Vol. 29. http://dx.doi.org/10.1016/B978-0-444-63527-3.00002-3

books on the subject would be useful for the reader who wants to go deeper into these methods. Monographic basic books are [1–3] and more theoretical ones [4,5]. Special mention deserves the three chapters devoted to the subject in the reference book *Comprehensive Chemometrics* [6–8]. A book devoted to applications about calibration with three-way data is [9].

In general, we will assume that we have a three-way data tensor **X** $(I \times J \times K)$ made up by real numbers, $x_{ijk}$, $i=1, \ldots, I$; $j=1, \ldots, J$; $k=1, \ldots, K$.

For example, with techniques based on gas chromatography-mass spectrometry (GC-MS), each value $x_{ijk}$ would be the abundance recorded at the $i$th retention time, from the $j$th $m/z$ ratio and the $k$th sample. However, with a HPLC-DAD instrument, each value $x_{ijk}$ would be, for sample $k$ and time $i$, the absorbance at the $j$th wavelength. Another example comes from data of excitation-emission molecular fluorescence spectroscopy, in which case the tensor **X** is built by concatenating together the $I$ matrices, one for each sample, $i=1, \ldots, I$, made up by the fluorescence intensity recorded at the $\lambda_j$, $j=1, \ldots, J$, emission wavelength and $\lambda_k$, $k=1, \ldots, K$, excitation wavelength. Nevertheless, the usefulness of PARAFAC is not limited to data generated in the laboratory by hyphenated instrumental techniques, but it also allows monitoring the evolution in $K$ temporal moments of $I$ objects characterized by $J$ physicochemical variables measured on each object.

In practice, each profile (way or mode) of the tensor will be identified by its meaning, for example, chromatographic, spectral, or sample profiles for GC-MS data; or sample, emission, and excitation profiles for fluorescence spectroscopy data. Occasionally, for brevity, we will only make reference to $I$, $J$, or $K$ profiles. Figure 1 is a graphical description of a three-way tensor for GC-MS data.

## 1.1 PARAFAC Model

A PARAFAC model of rank $F$ for tensor $\mathbf{X} = \left(x_{ijk}\right)$ is written [10,11] as

$$x_{ijk} = \sum_{f=1}^{F} a_{if} b_{jf} c_{kf} + e_{ijk}, \quad i=1,2,\ldots,I; j=1,2,\ldots,J; k=1,2,\ldots,K \tag{1}$$

where $e_{ijk}$ are the elements of tensor **E**, which is the one that contains the residuals of fitting the model.

Equation (1) shows that PARAFAC is a trilinear model, in the sense that it is linear in each of the three profiles.

The $f$th factor of the PARAFAC model is defined by three vectors, loadings, $\mathbf{a}_f = \left(a_{1f}, \ldots, a_{If}\right)$, $\mathbf{b}_f = \left(b_{1f}, \ldots, b_{Jf}\right)$, and $\mathbf{c}_f = \left(c_{1f}, \ldots, c_{Kf}\right)$ in the $I$, $J$, and $K$ profiles, respectively. It is important to highlight that all three vectors define each PARAFAC factor, their product being a three-way tensor, $\mathbf{X}_f$, of size $I \times J \times K$. Except for the residual, the sum of the $F$ tensors of rank one, $\mathbf{X}_1 + \mathbf{X}_2 + \cdots + \mathbf{X}_F$, reconstructs the original tensor **X**. This is also illustrated in Figure 1.

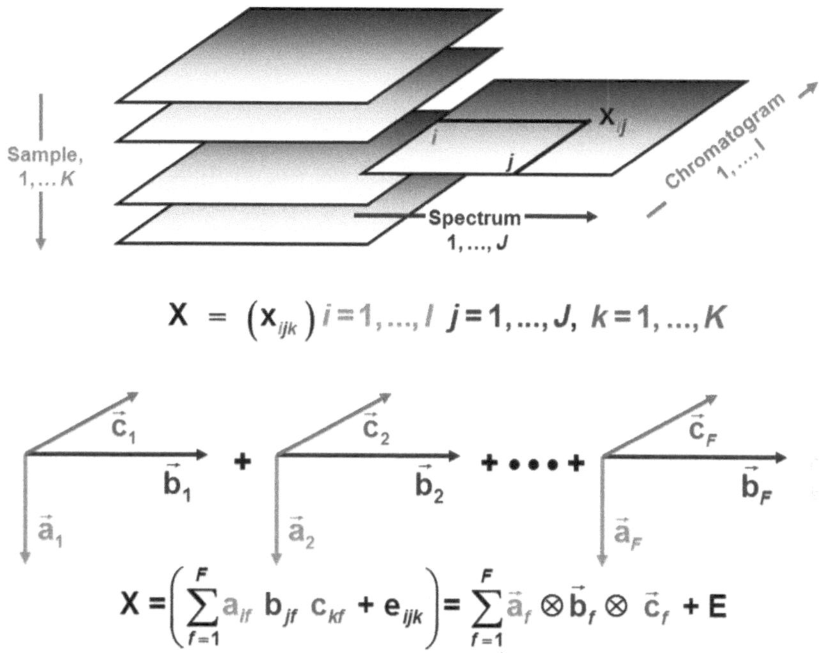


**FIGURE 1** Three-way data and the PARAFAC model.

Loadings $\mathbf{a}_f, \mathbf{b}_f,$ and $\mathbf{c}_f$ are the columns of matrices **A**, **B**, and **C** of size $I \times F$, $J \times F$, and $K \times F$, respectively. With these matrices, the PARAFAC model in Equation (1) can be written in terms of matrices $\mathbf{X}_k$, as in Equation (2), where each $\mathbf{X}_k$ is the $k$th slice of tensor **X** with size $I \times J$, also depicted in Figure 1 (sample slices)

$$\mathbf{X}_k = \mathbf{A}\mathbf{D}_k\mathbf{B}^{\mathrm{T}} + \mathbf{E}_k = c_{k1}\mathbf{a}_1\mathbf{b}_1^{\mathrm{T}} + \cdots + c_{kF}\mathbf{a}_F\mathbf{b}_F^{\mathrm{T}} + \mathbf{E}_k, \quad k = 1, 2, \ldots, K. \qquad (2)$$

In Equation (2), $\mathbf{D}_k$ is a diagonal square matrix ($F \times F$) whose non-null elements are the ones in the $k$th row of matrix **C**. Finally, $\mathbf{E}_k$ ($I \times J$) is the residual matrix. Consequently, each $\mathbf{X}_k$ is modeled by the same matrices of loadings **A** and **B** though with different weights, the ones in $\mathbf{D}_k$. As usual, the superscript "T" indicates transpose of the matrix or vector.

The equivalent way of expressing the PARAFAC model in Equation (2) describes very well its practical meaning. For example, with a tensor **X** of GC-MS data, each of the $F$ factors would correspond to a single analyte with chromatographic profile $\mathbf{a}_f$ and spectral profile $\mathbf{b}_f$. Except for a scale factor, these profiles are the chromatographic peak and mass spectrum of such analyte. The $F$ weights, $(c_{k1}, \ldots, c_{kF})$ in Equation (2) account for the different amount of each of the $F$ analytes in the $k$th sample. Now, Equation (2) is easily interpreted in each concrete case. Again, in the case of GC-MS data, Equation (2) means that in all the samples, each analyte has its own specific chromatographic peak and mass spectrum, but both common to all the samples, in such a way that the only difference from sample to sample is the amount of each individual analyte.

## 1.2 Uniqueness and Partial Uniqueness

Uniqueness means that PARAFAC model is the only possible combination of rank-one tensors that sums to **X** with the exception of the elementary indeterminacies of scaling and permutation. The permutation indeterminacy refers to the fact that the rank-one component tensors can be reordered arbitrarily, i.e., applying the same permutation to the columns of loadings matrices **A**, **B**, and **C**. The scaling indeterminacy refers to the fact that we can scale the individual loadings vectors, $\mathbf{a}_f$, $\mathbf{b}_f$, and $\mathbf{c}_f$ by factors $\alpha_f$, $\beta_f$, and $\gamma_f$ as long as their product $\alpha_f \times \beta_f \times \gamma_f$ is equal to 1.

The interest of the PARAFAC model is that the least squares solution is unique. That means that matrices **A**, **B**, and **C** computed for the *F* factors when fitting with least squares the model of Equation (1) is the only solution that gives the least sum of squares of residuals, so there is no other set of *F* factors (except for permutation and scaling) described with different loadings and with the same fit. In other words, any rotation of these factors cannot give the same residual sum of squares.

This uniqueness property is known as the "second-order advantage" [12] in chemical analysis and means that with a trilinear data tensor, built with the *K* slabs corresponding to *K–n* samples of known concentration of a target analyte together with the *n* slab of test samples, if one of the factors corresponds to the analyte, then the concentration of this analyte can be computed in the test sample even when there are interferents that were not in the calibration samples. This is so because the uniqueness property guarantees that there is only one sample profile linked to the analyte of interest, independently of the remaining factors.

With the *K–n* loadings corresponding to the samples, a calibration model is fitted, which can be used for computing the concentration corresponding to the loading of each test sample. From this point of view, Equation (2) of the PARAFAC model says that there will be a unique solution if all the matrices $\mathbf{X}_k$, $k = 1, 2, \ldots, K$, that make up tensor **X**, have the same pairs of profiles $\mathbf{a}_f, \mathbf{b}_f, f = 1, \ldots, F$ and differ from matrix to matrix in the amount of each analyte.

It should be noted that the uniqueness property is established in terms of the estimated values with the model of Equation (1); for example, in the case of GC-MS, there is no prior guarantee that necessarily one of the *F* factors corresponds to the target analyte, but if the PARAFAC model is correctly specified, with the appropriate number of factors and the trilinear structure is approximately valid, then it provides estimates of the underlying physical parameters generating data of tensor **X**, in the case of GC-MS data, the target analyte.

A sufficient condition for the PARAFAC model to provide unique estimates of the loadings **A**, **B**, and **C** is the one by Kruskal [13] that is based in the *k*-rank definition. Given a matrix **M**, its Kruskal rank is the largest *k* so that every set of *k* columns of **M** is linearly independent. Suppose that matrices **A**, **B**, and **C** have Kruskal ranks $k_{\mathbf{A}}$, $k_{\mathbf{B}}$, and $k_{\mathbf{C}}$ respectively. If

$$k_{\mathbf{A}} + k_{\mathbf{B}} + k_{\mathbf{C}} \geq 2F + 2 \tag{3}$$

then the PARAFAC model of Equation (1) is unique among all decompositions with at most $F$ factors. A model with a single factor is always unique regardless of the above. The Kruskal condition is not only sufficient but also necessary for $F = 2$ or 3 [14]. However, this is not so for other values of $F$.

In the specific situation where **A** and **B** have full column rank, the condition of Kruskal takes a particularly simple form. As, in this case, $k_{\mathbf{A}} = k_{\mathbf{B}} = F$, then the matrix **C** needs to have at least *k* rank equal to 2, which means that every pair of columns of **C** constitutes a matrix of rank two. Equivalently, **C** has no mutually proportional columns. This condition also rules out the possibility of having a zero column in **C**. For practical applications, this is a particularly useful result because having **A** and **B** of full column rank is satisfied in nearly all applications.

When some of the factors (triads of vectors) can be obtained in a unique way and other factors cannot, that is, they are solved with ambiguity, it is said that there is a solution with partial uniqueness. However, still the PARAFAC solution can be of interest, even it can have the second-order advantage. Suppose that we have a sample of an analyte and its excitation-emission matrix, EEM, has been recorded at 5 excitation wavelengths and 10 emission wavelengths and a test sample with the same analyte and two interferents [2]. Theoretically, tensor **X** of size $(5 \times 10 \times 2)$ has three factors. Columns of **A** $(5 \times 3)$ are the estimates of the excitation spectra of the three compounds, the ones of **B** $(10 \times 3)$ are those of emission, and finally matrix **C** $(2 \times 3)$ contains the three concentration profiles. Assuming that the excitation and emission spectra of the three fluorophores are not linearly related, that is, $\text{rank}(\mathbf{A}) = k_{\mathbf{A}} = \text{rank}(\mathbf{B}) = k_{\mathbf{B}} = 3$, then the structure of matrix **C** is

$$\begin{pmatrix} m & 0 & 0 \\ n & p & q \end{pmatrix}.$$

The first row has two zeros, because the two fluorophores (interferents) in the test sample are not in the calibration standards and, consequently $k_{\mathbf{C}} = 1$ because the second and third columns are linearly dependent. Therefore, the Kruskal condition, Equation (3), is not fulfilled because $3 + 3 + 1 = 7$ is less than $2 \times 3 + 2 = 8$. As for three factors the Kruskal condition is necessary and sufficient, there is not unique solution. Nevertheless, the three profiles associated to the calibrated analyte will be obtained without rotational ambiguity, although this is not so for the interferents. This partial uniqueness significantly expands the applicability of PARAFAC.

There is not a complete characterization of the conditions of uniqueness when the matrices of loadings of one or more modes do not have full rank. However, in Ref. [15], there is a list of six principles (or rules) that can guide researchers on full or partial uniqueness in their data. With the idea of evaluating partial uniqueness for specific data with three PARAFAC factors, the "bands" that include the effect of the lack of uniqueness on the estimation of the profiles can be visualized [16], an application of this strategy to analyze to what extent the uniqueness of the PARAFAC solution degrades as the profiles of two factors overlap is in Ref. [17].

## 1.3 Tucker3 Model

Tucker proposed several models for $n$-way data [18] that, later, have been widely analyzed [4]. Here, only Tucker3 model will be introduced since structurally PARAFAC is a particular case and, besides, Tucker3 has quite practical utility. Given the tensor $\mathbf{X}=\left(x_{ijk}\right)$, Tucker3 ($D$, $E$, $F$) model of $\mathbf{X}$, with $D$, $E$, and $F$ factors in the three different ways, is

$$x_{ijk}=\sum_{d=1}^{D}\sum_{e=1}^{E}\sum_{f=1}^{F}a_{id}b_{je}c_{kf}g_{def}+e_{ijk},\quad i=1,2,\ldots,I;j=1,2,\ldots,J;k=1,2,\ldots,K \tag{4}$$

where $e_{ijk}$ are the residuals of the model, $a_{id}$, $b_{je}$, and $c_{kf}$ are the elements of the matrices of loadings $\mathbf{A}$, $\mathbf{B}$, and $\mathbf{C}$ with size $I\times D$, $J\times E$, and $K\times F$, respectively, and $\mathbf{G}=\left(g_{def}\right)$ is the tensor core of size $D\times E\times F$. The PARAFAC model is a particular case of the model in Equation (4), the one obtained when the number of factors is the same in every way, (i.e. $D=E=F$) and the elements of tensor core $\mathbf{G}$ are equal to 1 when $d=e=f$ (the elements in the superdiagonal) and equal to 0 otherwise. A nondiagonal core tensor explicitly means that there are interactions between factors in the model. The Tucker3 model does not have the uniqueness property, and thus, it is usual to rotate the core matrix to achieve better interpretation of factors.

## 1.4 PARAFAC2 Model

While PARAFAC applies the same factors ($\mathbf{A}$, $\mathbf{B}$) across a parallel set of matrices $\mathbf{X}$, as shown in Equation (2), PARAFAC2 instead applies the same factor along one mode and allows the other factors in matrix to vary that. An advantage of PARAFAC2 is that, further to model a regular three-way tensor with fewer constraints than PARAFAC, it can also be applied to a collection of matrices with varying sizes in one mode, e.g., same column size but different row size.

The PARAFAC2 model for $F$ factors has the form in Equation (5):

$$\mathbf{X}_k=\mathbf{A}_k\mathbf{D}_k\mathbf{B}^{\mathrm{T}}+\mathbf{E}_k,\quad k=1,2,\ldots,K \tag{5}$$

where $\mathbf{A}_k$ is a $I_k \times F$ matrix, with $I_k$ rows not necessarily equal for all $k=1, \ldots, K$; $\mathbf{D}_k$ is a $F \times F$ diagonal matrix of weights with the same meaning as in Equation (2); and $\mathbf{B}$ is a $J \times F$ loadings matrix that does not vary with $k$. The model in Equation (5) does not have the uniqueness property. To obtain it, a restriction is added [19], which is imposing that the cross product $\boldsymbol{\phi} = \mathbf{A}_k^{\mathrm{T}}\mathbf{A}_k$ is the same for $k=1, \ldots, K$. With this restriction, matrix $\mathbf{A}_k$ in Equation (5) can be decomposed as $\mathbf{A}_k = \mathbf{Q}_k\mathbf{S}$. Therefore, the PARAFAC2 model in Equation (5) can be expressed as

$$\mathbf{X}_k = \mathbf{Q}_k\mathbf{S}\mathbf{D}_k\mathbf{B}^{\mathrm{T}} + \mathbf{E}_k, \quad k = 1,2,\ldots,K \tag{6}$$

where $\mathbf{Q}_k$ is an orthonormal $I_k \times F$ matrix and $\mathbf{S}$ is an $F \times F$ matrix that does not vary along slices. The matrix $\mathbf{Q}_k$ handles what is specific for each sample and $\mathbf{S}$ handles what is common to all of them. Algorithms for fitting a PARAFAC2 model work by either fitting the cross products of the covariance matrices, as proposed by Harshman in his pioneering work [19] (indirect fitting), or fitting Equation (6) to the original data itself (direct fitting) [20]. This last alternative offers several advantages.

The PARAFAC2 model has a unique solution if the loadings matrix $\mathbf{B}$ has full column rank, the cross product matrix $\boldsymbol{\phi}$ is positive semidefinite (has no negative eigenvalues), and at least one of the matrices $\mathbf{D}_k$, $k=1, 2, \ldots, K$ is nonsingular. In addition, Bro *et al.* [20] provide simulation results that suggest that the direct fitting PARAFAC2 solutions seem to be unique whenever $K \geq 4$ for any value of $F$. Details about uniqueness with indirect fit can be seen in Ref. [21].

The PARAFAC2 decomposition technique, as defined in Equation (6), overcomes the inequality in the chromatographic profiles and allows some deviation in them [22,23].

## 1.5 Model Complexity

The uniqueness property of PARAFAC and PARAFAC2 models and, consequently, the identification of the underlying phenomenon in the data tensor caused an increasing use of them, especially in Chemometrics. Nevertheless, it is still a challenge when using both of them the determination of the appropriate number of components. It is possible to use resampling techniques such as cross-validation [24] or split-half analysis [25], but both procedures are often unfeasible when the sample mode includes calibration standards, which are few and different to each other, making it difficult the division of the tensor in similar parts. Besides, they can be unattractive because of the intense computations involved and owing to the nonsequential fitting of the PARAFAC (PARAFAC2) model. Therefore, a diagnostic based on a single analysis, the Core Consistency Diagnostic index, CORCONDIA, has been developed for this purpose in Ref. [26] for PARAFAC and in Ref. [27] for PARAFAC2.

To understand the ideas under the index and its adaptation to PARAFAC2, let us assume that a PARAFAC model has been fitted to an experimental tensor **X**. The model is given by the loadings matrices **A**, **B**, and **C**. The PARAFAC core tensor is the core tensor constructed from matrices **A**, **B**, and **C** to see if the PARAFAC loadings vectors alone describe the data almost as well as the model involving interactions of these vector loadings does. For this task, the full Tucker3 model is fitted to the data tensor **X**, but using the loadings found by PARAFAC, obtaining the optimal tensor **G** (size $F \times F \times F$) that provides the best fit, by least squares, of the experimental data tensor.

The Tucker3 core tensor, **G**, can be considered as the regression of the experimental data tensor, **X**, on the subspace spanned by the loadings matrices, **A**, **B**, and **C**. Therefore, **G** is the optimal representation of the data tensor in this subspace whereas the PARAFAC model is a constrained regression in the same subspace, because specifically disregards any variation associated with off-super-diagonal core elements and, therefore, its core tensor is a $(F \times F \times F)$ tensor $\mathbf{T} = (t_{ijk})$ having ones in the superdiagonal ($t_{ijk} = 1$ if $i = j = k$) and zeros outside.

A way to assess whether **G** and **T** are similar is to quantify the Frobenius distance between one another. The CORCONDIA index in Equation (7) is this distance, computed as relative similarity (in percentage) against the norm of **T**.

$$\text{CORCONDIA} = 100\left(1 - \frac{\sum_{i=1}^{F}\sum_{j=1}^{F}\sum_{k=1}^{F}\left(g_{ijk} - t_{ijk}\right)^2}{\sum_{i=1}^{F}\sum_{j=1}^{F}\sum_{k=1}^{F}\left(t_{ijk}\right)^2}\right) \tag{7}$$

A CORCONDIA close to 100% implies an appropriate model, because **G** and **T** are similar, that is, the best fit (Tucker3) is similar to the PARAFAC fit and the dimension-wise trilinear combination of vector loadings are the only entities needed for describing the data tensor. Also, it says that the contribution of the interactions between one another is not appreciable. Moving away from 100%, there is not a threshold value accepted by all practitioners, but the decision about the closeness to 100% is to be made relative to the changes compared with models with fewer components.

In Ref. [1], Bro (p. 115–118) also proposed a core consistency plot with the values of the **G** tensor on the vertical axis and the core elements themselves on the horizontal axis with the superdiagonal elements plotted first.

It has already been shown in Refs. [26,28] that the core consistency greatly improves when the data that do not follow a trilinear structure are deleted from **X** (e.g., removing the Rayleigh scatter or the null fluorescent intensity at emission below the recorded excitation wavelength in an EEM tensor). That is, if an experimental tensor comprises a tensor with trilinear structure and another which is not trilinear, then it is possible to recover the trilinearity by removing the tensor with non-PARAFAC structure.

The CORCONDIA index has been extended in Ref. [27] for the PARAFAC2 model. With this purpose, Equations (5) and (6) are rewritten by considering that the first mode is the one with equal loadings for all samples, while the second changes from sample to sample. This implies no loss of generality because it is the researcher who decides the position of each mode in the experimental data. Now, PARAFAC2 equation would be

$$\mathbf{X}_k = \mathbf{A}\mathbf{D}_k\mathbf{B}_k^{\mathrm{T}} + \mathbf{E}_k,\quad k = 1,2,\ldots,K \tag{8}$$

where $\mathbf{A}$ is a $I \times F$ matrix, $\mathbf{D}_k$ is a $F \times F$ diagonal matrix of weights with the same meaning as in Equation (2), and $\mathbf{B}_k$ is a $J_k \times F$ loadings matrix that varies with $k$. The restriction to guarantee uniqueness in the solution is that the cross product $\mathbf{B}_k^{\mathrm{T}}\mathbf{B}_k$ is the same for $k=1, \ldots, K$. Similar to Equation (6), now $\mathbf{B}_k = \mathbf{Q}_k\mathbf{S}$ with $\mathbf{Q}_k$ orthonormal; therefore, $\mathbf{Q}_k^{\mathrm{T}}\mathbf{Q}_k = \mathbf{I}$.

As $\mathbf{X}_k \simeq \mathbf{A}\mathbf{D}_k\mathbf{B}_k^{\mathrm{T}}$ substituting $\mathbf{B}_k$ by $\mathbf{Q}_k\mathbf{S}$, we have $\mathbf{X}_k \simeq \mathbf{A}\mathbf{D}_k(\mathbf{Q}_k\mathbf{S})^{\mathrm{T}} = \mathbf{A}\mathbf{D}_k\mathbf{S}^{\mathrm{T}}\mathbf{Q}_k^{\mathrm{T}}$ and multiplying on the right by $\mathbf{Q}_k$ we have $\mathbf{X}_k\mathbf{Q}_k \simeq \mathbf{A}\mathbf{D}_k\mathbf{S}^{\mathrm{T}}\mathbf{Q}_k^{\mathrm{T}}\mathbf{Q}_k = \mathbf{A}\mathbf{D}_k\mathbf{S}^{\mathrm{T}}$. Therefore, the PARAFAC2 model is a PARAFAC model but for matrices $\mathbf{Y}_k = \mathbf{X}_k\mathbf{Q}_k$; $k=1, \ldots, K$, instead of $\mathbf{X}_k$. When the orthonormal $\mathbf{Q}_k$ matrices are known, we can rephrase the PARAFAC2 model as a PARAFAC model in terms of the frontal slabs of data, $\mathbf{X}_k$, "compressed" with their own specific $\mathbf{Q}_k$ matrix; hence, a PARAFAC model can be fitted on a data array of $\mathbf{Y}_k$ slabs. This is interesting in understanding how PARAFAC2 handles changes in the second mode, and it is also useful for the purpose of developing a core consistency measure for PARAFAC2 models.

## 2 QUANTIFICATION AND IDENTIFICATION WITH EXCITATION-EMISSION MOLECULAR FLUORESCENCE DATA

In fluorescence spectroscopy, for diluted solutions, the quantity of light emission measured is separately linear in the number of photons absorbed and in the fraction of photons absorbed that lead to emission at a specific wavelength. Therefore, when sample $i$th, $i=1, \ldots, I$ with $F$ fluorophores is illuminated with light at excitation wavelength $\lambda_k$, $k=1, \ldots, K$ and the subsequent emission of light is measured at wavelength $\lambda_j$, $j=1, \ldots, J$, then, in theory [29], the model for the corresponding fluorescence intensity $x_{ijk}$ is

$$x_{ijk} = \sum_{f=1}^{F} \alpha_{if}\beta_{jf}\delta_{kf},\quad i=1,2,\ldots,I; j=1,2,\ldots,J; k=1,2,\ldots,K \tag{9}$$

where $\alpha_{if}$ is linearly related with the concentration of fluorophore $f$ in the $i$th sample, $\beta_{jf}$ is the relative emission at wavelength $\lambda_j$, and $\delta_{kf}$ is the relative amount of light absorbed at the excitation $\lambda_k$, of fluorophore $f$. We will see that Equation (9) is the PARAFAC model of Equation (1). It must not be forgotten that there are some areas of the spectrum where this is not verified: emission is not possible at wavelengths below the excitation wavelength and Rayleigh and Raman scattering do not conform to the trilinear model. To obtain an appropriate PARAFAC model where the components primarily reflect (low-rank) trilinear variation in the data, these nontrilinear regions should be deleted (set to missing values) [30]. Other alternatives consist of modeling the first-order Rayleigh scatter as a separate component in the PARAFAC decomposition [31] by means of interpolation [32] or using Gaussian peaks [33].

Using EEMs with a PARAFAC model allows recovering the specificity needed to design analytical procedures that are cheap, fast, and simple, like in the case of the determination of enrofloxacin in poultry feeding water [34]. The model in Equation (9) is valid as long as there is not quenching effect. Due to the fact that many real matrices have native fluorescence, the determination of target analytes in these matrices has important difficulties. In Ref. [35], an experimental strategy is proposed that includes the use of a three-way PARAFAC model for the determination of tetracyclines in milk. However, the quenching problem can be handled by the addition of a fourth way to account for the variation of the amount of quencher [29]. A four-way model has also been used [36] for the determination of tetracycline in tea, which shows a strong quencher effect caused by other fluorophores present in the matrix, and again in Ref. [37] for the quantitative analysis of aromatic amino acids in human plasma using intrinsic fluorescence. The usefulness of four-way PARAFAC models with EEMs has been little explored, but it extends the possibilities of excitation-emission fluorescence, for example, hyphenating it with a photochemically induced kinetics [38]. On the other side, it also allows identification of the nontrilinear component so that its contribution can be subtracted from the signal [28] and then quantification is possible by using the (unique) factors associated to carbamate pesticides in dried lime tree.

The case study in this section is extracted from Ref. [39], and it is devoted to explain how to fit a global PARAFAC model with the data of excitation-emission molecular fluorescence recorded in a complex matrix. In particular, ciprofloxacin is determined in different samples of human urine, one of which has mesalazine, and the use of fluorescence signals allows avoiding the need of a prior extraction step.

Both the urine and the mesalazine have fluorophores which give fluorescence signal in the same range of wavelengths as the ciprofloxacin does. Thus, despite molecular fluorescence spectroscopy is a technique easy to use and nonexpensive with good sensitivity and low limit of detection, the signals might not be specific, like in this case where the matrix of the samples provides also fluorescence.

To highlight different characteristics of the problem as well as the way PARAFAC can be used to handle them, the study is developed as in three different cases.

## 2.1 Case I: Determination of Ciprofloxacin (Added in Human Urine)

The goal is to determine ciprofloxacin in human urine without the extraction step prior to the determination. With that purpose, excitation-emission fluorescence spectra are recorded for a total of 68 samples, all of them spiked with ciprofloxacin in the range 100–240 μg $L^{-1}$ (every 20 μg $L^{-1}$) and distributed as follows: 16 calibration standards of water spiked with ciprofloxacin and 48 samples of human urine spiked with ciprofloxacin. The urine samples corresponded to three different people, 17 samples of a young person, another 17 for urine of an old person, and the remaining 17 in urine of a woman. In each series, the first 8 samples were replicated twice and the last one is always a blank.

The excitation-emission matrices are stacked one above another, like in the illustrative image in Figure 2, to make up the tensor for computing a PARAFAC model according to Equation (1). The size of this tensor is $68 \times 201 \times 15$, meaning that the sample mode has 68 samples, emission

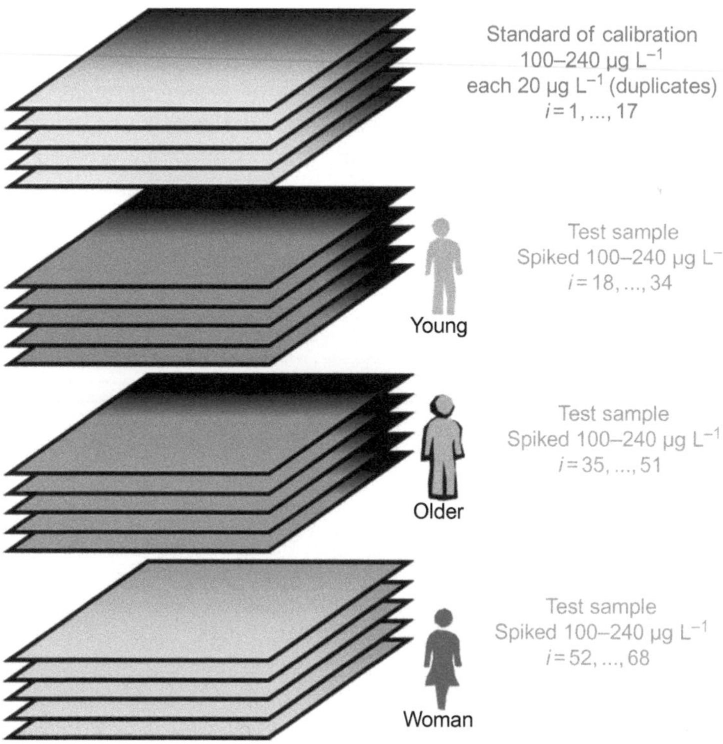


**FIGURE 2** Data tensor for the PARAFAC decomposition in case I.

spectra are recorded (every 1 nm) between 350 and 550 nm, and excitation spectra between 240 and 310 nm (each 5 nm).

A two-factor PARAFAC decomposition of the tensor explains 99.8% of variance with a CORCONDIA index of 99.99%. Identification of the factors is made by computing the correlation coefficient between the emission and excitation spectra extracted by PARAFAC and the spectra (whether excitation or emission spectra) of a standard of ciprofloxacin.

The loadings of the two factors on the emission and excitation modes are depicted in Figure 3A and B, respectively. The factor whose loadings are depicted in blue (dark gray in the print version) continuous line is related to ciprofloxacin. The correlation between the PARAFAC loadings and the experimental true spectra of ciprofloxacin is equal to 0.99 for both emission and excitation spectra. The other factor, in discontinuous green (gray in the print version) line, is related to urine and the correlation coefficients are also 0.99 for both modes.

The loadings on the sample mode for both factors are depicted in Figure 4: blue (dark gray in the print version) circles for the loadings of the factor identified as ciprofloxacin and green (gray in the print version) stars for the ones related to urine. The first loadings (samples 1–17, see Figure 2) are for the calibration standards in water, then the 17 samples of the urine of the young person, then the urine of the older man, and then the samples of the urine of the woman. The loadings corresponding to analytical blank samples recorded at the end of each series are also seen in Figure 4.

The pattern shown by the blue (dark gray in the print version) circles reproduces the design, samples in duplicate with increasing quantity of ciprofloxacin, and the blank samples with null value for the corresponding loading.

On the other hand, the green (gray in the print version) stars in Figure 4 show also a clear pattern: they are null for the calibration standards and non-null, although constant, in the rest of the series of samples. Clearly, there

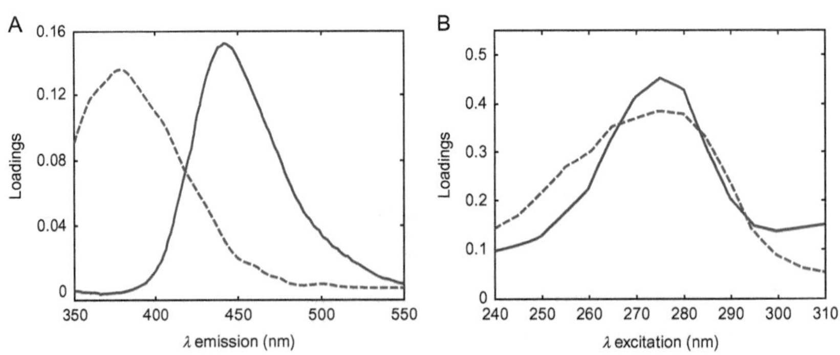


**FIGURE 3** Loadings on the (A) emission and (B) excitation profiles. Blue (dark gray in the print version) continuous line is for the factor identified as ciprofloxacin and green (gray in the print version) dashed lines are for the factor related to urine.

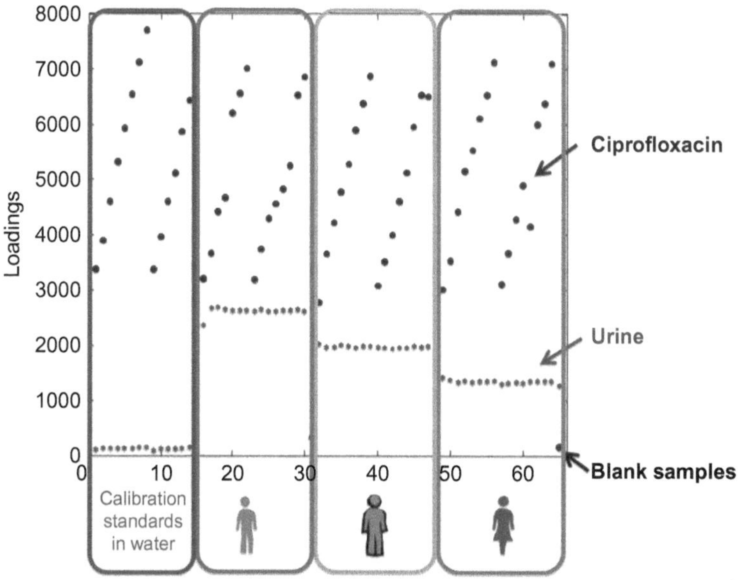


**FIGURE 4** Loadings on the sample mode. Blue (dark gray in the print version) circles are for the factor identified as ciprofloxacin and green (gray in the print version) stars are for the factor related to urine. Different frames are used to separate the different series of samples in the order of Figure 2.

is a compound which is present in the urine at the same concentration in each sample of urine but different from person to person.

Once the analyte of interest has been identified, its loadings on the sample mode for the calibration standards are used for quantification of ciprofloxacin, by fitting and validating a calibration line "sample mode loadings" ($y$) versus "true concentration" ($x$). The resulting calibration line is $y=206.2+31.4x$ with standard error of estimation of $s_{y|x}=62.8\ \mu g\ L^{-1}$ and coefficient of determination $R^2=0.998$. The mean of the absolute value of relative error (17 samples) in calibration is 0.9%.

With the calibration line and the loadings of the samples of urine, the concentration of ciprofloxacin in them is computed obtaining that the mean of the absolute values of the relative errors of test samples is 9.8% ($n=51$).

## 2.2 Case II: Determination of Ciprofloxacin (Added in Human Urine of a Patient Being Treated with Mesalazine)

In this case, the goal is still the determination of ciprofloxacin in human urine without the extraction step, but new samples of urine come from a person who is being treated with mesalazine, also fluorescent, that will act as an interferent in the determination.

Likewise, EEMs are recorded for additional 17 samples of urine: 16 of them spiked with ciprofloxacin in the same range of concentration 100–240 μg $L^{-1}$ (every 20 μg $L^{-1}$, two replicates in each level of concentration) plus a blank sample. At the bottom left part of Figure 5, the landscape of one of the samples spiked with 240 μg $L^{-1}$ of ciprofloxacin is plotted, showing huge differences with the other landscapes reproduced in Figure 5 (left column). The high overlap between the signal of the mesalazine and the rest is evident.

The excitation-emission fluorescence spectra of the new 17 samples are added to the tensor made up with the previous 68 samples, as it is illustrated in Figure 5 (right part). Consequently, the tensor now is $85 \times 201 \times 15$.

The PARAFAC model of this tensor has three factors, with CORCONDIA equal to 98.5% and explained variance equal to 99.9%. Figure 6 contains the loadings on the emission and excitation profiles of the three factors. The factors were identified, by using the correlation coefficient between the pure spectra and the PARAFAC loadings on the emission and excitation modes, as ciprofloxacin (in blue, dark gray in the print version, solid line), urine (in green, gray in the print version, dashed line), and mesalazine (in red, light

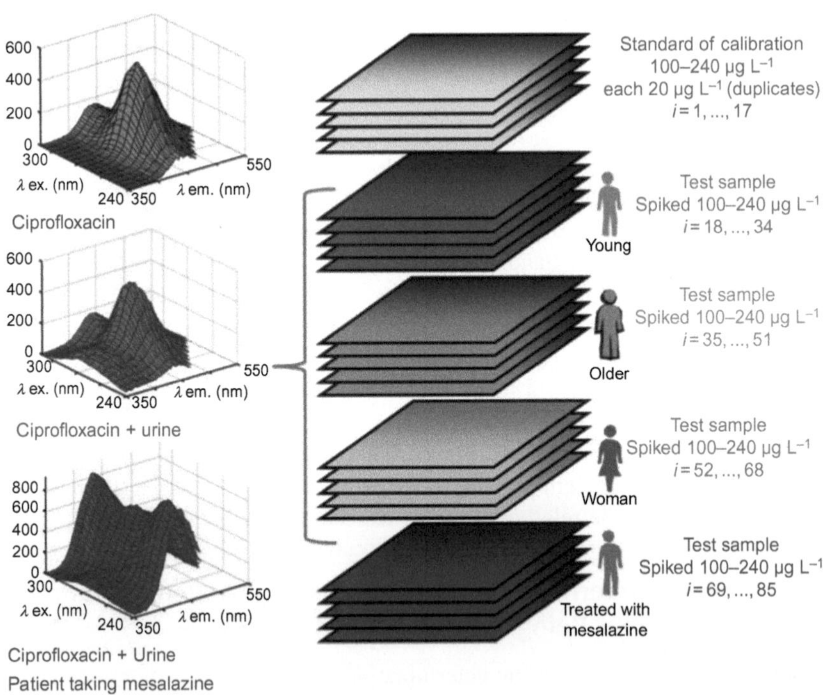


**FIGURE 5** Tensor for the PARAFAC decomposition in case II. On the left, different landscapes: samples in water, urine, and urine from the patient being treated with mesalazine, all of them spiked with 240 μg $L^{-1}$ of ciprofloxacin.

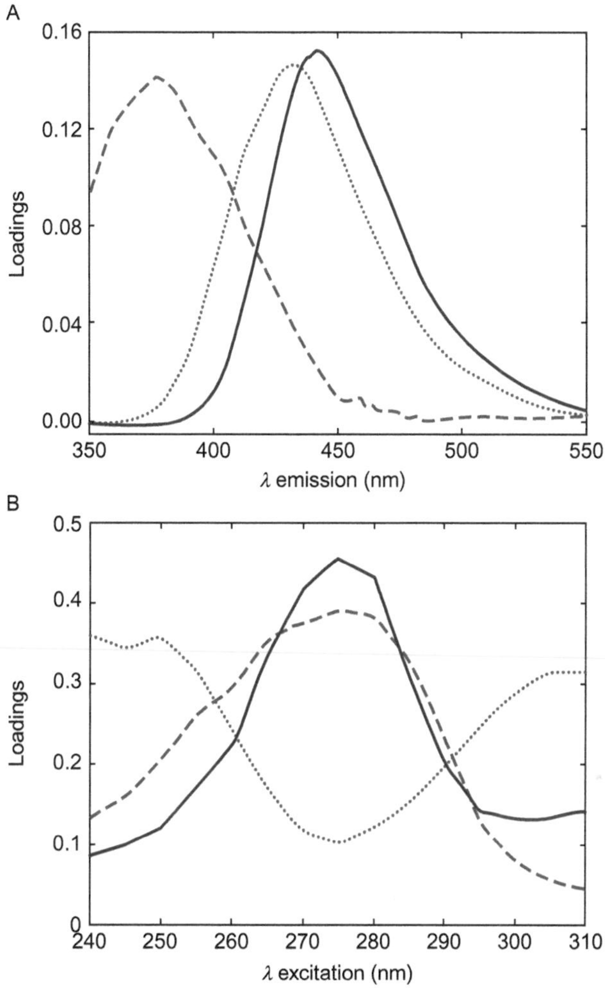


**FIGURE 6** Loadings on the (A) emission and (B) excitation profiles. Blue (dark gray in the print version) continuous line is for the factor identified as ciprofloxacin, green (gray in the print version) dashed line is for the factor related to urine, and red (light gray in the print version) dotted line is for the factor identified as mesalazine.

gray in the print version, dotted line), because when comparing emission spectra, the correlation coefficients were 0.999, 0.984, and 0.999 with the pure emission spectra of ciprofloxacin, urine, and mesalazine, respectively; whereas the correlation coefficients between the excitation spectra were 0.998, 0.999, and 0.997, again, for ciprofloxacin, urine, and mesalazine, respectively.

The graph of the loadings on the sample mode, Figure 7, shows the same pattern as in the previous case, blue (dark gray in the print version) circles

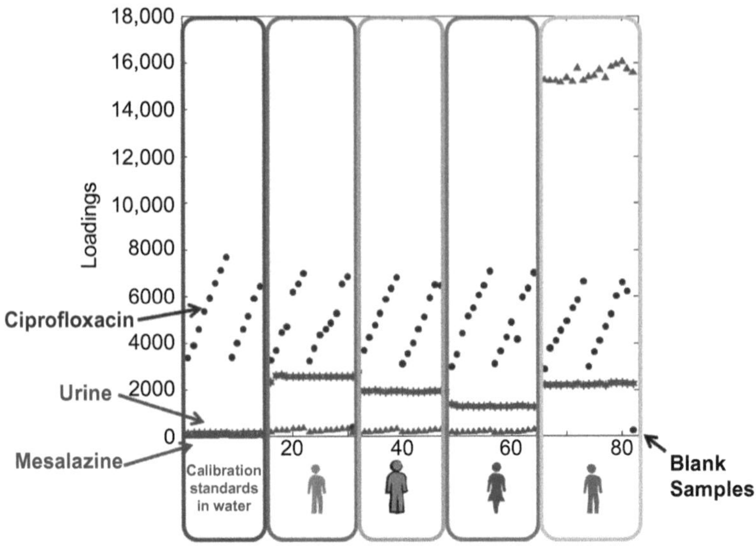


**FIGURE 7** Loadings on the sample mode. Blue (dark gray in the print version) circles are for the factor identified as ciprofloxacin, green (gray in the print version) stars are for the factor related to urine, and red (light gray in the print version) triangles are for the factor identified as mesalazine. Different frames are used to separate the different series of samples in the order of Figure 5.

related to ciprofloxacin increase in duplicates and similarly in all the series, except for the blank samples; the green (gray in the print version) stars related to the urine are also null in the calibration standards and non-null and different (though constant) in the urine of each person. The loadings of the third factor, identified as mesalazine (red, light gray in the print version, triangles), are clearly null in the calibration standards as well as in the urine samples of the three previous people, and non-null and large for the person who was taking it (samples 69–85).

With the loadings of the factor related to ciprofloxacin on the sample mode for the calibration standards, the calibration line "loadings" ($y$) versus "true ciprofloxacin concentration" ($x$) is $y=221.2+31.3x$, with $s_{y|x}=63.8$ μg L$^{-1}$, $R^2=0.998$, and gives a mean of the absolute values of relative error in calibration of 0.9%. When applied to the samples of urine, the mean of the absolute values of relative errors in the determination of ciprofloxacin in the 63 urine samples is 10.8%.

In this case study, mesalazine behaves as an interferent in the determination of ciprofloxacin, because the mesalazine is also fluorescent. Even so, the PARAFAC decomposition clearly separates the contribution of the ciprofloxacin in the EEM from the one of any other compound in the samples.

When comparing the performance of both methods for determining ciprofloxacin, first the accuracy line "concentration calculated with the PARAFAC-based calibration" ($y$) versus "true concentration" ($x$) should be fitted. The corresponding accuracy lines are in Table 1.

**TABLE 1** Performance Characteristics of the Methods in Cases I and II

| | Regression line | Correlation coefficient, $\rho$ | Standard error of estimation, $s_{y\|x}$ (precision) ($\mu g\ L^{-1}$) | Capability of discrimination at 200 $\mu g\ L^{-1}$ ($\mu g\ L^{-1}$) |
|---|---|---|---|---|
| Case I | $y = 0.0289 + 1.0001x$ | 0.999 | 1.97 | 8.2 |
| Case II | $y = 0.0128 + 1.0001x$ | 0.999 | 2.03 | 8.5 |

As consequence of the second-order property, the two calibration models based on PARAFAC are statistically equal, although each one is based on a different data cube. Also, the trueness has been statistically verified testing that the slope and intercept of both regression lines are 1 and 0, respectively (significance level equal to 5%).

The capability of discrimination, also known as analytical sensitivity, is evaluated with multivariate and/or multiway calibration models in Refs. [40,41]. It is defined as the quantity of analyte that can be distinguished from a reference concentration with a probability $\alpha$ of falsely stating that the concentration in sample is different to the reference concentration, and another probability $\beta$ of falsely stating that the concentration in the sample is equal to the one in the reference sample.

In our case, for 200 $\mu g\ L^{-1}$ as reference value, the obtained multivariate sensitivity is 8.2 and 8.5 $\mu g\ L^{-1}$ in case I and case II, respectively, when $\alpha$ and $\beta$ were both fixed at 0.05. Therefore, the inclusion of a new interferent does not significantly worsen the sensitivity of the procedure.

## 2.3 Case III: Determination of Ciprofloxacin in Human Urine of a Patient Who Is Taking Ciprofloxacin

This time the urine samples come from a person who is being treated with ciprofloxacin. To tackle the determination, 17 calibration standards are prepared in water in the range 100–240 $\mu g\ L^{-1}$ (every 20 $\mu g\ L^{-1}$ and replicated twice, plus a blank sample). Their EEMs are stacked together with the matrices of three replicates of the diluted urine (1:2000) and, besides, the new matrices obtained when measuring the same three samples 3 days later. Therefore, the tensor dimension is $23 \times 201 \times 15$.

The PARAFAC decomposition of the data tensor (two factors, CORCONDIA = 99.9% explained variance of 99.8%) correctly recovered the factor corresponding to ciprofloxacin with correlation coefficient between the spectra defined by the loadings and the spectra of ciprofloxacin of 0.99 for both emission and excitation spectra. The loadings of the other factor correlate with the spectra of urine with correlation coefficients of 0.80 and 0.94 for the emission and excitation spectra, respectively.

Figure 8 depicts the loadings on the sample mode, blue (dark gray in the print version) circles for the factor identified as ciprofloxacin and green (gray in the print version) stars for the loadings of the urine samples. The first 17 loadings in Figure 8 of the factor related to ciprofloxacin show the design of the calibration, i.e., increasing concentration in each of the two replicates per concentration and null loading for the blank. Meanwhile, the loadings of these same samples are all null for the factor related to urine (green, gray in the print version, stars) and are constant and non-null for the remaining six samples (indices 18–23).

With the loadings of the calibration standards for the factor identified as ciprofloxacin, a calibration line is fitted that allows the determination of the quantity of ciprofloxacin in the urine of the person taking it. The mean concentration of ciprofloxacin in the first three determinations is 383 and 412 μg L$^{-1}$ for the samples measured 3 days later.

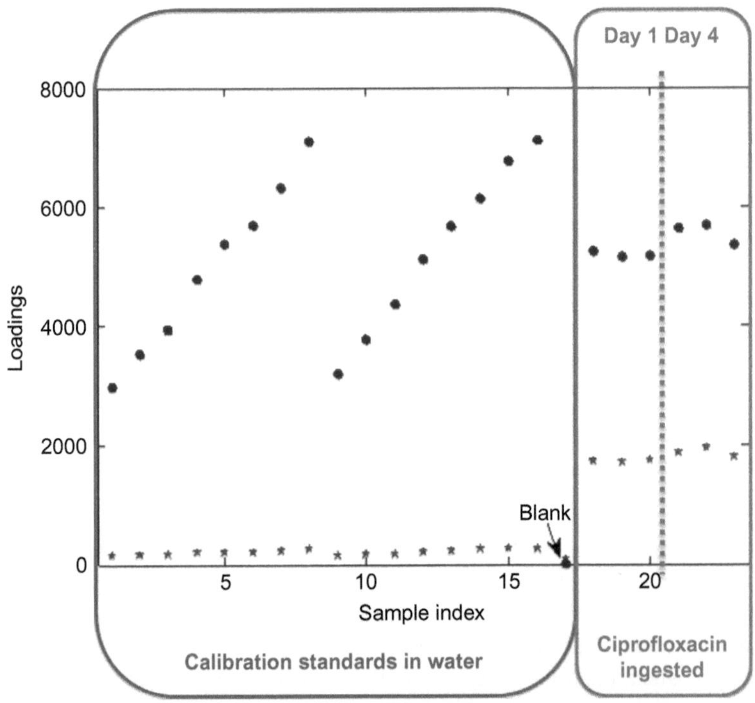


**FIGURE 8** Loadings on the sample mode versus the sample index. Blue (dark gray in the print version) circles are for the factor identified as ciprofloxacin, and green (gray in the print version) stars are for the factor related to urine. The frames separate the loadings of the calibration standards from the loadings of the test samples.

In summary, all three cases show that modeling second-order fluorescence signals through PARAFAC makes it possible the quick and precise identification and quantification of ciprofloxacin in human urine with independence of the person and other fluorescent substances in the matrix.

## 3 QUANTIFICATION AND IDENTIFICATION WITH PTV-GC-MS DATA IN THE CONTEXT OF REGULATED ANALYSIS

This section is intended for highlighting the usefulness of *n*-way techniques when there are legal requirements to comply with in the performance characteristics of methods, as it is the case of methods for determining hazardous to health substances such as pesticides or residues of veterinary substances. For these cases, the performance characteristics of the methods are defined in SANCO/12495/2011 [42] and in Decision 2002/657/EU [43] for pesticides and residues, respectively. Applications of PARAFAC and PARAFAC2 to chromatographic analysis (in particular, GC-MS) in this regulated context can be found in Refs. [44,45,46] about pesticides and in the determination of monomers and additives that migrate into food [47,48]; in all these cases, the performance of the analyses is regulated. Usefulness of three-way techniques is shown with other chromatographic techniques too, for example, to determine veterinary residues with HPLC-DAD in Ref. [49] or fungicides with LC-MS-MS in Ref. [50]. Also, Ref. [51] shows the use of these techniques for nontarget analyses.

We start by showing the similarity between the PARAFAC model and the physical model underlying the data acquisition by GC-MS. The data are arranged in a three-way array, **X**. In our case, we use a data tensor for each target analyte so that each tensor is rebuilt around the retention time of each target analyte.

In GC-MS, for each of the $K$ samples analyzed, the abundance of $J$ characteristic ions is recorded at $I$ times around the retention time of each compound, giving a mass spectrum at every time. The resulting experimental data obtained for every peak can thus be arranged in a cube or tensor **X** with dimension $I \times J \times K$.

Considering an analyte, its abundance $x_j$ recorded by an MS spectrometer at the $j$th $m/z$ ratio is

$$x_j = \varepsilon_j \chi, \quad j = 1,2,\ldots,J \tag{10}$$

where $\varepsilon_j$ is a coefficient of proportionality between the analyte concentration, $\chi$, and the abundance. $\varepsilon_j$ depends on the $j$th $m/z$ ratio; the vector of these coefficients constitutes the spectral profile (the mass spectrum).

As the mass spectrometer is coupled to a chromatograph, the abundance depends also on the elution time, because the fraction of analyte that is eluting from the chromatographic column to the mass spectrometer changes over time. So at the $i$th elution time, the recorded abundance becomes

$$x_{ij} = \tau_i \varepsilon_j \chi, \quad i = 1,2,\ldots,I; j = 1,2,\ldots,J \tag{11}$$

where $\tau_i$ can be considered as the fraction of analyte that is going through the mass spectrometer at time $i$. The vector of all $\tau_i$ forms the chromatographic profile (chromatographic peak). If $F$ spectrally active substances coelute, the recorded abundance is the sum of the contributions of these $F$ different compounds, thus

$$x_{ij} = \sum_{f=1}^{F} \tau_{if} \varepsilon_{jf} \chi_f, \quad i = 1,2,\ldots,I; j = 1,2,\ldots,J \tag{12}$$

Finally, assuming that the $f$th analyte has the same retention time in all chromatographic runs, the abundance recorded at the $j$th $m/z$ ratio, the $i$th retention time, and the $k$th sample can be expressed as

$$x_{ijk} = \sum_{f=1}^{F} \tau_{if} \varepsilon_{jf} \chi_{kf}, \quad i = 1,2,\ldots,I; j = 1,2,\ldots,J; k = 1,2,\ldots,K \tag{13}$$

The arrangement of all data forms the three-way array or tensor **X**. As can be observed, the PARAFAC model of Equation (1) of Section 1.1 corresponds to the physical model of Equation (13).

Assuming that the model of Equation (1) is adequate for the recorded data, then the parameters of the true underlying physical model of Equation (13) will also provide the best possible fit to the data. In this case, PARAFAC allows estimating the mass spectrum and the chromatographic profile (both normalized) of the analyte even if another nonmodeled analyte coelutes with it.

Under the above-mentioned conditions, a PARAFAC model of $F$ components can be used to estimate: (i) the mass spectral profile of each analyte for all $m/z$ ratios recorded, vector $\mathbf{b}_f = (b_{1f}, b_{2f}, \ldots, b_{Jf})$, (ii) the fraction of analyte that leaves the chromatographic column along the chromatographic run, that is, the chromatographic profile or chromatogram of the analyte in vector $\mathbf{a}_f = (a_{1f}, a_{2f}, \ldots, a_{If})$, and (iii) the relative concentration of every analyte in all $K$ samples, i.e., the sample profile of each analyte, by means of vector $\mathbf{c}_f = (c_{1f}, c_{2f}, \ldots, c_{Kf})$. The coordinates of vectors $\mathbf{b}_f$, $\mathbf{a}_f$, and $\mathbf{c}_f$ are referred to as spectral, chromatographic, and sample loadings, respectively. It is important to note that the procedure to fit by least squares the trilinear model provides two of the profiles normalized, which means that they are divided by its norm (usually, the chromatographic and spectral profiles). This normalization is needed because the PARAFAC decomposition is unique up to scaling and permutation as it is explained in Section 1.2.

In this section, we go into the determination of triazines in oranges (in the context of legislation) using a GC-MS system coupled to a programmable temperature vaporizing (PTV) inlet. The European Union establishes maximum residue levels, MRLs, for simazine (SZ), atrazine (AZ), terbuthylazine

(TZ), simetryn (ST), ametryn (AT), prometryn (PT), and terbutryn (TT) in oranges [52,53]. These MRLs are 0.10 mg $kg^{-1}$ for TZ, 0.05 mg $kg^{-1}$ for AZ, and 0.01 mg $kg^{-1}$ for SZ, ST, AT, PT, and TT.

The low concentration allowed for these pesticides in food [54], some of them are even forbidden, together with the need of controlling these residues in a large number of samples, make it necessary the use of analytical techniques for the pretreatment of samples, techniques that should be sensitive, fast, and efficient in the elimination of interferents, different from the target analytes. The latter is more difficult when performing multiresidue analysis in complex sample matrices. A commonly used technique for pretreatment of samples in multiresidue analyses is the use of QuEChERS (Quick Easy Cheap Effective Rugged and Safe) [55]. The QuEChERS approach typically involves an extraction with acetonitrile followed by a clean-up step which consists of a dispersive solid-phase extraction but even after this clean-up the extracts are usually relatively dirty because of the risk of removing pesticides along with other matrix compounds, so the extracts can still contain coextracted compounds. This implies in general the presence of coeluents that have a similar retention time and, besides, share some *m*/*z* ions with the target analyte. The presence of nontarget compounds can cause false negatives during pesticides identification [51] due to the fail in complying with the maximum permitted tolerances for relative ion abundances established in the regulations. Figure 9 depicts the total ion chromatogram from the injection of an extract obtained, after QuEChERS procedure, of a fortified sample of orange.

Analytical pretreatment using the QuEChERS procedure and injection with a PTV inlet requires the optimization of many parameters. In this situation, Ref. [56] reports an innovative procedure in the field of experimental optimization, where the responses to be optimized are the sample loadings of a PARAFAC2 model. Part of this section has been extracted from the mentioned work.

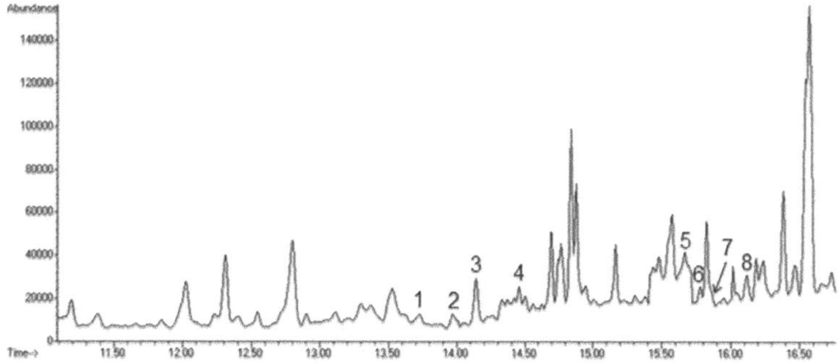


**FIGURE 9** Total ion chromatogram (TIC) from the injection of the extract of an orange sample fortified with 10 μg $L^{-1}$ of SZ, AZ, TZ, ST, AT, PT, and TT and extracted according to the experimental conditions in Ref. [56]. Peak labels: 1: SZ; 2: AZ; 3: PZ; 4: TZ; 5: ST; 6: AT; 7: PT; and 8: TT.

The use of the PARAFAC2 model allows unequivocal identification of target compounds according to document SANCO/12495/2011 [42]. For CG-MS, these requirements are (i) the relative retention time of the analyte should correspond to the one of a calibration solution with a tolerance of ±0.5% and (ii) the relative abundance of at least three diagnostic ions (*m*/*z*) must be within the corresponding tolerance intervals when the ions are recorded in SIM mode.

In the mentioned work [56], data tensors for each triazine were built up for PARAFAC decomposition in order to calculate responses for the experimental design (a D-optimal design picked for the problem). These data tensors contain the matrices recorded for both standards and the fortified samples of the experimental design, because, when complex matrices are analyzed, like in this case, the estimates of three-way models have proved [57] to be more precise when both, standards and fortified samples, are included in the decomposition step. In all the cases, 10 μg $L^{-1}$ of propazine (PZ) was used as internal standard. The second-order advantage of PARAFAC2 models guarantees that one of the factors is related to the target analyte, and its sample loadings are proportional to its quantity in each sample of the design. Consequently, there is no need of fitting a calibration model in each of the experimental conditions stated in the design.

Since the purpose of this section is illustration, only the model for simazine will be discussed. In this case, seven standards in the range 0–20 μg $L^{-1}$ were analyzed (the sample with 10 μg $L^{-1}$ was replicated 5 times). In the particular case of SZ, the signals recorded made up tensor **X** (31 × 5 × 25) where 31 refers to the number of scans, 5 refers to the number of ions, and 25 refers to the number of samples (7 standards plus 5 replicates plus 13 experiments of the D-optimal design). In Figure 10, a scheme of the QuEChERS extraction, clean-up procedure, and the tensor **X** built with signals are showed.

With the tensor, a PARAFAC model was built for the simazine (the nonnegativity constraint was enforced for the three ways), but the changes in the retention time between samples caused the PARAFAC model to fail. When studying the loadings on the chromatographic and spectral profiles, no factor matched the relative retention time of the reference standard and the relative ion abundances were not within the tolerance intervals established in Ref. [42]. To solve this problem, a PARAFAC2 model (unimodality and nonnegativity constraints in the chromatographic mode and nonnegativity constraint in spectral and sample modes) was built. In Section 1.4, it has already been explained that with PARAFAC2 each sample can have its own chromatographic profile.

The final PARAFAC2 model fitted for simazine has three factors that explain 96.53% of variance, with CORCONDIA index equal to 86%. No outlier data were detected by using $Q$ residuals and Hotelling's $T^2$ statistics.

The identification of pesticides according to the requirements established in document SANCO allows confirming the factor related to SZ in the PARAFAC2 model. The first three columns in Table 2 show the recorded ions

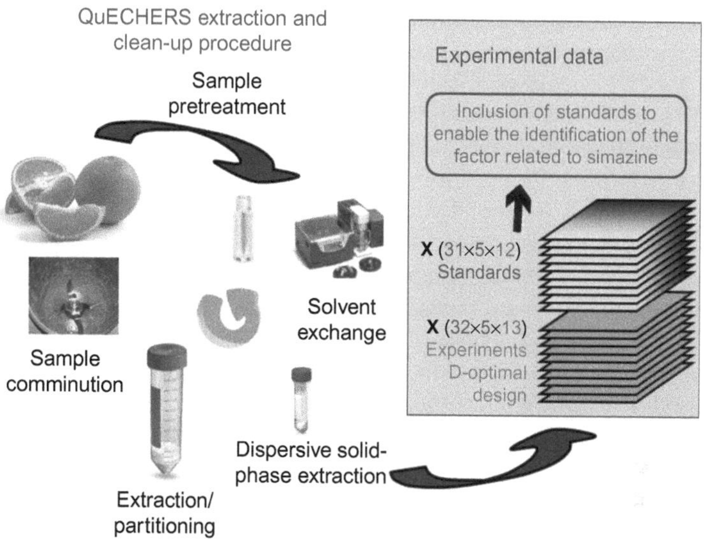


**FIGURE 10** Scheme of the QuEChERS procedure. **X** is the tensor built with standards plus the samples of the D-optimal design (experimental details in Ref. [56]).

**TABLE 2** Detected Ions (the Base Peak Is in Bold), Relative Abundance and Tolerance Intervals for the Reference Sample of 10 μg $L^{-1}$ of Simazine, and Relative Abundances Calculated with the Loadings of the Spectral Profile of the PARAFAC2 Model Built with the Samples Corresponding to the D-Optimal Design

| | Reference sample | | D-Optimal design samples (QuEChERS) | |
|---|---|---|---|---|
| *m/z* | *Relative abundance (%)* | *Tolerance interval (%)* | *Relative abundance (%)* | *Verified compliance* |
| 158 | 23.74 | 20.18–27.30 | 27.25 | Yes |
| 173 | 44.55 | 37.87–51.24 | 42.88 | Yes |
| 186 | 71.89 | 64.70–79.08 | 81.85 | No |
| **201** | **100.00** | – | **100.00** | – |
| 203 | 31.95 | 27.16–36.74 | 32.82 | Yes |

(*m/z* 201 is the most intense one, that is, the base peak), the relative abundance, and the tolerance intervals for the reference sample of 10 μg $L^{-1}$ of SZ. The last two columns refer to the relative abundances calculated with the loadings on the spectral profile of the samples corresponding to the

D-optimal design (using the QuEChERS procedure). The last column shows that three diagnostic ions verified the compliance as SANCO/12495/2011 establishes for unequivocal identification of a compound.

The loadings on chromatographic, spectral, and sample profiles obtained for SZ are showed in Figure 11. The loadings of the first factor, always in blue (dark gray in the print version) in Figure 11 (continuous line in Figure 11A, solid bar in Figure 11B, and circles in Figure 11C), were coherent with SZ, which enabled the unequivocal identification of this triazine. Figure 11A shows, through the chromatographic profile, that the shift in the retention time was prominent. However, taking into account that the retention time of simazine is 13.643 min and the one of the internal standard (PZ) is 14.130 min, the relative retention time is 0.9655 and, thus, the allowed variation according to Ref. [42] is (0.9607, 0.9704). The two extremes (left and right) chromatographic peaks in Figure 11A have relative retention times of 0.9684 and 0.9701 min, both inside the previous interval.

The loadings on the spectral mode of this factor (solid bars in Figure 11B, in blue or dark gray in the print version) matched the spectrum of a 10 μg $L^{-1}$ standard used as reference, because the relative abundance of each *m*/*z* ratio was within the tolerance interval (Table 2). It should be highlighted that the five recorded *m*/*z* ratios share at least one ion with one of the coeluting compounds (dashed and pointed bars, in blue and green, respectively, or in light gray and gray in the print version). This would have led to a false negative for simazine in these samples because, for each ion, the recorded abundance includes also the abundance of the coeluting compound, and thus, it is an overestimation of the one of simazine that would cause that the relative abundances would be outside the tolerance intervals.

Finally, in the sample mode, Figure 11C, the loadings of the first factor, blue circles (in blue or dark gray in the print version), increase as expected with the concentration of SZ for the first 7 standards while the other 5 (the replicates) have nearly equal loadings, all similar to the one with 10 μg $L^{-1}$. These 12 samples have loadings in the other two factors (squares and triangles in Figure 11C) that remain nearly constant and practically null. However, the loadings of these two factors increase for orange extracts (the last 13 samples corresponding to the experimental design).

The optimized procedure has been validated with a new tensor that includes standards and matrix-matched standards. With this purpose, the standardized loadings on the sample mode calculated for SZ (standardization is made by dividing each loading by the corresponding loading of the internal standard) are used for building the calibration line: $y=0.0459+0.0460x$ (being $y$ the standardized loadings and $x$ the true concentration), with standard error of estimation $s_{y|x}=0.0172$ μg $L^{-1}$ and correlation coefficient $\rho=0.999$.

Accuracy of the procedure is checked with the regression model "calculated concentration" versus "true concentration." The procedure has

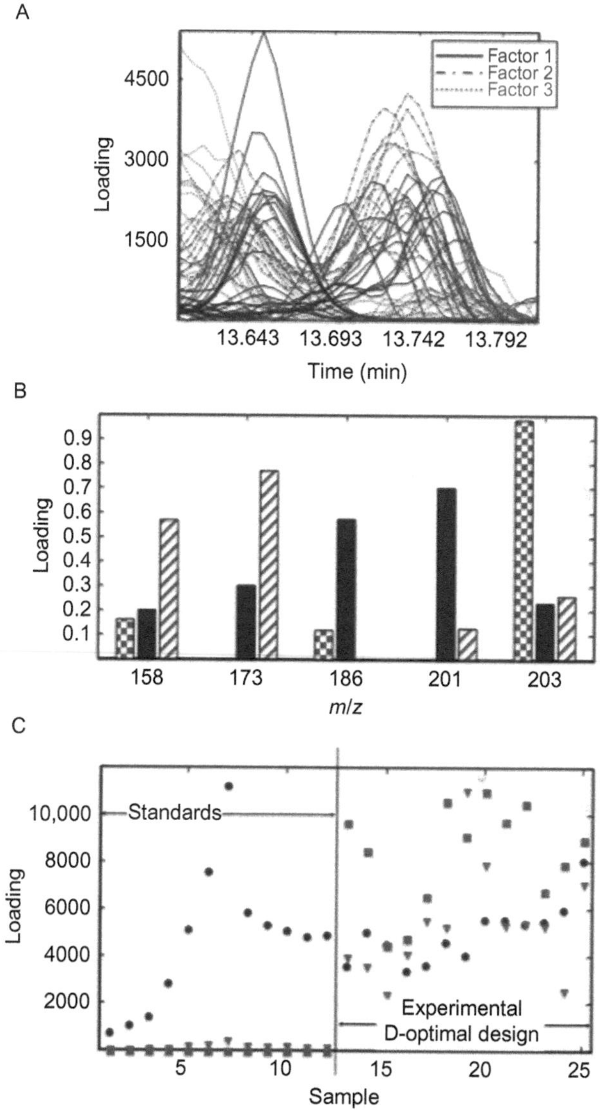


**FIGURE 11** Loadings of the PARAFAC2 model built for simazine with standards and samples pretreated with QuEChERS procedure. The first factor is always in blue (dark gray in the print version), the second in green (gray in the print version) and the third in red (light gray in the print version). (A) Chromatographic mode: continuous, dashed and dotted lines for factors 1, 2 and 3, respectively; (B) spectral mode: solid, pointed and dashed bars for factors 1, 2, and 3, respectively; (C) sample mode: circles, squares and triangles for factors 1, 2 and 3, respectively.

neither proportional nor constant bias according to the decisions made with the hypothesis tests for slope and intercept equal to 1 and 0, respectively. Therefore, trueness and precision have been assessed at a 5% significance level. These results are listed in Table 3 along with CCα and CCβ as required in Decision 657. In ISO 11843 [58] and IUPAC [59], these figures of merit are defined as decision limit and capability of detection, respectively, and both should be evaluated for null concentration ($x_0=0$) and also for the concentration in the MRL (MRL = 10 μg kg$^{-1}$ for simazine). For $n$-way data, these values can be computed with the procedure in Refs. [57,60]. For the case at hand, simazine, CCα varies from 0.81 to 1.15 μg kg$^{-1}$ and CCβ from 1.55 to 2.21 μg kg$^{-1}$ in $x_0=0$ in matrix matched and standards, respectively. These values are far below the MRL established by the European Union for simazine in oranges. Table 3 also shows decision limit and capability of detection in the MRL, with values between 10.78 and 11.08 μg kg$^{-1}$ for CCα and 11.51 and 12.08 μg kg$^{-1}$ for CCβ for matrix matched and standards, respectively. In all the cases, standard and matrix-matched standards and for both, null and MRL concentrations, the probabilities of false positive (or false no-compliance in MRL) and a false negative (or false compliance) were set to 0.05. No MRL violations were found in the commercial oranges examined for any of the seven triazines analyzed [56] (simazine, atrazine, terbuthylazine, ametryn, simetryn, prometryn, and terbutryn).

In summary, the "second-order advantage" of the PARAFAC2 decomposition allows the unequivocal identification according to document SANCO/12495/2011 (taking into account the tolerances for relative retention time and the relative abundance for the diagnostic ions), avoiding false negatives even in the presence of unknown coeluents in oranges that shared ions with simazine.

## 4 QUANTIFICATION WITH DATA OF AN ELECTRONIC NOSE BASED ON METAL OXIDE SENSORS

This section shows the usefulness of three-way models to predict, in this case, the storage time of small lobsters with data recorded with an electronic nose. The electronic nose is a device used for the detection of aromatic profiles of a sample. The sample can be analyzed practically without analytical pretreatment, which makes the electronic nose adequate for several in-line analyses in any kind of industry. In particular, the instrument (electronic nose) utilized for the present application is made up by electrochemical sensors with different sensitivity and specificity. This hybrid character of sensors makes this technique useful and adequate to discriminate and/or typify categories among samples in which the aroma plays an important role [61,62].

All the analyses were made using an electronic nose with a sensory array system consisting of 18 metal oxide semiconductor sensors placed in 3 cameras. The signals recorded for a sample of small lobster (the one identified

**TABLE 3** Parameters of Regression Models "$c_{calc}$ Versus $c_{true}$": Intercept, Slope, and Standard Deviation of Regression ($s_{y|x}$) and Number of Standards ($n$)

| | Concentration range (μg L$^{-1}$) | Intercept | Slope | $s_{y\|x}$ | Error (%)[a] | $x_0 = 0$ | | $x_0 = $ MRL | |
|---|---|---|---|---|---|---|---|---|---|
| | | | | | | CCα | CCβ | CCα | CCβ |
| | | | | | | (μg kg$^{-1}$)[b] | | (μg kg$^{-1}$)[b] | |
| Matrix-matched standards | 0–20 ($n$=8) | $6.26 \times 10^{-5}$ | 1 | 0.3736 | 5.94 | 0.81 | 1.55 | 10.78 | 11.51 |
| Standards | 0–20 ($n$=7) | $-9.09 \times 10^{-6}$ | 1 | 0.4998 | 3.51 | 1.15 | 2.21 | 11.08 | 12.08 |

Some figures of merit: errors, detection limit (CCα), and capability of detection (CCβ) at $x_0 = 0$ and at the maximum residue level ($x_0 = 10$ μg kg$^{-1}$) for both standards and matrix-matched standards.

[a]*Mean of the absolute value of relative errors in calibration.*

[b]*The extraction procedure gave final solution representing 1 mg of the commodity per mL of extract.*

**TABLE 4** Characteristics of the Samples of Small Lobster Used for the Analyses and Predictions with the Two Three-Way Models

| Sample | Stored with modified atmosphere | Days stored ($y$) | Days predicted with the PARAFAC2 model | Days predicted with the Tucker3 model |
|---|---|---|---|---|
| M1 | No | 1 | 1.78 | 1.94 |
| M2 | No | 1 | 1.54 | 1.61 |
| M3 | No | 4 | 3.01 | 2.91 |
| M4 | No | 4 | 3.82 | 3.68 |
| M5 | Yes | 4 | 3.43 | 3.89 |
| M6 | Yes | 4 | 4.79 | 4.87 |
| M7 | No | 7 | 5.68 | 5.68 |
| M8 | No | 7 | 7.33 | 7.31 |
| M9 | Yes | 7 | 7.78 | 7.79 |
| M10 | Yes | 7 | 8.16 | 8.18 |
| M11 | No | 10 | 8.76 | 8.79 |
| M12 | No | 10 | 10.62 | 10.57 |
| M13 | Yes | 10 | – | – |
| M14 | Yes | 10 | 9.30 | 9.29 |

in the first row of Table 4) and with a baseline correction using fractional approach are shown in Figure 12. There, the positive values correspond to the signals with the 6 sensors in the first camera, whereas the negative values are the signals recorded with the remaining 12 sensors in cameras two and three. Details can be consulted in Ref. [63].

Data obtained from an electronic nose are structurally three way (time × sample × sensor). However, in practice, the complete information related to the time profile is not used because usually only the signal at the maximum (in absolute value) of the time vector is recorded so that the structure of signals is reduced to two way (sample × sensor). In this work, the whole vector in the time profile is used; thus, the original three-way structure is preserved. Figure 12 also shows by red (light gray in the print version) filled circles some of the maxima in the time profile, which would be the data usually used for these analyses.

Data acquisition lasted 2 min, recording a single datum every second, thus 120 values per sensor and sample. Experimental data tensor, **X**, is then

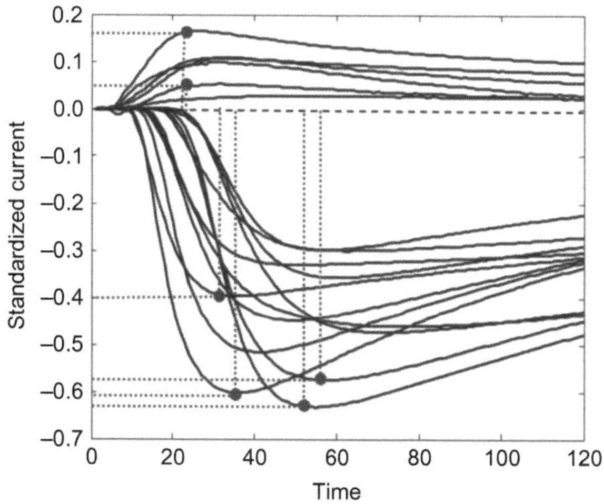


**FIGURE 12** Recorded signals for each sensor (120 s). The 6 sensors in camera 1 gave the positive values and the remaining 12 in cameras 2 and 3 gave the negative values. Red (light gray in the print version) circles mark the maximum (in magnitude) of the corresponding signal.

$120 \times 14 \times 18$ (120 recorded current, 14 samples of small lobster, and 18 sensors). The 14 samples are recorded in days 1, 4, 7, and 10 after their storage. Besides, except for samples in the first day, the remaining samples were stored with and without modified atmosphere to see if this has any effect on the storage time. Table 4 identifies the samples used for the analysis and the way they were stored.

Because the aim of the study is predicting, with the least possible error, the storage time of small lobsters, three-way methods are applied to look for the underlying factorial structure of tensor **X**. First, the unique factors were sought by fitting a PARAFAC model. Two factors gave a unique solution with CORCONDIA equal to 100% and explaining 75.70% of variance. However, this solution is not related to the storage time. The loadings on the sample mode are the projections of the samples in the space spanned by the two PARAFAC factors, but when they are used to predict the storage time by a multilinear regression model, the fitted model is not significant ($p$ value equal to 0.39 for the ANOVA) and it only explains 24% of the variability of the storage time, according to the coefficient of determination ($R^2$).

This is so because the PARAFAC model is fitted assuming that each factor in each mode is related in a unique way to the corresponding one in the remaining modes. That is, the first factor in the temporal mode is related with only one factor in the sample mode and with another sole factor in the mode "type of sensor," this is the second-order advantage. This property, which is decisive when the factors serve to identify a spectrum or a chromatogram and relate them with a reference sample (previous cases; Sections 2 and 3), turns out to be irrelevant in a case like this case where the found unique solution, despite the fact that the PARAFAC model is valid, is not related to the storage time.

With a PARAFAC2 model, factors are determined with fewer restrictions than for PARAFAC, in the sense that it is only required that the cross-product matrix of signals, instead of the raw matrix of signals, is equal from sample to sample. In this way, with PARAFAC2, each temporal profile is separately modeled, while PARAFAC estimates a single temporal profile, the same for all the experimental signals, that is, for each sensor and sample. The fitted PARAFAC2 model has also two factors. Table 5 shows the characteristics of the model that explains 99.9 % of the variance of tensor **X**, and has a CORCONDIA index of 100%. That means that this unique solution spans a projection space that is different from the one spanned with the PARAFAC model. In this case, the multilinear regression model that relates the sample loadings of the two factors with the storage time explains 92.3% of the variance of the response (storage time). The different time profiles estimated with the PARAFAC2 model are depicted in Figure 13A).

In Section 1.3, it is already explained that a Tucker3 model allows exploring the possible associations among factors in relation to the mode or way under consideration, although this implies the loss of uniqueness in solution. For the problem at hand, a Tucker3 (3,3,3) model with three factors for each mode explains 99.67% of tensor **X** and has a nondiagonal core array $\mathbf{G} = (g_{ijk})$. Table 6 contains some of the elements $g_{ijk}$ sorted (in absolute value) from largest to smallest according to the percentage of variation explained. For example, third row in Table 6 corresponds to term $g_{123}$ and collects the interaction between factor 1 in the time mode, factor 2 in the sample mode, and factor 3 in the sensor-type mode. This interaction represents 1.46% of the total variation in **G**. Looking at the first three terms, that together account for 99.70% of the variation, it is observed that only factors 1 and 2 are involved in the first two modes, though the three factors 1, 2, and 3 are intervening in the third mode. Consequently, a Tucker3 (2,2,3) model is used.

This new model explains 99.54 % of the variation of tensor **X**. The three most important elements of core tensor are the same as the ones in Table 6 with percentages of explained variance of 99.99, 2.39, and 1.47, respectively. Nevertheless, the new projection space spanned by the Tucker3 (2,2,3) model is certainly different from the spaces spanned by the factors of the previous PARAFAC and PARAFAC2 models. Its loadings on the sample mode will be used to predict the storage time of the small lobsters. With this goal, the resulting regression model fitted is significant with $R^2$ equal to 91.7% (Table 5). From the point of view of prediction of the storage time, the Tucker3 model includes a relation between the sample mode of the second factor and the sensor-type mode of the third factor, which is quantitatively important; on the other hand, there is no guarantee of uniqueness in the

**TABLE 5** Three-way Decomposition and related Regression Models

| | Three-way decomposition | | | Regression model | | Accuracy lines | | | |
|---|---|---|---|---|---|---|---|---|---|
| | *Factors number* | *Explained variance (%)* | *CCI*[a] *(%)* | *Equation* $y = b_0 + b_1x_1 + b_2x_2$ | $R^2$ (%) | *Intercept*[b] *(p value)* | *Slope*[c] *(p value)* | ρ | $s_{y\|x}$ ($\mu g\ L^{-1}$) |
| PARAFAC2 | 2 | 99.9 | 100 | $b_0 = -205.3$ <br> $b_1 = -55.7$ <br> $b_2 = 839.4$ | 92.3 | 0.45 (0.39) | 0.92 (0.34) | 0.96 | 0.87 |
| Tucker3 | (2, 2, 3) | 99.5 | – | $b_0 = -204.1$ <br> $b_1 = -781.3$ <br> $b_2 = 85.7$ | 91.7 | 0.49 (0.41) | 0.92 (0.36) | 0.96 | 0.99 |

Multilinear regression models "days of storage" as a function of "sample loadings" computed with PARAFAC2 and Tucker3 models. Accuracy lines "predicted response" versus "true values."
[a]*CCI means CORCONDIA index of model (Equation 7).*
[b]*Null hypothesis: intercept equal to 0.*
[c]*Null hypothesis: slope equal to 1.*

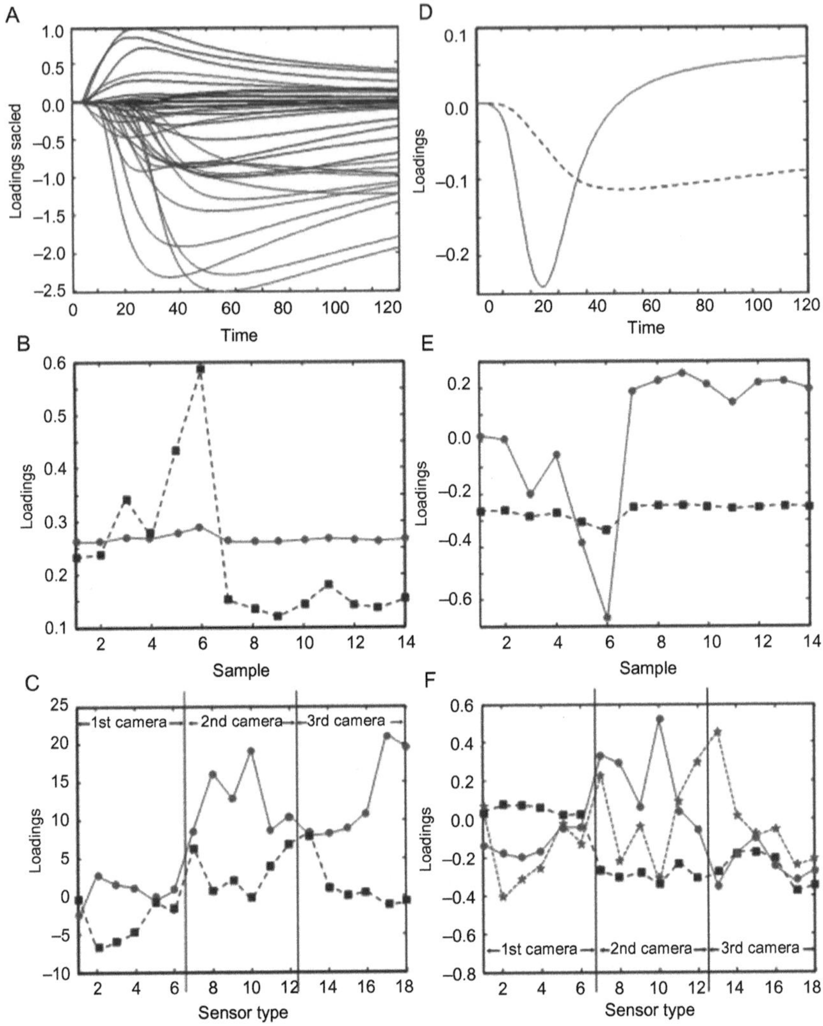


**FIGURE 13** On the left, loadings of the PARAFAC2 model with two factors: loadings on (A) time, (B) sample, and (C) sensor profiles. On the right, loadings of the Tucker3 model on (D) time, (E) sample, and (F) sensor profiles. First factor in blue (dark gray in the print version; dashed lines and squares), second in green (gray in the print version; continuous line and circles), and third in red (light gray in the print version; dotted lines and stars).

solution. This analysis suggests that storage causes the emergence of compounds that give signals reacting to various effects of decomposition of the small lobsters which are reflected in the responses of the sensors.

Figure 13 shows the found factors in each mode for both PARAFAC2 and Tucker3 models. Left part of the figure is for the PARAFAC2 model and

**TABLE 6** Analysis of 3×3×3 Core Array of Tucker3 (3,3,3) Model

| Component | Value | Squared value | Fractional variation (%) | Summed fraction (%) |
|---|---|---|---|---|
| (1,1,1) | 52.49 | 2755.00 | 95.86 | 95.86 |
| (2,1,2) | −8.26 | 68.24 | 2.37 | 98.24 |
| (1,2,3) | −6.49 | 42.09 | 1.46 | 99.70 |
| (3,1,3) | −1.82 | 3.33 | 0.12 | 99.87 |
| (1,2,2) | −1.19 | 1.42 | 0.05 | 99.90 |
| (2,1,3) | 1.06 | 1.12 | 0.04 | 99.94 |
| (2,2,3) | −1.03 | 1.06 | 0.04 | 99.97 |

$(i,j,k)$ refer to the subscripts of $g_{ijk}$ (Equation 4 of Section 1.3).

contains loadings of the two factors on the three modes: temporal (Figure 13A), sample (Figure 13B), and sensor type (Figure 13C). In the first mode, time profile, there is a different estimate for each sample and sensor. Right part of Figure 13 shows the modes and different factors of the Tucker3 model. Loadings of the two factors on the time and sample profiles are depicted in Figure 13D and E, respectively, and loadings of the three factors on the sensor-type mode are in Figure 13F. These last loadings show that the sensors behave differently depending on whether the sensors are in the first, second, or third camera; for example, the first factor (in blue, dark gray in the print version, squares) has loadings practically null for the sensors in the first camera (the first six).

The validity of the PARAFAC2 and Tucker3 (2,2,3) models to predict storage time is evaluated by fitting the accuracy line. This is the regression model that relates the predicted values (with the corresponding model) and the true value of "storage days." The accuracy lines fitted are also in Table 5. *P* values of the hypothesis tests to test whether the slope is 1 and the intercept is 0 lead to the conclusion (5% significance level) that there is neither proportional nor constant errors when predicting storage time of the small lobsters. The last two columns of Table 4 contain the predicted values with the two decompositions of the three-way data coming from an electronic nose. It is observed that the predictions with the two decomposition models differ from the real ones at most 1.3 storage days. Sample M13 was declared outlier and was consequently removed. Finally, in this study, no difference between small lobsters stored with or without modified atmosphere was observed.

## 5 DESCRIBING THE GRAPE MATURITY AT HARVEST BY MEANS OF PHYSICOCHEMICAL VARIABLES

The content in this section was partially published in Refs. [64,65]. Grape quality at harvest is the main factor which will determine wine quality. Nevertheless, grape maturation is a continuous process that evolves through various ripening stages. The process "grape maturation" in a geographical zone, such as the *Denominación de Origen* (designation of origin) D.O. Rioja, is complex and multidimensional. It is complex because each parcel has its own geographical conditions and it is planted with a determined grape variety (and not with any other). It is multidimensional because it is not possible to define grape maturity from a single physicochemical parameter. This multidimensional and evolving character makes the Regulating Board of the Qualified D.O. Rioja (for now on, DOC[1] Rioja) to conduct, every year, a control on the maturation in the representative parcels along the different wine-growing zones and the most important grape varieties of Rioja. Weekly, approximately 1 month before harvesting, samples of grapes are collected in fixed plots; these grapes are analyzed at the laboratory of the Estación Enológica de Haro [66] (which is the official laboratory of DOC Rioja, accredited by UNE-EN ISO/IEC 17025:2005 norm), and the results are delivered to growers to facilitate their decision to harvest.

Traditionally, the decision of harvesting is taken after analyzing only sugar concentration, titratable acidity, and pH of the grape juice (technological maturity). However, these parameters merely provide information about the pulp ripeness and overlook the real degree of maturity of both skins and seeds (phenolic maturity). Nevertheless, technological and phenolic maturities are not simultaneously reached; on the contrary, they tend to separate depending on several factors: grape variety, cultivar, adverse weather conditions, soil, water availability, and cultural practices. Besides, this divergence is increasing as a consequence of the climate changes (larger quantities of $CO_2$, less rain, and higher temperatures). Therefore, the complexity of the maturity process, which is described in Ref. [67], is clear. In summary, in the DOC Rioja, the decision to harvest is based on the analyses of the 12 physicochemical parameters listed in Table 7, with all the analyses carried out at the laboratory of the Estación Enológica de Haro.

The samples for this study correspond to the harvests of 2009 and were collected during the last four consecutive weeks before harvesting. For the reasons mentioned in the first paragraph, grapes were collected in plots representative of the DOC Rioja, as well as from different varieties of grape: *tempranillo*, *garnacha*, *mazuelo*, and *graciano*. Table 8 shows the parcels and the grape variety in each zone, while Figure 14 depicts the geographical location of the parcels in Table 8.

1. The acronym refers to the Spanish "Denominación de origen calificada."

**TABLE 7** Physicochemical Parameters Determined in Grapes

| Index | Physicochemical parameters |
|---|---|
| 1 | Probable alcoholic degree |
| 2 | Total acidity |
| 3 | pH |
| 4 | Malic acid |
| 5 | Potassium |
| 6 | Total polyphenols index |
| 7 | Anthocyans |
| 8 | Absorbance at 420 nm |
| 9 | Absorbance at 520 nm |
| 10 | Absorbance at 620 nm |
| 11 | Color index |
| 12 | Tartaric acid |

The magnitude of the 12 physicochemical parameters under control and the correlation between one another change during ripening of grapes, that is, ripening changes the internal structure of the space of the chemical variables representing maturation. These changes are clearly observed in Figure 15 that contains a colored representation of the correlation matrices in the 4 weeks of collection.

For example, variable 12, tartaric acid, is either not correlated or slightly negatively correlated with the variables of color (number 8, 9, 10, and 11, that is, the absorbance at 420, 520, and 620 nm, and color index) in the first 2 weeks (Figure 15A and B), and then in the third and fourth weeks (Figure 15C and D), the tartaric acid becomes positively correlated with the variables of color.

However, all these correlation coefficients cannot reveal the fact that each plot/variety may have a specific behavior. Therefore, it seems evident that all the physicochemical variables must be jointly handled respecting the three-way structure of data.

The obtained data, in a natural way, make up a three-way tensor, **X**, of size $(16 \times 12 \times 4)$, corresponding to the sample profile (16 parcel/variety), physicochemical profile (12 parameters), and temporal profile (4 weeks). The main goal when using three-way methods to analyze three-way data is to unravel the complex patterns of dependencies among observations, without the need of explaining any other external variable, like in sections 2–4.

**TABLE 8** Geographical Distribution of Samples by Zones, Variety, and Altitude Above the Sea Level

| | Zone | # Sample | Location | Variety | Altitude (m) |
|---|---|---|---|---|---|
| Rioja Alta | I—Obarenes | 1 | Haro | Tempranillo (t) | 438 |
| | III—Sonsierra | 2 | San Vicente | Tempranillo (t) | 440 |
| | IV—Valpierre | 3 | San Asensio | Tempranillo (t) | 457 |
| | V—Bajo Najerilla | 4 | Cenicero | Tempranillo (t) | 434 |
| | VI—Centro | 5 | Fuenmayor | Tempranillo (t) | 428 |
| Rioja Baja | X—Iregua-Leza | 6 | Murillo | Mazuelo (m) | 460 |
| | | 7 | Murillo | Tempranillo (t) | 460 |
| | | 8 | Murillo | Garnacha (g) | 460 |
| | XI—Valle de Ocón | 9 | Alcanadre | Tempranillo (t) | 400 |
| | | 10 | Alcanadre | Garnacha (g) | 400 |
| | | 11 | Ausejo | Tempranillo (t) | 565 |
| | | 12 | Ausejo | Garnacha (g) | 565 |
| | | 13 | Ausejo | Graciano (r) | 565 |
| | XIV—Alhama-Aldeanueva | 14 | Aldeanueva | Tempranillo (t) | 397 |
| | | 15 | Aldeanueva | Garnacha (g) | 397 |
| | | 16 | Aldeanueva | Graciano (r) | 350 |

In any case, the tensor **X** ($16 \times 12 \times 4$) just defined is used to fit a PARAFAC model. The one with two factors has a CORCONDIA index equal to 99.99%; thus, the solution is unique. The model explains 80.13% of the variance of the data (although strictly speaking, since the data are not mean-centered, it is not variance but just sum of squares data). 90.6% of this "variance" is explained by the first factor and the remaining 9.4% by the second one. A PARAFAC model assigns the same factor in each profile, so that it suffices to identify the factor related to grape maturity to know how this maturity is expressed in each sample (parcel/variety) and each week.

Taking into account the well-known behavior or evolution of the physicochemical variables during ripening [68,69], their profile will be the first one to be interpreted. It is known that the probable alcoholic degree increases during maturity, as well as pH and color components (A420, A520, A620, total

**FIGURE 14** Map of La Rioja in Spain. The coding of zones is the same as in Table 8.

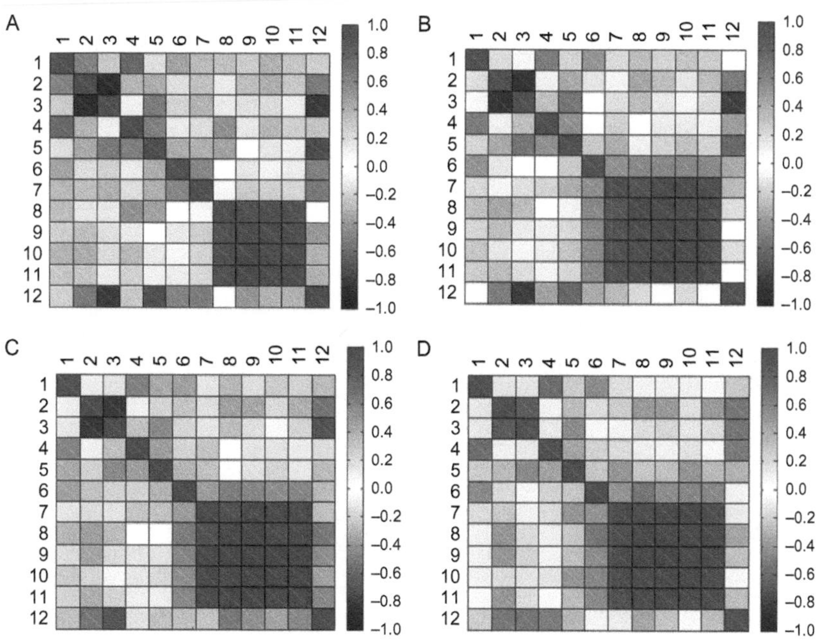


**FIGURE 15** Correlation matrices among the physicochemical parameters listed in Table 7. (A) First week, September 1. (B) Second week, September 8. (C) Third week, September 15. (D) Fourth week, September 24.

polyphenols index, and anthocyans), whereas total acidity, malic acid, and tartaric acid decrease [70].

The loadings of the two PARAFAC factors on the physicochemical profile are plotted in Figure 16. Clearly, the loadings of the second factor, depicted as red (light gray in the print version) stars, are the ones describing the chemical evolution of maturity. It is observed that the loadings on the variables that increase during ripening are larger than the ones for the variables that decrease; this opposition is greatly patent, even with a change in sign, in the case of the color components (phenolic evolution) and malic acid (technological maturity).

Once identified the second factor as grape maturation, the first factor, blue (dark gray in the print version) circles in Figure 16 (with loadings practically equal in every variable), collects what is common to all parcels, varieties, and weeks. One can say that this is the factor related to the own characteristics that constitute grape (in the case of instrumental signals, that would be the "background" of signals). From the variance explained by the model, this factor explains 10 times more (90.6%) than the factor related to maturity (9.4%), the ripening process, which is a kind of "form factor," different from sample to sample, but responsible for much less variability than the "background factor" that behaves nearly like a constant factor.

Figure 17 shows the temporal profile and, indeed, the loadings of the first factor in blue (dark gray in the print version) circles, are almost constant during the 4 weeks, whereas the loadings of the second factor in red (light gray in

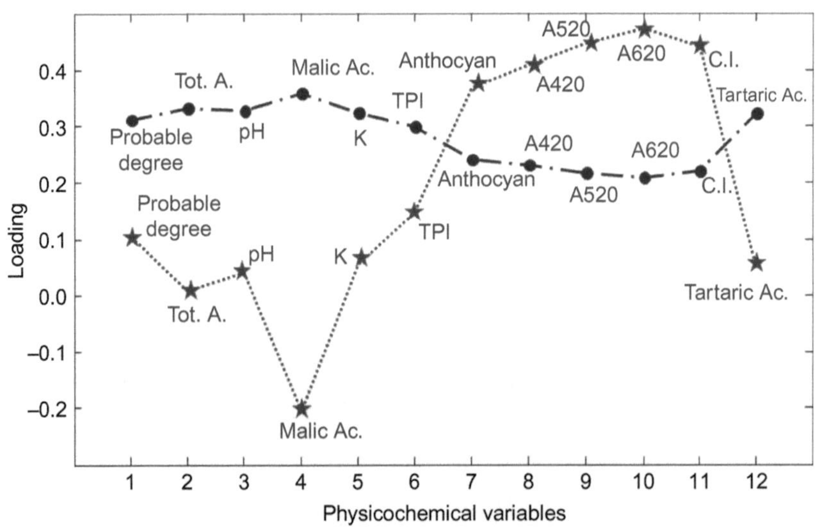


**FIGURE 16** PARAFAC model. Loadings on mode 2, physicochemical variables. First factor in blue (dark gray in the print version; dashed lines and circles) and second in red (gray in the print version; dotted lines and stars). Variables in the order of Table 7.

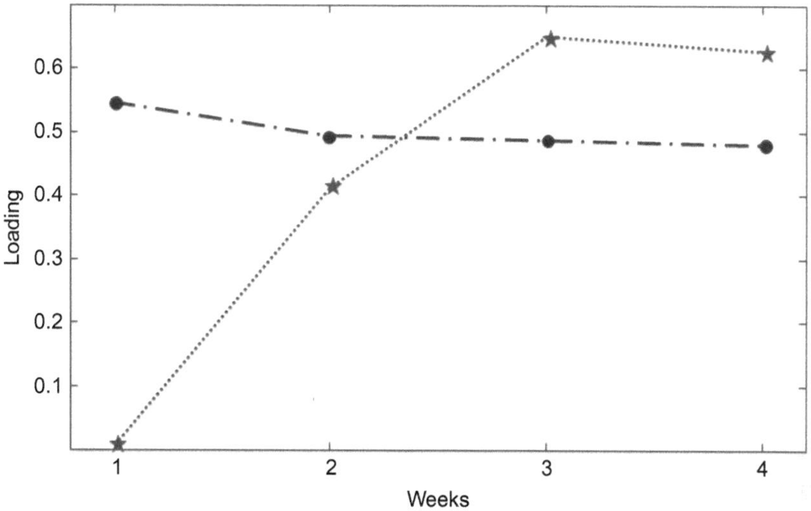


**FIGURE 17** PARAFAC model. Loadings on mode 3, temporal profile. First factor in blue (dark gray in the print version) dashed lines and circles, and second in red (gray in the print version) pointed lines and stars.

the print version) stars, which was identified as grape maturity, are coherent with this evolution. Note how, from the first to the second week, the evolution measured as the slope of the line that joints both loadings is more pronounced than the slope when moving from the second to the third week, and that the transition from the third to the fourth shows a stagnation or stopping in the evolution of maturation.

Figure 18 contains the loadings of the two factors on the sample profile. They represent the description of the ripening process according to parcel/variety. The loadings in the figure have been plotted with the code of parcel and variety (according to Table 8) and with blue (black in the print version) circles for the loadings of the first factor and red (light gray in the print version) stars for those of the second, related to grape maturity. Like in the two other profiles, the "background" factor practically is constant, that is, the rank-one tensor that defines this factor is nearly constant.

The observed behavior of maturation (second factor in pointed line, red or light gray in the print version) in relation to the sample is interesting. As regards the color properties, the grape varieties are arranged by intensity from small to large in this way: *garnacha* < *tempranillo* < *graciano*. Looking at one particular zone, for instance, XIV, the loadings of the three varieties (samples 14, 15, and 16, surrounded by a purple, dark gray in the print version, dashed circumference) are exactly in this order. The same is observed with the five samples (from 9 to 14, green, gray in the print version, continuous circumference) that correspond to zone XI, the *garnacha* variety (samples 10 and 12) has loadings less than the ones for *tempranillo* (samples 9 and 11) and the

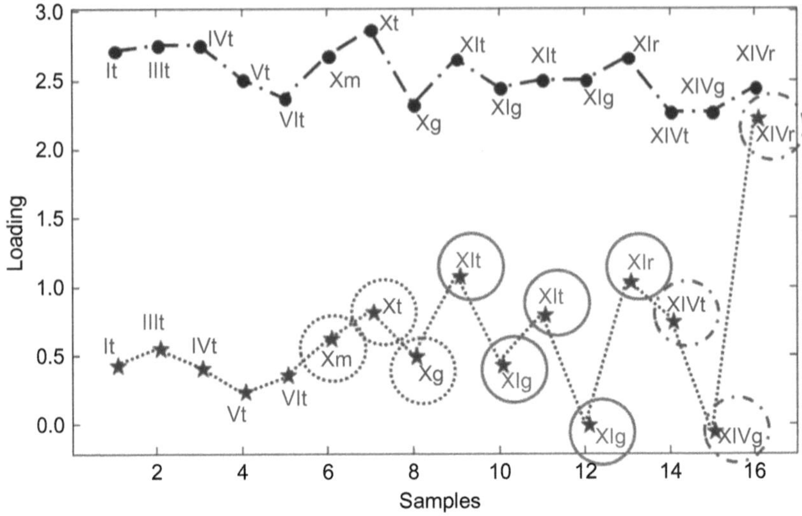


**FIGURE 18** PARAFAC model. Loadings on mode 1, sample (parcel/variety) profile. First factor in blue (black in the print version) dashed lines and circles, and second in red (light gray in the print version) pointed lines and stars. Circumferences denoted "Rioja Baja," different line type for the different parcels in Table 8.

latter less than those for *graciano* (sample 13). The same happens in zone X, with samples surrounded by a red (light gray in the print version) pointed circumference, *garnacha* variety (sample 8) has a loading less than the one of *tempranillo* (sample 7).

If now the loadings on the factor "maturity" are analyzed, by keeping variety and zone (both in "Rioja Baja," Table 8) constant, one discovers the pattern related to altitude. It is well known that altitude directly influences on the maturity, since the lower the altitude, the higher the temperature and the accumulation of color substances. This effect is shown in Figure 19, *graciano* variety has a larger loading in zone XIV (sample 16) than in zone XI (sample 13), and Table 8 shows that sample 16 is 350 m above the sea level while the sample 13, even being the same grape variety, raises higher, 565 m above the sea level. In zone XI, samples 10 and 12 are both *garnacha* and the loading is larger in sample 10, which is 400 m, than in sample 12, which is 565 m above the sea level. The same happens for *tempranillo*, samples 9 and 11, the former has a larger loading and is at lower height, 400 m compared to 565 m in the latter. In "Rioja Alta," samples 1–5, altitude is very similar in all zones and there is only *tempranillo* variety there.

Finally, in each zone, the loadings of the maturity factor follow a decreasing range in the following order: zone XIV > zone XI > zone X. This range is more pronounced than the one in zones I, III, IV, V, and VI of "Rioja Alta." This effect is related to the progress in maturation, which in La Rioja runs from east to west according to the map of Figure 14, it appears that the eastern

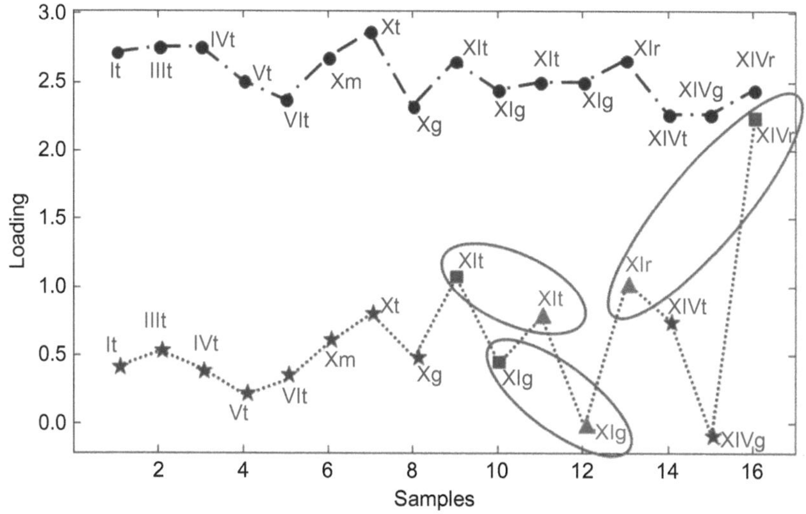


**FIGURE 19** PARAFAC model. Loadings on mode 1, sample (parcel/variety) profile. First factor in blue (black in the print version) dashed lines and circles. Second factor in red (light gray in the print version) pointed lines, purple (dark gray in the print version) squares for the parcels with lower altitude, and green (gray in the print version) triangles for those with higher altitude when different parcels have the same grape variety (grouped with an ellipse).

zones have more time to mature, and this causes more pronounced differences among varieties.

In summary, the PARAFAC decomposition of the physicochemical variables, with unique solution, allows describing with a great detail a complex process, such as the ripening of grapes, in a territory and a broad time span, accurately modeling the effect of plot and variety.

## ACKNOWLEDGMENTS

The authors thank the financial support provided by projects of Ministerio de Economía y Competitividad (CTQ2011-26022) and Junta de Castilla y León (BU108A11-2).

## REFERENCES


[1] Bro R. Multi-way analysis in the food industry. Models, algorithms and applications. PhD thesis, Amsterdam, The Netherlands: University of Amsterdam; 1998.

[2] Smilde A, Bro R, Geladi P. Multi-way analysis applications in the chemical sciences. Chichester: John Wiley & Sons; 2004.

[3] Kroonenberg PM. Applied multiway data analysis. Hoboken: John Wiley & Sons, Inc.; 2008.

[4] Kroonenberg PM. Three-mode principal component analysis: theory and applications. Leiden: DSWO Press; 1983. Available in, http://three-mode.leidenuniv.nl/index.html (last visit: 25/11/2014).

[5] Krijnen WP. The analysis of three-way arrays by constrained PARAFAC methods. Leiden: DSWO Press; 1993.
[6] Ferré J, Boqué R, Faber NM. Multiway data analysis: eigenvector-based methods. In: Brown S, Tauler R, Walczak R, editors. Comprehensive chemometrics, vol. 2. Oxford: Elsevier; 2009. p. 365–409.
[7] Tomasi G, Bro R. Multilinear models: iterative methods. In: Brown S, Tauler R, Walczak R, editors. Comprehensive chemometrics, vol. 2. Oxford: Elsevier; 2009. p. 411–51.
[8] Booksh KS, Bronk B, Czege J. Three-way calibration. In: Brown S, Tauler R, Walczak R, editors. Comprehensive chemometrics, vol. 3. Oxford: Elsevier; 2009. p. 379–412.
[9] Olivieri AC, Escandar GM. Practical three-way calibration. Amsterdam: Elsevier; 2014.
[10] Anderson CA, Bro R. The *N*-way toolbox for MATLAB. Chemometr Intell Lab Syst 2000;52:1–4.
[11] Bro R. Exploratory study of sugar production using fluorescence spectroscopy and multi-way analysis. Chemometr Intell Lab Syst 1999;46:133–47.
[12] Booksh KS, Kowalski BR. Theory of analytical chemistry. Anal Chem 1994;66:782A–791A.
[13] Kruskal JB. Rank, decomposition, and uniqueness for 3-way and N-way arrays. In: Coppi R, Bolasco S, editors. Multiway data analysis. Amsterdam: Elsevier; 1989.
[14] ten Berge JMF, Sidiriopolous ND. On uniqueness in CANDECOMP/PARAFAC. Psychometrika 2002;67:399–409.
[15] Bro R, Harshman RA, Sidiropoulos ND, Lundy ME. Modeling multi-way data with linearly dependent loadings. J Chemometr 2009;23:324–40.
[16] Omikidia N, Abdollahi H, Kompany-Zareh M. Visualization and establishment of partial uniqueness rules in parallel factor analysis. J Chemometr 2013;27:330–40.
[17] Omikidia N, Abdollahi H, Kompany-Zareh M. On uniqueness and selectivity in three-component parallel factor analysis. Anal Chim Acta 2013;782:12–20.
[18] Tucker L. The extension of factor analysis to three-dimensional matrices. In: Frederiksen N, Gulliksen H, editors. Contributions to mathematical psychology. New York: Holt, Rinehart Winston; 1964. p. 110–82.
[19] Harshman RA. PARAFAC2: mathematical and technical notes. UCLA working papers in phonetics, vol. 22; 1972. p. 30–47.
[20] Bro R, Anderson CA, Kiers HAL. PARAFAC2—part I. A direct fitting algorithm for the PARAFAC2 model. J Chemometr 1999;13:275–94.
[21] ten Berge JMF, Kiers HAL. Some uniqueness results for PARAFAC2. Psychometrika 1996;61:123–32.
[22] Bro R, Andersson CA, Kiers HAL. PARAFAC2—part II. Modelling chromatographic data with retention time shifts. J Chemometr 1999;13:295–309.
[23] Amigo JM, Popielarz MJ, Callejón RM, Morales ML, Troncoso AM, Petersen MA, et al. Comprehensive analysis of chromatographic data by using PARAFAC2 and principal components analysis. J Chromatogr A 2010;1217:4422–9.
[24] Gourvénec S, Tomasi G, Durville C, Di Crescenzo E, Saby CA, Massart DL, et al. CuBatch, a MATLAB® interface for *n*-mode data analysis. Chemometr Intell Lab Syst 2005;77:122–30.
[25] Harshman RA, de Sarbo WS. An application of PARAFAC to a small sample problem, demonstrating preprocessing, orthogonality constraints, and split-half diagnostic techniques. In: Law HG, Snyder CW, Hattie JA, McDonald RP, editors. Research methods for multimode data analysis. New York: Praeger; 1984. p. 602–42.
[26] Bro R, Kiers HAL. A new efficient method for determining the number of components in PARAFAC models. J Chemometr 2003;17:274–86.

[27] Kamstrup-Nielsen MH, Johnsen LG, Bro R. Core consistency diagnostic in PARAFAC2. J Chemometr 2013;27:99–105.
[28] Rubio L, Ortiz MC, Sarabia LA. Identification and quantification of carbamate pesticides in dried lime tree flowers by means of excitation-emission molecular fluorescence and parallel factor analysis when quenching effect exists. Anal Chim Acta 2014;820:9–22.
[29] Leurgans S, Ross RT. Multilinear models: applications in spectroscopy. Stat Sci 1992;7:289–319.
[30] Andersen CM, Bro R. Practical aspects of PARAFAC modeling of fluorescence excitation-emission data. J Chemometr 2003;17:200–15.
[31] Rinnan Å, Booksh KS, Bro R. First order Rayleigh scatter as a separate component in the decomposition of fluorescence landscapes. Anal Chim Acta 2005;537:349–58.
[32] Bahram M, Bro R, Stedmon C, Afkami A. Handling of Rayleigh and Raman scatter for PARAFAC modeling of fluorescence data using interpolation. J Chemometr 2006;20:99–105.
[33] Eilers PHC, Kroonenberg PM. Modeling and correction of Raman and Rayleigh scatter in fluorescence landscapes. Chemometr Intell Lab Syst 2014;130:1–5.
[34] Giménez D, Sarabia LA, Ortiz MC. Identification and quantification of enrofloxacin in poultry feeding water through excitation emission fluorescence and three-way PARAFAC calibration. Analyst 2005;130:1639–47.
[35] Rodríguez N, Ortiz MC, Sarabia LA. Study of robustness based on *n*-way models in the spectrofluorimetric determination of tetracyclines in milk when quenching exists. Anal Chim Acta 2009;651:149–58.
[36] Rodríguez N, Ortiz MC, Sarabia LA. Fluorescence quantification of tetracycline in the presence of quenching matrix effect by means of a four-way model. Talanta 2009;77:1129–36.
[37] Kang C, Wu H-L, Xie L-X, Xiang S-X, Yu R-Q. Direct quantitative analysis of aromatic aminoacids in human plasma by four-way calibration using intrinsic fluorescence: exploration of third-order advantages. Talanta 2014;122:293–301.
[38] Jiménez Girón A, Durán-Merás I, Espinosa-Mansilla A, Muñoz de la Peña A, Cañada Cañada F, Olivieri AC. On line photochemically induced excitation-emission-kinetic four-way data: analytical application for the determination of folic acid and its two main metabolites in serum by U-PLS and N-PLS/residual trilinearization (RTL) calibration. Anal Chim Acta 2008;622:94–103.
[39] Ortiz MC, Sarabia LA, Sánchez MS, Giménez D. Identification and quantification of ciprofloxacin in urine through excitation-emission fluorescence and three-way PARAFAC calibration. Anal Chim Acta 2009;642:193–205.
[40] Sanz MB, Sarabia LA, Herrero A, Ortiz MC. The capability of discrimination: application to soft calibration methods. Anal Chim Acta 2001;446:297–311.
[41] Ortiz MC, Sarabia LA, Sánchez MS. Tutorial on evaluation of type I and type II errors in chemical analyses: from the analytical detection to authentication of products and process control. Anal Chim Acta 2010;674:123–42.
[42] Document No SANCO/12495/2011: Method validation and quality control procedures for pesticide residues analysis in food and feed. EU, Brussels; 2011.
[43] Commission Decision 2002/657/EC of 12 August 2002 implementing Council Directive 96/23/EC concerning the performance of analytical methods and the interpretation of results. Off J Eur Commun L221 [17.08.2002].
[44] Real BD, Ortiz MC, Sarabia LA. Develop a multiway chemometric-based analytical method fulfilling regulatory identification criteria: applications to GC-MS pesticide residue analysis. J Chromatogr B 2012;910:122–37.

[45] Morales R, Ortiz MC, Sarabia LA. Usefulness of a PARAFAC decomposition in the fiber selection procedure to determine chlorophenols by means SPME-GC-MS. Anal Bioanal Chem 2012;403(4):1095–107.
[46] Herrero A, Reguera C, Ortiz MC, Sarabia LA. Determination of dichlobenil and its major metabolite (BAM) in onions by PTV-GC-MS using PARAFAC2 and experimental design methodology. Chemometr Intell Lab Syst 2014;133:92–108.
[47] Oca ML, Ortiz MC, Herrero A, Sarabia LA. Optimization of a GC-MS procedure that uses parallel factor analysis for the determination of bisphenols and their diglycidyl ethers after migration from polycarbonate tableware. Talanta 2013;106:266–80.
[48] Rubio L, Sanllorente S, Sarabia LA, Ortiz MC. Optimization of a headspace solid-phase microextraction and gas chromatography/mass spectrometry procedure for the determination of aromatic amines in water and in polyamide spoons. Chemometr Intell Lab Syst 2014;133:121–35.
[49] García I, Sarabia L, Ortiz MC, Aldama JM. Robustness of the extraction step when parallel factor analysis (PARAFAC) is used to quantify sulfonamides in kidney by high performance liquid chromatography-diode array detection (HPLC-DAD). Analyst 2004;129(8):766–71.
[50] Arroyo D, Ortiz MC, Sarabia LA, Palacios F. Determination and identification, according to European Union Decision 2002/657/EC, of malachite green and its metabolite in fish by liquid chromatography-tandem mass spectrometry using a optimized extraction procedure and three-way calibration. J Chromatogr A 2009;1216:5472–82.
[51] Rubio L, Sarabia LA, Herrero A, Ortiz MC. Advantages of programmed temperature vaporizer inlet and parallel factor analysis in the determination of triazines in the presence of non-intentionally added substances by gas chromatography. Anal Bioanal Chem 2012;403(4):1131–43.
[52] Commission Regulation (EU) No. 310/2011 of 28 March 2011 amending Annexes II and III to Regulation (EC) No 396/2005 of the European Parliament and of the Council as regards maximum residue levels for aldicarb, bromopropylate, chlorfenvinphos, endosulfan, EPTC, ethion, fenthion, fomesafen, methabenzthiazuron, methidathion, simazine, tetradifon and triforine in or on certain products. Off J Eur Union L86:1 [1.4.2011].
[53] Regulation (EC) No. 396/2005 of the European Parliament and of the Council of 23 February 2005 on maximum residue levels of pesticides in or on food and feed of plant and animal origin and amending Council Directive 91/414/EEC. Off J Eur Union L70:1 [16.3.2005].
[54] Commission Regulation (EC) No. 149/2008 of 29 January 2008 amending Regulation (EC) No. 396/2005 of the European Parliament and of the Council by establishing Annexes II, III and IV setting maximum residue levels for products covered by Annex I thereto. Off J Eur Union L58:1 [1.3.2008].
[55] Anastassiades M, Lehotay SJ, Stajnbaher D, Schenck FJ. Fast and easy multiresidue method employing acetonitrile extraction/partitioning and "dispersive solid-phase extraction" for the determination of pesticide residues in produce. J AOAC Int 2003;86(2):412–31.
[56] Herrero A, Ortiz MC, Sarabia LA. D-optimal experimental design coupled with parallel factor analysis 2 decomposition: a useful tool in the determination of triazines in oranges by programmed temperature vaporization-gas chromatography–mass spectrometry when using dispersive-solid phase extraction. J Chromatogr A 2013;1288:111–26.
[57] Ortiz MC, Sarabia LA, García I, Giménez D, Meléndez E. Capability of detection and three-way data. Anal Chim Acta 2006;559:124–36.
[58] International Standard ISO 11843. Capability of detection. Part 1. Terms and definitions. Part 2. Methodology in the linear calibration case. Geneva, Switzerland: International Organisation for Standardisation; 1997 and 2000.

[59] Inczédy J, Lengyel T, Ure AM, Gelencsér A, Hulanicki A. Compendium of analytical nomenclature. 3rd ed. Baltimore: International Union of Pure and Applied Chemistry, Port City Press, Inc.; 2003, p. 74.

[60] Ortiz MC, Sánchez MS, Sarabia LA. Quality of analytical measurements: univariate regression. In: Brown S, Tauler R, Walczak R, editors. Comprehensive chemometrics, vol. 1. Oxford: Elsevier; 2009. p. 127–69.

[61] Skov T, Bro R. A new approach for modelling sensor based data. Sens Actuator B Chem 2005;106:719–29.

[62] Berna A. Metal oxide sensors for electronic noses and their application to food analysis. Sensors 2010;10(4):3882–910.

[63] Sarabia LA, Ortiz MC, González J, Jaime I, Rovira J. Avances de la Ciencia y Tecnología de los Alimentos en los inicios del siglo XXI. Burgos: Servicio de Publicaciones de la Universidad de Burgos; 2005. p.317–20.

[64] Meléndez E, Íñiguez M, Puras P, Ortiz MC, Sarabia LA, Dulau L. Cuadernos de Campo 46. Ed. por Gobierno de la Rioja, Consejería de Agricultura y Ganadería y Desarrollo Rural, Logroño (Spain); 2011. p. 34–9. ISSN:1137-2095. Available in http://www.larioja.org/npRioja/cache/documents/543616_campo_46.pdf?idtab=44508 [25.11.14].

[65] Meléndez E, Sarabia LA, Ortiz MC. PARAFAC model for monitoring data from a grape harvest in DOC Rioja. Chemometrics Intell Lab Syst, accepted for publication.

[66] Estación Enológica de Haro, Gobierno de la Rioja, Spain, http://www.larioja.org/npRioja/default/defaultpage.jsp?idtab=438528.

[67] Meléndez E, Ortiz MC, Sarabia LA, Íñiguez M, Puras P. Modelling phenolic and technological maturities of grapes by means of the multivariate relation between organoleptic and physicochemical properties. Anal Chim Acta 2013;761:53–61.

[68] Blouin J, Guimberteau G. Maturation et maturité des raisins. Bordeaux: Éditions Féret; 2000.

[69] Hidalgo Togores J. La calidad del vino desde el viñedo. Madrid: Muni-Prensa; 2006.

[70] Flanzy C. Enología: fundamentos científicos y tecnológicos. 2ª ed. Paris: Editorial Technique et documentation; 2003.

Chapter 3

# Multiway Calibration Based on Alternating Multilinear Decomposition

**Hai-Long Wu[1], Yong Li, Chao Kang and Ru-Qin Yu[1]**
*State Key Laboratory of Chemo/Biosensing and Chemometrics, College of Chemistry and Chemical Engineering, Hunan University, Changsha, Hunan, China*
*[1]Corresponding authors: e-mail: hlwu@hnu.edu.cn; rqyu@hnu.edu.cn*

## Chapter Outline



Data Handling in Science and Technology, Vol. 29. http://dx.doi.org/10.1016/B978-0-444-63527-3.00003-5

# 1 INTRODUCTION

With the increasing number of available analytical instruments capable of generating multidimensional arrays of experimental data per sample, the multiway data analysis has attracted more and more attention and gained widespread acceptance in many scientific fields, such as chemistry, medicine, food safety, environmental, and single-cell sciences, as could be seen from an explosion of relevant literature [1–17]. Multiway or higher-order (second-order, third-order, fourth-order,...) calibration can compensate for potential interferences not included in the calibration set and accurately estimate the concentrations of individual components of interest. This property, usually referred to as the "second-order advantage" and "higher-order advantage," bestows an immense potential in multicomponent analysis and becomes a hotspot of theoretical interest and of intensive experimental research.

In this chapter, some aspects of multiway calibration such as the multilinear models, the multiway cyclic symmetry property, the algorithms for multiway calibration, the estimation of the chemical rank, the MATLAB toolbox for multiway calibration, and other fundamental issues, will be briefly described. The discussion will be mainly focused on second-order and third-order as well as fourth-order calibration based on alternating multilinear decomposition.

# 2 TERMINOLOGY AND NOMENCLATURE IN MULTIWAY DATA ANALYSIS

Some confusion and misunderstandings may arise in multivariate analysis on multiway data arrays. Therefore, a prior definition of the terminology and nomenclature adopted is necessary.

## 2.1 Terminology

The relationship and distinction should firstly be made between the concepts of "data order" and "data way." The terms "order" and "way" derived from tensor algebra are employed to refer to the dimensions for data of a single sample and data arrays stacked by a series of samples with similar properties, respectively. As illustrated in Figure 1, the instrumental response per sample can be zeroth-order (a scalar), first-order (a vector), second-order (a matrix), third-order (a three-dimensional array), and higher-order tensors, which yield one-way, two-way, three-way, four-way, and $N$-way data sets, respectively, when a group of samples are joined together. The procedure performed on zeroth-order tensor is known as univariate calibration, which requires full selectivity of the signal with respect to the analyte of interest. Whereas all the methodologies except

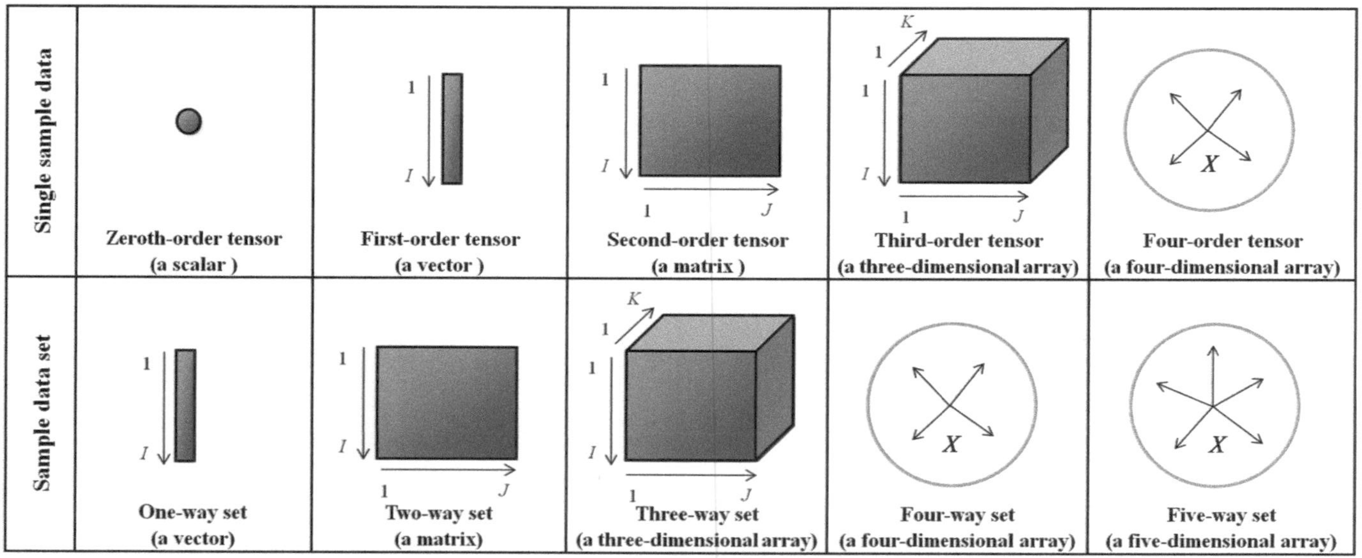


**FIGURE 1** Relationship and distinction between "data order" and "data way."

zeroth-order tensor for the analysis of data are classified into multivariate calibration, the analysis of second-order tensor and beyond is denoted as multiway multivariate or multicomponent calibration in the sense that it is a multivariate calibration referring to multiway data. The experimental data analyzed by our group are primarily second-order, third-order, and fourth-order as well as fifth-order tensors, among them the second-order tensor is mostly recorded by excitation–emission matrix (EEM) fluorescence spectroscopy, high-performance liquid chromatography-photodiode array detection (HPLC-DAD) as well as by fluorescence or UV–visible spectrophotometry following the chemical reaction kinetics in time. In addition, through introducing extra dimensions, third-order, fourth-order, or fifth-order tensors can also be achieved, such as EEM as a function of pH or/and reaction time.

## 2.2 Nomenclature

Scalars are denoted by lowercase italics; bold lowercase characters mean vectors; bold capitals designate two-way matrices; italic bold capitals represent three-way or higher-way arrays, and the superscript T symbolizes the transpose of a matrix, the superscript "+" means the Moore–Penrose generalized inverse of a matrix. $\|\cdot\|_F$ denotes the Frobenius matrix norm. Tables 1 and 2 list Terminologies in second- and third-order calibrations. Theoretically, the four-and five-way cyclic symmetries for quadrilinear and quinquelinear decomposition, respectively, are similar to the three-way cyclic symmetry and easy obtainable by simple mathematical manipulation of exchanging the symbols similar to three-way cyclic symmetry. These regularities provide helpful guide for the standardization of symbol systems in multiway data analysis, for deeper understanding of the essence of multiway multilinear decomposition, and for developing new multiway calibration algorithms and exploring multilinear algebra in mathematics.

# 3 MULTILINEAR MODELS

## 3.1 Trilinear Model

The PARAFAC (PARallel FACtor analysis) model was introduced in 1970 by Harshman [18] and simultaneously by Carroll and Chang [19] under the name CANDECOMP. For brevity, it is often referred as the PARAFAC model in the chemometric literature. In this trilinear model, the elements $x_{ijk}$ of a three-way array $\boldsymbol{X}$ $(I \times J \times K)$ are expressed as sums of products of elements from the matrices, **A**, **B**, and **C**, according to the following equation:

$$x_{ijk} = \sum_{n=1}^{N} a_{in} b_{jn} c_{kn} + e_{ijk}, \quad \text{for } i = 1,2,\ldots,I;\ j = 1,2,\ldots,J;\ k = 1,2,\ldots,K, \qquad (1)$$

where $N$ denotes the number of factors, which should correspond to the total number of detectable species, consisting of the component(s) of interest and

**TABLE 1 Terminology in Second-Order Calibration**

| | |
|---|---|
| $\boldsymbol{X}$ $(I \times J \times K)$ | Three-way data array |
| $\boldsymbol{E}$ $(I \times J \times K)$ | Three-way residue array |
| $x_{ijk}$, | The $ijk$th element of $\boldsymbol{X}$ $(I \times J \times K)$ |
| $e_{ijk}$ | The $ijk$th element of $\boldsymbol{E}$ $(I \times J \times K)$ |
| $I$, $J$, $K$ | The dimensions of three modes of $\boldsymbol{X}$ $(I \times J \times K)$ |
| $\mathbf{A}_{I\times N}$, $\mathbf{B}_{J\times N}$, $\mathbf{C}_{K\times N}$ | The three underlying profile matrices of $\boldsymbol{X}$ $(I \times J \times K)$ with $I \times N$, $J \times N$, $K \times N$, respectively |
| $a_{in}$, $b_{jn}$, $c_{kn}$ | The $in$th, $jn$th, and $kn$th elements of the three underlying profile matrices **A**, **B**, and **C**, respectively |
| $\mathbf{a}_{(i)}$, $\mathbf{b}_{(j)}$, $\mathbf{c}_{(k)}$ | The $i$th, $j$th, $k$th row vectors of profile matrices **A**, **B**, and **C**, respectively |
| diag($\mathbf{a}_{(i)}$), diag($\mathbf{b}_{(j)}$), diag($\mathbf{c}_{(k)}$) | Diagonal matrices with elements equal to the elements of $\mathbf{a}_{(i)}$, $\mathbf{b}_{(j)}$, and $\mathbf{c}_{(k)}$, respectively |
| $\mathbf{X}_{i..}$, $\mathbf{X}_{.j.}$, $\mathbf{X}_{..k}$ | The $i$th horizontal, $j$th lateral, and $k$th frontal slices of $\boldsymbol{X}$ $(I \times J \times K)$, respectively |
| $\mathbf{E}_{i..}$, $\mathbf{E}_{.j.}$, $\mathbf{E}_{..k}$ | The $i$th horizontal, $j$th lateral, and $k$th frontal slices of the three-way residue array $\boldsymbol{E}$ $(I \times J \times K)$, respectively |

the background as well as unknown interferences. Figure 2 shows the graphical representation of trilinear model of three-way data array $\boldsymbol{X}$. To further illustrate the mathematic meaning of the trilinear model expressed with graphics, the response data array $\boldsymbol{X}$ is returned with three inverse steps as described in Figure 3.

The trilinear model could be expressed in vector forms as:

$$\mathbf{X}_{ij.} = \mathbf{b}_{(j)} diag\left(\mathbf{a}_{(i)}\right)\mathbf{C}^{\mathrm{T}} + \mathbf{E}_{ij.}, \quad \text{for } i = 1,2,\ldots,I;\ j = 1,2,\ldots,J, \tag{2}$$

$$\mathbf{X}_{.jk} = \mathbf{c}_{(k)} diag\left(\mathbf{b}_{(j)}\right)\mathbf{A}^{\mathrm{T}} + \mathbf{E}_{.jk}, \quad \text{for } j = 1,2,\ldots,J;\ k = 1,2,\ldots,K, \tag{3}$$

$$\mathbf{X}_{i.k} = \mathbf{a}_{(i)} diag\left(\mathbf{c}_{(k)}\right)\mathbf{B}^{\mathrm{T}} + \mathbf{E}_{i.k}, \quad \text{for } i = 1,2,\ldots,I;\ k = 1,2,\ldots,K. \tag{4}$$

The trilinear model could be expressed in slice forms as:

$$\mathbf{X}_{i..} = \mathbf{B} diag\left(\mathbf{a}_{(i)}\right)\mathbf{C}^{\mathrm{T}} + \mathbf{E}_{i..}, \quad \text{for } i = 1,2,\ldots,I, \tag{5}$$

$$\mathbf{X}_{.j.} = \mathbf{C} diag\left(\mathbf{b}_{(j)}\right)\mathbf{A}^{\mathrm{T}} + \mathbf{E}_{.j.}, \quad \text{for } j = 1,2,\ldots,J, \tag{6}$$

$$\mathbf{X}_{..k} = \mathbf{A} diag\left(\mathbf{c}_{(k)}\right)\mathbf{B}^{\mathrm{T}} + \mathbf{E}_{..k}, \quad \text{for } k = 1,2,\ldots,K. \tag{7}$$

**TABLE 2** Terminology in Third-Order Calibration

| | |
|---|---|
| $\boldsymbol{X}$ $(I \times J \times K \times L)$ | Four-way data array |
| $\boldsymbol{E}$ $(I \times J \times K \times L)$ | Four-way residue array |
| $x_{ijkl}$ | The *ijkl*th element of $\boldsymbol{X}$ $(I \times J \times K \times L)$ |
| $e_{ijkl}$ | The *ijkl*th element of $\boldsymbol{E}$ $(I \times J \times K \times L)$ |
| $I, J, K, L$ | The dimensions of four modes of $\boldsymbol{X}$ $(I \times J \times K \times L)$ |
| $\mathbf{A}_{I \times N}, \mathbf{B}_{J \times N}, \mathbf{C}_{K \times N}, \mathbf{D}_{L \times N}$ | The four underlying profile matrices of $\boldsymbol{X}$ $(I \times J \times K \times L)$ with $I \times N$, $J \times N$, $K \times N$, $L \times N$, respectively |
| $a_{in}, b_{jn}, c_{kn}, d_{ln}$ | The *in*th, *jn*th, *kn*th, and *ln*th elements of the four underlying profile matrices **A**, **B**, **C**, and **D**, respectively |
| $\mathbf{a}_{(i)}, \mathbf{b}_{(j)}, \mathbf{c}_{(k)}, \mathbf{d}_{(l)}$ | The *in*th, *jn*th, *kn*th, and *ln*th row vectors of profile matrices **A**, **B**, **C**, and **D**, respectively |
| $\text{diag}(\mathbf{a}_{(i)})$, $\text{diag}(\mathbf{b}_{(j)})$, $\text{diag}(\mathbf{c}_{(k)})$, $\text{diag}(\mathbf{d}_{(l)})$ | Diagonal matrices with elements equal to the elements of $\mathbf{a}_{(i)}$, $\mathbf{b}_{(j)}$, $\mathbf{c}_{(k)}$, and $\mathbf{d}_{(l)}$, respectively |
| $\mathbf{X}_{ij..}, \mathbf{X}_{.jk.}, \mathbf{X}_{..kl}, \mathbf{X}_{i..l}$ | The slices of $\boldsymbol{X}$ $(I \times J \times K \times L)$ |
| $\mathbf{X}_{I \times JKL}, \mathbf{X}_{J \times KLI}, \mathbf{X}_{K \times LIJ}, \mathbf{X}_{L \times IJK}$ | The fully stretched matrices of $\boldsymbol{X}$ $(I \times J \times K \times L)$ |
| $\mathbf{X}_{i...}, \mathbf{X}_{.j..}, \mathbf{X}_{..k.}, \mathbf{X}_{...l}$ | Not fully stretched matrices of $\boldsymbol{X}$ $(I \times J \times K \times L)$ |
| $\mathbf{E}_{ij..}, \mathbf{E}_{.jk.}, \mathbf{E}_{..kl}, \mathbf{E}_{i..l}$ | The slices of $\boldsymbol{E}$ $(I \times J \times K \times L)$ |

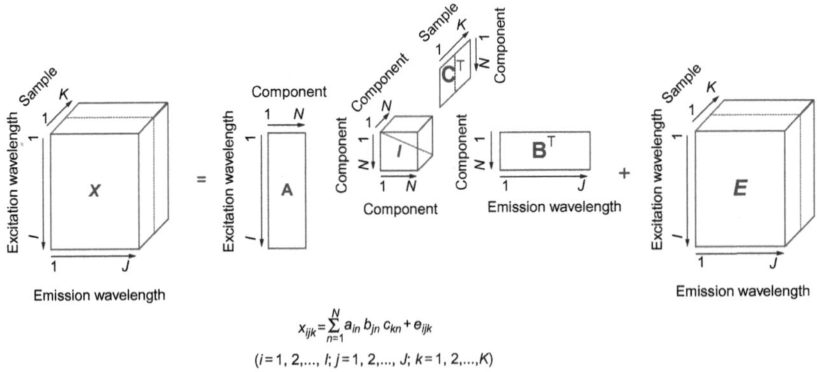


**FIGURE 2** Graphical representation of the trilinear model of three-way fluorescence data array $X$ $(I \times J \times K)$. *See Ref. [2], [3], [15].*

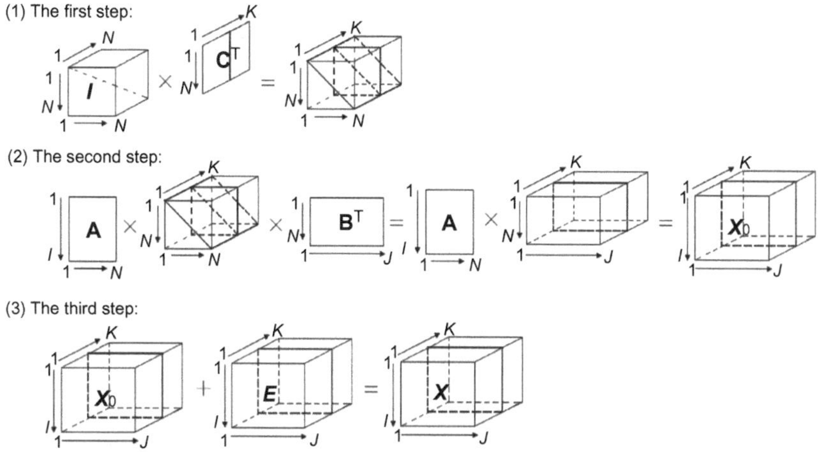


FIGURE 3 Inverse procedure to return the three-way data array $X$ $(I \times J \times K)$.

The trilinear model could be expressed in three stretched matrices:

$$\mathbf{X}_{I\times JK} = \mathbf{A}(\mathbf{C}\odot\mathbf{B})^{\mathrm{T}} + \mathbf{E}_{I\times JK}, \tag{8}$$

$$\mathbf{X}_{J\times KI} = \mathbf{B}(\mathbf{A}\odot\mathbf{C})^{\mathrm{T}} + \mathbf{E}_{J\times KI}, \tag{9}$$

$$\mathbf{X}_{K\times IJ} = \mathbf{C}(\mathbf{B}\odot\mathbf{A})^{\mathrm{T}} + \mathbf{E}_{K\times IJ}, \tag{10}$$

here, $\odot$ denotes the Khatri-Rao product.

## 3.2 Quadrilinear Model

Suppose that third-order data are collected in an $I \times J \times K \times L$ four-way data array $X$ $(I \times J \times K \times L)$, each element $x_{ijkl}$ of the four-way data array can be represented as the following form [2,3]:

$$x_{ijkl} = \sum_{n=1}^{N} a_{in} b_{jn} c_{kn} d_{ln} + e_{ijkl}, \quad \text{for } i = 1,2,\ldots,I;\ j = 1,2,\ldots,J;\ k = 1,2,\ldots,K;\ l = 1,2,\ldots,L. \tag{11}$$

Figure 4 shows the graphical representation of the quadrilinear model of four-way data array.

The quadrilinear model can be, respectively, expressed in four fully equivalent slice-wise matrices as:

$$\mathbf{X}_{..kl} = \mathbf{A}\,diag\left(\mathbf{d}_{(l)}\right)diag\left(\mathbf{c}_{(k)}\right)\mathbf{B}^{\mathrm{T}} + \mathbf{E}_{..kl}, \quad \text{for } k = 1,2,\ldots,K;\ l = 1,2,\ldots,L, \tag{12}$$

$$\mathbf{X}_{i..l} = \mathbf{B}\,diag\left(\mathbf{a}_{(i)}\right)diag\left(\mathbf{d}_{(l)}\right)\mathbf{C}^{\mathrm{T}} + \mathbf{E}_{i..l}, \quad \text{for } l = 1,2,\ldots,L;\ i = 1,2,\ldots,I, \tag{13}$$

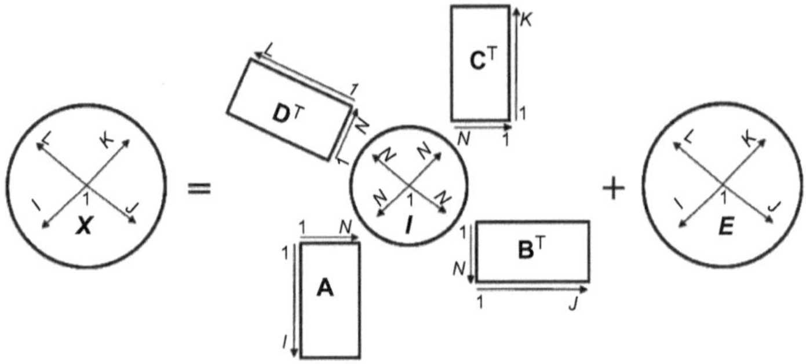


$$x_{ijkl} = \sum_{n=1}^{N} a_{in} b_{jn} c_{kn} d_{ln} + e_{ijkl}$$

$(i = 1, 2, \ldots, I;\ j = 1, 2, \ldots, J;\ k = 1, 2, \ldots, K;\ l = 1, 2, \ldots, L)$

**FIGURE 4** Graphical representation of the quadrilinear model of four-way data array $\boldsymbol{X}$ $(I \times J \times K \times L)$.

$$\mathbf{X}_{ij..} = \mathbf{C}\,diag\left(\mathbf{b}_{(j)}\right)diag\left(\mathbf{a}_{(i)}\right)\mathbf{D}^{\mathrm{T}} + \mathbf{E}_{ij..}, \quad \text{for } i = 1,2,\ldots,I;\ j = 1,2,\ldots,J, \tag{14}$$

$$\mathbf{X}_{.jk.} = \mathbf{D}\,diag\left(\mathbf{c}_{(k)}\right)diag\left(\mathbf{b}_{(j)}\right)\mathbf{A}^{\mathrm{T}} + \mathbf{E}_{.jk.}, \quad \text{for } j = 1,2,\ldots,J;\ k = 1,2,\ldots,K, \tag{15}$$

and four fully equivalent stretched matrices as:

$$\mathbf{X}_{I \times JKL} = \mathbf{A}(\mathbf{D} \odot \mathbf{C} \odot \mathbf{B})^{\mathrm{T}} + \mathbf{E}_{I \times JKL}, \tag{16}$$

$$\mathbf{X}_{J \times KLI} = \mathbf{B}(\mathbf{A} \odot \mathbf{D} \odot \mathbf{C})^{\mathrm{T}} + \mathbf{E}_{J \times KLI}, \tag{17}$$

$$\mathbf{X}_{K \times LIJ} = \mathbf{C}(\mathbf{B} \odot \mathbf{A} \odot \mathbf{D})^{\mathrm{T}} + \mathbf{E}_{K \times LIJ}, \tag{18}$$

$$\mathbf{X}_{L \times IJK} = \mathbf{D}(\mathbf{C} \odot \mathbf{B} \odot \mathbf{A})^{\mathrm{T}} + \mathbf{E}_{L \times IJK}. \tag{19}$$

For the quadrilinear model, not fully stretched matrix form can be expressed as:

$$\mathbf{X}_{...l} = \mathbf{A}\,diag\left(\mathbf{d}_{(l)}\right)(\mathbf{C} \odot \mathbf{B})^{\mathrm{T}} + \mathbf{E}_{...l}, \text{ for } l = 1,2,\ldots,L, \tag{20}$$

$$\mathbf{X}_{i...} = \mathbf{B}\,diag\left(\mathbf{a}_{(i)}\right)(\mathbf{D} \odot \mathbf{C})^{\mathrm{T}} + \mathbf{E}_{i...}, \text{ for } i = 1,2,\ldots,I, \tag{21}$$

$$\mathbf{X}_{.j..} = \mathbf{C}\,diag\left(\mathbf{b}_{(j)}\right)(\mathbf{A} \odot \mathbf{D})^{\mathrm{T}} + \mathbf{E}_{.j..}, \text{ for } j = 1,2,\ldots,J, \tag{22}$$

$$\mathbf{X}_{..k.} = \mathbf{D}\,diag\left(\mathbf{c}_{(k)}\right)(\mathbf{B} \odot \mathbf{A})^{\mathrm{T}} + \mathbf{E}_{..k.} \text{ for } k = 1,2,\ldots,K. \tag{23}$$

## 3.3 Quinquelinear Model

Similar to the trilinear and quadrilinear models, the quinquelinear model can be obtained as:

$$x_{ijklm} = \sum_{n=1}^{N} a_{in} b_{jn} c_{kn} d_{ln} f_{mn} + e_{ijklm}, \quad \text{for } i = 1,2,\ldots,I;\ j = 1,2,\ldots,J; \\ k = 1,2,\ldots,K;\ l = 1,2,\ldots,L;\ m = 1,2,\ldots,M, \tag{24}$$

where $x_{ijklm}$ is the element of one five-way data array $\boldsymbol{X}$ $(I \times J \times K \times L \times M)$.

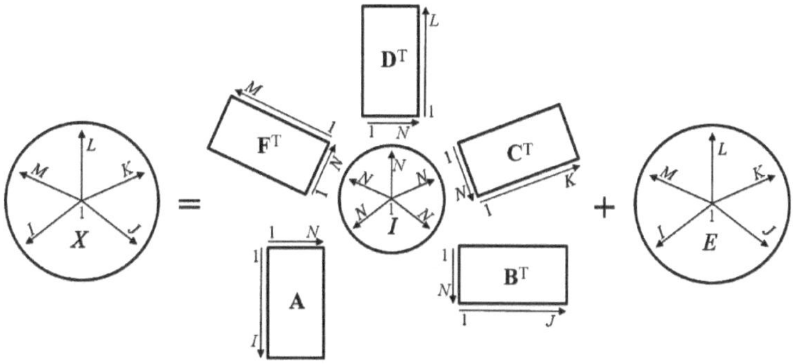


$$x_{ijklm} = \sum_{n=1}^{N} a_{in} b_{jn} c_{kn} d_{ln} f_{mn} + e_{ijklm}$$

$$(i = 1,2,...,I; j = 1,2,...,J; k = 1,2,...,K; l = 1,2,...,L; m = 1,2,...,M)$$

**FIGURE 5** Graphical representation of the quinquelinear model of five-way data array $X$ $(I \times J \times K \times L \times M)$.

Figure 5 shows the graphical representation of quinquelinear model of five-way data array.

The slice, stretched matrix, and not-fully stretched matrix forms can also be acquired.

$$\mathbf{X}_{..klm} = \mathbf{A}\,diag(\mathbf{f}_{(m)})\,diag(\mathbf{d}_{(l)})\,diag(\mathbf{c}_{(k)})\mathbf{B}^{\mathrm{T}} + \mathbf{E}_{..klm}, \quad \text{for } k = 1,2,\ldots,K; \; l = 1,2,\ldots,L; \; m = 1,2,\ldots,M, \tag{25}$$

$$\mathbf{X}_{i..lm} = \mathbf{B}\,diag(\mathbf{a}_{(i)})\,diag(\mathbf{f}_{(m)})\,diag(\mathbf{d}_{(l)})\mathbf{C}^{\mathrm{T}} + \mathbf{E}_{i..lm}, \quad \text{for } l = 1,2,\ldots,L; \; m = 1,2,\ldots,M; \; i = 1,2,\ldots,I, \tag{26}$$

$$\mathbf{X}_{ij..m} = \mathbf{C}\,diag(\mathbf{b}_{(j)})\,diag(\mathbf{a}_{(i)})\,diag(\mathbf{f}_{(m)})\mathbf{D}^{\mathrm{T}} + \mathbf{E}_{ij..m}, \quad \text{for } m = 1,2,\ldots,M; \; i = 1,2,\ldots,I; \; j = 1,2,\ldots,J, \tag{27}$$

$$\mathbf{X}_{ijk..} = \mathbf{D}\,diag(\mathbf{c}_{(k)})\,diag(\mathbf{b}_{(j)})\,diag(\mathbf{a}_{(i)})\mathbf{F}^{\mathrm{T}} + \mathbf{E}_{ijk..}, \quad \text{for } i = 1,2,\ldots,I; \; j = 1,2,\ldots,J; \; k = 1,2,\ldots,K, \tag{28}$$

$$\mathbf{X}_{.jkl.} = \mathbf{F}\,diag(\mathbf{d}_{(l)})\,diag(\mathbf{c}_{(k)})\,diag(\mathbf{b}_{(j)})\mathbf{A}^{\mathrm{T}} + \mathbf{E}_{.jkl.}, \quad \text{for } j = 1,2,\ldots,J; \; k = 1,2,\ldots,K; \; l = 1,2,\ldots,L. \tag{29}$$

The fully equivalent stretched matrices can be expressed as:

$$\mathbf{X}_{I \times JKLM} = \mathbf{A}(\mathbf{F} \odot \mathbf{D} \odot \mathbf{C} \odot \mathbf{B})^{\mathrm{T}} + \mathbf{E}_{I \times JKLM}, \tag{30}$$

$$\mathbf{X}_{J \times KLMI} = \mathbf{B}(\mathbf{A} \odot \mathbf{F} \odot \mathbf{D} \odot \mathbf{C})^{\mathrm{T}} + \mathbf{E}_{J \times KLMI}, \tag{31}$$

$$\mathbf{X}_{K \times LMIJ} = \mathbf{C}(\mathbf{B} \odot \mathbf{A} \odot \mathbf{F} \odot \mathbf{D})^{\mathrm{T}} + \mathbf{E}_{K \times LMIJ}, \tag{32}$$

$$\mathbf{X}_{L\times MIJK} = \mathbf{D}(\mathbf{C}\odot\mathbf{B}\odot\mathbf{A}\odot\mathbf{F})^{\mathrm{T}} + \mathbf{E}_{L\times MIJK}, \tag{33}$$

$$\mathbf{X}_{M\times IJKL} = \mathbf{F}(\mathbf{D}\odot\mathbf{C}\odot\mathbf{B}\odot\mathbf{A})^{\mathrm{T}} + \mathbf{E}_{M\times IJKL}. \tag{34}$$

Not-fully stretched matrix form can be expressed as:

$$\mathbf{X}_{....m} = \mathbf{A}diag\left(\mathbf{f}_{(m)}\right)(\mathbf{D}\odot\mathbf{C}\odot\mathbf{B})^{\mathrm{T}} + \mathbf{E}_{....m}, \text{ for } m = 1,2,\ldots,M, \tag{35}$$

$$\mathbf{X}_{i...} = \mathbf{B}diag\left(\mathbf{a}_{(i)}\right)(\mathbf{F}\odot\mathbf{D}\odot\mathbf{C})^{\mathrm{T}} + \mathbf{E}_{i....}, \text{ for } i = 1,2,\ldots,I, \tag{36}$$

$$\mathbf{X}_{.j...} = \mathbf{C}diag\left(\mathbf{b}_{(j)}\right)(\mathbf{A}\odot\mathbf{F}\odot\mathbf{D})^{\mathrm{T}} + \mathbf{E}_{.j...}, \text{ for } j = 1,2,\ldots,J, \tag{37}$$

$$\mathbf{X}_{..k...} = \mathbf{D}diag\left(\mathbf{c}_{(k)}\right)(\mathbf{B}\odot\mathbf{A}\odot\mathbf{F})^{\mathrm{T}} + \mathbf{E}_{..k..}, \text{ for } k = 1,2,\ldots,K, \tag{38}$$

$$\mathbf{X}_{...l.} = \mathbf{F}diag\left(\mathbf{d}_{(l)}\right)(\mathbf{C}\odot\mathbf{B}\odot\mathbf{A})^{\mathrm{T}} + \mathbf{E}_{...l.}, \text{ for } l = 1,2,\ldots,L. \tag{39}$$

# 4 ADVANTAGES OF MULTIWAY CALIBRATION

## 4.1 Uniqueness Property

The uniqueness property in multilinear decomposition means that only permutations and scaling transformations are allowed without altering the residuals of the fit [20–22]. Wu and coworkers studied the uniqueness of trilinear decomposition in a true trilinear sense, i.e., from the aspect of space [23]. To express this algebraically, an alternative solution is represented as:

$$\mathbf{A}^* = \mathbf{A}\times\mathbf{D}_a\times\mathbf{P};\ \mathbf{B}^* = \mathbf{B}\times\mathbf{D}_b\times\mathbf{P};\ \mathbf{C}^* = \mathbf{C}\times\mathbf{D}_c\times\mathbf{P}. \tag{40}$$

Here the $\mathbf{D}_a$, $\mathbf{D}_b$, and $\mathbf{D}_c$ must conform to the requisite condition, which is shown in Figure 6. Since the effect of any internal rescaling (described by the **D** matrices) must "cancel out." Here ***I*** is a three-way (superdiagonal) core array of size $N\times N\times N$ with ones on the superdiagonal and zeros elsewhere; **P** is a matrix for elementary transformation. These indeterminacies can be

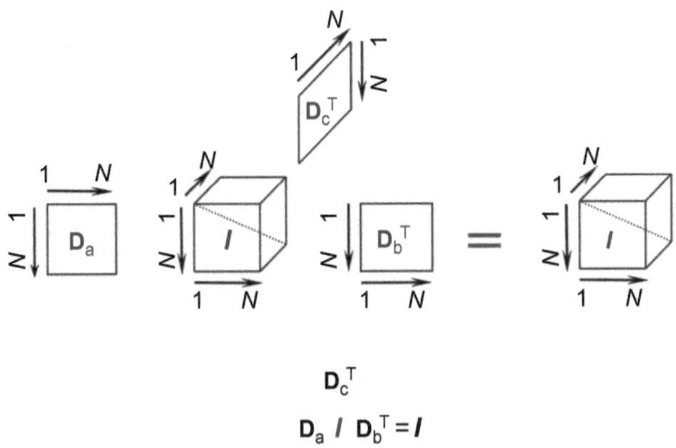


**FIGURE 6** Requisite condition which $\mathbf{D}_a$, $\mathbf{D}_b$ and $\mathbf{D}_c$ must conform to in the trilinear model. *Adapted from Ref. [3], figure 5, p. 482.*

considered trivial because they do not affect the interpretation of the solutions. Geometrically, the effect of the permutation matrix **P** is simply to "renumber" the axes without moving them, and the effect of the rescaling $\mathbf{D}_a$ and so forth is simply to stretch or contract the axis in one space with compensatory contraction or stretch of the corresponding axis in the space of another mode.

## 4.2 Second- and Third-Order Advantages

The trilinear data analysis possesses so-called second-order advantage, e.g., several components of interest can be quantified even in the presence of unknown interferents by using such calibration methodology. The first experimental demonstration of second-order advantage was reported in 1978, when perylene was quantified by the rank annihilation method from multicomponent fluorescence data [24]. However, the term "second-order advantage" was not coined until 1994 by the group of Kowalski [1]. Based on this advantage, second-order calibration coupled with some available analytical instruments, such as EEM fluorescence and HPLC-DAD, has been applied in many scientific fields.

Theoretically, the "third-order advantage," held by third-order calibration in theory, not only retains the second-order advantage (inherent in second-order calibration) but can also have additional advantages [9]:

**(1)** It offers the possibility of decomposing the data array for each sample exclusively, independent of other samples;
**(2)** It offers enhancement in sensitivity and selectivity, which is a characteristic feature that accompanies multiway data with increase in its dimension;
**(3)** The introduction of fourth mode can relieve the serious problem of collinearity. However, there is a lack of general consensus among the chemometrics community on the existence of third- or higher-order advantage.

## 4.3 Multiway Cyclic Symmetry

Before perusing the following parts of this chapter, readers are recommended to refer an inner law for trilinear decomposition which is firstly discovered by our laboratory and named as three-way cyclic symmetry (see Figure 7) [2,3,15]. As depicted in Figure 7, physical modes, subscripts, elements, and vectors in resolved matrices, sliced matrices, and unfolded matrices, as well as residue and resolution formulas, all of them obey the inner law, circumrotating along the same way. The component matrices are equivalent from the calculation point of view, in spite of their differences in physical/chemical meaning. When the three-way data set $\boldsymbol{X}$ rotates every 120° on its rotation axis (Figure 8), the component matrices will be exchanged circularly and equivalently. Consequently, the mathematical derivation of one component matrix manipulation can be taken as a demonstration, while those for others can be obtained just by exchanging the symbols according to three-way cyclic

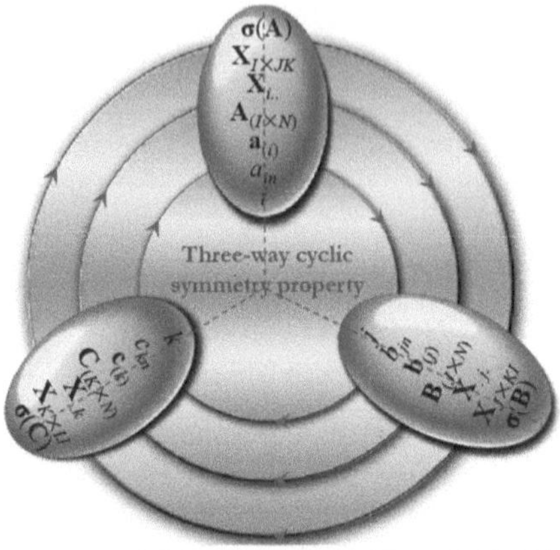


$$\mathbf{a}_{(i)}^{\mathrm{T}} = diagm(\mathbf{B}^{+}\mathbf{X}_{i..}(\mathbf{C}^{\mathrm{T}})^{+}),\quad (i=1,2,\ldots,I)$$
$$\mathbf{X}_{I\times JK} = [\mathbf{X}_{..1}, \mathbf{X}_{..2}, \ldots, \mathbf{X}_{..K}]$$
$$\sigma(\mathbf{A}) = \sum_{i=1}^{I} \left\| \mathbf{X}_{i..} - \mathbf{B}diag(\mathbf{a}_{(i)})\mathbf{C}^{\mathrm{T}} \right\|_{\mathrm{F}}^{2}$$

$$\mathbf{b}_{(j)}^{\mathrm{T}} = diagm(\mathbf{C}^{+}\mathbf{X}_{.j.}(\mathbf{A}^{\mathrm{T}})^{+}),\quad (j=1,2,\ldots,J)$$
$$\mathbf{X}_{J\times KI} = [\mathbf{X}_{1..}, \mathbf{X}_{2..}, \ldots, \mathbf{X}_{I..}]$$
$$\sigma(\mathbf{B}) = \sum_{j=1}^{J} \left\| \mathbf{X}_{.j.} - \mathbf{C}diag(\mathbf{b}_{(j)})\mathbf{A}^{\mathrm{T}} \right\|_{\mathrm{F}}^{2}$$

$$\mathbf{c}_{(k)}^{\mathrm{T}} = diagm(\mathbf{A}^{+}\mathbf{X}_{..k}(\mathbf{B}^{\mathrm{T}})^{+}),\quad (k=1,2,\ldots,K)$$
$$\mathbf{X}_{K\times IJ} = [\mathbf{X}_{.1.}, \mathbf{X}_{.2.}, \ldots, \mathbf{X}_{.J.}]$$
$$\sigma(\mathbf{C}) = \sum_{k=1}^{K} \left\| \mathbf{X}_{..k} - \mathbf{A}diag(\mathbf{c}_{(k)})\mathbf{B}^{\mathrm{T}} \right\|_{\mathrm{F}}^{2}$$

**FIGURE 7** Visualization of the three-way cyclic symmetry property for the trilinear component model.

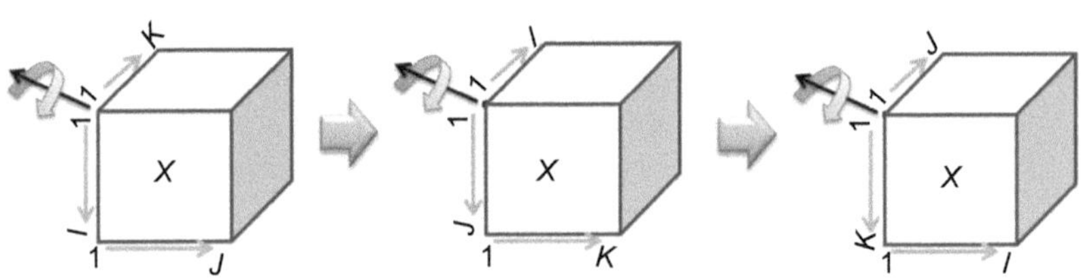


**FIGURE 8** Schema for rotation of a three-way data set.

symmetry. These regularities provide a sound theoretical basis for the standardization of symbols used in multiway data analysis, which is very helpful for deeper understanding the essence of multiway multilinear decomposition and developing new multiway calibration methods and algorithms by exploiting the fundamentals of multilinear algebra. These observations can also be extended to higher-order calibration, such as third-order calibration. Similar to three-way cyclic symmetry, four-way cyclic symmetry has been given in Ref. [3], which is shown in Figure 9.

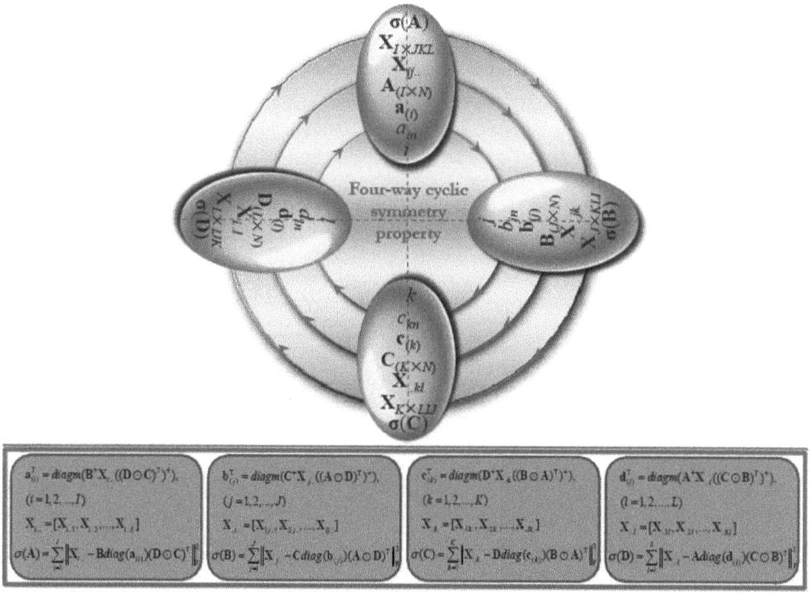


**FIGURE 9** Visualization of the four-way cyclic symmetry property for the quadrilinear component model.

# 5 ALGORITHMS FOR MULTIWAY CALIBRATION

## 5.1 Second-Order Calibration

Algorithms for second-order calibration can be classified into three groups and are described in the following sections (Table 3). The first group includes noniterative methods. The earliest one is the rank annihilation factor analysis (RAFA) [25]. It makes use of relationship that equates the component number in sample signals to the rank of data matrix. RAFA can quantify a single component in a complex mixture at a time without accounting for other components. In 1986, RAFA was improved by Kowalski and coworkers and was renamed generalized rank annihilation method (GRAM) [26]. The enhancements in GRAM convert the computational problem of RAFA to the resolution of characteristic equations, and with this modification, GRAM can determine the concentration of several analytes in a single sample at a time. A related variation of GRAM, presented by Gemperline and coworkers in 1992, can transform complex eigenvalues and eigenvectors into real eigenvalues and eigenvectors and allow the estimation of the spectra and profiles of pure constituents [49]. Following this, Kowalski and coworkers developed a new method, direct trilinear decomposition (DTLD) [27], which, like the GRAM method, estimates the intrinsic profiles of each component present in each sample by solving an eigenvector/eigenvalue problem. The relative concentration of each component between the samples is found by the least

**TABLE 3 Algorithms for Second-Order Calibration**

| Group | Algorithm | Acronym | References |
|---|---|---|---|
| Noniterative | (1) Rank annihilation factor analysis | RAFA | [25] |
| | (2) Generalized rank annihilation method | GRAM | [26] |
| | (3) Direct Trilinear Decomposition | DTLD | [27] |
| Residual bilinearization | (1) Bilinear least squares/ residual bilinearization | BLLS/RBL | [28] |
| | (2) Unfold partial least squares/ residual bilinearization | U-PLS/RBL | [29] |
| | (3)Multiway partial least squares/ residual bilinearization | N-PLS/RBL | [29] |
| | (4) Unfold principal component analysis/residual bilinearization | U-PCA/RBL | [9] |
| | (5) artificial neural network/ residual bilinearization | ANN/RBL | [9] |
| Iterative | (1) Parallel factor analysis | PARAFAC | [30] |
| | (2) Multivariate curve resolution with alternating least squares | MCR-ALS | [31] |
| | (3) Alternating trilinear decomposition | ATLD | [32] |
| | (4) Self-weighted alternating trilinear decomposition | SWATLD | [33] |
| | (5) Alternating penalty trilinear decomposition | APTLD | [34] |
| | (6) Alternating asymmetric trilinear decomposition | AATLD | [35] |
| | (7) Coupled vectors resolution | COVER | [36] |
| | (8) Alternating coupled vectors resolution | ACOVER | [37] |
| | (9) Alternating slice-wise diagonalization | ASD | [38] |
| | (10) Alternating coupled matrix resolution | ACOMAR | [39] |
| | (11) Pseudo alternating least squares | PALS | [40] |

**TABLE 3 Algorithms for Second-Order Calibration—Cont'd**

| Group | Algorithm | Acronym | References |
|---|---|---|---|
| | (12) Constrained PARAFAC | CPARAFAC | [41] |
| | (13) Full-rank parallel factor analysis | FRA-PARAFAC | [42] |
| | (14) Penalty diagonalization error | PDE | [43] |
| | (15) Alternating fitting residue | AFR | [44] |
| | (16) Alternating normalization-weighted error | ANWE | [45] |
| | (17) Self-weighted alternating normalized residue fitting algorithm | SWANRF | [46] |
| | (18) Alternating coupled two-unequal functions | ACTUF | [47] |
| | (19) Algorithm combination methodology | ACM | [48] |

squares fitting of the intrinsic profiles to the instrumental response of the samples. When compared with GRAM, DTLD can more efficiently analyze data from multiple samples simultaneously; however, DTLD requires the construction of two pseudosamples, which unavoidably results in the loss of sample information. Moreover, DTLD can occasionally yield imaginary solutions and exhibit inflated variance, especially when the data array does not strictly obey the trilinear model or when the noise level is high.

The second group of algorithms that handle second-order data mainly include the bilinear least-squares (BLLS) [28], multiway PLS (U-PLS and N-PLS) [29], artificial neural network (ANN) [9], and unfold principal component analysis (U-PCA) [9]. Notice that these methods are two-block methods, while the remaining algorithms involve only one block, i.e., they are directly applied to a multiway data array without *a priori* knowledge of the concentration. However, these algorithms preserve the second-order advantage only by combining with the residual bilinearization (RBL) [50], which separates the calibration derived portion of the signal from the contributions to the signal from potential interferents. Unlike BLLS/RBL, which deals only with trilinear models, U-PLS/RBL and N-PLS/RBL can process second-order data, irrespective of that whether they conform to or deviate from the trilinear data structure.

The last group of algorithms used in second-order data analysis is based on the iterative least-squares principle, which is the most efficient method for three-way data decomposition. Some algorithms of this group, such as

PARAFAC [30] and multivariate curve resolution-alternating least squares method (MCR-ALS) [31], are sensitive to the preestimated component number. The application of MCR-ALS in quantitative analysis can be interpreted as the unfolding of a three-way data into a bilinear two-way data matrix. The basis for the successful data resolution is the building of an augmented data matrix created from calibration and unknown matrices. Decomposition is achieved through ALS, aided by a series of sensible constraints (such as nonnegativity, selectivity and zero concentration windows, unimodality and closure), which provide a solution with physical interpretability and help to limit the number of possible solutions. Based on the advantages of MCR-ALS, it appears that the three-way data that conform to the trilinearity condition or even mixed bilinear/trilinear case can be handled by this algorithm, thereby showing general applicability, especially in GC-MS, LC-MS, and LC-DAD-generated data.

PARAFAC, based on strict minimization of residual sum of squares (SSR), provides a nice fit to three-way trilinear data. Initially, its use was restricted to qualitative analysis. However, in 1998, Wu and coworkers presented the first application of PARAFAC in the simultaneous quantification of multiple target components in several complex samples even in the presence of unknown interferents [32]. The practical deployment of the classical PARAFAC algorithm can run into some problems, such as sensitivity to preestimated component number, initialization values and collinearity, in addition to exhibiting slow convergence. Furthermore, when the three-way responsive data does not strictly follow the trilinear model, the analysis can easily drop into the "swamp" of calculation and produce degenerate solutions without chemical significance. These defects reveal that the classical PARAFAC algorithm only works well for systems where the component number in measuring systems can be accurately estimated. To circumvent the disadvantages of the traditional PARAFAC algorithm and other noniterative methods, our group developed a series of second-order calibration methods and algorithms for analyzing three-way trilinear data. The earliest one is the alternating trilinear decomposition (ATLD) method [32] that is based on the alternating least-squares principle and extracts diagonal elements, thereby reducing the demand in computational memory. Moreover, performing Moor–Penrose generalized inverse calculations based on truncated singular value decomposition (SVD) makes ATLD insensitive to the overestimated component number and allows for faster convergence. The convergence rate of ATLD is the fastest one among all iteration algorithms reported so far, and a rather low memory is required for the deployment of ATLD. The self-weighted alternating trilinear decomposition (SWATLD) [33], which employs a similar ideology, shows an improved performance. It retains the merits of fast convergence and insensitivity to excess factors used in calculation. In addition, SWATLD offers more satisfactory results when compared with those from ATLD under higher noise and collinear levels. The alternating penalty trilinear decomposition (APTLD) [34] brings penalty functions into three objective functions, consequently, transforming the process of target minimization into a constrained optimization problem. With the application of large penalty factors, the constrained problem can further be converted into an unconstrained problem. Besides, we have developed several algorithms with some enhanced features, such as alternating asymmetric trilinear decomposition (AATLD) [35], coupled vectors resolution (COVER) [36], alternating coupled vectors resolution (ACOVER) [37], alternating slice-wise diagonalization (ASD) [38], alternating coupled matrices resolution (ACOMAR) [39], pseudo alternating least-squares (PALS) [40], constrained parallel factor analysis (CPARAFAC) [41], full-rank parallel factor analysis (FRA-PARAFAC) [42], alternating normalization-weighted error (ANWE) [45], self-weighted alternating normalized residue fitting (SWANRF) [46], alternating coupled two-unequal functions (ACTUF) [47], and algorithm combination methodology (ACM) [48]. Here, a short description will be given for some of them together with PARAFAC, ATLD, and ATLD's variants that are based on ATLD.

### 5.1.1 Parallel Factor Analysis

One way for PARAFAC based on the slice forms:

$$\sigma(\mathbf{A},\mathbf{B},\mathbf{C}) = \sum_{k=1}^{K} \|\mathbf{E}_{..k}\|_{\mathrm{F}}^{2} = \sum_{k=1}^{K} \left\| \left(\mathbf{X}_{..k} - \mathbf{A}\,diag\left(\mathbf{c}_{(k)}\right)\mathbf{B}^{\mathrm{T}}\right) \right\|_{\mathrm{F}}^{2}, \tag{41}$$

Then we can obtain **A**, **B**, and **C** as:

$$\mathbf{A} = \left(\sum_{k=1}^{K} \mathbf{X}_{..k}\mathbf{B}\,diag(\mathbf{c}_k)\right)\left[\sum_{k=1}^{K} diag(\mathbf{c}_k)\mathbf{B}^{\mathrm{T}}\mathbf{B}\,diag(\mathbf{c}_k)\right]^{-1}, \tag{42}$$

$$\mathbf{B} = \left(\sum_{k=1}^{K} \mathbf{X}_{..k}^{\mathrm{T}}\mathbf{A}\,diag(\mathbf{c}_k)\right)\left[\sum_{k=1}^{K} diag(\mathbf{c}_k)\mathbf{A}^{\mathrm{T}}\mathbf{A}\,diag(\mathbf{c}_k)\right]^{-1}, \tag{43}$$

$$\mathbf{c}_{(k)} = \left(\left((\mathbf{A}^{\mathrm{T}}\mathbf{A}) * (\mathbf{B}^{\mathrm{T}}\mathbf{B})\right)^{-1} diagm\left(\mathbf{A}^{\mathrm{T}}\mathbf{X}_{..k}\mathbf{B}\right)\right)^{\mathrm{T}}. \tag{44}$$

Another set of loss functions for PARAFAC can be expressed in stretched matrices:

$$\sigma(\mathbf{A}) = \|\mathbf{E}_{I\times JK}\|_{\mathrm{F}}^{2} = \left\|\mathbf{X}_{I\times JK} - \mathbf{A}(\mathbf{C}\odot\mathbf{B})^{\mathrm{T}}\right\|_{\mathrm{F}}^{2}, \tag{45}$$

$$\sigma(\mathbf{B}) = \|\mathbf{E}_{I\times JK}\|_{\mathrm{F}}^{2} = \|\mathbf{E}_{J\times KI}\|_{\mathrm{F}}^{2} = \left\|\mathbf{X}_{J\times KI} - \mathbf{B}(\mathbf{A}\odot\mathbf{C})^{\mathrm{T}}\right\|_{\mathrm{F}}^{2}, \tag{46}$$

$$\sigma(\mathbf{C}) = \|\mathbf{E}_{I\times JK}\|_{\mathrm{F}}^{2} = \|\mathbf{E}_{K\times IJ}\|_{\mathrm{F}}^{2} = \left\|\mathbf{X}_{K\times IJ} - \mathbf{C}(\mathbf{B}\odot\mathbf{A})^{\mathrm{T}}\right\|_{\mathrm{F}}^{2}. \tag{47}$$

According to Equations (45)–(47), the PARAFAC-ALS algorithm has found out a solution of **A** for fixed **B** and **C**, or **B** for fixed **C** and **A**, or **C** for fixed **A** and **B**, then **A**, **B** and **C** are updated by using the strictly least-squares principle as follows:

$$\mathbf{A} = \mathbf{X}_{I \times JK}\left((\mathbf{C} \odot \mathbf{B})^{\mathrm{T}}\right)^{+}, \tag{48}$$

$$\mathbf{B} = \mathbf{X}_{J \times KI}\left((\mathbf{A} \odot \mathbf{C})^{\mathrm{T}}\right)^{+}, \tag{49}$$

$$\mathbf{C} = \mathbf{X}_{K \times IJ}\left((\mathbf{B} \odot \mathbf{A})^{\mathrm{T}}\right)^{+}. \tag{50}$$

Evidently, PARAFAC-ALS holds strictly least-squares solutions in mathematical sense. However, by minimizing the only one loss function, i.e., the whole residuals, this algorithm would meet the problems of slow convergence, sensitivity to the component number, and easily dropping into locally optimal solutions. Initializing PARAFAC for the study of three-way arrays can be done using several options: (I) SVD vectors; (II) random orthogonalized values; and (III) the best-fitting model of several models fitted using a few iterations.

### 5.1.2 *Alternating Trilinear Decomposition*

The ATLD algorithm [32] is an alternative one for decomposition of three-way data arrays, which is an improvement of the traditional PARAFAC algorithm without any constraints. It is based on an alternating least squares principle and an improved iterative procedure that utilizes the Moore–Penrose generalized inverse obtained by SVD. When the data follow the trilinear model, ATLD can make the calibration possible even in the presence of interferences that are not present in the calibration samples.

The three-way data array can be alternately decomposed based on sliced matrices. Due to the property named as cyclic symmetry of the trilinear model, the three expressions are mathematically equal to each other. According to Equations (5)–(7), the loss functions to be minimized is the sum of the squares of the elements of the residual matrices, which can be represented as:

$$\sigma\left(\mathbf{a}_{(i)}\right) = \sum_{i=1}^{I} \left\| \mathbf{X}_{i..} - \mathbf{B}\,diag\left(\mathbf{a}_{(i)}\right)\mathbf{C}^{\mathrm{T}} \right\|_{\mathrm{F}}^{2}, \tag{51}$$

$$\sigma\left(\mathbf{b}_{(j)}\right) = \sum_{j=1}^{J} \left\| \mathbf{X}_{.j.} - \mathbf{C}\,diag\left(\mathbf{b}_{(j)}\right)\mathbf{A}^{\mathrm{T}} \right\|_{\mathrm{F}}^{2}, \tag{52}$$

$$\sigma\left(\mathbf{c}_{(k)}\right) = \sum_{k=1}^{K} \left\| \mathbf{X}_{..k} - \mathbf{A}\,diag\left(\mathbf{c}_{(k)}\right)\mathbf{B}^{\mathrm{T}} \right\|_{\mathrm{F}}^{2}. \tag{53}$$

Using the loss functions mentioned above, the ATLD algorithm minimizes alternately one of the three objective functions over **C** on fixed **A** and **B**, then over **A** on fixed **B** and **C**, and then over **B** on fixed **C** and **A**. The updates for **A**, **B**, and **C** based on the least squares principle, are as follows:

$$\mathbf{a}_{(i)}^{\mathrm{T}} = diagm\left(\mathbf{B}^{+}\mathbf{X}_{i..}\left(\mathbf{C}^{\mathrm{T}}\right)^{+}\right), \quad \text{for } i=1,2,\ldots,I, \tag{54}$$

$$\mathbf{b}_{(j)}^{\mathrm{T}} = diagm\left(\mathbf{C}^{+}\mathbf{X}_{.j.}\left(\mathbf{A}^{\mathrm{T}}\right)^{+}\right), \quad \text{for } j=1,2,\ldots,J, \tag{55}$$

$$\mathbf{c}_{(k)}^{\mathrm{T}} = diagm\left(\mathbf{A}^{+}\mathbf{X}_{..k}\left(\mathbf{B}^{\mathrm{T}}\right)^{+}\right), \quad \text{for } k=1,2,\ldots,K. \tag{56}$$

where $diagm(\cdot)$ denotes a column $N$-vector whose elements are diagonal elements of a square matrix. In each iteration cycle, **A** and **B** are normalized columnwise to unit length. With the aid of the resolved profile matrices, the analyte concentrations can be obtained by regression of the appropriate column of **C** corresponding to each analyte against its standard concentrations.

ATLD holds the fastest convergence, which is attributed to the operation based on sliced matrices with less size and two other major strategies. One is the truncated least-squares method, which uses the tolerance to truncate the small singular values in the SVD, the other is the operation of selecting diagonal elements, which also makes ATLD retaining trilinearity property indeed and be insensitive to the estimated component number. One example is shown that a simulated four-component fluorescent data set was analyzed by ATLD with different component numbers. When the chosen factor number is equal to the real one (i.e. $N=4$), ATLD can acquire perfect excitation and emission spectra and the concentration profiles (Figure 10A–C). Figure 10D shows the iteration process of ATLD, suggesting that ATLD only needs six iterations to reach convergence. Figure 11 displays the results of ATLD with the factor number equal to 5, which is larger than the real one. It can be seen

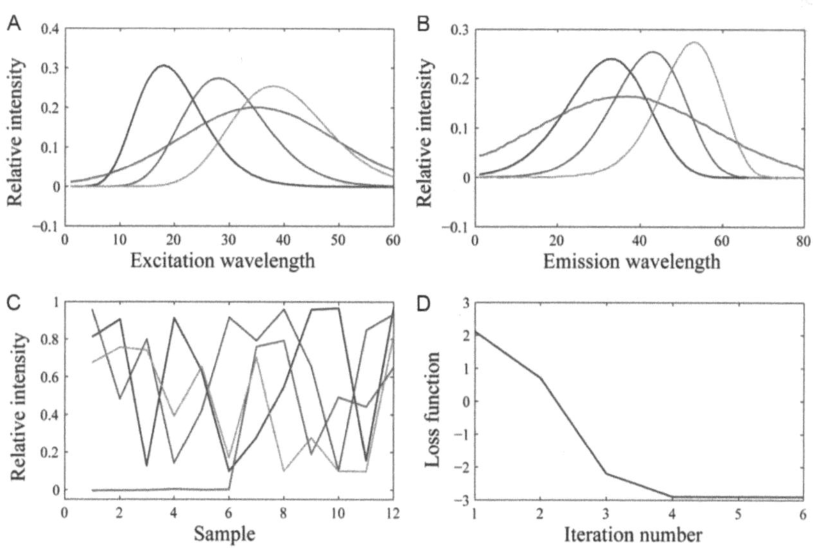


**FIGURE 10** Results of ATLD with the correctly estimated factor number equal to 4.

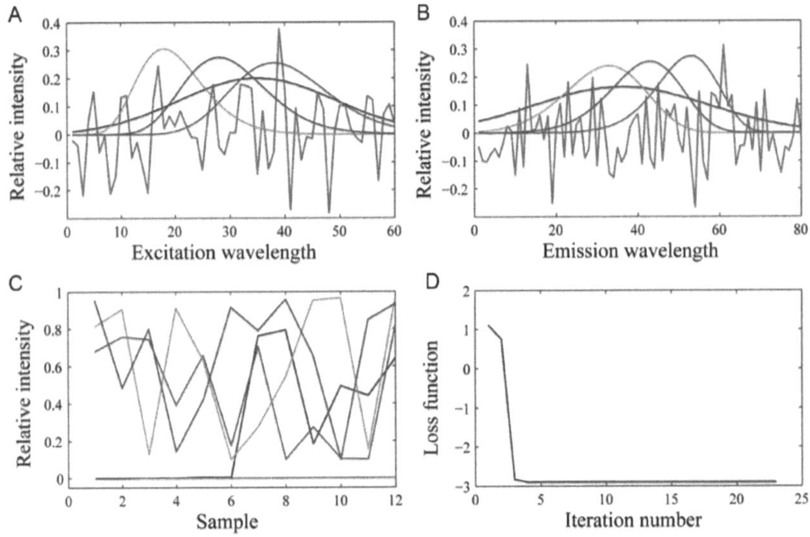


**FIGURE 11** Results of ATLD with the overestimated factor number equal to 5.

that satisfactory results are also obtained by ATLD and the excessive columns represent noise which can be easily discriminated from the desired ones in resolved matrices.

Having derived the updating equations for parameter matrices, the general process (Figure 12) for ATLD and its variants is as follows:

**(1)** Input a pre-estimated number of chemical rank, and then initialize **A** and **B**.
**(2)** Compute **C**.
**(3)** Compute **A** and scale **A** to be columnwisely normalized.
**(4)** Compute **B** and scale **B** to be columnwisely normalized.
**(5)** Compute **C**.
**(6)** Repeat steps 3–5 until a stopping criterion is satisfied.
**(7)** Postprocess **A**, **B**, and **C** according to the uniqueness property and predict concentration of each component of interest in unknown sample(s) by regression of the appropriate column of **C** corresponding to each analyte of interest against its standard concentrations or by regression from the corresponding standard concentrations.

The optimizing procedure of the improved method is terminated when the following criterion reaches a certain threshold $\varepsilon = 10^{-6}$. A maximal number of 3000 iterations are adopted to avoid unduly slow convergence.

$$\mathrm{SSR}^{(m)} = \sum_{k=1}^{K} \left\| \mathbf{X}_{..k} - \mathbf{A}^{(m)} diag\left(\mathbf{c}_{(k)}\right)^{(m)} \left(\mathbf{B}^{\mathrm{T}}\right)^{(m)} \right\|_{\mathrm{F}}^{2}, \tag{57}$$

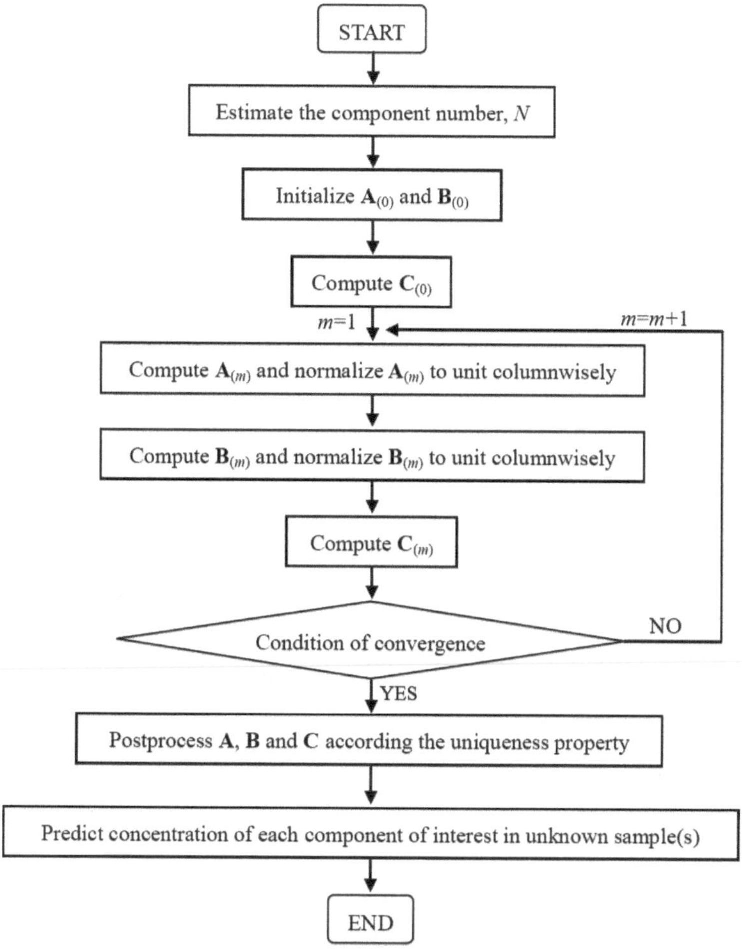


**FIGURE 12** Scheme illustrating the procedure of ATLD and its variants.

$$\left|\frac{\mathrm{SSR}^{(m)} - \mathrm{SSR}^{(m-1)}}{\mathrm{SSR}^{(m-1)}}\right| \le \varepsilon, \tag{58}$$

where SSR is the residual sum of squares and $m$ is the current iteration number.

The advantages of ATLD [32] and its variants, such as insensitivity to excessive component number, fast convergence, low physical memory needed, and the ability to circumvent the difficulty caused by slight deviation of linearity, are important features in applications of second-order calibration. First, ATLD and its variants coupled with some commonly used instruments including EEM fluorescence spectrophotometry, HPLC-DAD and high-performance liquid chromatography-mass spectrometry can be used to quantify some components of interest in many complex systems, such as biological samples, food, environmental, pharmaceuticals, and cosmetics [2,3,51–67]. Second, ATLD and its variants enable one or more analyte(s) of interest to be determined simultaneously in the samples with various types of matrices. This strategy can save more time compared to the approach that analyzes samples with various types of matrices separately. Reference [68] has reported the second-order calibration coupled with EEM fluorescence for the determination of napropamide in soil, river sediment, and wastewater as well as river water samples. Based on the global model based on the ATLD's variant, APTLD, good results were obtained. In addition, the ATLD method coupled with some instruments with fast measurement can be used for industrial process analysis, which is a great challenge in industry due to the complexity of industrial production [69,70]. Reference [69] has reported an approach based on online near-infrared spectroscopy and ATLD for industrial process analysis of tobacco production, whose result has shown that the coupled method is able to capture the intrinsic information of the products and performs well in process analysis and quality assurance.

### 5.1.3 Self-Weighted Alternating Trilinear Decomposition

The SWATLD algorithm as a version derived from ATLD can yield better results in many cases where it is widely employed. It alternately minimizes three objective functions with intrinsic relationship and provides the features of fast convergence and insensitivity to the excess factors used in calculation. Mathematical explanations of these valuable properties are presented by authors in original paper [33]. New residues can be obtained:

$$\begin{aligned} \mathbf{B}^{+}\mathbf{X}_{i..} &= diag(\mathbf{a}_{(i)})\mathbf{C}^{\mathrm{T}} + \mathbf{B}^{+}\mathbf{E}_{i..}, \\ \mathbf{X}_{i..}(\mathbf{C}^{+})^{\mathrm{T}} &= \mathbf{B}diag(\mathbf{a}_{(i)}) + \mathbf{E}_{i..}(\mathbf{C}^{+})^{\mathrm{T}}, \quad \text{for } i = 1,2,\ldots,I, \end{aligned} \tag{59}$$

$$\begin{aligned} \mathbf{C}^{+}\mathbf{X}_{.j.} &= diag(\mathbf{b}_{(j)})\mathbf{A}^{\mathrm{T}} + \mathbf{C}^{+}\mathbf{E}_{.j.}, \\ \mathbf{X}_{.j.}(\mathbf{A}^{+})^{\mathrm{T}} &= \mathbf{C}diag(\mathbf{b}_{(j)}) + \mathbf{E}_{.j.}(\mathbf{A}^{+})^{\mathrm{T}}, \quad \text{for } j = 1,2,\ldots,J, \end{aligned} \tag{60}$$

$$\begin{aligned} \mathbf{A}^{+}\mathbf{X}_{..k} &= diag(\mathbf{c}_{(k)})\mathbf{B}^{\mathrm{T}} + \mathbf{A}^{+}\mathbf{E}_{..k}, \\ \mathbf{X}_{..k}(\mathbf{B}^{+})^{\mathrm{T}} &= \mathbf{A}diag(\mathbf{c}_{(k)}) + \mathbf{E}_{..k}(\mathbf{B}^{+})^{\mathrm{T}}, \quad \text{for } k = 1,2,\ldots,K. \end{aligned} \tag{61}$$

Three new objective functions for SWATLD through introducing some reasonable weight terms are established as follows:

$$\begin{aligned} \mathrm{S}(\mathbf{A}) = &\sum_{i=1}^{I} \left\| (\mathbf{B}^{+}\mathbf{X}_{i..} - diag(\mathbf{a}_{(i)})\mathbf{C}^{\mathrm{T}})^{\mathrm{T}} \times diag(sqrt(\mathbf{1}./diagm(\mathbf{C}^{\mathrm{T}}\mathbf{C}))) \right\|_{\mathrm{F}}^{2} \\ &+ \sum_{i=1}^{I} \left\| \left(\mathbf{X}_{i..}(\mathbf{C}^{\mathrm{T}})^{+} - \mathbf{B}diag(\mathbf{a}_{(i)})\right) \times diag(sqrt(\mathbf{1}./diagm(\mathbf{B}^{\mathrm{T}}\mathbf{B}))) \right\|_{\mathrm{F}}^{2}, \end{aligned} \tag{62}$$

$$S(\mathbf{B}) = \sum_{j=1}^{J} \|(\mathbf{C}^{+}\mathbf{X}_{.j.} - diag(\mathbf{b}_{(j)})\mathbf{A}^{\mathrm{T}})^{\mathrm{T}} \times diag(sqrt(\mathbf{1}./diagm(\mathbf{A}^{\mathrm{T}}\mathbf{A})))\|_{\mathrm{F}}^{2} + \sum_{j=1}^{J} \left\| \left(\mathbf{X}_{.j.}(\mathbf{A}^{\mathrm{T}})^{+} - \mathbf{C}diag(\mathbf{b}_{(j)})\right) \times diag(sqrt(\mathbf{1}./diagm(\mathbf{C}^{\mathrm{T}}\mathbf{C}))) \right\|_{\mathrm{F}}^{2}, \tag{63}$$

$$S(\mathbf{C}) = \sum_{k=1}^{K} \|(\mathbf{A}^{+}\mathbf{X}_{..k} - diag(\mathbf{c}_{(k)})\mathbf{B}^{\mathrm{T}})^{\mathrm{T}} \times diag(sqrt(\mathbf{1}./diagm(\mathbf{B}^{\mathrm{T}}\mathbf{B})))\|_{\mathrm{F}}^{2} + \sum_{k=1}^{K} \left\| \left(\mathbf{X}_{..k}(\mathbf{B}^{+})^{\mathrm{T}} - \mathbf{A}diag(\mathbf{c}_{(k)})\right) \times diag(sqrt(\mathbf{1}./diagm(\mathbf{A}^{\mathrm{T}}\mathbf{A}))) \right\|_{\mathrm{F}}^{2}. \tag{64}$$

Based on the least-squares principle, **A**, **B**, and **C** can be acquired as follows:

$$\mathbf{a}_{(i)} = \frac{1}{2}(diagm(\mathbf{B}^{+}\mathbf{X}_{i..}\mathbf{C})./diagm(\mathbf{C}^{\mathrm{T}}\mathbf{C})) + \frac{1}{2}(diagm(\mathbf{C}^{+}\mathbf{X}_{i..}{}^{\mathrm{T}}\mathbf{B})/diagm(\mathbf{B}^{\mathrm{T}}\mathbf{B})), \text{ for } i = 1,2,\ldots,I, \tag{65}$$

$$\mathbf{b}_{(j)} = \frac{1}{2}(diagm(\mathbf{C}^{+}\mathbf{X}_{.j.}\mathbf{A})./diagm(\mathbf{A}^{\mathrm{T}}\mathbf{A})) + \frac{1}{2}(diagm(\mathbf{A}^{+}\mathbf{X}_{.j.}{}^{\mathrm{T}}\mathbf{C})./diagm(\mathbf{C}^{\mathrm{T}}\mathbf{C})), \text{ for } j = 1,2,\ldots,J, \tag{66}$$

$$\mathbf{c}_{(k)} = \frac{1}{2}(diagm(\mathbf{A}^{+}\mathbf{X}_{..k}\mathbf{B})./diagm(\mathbf{B}^{\mathrm{T}}\mathbf{B})) + \frac{1}{2}(diagm(\mathbf{B}^{+}\mathbf{X}_{..k}{}^{\mathrm{T}}\mathbf{A})./diagm(\mathbf{A}^{\mathrm{T}}\mathbf{A})), \text{ for } k = 1,2,\ldots,K. \tag{67}$$

Due to the unique optimizing scheme, the algorithm is more efficient than the other algorithms. It not only holds the properties of fast convergence and insensitivity to excess factors but also can offer more satisfactory results than ATLD under moderate noise levels. In addition, this algorithm can cope with the problem of moderate collinearity; but it is not so effective for severely collinear data.

### *5.1.4 Alternating Penalty Trilinear Decomposition*

The APTLD algorithm developed by Xia *et al.* provides some improved properties. It aims at using alternately the constraint functions as penalty terms of the PARAFAC error, which minimizes three new least-squares-based objective functions. APTLD transforms these constrained problems into nonconstrained ones by introducing penalty terms and combines them with residue functions (Equations 68–70) to construct three objective functions. Then it minimizes the following three objective functions in an alternating manner to establish the model:

$$
\begin{aligned}
S(\mathbf{A}) = & \sum_{k=1}^{K} \| \mathbf{X}_{..k} - \mathbf{A}\, diag(\mathbf{c}_{(k)}) \mathbf{B}^{\mathrm{T}} \|_{\mathrm{F}}^{2} \\
& + q \left( \sum_{j=1}^{J} \| diag(sqrt(\mathbf{1}./diagm(\mathbf{B}^{\mathrm{T}}\mathbf{B}))) (\mathbf{C}^{+}\mathbf{X}_{.j.} - diag(\mathbf{b}_{(j)})\mathbf{A}^{\mathrm{T}}) \|_{\mathrm{F}}^{2} \right. \\
& \left. + \sum_{k=1}^{K} \left\| \left( \mathbf{X}_{..k} (\mathbf{B}^{\mathrm{T}})^{+} - \mathbf{A}\, diag(\mathbf{c}_{(k)}) \right) diag(sqrt(\mathbf{1}./diagm(\mathbf{C}^{\mathrm{T}}\mathbf{C}))) \right\|_{\mathrm{F}}^{2} \right),
\end{aligned} \tag{68}
$$

$$
\begin{aligned}
S(\mathbf{B}) = & \sum_{i=1}^{I} \| \mathbf{X}_{i..} - \mathbf{B}\, diag(\mathbf{a}_{(i)}) \mathbf{C}^{\mathrm{T}} \|_{\mathrm{F}}^{2} \\
& + r \left( \sum_{k=1}^{K} \| diag(sqrt(\mathbf{1}./diagm(\mathbf{C}^{\mathrm{T}}\mathbf{C}))) (\mathbf{A}^{+}\mathbf{X}_{..k} - diag(\mathbf{c}_{(k)})\mathbf{B}^{\mathrm{T}}) \|_{\mathrm{F}}^{2} \right. \\
& \left. + \sum_{i=1}^{I} \left\| \left( \mathbf{X}_{i..} (\mathbf{C}^{\mathrm{T}})^{+} - \mathbf{B}\, diag(\mathbf{a}_{(i)}) \right) diag(sqrt(\mathbf{1}./diagm(\mathbf{A}^{\mathrm{T}}\mathbf{A}))) \right\|_{\mathrm{F}}^{2} \right),
\end{aligned} \tag{69}
$$

$$
\begin{aligned}
S(\mathbf{C}) = & \sum_{j=1}^{J} \| \mathbf{X}_{.j.} - \mathbf{C}\, diag(\mathbf{b}_{(j)}) \mathbf{A}^{\mathrm{T}} \|_{\mathrm{F}}^{2} \\
& + p \left( \sum_{i=1}^{I} \| diag(sqrt(\mathbf{1}./diagm(\mathbf{A}^{\mathrm{T}}\mathbf{A}))) (\mathbf{B}^{+}\mathbf{X}_{i..} - diag(\mathbf{a}_{(i)})\mathbf{C}^{\mathrm{T}}) \|_{\mathrm{F}}^{2} \right. \\
& \left. + \sum_{j=1}^{J} \left\| \left( \mathbf{X}_{.j.} (\mathbf{A}^{\mathrm{T}})^{+} - \mathbf{C}\, diag(\mathbf{b}_{(j)}) \right) diag(sqrt(\mathbf{1}./diagm(\mathbf{B}^{\mathrm{T}}\mathbf{B}))) \right\|_{\mathrm{F}}^{2} \right),
\end{aligned} \tag{70}
$$

where $p$, $q$, and $r$ are penalty factors. The performance of APTLD is related to the choice of the value of these penalty factors. When $p$, $q$, and $r$ are very small, it will cause a large number of iterations and sensitivity to excess factors used in calculations for the final results of APTLD, which is similar to the situation of the PARAFAC algorithm, particularly, when $p=q=r=0$, APTLD can be considered as a variant of PARAFAC. However, selecting larger $p$, $q$, and $r$ will make APTLD insensitive to excess factors and speed up convergence of the algorithm. A further increase in $p$, $q$, and $r$ will make APTLD to perform theoretically better according to the variance among different trials and computational burdens. Hence, the performance of APTLD can be skillfully improved by adjusting the penalty factors $p$, $q$, and $r$ according to particular circumstances and special needs.

### *5.1.5 Alternating Coupled Two-Unequal Residual Functions*

The ACTUF algorithm was developed to analyze three-way data based on two loss functions obtained from two kinds of residuals. The process of the new algorithm is shown as follows.

First, two kinds of residuals are acquired. Measurement residuals can be easily gained from Equation (6) as:

$$\mathbf{X}_{.j.} - \mathbf{C}diag(\mathbf{b}_{(j)})\mathbf{A}^{\mathrm{T}} = \mathbf{E}_{.j.} \quad \text{for } j = 1,2,\ldots,J. \tag{71}$$

Similarly, approximate residuals can be gained as:

$$\mathbf{X}_{.j.}^{\mathrm{T}}(\mathbf{C}^{\mathrm{T}})^{+} - \mathbf{A}diag(\mathbf{b}_{(j)}) = \mathbf{E}_{.j.}^{\mathrm{T}}(\mathbf{C}^{\mathrm{T}})^{+} = \widetilde{\mathbf{E}}_{.j.}^{\mathrm{T}} \quad \text{for } j = 1,2,\ldots,J. \tag{72}$$

Then, according to the two kinds of residuals, two unequal residual functions can be constructed by the principle of SSR, respectively:

$$\sigma(\mathbf{A},\mathbf{B}) = \sum_{j=1}^{J} \left\| \widetilde{\mathbf{E}}_{.j.}^{\mathrm{T}} \right\|_{\mathrm{F}}^{2} = \sum_{j=1}^{J} \left\| \mathbf{X}_{.j.}^{\mathrm{T}}(\mathbf{C}^{\mathrm{T}})^{+} - \mathbf{A}diag(\mathbf{b}_{(j)}) \right\|_{\mathrm{F}}^{2}, \tag{73}$$

$$\sigma(\mathbf{C}) = \sum_{j=1}^{J} \| \mathbf{E}_{.j.} \|_{\mathrm{F}}^{2} = \sum_{j=1}^{J} \| (\mathbf{X}_{.j.} - \mathbf{C}diag(\mathbf{b}_{(j)})\mathbf{A}^{\mathrm{T}}) \|_{\mathrm{F}}^{2}. \tag{74}$$

where $\|\cdot\|_{\mathrm{F}}$ is the Frobenius matrix norm.

By minimizing SSR, **A** and **C** can be easily resolved from Equation (73) and (74), respectively:

$$\mathbf{A} = \left[ \sum_{j=1}^{J} \mathbf{X}_{.j.}^{\mathrm{T}}(\mathbf{C}^{\mathrm{T}})^{+} diag(\mathbf{b}_{(j)}) \right] \left[ \sum_{j=1}^{J} diag(\mathbf{b}_{(j)}) diag(\mathbf{b}_{(j)}) \right]^{+}, \tag{75}$$

$$\mathbf{C} = \left[ \sum_{j=1}^{J} \mathbf{X}_{.j.}\mathbf{A}diag(\mathbf{b}_{(j)}) \right] \left[ \sum_{j=1}^{J} diag(\mathbf{b}_{(j)})\mathbf{A}^{\mathrm{T}}\mathbf{A}diag(\mathbf{b}_{(j)}) \right]^{+}. \tag{76}$$

where superscript + is the Moore–Penrose generalized inverse of a matrix.

In order to facilitate solving **B**, Equation (73) can be rewritten as:

$$\sigma(\mathbf{A},\mathbf{B}) = \sum_{j=1}^{J} \left\| \widetilde{\mathbf{E}}_{.j.}^{\mathrm{T}} \right\|_{\mathrm{F}}^{2} = \sum_{i=1}^{I} \left\| \mathbf{X}_{i..}(\mathbf{C}^{\mathrm{T}})^{+} - \mathbf{B}diag(\mathbf{a}_{(i)}) \right\|_{\mathrm{F}}^{2}. \tag{77}$$

Similarly, **B** can be resolved from Equation (77) as:

$$\mathbf{B} = \left[ \sum_{i=1}^{I} \mathbf{X}_{i..}(\mathbf{C}^{\mathrm{T}})^{+} diag(\mathbf{a}_{(i)}) \right] \left[ \sum_{i=1}^{I} diag(\mathbf{a}_{(i)}) diag(\mathbf{a}_{(i)}) \right]^{+}. \tag{78}$$

For comparison of PARAFAC with one loss function as well as SWATLD with three loss functions, ACTUF, based on two unequal loss functions to combine the merits of PARAFAC and SWATLD, is of particular significance.

First, similar to SWATLD, ACTUF is based on the idea of two unequal loss functions that makes it insensitive to excessive factors and provides fast convergence. Second, similar to PARAFAC, this new method provides a nice fitting to three-way data and resistance to high level noise. In addition, superior to PARAFAC and SWATLD, this new method can overcome the problems due to severe collinearity. The reason for this is that computing **A**, **B**, and **C** according to two unequal loss functions makes this new method easily jump out of the "swamp" and pay more attention to extracting the useful information from data.

One example is shown that a four-component-simulated fluorescent data set with the problem of severe collinearity is analyzed by ACTUF. In this data set, the condition numbers of the excitation and emission profiles are 59.92 and 61.67, respectively. Some commonly used second-order calibration methods, such as PARAFAC and SWATLD, cannot obtain reasonable results for severe collinearity. On the contrary, ACTUF provides satisfying results (Figure 13), which shows the favorable performance of ACTUF to deal with the problem due to severe collinearity.

### *5.1.6 Algorithm Combination Methodology*

Yu and coworkers suggested that the complete process of ATLD procedure for second-order calibration (Figure 14) can be roughly divided into three stages according to the changes of loss of fitness. In the first stage, random initial values are rapidly optimized, whereas in the second stage, the estimators are optimized in a very gentle manner, while in the final stage, methodology converges precisely to provide meaningful results. The three stages

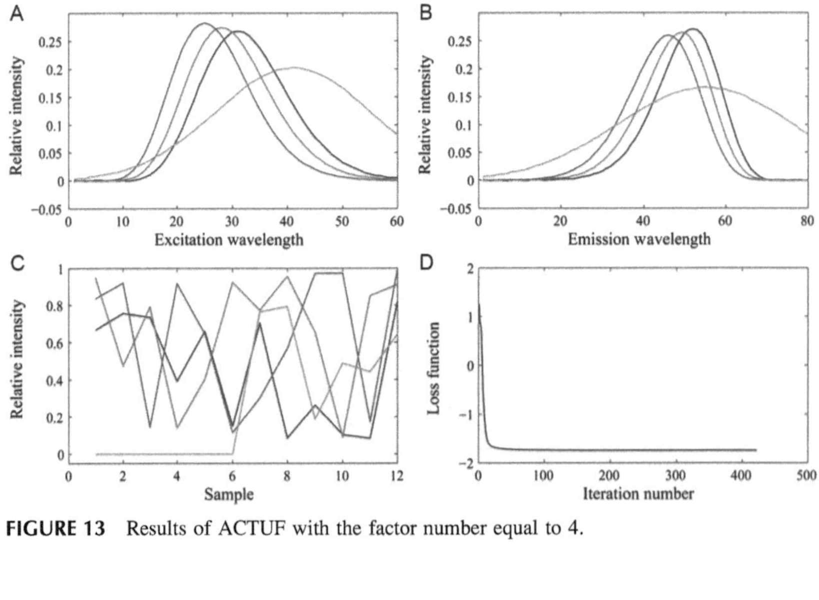


**FIGURE 13** Results of ACTUF with the factor number equal to 4.

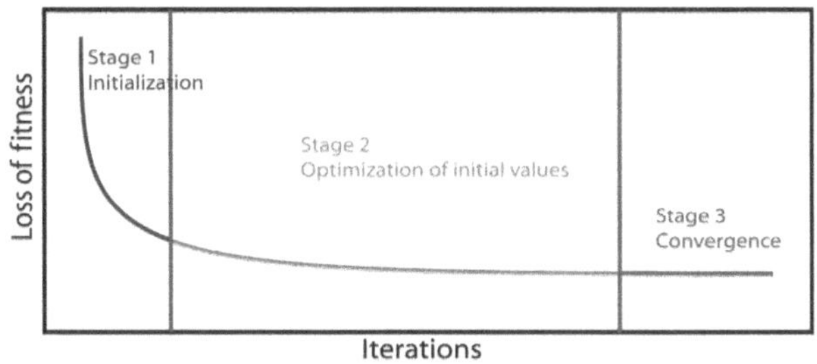


**FIGURE 14** Graphical illustration of the optimization procedure of alternating trilinear decomposition. *Adapted from Ref. [48], figure 1, p. 199.*

were named initialization stage, optimization stage, and convergence stage for the sake of simplicity in this study. Then, they proposed the ACM algorithm: in the first stage, ATLD is used to rapidly optimize random initial values; in the second optimization stage, SWATLD is used to effectively optimize initialization from ATLD; and in the third stage, steady convergence is achieved by PARAFAC to provide an accurate solution. The scheme of the combined algorithm is shown in Figure 15. In theory, ACM has several additional advantages, which can be summarized as following:

**(1)** Solutions can be uniquely obtained, yielding accurate estimates for the profiles of analytes of interest.
**(2)** The combination strategy, ACM can provide further advantages over any one of ATLD, SWATLD. or PARAFAC individually.
**(3)** ACM is very stable with respect to noise level, collinearity, and overestimations of the number of components.
**(4)** ACM can automatically converge in ATLD, SWATLD, or PARAFAC according to the natural structure of the analyzed data array, making the iterative process more efficient.

All of the characteristics of ACM offer the advantage that in second-order calibration one can circumvent the difficulty of selecting suitable methodology for analyzing the data at hand and the problem of determining a proper number of components presented in the samples.

### *5.1.7 Comparison of Algorithms for Second-Order Calibration*

Faber and coworkers reviewed several proposed algorithms for fitting the PARAFAC model [71]. Some of these include PARAFAC-ALS, DTLD, ATLD, SWATLD, PALS, ACOVER, ASD, and ACOMAR. As reported, PARAFAC-ALS provided solutions of a better quality when compared with those from the alternatives, but unfortunately, it is computationally demanding (slow) when overfactoring. Based on the reported results (in particular the poor performance of DTLD), (a slightly modified) ASD is a good alternative to ALS when a faster algorithm is designed. While this work only aimed

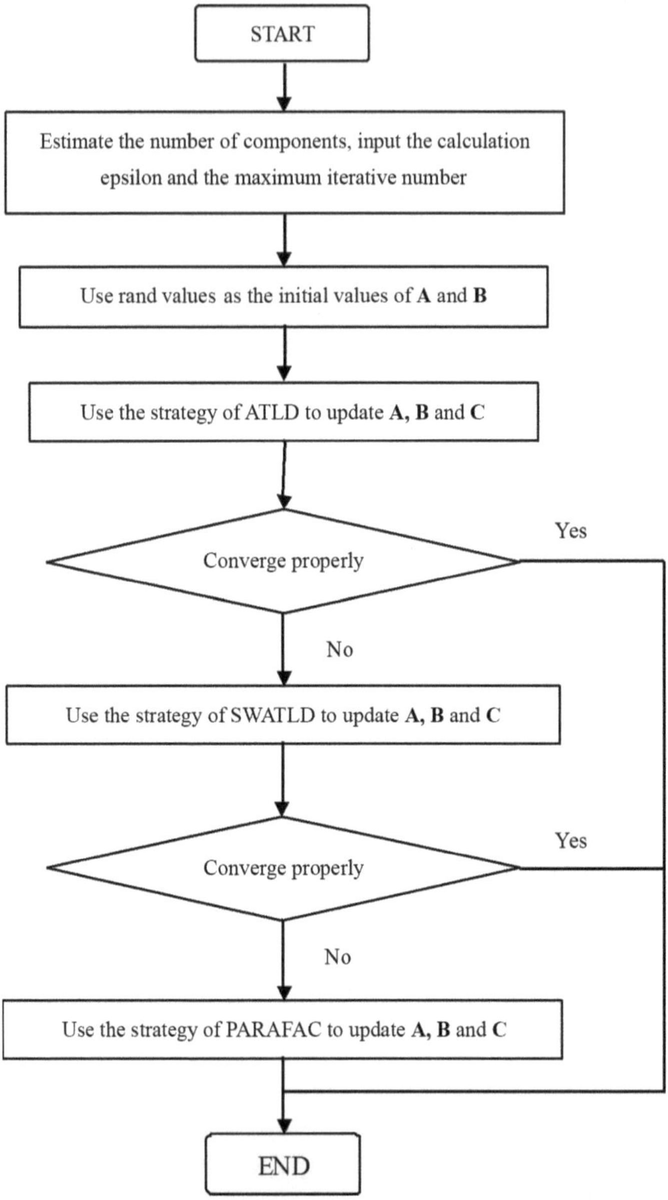


**FIGURE 15** Scheme illustrating the procedure of combination of ATLD, SWATLD and PARAFAC algorithms *(Adapted from Ref. [48], figure 2, p. 199).*

at fitting the PARAFAC model by using different algorithms, and the second-order advantage was not taken into consideration. This is a situation similar to an early discussion about both unbiased and biased fitting problem of PCR and PLS methods in the field of statistics [6].

Wu *et al.* compared four algorithms in 1990s, including GRAM, DTLD, PARAFAC, and ATLD [15]. The relationship among the ATLD method and the GRAM, DTLD, and PARAFAC methods are diagrammatized in Figure 16. This figure showed that: GRAM may be applied to multiple component systems but constrained within only one standard and one mixture sample at a time; DTLD may be used to analyze data from multiple samples simultaneously but only useful when the cumulative percent variance of the first principal components presented in the sample space is artificially restricted to two; PARAFAC may be used in a multiple component system from multiple samples, but lacks the true trilinear sense; the ATLD method includes the advantages of the GRAM, DTLD, and PARAFAC methods and has effectively overcome their disadvantages. Hence, ATLD may be considered as a general trilinear decomposition method for three-way trilinear data arrays and general second-order calibration method. It is suitable for simultaneous multicomponent analysis of multiple unknown samples even in the presence of uncalibrated interferent(s).

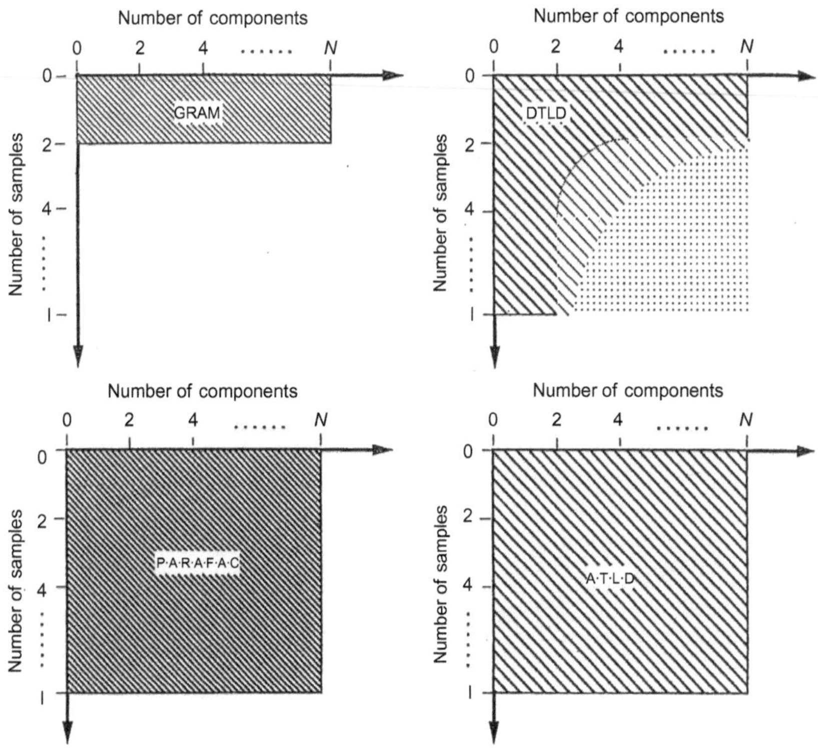


**FIGURE 16** Applicability of four second-order calibration methods. In the case of DTLD, the position of the blend part depends on the correlation between the spectral profiles of the components and their degree of overlapping in the time domain since the number of principal components in sample space is artificially restricted to two. *Adapted from Ref. [15], figure 6, p.172.*

Yu and coworkers [72] compared several established second-order calibration methods, i.e., PARAFAC, ATLD, SWATLD, and APTLD, and suggested a protocol for solving complex quantitative analysis problems. Based on different input parameters including noise level, initial value, number of estimated components, and collinearity in simulated and real data, the performances of these methods were evaluated in terms of recovery, consistency of resolved and real profiles, fitness obtained by selected components and speed of convergence. The performance of four algorithms on a data is like a seesaw (Figure 17). The position of ATLD and PARAFAC is at two end points. According to our experience, if ATLD performs better, PARAFAC will be worse and vice versa. When SWATLD and APTLD are in the middle position, and they almost in between ATLD and PARAFAC and their results are very close. Several advisements can be shown as:

**(1)** All of algorithms can give acceptable and reasonable results in easy cases. And their solutions have no obvious differences.
**(2)** For a three-way data with low signal-to-noise ratio, one should be careful to use ATLD because the update procedure of ATLD is influenced more heavily by the noise than the other three algorithms. And if only slight nonlinearity exists in the data (such as Rayleigh scatter in EEM), the sequences of preference of the four algorithms are PARAFAC, APTLD, SWATLD, and ATLD.
**(3)** For a three-way data with high collinearity, one should be careful to employ PARAFAC because PARAFAC can be easily trapped into swamps, while other three algorithms can more efficiently overcome this problem.
**(4)** Methods used for pre-estimation of the number of components only provide a reference value for the decomposition. Higher component number should be used to check whether small components are missing.
**(5)** PARAFAC is sensitive to the pre-estimated component number, while the other three algorithms are not. If the excess component number is employed, DTLD can be used as an initial value for simple structure data.

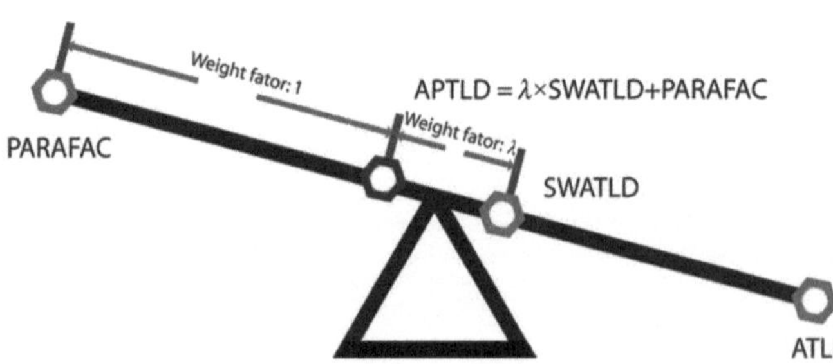


**FIGURE 17** A graphical representation of the relationship among four trilinear decomposition algorithms for second-order calibration.

(6) The manner of the calculation of fitness can be used to indicate whether an algorithm is trapped into swamps and can also be used to indicate whether the component number used for PARAFAC is more than required.

Zhang and coworkers [73] presented an in-depth discussion of the differential properties of various iterative trilinear decomposition algorithms, including PARAFAC-ALS, ATLD, SWATLD, and APTLD. The shape of each algorithm's objective function ("convex" or "strictly convex") is related to the algorithm's sensitivity to the pre-estimated component number of the trilinear system. Different situations near the objective solution are analyzed both theoretically and numerically. The wall of perturbation generated by deviations in the iterative steps prevents the PARAFAC algorithm from achieving the objective solution when the component number is overestimated. This may explain, from a calculational perspective, why the PARAFAC algorithm could not obtain the objective solution or any equivalent thereto (although equivalents might still be chemically meaningful optimal solutions). The different effects of deviation and residual on the algorithms are demonstrated by numerical analysis by the authors. The convergence rate can be improved by the use of high-performance computing strategy of the specific algorithm.

## 5.2 Third-Order Calibration

As the modern analytical instrumentation continues to develop toward more complex arrangements, third-order data can be produced in a relatively easy way. One of the common strategies is to record full EEM as a function of reaction time, pH, volume of quencher, or decay time using a single instrument, e.g., spectrofluorometer. The alternative strategy is by resorting to hyphenated instruments, such as two-dimensional liquid or gas chromatography (LC or GC) with time-of-flight mass spectrometry (TOFMS) or diode array detector (DAD) and LC-DAD as a function of reaction time. Methods that quantitatively analyze four-way responsive data based on the quadrilinear model are known as "third-order calibration."

Up to now, some third-order calibration methods for the analysis of four-way data arrays have been developed and can be generally classified into two types (Table 4). The first type is based on the iterative least-squares principle, including four-way PARAFAC (FPARAFAC) [74], such as the alternating quadrilinear decomposition (AQLD) [75], the APQLD [76], the alternating weighted residual constraint quadrilinear decomposition (AWRCQLD) [77], the four-way self-weighted alternating normalized residue fitting (FSWANRF) algorithm [78], and the regularized self-weighted alternating quadrilinear decomposition (RSWAQLD) [79]. The second type is built on a direct least-squares procedure, for example, trilinear least-squares algorithm combined with the separate procedure known as U-PLS/residual trilinearization (U-PLS/RTL) [80], multiway partial least-squares/RTL (N-PLS/RTL) [80],

**TABLE 4 Algorithms for Third-Order Calibration**

| Algorithm | Acronym | References |
|---|---|---|
| (1) Four-way PARAFAC | FPARAFAC | [74] |
| (2) Alternating quadrilinear decomposition | AQLD | [75] |
| (3) Alternating penalty quadrilinear decomposition | APQLD | [76] |
| (4) Alternating weighted residual constraint quadrilinear decomposition | AWRCQLD | [77] |
| (5) Four-way self-weighted alternating normalized residue fitting algorithm | FSWANRF | [78] |
| (6) Regularized self-weighted alternating quadrilinear decomposition | RSWAQLD | [79] |
| (7) Unfolded partial least-squares/residual trilinearization | U-PLS/RTL | [80] |
| (8) Multiway partial least-squares/residual trilinearization | N-PLS/RTL | [80] |
| (9) Trilinear least-squares/ residual trilinearization | TLLS/RTL | [81] |

and hybrid technique trilinear least-squares/RTL (TLLS/RTL) [81]. These methodologies have been successfully applied to extracting the qualitative and quantitative information of single component or mixture components in complex practical samples. Two algorithms, TLLS/RTL and UPLS/RTL, have been employed by Alejandro *et al.* for the quantitative analysis of methotrexate and leucovorin [81]. Xia *et al.* has proposed APQLD for the study of the hydrolyzed reaction of procaine hydrochloride embedded in a human plasma background [76]. A genuine multidimensional tool, N-PLS/RTL, has been developed by Damiani *et al.* for simultaneous analysis of procaine and its metabolite p-aminobenzoic acid in equine serum [80]. AWRCQLD proposed by Fu *et al.* has been employed to predict the oxprozin concentration in the presence of high collinearity and matrix effect between the target analyte and plasma background [77]. The two latest third-order algorithms, FSWANRF and RSWAQLD, have been applied to process four-way fluorescence excitation–emission–pH data [78,79].

### 5.2.1 Four-way PARAFAC

The four-way PARAFAC algorithm is commonly carried out through an alternating least-squares minimization scheme. According to Equations (16)–(19), four loss functions can be obtained as:

$$\sigma(\mathbf{A}) = \left\|\mathbf{X}_{I\times JKL} - \mathbf{A}(\mathbf{D}\odot\mathbf{C}\odot\mathbf{B})^{\mathrm{T}}\right\|_{\mathrm{F}}^{2}, \quad (79)$$

$$\sigma(\mathbf{B}) = \left\|\mathbf{X}_{I\times JKL} - \mathbf{B}(\mathbf{A}\odot\mathbf{D}\odot\mathbf{C})^{\mathrm{T}}\right\|_{\mathrm{F}}^{2}, \quad (80)$$

$$\sigma(\mathbf{C}) = \left\|\mathbf{X}_{K\times LIJ} - \mathbf{C}(\mathbf{B}\odot\mathbf{A}\odot\mathbf{D})^{\mathrm{T}}\right\|_{\mathrm{F}}^{2}, \quad (81)$$

$$\sigma(\mathbf{D}) = \left\|\mathbf{X}_{L\times IJK} - \mathbf{D}(\mathbf{C}\odot\mathbf{B}\odot\mathbf{A})^{\mathrm{T}}\right\|_{\mathrm{F}}^{2}. \quad (82)$$

The solutions are as follows:

$$\mathbf{A} = \mathbf{X}_{I\times JKL}\left[(\mathbf{D}\odot\mathbf{C}\odot\mathbf{B})^{\mathrm{T}}\right]^{+}, \quad (83)$$

$$\mathbf{B} = \mathbf{X}_{J\times KLI}\left[(\mathbf{A}\odot\mathbf{D}\odot\mathbf{C})^{\mathrm{T}}\right]^{+}, \quad (84)$$

$$\mathbf{C} = \mathbf{X}_{K\times LIJ}\left[(\mathbf{B}\odot\mathbf{A}\odot\mathbf{D})^{\mathrm{T}}\right]^{+}, \quad (85)$$

$$\mathbf{D} = \mathbf{X}_{L\times IJK}\left[(\mathbf{C}\odot\mathbf{B}\odot\mathbf{A})^{\mathrm{T}}\right]^{+}. \quad (86)$$

A typical iterative procedure for four-way PARAFAC can be described as follows: Estimate the number of factors $N$ and randomly initialize **A**, **B**, and **C**, then alternately compute **A**, **B**, **C**, and **D** using above equations until a stopping criterion is satisfied.

The excellent characteristics of four-way PARAFAC are similar to those of trilinear PARAFAC. Theoretically, the four-way PARAFAC will be insensitive to the low value of signal-to-noise ratio and be steady to retrieved solutions as long as the accurate number of components was employed. Since the update of **A**, **B**, **C**, and **D** were calculated according to a strict least-squares manner, four-way PARAFAC will provide the best fitness to the analyzed data, provided that it converges properly. However, similar to PARAFAC, FPARAFAC is often affected by the problem of severe collinearity.

### 5.2.2 Alternating Quadrilinear Decomposition

AQLD has been extended from the ATLD method, whose four objective functions are shown as follows:

$$\sigma(\mathbf{D}) = \sum_{l=1}^{L}\left\|\mathbf{X}_{\ldots l} - \mathbf{A}\,diag\left(\mathbf{d}_{(l)}\right)(\mathbf{C}\odot\mathbf{B})^{\mathrm{T}}\right\|_{\mathrm{F}}^{2}, \quad (87)$$

$$\sigma(\mathbf{A}) = \sum_{i=1}^{I}\left\|\mathbf{X}_{i\ldots} - \mathbf{B}\,diag\left(\mathbf{a}_{(i)}\right)(\mathbf{D}\odot\mathbf{C})^{\mathrm{T}}\right\|_{\mathrm{F}}^{2}, \quad (88)$$

$$\sigma(\mathbf{B}) = \sum_{j=1}^{J}\left\|\mathbf{X}_{.j..} - \mathbf{C}\,diag\left(\mathbf{b}_{(j)}\right)(\mathbf{A}\odot\mathbf{D})^{\mathrm{T}}\right\|_{\mathrm{F}}^{2}, \quad (89)$$

$$\sigma(\mathbf{C}) = \sum_{k=1}^{K} \left\| \mathbf{X}_{..k.} - \mathbf{D} diag\left(\mathbf{c}_{(k)}\right)(\mathbf{B} \odot \mathbf{A})^{\mathrm{T}} \right\|_{\mathrm{F}}^{2}. \tag{90}$$

According to the least-squares principle, **A**, **B**, C, and **D** can be updated by using Equations (87)–(90). Therefore, the least-square solutions of AQLD are described as follows:

$$\mathbf{d}_{(l)}^{\mathrm{T}} = diagm\left(\mathbf{A}^{+}\mathbf{X}_{...l}\left(\left((\mathbf{C}\odot\mathbf{B})^{\mathrm{T}}\right)^{+}\right)\right), \text{ for } l = 1,2,\ldots,L, \tag{91}$$

$$\mathbf{a}_{(i)}^{\mathrm{T}} = diagm\left(\mathbf{B}^{+}\mathbf{X}_{i...}\left(\left((\mathbf{D}\odot\mathbf{C})^{\mathrm{T}}\right)^{+}\right)\right), \text{ for } i = 1,2,\ldots,I, \tag{92}$$

$$\mathbf{b}_{(j)}^{\mathrm{T}} = diagm\left(\mathbf{C}\mathbf{X}_{.j..}\left(\left((\mathbf{A}\odot\mathbf{D})^{\mathrm{T}}\right)^{+}\right)\right), \text{ for } j = 1,2,\ldots,J, \tag{93}$$

$$\mathbf{c}_{(k)}^{\mathrm{T}} = diagm\left(\mathbf{D}^{+}\mathbf{X}_{..k.}\left(\left((\mathbf{B}\odot\mathbf{A})^{\mathrm{T}}\right)^{+}\right)\right), \text{ for } k = 1,2,\ldots,K. \tag{94}$$

The AQLD algorithm not only holds the positive features of ATLD, such as fast convergence, being insensitive to the excess factors, and holding the second-order advantage, but also can improve resolution and quantitative analysis capability and provide more information.

The steps of the AQLD algorithm have been described as follows:

**(1)** Choose the number of factors, *N*;
**(2)** Randomly initialize the matrices **A**, **B**, and **C**;
**(3)** Compute **D** using Equation (91);
**(4)** Compute **A** using Equation (92) and scale **A** to be columnwisely normalized;
**(5)** Compute **B** using Equation (93) and scale **B** to be columnwisely normalized;
**(6)** Compute **C** using Equation (94) and scale **C** to be column-wisely normalized;
**(7)** Compute **D** using Equation (91);
**(8)** Update **A**, **B**, **C** and **D** according to steps 4–7, until the following stop criterion is reached.
**(9)** Postprocess **A**, **B**, **C** and **D** according to the uniqueness property and predict concentration of each component of interest in unknown sample(s) by regression of the appropriate column of **D** corresponding to each analyte of interest against its standard concentrations or by regression from the corresponding standard concentrations.

The optimizing procedure of the improved method is terminated when the following criterion reaches a certain threshold $\varepsilon = 10^{-6}$. A maximal number of 3000 iterations are adopted to avoid unduly slow convergence.

$$\mathrm{SSR}^{(m)} = \sum_{l=1}^{L} \left\| \mathbf{X}_{...l} - \mathbf{A}^{(m)} diag\left(\mathbf{d}_{(l)}\right)^{(m)} \left(\mathbf{C}^{(m)} \odot \mathbf{B}^{(m)}\right)^{\mathrm{T}} \right\|_{\mathrm{F}}^{2}, \tag{95}$$

$$\left|\frac{\mathrm{SSR}^{(m)}-\mathrm{SSR}^{(m-1)}}{\mathrm{SSR}^{(m-1)}}\right|\le\varepsilon. \tag{96}$$

where SSR is the residual sum of squares and $m$ is the current iteration number.

### *5.2.3 Alternating Penalty Quadrilinear Decomposition*

The APQLD algorithm as an extension of APTLD for decomposition of quadrilinear data is applied to third-order calibration. Similar to APTLD, four objective functions can be obtained as follows:

$$\begin{aligned}\mathrm{S}(\mathbf{A})=&\sum_{k=1}^{K}\sum_{l=1}^{L}\left\|\mathbf{X}_{..kl}-\mathbf{A}\,diag\left(\mathbf{d}_{(l)}\right)diag\left(\mathbf{c}_{(k)}\right)\mathbf{B}^{\mathrm{T}}\right\|_{\mathrm{F}}^{2}\\&+q\left(\sum_{k=1}^{K}\sum_{l=1}^{L}\left\|\left(\mathbf{X}_{..kl}\left(\mathbf{B}^{\mathrm{T}}\right)^{+}-\mathbf{A}\,diag\left(\mathbf{d}_{(l)}\right)diag\left(\mathbf{c}_{(k)}\right)\right)sqrt(\mathbf{W}_{\mathrm{B}})\right\|_{\mathrm{F}}^{2}\right.\\&\left.+\sum_{j=1}^{J}\sum_{k=1}^{K}\left\|sqrt(\mathbf{W}_{\mathrm{D}})\left(\mathbf{D}^{+}\mathbf{X}_{.jk.}-diag\left(\mathbf{c}_{(k)}\right)diag\left(\mathbf{b}_{(j)}\right)\mathbf{A}^{\mathrm{T}}\right)\right\|_{\mathrm{F}}^{2}\right),\end{aligned} \tag{97}$$

$$\begin{aligned}\mathrm{S}(\mathbf{B})=&\sum_{l=1}^{L}\sum_{i=1}^{I}\left\|\mathbf{X}_{i..l}-\mathbf{B}\,diag\left(\mathbf{a}_{(i)}\right)diag\left(\mathbf{d}_{(l)}\right)\mathbf{C}^{\mathrm{T}}\right\|_{\mathrm{F}}^{2}\\&+r\left(\sum_{l=1}^{L}\sum_{i=1}^{I}\left\|\left(\mathbf{X}_{i..l}\left(\mathbf{C}^{\mathrm{T}}\right)^{+}-\mathbf{B}\,diag\left(\mathbf{a}_{(i)}\right)diag\left(\mathbf{d}_{(l)}\right)\right)sqrt(\mathbf{W}_{\mathrm{C}})\right\|_{\mathrm{F}}^{2}\right.\\&\left.+\sum_{k=1}^{K}\sum_{l=1}^{L}\left\|sqrt(\mathbf{W}_{\mathrm{A}})\left(\mathbf{A}^{+}\mathbf{X}_{..kl}-diag\left(\mathbf{d}_{(l)}\right)diag\left(\mathbf{c}_{(k)}\right)\mathbf{B}^{\mathrm{T}}\right)\right\|_{\mathrm{F}}^{2}\right),\end{aligned} \tag{98}$$

$$\begin{aligned}\mathrm{S}(\mathbf{C})=&\sum_{i=1}^{I}\sum_{j=1}^{J}\left\|\mathbf{X}_{ij..}-\mathbf{C}\,diag\left(\mathbf{b}_{(j)}\right)diag\left(\mathbf{a}_{(i)}\right)\mathbf{D}^{\mathrm{T}}\right\|_{\mathrm{F}}^{2}\\&+s\left(\sum_{i=1}^{I}\sum_{j=1}^{J}\left\|\left(\mathbf{X}_{ij..}\left(\mathbf{D}^{\mathrm{T}}\right)^{+}-\mathbf{C}\,diag\left(\mathbf{b}_{(j)}\right)diag\left(\mathbf{a}_{(i)}\right)\right)sqrt(\mathbf{W}_{\mathrm{D}})\right\|_{\mathrm{F}}^{2}\right.\\&\left.+\sum_{l=1}^{L}\sum_{i=1}^{I}\left\|sqrt(\mathbf{W}_{\mathrm{B}})\left(\mathbf{B}^{+}\mathbf{X}_{i..l}-diag\left(\mathbf{a}_{(i)}\right)diag\left(\mathbf{d}_{(l)}\right)\mathbf{C}^{\mathrm{T}}\right)\right\|_{\mathrm{F}}^{2}\right),\end{aligned} \tag{99}$$

$$\begin{aligned}\mathrm{S}(\mathbf{D})=&\sum_{j=1}^{J}\sum_{k=1}^{K}\left\|\mathbf{X}_{.jk.}-\mathbf{D}\,diag\left(\mathbf{c}_{(k)}\right)diag\left(\mathbf{b}_{(j)}\right)\mathbf{A}^{\mathrm{T}}\right\|_{\mathrm{F}}^{2}\\&+p\left(\sum_{j=1}^{J}\sum_{k=1}^{K}\left\|\left(\mathbf{X}_{.jk.}\left(\mathbf{A}^{\mathrm{T}}\right)^{+}-\mathbf{D}\,diag\left(\mathbf{c}_{(k)}\right)diag\left(\mathbf{b}_{(j)}\right)\right)sqrt(\mathbf{W}_{\mathrm{A}})\right\|_{\mathrm{F}}^{2}\right.\\&\left.+\sum_{i=1}^{I}\sum_{j=1}^{J}\left\|sqrt(\mathbf{W}_{\mathrm{C}})\left(\mathbf{C}^{+}\mathbf{X}_{ij..}-diag\left(\mathbf{b}_{(j)}\right)diag\left(\mathbf{a}_{(i)}\right)\mathbf{D}^{\mathrm{T}}\right)\right\|_{\mathrm{F}}^{2}\right),\end{aligned} \tag{100}$$

where $\mathbf{W}_A = diag(\mathbf{1}./diagm(\mathbf{A}^T\mathbf{A}))$, $\mathbf{W}_B = diag(\mathbf{1}./diagm(\mathbf{B}^T\mathbf{B}))$, $\mathbf{W}_C = diag(\mathbf{1}./diagm(\mathbf{C}^T\mathbf{C}))$ and $\mathbf{W}_D = diag(\mathbf{1}./diagm(\mathbf{D}^T\mathbf{D}))$. The APQLD algorithm solves the quadrilinear model by alternatively optimizing the above four objective functions. The performance of APQLD can be also improved through adjusting the penalty factors $p$, $q$, $r$, and $s$, which should be theoretically close to positive infinite. Apparently, when $p = q = r = s = 0$, APQLD can be regarded as a variant of the four-way PARAFAC, that is, the four-way PARAFAC is just a special form of APQLD. APQLD not only retains the second-order advantage possessed by second-order calibration but also holds additional advantage. The introduction of fourth mode can relieve the serious problem of collinearity, which cannot be handled by three-way algorithms.

### 5.2.4 Alternating Weighted Residual Constraint Quadrilinear Decomposition

The design of AWRCQLD is based on such a way that the new residue functions are, respectively, combined with weighted factors $\mathbf{W}_A$, $\mathbf{W}_B$, $\mathbf{W}_C$, and $\mathbf{W}_D$ as constraint items to fit the loss function of PARAFAC and minimized simultaneously to construct the four new objective functions for decomposition of the quadrilinear model:

$$\sigma(\mathbf{A}) = \sum_{l=1}^{L} \left\| \mathbf{X}_{...l} - \mathbf{A}diag\left(\mathbf{d}_{(l)}\right)(\mathbf{C}\odot\mathbf{D})^{\mathrm{T}} \right\|_{\mathrm{F}}^{2} + \lambda_{\mathbf{A}} \sum_{l=1}^{L} \left\| \left( \mathbf{X}_{...l}\left((\mathbf{C}\odot\mathbf{B})^{\mathrm{T}}\right)^{+} - \mathbf{A}diag\left(\mathbf{d}_{(l)}\right)\right)\mathbf{W}_{\mathbf{D}} \right\|_{\mathrm{F}}^{2}, \quad (101)$$

$$\sigma(\mathbf{B}) = \sum_{i=1}^{I} \left\| \mathbf{X}_{i...} - \mathbf{B}diag\left(\mathbf{a}_{(i)}\right)(\mathbf{D}\odot\mathbf{C})^{\mathrm{T}} \right\|_{\mathrm{F}}^{2} + \lambda_{\mathbf{B}} \sum_{i=1}^{I} \left\| \left( \mathbf{X}_{i...}\left((\mathbf{D}\odot\mathbf{C})^{\mathrm{T}}\right)^{+} - \mathbf{B}diag\left(\mathbf{a}_{(i)}\right)\right)\mathbf{W}_{\mathbf{A}} \right\|_{\mathrm{F}}^{2}, \quad (102)$$

$$\sigma(\mathbf{C}) = \sum_{j=1}^{J} \left\| \mathbf{X}_{.j..} - \mathbf{C}diag\left(\mathbf{b}_{(j)}\right)(\mathbf{A}\odot\mathbf{D})^{\mathrm{T}} \right\|_{\mathrm{F}}^{2} + \lambda_{\mathbf{C}} \sum_{j=1}^{J} \left\| \left( \mathbf{X}_{.j..}\left((\mathbf{A}\odot\mathbf{D})^{\mathrm{T}}\right)^{+} - \mathbf{C}diag\left(\mathbf{b}_{(j)}\right)\right)\mathbf{W}_{\mathbf{B}} \right\|_{\mathrm{F}}^{2}, \quad (103)$$

$$\sigma(\mathbf{D}) = \sum_{k=1}^{K} \left\| \mathbf{X}_{..k.} - \mathbf{D}diag\left(\mathbf{c}_{(k)}\right)(\mathbf{B}\odot\mathbf{A})^{\mathrm{T}} \right\|_{\mathrm{F}}^{2} + \lambda_{\mathbf{D}} \sum_{k=1}^{K} \left\| \left( \mathrm{X}_{..k.}\left((\mathbf{B}\odot\mathbf{A})^{\mathrm{T}}\right)^{+} - \mathbf{D}diag\left(\mathbf{c}_{(k)}\right)\right)\mathbf{W}_{\mathbf{C}} \right\|_{\mathrm{F}}^{2}, \quad (104)$$

where $\mathbf{W_D} = diag\left(sqrt\left(\mathbf{1}./diagm\left(\mathbf{D}^T\mathbf{D}\right)\right)\right)$, $\mathbf{W_A} = diag\left(sqrt\left(\mathbf{1}./diagm\left(\mathbf{A}^T\mathbf{A}\right)\right)\right)$ $\mathbf{W_B} = diag\left(sqrt\left(\mathbf{1}./diagm\left(\mathbf{B}^T\mathbf{B}\right)\right)\right)$, $\mathbf{W_C} = diag\left(sqrt\left(\mathbf{1}./diagm\left(\mathbf{C}^T\mathbf{C}\right)\right)\right)$. $\lambda_\mathbf{A}$, $\lambda_\mathbf{B}$, $\lambda_\mathbf{C}$, and $\lambda_\mathbf{D}$ are coefficients of constraint items. Based on the above four objective functions, a new algorithm called AWRCQLD is developed to exploit the solution. This minimizes the four objective functions in an alternating manner, i.e., it minimizes the AWRCQLD error Equation (101) over **A** for fixed **B**, **C**, and **D**, Equation (102) over **B** for fixed **A**, **C**, and **D**, Equation (103) over **C** for fixed **A**, **B**, and **D**, and Equation (104) over **D** for fixed **A**, **B**, and **C**.

$$\mathbf{A} = \left(\sum_{l=1}^{L}\mathbf{X}_{...l}\left((\mathbf{C}\odot\mathbf{B}) + \lambda_\mathbf{A}\left((\mathbf{C}\odot\mathbf{B})^T\right)^+\mathbf{W_D}\mathbf{W_D}\right)diag\left(\mathbf{d}_{(l)}\right)\right)\cdot \left(\sum_{l=1}^{L}diag\left(\mathbf{d}_{(l)}\right)\left((\mathbf{C}\odot\mathbf{B})^T(\mathbf{C}\odot\mathbf{B}) + \lambda_\mathbf{A}\mathbf{W_D}\mathbf{W_D}\right)diag\left(\mathbf{d}_{(l)}\right)\right)^+, \quad (105)$$

$$\mathbf{B} = \left(\sum_{i=1}^{I}\mathbf{X}_{i...}\left((\mathbf{D}\odot\mathbf{C}) + \lambda_\mathbf{B}\left((\mathbf{D}\odot\mathbf{C})^T\right)^+\mathbf{W_A}\mathbf{W_A}\right)diag\left(\mathbf{a}_{(i)}\right)\right)\cdot \left(\sum_{i=1}^{I}diag\left(\mathbf{a}_{(i)}\right)\left((\mathbf{D}\odot\mathbf{C})^T(\mathbf{D}\odot\mathbf{C}) + \lambda_\mathbf{B}\mathrm{W_A}\mathrm{W_A}\right)diag\left(\mathbf{a}_{(i)}\right)\right)^+, \quad (106)$$

$$\mathbf{C} = \left(\sum_{j=1}^{J}\mathbf{X}_{.j..}\left((\mathbf{A}\odot\mathbf{D}) + \lambda_\mathbf{C}\left((\mathbf{A}\odot\mathbf{D})^T\right)^+\mathbf{W_B}\mathbf{W_B}\right)diag\left(\mathbf{b}_{(j)}\right)\right)\cdot \left(\sum_{j=1}^{J}diag\left(\mathbf{b}_{(j)}\right)\left((\mathbf{A}\odot\mathbf{D})^T(\mathbf{A}\odot\mathbf{D}) + \lambda_\mathbf{C}\mathbf{W_B}\mathbf{W_B}\right)diag\left(\mathbf{b}_{(j)}\right)\right)^+, \quad (107)$$

$$\mathbf{D} = \left(\sum_{k=1}^{K}\mathbf{X}_{..k.}\left((\mathbf{B}\odot\mathbf{A}) + \lambda_\mathbf{D}\left((\mathbf{B}\odot\mathbf{A})^T\right)^+\mathrm{W_C}\mathrm{W_C}\right)diag\left(\mathbf{c}_{(k)}\right)\right)\cdot \left(\sum_{k=1}^{K}diag\left(\mathbf{c}_{(k)}\right)\left((\mathbf{B}\odot\mathbf{A})^T(\mathbf{B}\odot\mathbf{A}) + \lambda_\mathbf{D}\mathbf{W_C}\mathbf{W_C}\right)diag\left(\mathbf{c}_{(k)}\right)\right)^+. \quad (108)$$

In addition, compared with four-way PARAFAC, the proposed method not only can resist the influence of the excess component number but also largely reduce the number of iterations required.

### 5.2.5 *Four-Way Self-Weighted Alternating Normalized Residue Fitting Algorithm*

According to Equations (20)–(23), the following equations hold for a quadrilinear model:

$$\mathbf{A}^{+}\mathbf{X}_{...l}\left((\mathbf{C}\odot\mathbf{B})^{\mathrm{T}}\right)^{+} = diag\left(\mathbf{d}_{(l)}\right) + \mathbf{A}^{+}\mathbf{E}_{....l}\left((\mathbf{C}\odot\mathbf{B})^{\mathrm{T}}\right)^{+},$$
$$\mathbf{A}^{+}\mathbf{X}_{...l} = diag\left(\mathbf{d}_{(l)}\right)(\mathbf{C}\odot\mathbf{B})^{\mathrm{T}} + \mathbf{A}^{+}\mathbf{E}_{...l},$$
$$\mathbf{X}_{...l}\left((\mathbf{C}\odot\mathbf{B})^{\mathrm{T}}\right)^{+} = \mathbf{A}diag\left(\mathbf{d}_{(l)}\right) + \mathbf{E}_{...l}\left((\mathbf{C}\odot\mathbf{B})^{\mathrm{T}}\right)^{+} \quad \text{for } l = 1,2,\ldots,L, \tag{109}$$

$$\mathbf{B}^{+}\mathbf{X}_{i...}\left((\mathbf{D}\odot\mathbf{C})^{\mathrm{T}}\right)^{+} = diag\left(\mathbf{a}_{(i)}\right) + \mathbf{B}^{+}\mathbf{E}_{i...}\left((\mathbf{D}\odot\mathbf{C})^{\mathrm{T}}\right)^{+},$$
$$\mathbf{B}^{+}\mathbf{X}_{i...} = diag\left(\mathbf{a}_{(i)}\right)(\mathbf{D}\odot\mathbf{C})^{\mathrm{T}} + \mathbf{B}^{+}\mathbf{E}_{i...},$$
$$\mathbf{X}_{i...}\left((\mathbf{D}\odot\mathbf{C})^{\mathrm{T}}\right)^{+} = \mathbf{B}diag\left(\mathbf{a}_{(i)}\right) + \mathbf{E}_{i...}\left((\mathbf{D}\odot\mathbf{C})^{\mathrm{T}}\right)^{+} \quad \text{for } i = 1,2,\ldots,I, \tag{110}$$

$$\mathbf{C}^{+}\mathbf{X}_{.j..}\left((\mathbf{A}\odot\mathbf{D})^{\mathrm{T}}\right)^{+} = diag\left(\mathbf{b}_{(j)}\right) + \mathbf{C}^{+}\mathbf{E}_{.j..}\left((\mathbf{A}\odot\mathbf{D})^{\mathrm{T}}\right)^{+},$$
$$\mathbf{C}^{+}\mathbf{X}_{.j..} = diag\left(\mathbf{b}_{(j)}\right)(\mathbf{A}\odot\mathbf{D})^{\mathrm{T}} + \mathbf{C}^{+}\mathbf{E}_{.j..},$$
$$\mathbf{X}_{.j..}\left((\mathbf{A}\odot\mathbf{D})^{\mathrm{T}}\right)^{+} = \mathbf{C}diag\left(\mathbf{b}_{(j)}\right) + \mathbf{E}_{.j..}\left((\mathbf{A}\odot\mathbf{D})^{\mathrm{T}}\right)^{+} \quad \text{for } j = 1,2,\ldots,J, \tag{111}$$

$$\mathbf{D}^{+}\mathbf{X}_{..k.}\left((\mathbf{B}\odot\mathbf{A})^{\mathrm{T}}\right)^{+} = diag\left(\mathbf{c}_{(k)}\right) + \mathbf{D}^{+}\mathbf{E}_{..k.}\left((\mathbf{B}\odot\mathbf{A})^{\mathrm{T}}\right)^{+},$$
$$\mathbf{D}^{+}\mathbf{X}_{..k.} = diag\left(\mathbf{c}_{(k)}\right)(\mathbf{B}\odot\mathbf{A})^{\mathrm{T}} + \mathbf{D}^{+}\mathbf{E}_{..k.},$$
$$\mathbf{X}_{..k.}\left((\mathbf{B}\odot\mathbf{A})^{\mathrm{T}}\right)^{+} = \mathbf{D}diag\left(\mathbf{c}_{(k)}\right) + \mathbf{E}_{..k.}\left((\mathbf{B}\odot\mathbf{A})^{\mathrm{T}}\right)^{+} \quad \text{for } k = 1,2,\ldots,K, \tag{112}$$

To develop an algorithm with high convergence rate and being insensitive to the excess factors used in calculation, we consider the possibility to decompose the quadrilinear component model by alternatively solving the following four nearby least-squares problems:

$$\begin{aligned} S(\mathbf{A}) = & \sum_{i=1}^{I}\left\|\mathbf{B}^{+}\mathbf{X}_{i...}\left((\mathbf{D}\odot\mathbf{C})^{\mathrm{T}}\right)^{+} - diag\left(\mathbf{a}_{(i)}\right)\right\|_{\mathrm{F}}^{2} \\ & + \lambda_{\mathbf{A}}\sum_{i=1}^{I}\left\|\left(\mathbf{B}^{+}\mathbf{X}_{i...} - diag\left(\mathbf{a}_{(i)}\right)(\mathbf{D}\odot\mathbf{C})^{\mathrm{T}}\right)^{\mathrm{T}}\right. \\ & \left. .^{*}diag\left(sqrt\left(\mathbf{1}./diagm\left((\mathbf{D}\odot\mathbf{C})^{\mathrm{T}}(\mathbf{D}\odot\mathbf{C})\right)\right)\right)\right\|_{\mathrm{F}}^{2} \\ & + \mu_{\mathbf{A}}\sum_{i=1}^{I}\left\|\left(\mathbf{X}_{i...}\left((\mathbf{D}\odot\mathbf{C})^{\mathrm{T}}\right)^{+} - \mathbf{B}diag\left(\mathbf{a}_{(i)}\right)\right)\right. \\ & \left. .^{*}diag\left(sqrt\left(\mathbf{1}./diagm\left(\mathbf{B}^{\mathrm{T}}\mathbf{B}\right)\right)\right)\right\|_{\mathrm{F}}^{2}, \end{aligned} \tag{113}$$

$$
\begin{aligned}
S(\mathbf{B}) = & \sum_{j=1}^{J} \left\| \mathbf{C}^{+}\mathbf{X}_{.j..}\left( (\mathbf{A} \odot \mathbf{D})^{\mathrm{T}} \right)^{+} - diag\left(\mathbf{b}_{(j)}\right) \right\|_{\mathrm{F}}^{2} \\
& + \lambda_{\mathbf{B}} \sum_{j=1}^{J} \left\| \left( \mathbf{C}^{+}\mathbf{X}_{.j..} - diag\left(\mathbf{b}_{(j)}\right)(\mathbf{A} \odot \mathbf{D})^{\mathrm{T}} \right)^{\mathrm{T}} \right. \\
& \left. .^{*} diag\left( sqrt\left( \mathbf{1}./diagm\left( (\mathbf{A} \odot \mathbf{D})^{\mathrm{T}} (\mathbf{A} \odot \mathbf{D}) \right) \right) \right) \right\|_{\mathrm{F}}^{2} \\
& + \mu_{\mathbf{B}} \sum_{j=1}^{J} \left\| \left( \mathbf{X}_{.j.}\left( (\mathbf{A} \odot \mathbf{D})^{\mathrm{T}} \right)^{+} - \mathbf{C} diag\left(\mathbf{b}_{(j)}\right) \right) \right. \\
& \left. .^{*} diag\left( sqrt\left( \mathbf{1}./diagm\left( \mathbf{C}^{\mathrm{T}}\mathbf{C} \right) \right) \right) \right\|_{\mathrm{F}}^{2},
\end{aligned}
\tag{114}
$$

$$
\begin{aligned}
S(\mathbf{C}) = & \sum_{k=1}^{K} \left\| \mathbf{D}^{+}\mathbf{X}_{..k.}\left( (\mathbf{B} \odot \mathbf{A})^{\mathrm{T}} \right)^{+} - diag\left(\mathbf{c}_{(k)}\right) \right\|_{\mathrm{F}}^{2} \\
& + \lambda_{\mathbf{C}} \sum_{k=1}^{K} \left\| \left( \mathbf{D}^{+}\mathbf{X}_{..k.} - diag\left(\mathbf{c}_{(k)}\right)(\mathbf{B} \odot \mathbf{A})^{\mathrm{T}} \right)^{\mathrm{T}} \right. \\
& \left. .^{*} diag\left( sqrt\left( \mathbf{1}./diagm\left( (\mathbf{B} \odot \mathbf{A})^{\mathrm{T}} (\mathbf{B} \odot \mathbf{A}) \right) \right) \right) \right\|_{\mathrm{F}}^{2} \\
& + \mu_{\mathbf{C}} \sum_{k=1}^{K} \left\| \left( \mathbf{X}_{..k.}\left( (\mathbf{B} \odot \mathbf{A})^{\mathrm{T}} \right)^{+} - \mathbf{D} diag\left(\mathbf{c}_{(j)}\right) \right) \right. \\
& \left. .^{*} diag\left( sqrt\left( \mathbf{1}./diagm\left( \mathbf{D}^{\mathrm{T}}\mathbf{D} \right) \right) \right) \right\|_{\mathrm{F}}^{2},
\end{aligned}
\tag{115}
$$

$$
\begin{aligned}
S(\mathbf{D}) = & \sum_{l=1}^{L} \left\| \mathbf{A}^{+}\mathbf{X}_{...l}\left( (\mathbf{C} \odot \mathbf{B})^{\mathrm{T}} \right)^{+} - diag\left(\mathbf{d}_{(l)}\right) \right\|_{\mathrm{F}}^{2} \\
& + \lambda_{\mathbf{D}} \sum_{l=1}^{L} \left\| \left( \mathbf{A}^{+}\mathbf{X}_{...l} - diag\left(\mathbf{d}_{(l)}\right)(\mathbf{C} \odot \mathbf{B})^{\mathrm{T}} \right)^{\mathrm{T}} \right. \\
& \left. .^{*} diag\left( sqrt\left( \mathbf{1}./diagm\left( (\mathbf{C} \odot \mathbf{B})^{\mathrm{T}} (\mathbf{C} \odot \mathbf{B}) \right) \right) \right) \right\|_{\mathrm{F}}^{2} \\
& + \mu_{\mathbf{D}} \sum_{l=1}^{L} \left\| \left( \mathbf{X}_{...l}\left( (\mathbf{C} \odot \mathbf{B})^{\mathrm{T}} \right)^{+} - \mathbf{A} diag\left(\mathbf{d}_{(l)}\right) \right) \right. \\
& \left. .^{*} diag\left( sqrt\left( \mathbf{1}./diagm\left( \mathbf{A}^{\mathrm{T}}\mathbf{A} \right) \right) \right) \right\|_{\mathrm{F}}^{2},
\end{aligned}
\tag{116}
$$

It is worthwhile to note that, in the solving process, the Moore–Penrose generalized inverse based on SVD is employed. This approach could make the numerical solution more stable with similarities in profiles.

The four above-described objective functions have strong intrinsic relationships, and their solutions are identical to the real underlying profile matrices of error-free quadrilinear component model, which is the reason one needs to minimize them alternatively. In the presence of noise, the four objective functions have different responses to hypersurfaces. Therefore, it can be expected that alternatively minimizing the four objective functions may avoid possible "swamp."

According to the above-mentioned four objective functions, a new algorithm called the RSWAQLD can be proposed as follows. It alternately minimizes $S(\mathbf{D})$ to estimate $\mathbf{D}$ with $\mathbf{A}$, $\mathbf{B}$, and $\mathbf{C}$ fixed, $S(\mathbf{A})$ to estimate $\mathbf{A}$ with $\mathbf{B}$, $\mathbf{C}$ and $\mathbf{D}$ fixed, $S(\mathbf{B})$ to estimate $\mathbf{B}$ with $\mathbf{C}$, $\mathbf{D}$ and $\mathbf{A}$ fixed, $S(\mathbf{C})$ to estimate $\mathbf{C}$ with $\mathbf{D}$, $\mathbf{A}$ and $\mathbf{B}$ fixed. The updates for $\mathbf{D}$, $\mathbf{A}$, $\mathbf{B}$, and $\mathbf{C}$ from Equations (113)–(116), based on the least-squares principle, are as follows:

$$\begin{aligned}\mathbf{a}_{(i)} = &\frac{1}{1+\lambda_{\mathbf{A}}+\mu_{\mathbf{A}}} diagm\Big[\mathbf{B}^{+}\mathbf{X}_{i...}\left((\mathbf{D}\odot\mathbf{C})^{\mathrm{T}}\right)^{+}\\ &+\lambda_{\mathbf{A}}(\mathbf{B}^{+}\mathbf{X}_{i...}(\mathbf{D}\odot\mathbf{C}))./((\mathbf{D}^{\mathrm{T}}\mathbf{D})\cdot*(\mathbf{C}^{\mathrm{T}}\mathbf{C}))\\ &+\mu_{\mathbf{A}}\left((\mathbf{D}\odot\mathbf{C})^{+}(\mathbf{X}_{i...})^{\mathrm{T}}\mathbf{B}\right)./(\mathbf{B}^{\mathrm{T}}\mathbf{B})\Big]^{\mathrm{T}} \quad \text{for } i=1,2,\ldots,I,\end{aligned} \tag{117}$$

$$\begin{aligned}\mathbf{b}_{(j)} = &\frac{1}{1+\lambda_{\mathbf{B}}+\mu_{\mathbf{B}}} diagm\Big[\mathbf{C}^{+}\mathbf{X}_{.j..}\left((\mathbf{A}\odot\mathbf{D})^{\mathrm{T}}\right)^{+}\\ &+\lambda_{\mathbf{B}}(\mathbf{C}^{+}\mathbf{X}_{.j..}(\mathbf{A}\odot\mathbf{D}))./((\mathbf{A}^{\mathrm{T}}\mathbf{A})\cdot*(\mathbf{D}^{\mathrm{T}}\mathbf{D}))\\ &+\mu_{\mathbf{B}}\left((\mathbf{A}\odot\mathbf{D})^{+}(\mathbf{X}_{.j..})^{\mathrm{T}}\mathbf{C}\right)./(\mathbf{C}^{\mathrm{T}}\mathbf{C})\Big]^{\mathrm{T}} \quad \text{for } j=1,2,\ldots,J,\end{aligned} \tag{118}$$

$$\begin{aligned}\mathbf{c}_{(k)} = &\frac{1}{1+\lambda_{\mathbf{C}}+\mu_{\mathbf{C}}} diagm\Big[\mathbf{D}^{+}\mathbf{X}_{..k.}\left((\mathbf{B}\odot\mathbf{A})^{\mathrm{T}}\right)^{+}\\ &+\lambda_{\mathbf{C}}(\mathbf{D}^{+}\mathbf{X}_{..k.}(\mathbf{B}\odot\mathbf{A}))\cdot/((\mathbf{B}^{\mathrm{T}}\mathbf{B})\cdot*(\mathbf{A}^{\mathrm{T}}\mathbf{A}))\\ &+\mu_{\mathbf{C}}\left((\mathbf{B}\odot\mathbf{A})^{+}(\mathbf{X}_{..k.})^{\mathrm{T}}\mathbf{D}\right)./(\mathbf{D}^{\mathrm{T}}\mathbf{D})\Big]^{\mathrm{T}} \quad \text{for } k=1,2,\ldots,K,\end{aligned} \tag{119}$$

$$\begin{aligned}\mathbf{d}_{(l)} = &\frac{1}{1+\lambda_{\mathbf{D}}+\mu_{\mathbf{D}}} diagm\Big[\mathbf{A}^{+}\mathbf{X}_{..l}\left((\mathbf{C}\odot\mathbf{B})^{\mathrm{T}}\right)^{+}\\ &+\lambda_{\mathbf{D}}(\mathbf{A}^{+}\mathbf{X}_{...l}(\mathbf{C}\odot\mathbf{B}))\cdot/((\mathbf{C}^{\mathrm{T}}\mathbf{C})\cdot*(\mathbf{B}^{\mathrm{T}}\mathbf{B}))\\ &+\mu_{\mathbf{D}}\left((\mathbf{C}\odot\mathbf{B})^{+}(\mathbf{X}_{...l})^{\mathrm{T}}\mathbf{A}\right)\cdot/(\mathbf{A}^{\mathrm{T}}\mathbf{A})\Big]^{\mathrm{T}} \quad \text{for } l=1,2,\ldots,L.\end{aligned} \tag{120}$$

RSWAQLD has the properties of fast convergence and being insensitive to the excess factors used in calculation.

## 5.3 Fourth-Order Calibration

Fourth-order data would display the obvious advantage of providing richer analytical information, implying more stable methods toward interferences and matrix effects and less prone to minor changes in reaction conditions. This should allow for an improvement in predictive ability. In recent years, U-PLS/residual quadrilinearization has also been proposed by Olivieri and coworkers to decompose a fourth-order data array (excitation–emission–kinetic–pH), and identify its higher-order advantage [9].

Very recently, the alternating quinquelinear decomposition (AQQLD) algorithm for fourth-order calibration was developed by Qing and coworkers for the processing of five-way data, which were obtained by recording the HPLC-DAD signals of samples as a function of the reaction time and pH values [16]. Actually, AQQLD was extended from ATLD based on pseudo-fully stretched matrix forms of the quinquelinear model, which has the advantage of faster convergence rate and being insensitive to the excess component number adopted in the model. AQQLD and five-way PARAFAC have been successfully applied to investigate quantitatively the kinetics of naptalam hydrolysis in two practical sample systems. Satisfactory results were obtained for the determination of NAP in these samples by the both algorithms.

## 6 ESTIMATION OF THE CHEMICAL RANK IN MULTIWAY DATA

Before decomposing three-way responsive data, estimating chemical rank (the number of factors or components) responsible for the trilinear model is always an important and intractable problem. It can be seemingly solved by the algorithms which are insensitive to the overestimation of the chemical rank. However, these algorithms still require that the chemical rank chosen is no less than the underlying one. According to our practical experience, if the chosen model dimensionality is much more than the actual number of components, generally, a model fitting error and a deviation between the predicted and actual results will arise. On the contrary, the algorithm can perform well with more acceptable resolution and calibration when the factor number chosen is the same as the chemical rank of the analytical system.

With the above beliefs, a number of suitable methods for estimating the chemical rank have been developed in the literature. Generally, they can be classified into two main groups. The first group is based on the trilinear model, such as Wu's maximum rank method [32], split-half analysis [82], ADD-ONE-UP [83], core consistency diagnostic (CORCONDIA) [84] and self-weighted alternating trilinear decomposition and Monte Carlo simulation (SWATLD-MCS) [85]. Split-half analysis is a type of jack-knife analysis. It involves a relatively complex splitting skill and a poor splitting scheme that will hinder the correct result. ADD-ONE-UP fits two reconstructed three-way data arrays by PARAFAC with a gradually increasing number of components and determines the chemical rank by examining their SSR. It can cope with nonideal experimental conditions, such as slight collinearity and varying backgrounds. CORCONDIA, as one commonly used method, determines the chemical rank by comparing the core matrices of PARAFAC and Tucker3 models. However, ADD-ONE-UP and CORCONDIA are very time-consuming for requiring PARAFAC to run many times. Moreover, the data with severe collinearity may make them suffer from heavy computation burden and acquiring inaccurate results. The method, SWATLD-MCS estimates the chemical rank by comparing the values of sorted mean relative-concentration (SMRC), which are obtained from SWATLD by decomposing one pseudo three-way data array created by Monte Carlo simulation. The other group includes non-model methods, such as factor indicator function (IND) [86], orthogonal projection approach (OPA) [87], two-mode subspace comparison (TMSC) [88], linear transform method incorporating Monte Carlo simulation (LTMC) [89], subspace projection of pseudo high-way array (SPPH) [90], and region-based on moving windows subspace projection technique (RMWSPT) [91]. Though all of above methods can be used for rank-estimation, it is a pity that none of them can guarantee the correct results under all circumstances. In practice, two or more methods are often adopted to confirm the analytical results [2,3].

## 6.1 Maximum Rank Method

The maximum rank method was firstly proposed by Wu *et al.* to estimate the chemical rank for ATLD and ATLD's variants, as the following form shows:

$$rank\underline{\mathbf{X}} = max\{rank(\mathbf{X}_{I\times JK}), rank(\mathbf{X}_{J\times KI}), rank(\mathbf{X}_{K\times IJ})\}, \tag{121}$$

In practice, the number of factors will be also determined as follows:

$$rank\underline{\mathbf{X}} = max\left\{rank\left(\sum_{i=1}^{I}\mathbf{X}_{i..}\right), rank\left(\sum_{j=1}^{J}\mathbf{X}_{.j.}\right), rank\left(\sum_{k=1}^{K}\mathbf{X}_{..k}\right)\right\}, \tag{122}$$

or

$$rank\underline{\mathbf{X}} = max\{rank(\mathbf{X}_{I\times JK}\mathbf{X}_{I\times JK}^{\mathrm{T}}), rank(\mathbf{X}_{J\times KI}\mathbf{X}_{J\times KI}^{\mathrm{T}}), rank(\mathbf{X}_{K\times IJ}\mathbf{X}_{K\times IJ}^{\mathrm{T}})\}, \tag{123}$$

where rank (.) denotes the numerical rank estimate of a matrix based on an SVD procedure with a default tolerance. This method is conformable to be used in any instance and estimates the chemical rank of the three-way data with satisfactory results.

## 6.2 ADD-ONE-UP

ADD-ONE-UP, proposed by Chen *et al.* in 2001 for determining the chemical rank, fits two reconstructed three-way data arrays by PARAFAC with a gradually increasing number of components and then determines the chemical rank by examining the SSR. This method can cope with nonideal experimental conditions, such as slight collinearity and varying background contributions.

**(1)** For a trilinear data array $\underline{\mathbf{X}}$, unfold it into a two-way data set $\mathbf{X}_{I\times JK}$.
**(2)** Decompose $\mathbf{X}_{I\times JK}$ by SVD: $\mathbf{X}_{I\times JK} = \mathbf{USV}^{\mathrm{T}}$.
**(3)** Define $\mathbf{X}_c = \mathbf{U}_c\mathbf{S}_c\mathbf{V}_c^{\mathrm{T}}$, where $\mathbf{U}_c$ and $\mathbf{V}_c$ consist of the first $c$ columns of $\mathbf{U}$ and $\mathbf{V}$, respectively, $\mathbf{S}_c$ is a diagonal matrix with diagonal elements equal to the first c diagonal elements of $\mathbf{S}$.

(4) Fold $\mathbf{X}_c$ into a three-way data array $\mathbf{X}c$ and decompose it by PARAFAC with $N=c$. The residual sum of squares is represented by $\text{SSR}_c$.
(5) Repeat steps 3 and 4, for $c=1, 2, 3,\ldots,$ until $\text{SSR}_{c1}$ reaches its minimum or satisfies the following in equations: $\text{SSR}_{c1} < \text{s}_{c1}^2$ and $\text{SSR}_{c1+1} > \text{s}_{c1+1}^2$ and $\text{SSR}_{c1+2} > \text{s}_{c1+2}^2$ ($\text{s}_i$ is the $i$th diagonal element of matrix $\mathbf{S}$, $\text{s}_{c1}^2$ is the variance introduced by the inclusion of c1-th component in the truncating step).
(6) Unfold $\boldsymbol{X}$ in another mode to get $\mathbf{X}_{IK\times J}$, and then perform the same operations from steps 2 to 5 to get $c2$, which satisfies similar relationships as for $c1$.
(7) The number of factors needed in decomposing the trilinear data array $\boldsymbol{X}$ should be the smaller one between $c1$ and $c2$, i.e., $F=\min(c1, c2)$.

This method takes advantage of the eigenvalues of factor analysis and the residuals of trilinear decomposition. It can cope with nonideal experimental conditions, such as moderate collinearity and varying backgrounds. But, because it is based on the PARAFAC algorithm, ADD-ONE-UP has some drawbacks. It is very time-consuming, because of the need to run PARAFAC for many times. In addition, two-factor degeneracies may make this method suffer from a heavy computational burden and may yield inaccurate results. Table 5 lists the common performance of ADD-ONE-UP for data arrays with random noise, and, as expected in the design stage, $\text{SSR}_i$ attains its minimum when the number of retained factors is equal to 4, which is exactly the number of actual underlying factors. It also confirms that the variances ($\text{s}_i^2$) brought in by the addition of noise factors in truncating procedure are smaller than the corresponding residuals sum of square of PARAFAC with $N$ equal to the number of factors retained. The two criteria employed in ADD-ONE-UP method have been simultaneously satisfied.

**TABLE 5** General Behavior of ADD-ONE-UP Method for a Randomly Simulated Four-Component Data Array with Small Homoscedastic Noise Level

| $i$ | $s_i^2$ | $\text{SSR}_i$ |
|---|---|---|
| 1 | 23.48 | 0.3703 |
| 2 | 0.4526 | 0.2598 |
| 3 | 0.2162 | $8.337\times10^{-2}$ |
| 4 | 0.1081 | $2.392\times10^{-3}$ |
| 5 | $1.391\times10^{-3}$ | $3.630\times10^{-3}$ |
| 6 | $1.293\times10^{-3}$ | $4.796\times10^{-3}$ |

## 6.3 CORCONDIA

The principle of CORCONDIA is to assess the similarity between the superdiagonal array $\boldsymbol{T}$ and the least-squares-fitted $\boldsymbol{G}$ with a gradually increasing number of components. CORCONDIA is defined as:

$$\text{core consistency} = 100 \times \left(1 - \frac{\sum_{d=1}^{N}\sum_{e=1}^{N}\sum_{f=1}^{N}\left(g_{def} - t_{def}\right)^2}{\sum_{d=1}^{N}\sum_{e=1}^{N}\sum_{f=1}^{N} t_{def}^2}\right), \quad (124)$$

where $g_{def}$ is the element of $\boldsymbol{G}$; $t_{def}$ is the elements of $\boldsymbol{T}$; $N$ is the number of factors in the model.

For an ideal trilinear model, $g_{def}$ are equal to $t_{def}$ and the value of core consistency will be equal to 100%. Usually, a value of the core consistency above 90% can be regarded as "very trilinear," whereas a value in the neighborhood of 50% will imply a problematic model, which contains both trilinear and nontrilinear variations. A value close to zero or even negative means the model is not valid. Though it is an effective method, it suffers from the drawbacks of PARAFAC. Figure 18 shows the result of CORCONDIA to estimate one simulated data with three components.

## 6.4 Self-Weighted Alternating Trilinear Decomposition and Monte Carlo Simulation

The method, SWATLD-MCS, estimates the chemical rank by comparing the values of SMRC, which are obtained from SWATLD by decomposing one pseudo three-way data array created by Monte Carlo simulation.

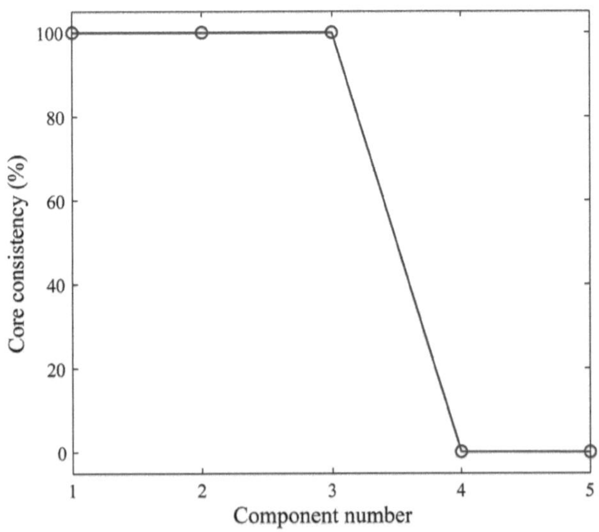


**FIGURE 18** Result of CORCONDIA to estimate one simulated data with three components.

The process of this new method is demonstrated as follows:

**(1)** $M$ pseudo-sample matrices are obtained by Monte Carlo simulation as follows:

$$\mathbf{R}_{..m} = \left(\sum_{k=1}^{K} w_{km}\mathbf{X}_{..k}\right)\left(\sum_{k=1}^{K} w_{km}\right)^{-1}, \quad \text{for } m = 1, 2, \cdots, M. \tag{125}$$

where $\mathbf{R}_{..m}$ is a pseudo sample and the size is $I \times J$; $M$, $(K \leq M \leq K(K-1)/2)$, is the number of pseudo samples; $w_{km}$, $(0 < w_{km} < 1,\ k = 1, 2, \ldots, K;\ m = 1, 2, \ldots, M)$, is a stochastic number, which is produced by Monte Carlo simulation.

**(2)** The pseudo three-way data, stacked by **M** pseudo-sample matrices, is decomposed by SWATLD with $N$ as the factor number ($N$ is larger than the underlying one), and the relative-concentration matrix **C** is obtained.

**(3)** The sorted mean relative-concentration (SMRC) is computed and the chemical rank is estimated based on the following criteria: the value of SMRC will decrease obviously from a relatively large value to a very small one when the number of retained components exceeds the real one, and then vary over a small range for the latter factors.

This new method can accurately estimate chemical ranks of complex systems even when heavy collinearity and high-intensity noise are presented. Also the method has a lower computational burden than competitive methods, which saves overall analysis time. In addition, the application of this new methodology can be further extended, in particular, (I) other second-order calibration algorithms, being insensitive to excessive factors, can be used with this method; (II) the chemical rank of higher-order data can be determined by this method, using higher-order calibration algorithms.

An example showing its behavior is illustrated in Figure 19, concerning one simulated three-way data with four underlying components. The resolved

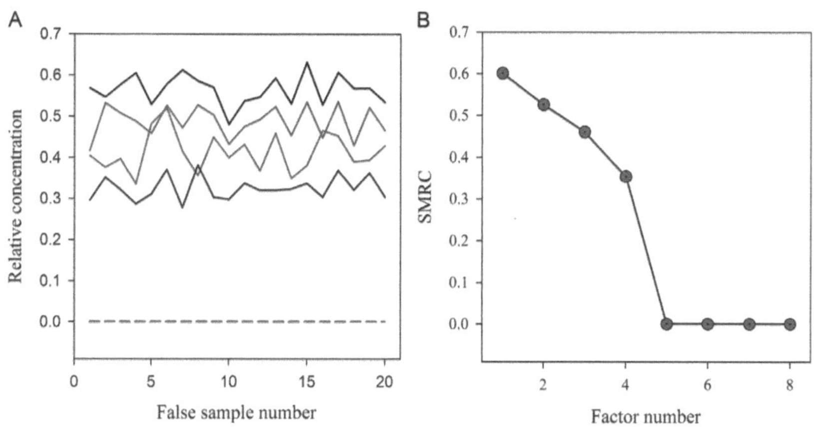


**FIGURE 19** Results from SWATLD-MCS on simulated data with 8 as the factor number: (A) the profile of relative concentration matrix, **C**; (B) the SMRC profile.

matrix **C** from the pseudo three-way data is plotted in Figure 19A. As can be seen, there exist four lines which are obviously higher than other ones, and they should represent the information of underlying components. The SMRC is plotted in Figure 19B, suggesting that when the number of retained components exceeds four-factors, the corresponding value of SMRC decreases from a relatively large value to a very small one and varies over a small range for latter factors. Consequently, the suggested number of components is four, which is the same as the underlying one.

## 6.5 Factor Indicator Function

An empirical function, called IND, is discovered by Malinowski [86]. The factor indicator function is defined as

$$\mathrm{IND} = \left(\sum_{j=n+1}^{c} \lambda_j^0\right)^{1/2} r^{-1/2}(c-n)^{-5/2}. \tag{126}$$

Here, $r$ and $c$ are the number of rows and columns of data matrix **M**, respectively (assume $c < r$); $n$ denotes the number of factors; $\lambda_j^0$ is the eigenvalue of the covariance matrix $\mathbf{M}^{\mathrm{T}}\mathbf{M}$. The IND function reaches minimum when the correct number of factors is employed. Figure 20 shows the result of IND to estimate one simulated data with three components.

## 6.6 Subspace Projection of Pseudo High-way Data Array

SPPH was developed by Xia and coworkers for estimating the chemical rank of high-way data arrays. The proposed method determines the chemical rank

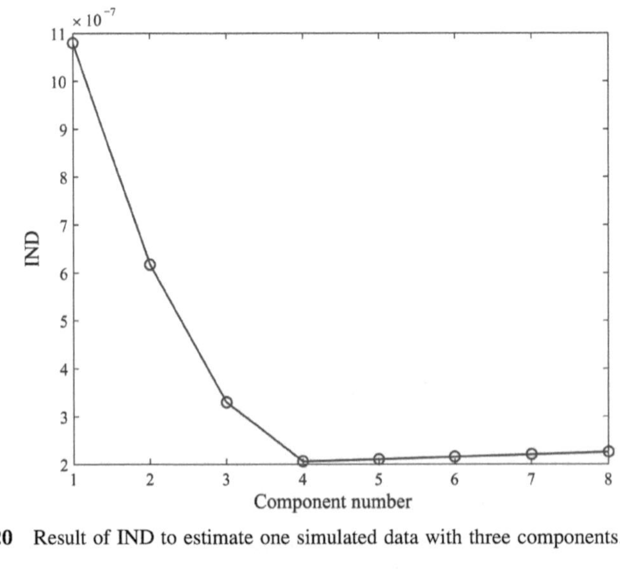


**FIGURE 20** Result of IND to estimate one simulated data with three components.

through performing SVD on the slice matrices of original high-way data array to produce a pseudo high-way data array and employing the idea of the difference of the original truncated data set and the pseudo one.

The steps for SPPH are as follows:

**(1)** For a trilinear data array $\boldsymbol{X}$, unfold it into two-way data matrices $\mathbf{X}_{I\times JK}$ and $\mathbf{X}_{J\times KI}$.

**(2)** Decompose $\mathbf{X}_{I\times JK}$ and $\mathbf{X}_{J\times KI}$ by SVD, respectively, $\mathbf{X}_{I\times JK}=\mathbf{U}_I{}^*\mathbf{S}_I{}^*(\mathbf{V}_I{}^*)^{\mathrm{T}}$ and $\mathbf{X}_{J\times KI}=\mathbf{U}_J{}^*\mathbf{S}_J{}^*(\mathbf{V}_J{}^*)^{\mathrm{T}}$.

**(3)** Decompose block matrix $\mathbf{X}_{..k}$ by SVD, $\mathbf{X}_{..k}=\mathbf{U}^k\mathbf{S}^k(\mathbf{V}^k)^{\mathrm{T}}$ ($k=1, 2, \ldots, K$).

**(4)** Define $\mathbf{X}_f^k=\mathbf{U}_f^k\mathbf{S}_f^k\left(\mathbf{V}_f^k\right)^{\mathrm{T}}$, where $\mathbf{U}_f^k$ and $\mathbf{V}_f^k$ consist of the first $f$ columns of $\mathbf{U}^k$ and $\mathbf{V}^k$, respectively, $\mathbf{S}_f^k$ is a diagonal matrix with diagonal elements equal to the first $f$ diagonal elements of $\mathbf{S}^k$. Then obtain a new three-way $\underline{\mathbf{X}}_f$.

**(5)** Unfold $\underline{\mathbf{X}}_f$ into a two-way data matrix $\mathbf{X}_{(f)I\times JK}=\left[\mathbf{X}_f^1, \mathbf{X}_f^2, \ldots, \mathbf{X}_f^k\right]$ and decompose $\mathbf{X}_{(f)I\times JK}$ by SVD, $\mathbf{X}_{(f)I\times JK}=\mathbf{U}_I\mathbf{S}_I\mathbf{V}_I^{\mathrm{T}}$.

**(6)** Select factor number $f=1, 2, \ldots, F$ to calculate the projection residual ps*i*. The process will stop when the projection residuals decrease remarkably from a very large value to a very small one.

**(7)** Two criteria are used. One criterion: the residual will increase rapidly and obviously from a very small value to a relatively large one when the selected factor number $f=N_1+1$, where $N_1$ denotes the *I*-mode chemical rank of the three-way data array. One can directly observe and determine the appropriate component to fit the model by the figure of projection residuals. The other criterion: Define $m=\min\ (I, J)$ and $ss=S^*(\mathrm{m})/\|\mathbf{S}^*\|\mathrm{F}$; the residual is larger than $ss$ when $f>N_1$, where $S^*(m)$ denotes the $m$th singular value of $\mathbf{X}_{I\times JK}$.

**(8)** Unfold $\mathbf{X}_f$ into another two-way data matrix $\mathbf{X}_{(f)J\times KI}=\left[\left(\mathbf{X}_f^1\right)^{\mathrm{T}}, \left(\mathbf{X}_f^2\right)^{\mathrm{T}}, \ldots, \left(\mathbf{X}_f^K\right)^{\mathrm{T}}\right]$ and decompose $\mathbf{X}_{(f)J\times IK}$ by SVD, and then perform the same operations from step 6 to 7 to get $N_J$, which satisfies similar relationships as for $N_1$.

**(9)** The number of factors needed in decomposing the three-way data array $\boldsymbol{X}$ is equal to $N_1$ or $N_J$ when $N_1$ is equal to $N_J$, i.e. $N=N_1=N_J$. However, when $N_1$ is not equal to $N_J$, a more conservative rule would be examining the possible sources of this difference, such as, $N=\max\ (N_1, N_J)$ for having a heavy situation of collinearity and $N=\min\ (N_1, N_J)$ for having a nontrilinear system.

This method uses the information from eigenvectors combined with the projection residual to estimate the rank of the three-way data array instead of using the eigenvalue. It can accurately and quickly determine the chemical rank to fit the trilinear model and deal with more complex situations with existence of severe collinearity and trace concentration than many other methods can and performs well in practical applications.

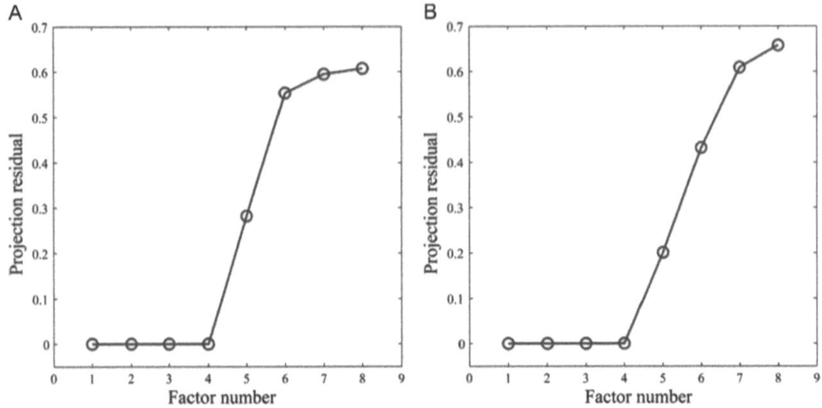


**FIGURE 21** Projection residuals of VSPMCS for one simulated data with four underlying components: (A) for *I* direction; (B) for *J* direction.

## 6.7 Vector Subspace Projection with Monte Carlo Simulation

Vector subspace projection with Monte Carlo simulation (VSPMCS) is proposed for three-way fluorescence data to achieve this goal. This new method estimates an appropriate chemical rank by comparing the projection residuals which are obtained from vector subspace projection analysis of two similar pseudo matrices constructed by the technology of Monte Carlo simulation. Here, the technology, vector subspace projection, is used to distinguish the difference between two corresponding vectors, each of which is from the orthonormal matrix acquired by SVD. It can be shown that the vector subspace projection is a "constrained" version of the subspace projection. Compared with some methods based on the technology of subspace projection, such as SPPH, this new method can resist the influence of nontrilinear contributions to some degrees and is more suitable for the task of estimating the chemical rank of three-way fluorescence data in complex systems. Generally, the new method contains two main steps: (I) Monte Carlo simulation is applied to create two similar pseudo matrices from original three-way data; (II) Vector subspace projection analysis is performed on the two pseudo matrices. An example showing its behavior is illustrated in Figure 21, concerning one simulated three-way data with four underlying components. This figure displays the projection residuals as a function of the number of factors. It shows that along both *I* and *J* directions, the projection residuals for the first four factors are near zero and the sudden increase occurs for the fifth factor, giving evidence for four chemical factors clearly existing in this data set.

# 7 SOFTWARE AND TOOLBOX FOR MULTIWAY CALIBRATION

Together with collaborators from the Olivieri's laboratory, we have jointly released a MATLAB graphical interface toolbox for second-order multivariate calibration called MVC2 [92]. This toolbox based on graphical user interfaces allows the execution of the calculations in an easy-to-use graphical environment. MVC2 involves a single main window (see Figure 22), from which all steps required to implement the different second-order multivariate calibration strategies can be carried out. MVC2 implements several second-order multivariate calibration methodologies, including PARAFAC, ATLD, APTLD, SWATLD, BLLS/RBL, U-PLS/RBL, and N-PLS/RBL. The toolbox accepts a variety of input data formats, arranged in either matrices or vectors (i.e., unfolded matrices), and contained in ASCII files. It allows one to manually select working sensor regions and plot landscapes for selected samples. The development of each model and its subsequent application to unknown samples are straightforward. Prediction results are revealed along with analytical figures of merit and standard concentration errors, as calculated by modern concepts of uncertainty propagation.

Recently, the MVC3 toolbox for third-order multivariate calibration was released [93]. It is a sequel to the already described MVC2 toolbox. Quadrilinear decomposition techniques and latent structured models based on partial least-squares regression and residual trilinearization were also included in the software, such as four-way PARAFAC, AQTLD, AWRCQLD, TLLS/RBL, U-PLS/RTL and N-PLS/RTL. The release of MVC2 and MVC3 greatly promotes the application of multiway calibration methods in qualitative and quantitative analysis of the component(s) of interest in complicated systems. The two toolboxes are freely available and can be downloaded from www.chemometry.com.

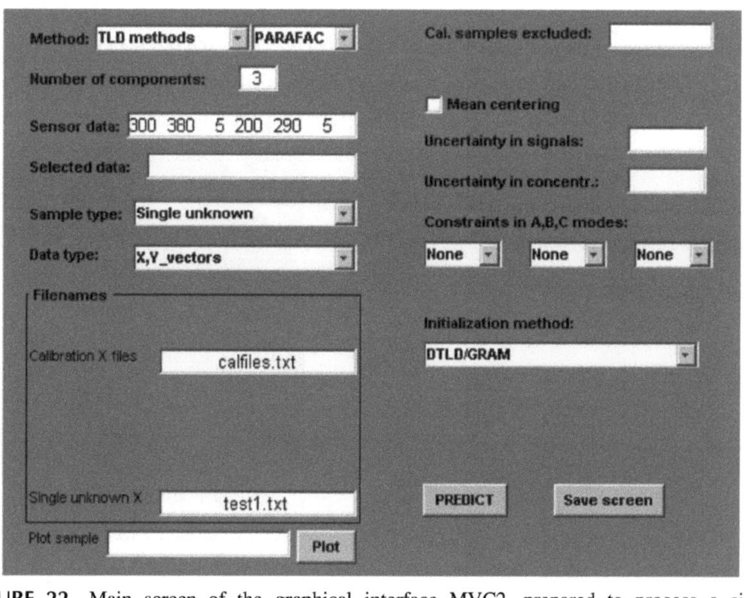


**FIGURE 22** Main screen of the graphical interface MVC2, prepared to process a single unknown sample with PARAFAC using three responsive components. *Adapted from Ref. [92], figure 2, p. 247.*

# 8 SOME RELATED FUNDAMENTAL ISSUES

## 8.1 Dealing of the Scattering in Three-way Fluorescence

Rayleigh scattering, which is often contained in the excitation–emission fluorescence data (Figure 23A), does not conform to the trilinear structure, and then its existence complicates the decomposition using second-order calibration methods. Analysts are familiar with the strategy of selecting a rectangular scope (abbreviated as selecting method) that selects a region of the signals of analytes and avoids Rayleigh scattering at the same time (see the black rectangular scope shown in Figure 23B). Obviously, this method may lose part of valuable information of analytes. In addition, this method could not be suitable for simultaneously analyzing multiple analytes of interest, such as the situation of Figure 23B.

By contraries, chemometricians prefer to employ several other ways to handle Rayleigh scattering in relation to the trilinear modeling, such as down-weighting of the scatter region [95], inserting missing values [96], inserting zeros outside the data area [97], substituting scattering by interpolation [98], and specific modeling of scatter [99]. The successful applications of these methods have been demonstrated in the published works. Nie and coworkers have developed one method based on a trilinear decomposition algorithm, which is not sensitive to the excess factor number [100]. It constructs two 3-way data arrays containing the Rayleigh scattering along both the *I*-mode and *J*-mode directions, then uses a trilinear decomposition algorithm to model them, and finally removes the scattering as a responsive component or factor from the three-way data array. Li and coworker developed a method named as two-direction resection PARAFAC (TDR-PARAFAC) for dealing with the Rayleigh scattering problem [94]. It aims at directly decomposing

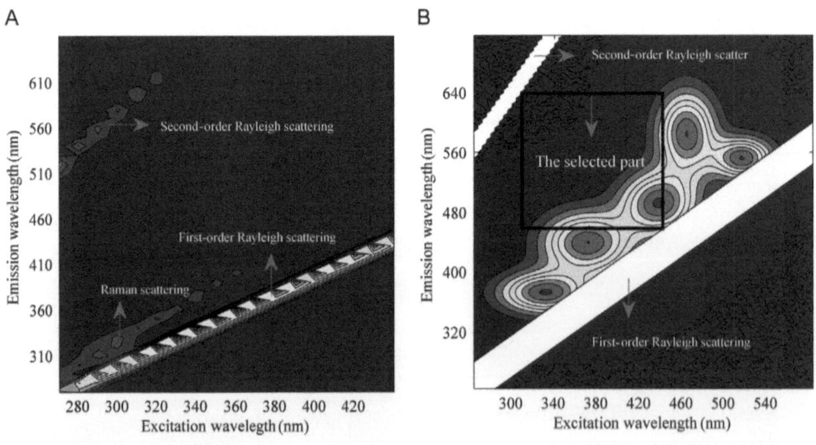


**FIGURE 23** (A) The EEM spectra of a water sample, where the light scatterings are pointed out. (B) A simulated EEM where Rayleigh scatterings are substituted by missing data: the black rectangular scope shows the selected data, suggesting that it is unavoidable to lose massive information by the selecting method. *Adapted from Ref. [94], figure 1, p. 3988.*

the treated data after the scattering regions are removed from the original data. Another method based on background constraining and iterative correcting techniques was proposed to eliminate the effect of model deviation on the estimated chemical loading parameters [101].

### 8.1.1 Substituting Scattering by Missing Values

The traditional PARAFAC method combined with the strategy of inserting missing values can also be employed to deal with the problem of Rayleigh scattering [96]. In the beginning, the missing data are initially replaced with either sensible or random elements. A standard algorithm is used for estimating the model parameters using the complete, but incorrect, data set. After each iteration, the model is calculated and the missing elements are replaced with the model estimates. The iterations and replacements are continued until no changes occur in the estimates of the missing elements and the overall convergence criterion is fulfilled. When the algorithm has converged, the elements replacing the missing elements have zero residual.

### 8.1.2 Inserting Zeros Outside the Data Area

Inserting zeros outside the data area can be seen as a pretreatment, where zeros are inserted into the matrix [97]. PARAFAC will then converge faster and give better estimates of the spectra. This method can be used on any type of fluorescence landscapes, but it does not have a significant effect on the resolved spectra unless one or more of the peaks lie close to or on the Rayleigh scatter lines.

### 8.1.3 Substituting Scattering by Interpolation

Bahram and coworkers developed a method for handling scattering using interpolation in the areas affected by first- and second-order Rayleigh and Raman scattering in such a way that the interfering signal is, at best, removed. The suggested method is fast and requires no additional input other than specifying the scattering region [98]. The main advantage of the interpolation method is in its lack of additional metaparameters, its algorithmic speed and subsequent speed-up of PARAFAC modeling. In fact, this method can be seen as a pretreatment, where interpolation values are inserted into the data set. Then all algorithms for second-order calibration can be the three-way data after treatment.

### 8.1.4 Two-Directional Resection PARAFAC

TDR-PARAFAC was proposed for analyzing the three-way fluorescence data in the presence of Rayleigh scattering [94]. First, the data after resecting the region of Rayleigh scattering follows the trilinear structure. Then TDR-PARAFAC is applied to directly decompose the incomplete data based on the principle of trilinear model. Figure 24 shows the graphical representation of procedure of the strategy, two-directional resection for trilinear model. The new method possesses certain advantages: (1) It is a convenient method where

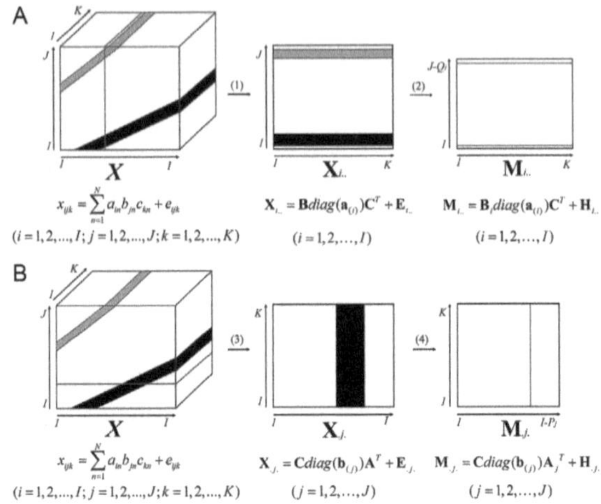


**FIGURE 24** Graphical representation of procedure of the strategy, two-direction resection for trilinear model: (A) is the process of obtaining the rest matrix $\mathbf{M}_{i..}$ by resecting the Rayleigh scattering region from $\mathbf{X}_{i..}$; similarly, the rest matrix $\mathbf{M}_{.j.}$ could be obtained from (B). The black and gray regions indicate first- and second-order Rayleigh scatterings, respectively. Steps (1), (2), (3) and (4) represent the processes of getting $\mathbf{X}_{i..}$, $\mathbf{M}_{i..}$, $\mathbf{X}_{.j.}$ and $\mathbf{M}_{.j.}$ respectively. *Adapted from Ref. [94], figure 2, p. 3988.*

only two parameters, i.e., the scattering regions and the number of components, are required; (2) the effects of Rayleigh scattering can be successfully overcome; (3) it holds fast convergence for avoiding the trivial procedure of solving or substituting the data in the Rayleigh scattering region; and (4) Valuable information involving normalized spectra and relative concentrations for analytes of interest could be satisfactorily retrieved, regardless of the presence of interferents (i.e. preserving the second-order advantage). Moreover, the ideology behind this method could be extended into other second-order calibration methods, such as ATLD and SWATLD. In addition, for severe Raman scattering which is in the same position of each sample matrix as Rayleigh scattering, it can also be handled by this new method.

## 8.2 Chromatographic Peak Alignment and Background Drift in the Chromatographic Analysis

### 8.2.1 Chromatographic Peak Alignment

Chromatographic peak alignment remains a challenge in the field of complex sample analysis by multiway calibration methods. Some methods for peak alignment are built upon the second-order instruments, which generate a matrix of data per sample. These methods [102], typically exemplified by rank minimization (RM), iterative target factor analysis coupled to COW (ITTFA-COW), parallel factor analysis alignment, and other recently proposed methods that are based on MCR-ALS with the advantage of two-dimensional structure of signals for aligning chromatographic peaks shifts in much more complicated samples. In principle, these methods are built on the alignment of local chromatographic domains and thus are capable of providing reasonable results for the time shifts present in the whole chromatographic data. The significant advantage of these methods is that accurate time alignment can be obtained regardless of the presence of the coelution interferences. Recently, Yu and coworkers proposed a new time shift alignment algorithm deriving from the well-known RM method. It aligns chromatographic peak shift among samples and then uses trilinear decomposition methodology to interpret the overlapping chromatographic peaks to quantify analytes of interest [102].

The graphical representation of the RM method has been depicted in Figure 25A. A remarkable advantage of this method lies in the fact that alignment can be successfully performed even when the potential interferences coelute with the analyte of interest. To provide a more clear view on this method, a series of fixed-size time window (rectangles) along the elution directions were adopted in Figure 25A. Specifically, the red (light gray in the print version) rectangle $\mathbf{M}_0$ denotes the elution range of analyte in the response of reference sample, while the elution range between blue (gray in the print version) and green (gray in the print version) rectangles in the response of a test sample points out the underlying time shift ranges of the analyte (see Figure 25A). By rowwisely moving the fixed-size time window on the test sample along the elution direction, a number of rectangles, $\mathbf{M}_1$ to $\mathbf{M}_n$ as shown in Figure 25A, can be extracted from the response of the test sample, then $n$-augmented matrices in the elution direction can be obtained which were defined as $[\mathbf{M}_0 \mid \mathbf{M}_1],\ldots, [\mathbf{M}_0 \mid \mathbf{M}_n]$ (Stage 2 in Figure 25A). Finally, the SVD is performed on these augmented matrices and leads to a list of residual variance. Consequently, the percent of the residual variance plotted against each chromatographic time shift will mark clearly the time shift point correction corresponding to the minimum residual variance.

The abstract subspace difference (ASSD) method makes use of abstract chromatographic profiles instead of the original measured data for alignment. Mathematically, the response matrix $\mathbf{X}$ can be represented in SVD notation as follows:

$$\mathbf{X} = \mathbf{U}\mathbf{S}\mathbf{V}^{\mathrm{T}} + \mathbf{E}. \tag{127}$$

For the chromatographic analysis illustrated here, the columns of $\mathbf{U}$ and $\mathbf{V}$ are frequently described as the abstract chromatographic profiles and the abstract spectra, respectively, to emphasize that these profiles are not necessarily correspond to the real chromatographic profiles and spectra. Suppose that two data matrices have been collected: a reference data, $\mathbf{X}_{\mathrm{ref}}$, which contains a single analyte, and a test data, $\mathbf{X}_{\mathrm{test}}$, which collects the same analyte as well as several coeluted interferences. Hereby, the abstract chromatographic profiles for reference and test samples can be obtained independently according to the SVD:

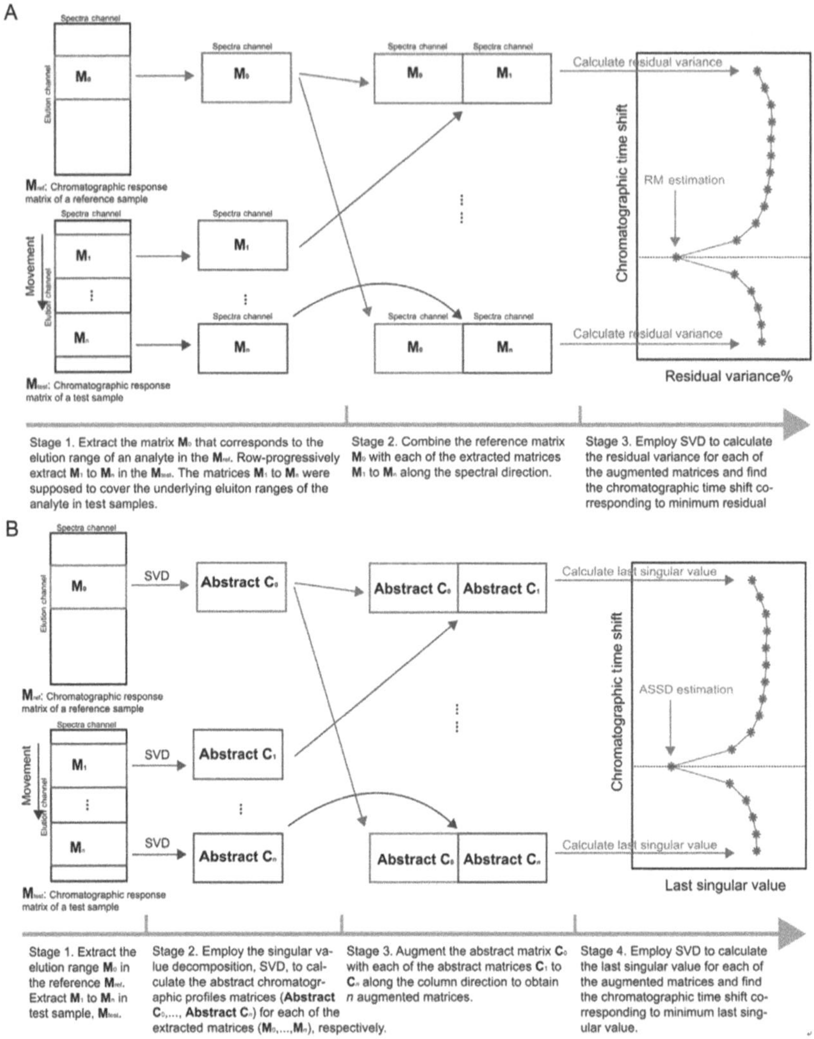


**FIGURE 25** Graphical illustration of RM and ASSD.

$$\mathbf{X}_{\text{ref}} = \mathbf{U}_{\text{ref}}\mathbf{S}_{\text{ref}}\mathbf{V}_{\text{ref}}^{\text{T}} + \mathbf{E}_{\text{ref}}, \tag{128}$$

$$\mathbf{X}_{\text{test}} = \mathbf{U}_{\text{test}}\mathbf{S}_{\text{test}}\mathbf{V}_{\text{test}}^{\text{T}} + \mathbf{E}_{\text{test}}. \tag{129}$$

In the ideal situation, no noise is present, and there is no time shift between reference and test samples. In this case, the mathematical rank of the augmented matrix $[\mathbf{U}_{\text{ref}} \mid \mathbf{U}_{\text{test}}]$ will be identical to that of $\mathbf{U}_{\text{test}}$. However, in the situations where the chromatographic retention time of the analyte changed between the time, the reference, and test samples were collected, the mathematical rank of the augmented matrix, $[\mathbf{U}_{\text{ref}} \mid \mathbf{U}_{\text{test}}]$, will be greater than the number of components. Consequently, the core ideology of ASSD method is in search of the augmented matrix with minimum mathematical rank for alignment, which is identical to the well-known RM method except that the abstract chromatographic profiles instead of the underlying ones were employed for peak correction.

The graphical illustration of the ASSD method is shown in Figure 25B. The comparison with the procedure of RM method indicates obviously that an additional step, SVD, has been introduced in the Stage 1 of Figure 25B, in order to calculate the abstract chromatographic profiles for each of the extracted matrices $\mathbf{M}_1$ to $\mathbf{M}_n$. In addition, the last singular value instead of the percent of residual variance was adopted in the last stage by this new method to indicate time shift correction to be made. Due to the fact of practical measurements, it is impracticable to align time shift of an analyte between the reference and a test sample according to the critical criterion of the mathematical rank of the augmented matrix. However, the augmented matrix, $[\mathbf{U}_{\text{ref}} \mid \mathbf{U}_{\text{test}}]$, will be a seriously ill-conditioned matrix provided that the time shift has been successfully aligned. Hereby, time shift alignment can be transformed to an optimization problem that finds the most ill-conditioned augmented matrix among the augmented matrices as shown in the Stage 3 of Figure 25B. Since the total variance is the sum of the squared elements of the augment matrix, $[\mathbf{U}_{\text{ref}} \mid \mathbf{U}_{\text{test}}]$, and will be a constant value which is equal to the number of columns, a smaller last singular value will definitely correspond to a more ill-conditioned matrix.

### 8.2.2 Background Drift

The quality of measured data from hyphenated (chromatographic/spectroscopic) techniques is influenced by many artifacts, such as the background drift attributed to the nature of complicated matrices; these artifacts might produce invalid qualitative or quantitative results of the aforementioned chemometric methods. Amigo and coworkers have reviewed the mathematical models and intuitive graphics used in solving chromatographic data issues [103]. Typical examples are multivariate curve resolution (MCR) methods.

In 2007, a chromatographic background drift correction strategy [104] was proposed by our group for LC × LC × DAD data. The basic idea is to perform trilinear decomposition on the instrumental response data, which is based on the ATLD algorithm. In model construction, the background drift is modeled as one component or factor as well as the analytes of interest; hence, the drift is explicitly included into the calibration. The method involves performing trilinear decomposition on the raw data, then extracting the background component and subtracting this background data from the raw data, leaving the analytes' signal on a flat baseline. Figure 26 is a schematic description of this method to remove the background drift from three-dimensional instrumental data.

Very recently, an approach that employs orthogonal spectral signal projection (OSSP) to simultaneously treat various types of chromatographic background drift was studied [105]. The experimental results demonstrating that the combination of OSSP with PARAFAC can be used to resolve coelution

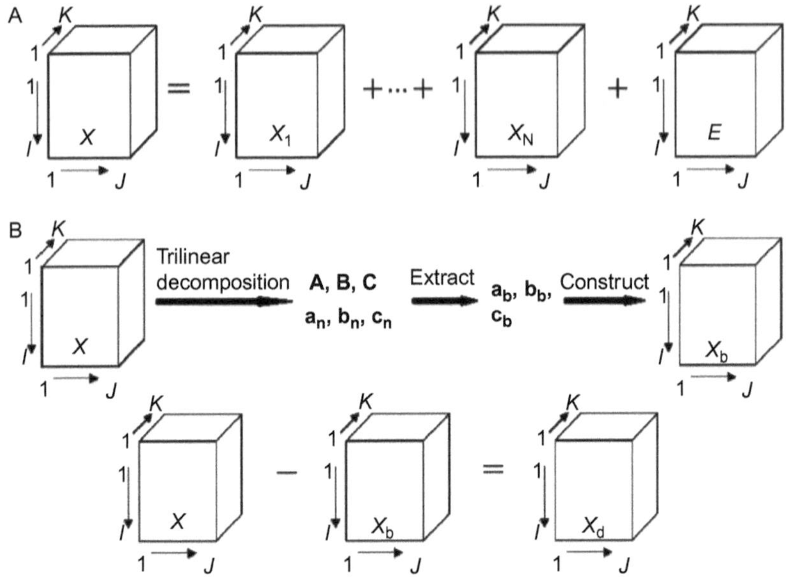


**FIGURE 26** Schematic description of the proposed method to remove background drift from 3D data. $N$ is the total number of responsive components; $\mathbf{X}$ represents the instrument response signal; $\mathbf{X}_N$ represents the signal of individual components; $\mathbf{X}_b$ represents the signal of background drift; $\mathbf{X}_d$ represents the response signal after background removal; $\mathbf{E}$ represents the residual error. $I$, $K$, $J$ are the data points in the second-dimension column, first-dimension column, and spectral dimension, respectively.

problems in chromatographic analysis; therefore, more accurate quantitative results for analytes of interest can be achieved, regardless of the presence of background drift and unknown interferences. One example of OSSP dealing with the problem due to background drift is shown in Figure 27.

## 8.3 Second-Order Standard Addition Method

The standard addition method (SAM) as a means of overcoming matrix or background effects is usually applied to zeroth-order instrumentation (instruments that return only a scalar quantity per sample analyzed) [106–108]. For a successful calibration, SAM requires two assumptions to be fulfilled: (I) there is a linear change in the instrument response with increasing analyte concentration; (II) for zero concentration of an analyte the instrument response must be zero. A plot of the instrument response (ordinate) against the amount of standard added (abscissa) estimates the analyte concentration in the sample by fitting a line to the data and finding the intercept of the line on the abscissa. Ideally, a second-order calibration can be performed in the presence of interfering species unaccounted by the calibration model. However, if the interfering species change the instrument response of the analyte (in scale or shape), standard additions must be employed to ensure an accurate analyte

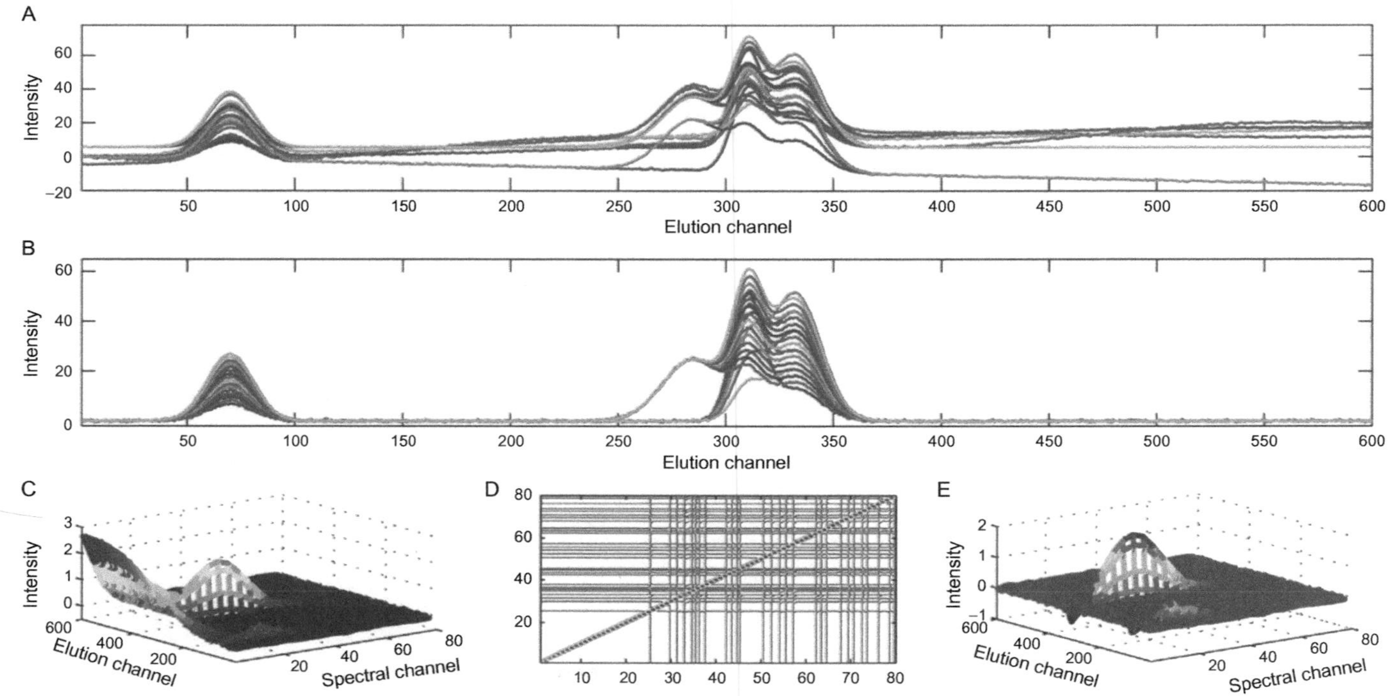


**FIGURE 27** Total chromatograms of 20 samples (A), and results obtained with background drift corrected by OSSP (B). For the sixth sample: (C) Original three-dimensional plots, (D) contour plot of the projection matrix, and (E) three-dimension plots after background drift correction. Note that the total chromatogram of each sample was calculated as the summation of the chromatograms under each spectral channel.

concentration estimation. Wu and coworkers developed a second-order SAM based on the alternating trilinear decomposition algorithm [32] (ATLD-SOSAM), which preserved the second-order advantage [109]. They outlined three major steps of ATLD-SOSAM:

**(1)** Decomposition of one three-way response data array $\boldsymbol{X}$ with an appropriate number of factors by ATLD;
**(2)** Location of the column of N relative concentration estimates corresponding to the analyte of interest;
**(3)** Finding the original relative concentration of each component of interest by regression from the corresponding added standard concentrations.

They used DTLD-SOSAM, PARAFAC-SOSAM, and ATLD-SOSAM algorithms for the quantitative analysis of *o*-chlorotoluene in the presence of *p*-chlorotoluene. The three matrices, that is, the relative concentration, spectrum, time profile matrices, and a loss function vector in variation with the iteration number could be obtained for each trilinear decomposition. Figure 28 shows one set of the results based on using the three-factor ATLD-SOSAM algorithm. Based on the corresponding relative concentration matrices, the regression curves were obtained by plotting the relative concentration values corresponding to *o*-chlorotoluene versus their added standards. The concentrations of *o*-chlorotoluene obtained from the slope and intercept were 26.37 μg/ml for the two-factor DTLD-SOSAM, 26.44 μg/ml for the two-factor PARAFAC-SOSAM, 24.96 μg/ml for the two-factor ATLD-SOSAM, and 20.42 μg/ml for the three-factor ATLD-SOSAM. Their corresponding recoveries were 117.7%, 118.0%, 111.4%, and 91.2%, respectively. The results show that ATLD-SOSAM method is slightly superior to both DTLD-SOSAM, and PARAFAC-SOSAM methods.

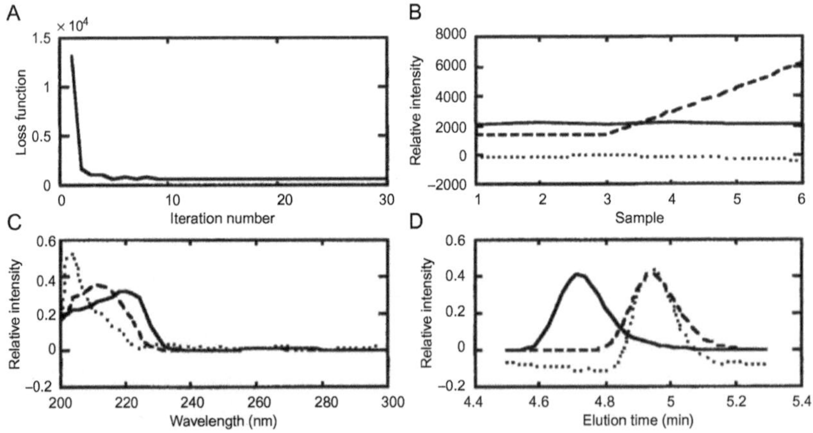


**FIGURE 28** Results obtained for HPLC-DAD experimental data using the three-factor ATLD-SOSAM method. The average predicted concentration of *o*-chlorotoluene in samples #1 to #3 was 20.40 ± 0.4 mg/ml and their average recovery was 91.1 ± 1.8%. *Adapted from Ref. [109], figure 2, p. 218.*

## 9 PERSPECTIVES

This chapter scientifically summarizes in details the multiway chemometric methodologies mainly based on alternating multilinear decomposition reported by our group, including multilinear model, multiway cyclic symmetry property, algorithms for multiway calibration, estimation of the chemical rank, software and toolbox for multiway calibration, and other fundamental issues. Multiway data analysis and tensorial calibration become research hot spots in many scientific fields with the help of the abundance of instruments, which can automatically yield precise multiway data arrays in a short period of time. Many fundamental issues, such as higher-order advantages, figures of merits, and rank-estimation for higher-order data, are still immature. Therefore, related works on alternating multilinear decomposition should be studied in more depth.

## ACKNOWLEDGMENTS

The authors gratefully acknowledge the Sasakawa Scientific Research Grant from The Japan Science Society, the National Natural Science Foundation of China (Grant Nos. 29735150, 29975007, 20475014, 20775025, 21175041, and 21221003), and the National Basic Research Program (Nos. 2007CB216404 and 2012CB910602) as well as the Grants from the Ministry of Education of China for financial support. The authors would also like to thank Mr. Zhao-Yang Ai and Ms. Ding Zeng for their English language review. Finally, the authors would like to thank our friends for their continuing support.

## APPENDIXES ALGORITHM'S SOLUTIONS AND MATLAB PROGRAMS FOR SOME MULTIWAY CALIBRATION METHODS

### A.1 Solutions for Some Algorithms

#### A.1.1 How to Get the Final Solutions of ATLD

*Theorem 1*

Assume that $\mathbf{AXB}$ and $\mathbf{AX}^{\mathrm{T}}\mathbf{B}$ are square matrices, then the following expressions hold:

$$\frac{trace(\mathbf{AXB})}{\partial \mathbf{X}} = (\mathbf{BA})^{\mathrm{T}} = \mathbf{A}^{\mathrm{T}}\mathbf{B}^{\mathrm{T}},$$
$$\frac{trace(\mathbf{AX}^{\mathrm{T}}\mathbf{B})}{\partial \mathbf{X}} = \mathbf{BA}.$$

*Proof*

Suppose that $\mathbf{D} = \mathbf{AXB}$, $\mathbf{A}$, $\mathbf{X}$, and $\mathbf{B}$ with the size of $\mathrm{L} \times \mathrm{P}$, $\mathrm{P} \times \mathrm{Q}$, and $\mathrm{Q} \times \mathrm{L}$, respectively. The $ij$th element of $\mathbf{D}$ can be expressed as

$$d_{ij} = \sum_{p=1}^{P} \sum_{q=1}^{Q} a_{ip} x_{pq} b_{qj}.$$

and the trace of **D** is

$$trace(\mathbf{D}) = \sum_{l=1}^{L}\sum_{p=1}^{P}\sum_{q=1}^{Q} a_{lp}x_{pq}b_{ql}.$$

Then, one can get:

$$\frac{\partial trace(\mathbf{D})}{\partial \mathbf{X}} = \frac{\partial\left(\sum_{l=1}^{L}\sum_{p=1}^{P}\sum_{q=1}^{Q} a_{lp}x_{pq}b_{ql}\right)}{\partial x_{ij}}.$$

Since

$$\frac{\partial\left(a_{lp}x_{pq}b_{ql}\right)}{\partial x_{ij}} = \begin{cases} a_{lp}b_{ql} & p=i \,\text{and}\, q=j \\ 0 & \text{others} \end{cases}.$$

And one can obtain:

$$\frac{\partial trace(\mathbf{D})}{\partial x_{ij}} = \sum_{l=1}^{L} a_{li}b_{jl} = \sum_{l=1}^{L} b_{jl}a_{li}.$$

Proof is complete.

*Theorem 2*

The expressions of Equation (54)–(56) are equal to the three following equations, respectively:

$$diag\left(\mathbf{a}_{(i)^{\mathrm{T}}}\right) = \mathbf{B}^{+}\mathbf{X}_{i..}\left(\mathbf{C}^{\mathrm{T}}\right)^{+}, \quad \text{for } i=1,2,\ldots,I,$$
$$diag\left(\mathbf{b}_{(j)^{\mathrm{T}}}\right) = \mathbf{B}^{+}\mathbf{X}_{i..}\left(\mathbf{C}^{\mathrm{T}}\right)^{+}, \quad \text{for } i=1,2,\ldots,J,$$
$$diag\left(\mathbf{c}_{(k)^{\mathrm{T}}}\right) = \mathbf{A}^{+}\mathbf{X}_{..k}\left(\mathbf{B}^{\mathrm{T}}\right)^{+}, \quad \text{for } k=1,2,\ldots,K,$$

*Proof*

Here, we only provide the proof for the first equation, as the others can be obtained similarly.

$$\frac{\partial\|\mathbf{E}_{i..}\|_{\mathrm{F}}^{2}}{\partial diag(\mathbf{a}_{(i)})} = \frac{\partial\left(trace\left(\left(\mathbf{X}_{i..} - \mathbf{B}diag(a_{(i)})\mathbf{C}^{\mathrm{T}}\right)^{\mathrm{T}}\left(\mathbf{X}_{i..} - \mathbf{B}diag(a_{(i)})\mathbf{C}^{\mathrm{T}}\right)\right)\right)}{\partial diag(a_{(i)})}$$
$$= \frac{\partial\left(trace\left(\left(\mathbf{X}_{i..}^{\mathrm{T}} - \mathbf{C}diag(\mathbf{a}_{(i)})\mathbf{B}^{\mathrm{T}}\right)\left(\mathbf{X}_{i..} - \mathbf{B}diag(a_{(i)})\mathbf{C}^{\mathrm{T}}\right)\right)\right)}{\partial diag(\mathbf{a}_{(i)})}$$
$$= \frac{\partial\left(trace\left(\mathbf{X}_{i..}^{\mathrm{T}}\mathbf{X}_{i..} - \mathbf{C}diag(\mathbf{a}_{(i)})\mathbf{B}^{\mathrm{T}}\mathbf{X}_{i..} - \mathbf{X}_{i..}^{\mathrm{T}}\mathbf{B}diag(\mathbf{a}_{(i)})\mathbf{C}^{\mathrm{T}} + \mathbf{C}diag(\mathbf{a}_{(i)})\mathbf{B}^{\mathrm{T}}\mathbf{B}diag(\mathbf{a}_{(i)})\mathbf{C}^{\mathrm{T}}\right)\right)}{\partial diag(\mathbf{a}_{(i)})}$$
$$= -\mathbf{B}^{\mathrm{T}}\mathbf{X}_{i..}\mathbf{C} - \left(\mathbf{C}^{\mathrm{T}}\mathbf{X}_{i..}{}^{\mathrm{T}}\mathbf{B}\right)^{\mathrm{T}} + \mathbf{B}^{\mathrm{T}}\mathbf{B}diag(\mathbf{a}_{(i)})\mathbf{C}^{\mathrm{T}}\mathbf{C} + \left(\mathbf{C}^{\mathrm{T}}\mathbf{C}diag(\mathbf{a}_{(i)})\mathbf{B}^{\mathrm{T}}\mathbf{B}\right)^{\mathrm{T}}$$
$$= -2\mathbf{B}^{\mathrm{T}}\mathbf{X}_{i..}\mathbf{C} + 2\mathbf{B}^{\mathrm{T}}\mathbf{B}diag(\mathbf{a}_{(i)})\mathbf{C}^{\mathrm{T}}\mathbf{C}.$$

So $\mathbf{B}^{\mathrm{T}}\mathbf{B}diag(\mathbf{a}_{(i)})\mathbf{C}^{\mathrm{T}}\mathbf{C}=\mathbf{B}^{\mathrm{T}}\mathbf{X}_{i..}\mathbf{C}$.

Then, $diag(\mathbf{a}_{(i)})=(\mathbf{B}^{\mathrm{T}}\mathbf{B})^{-1}\mathbf{B}^{\mathrm{T}}\mathbf{X}_{i..}\mathbf{C}(\mathbf{C}^{\mathrm{T}}\mathbf{C})^{-1}=\mathbf{B}^{+}\mathbf{X}_{i..}(\mathbf{C}^{\mathrm{T}})^{+}$.

After that, $\mathbf{a}_{(i)}^{\mathrm{T}}=diagm\left(\mathbf{B}^{+}\mathbf{X}_{i..}(\mathbf{C}^{\mathrm{T}})^{+}\right)$.

Proof is complete.

### A.1.2 How to Get the Final Solution of SWATLD

*Theorem 3*

For any equation with the following form:

$$\mathbf{X}=\mathbf{A}diag(\mathbf{c})+\mathbf{E}.$$

The least-squares solutions of diag (**c**) and **c** are

$$diag(\mathbf{c})=\mathbf{A}^{+}\mathbf{X},$$

$$\mathbf{c}=diagm(\mathbf{A}^{\mathrm{T}}\mathbf{X})./diagm(\mathbf{A}^{\mathrm{T}}\mathbf{A}).$$

where"./" denotes array division.

*Proof*

$$\frac{\partial\|\mathbf{E}\|_F^2}{\partial c_m}=\frac{\partial\left(trace\left((\mathbf{X}-\mathbf{A}diag(\mathbf{c}))^{\mathrm{T}}(\mathbf{X}-\mathbf{A}diag(\mathbf{c}))\right)\right)}{\partial c_m}$$

$$=\frac{\partial\left(\sum_{n=1}^{N}(\mathbf{x}_n-\mathbf{a}_n c_n)^{\mathrm{T}}(\mathbf{x}_n-\mathbf{a}_n c_n)\right)}{\partial c_m}$$

$$=\frac{\partial\left((\mathbf{x}_m-\mathbf{a}_m c_m)^{\mathrm{T}}(\mathbf{x}_m-\mathbf{a}_m c_m)\right)}{\partial c_m}$$

$$=\frac{\partial\left((\mathbf{x}_m^{\mathrm{T}}\mathbf{x}_m-c_m\mathbf{a}_m^{\mathrm{T}}\mathbf{x}_m-\mathbf{x}_m^{\mathrm{T}}\mathbf{a}_m c_m+c_m\mathbf{a}_m^{\mathrm{T}}\mathbf{a}_m c_m)\right)}{\partial c_m}$$

$$=-2\mathbf{a}_m^{\mathrm{T}}\mathbf{x}_m+2c_m\mathbf{a}_m^{\mathrm{T}}\mathbf{a}_m.$$

And one can get $c_m$ as

$$c_m=\mathbf{a}_m^{\mathrm{T}}\mathbf{x}_m/\mathbf{a}_m^{\mathrm{T}}\mathbf{a}_m.$$

So the **c** is

$$\mathbf{c}=\begin{bmatrix}c_1\\c_2\\\vdots\\c_N\end{bmatrix}=\begin{bmatrix}\mathbf{a}_1^{\mathrm{T}}\mathbf{x}_1/\mathbf{a}_1^{\mathrm{T}}\mathbf{a}_1\\\mathbf{a}_2^{\mathrm{T}}\mathbf{x}_2/\mathbf{a}_2^{\mathrm{T}}\mathbf{a}_2\\\vdots\\\mathbf{a}_N^{\mathrm{T}}\mathbf{x}_N/\mathbf{a}_N^{\mathrm{T}}\mathbf{a}_N\end{bmatrix}=diagm(\mathbf{A}^{\mathrm{T}}\mathbf{X})./diagm(\mathbf{A}^{\mathrm{T}}\mathbf{A}).$$

Proof is complete.

In SWATLD, the loss function of **A** is

$$\sigma(\mathbf{A}) = \sum_{i=1}^{I} \left\| \left( \mathbf{X}_{i..}{}^{\mathrm{T}} (\mathbf{B}^{\mathrm{T}})^{+} - \mathbf{C}\,diag(\mathbf{a}_{(i)}) \right) diag\left(sqrt\left(\mathbf{1}./diagm(\mathbf{C}^{\mathrm{T}}\mathbf{C})\right)\right) \right\|_{\mathrm{F}}^{2} + \sum_{i=1}^{I} \left\| \left( \mathbf{X}_{i..} (\mathbf{C}^{\mathrm{T}})^{+} - \mathbf{B}\,diag(\mathbf{a}_{(i)}) \right) diag\left(sqrt\left(\mathbf{1}./diagm(\mathbf{B}^{\mathrm{T}}\mathbf{B})\right)\right) \right\|_{\mathrm{F}}^{2}.$$

Let $L_A 1 = \sum_{i=1}^{I} \left\| \left( \mathbf{X}_{i..}{}^{\mathrm{T}} (\mathbf{B}^{\mathrm{T}})^{+} - \mathbf{C}\,diag(\mathbf{a}_{(i)}) \right) diag\left(sqrt\left(\mathbf{1}./diagm(\mathbf{C}^{\mathrm{T}}\mathbf{C})\right)\right) \right\|_{\mathrm{F}}^{2}.$

And $L_A 2 = \sum_{i=1}^{I} \left\| \left( \mathbf{X}_{i..} (\mathbf{C}^{\mathrm{T}})^{+} - \mathbf{B}\,diag(\mathbf{a}_{(i)}) \right) diag\left(sqrt\left(\mathbf{1}./diagm(\mathbf{B}^{\mathrm{T}}\mathbf{B})\right)\right) \right\|_{\mathrm{F}}^{2}.$

According to Theorem 3, the least-squares solution of $L_A 1$ and $L_A 2$ are

$$\mathbf{a}_{(i)} = \left( diagm\left( \mathbf{C}^{\mathrm{T}} \mathbf{X}_{i..}{}^{\mathrm{T}} (\mathbf{B}^{\mathrm{T}})^{+} \right) ./diagm(\mathbf{C}^{\mathrm{T}}\mathbf{C}) \right),$$
$$\mathbf{a}_{(i)} = \left( diagm\left( \mathbf{B}^{\mathrm{T}} \mathbf{X}_{i..} (\mathbf{C}^{\mathrm{T}})^{+} \right) ./diagm(\mathbf{B}^{\mathrm{T}}\mathbf{B}) \right).$$

So one can get:

$$\mathbf{a}_{(i)} = \frac{1}{2}\left(diagm(\mathbf{B}^{+}\mathbf{X}_{i..}\mathbf{C})./diagm(\mathbf{C}^{\mathrm{T}}\mathbf{C})\right) + \frac{1}{2}\left(diagm(\mathbf{C}^{+}\mathbf{X}_{i..}^{\mathrm{T}}\mathbf{B})/diagm(\mathbf{B}^{\mathrm{T}}\mathbf{B})\right).$$

It is noted that **A** is calculated as the combination of minimization of both $L_A 1$ and $L_A 2$. One can also use Theorem 3 to calculate the actual least-squares solution of **B** and **C**:

$$\mathbf{b}_{(j)} = \frac{1}{2}\left(diagm(\mathbf{C}^{+}\mathbf{X}_{.j.}\mathbf{A})./diagm(\mathbf{A}^{\mathrm{T}}\mathbf{A})\right) + \frac{1}{2}\left(diagm(\mathbf{A}^{+}\mathbf{X}_{.j.}{}^{\mathrm{T}}\mathbf{C})./diagm(\mathbf{C}^{\mathrm{T}}\mathbf{C})\right),$$
$$\mathbf{c}_{(k)} = \frac{1}{2}\left(diagm(\mathbf{A}^{+}\mathbf{X}_{..k}\mathbf{B})./diagm(\mathbf{B}^{\mathrm{T}}\mathbf{B})\right) + \frac{1}{2}\left(diagm(\mathbf{B}^{+}\mathbf{X}_{..k}{}^{\mathrm{T}}\mathbf{A})./diagm(\mathbf{A}^{\mathrm{T}}\mathbf{A})\right).$$

### *A.1.3 How to Get the Final Solution of AWRCQLD*

The proofs of Equations (105)-(108) are similar. Here, only the proof of equation (105) is provided.

*Proof*

Equation (101) can be divided into two parts:

$$\sigma(\mathbf{A}) = L1 + L2,$$
$$L1 = \sum_{l=1}^{L} \left\| \mathbf{X}_{...l} - \mathbf{A}\,diag(\mathbf{d}_{(l)})(\mathbf{C}\odot\mathbf{B})^{\mathrm{T}} \right\|_{\mathrm{F}}^{2},$$
$$L2 = \lambda_{\mathbf{A}} \sum_{l=1}^{L} \left\| \left( \mathbf{X}_{...l} \left( (\mathbf{C}\odot\mathbf{B})^{\mathrm{T}} \right)^{+} - \mathbf{A}\,diag(\mathbf{d}_{(l)}) \right) \mathbf{W_D} \right\|_{\mathrm{F}}^{2}.$$

The derivation of **A** in equation ($L1$) can be written as:

$$\frac{\partial(L1)}{\partial \mathbf{A}} = \sum_{l=1}^{L}\left(2\mathbf{X}_{\dots l}(\mathbf{C}\odot\mathbf{B})diag\left(\mathbf{d}_{(l)}\right) - 2\mathbf{A}\,diag\left(\mathbf{d}_{(l)}\right)(\mathbf{C}\odot\mathbf{B})(\mathbf{C}\odot\mathbf{B})diag\left(\mathbf{d}_{(l)}\right)\right)$$

$$= \sum_{l=1}^{L}\left(2\mathbf{X}_{\dots l}(\mathbf{C}\odot\mathbf{B})diag\left(\mathbf{d}_{(l)}\right) - 2\mathbf{A}\,diag\left(\mathbf{d}_{(l)}\right)\left(\left(\left(\mathbf{C}^{\mathrm{T}}\mathbf{C}\right)\right)*\left(\mathbf{B}^{\mathrm{T}}\mathbf{B}\right)\right)diag\left(\mathbf{d}_{(l)}\right)\right).$$

Likewise, the derivation of **A** in equation ($L2$) is:

$$\frac{\partial(L2)}{\partial \mathbf{A}} = \sum_{l=1}^{L}\left(2\lambda_{\mathbf{A}}\mathbf{X}_{\dots l}\left(\left((\mathbf{C}\odot\mathbf{B})^{\mathrm{T}}\right)\right)^{+}\mathbf{W}_{\mathbf{D}}\mathbf{W}_{\mathbf{D}}diag\left(\mathbf{d}_{(l)}\right)\right.$$

$$\left. + 2\lambda_{\mathbf{A}}\mathbf{A}\,diag\left(\mathbf{d}_{(l)}\right)\mathbf{W}_{\mathbf{D}}\mathbf{W}_{\mathbf{D}}diag\left(\mathbf{d}_{(l)}\right)\right).$$

Hence, solution of **A** can be deduced as follows:

$$\frac{\partial(\sigma(\mathbf{A}))}{\partial \mathbf{A}} = \frac{\partial(L1)}{\partial \mathbf{A}} + \frac{\partial(L2)}{\partial \mathbf{A}}$$

$$= \sum_{l=1}^{L}\left(2\mathbf{X}_{\dots l}(\mathbf{C}\odot\mathbf{B})diag\left(\mathbf{d}_{(l)}\right) - 2\mathbf{A}\,diag\left(\mathbf{d}_{(l)}\right)\left(\left(\mathbf{C}^{\mathrm{T}}\mathbf{C}\right)*\left(\mathbf{B}^{\mathrm{T}}\mathbf{B}\right)\right)diag\left(\mathbf{d}_{(l)}\right)\right)$$

$$+ \sum_{l=1}^{L}\left(2\lambda_{\mathbf{A}}\mathbf{X}_{\dots l}\left((\mathbf{C}\odot\mathbf{B})^{\mathrm{T}}\right)^{+}\mathbf{W}_{\mathbf{D}}\mathbf{W}_{\mathbf{D}}diag\left(\mathbf{d}_{(l)}\right)\right.$$

$$\left. + 2\lambda_{\mathbf{A}}\mathbf{A}diag\left(\mathbf{d}_{(l)}\right)\mathbf{W}_{\mathbf{D}}\mathbf{W}_{\mathbf{D}}diag\left(\mathbf{d}_{(l)}\right)\right) = 0$$

$$\Rightarrow \mathbf{A}\sum_{l=1}^{L}\left(diag\left(\mathbf{d}_{(l)}\right)\left(\left(\mathbf{C}^{\mathrm{T}}\mathbf{C}\right)*\left(\mathbf{B}^{\mathrm{T}}\mathbf{B}\right)\right)diag\left(\mathbf{d}_{(l)}\right) + \lambda_{\mathbf{A}}diag\left(\mathbf{d}_{(l)}\right)\mathbf{W}_{\mathbf{D}}\mathbf{W}_{\mathbf{D}}diag\left(\mathbf{d}_{(l)}\right)\right)$$

$$= \sum_{l=1}^{L}\left(diag\left(\mathbf{d}_{(l)}\right)\left(\left(\mathbf{C}^{\mathrm{T}}\mathbf{C}\right)*\left(\mathbf{B}^{\mathrm{T}}\mathbf{B}\right)\right)diag\left(\mathbf{d}_{(l)}\right) + \lambda_{\mathbf{A}}diag\left(\mathbf{d}_{(l)}\right)\mathbf{W}_{\mathbf{D}}\mathbf{W}_{\mathbf{D}}diag\left(\mathbf{d}_{(l)}\right)\right)$$

$$\Rightarrow \mathbf{A} = \left(\sum_{l=1}^{L}\mathbf{X}_{\dots l}\left((\mathbf{C}\odot\mathbf{B}) + \lambda_{\mathbf{A}}\left((\mathbf{C}\odot\mathbf{B})^{\mathrm{T}}\right)^{+}\mathbf{W}_{\mathbf{D}}\mathbf{W}_{\mathbf{D}}\right)diag\left(\mathbf{d}_{(l)}\right)\right)\cdot$$

$$\left(\sum_{l=1}^{L}\left(diag\left(\mathbf{d}_{(l)}\right)\left(\left(\mathbf{C}^{\mathrm{T}}\mathbf{C}\right)*\left(\mathbf{B}^{\mathrm{T}}\mathbf{B}\right) + \lambda_{\mathbf{A}}\mathbf{W}_{\mathbf{D}}\mathbf{W}_{\mathbf{D}}\right)diag\left(\mathbf{d}_{(l)}\right)\right)\right)^{+}.$$

Proof is complete.

### A.1.4 How to Get the Final Solution of RSWAQLD

The proofs of Equations (117)–(120) are similar. Only the proof of equation (120) is provided.

*Proof*

For Equation (116), let

$$S1(\mathbf{D}) = \sum_{l=1}^{L} \left\| \mathbf{A}^{+}\mathbf{X}_{...l}\left((\mathbf{C}\odot\mathbf{B})^{\mathrm{T}}\right)^{+} - diag\left(\mathbf{d}_{(l)}\right)\right\|_{\mathrm{F}}^{2},$$

$$S2(\mathbf{D}) = \sum_{l=1}^{L} \left\| \left(\mathbf{A}^{+}\mathbf{X}_{...l} - diag\left(\mathbf{d}_{(l)}\right)(\mathbf{C}\odot\mathbf{B})^{\mathrm{T}}\right)^{\mathrm{T}} .* \mathbf{W}_{\mathbf{CB}}\right\|_{\mathrm{F}}^{2},$$

$$S3(\mathbf{D}) = \sum_{l=1}^{L} \left\| \left(\mathbf{X}_{...l}\left((\mathbf{C}\odot\mathbf{B})^{\mathrm{T}}\right)^{+} - \mathbf{A}\,diag\left(\mathbf{d}_{(l)}\right)\right) .* \mathbf{W}_{\mathbf{A}}\right\|_{\mathrm{F}}^{2},$$

where

$$\mathbf{W}_{\mathbf{CB}} = diag\left(sqrt\left(\mathbf{1}./diagm(\mathbf{C}\odot\mathbf{B})^{\mathrm{T}}(\mathbf{C}\odot\mathbf{B})\right)\right),$$

$$\mathbf{W}_{\mathbf{A}} = diag\left(sqrt\left(\mathbf{1}./diagm\left(\mathbf{A}^{\mathrm{T}}\mathbf{A}\right)\right)\right).$$

Then, Equation (116) is divided into three parts, i.e., $S(\mathbf{D}) = S1(\mathbf{D}) + S2(\mathbf{D}) + S3(\mathbf{D})$. The derivation of $diag(\mathbf{d}_{(l)})$ in Equation $S1(\mathbf{D})$ is

$$\frac{\partial(S1(\mathbf{D}))}{\partial\left(diag\left(\mathbf{d}_{(l)}\right)\right)} = \frac{\partial\left(\left\|\mathbf{A}^{+}\mathbf{X}_{...l}\left((\mathbf{C}\odot\mathbf{B})^{\mathrm{T}}\right)^{+} - diag\left(\mathbf{d}_{(l)}\right)\right\|_{\mathrm{F}}^{2}\right)}{\partial\left(diag\left(\mathbf{d}_{(l)}\right)\right)}$$

$$= \frac{\partial\left(trace\left(\left(\mathbf{A}^{+}\mathbf{X}_{...l}\left((\mathbf{C}\odot\mathbf{B})^{\mathrm{T}}\right)^{+} - diag\left(\mathbf{d}_{(l)}\right)\right)^{\mathrm{T}}\left(\mathbf{A}^{+}\mathbf{X}_{...l}\left((\mathbf{C}\odot\mathbf{B})^{\mathrm{T}}\right)^{+} - diag\left(\mathbf{d}_{(l)}\right)\right)\right)\right)}{\partial\left(diag\left(\mathbf{d}_{(l)}\right)\right)}$$

$$= -2\mathbf{A}^{+}\mathbf{X}_{...l}\left((\mathbf{C}\odot\mathbf{B})^{\mathrm{T}}\right)^{+} + 2diag\left(\mathbf{d}_{(l)}\right).$$

Analogously, the derivation of $diag(\mathbf{d}_{(l)})$ in Equation $S2(\mathbf{D})$ is

$$\frac{\partial(S2(\mathbf{D}))}{\partial\left(diag\left(\mathbf{d}_{(l)}\right)\right)} = \frac{\partial\left(\left\|(\mathbf{X}_{...l})^{\mathrm{T}}(\mathbf{A}^{+})^{\mathrm{T}} .* \mathbf{W}_{\mathbf{CB}} - (\mathbf{C}\odot\mathbf{B})\left(diag\left(\mathbf{d}_{(l)}\right)\right)^{\mathrm{T}} .* \mathbf{W}_{\mathbf{CB}}\right\|_{\mathrm{F}}^{2}\right)}{\partial\left(diag\left(\mathbf{d}_{(l)}\right)\right)}$$

$$= \frac{\partial\left(trace\left(\begin{array}{c}\left((\mathbf{X}_{...l})^{\mathrm{T}}(\mathbf{A}^{+})^{\mathrm{T}}.*\mathbf{W}_{\mathbf{CB}} - (\mathbf{C}\odot\mathbf{B})\left(diag\left(\mathbf{d}_{(l)}\right)\right)^{\mathrm{T}}.*\mathbf{W}_{\mathbf{CB}}\right)^{\mathrm{T}} \\ \left((\mathbf{X}_{...l})^{\mathrm{T}}(\mathbf{A}^{+}).*\mathbf{W}_{\mathbf{CB}} - (\mathbf{C}\odot\mathbf{B})\left(diag\left(\mathbf{d}_{(l)}\right)\right)^{\mathrm{T}}.*\mathbf{W}_{\mathbf{CB}}\right)\end{array}\right)\right)}{\partial\left(diag\left(\mathbf{d}_{(l)}\right)\right)}$$

$$= -2(\mathbf{C}\odot\mathbf{B})^{\mathrm{T}}(\mathbf{X}_{...l})^{\mathrm{T}}(\mathbf{A}^{+})^{\mathrm{T}}.*\mathbf{W}_{\mathbf{CB}}.*(\mathbf{W}_{\mathbf{CB}})^{\mathrm{T}} + 2diag\left(\mathbf{d}_{(l)}\right).$$

The derivation of $diag(\mathbf{d}_{(l)})$ in Equation $S3(\mathbf{D})$ is

$$\frac{\partial(S3(\mathbf{D}))}{\partial(diag(\mathbf{d}_{(l)}))}=\frac{\partial\left(\left\|\mathbf{X}_{...l}\left((\mathbf{C}\odot\mathbf{B})^{\mathrm{T}}\right)^{+}.*\mathbf{W}_{\mathbf{A}}-\mathbf{A}diag(\mathbf{d}_{(l)}).*\mathbf{W}_{\mathbf{A}}\right\|_{\mathrm{F}}^{2}\right)}{\partial(diag(\mathbf{d}_{(l)}))}$$

$$=\frac{\partial\left(trace\left(\begin{array}{c}\left(\mathbf{X}_{...l}\left((\mathbf{C}\odot\mathbf{B})^{\mathrm{T}}\right)^{+}.*\mathbf{W}_{\mathbf{A}}-\mathbf{A}diag(\mathbf{d}_{(l)}).*\mathbf{W}_{\mathbf{A}}\right)^{\mathrm{T}}\\ \left(\mathbf{X}_{...l}\left((\mathbf{C}\odot\mathbf{B})^{\mathrm{T}}\right)^{+}.*\mathbf{W}_{\mathbf{A}}-\mathbf{A}diag(\mathbf{d}_{(l)}).*\mathbf{W}_{\mathbf{A}}\right)\end{array}\right)\right)}{\partial(diag(\mathbf{d}_{(l)}))}$$

$$=-2\mathbf{A}^{\mathrm{T}}\mathbf{X}_{...l}\left((\mathbf{C}\odot\mathbf{B})^{\mathrm{T}}\right)^{+}.*\mathbf{W}_{\mathbf{A}}.*\mathbf{W}_{\mathbf{A}}^{\mathrm{T}}+2diag(\mathbf{d}_{(l)}).$$

Therefore:

$$\frac{\partial(S(\mathbf{D}))}{\partial(diag(\mathbf{d}_{(l)}))}=\frac{\partial(S1(\mathbf{D}))}{\partial(diag(\mathbf{d}_{(l)}))}+\lambda_{\mathbf{D}}\frac{\partial(S2(\mathbf{D}))}{\partial(diag(\mathbf{d}_{(l)}))}+\mu_{\mathbf{D}}\frac{\partial(S3(\mathbf{D}))}{\partial(diag(\mathbf{d}_{(l)}))}$$

$$=-2\mathbf{A}^{+}\mathbf{X}_{...l}\left((\mathbf{C}\odot\mathbf{B})^{\mathrm{T}}\right)^{+}+2diag(\mathbf{d}_{(l)})$$

$$+\lambda_{\mathbf{D}}\left(-2(\mathbf{C}\odot\mathbf{B})^{\mathrm{T}}(\mathbf{X}_{...l})^{\mathrm{T}}(\mathbf{A}^{+})^{\mathrm{T}}\cdot*\mathbf{W}_{\mathbf{CB}}\cdot*(\mathbf{W}_{\mathbf{CB}})^{\mathrm{T}}+2diag(\mathbf{d}_{(l)})\right)$$

$$+\mu_{\mathbf{D}}\left(-2\mathbf{A}^{\mathrm{T}}\mathbf{X}_{...l}\left((\mathbf{C}\odot\mathbf{B})^{\mathrm{T}}\right)^{+}\cdot*\mathbf{W}_{\mathbf{A}}.*\mathbf{W}_{\mathbf{A}}^{\mathrm{T}}+2diag(\mathbf{d}_{(l)})\right)$$

$$=0$$

$$\Rightarrow diag(\mathbf{d}_{(l)})+\lambda_{\mathbf{D}}diag(\mathbf{d}_{(l)})+\mu_{\mathbf{D}}diag(\mathbf{d}_{(l)})$$

$$=\mathbf{A}^{+}\mathbf{X}_{..l}\left((\mathbf{C}\odot\mathbf{B})^{\mathrm{T}}\right)^{+}+\lambda_{\mathbf{D}}(\mathbf{C}\odot\mathbf{B})^{\mathrm{T}}(\mathbf{X}_{..l})^{\mathrm{T}}(\mathbf{A}^{+})^{\mathrm{T}}\cdot*\mathbf{W}_{\mathbf{CB}}$$

$$\cdot*(\mathbf{W}_{\mathbf{CB}})^{\mathrm{T}}+\mu_{\mathbf{D}}\mathbf{A}^{\mathrm{T}}\mathbf{X}_{..l}\left((\mathbf{C}\odot\mathbf{B})^{\mathrm{T}}\right)^{+}\cdot*\mathbf{W}_{\mathbf{A}}\cdot*\mathbf{W}_{\mathbf{A}}^{\mathrm{T}}$$

$$\Rightarrow diag(\mathbf{d}_{(l)})=\frac{1}{1+\lambda_{\mathbf{D}}+\mu_{\mathbf{D}}}\left[\mathbf{A}^{+}\mathbf{X}_{...l}\left((\mathbf{C}\odot\mathbf{B})^{\mathrm{T}}\right)^{+}\right.$$

$$+\lambda_{\mathbf{D}}(\mathbf{C}\odot\mathbf{B})^{\mathrm{T}}(\mathbf{X}_{..l})^{\mathrm{T}}(\mathbf{A}^{+})^{\mathrm{T}}\cdot*\mathbf{W}_{\mathbf{CB}}\cdot*\mathbf{W}_{\mathbf{CB}}^{\mathrm{T}}$$

$$\left.+\mu_{\mathbf{D}}\mathbf{A}^{\mathrm{T}}\mathbf{X}_{...l}\left((\mathbf{C}\odot\mathbf{B})^{\mathrm{T}}\right)^{+}\cdot*\mathbf{W}_{\mathbf{A}}\cdot*\mathbf{W}_{\mathbf{A}}^{\mathrm{T}}\right].$$

Thus, one has

$$\mathbf{d}_{(l)}=\frac{1}{1+\lambda_{\mathbf{D}}+\mu_{\mathbf{D}}}\mathrm{diagm}\left[\mathbf{A}^{+}\mathbf{X}_{...l}\left((\mathbf{C}\odot\mathbf{B})^{\mathrm{T}}\right)^{+}\right.$$

$$+\lambda_{\mathbf{D}}(\mathbf{A}^{+}\mathbf{X}_{...l}(\mathbf{C}\odot\mathbf{B}))\cdot/((\mathbf{C}^{\mathrm{T}}\mathbf{C})\cdot*(\mathbf{B}^{\mathrm{T}}\mathbf{B}))$$

$$\left.+\mu_{\mathbf{D}}\left((\mathbf{C}\odot\mathbf{B})^{+}(\mathbf{X}_{...l})^{\mathrm{T}}\mathbf{A}\right)\cdot/(\mathbf{A}^{\mathrm{T}}\mathbf{A})\right]^{\mathrm{T}}\quad\text{for } l=1,2,\ldots,L,$$

## A.2 MATLAB Programs for Some Algorithms

### A.2.1 MATLAB Program for ATLD

```
function [A,B,C,LFT,M]=ATLD(XIJK,N,epsilon)
%_______________________________________________________________
% XIJK is one three-way data array with the size of I*J*K
% epsilon is tolerance.
% I is the number of rows
% J is the number of columns
% K is the number of channels
% N is the number of components
% LFT is the loss function
% A(I,N),B(J,N),C(K,N)
% State Key Laboratory of Chemo/Biosensing and Chemometrics,
% College of Chemistry and Chemical Engineering,
% Hunan University, Changsha 410082, China.
% ® Hai–Long Wu (hlwu@hnu.edu.cn)
% ----------STEP 0------------
% decompose the cube X along I,J,K direction, respectively
if nargin < 3
epsilon=10*eps*norm(XK,1)*max(size(XK));
end
[I,J,K]=size(XIJK);
XK=reshape(XIJK,I,J*K);
XJKI=shiftdim(XIJK,1); % cut cube X along I direction
XKIJ=shiftdim(XJKI,1); % cut cube X along J direction
% ---------STEP 1-------------
% initialize A & B and compute C
%[A,B,C]=ATLD(XIJK,N);
A=randn(I,N);
B=randn(J,N);
%[A,B,C]= ATLD(XIJK,N);
  PA=pinv(A,epsilon);
  PB=pinv(B,epsilon);
for k=1:K
C(k,:)=diag(PA*XIJK(:,:,k)*PB')';
end
 % -----------STEP 2-------------
% start to caculate LFT and do iteration
TOL=10;
M=0;
LFT=[];
LF=0.01;
while TOL > epsilon && M < 500
       % compute A according to B and C
```

```
        PC=pinv(C,epsilon);
for i=1:I
A(i,:)=diag(PB*XJKI(:,:,i)*PC');
end
        % normalize A columnwisely
        A=A*diag(1./diag(sqrt(A'*A)));
        % compute B according to A and C
        PA=pinv(A,epsilon);
for j=1:J
B(j,:)=diag(PC*XKIJ(:,:,j)*PA')';
end
        % normalize B columnwisely
        B=B*diag(1./diag(sqrt(B'*B)));
        % compute C according to A and B
        PA=pinv(A,epsilon);
        PB=pinv(B,epsilon);
for k=1:K
C(k,:)=diag(PA*XIJK(:,:,k)*PB')';
end
         % calculate the loss function LFT
         LFTT=0;
for k=1:K
XXX(:,:,k)=A*diag(C(k,:))*B';
 LFTT=LFTT+trace((XIJK(:,:,k)-XXX(:,:,k))'*(XIJK(:,:,k)-XXX(:,:,
        k)));
end
      TOL=abs((LFTT-LF)/LF);
      LFT=[LFT,LFTT];
      LF=LFTT;
      M=M+1;
end;
% ------------STEP 3--------------
% post-processing to keep sign convention
[maxa,inda]=max(abs(A));
[maxb,indb]=max(abs(B));
asign=ones(N,1);
bsign=ones(N,1);
for n=1:N
asign(n)=sign(A(inda(n),n));
bsign(n)=sign(B(indb(n),n));
end
A=A*diag(asign);
B=B*diag(bsign);
C=C*diag(asign)*diag(bsign);
```

### A.2.2 MATLAB Program for SWATLD

```
function [A,B,C,LFT,M]=SWATLD(XIJK,N,epsilon)
% ____________________________________________________________
% XIJK is one three-way data array with the size of I*J*K
% epsilon is tolerance.
% I is the number of rows
% J is the number of columns
% K is the number of channels
% N is the number of components
% LFT is the loss function
% M is the iterative number
% A(I,N),B(J,N),C(K,N)
% ____________________________________________________________
% State Key Laboratory of Chemo/Biosensing and Chemometrics,
% College of Chemistry and Chemical Engineering,
% Hunan University, Changsha 410082, China.
% ®Hai-Long Wu (hlwu@hnu.edu.cn)& Ru-Qin Yu (rqyu@hnu.edu.cn)
% ---------STEP 0-------------
% decompose the cube X along I,J,K direction, respectively
if nargin < 3
epsilon=10*eps*norm(XK,1)*max(size(XK));
end
[I,J,K]=size(XIJK);
XK=reshape(XIJK,I,J*K);
XJKI=shiftdim(XIJK,1);      % cut X along I direction
XKIJ=shiftdim(XJKI,1);      % cut X along J direction
% ---------STEP 1--------------
% initialize A & B and compute C
A=rand(I,N);
B=rand(J,N);
%[A,B,C]=SWATLD(XIJK,N);
PA=pinv(A,epsilon);
PB=pinv(B,epsilon);
for k=1:K
CC=0.5*(diag(PB*XIJK(:,:,k)'*A)./diag(A'*A)+diag(PA*XIJK(:,:,k)
*B)./diag(B'*B));
C(k,:)=CC';
end
% -----------STEP 2-------------
% start to calculate LFT and do iteration
TOL=10;
M=0;
LFT=[];
```

```
LF=0.01;
while TOL > epsilon && M < 500
     % compute A according to B and C
     PC=pinv(C,epsilon);
for i=1:I
AA=0.5*(diag(PC*XJKI(:,:,i)'*B)./diag(B'*B)+diag(PB*XJKI(:,:,i)
    *C)./diag(C'*C));
A(i,:)=AA';
end
     % normalize A columnwisely
     A=A*diag(1./diag(sqrt(A'*A)));
     % compute B according to A and C
    PA=pinv(A,epsilon);
for j=1:J
BB=0.5*(diag(PA*XKIJ(:,:,j)'*C)./diag(C'*C)+diag(PC*XKIJ(:,:,j)
    *A)./diag(A'*A));
B(j,:)=BB';
end
     % normalize B columnwisely
     B=B*diag(1./diag(sqrt(B'*B)));
     % compute C according to A and B
     PA=pinv(A,epsilon);
     PB=pinv(B,epsilon);
for k=1:K
CC=0.5*(diag(PB*XIJK(:,:,k)'*A)./diag(A'*A)+diag(PA*XIJK(:,:,k)
    *B)./diag(B'*B));
C(k,:)=CC';
end
     %calculate the loss function
    LFTT=0;
for k=1:K
XXX(:,:,k)=A*diag(C(k,:))*B';
LFTT=LFTT+trace((XIJK(:,:,k)-XXX(:,:,k))'*(XIJK(:,:,k)-XXX(:,:,
k)));
end
     TOL=abs((LFTT-LF)/LF);
     LFT=[LFT,LFTT];
     LF=LFTT;
     M=M+1 ;
end
% -----------STEP 3-------------
% post-processing to keep sign convention
[maxa,inda]=max(abs(A));
[maxb,indb]=max(abs(B));
```

```
asign=ones(N,1);
bsign=ones(N,1);
for n=1:N
asign(n)=sign(A(inda(n),n));
bsign(n)=sign(B(indb(n),n));
end
A=A*diag(asign);
B=B*diag(bsign);
C=C*diag(asign)*diag(bsign);
```

### *A.2.3 MATLAB Program for APTLD*

```
10.2.3function [A,B,C,LFT,M]=APTLD(XIJK,N,epsilon)
% ___________________________________________________________
% XIJK is one three-way data array with the size of I*J*K
% epsilon is tolerance.
% I is the number of rows
% J is the number of columns
% K is the number of channels
% N is the number of components
% LFT is the loss function
% M is the iterative number
% A(I,N),B(J,N),C(K,N)
% State Key Laboratory of Chemo/Biosensing and Chemometrics,
% College of Chemistry and Chemical Engineering,
% Hunan University, Changsha 410082, China.
% ® Hai-Long Wu (hlwu@hnu.edu.cn)
% ___________________________________________________________
% ---------STEP 0-------------
% decompose the cube X along I,J,K direction, respectively
if nargin < 3
epsilon=10*eps*norm(XK,1)*max(size(XK));
end
[I,J,K]=size(XIJK);
XK=reshape(XIJK,I,J*K);
XJKI=shiftdim(XIJK,1); %cut X along I direction
XKIJ=shiftdim(XJKI,1); %cut X along J direction
% ---------STEP 1-------------
% initialize A & B and compute C
A=rand(I,N);
B=rand(J,N);
%[A,B,C]=dtld(XIJK,N);
CD1=0; CD2=0; CD3=0; CD4=0; aa=0; bb=0;
p=10^20; q=10^20; r=10^20;
Wa=diag(1./diag(A'*A));
```

```
Wb=diag(1./diag(B'*B));
PA=pinv(A,epsilon);
PB=pinv(B,epsilon);
for i=1:I
     CD1=CD1+XJKI(:,:,i)'*PB'*Wa*diag(A(i,:));
aa=aa+diag(A(i,:))*Wa*diag(A(i,:));
end
for j=1:J
    CD2=CD2+XKIJ(:,:,j)*PA'*Wb*diag(B(j,:));
    CD3=CD3+XKIJ(:,:,j)*A*diag(B(j,:));
    CD4=CD4+diag(B(j,:))*A'*A*diag(B(j,:));
    bb=bb+diag(B(j,:))*Wb*diag(B(j,:));
end
C=(p*(CD1+CD2)+CD3)*pinv(p*aa+p*bb+CD4,epsilon);
% -----------STEP 2-------------
% start to calculate LFT and do iteration
TOL=10;
M=0;
LFT=[];
LF=0.01;

while TOL > epsilon && M < 500
     % compute A according to B and C
     AD1=0; AD2=0; AD3=0; AD4=0; bb=0; cc=0;
     Wc=diag(1./diag(C'*C));
     PC=pinv(C,epsilon);
for j=1:J
        AD1=AD1+XKIJ(:,:,j)'*PC'*Wb*diag(B(j,:));
bb=bb+diag(B(j,:))*Wb*diag(B(j,:));
end
for k=1:K
     AD2=AD2+XIJK(:,:,k)*PB'*Wc*diag(C(k,:));
     AD3=AD3+XIJK(:,:,k)*B*diag(C(k,:));
     AD4=AD4+diag(C(k,:))*B'*B*diag(C(k,:));
     cc=cc+diag(C(k,:))*Wc*diag(C(k,:));
end
     A=(r*(AD1+AD2)+AD3)*pinv(r*bb+r*cc+AD4,epsilon);
     % normalize A columnwisely
     A=A*diag(1./diag(sqrt(A'*A)));
     % compute B according to A and C
     BD1=0; BD2=0; BD3=0; BD4=0; cc=0;aa=0;
     Wa=diag(1./diag(A'*A));
     PA=pinv(A,epsilon);
for k=1:K
        BD1=BD1+XIJK(:,:,k)'*PA'*Wc*diag(C(k,:));
```

```
        cc=cc+diag(C(k,:))*Wc*diag(C(k,:));
end
for i=1:I
        BD2=BD2+XJKI(:,:,i)*PC'*Wa*diag(A(i,:));
        BD3=BD3+XJKI(:,:,i)*C*diag(A(i,:));
        BD4=BD4+diag(A(i,:))*C'*C*diag(A(i,:));
aa=aa+diag(A(i,:))*Wa*diag(A(i,:));
end
    B=(q*(BD1+BD2)+BD3)*pinv(q*cc+q*aa+BD4,epsilon);
    % normalize B columnwisely
    B=B*diag(1./diag(sqrt(B'*B)));
    % compute C according to A and B
    CD1=0; CD2=0; CD3=0; CD4=0; aa=0; bb=0;
    Wa=diag(1./diag(A'*A));
    Wb=diag(1./diag(B'*B));
    PA=pinv(A,epsilon);
    PB=pinv(B,epsilon);
for i=1:I
        CD1=CD1+XJKI(:,:,i)'*PB'*Wa*diag(A(i,:));
aa=aa+diag(A(i,:))*Wa*diag(A(i,:));
end
for j=1:J
        CD2=CD2+XKIJ(:,:,j)*PA'*Wb*diag(B(j,:));
        CD3=CD3+XKIJ(:,:,j)*A*diag(B(j,:));
        CD4=CD4+diag(B(j,:))*A'*A*diag(B(j,:));
bb=bb+diag(B(j,:))*Wb*diag(B(j,:));
end
   C=(p*(CD1+CD2)+CD3)*pinv(p*aa+p*bb+CD4,epsilon);
   % calculate the loss function
   LFTT=0;
for k=1:K
XXX(:,:,k)=A*diag(C(k,:))*B';
LFTT=LFTT+trace((XIJK(:,:,k)-XXX(:,:,k))'*(XIJK(:,:,k)-XXX(:,:,
     k)));
end
   TOL=abs((LFTT-LF)/LF);
   LFT=[LFT,LFTT];
   LF=LFTT;
   M=M+1;
end
% -----------STEP 3--------------
%post-processing to keep sign convention
[maxa,inda]=max(abs(A));
[maxb,indb]=max(abs(B));
```

```
asign=ones(N,1);
bsign=ones(N,1);
for n=1:N
asign(n)=sign(A(inda(n),n));
bsign(n)=sign(B(indb(n),n));
end
A=A*diag(asign);
B=B*diag(bsign);
C=C*diag(asign)*diag(bsign);
```

### *A.2.4 MATLAB Program for APQLD*

```
function [A,B,C,D,LFT,M]=APQLD(XIJKL,K,L,N,epsilon)
% ________________________________________________________________
% XIJKL is one four-way data array with the size of I*J*K*L
% epsilon is tolerance.
% I is the row
% J is the column
% K is the channel
% L is the second channel
% N is the estimated component number
% LFT is the loss function
% M is the iterative number
% State Key Laboratory of Chemo/Biosensing and Chemometrics,
% College of Chemistry and Chemical Engineering,
% Hunan University, Changsha 410082, China.
% ® Hai-Long Wu (hlwu@hnu.edu.cn)
% ________________________________________________________________
% ----------STEP 0------------
% decompose the cube X along I,J,K,L direction, respectively
if nargin < 5
epsilon=10*eps*norm(XL,1)*max(size(XIJKL));
end
[I,J,K,L]=size(XIJKL);
XJKLI=shiftdim(XIJKL,1);
XKLIJ=shiftdim(XJKLI,1);
XLIJK=shiftdim(XKLIJ,1);
% ---------STEP 1-------------
% initialize A & B & C and compute D
A=rand(I,N);
B=rand(J,N);
C=rand(K,N);
FD1=0; FD2=0; FD3=0; FD4=0;
p=10^20; q=10^20; r=10^20; s=10^20;
```

```
Wa=diag(1./diag(A'*A));
Wb=diag(1./diag(B'*B));
Wc=diag(1./diag(C'*C));
PA=pinv(A,epsilon);
PC=pinv(C,epsilon);
for j=1:J
for k=1:K
    FD1=FD1+XLIJK(:,:,j,k)*(A+p*PA'*Wa^2)*diag(B(j,:)*diag(C
        (k,:)));
    FD2=FD2+diag(C(k,:))*diag(B(j,:))*(A'*A+p*Wa)*diag(C(k,:))
        *diag(B(j,:));
end
end
for i=1:I
for j=1:J
    FD3=FD3+XKLIJ(:,:,i,j)'*PC'*Wc*diag(A(i,:))*diag(B(j,:));
    FD4=FD4+diag(A(i,:))*diag(B(j,:))*Wc*diag(A(i,:))*diag(B
        (j,:));
end
end
D=(FD1+p*FD3)*pinv(FD2+p*FD4,epsilon);
% -----------STEP 2--------------
% start to calculate LFT and do iteration
TOL=10;
M=0;
LFT=[];
LF=0.01;
while TOL > epsilon && M < 3000
    % compute A according to B, C and D
    FA1=0; FA2=0; FA3=0; FA4=0;
    Wd=diag(1./diag(D'*D));
    PB=pinv(B,epsilon);
    PD=pinv(D,epsilon);
for k=1:K
for l=1:L
            FA1=FA1+XIJKL(:,:,k,l)*(B+q*PB'*Wb)*diag(D(l,:))
                *diag(C(k,:));
            FA2=FA2+diag(D(l,:))*diag(C(k,:))*(B'*B+q*Wb)*diag(D
                (l,:))*diag(C(k,:));
end
end
for j=1:J
for k=1:K
```

```
        FA3=FA3+XLIJK(:,:,j,k)'*PD'*Wd*diag(B(j,:))*diag(C
            (k,:));
        FA4=FA4+diag(B(j,:))*diag(C(k,:))*Wd*diag(B(j,:))
            *diag(C(k,:));
end
end
  A=(FA1+q*FA3)*pinv(FA2+q*FA4,epsilon);
  % normalize A columnwisely
   A=A*diag(1./diag(sqrt(A'*A)));
  % compute B according to A, C and D;
  FB1=0; FB2=0; FB3=0; FB4=0;
  Wa=diag(1./diag(A'*A));
  PA=pinv(A,epsilon);
for l=1:L
for i=1:I
       FB1=FB1+XJKLI(:,:,l,i)*(C+r*PC'*Wc)*diag(A(i,:))*diag(D
           (l,:));
       FB2=FB2+diag(A(i,:))*diag(D(l,:))*(C'*C+r*Wc)*diag(D
           (l,:))*diag(A(i,:));
end
end
for k=1:K
for l=1:L
      FB3=FB3+XIJKL(:,:,k,l)'*PA'*Wa*diag(C(k,:))*diag(D
          (l,:));
      FB4=FB4+diag(C(k,:))*diag(D(l,:))*Wa*diag(D(l,:))*diag(C
          (k,:));
end
end
  B=(FB1+r*FB3)*pinv(FB2+r*FB4,epsilon);
   % normalize B columnwisely
   B=B*diag(1./diag(sqrt(B'*B)));
  % compute C according to A, B and D
  FC1=0; FC2=0; FC3=0; FC4=0;
  Wb=diag(1./diag(B'*B));
  PB=pinv(B,epsilon);
for i=1:I
for j=1:J
       FC1=FC1+XKLIJ(:,:,i,j)*(D+s*PD'*Wd)*diag(B(j,:))
           *diag(A(i,:));
       FC2=FC2+diag(A(i,:))*diag(B(j,:))*(D'*D+s*Wd)*diag(A
           (i,:))*diag(B(j,:));
end
end
```

```
for l=1:L
for i=1:I
       FC3=FC3+XJKLI(:,:,l,i)'*PB'*Wb*diag(D(l,:))*diag(A
           (i,:));
       FC4=FC4+diag(D(l,:))*diag(A(i,:))*Wb*diag(D(l,:))*diag(A
           (i,:));
end
end
  C=(FC1+s*FC3)*pinv(FC2+s*FC4,epsilon);
   % normalize C columnwisely
   C=C*diag(1./diag(sqrt(C'*C)));
  % compute D according to A, B and C
  FD1=0; FD2=0; FD3=0; FD4=0;
  Wa=diag(1./diag(A'*A));
  Wb=diag(1./diag(B'*B));
  Wc=diag(1./diag(C'*C));
  PA=pinv(A,epsilon);
  PC=pinv(C,epsilon);
for j=1:J
for k=1:K
         FD1=FD1+XLIJK(:,:,j,k)*(A+p*PA'*Wa)*diag(B(j,:)*diag(C
             (k,:)));
         FD2=FD2+diag(C(k,:))*diag(B(j,:))*(A'*A+p*Wa)*diag(C
             (k,:))*diag(B(j,:));
end
end
for i=1:I
for j=1:J
         FD3=FD3+XKLIJ(:,:,i,j)'*PC'*Wc*diag(A(i,:))*diag(B
             (j,:));
         FD4=FD4+diag(A(i,:))*diag(B(j,:))*Wc*diag(A(i,:))
             *diag(B(j,:));
end
end
  D=(FD1+p*FD3)*pinv(FD2+p*FD4,epsilon);
  % compute the loss function
  LFTT=0;
for k=1:K
for l=1:L
          XXX(:,:,k,l)=A*diag(D(l,:))*diag(C(k,:))*B';
          LFTT=LFTT+trace((XIJKL(:,:,k,l)-XXX(:,:,k,l))'*(XIJKL
              (:,:,k,l)-XXX(:,:,k,l)));
end
end
```

```
    TOL=abs((LFTT-LF)/LF);
    LFT=[LFT,LFTT];
    LF=LFTT;
    M=M+1;
end
% -----------STEP 3-------------
%post-processing to keep sign convention
[maxa,inda]=max(abs(A));
[maxb,indb]=max(abs(B));
[maxc,indc]=max(abs(C));

asign=ones(N,1);
bsign=ones(N,1);
csign=ones(N,1);
for n=1:N
asign(n)=sign(A(inda(n),n));
bsign(n)=sign(B(indb(n),n));
csign(n)=sign(C(indc(n),n));
end
A=A*diag(asign);
B=B*diag(bsign);
C=C*diag(csign);
D=D*diag(asign)*diag(bsign)*diag(csign);
```

## REFERENCES


[1] Booksh KS, Kowalski BR. Theory of analytical chemistry. Anal Chem 1994;66:782A–791A.

[2] Wu HL, Nie JF, Yu YJ, Yu RQ. Multiway chemometric methodologies and applications: a central summary of our research work. Anal Chim Acta 2009;650:131–42.

[3] Wu HL, Li Y, Yu RQ. Recent developments of chemical multiway calibration methodologies with second-order or higher-order advantages. J Chemometr 2014;28:476–89.

[4] Brown SD, Tauler R, Walzak B. Comprehensive chemometrics: chemical and biochemical data analysis. Amsterdam: Elsevier; 2009.

[5] Meyers RA. Encyclopedia of analytical chemistry: applications, theory, and instrumentation. New York: John Wiley & Sons Ltd; 2002.

[6] Gemperline P. Practical guide to chemometrics. New York: Taylor & Francis; 2006. p. 30–80.

[7] Wold S. Chemometrics: what do we mean with it, and what do we want from it? Chemom Intell Lab Syst 1995;3:109–15.

[8] Hierlemann A, Gutierrez-Osuna R. Higher-order chemical sensing. Chem Rev 2008;108:63–613.

[9] Olivieri AC. Recent advances in analytical calibration with multi-way data. Anal Methods 2012;4:1876–86.

[10] Gómez V, Callao MP. Analytical applications of second-order calibration methods. Anal Chim Acta 2008;627:169–83.

[11] Andrade-Eiroa A, Canle M, Cerdá V. Environmental applications of excitation-emission spectrofluorimetry: an in-depth review I. Appl Spectrosc Rev 2013;48:1–49.

[12] Arancibia JA, Damiani PC, Escandar GM, Ibañez GA, Olivieri AC. A review on second- and third-order multivariate calibration applied to chromatographic data. J Chromatogr B 2012;910:22–30.

[13] Olivieri AC. Analytical advantages of multivariate data processing. One, two, three, infinity? Anal Chem 2008;80:5713–20.

[14] Escandar GM, Damiani PC, Goicoechea HC, Olivieri AC. A review of multivariate calibration methods applied to biomedical analysis. Microchem J 2006;82:29–42.

[15] Wu HL. Studies on chemometric algorithms for multivariate calibration based on singular value decomposition. [Dissertation] (supervisor: Koichi Oguma). Chiba University; 1998; Collected in Tokyo: NIPPON Kokuritsu Kokka Toshukan; National Diet Library.

[16] Qing XD. The research on multi-way chemometric methodologies and their applications in soil and water analysis. [Dissertation] (supervisor: Hai-Long Wu). Hunan University; 2014; Collected in National Digital Library of China.

[17] Escandar GM, Faber NM, Goicoechea HC, de la Peña AM, Olivieri AC, Poppi RJ. Second- and third-order multivariate calibration: data, algorithms and applications. Trends Anal Chem 2007;26:752–65.

[18] Harshman RA. Foundations of the PARAFAC procedure: models and conditions for an explanatory "multimodal factor analysis" UCLA Work Pap Phon 1970;16:1–84.

[19] Carroll JD, Chang JJ. Analysis of individual differences in multidimensional scaling via an N-way generalization of "Eckart-Young" decomposition. Psychometrika 1970;35:283–319.

[20] Kruskal JB. Rank, decomposition, and uniqueness for 3-way and N-way arrays. In: Coppi R, Bolasco S, editors. Multiway data analysis. Amsterdam: Elsevier; 1989. p. 8–18.

[21] Kruskal JB. Three-way arrays: rank and uniqueness of trilinear decompositions, with applications to arithmetic complexity and statistics. Linear Algebra Appl 1997;18:95–138.

[22] Sidiropoulos ND, Bro R. On the uniqueness of multilinear decomposition of N-way arrays. J Chemometr 2000;14:229–39.

[23] Wu HL, Yu RQ, Oguma K. Trilinear component analysis in modern analytical chemistry. Anal Sci 2001;17:i483–6.

[24] Ho CN, Christian GD, Davidson ER. Application of the method of rank annihilation to quantitative analyses of multicomponent fluorescence data from the video fluorometer. Anal Chem 1978;50:1108–13.

[25] Ho CN, Christian GD, Davidson ER. Simultaneous multicomponent rank annihilation and applications to multicomponent fluorescent data acquired by the video fluorometer. Anal Chem 1981;53:92–8.

[26] Sánchez E, Kowalski BR. Generalized rank annihilation factor analysis. Anal Chem 1986;58:496–9.

[27] Booksh KS, Lin Z, Wang Z, Kowalski BR. Extension of trilinear decomposition method with an application to the flow probe sensor. Anal Chem 1994;66:2561–9.

[28] Linder M, Sundberg R. Second-order calibration: bilinear least squares regression and a simple alternative. Chemom Intell Lab Syst 1998;42:159–78.

[29] Olivieri AC. On a versatile second-order multivariate calibration method based on partial least-squares and residual bilinearization: second-order advantage and precision properties. J Chemometr 2005;19:253–65.

[30] Bro R. PARAFAC. Tutorial and applications. Chemom Intell Lab Syst 1997;38:149–71.

[31] Tauler R, Smilde A, Kowalski BR. Selectivity, local rank, three-way data analysis and ambiguity in multivariate curve resolution. J Chemometr 1995;9:31–58.

[32] Wu HL, Shibukawa M, Oguma K. An alternating trilinear decomposition algorithm with application to calibration of HPLC-DAD for simultaneous determination of overlapped chlorinated aromatic hydrocarbons. J Chemometr 1998;12:1–26.

[33] Chen ZP, Wu HL, Jiang JH, Li Y, Yu RQ. A novel trilinear decomposition algorithm for second-order linear calibration. Chemom Intell Lab Syst 2000;52:75–86.
[34] Xia AL, Wu HL, Fang DM, Ding YJ, Hu LQ, Yu RQ. Alternating penalty trilinear decomposition algorithm for second-order calibration with application to interference-free analysis of excitation–emission matrix fluorescence data. J Chemometr 2005;19:65–76.
[35] Hu LQ, Wu HL, Ding YJ, Fang DM, Xia AL, Yu RQ. Alternating asymmetric trilinear decomposition for three-way data arrays analysis. Chemom Intell Lab Syst 2006;82:145–53.
[36] Jiang JH, Wu HL, Chen ZP, Yu RQ. Coupled vectors resolution method for chemometric calibration with three-way data. Anal Chem 1999;71:4254–62.
[37] Jiang JH, Wu HL, Li Y, Yu RQ. Alternating coupled vectors resolution (ACOVER) method for trilinear analysis of three-way data. J Chemometr 1999;13:557–78.
[38] Jiang JH, Wu HL, Li Y, Yu RQ. Three-way data resolution by alternating slice-wise diagonalization (ASD) method. J Chemometr 2000;14:15–36.
[39] Li Y, Jiang JH, Wu HL, Chen ZP, Yu RQ. Alternating coupled matrices resolution method for three-way arrays analysis. Chemom Intell Lab Syst 2000;52:33–43.
[40] Chen ZP, Li Y, Yu RQ. Pseudo alternating least squares algorithm for trilinear decomposition. J Chemometr 2001;15:149–67.
[41] Chen ZP, Wu HL, Li Y, Yu RQ. Novel constrained PARAFAC algorithm for second-order linear calibration. Anal Chim Acta 2000;423:187–96.
[42] Xie HP, Chu X, Jiang JH, Cui H, Shen GL, Yu RQ. Competitive interactions of adriamycin and ethidium bromide with DNA as studied by full rank parallel factor analysis of fluorescence three-way array data. Spectrochim Acta A Mol Biomol Spectrosc 2003;59:743–9.
[43] Cao YZ, Chen ZP, Mo CY, Wu HL, Yu RQ. A PARAFAC algorithm using penalty diagonalization error (PDE) for three-way data array resolution. Analyst 2000;125:2303–10.
[44] Xia AL, Wu HL, Fang DM, Ding YJ, Hu LQ, Yu RQ. Determination of daunomycin in human plasma and urine using an interference-free analysis of excitation-emission matrix fluorescence data with second-order calibration. Anal Sci 2006;22:1189–95.
[45] Xia AL, Wu HL, Zhu SH, Han QJ, Zhang Y, Yu RQ. Determination of psoralen in human plasma using excitation-emission matrix fluorescence coupled to second-order calibration. Anal Sci 2008;24:1171–5.
[46] Nie JF, Wu HL, Zhang SR, Yu YJ, Yu RQ. Self-weighted alternating normalized residue fitting algorithm with application to quantitative analysis of excitation-emission matrix fluorescence data. Anal Methods 2010;2:1918–26.
[47] Li Y, Wu HL, Nie CC, Gu HW, Qing XD, Zuo Q, et al. Alternating coupled two-unequal residual functions algorithm for second-order calibration. Anal Methods 2014;6:6322–31.
[48] Yu YJ, Wu HL, Kang C, Wang Y, Zhao J, Li YN, et al. Algorithm combination strategy to obtain the second-order advantage: simultaneous determination of target analytes in plasma using three-dimensional fluorescence spectroscopy. J Chemometr 2012;26:197–208.
[49] Li S, Hamilton JC, Gemperline PJ. Generalized rank annihilation method using similarity transformations. Anal Chem 1992;64:599–607.
[50] Öhman J, Geladi P, Wold S. Residual bilinearization. Part1: theory and algorithms. J Chemometr 1990;4:79–90.
[51] Zou HY, Wu HL, Zhang Y, Li SF, Nie JF, Fu HY, et al. Studying the interaction of pirarubicin with DNA and determining pirarubicin in human urine samples: combining excitation-emission fluorescence matrices with second-order calibration methods. J Fluoresc 2009;6:955–66.
[52] Li SF, Wu HL, Xia AL, Zhu SH, Nie JF, Yu YJ, et al. Quantitative analysis of epinephrine in human plasma samples using kinetic fluorometric method combined with second-order calibration. Anal Sci 2009;25:1231–6.

[53] Gu HW, Wu HL, Liu YJ, Yin XL, Su ZY, Li SS, et al. Simultaneous determination of metoprolol and α-hydroxymetoprolol in human plasma using excitation–emission matrix fluorescence coupled with second-order calibration methods. Bioanalysis 2012;4:2781–93.

[54] Zhang SR, Wu HL, Chen Y, Zhang XH, Wang JY, Li Y, et al. An investigation on hydrogen bonding between 3-methylindole and ethanol using trilinear decomposition of fluorescence excitation-emission matrix. Chemom Intell Lab Syst 2013;121:9–14.

[55] Tan F, Tan C, Zhao A, Li ML. Simultaneous determination of free amino acid content in tea infusions by using high-performance liquid chromatography with fluorescence detection coupled with alternating penalty trilinear decomposition algorithm. J Agric Food Chem 2011;59:10839–47.

[56] Zhu SH, Wu HL, Li BR, Xia AL, Han QJ, Zhang Y, et al. Determination of pesticides in honey using excitation emission matrix fluorescence coupled with second-order calibration and second-order standard addition methods. Anal Chim Acta 2008;619:165–72.

[57] Yu YJ, Wu HL, Shao SZ, Kang C, Zhao J, Wang Y, et al. Using second-order calibration method based on trilinear decomposition algorithms coupled with high performance liquid chromatography with diode array detector for determination of quinolones in honey samples. Talanta 2011;85:1549–59.

[58] Zhang XH, Wu HL, Yin XL, Li LH, Wang JY, Chen Y, et al. A combined theoretical and experimental study for the chiral discrimination of naproxen enantiomers by molecular modeling and second-order standard addition method. Food Chem 2013;138:62–9.

[59] Li YN, Wu HL, Qing XD, Li Q, Li SF, Fu HY, et al. Quantitative analysis of triazine herbicides in environmental samples by using high performance liquid chromatography and diode array detection combined with second-order calibration based on an alternating penalty trilinear decomposition algorithm. Anal Chim Acta 2010;678:26–33.

[60] Wang JY, Wu HL, Chen Y, Sun YM, Yu YJ, Zhang XH, et al. Fast analysis of synthetic antioxidants in edible vegetable oil using trilinear component modeling of liquid chromatography–diode array detection data. J Chromatogr A 2012;1264:63–71.

[61] Zhu SH, Wu HL, Xia AL, Nie JF, Bian YC, Cai CB, et al. Excitation-emission-kinetic fluorescence coupled with third-order calibration for quantifying carbaryl and investigating the hydrolysis in effluent water. Talanta 2009;77:1640–6.

[62] Ding YJ, Wu HL, Xia AL, Cui H, Yu RQ. Simultaneous determination of two antituberculosis drugs using alternating fitting residue algorithm combined with HPLC-DAD. J Anal Sci 2008;24:1–4.

[63] Ouyang LQ, Wu HL, Nie JF, Zhang Y, Zou HY, Fu HY, et al. Simultaneous determination of psoralen and isopsoralen in plasma and Chinese medicine Xian Ling Gu Bao capsule by using HPLC–DAD coupled with alternating trilinear decomposition algorithm. Anal Chim Acta 2009;650:160–6.

[64] Zou HY, Wu HL, Ouyang LQ, Zhang Y, Nie JF, Fu HY, et al. Fluorescent quantification of terazosin hydrochloride content in human plasma and tablets using second-order calibration based on both parallel factor analysis and alternating penalty trilinear decomposition. Anal Chim Acta 2009;650:143–9.

[65] Nie JF, Wu HL, Zhu SH, Han QJ, Fu HY, Li SF, et al. Simultaneous determination of 6-methylcoumarin and 7-methoxycoumarin in cosmetics by three-dimensional excitation-emission matrix fluorescence coupled with second-order calibration methods. Talanta 2008;75:1260–9.

[66] Nie JF, Wu HL, Wang XM, Zhang Y, Zhu SH, Yu RQ. Determination of testosterone propionate in cosmetics using excitation-emission matrix fluorescence based on oxidation derivatization with the aid of second-order calibration methods. Anal Chim Acta 2008;628:24–32.

[67] Tu DZ, Wu HL, Li YN, Zhang J, Li Y, Nie CC, et al. Measuring estriol and estrone simultaneously in liquid cosmetic samples using second-order calibration coupled with excitation–emission matrix fluorescence based on region selection. Anal Methods 2012;4:222–9.

[68] Li YN, Wu HL, Qing XD, Nie CC, Li SF, Yu YJ, et al. The maintenance of the second-order advantage: second-order calibration of excitation-emission matrix fluorescence for quantitative analysis of herbicide napropamide in various environmental samples. Talanta 2011;85:325–32.

[69] Liu JJ, Ma X, Wen YD, Wang Y, Cai WS, Shao XG. Online near-infrared spectroscopy combined with alternating trilinear decomposition for process analysis of industrial production and quality assurance. Ind Eng Chem Res 2011;50:7677–81.

[70] Liu Y, Cai WS, Shao XG. Standardization of near infrared spectra measured on multiinstrument. Anal Chim Acta 2014;836:18–23.

[71] Faber NM, Bro R, Hopke PK. Recent developments in CANDECOMP/PARAFAC algorithms: a critical review. Chemom Intell Lab Syst 2003;65:119–37.

[72] Yu YJ, Wu HL, Nie JF, Zhang SR, Li SF, Li YN, et al. A comparison of several trilinear second-order calibration algorithms. Chemom Intell Lab Syst 2011;106:93–107.

[73] Zhang SR, Wu HL, Yu RQ. A study on the differential strategy of some iterative trilinear decomposition algorithms: PARAFAC-ALS, ATLD, SWATLD and APTLD. J Chemometr 2014;29:179–92.

[74] Olivieri AC, Arancibia JA, de la Peña AM, Durán-Merás I, Eansilla A. Second-order advantage achieved with four-way fluorescence excitation-emission-kinetic data processed by parallel factor analysis and trilinear least-squares. Determination of methotrexate and leucovorin in human urine. Anal Chem 2004;76:5657–66.

[75] Qing XD, Wu HL, Yan XF, Li Y, Ouyang LQ, Nie CC, et al. Development of a novel alternating quadrilinear decomposition algorithm for the kinetic analysis of four-way room-temperature phosphorescence data. Chemom Intell Lab Syst 2014;132:8–17.

[76] Xia AL, Wu HL, Li SF, Zhu SH, Hu LQ, Yu RQ. Alternating penalty quadrilinear decomposition algorithm for an analysis of four-way data arrays. J Chemometr 2007;21:133–44.

[77] Fu HY, Wu HL, Yu YJ, Yu LL, Zhang SR, Nie JF, et al. A new third-order calibration method with application for analysis of four-way data arrays. J Chemometr 2011;25:408–29.

[78] Liu YJ, Wu HL, Kang C, Gu HW, Nie JF, Li SS, et al. Four-way self-weighted alternating normalized residue fitting algorithm with application for the analysis of serotonin in human plasma. Anal Sci 2012;28:1097–104.

[79] Kang C, Wu HL, Yu YJ, Liu YJ, Zhang SR, Zhang XH, et al. An alternative quadrilinear decomposition algorithm for four-way calibration with application to analysis of four-way fluorescence excitation-emission-pH data array. Anal Chim Acta 2013;758:45–57.

[80] Damiani PC, Durán-Merás I, García-Reiriz A, Jiménez-Girón A, de la Peña AM, Olivieri AC. Multiway partial least-squares coupled to residual trilinearization: a genuine multidimensional tool for the study of third-order data. Simultaneous analysis of procaine and its metabolite p-aminobenzoic acid in equine serum. Anal Chem 2007;79:6949–58.

[81] Alejandro JA, Olivieri AC, Gil DB, Mansilla AE, Durán-Merás I, de la Peña AM. Trilinear least-squares and unfolded-PLS coupled to residual trilinearization: new chemometric tools for the analysis of four-way instrumental data. Chemom Intell Lab Syst 2006;80:77–86.

[82] Harshman RA, Lundy ME. Research methods for multimode data analysis. New York: Praeger; 1984.

[83] Chen ZP, Liu Z, Cao YZ, Yu RQ. Efficient way to estimate the optimum number of factors for trilinear decomposition. Anal Chim Acta 2001;444:295–307.

[84] Bro R, Kiers HHL. A new efficient method for determining the number of components in PARAFAC modes. J Chemometr 2003;17:274–86.
[85] Li Y, Wu HL, Qing XD, Zuo Q, Chen Y, Yu RQ. A novel method to estimate the chemical rank of three-way data for second-order calibration. Chemom Intell Lab Syst 2013;127:177–84.
[86] Malinowski ER. Determination of the number of factors and the experimental error in a data matrix. Anal Chem 1977;49:612–7.
[87] Sanchez FC, Toft J, van den Bogaert B, Massart DL. Orthogonal projection approach applied to peak purity assessment. Anal Chem 1996;68:79–85.
[88] Xie HP, Jiang JH, Long N, Shen GL, Wu HL, Yu RQ. Estimation of chemical rank of a three-way array using a two-mode subspace comparison approach. Chemom Intell Lab Syst 2003;66:101–15.
[89] Hu LQ, Wu HL, Jiang JH, Han QJ, Xia AL, Yu RQ. Estimating the chemical rank of three-way data arrays by a simple linear transform incorporating Monte Carlo simulation. Talanta 2007;71:373–80.
[90] Xia AL, Wu HL, Zhu SH, Han QJ, Yu RQ. A novel efficient way to estimate the chemical rank of high-way data arrays. Anal Chim Acta 2007;598:1–11.
[91] Nie JF, Wu HL, Wang JY, Liu YJ, Yu RQ. The chemical rank estimation for excitation-emission matrix fluorescence data by region-based moving window subspace projection technique and Monte Carlo simulation. Chemom Intell Lab Syst 2010;104:271–80.
[92] Olivieri AC, Wu HL, Yu RQ. MVC2: a MATLAB graphical interface toolbox for second-order multivariate calibration. Chemom Intell Lab Syst 2009;96:246–51.
[93] Olivieri AC, Wu HL, Yu RQ. MVC3: a MATLAB graphical interface toolbox for third-order multivariate calibration. Chemom Intell Lab Syst 2012;116:9–16.
[94] Li Y, Wu HL, Yu YJ, Zhang SR, Chen Y, Tu DZ, et al. A novel method to handle Rayleigh scattering in three-way excitation-emission fluorescence data. Anal Methods 2012;4:3987–96.
[95] Bro R, Sidiropoulos ND, Smilde AK. Maximum likelihood fitting using ordinary least squares algorithms. J Chemometr 2002;16:387–400.
[96] Tomasi G, Bro R. PARAFAC and missing values. Chemom Intell Lab Syst 2005;75:163–80.
[97] Thygesen LG, Rinnan A, Barsberg S, Moller JKS. Stabilizing the PARAFAC decomposition of fluorescence spectra by insertion of zeros outside the data area. Chemom Intell Lab Syst 2004;71:97–106.
[98] Bahram M, Bro R, Stedmon C, Afkhami A. Handling of Rayleigh and Raman scatter for PARAFAC modeling of fluorescence data using interpolation. J Chemometr 2006;20:99–105.
[99] Rinnan A, Booksh KS, Bro R. First order Rayleigh scatter as a separate component in the decomposition of fluorescence landscapes. Anal Chim Acta 2005;537:349–58.
[100] Nie JF, Wu HL, Zhang Y, Liu YJ, Wang JY, Li Y, et al. Removal of first-order Rayleigh scatter nonlinear interference in multiway calibration of excitation-emission matrix fluorescence spectra with the use of trilinear decomposition methods. Sci China Chem 2011;41:1740–9.
[101] Chen ZP, Yu RQ. Mitigating model deficiency in three-way data analysis by the combination of background constraining and iterative correcting techniques. Anal Chim Acta 2003;487:171–80.
[102] Yu YJ, Wu HL, Niu JF, Zhao J, Li YN, Kang C, et al. A novel chromatographic peak alignment method coupled with trilinear decomposition for three dimensional chromatographic data analysis to obtain the second-order advantage. Analyst 2013;138:627–34.

[103] Amigo JM, Skov T, Bro R. ChroMATHography: solving chromatographic issues with mathematical models and intuitive graphics. Chem Rev 2010;110:4582–605.

[104] Zhang Y, Wu HL, Xia AL, Hu LH, Zou HF, Yu RQ. Trilinear decomposition method applied to removal of three-dimensional background drift in comprehensive two-dimensional separation data. J Chromatogr A 2007;1167:178–83.

[105] Yu YJ, Wu HL, Fu HY, Zhao J, Li YN, Li SF, et al. Chromatographic background drift correction coupled with parallel factor analysis to resolve coelution problems in three-dimensional chromatographic data: quantification of eleven antibiotics in tap water samples by high-performance liquid chromatography coupled with a diode array detector. J Chromatogr A 2013;1302:72–80.

[106] Skoog DA, West DM. Analytical chemistry, an introduction. 4th ed. Philadephia: Saunders College Publishing; 1986.

[107] Saxberg B, Kowalski BR. Generalized standard addition method. Anal Chem 1979;51:1031–8.

[108] Booksh KS, Henshaw JM, Burgess LW, Kowalski BR. A second-order standard addition method with application to calibration of a kinetics-spectroscopic sensor for quantitation of trichloroethylene. J Chemometr 1995;9:263–82.

[109] Wu HL, Yu RQ, Shibukawa M, Öguma K. Second-order standard addition method based on alternating trilinear decomposition. Anal Sci 2000;16:217–20.

Chapter 4

# Practical Analytical Applications of Multiway Calibration Methods Based on Alternating Multilinear Decomposition

**Hai-Long Wu[1], Chao Kang, Yong Li and Ru-Qin Yu[1]**

*State Key Laboratory of Chemo/Biosensing and Chemometrics, College of Chemistry and Chemical Engineering, Hunan University, Changsha, Hunan, China*

*[1]Corresponding authors: e-mail: hlwu@hnu.edu.cn; rqyu@hnu.edu.cn*

## Chapter Outline



## 1 INTRODUCTION

In quantitative chemical analysis of analytes of interest in complex systems, the calibration of an analyte of interest is traditionally done by the standard curve method, which is also known as one-way calibration (or zeroth-order tensor calibration) in chemometrics. In one-way calibration (i.e., zeroth-order calibration), the measurement data for each sample is a zeroth-order tensor (e.g., ultraviolet absorbance at the maximum absorption wavelength).

Data Handling in Science and Technology, Vol. 29. http://dx.doi.org/10.1016/B978-0-444-63527-3.00004-7

The concentration of an analyte of interest in an unknown sample may be predicted by the use of a standard calibration curve, based on the linear model $y = m_0 + m_1 x$. However, only one analyte of interest can be analyzed at a time, and a fundamental limitation of one-way calibration is that the signal must be fully selective for the analyte of interest. This has led to the development of two-way calibration methods [1–15].

In two-way calibration (i.e., first-order tensor calibration), the measurement data for each sample is a first-order tensor (such as a chromatogram or spectrum). With two-way calibration methods based on the bilinear model $x_{ij} = \sum_{n=1}^{N} a_{in} b_{jn}$, multianalyte quantitative analysis becomes possible, and signals no longer have to be fully selective only for the analyte of interest but can now be responding to known interfering species [2]. These types of methods include multivariate linear regression [1–3], principal component regression [5], and partial least squares [4], which have become popular in chemometrics literature. It therefore follows that, if the measurement data for each sample is a second-order tensor (such as excitation–emission matrix (EEM) fluorescence data), more advantages could be obtained using three-way calibration methods [16–54].

In three-way calibration (i.e., second-order calibration), the measurement data for each sample is a second-order tensor (such as an EEM data matrix). The trilinear model (also called the PARAFAC model) is one of the most popular three-way calibration models, independently proposed by Carroll and Chang [16] and Harshman [17] in 1970 in behavioral science. Thereafter, the trilinear model was introduced into the field of three-way (second-order tensor) calibration in chemometrics [19,26,55,56]. With three-way calibration methods based on the trilinear model $x_{ijk} = \sum_{n=1}^{N} a_{in} b_{jn} c_{kn} + e_{ijk}$, $(i = 1,\ldots,I; j = 1,\ldots,J; k = 1,\ldots,K)$, analysts can achieve direct quantitative analysis of multiple analytes of interest in complex systems even in the presence of unknown, uncalibrated interferents, which is known as the "second-order advantage" [23,57]. In addition, three-way calibration can provide some other advantages, such as auto-correction of varying background signals and higher sensitivity. While, for much more collinear and complex systems, even three-way calibration may not correctly resolve the signals of analytes of interest, which has led to the development of four-way calibration methods.

In four-way calibration (i.e., third-order calibration), the measurement data for each sample is a third-order tensor (such as the EEM data as a function of reaction time). With four-way calibration methods based on the quadrilinear model $x_{ijkl} = \sum_{n=1}^{N} a_{in} b_{jn} c_{kn} d_{ln} + e_{ijkl}$, $(i = 1,\ldots,I; j = 1,\ldots,J; k = 1,\ldots,K; l = 1,\ldots,L)$, analysts can not only obtain the "second-order advantage" but also obtain additional advantages such as autocorrecting varying background signals and providing more resolving power as well as higher sensitivity [35,37,38,58,59]. Analysts could expect that, multiway calibration (i.e., calibration method that has three or more than three ways, also called higher-order calibration) based on the multilinear model can not only maintain the "second-order advantage" but also provide additional advantages such as auto-correction of varying background signals, possessing more resolving power and higher sensitivity [35,37,38,59].

Multiway calibration methods based on multilinear models are gaining more and more attention in the field of analytical science. Based on the "mathematical separation," multiway calibration methods enable one to achieve direct quantitative analysis of analytes of interest in complex systems even in the presence of unknown, uncalibrated interferents and varying background signals. When combined with advanced analytical instruments capable of generating multidimensional arrays (e.g., EEMs, high-performance liquid chromatography-diode array detection (HPLC-DAD), and LC-MS), multiway calibration can be a smart and green quantitative analysis strategy for complex chemical systems, based on the "mathematical separation" instead of chemical and physical separations.

First, this chapter briefly introduces multiway calibration methods based on alternating multilinear decompositions and reviews the practical analytical applications of these methods. Second, this chapter illustrates the general procedure of multiway calibration methods based on alternating multilinear decompositions, by the use of a simulated EEMs data array. Third, this chapter uses practical analytical applications as case studies to further illustrate the practical analytical applications of three- and four-way calibration methods.

## 2 GENERAL PROCEDURE OF MULTIWAY CALIBRATION

Multiway calibration methods based on the multilinear models are natural extensions of the one-way calibration method based on the linear model. The general procedure of multiway calibration methods based on alternating multilinear decomposition is similar to that of one-way calibration. However, the procedure of multiway calibration is some more complex. In this section, we will explain the general procedure of multiway calibration methods based on the multilinear models step by step, accompanied by the analysis of a simulated EEMs three-way data array by the three-way calibration method based on the alternating trilinear decomposition (ATLD) algorithm [26] as an example. The detailed flow chart of the three-way calibration method based on the trilinear model is shown in Figure 1.

Taking the analysis of the three-way EEMs data array by using three-way calibration method based on the ATLD algorithm as an example, the general procedure could be illustrated as shown in Figure 2.

The MATLAB demo on how to generate the simulated EEMs data and how to use the three-way calibration method and some related functions are provided in Appendix.

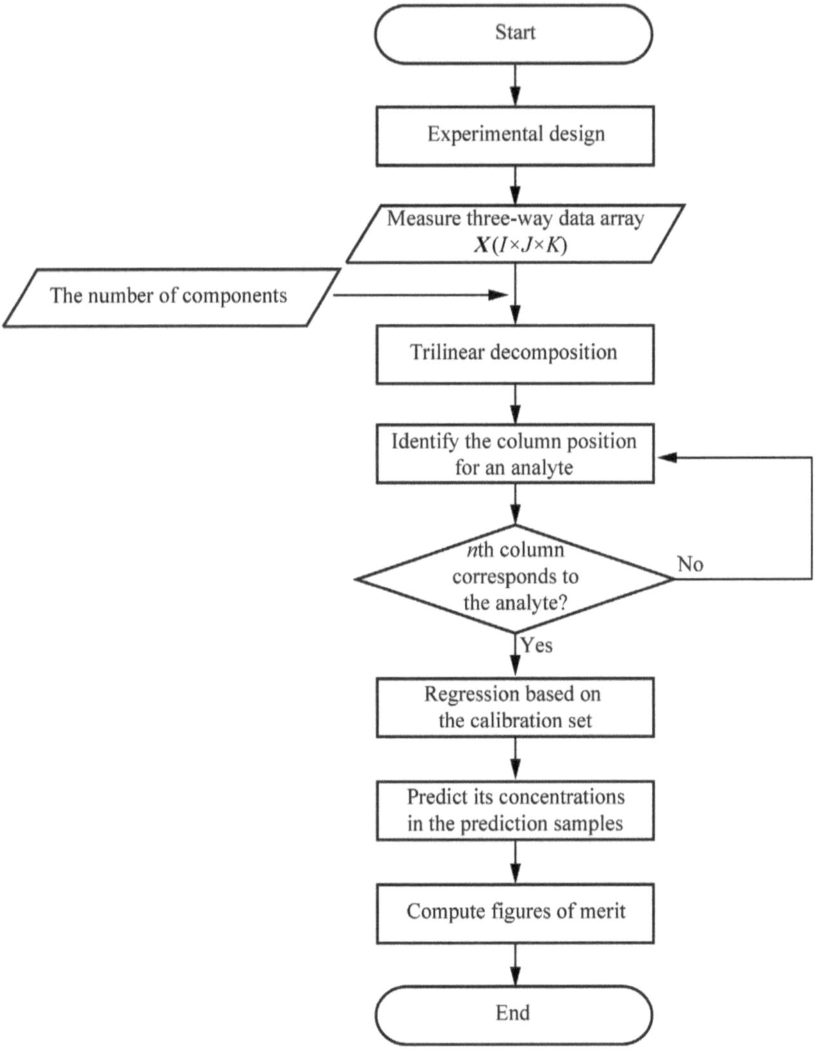


**FIGURE 1** Flow chart of the three-way calibration method based on the trilinear model.

## 2.1 Experimental Design

In a multiway calibration method, we often need to prepare the calibration set, the validation set, the prediction set, and the method blank set. The calibration set is often prepared in a simple buffer solution, without need to contain the responsive components presented in a future prediction sample. This property originated from the second-order advantage makes these multiway calibration methods very simple and flexible.

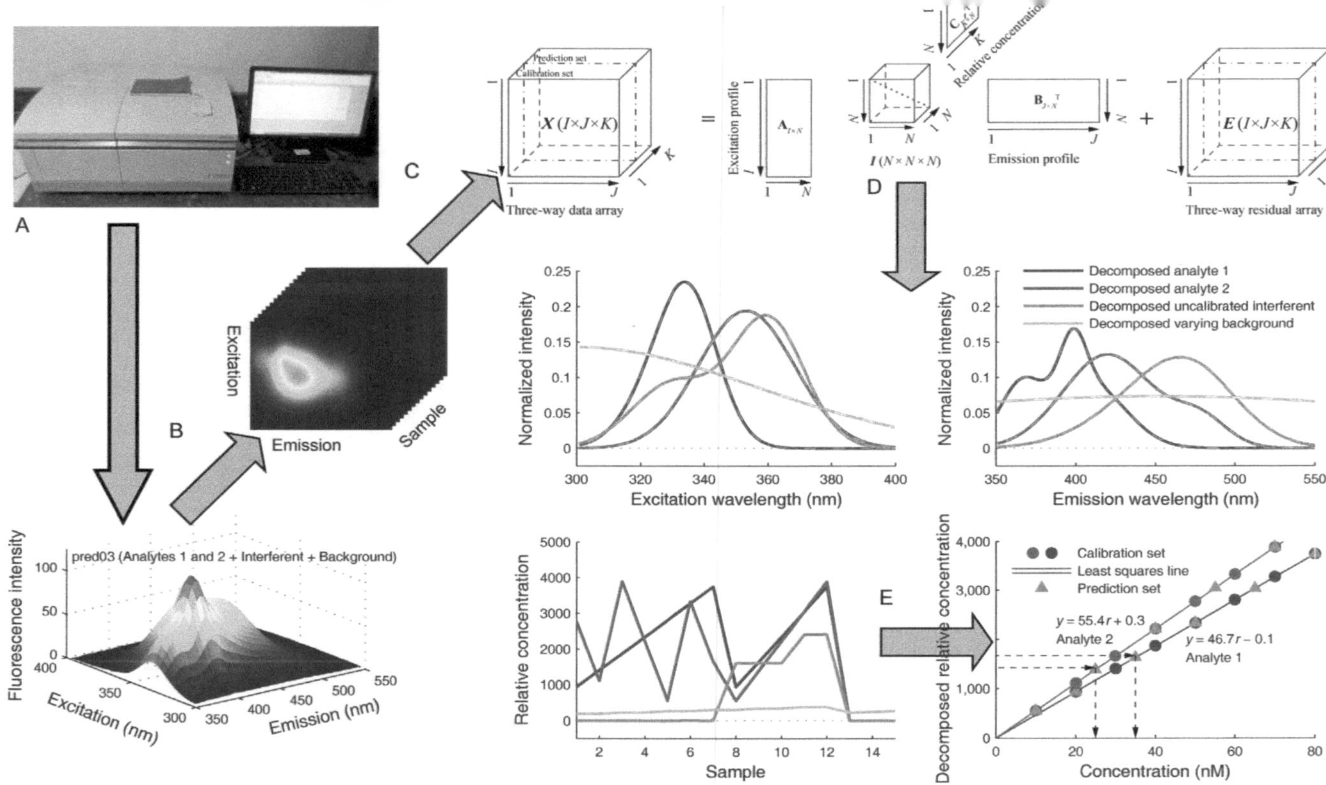


**FIGURE 2** Illustration of the three-way calibration method based on the trilinear model (Taking the analysis of the EEMs three-way data array by the three-way calibration method based on the ATLD algorithm as an example). (A) Measuring a two-way data matrix per sample by EEM fluorescence; (B) arranging two-way data matrices of calibration samples and prediction samples into three-way data array $\boldsymbol{X}$ $(I \times J \times K)$; (C) building the trilinear model for three-way calibration; (D) decomposing the three-way data array into underlying contributions of each analytes of interest with a proper trilinear decomposition algorithm; (E) predicting the concentration of an analyte of interest in prediction samples, by the regression equation obtained in a univariate regression of the decomposed relative concentration against the concentration based on the calibration set.

The validation set here represents the spiked validation samples, in which appropriate amounts of all the analytes of interest are spiked with the system studied. The validation set is used to validate whether or not a multiway calibration method can accurately predict the concentrations of all the analytes of interest in the system studied, even in the presence of uncalibrated interferents. The validation set is also used to test whether the response to a sample is the same as that expected from a multiway calibration. The spiked validation samples should be analyzed in the same manner as the prediction samples [60].

The prediction set here represents the real samples of the system studied. The prediction set is used to determine the concentration of an analyte of interest in the system studied. Accuracy is often demonstrated by analyzing a standard reference material, analyzing a sample spiked with a known addition of analyte, or comparing results by an approved analytical method. Although the last way is the most preferred, spiking is the most common method to evaluate accuracy because reference materials are not usually available and a second analytical method may not be readily available [60].

The method blanks are samples containing all components except the analytes of interest, and they have gone through all steps of the analytical procedure [60]. The method blank set is used to subtract the response of the blank sample from the response of a real sample prior to predicting the concentration of an analyte of interest in the sample. Meanwhile, multiway calibration methods based on the multilinear models can use an additional component to model the background signal, even a varying background signal [56]. Therefore, in some applications of multiway calibration, the background signal is modeled by an additional component, instead of being subtracted by the method blank set.

The design of the concentration levels of analytes of interest in the calibration set is very important. The calibration set often consists of five to seven samples, and each sample provides a concentration level of each analyte. Given the quantitative analysis of two analytes of interest in a complex system by multiway calibration, if we want to design seven concentration levels for each analyte, the uniform design is a very promising design for the concentration level [61,62], since the uniform design could provide an appropriate $U_7(7^2)$ uniform design for this seven-level, two-factor concentration design in which only seven experiments (i.e., seven calibration samples) are needed.

In the example of the analysis of the simulated EEMs three-way data array by the three-way calibration method based on the ATLD algorithm, there are seven calibration samples only containing the two analytes of interest, five prediction samples, and three method blank samples. The sample design and the concentration design are given in Table 1. The seven calibration samples only contain the two analytes of interest, while the five prediction samples not only contain the two analytes of interest but also contain a spectral interferent, and there exists a varying background signal in all samples.

**TABLE 1** Sample Design and Concentration Design for the Simulated EEMs Three-Way Data Array

| No. | Sample | Concentration (nM) | | Uncalibrated interferent (nM) | Varying background |
|---|---|---|---|---|---|
| | | *Analyte 1* | *Analyte 2* | | |
| **Calibration set—$U_7(7^2)$ *uniform design*** | | | | | |
| 1 | cali01 | 20.0 | 50.0 | 0 | 5.0 |
| 2 | cali02 | 30.0 | 20.0 | 0 | 5.0 |
| 3 | cali03 | 40.0 | 70.0 | 0 | 6.0 |
| 4 | cali04 | 50.0 | 40.0 | 0 | 6.0 |
| 5 | cali05 | 60.0 | 10.0 | 0 | 7.0 |
| 6 | cali06 | 70.0 | 60.0 | 0 | 7.0 |
| 7 | cali07 | 80.0 | 30.0 | 0 | 8.0 |
| **Prediction set** | | | | | |
| 8 | pred01 | 20.0 | 10.0 | 20.0 | 8.0 |
| 9 | pred02 | 35.0 | 25.0 | 20.0 | 9.0 |
| 10 | pred03 | 50.0 | 40.0 | 20.0 | 9.0 |
| 11 | pred04 | 65.0 | 55.0 | 30.0 | 10.0 |
| 12 | pred05 | 80.0 | 70.0 | 30.0 | 10.0 |
| **Method blank set** | | | | | |
| 13 | mblank01 | 0 | 0 | 0 | 6.0 |
| 14 | mblank02 | 0 | 0 | 0 | 6.5 |
| 15 | mblank03 | 0 | 0 | 0 | 7.0 |

The actual relative profiles in three modes, respectively, for the simulated EEMs three-way data array are given in Figure 3. The simulated spectra of analyte 1, analyte 2, and the uncalibrated interferent in each mode overlap seriously. And the three-dimensional landscapes of EEM for cali04 and pred03 are also given in Figure 3.

## 2.2 Multilinear Model

In this section, we present two commonly used multilinear models, i.e., the trilinear model and the quadrilinear model.

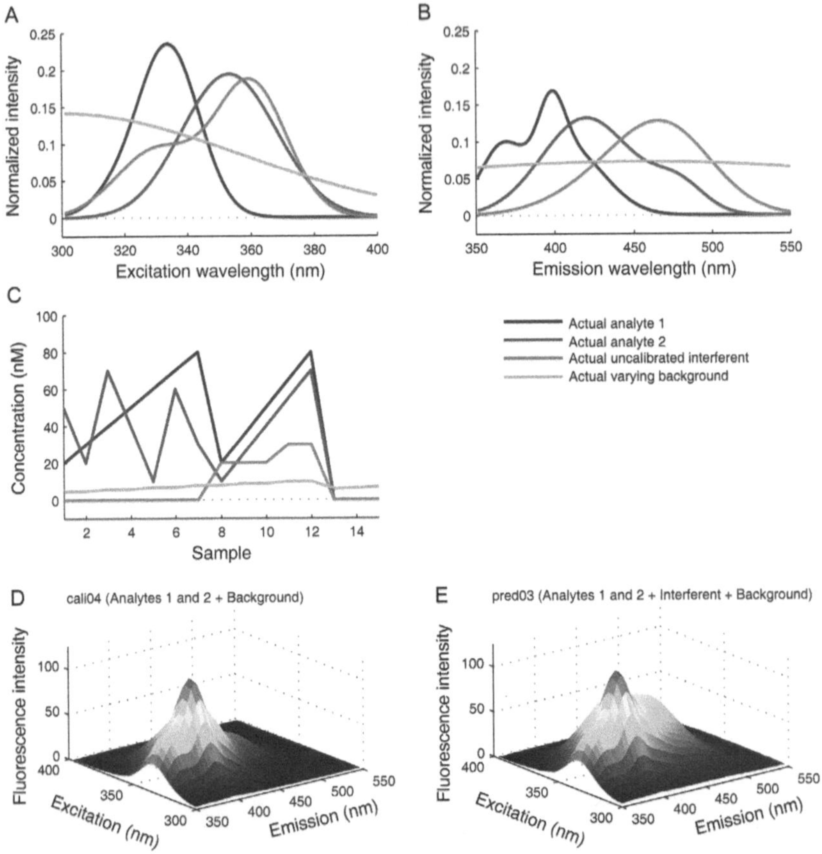


**FIGURE 3** Actual relative profiles in three modes and three-dimensional landscapes of EEM fluorescence for cali04 and pred03, in the simulated EEMs three-way data array.

### 2.2.1 *Trilinear Model*

Considering a model of the real-valued three-way data array $\boldsymbol{X}$ $(I \times J \times K)$, in which each element $x_{ijk}$ can be expressed as follows:

$$x_{ijk} = \sum_{n=1}^{N} a_{in} b_{jn} c_{kn} + e_{ijk}, \quad \text{for } i = 1, 2, \ldots, I; \; j = 1, 2, \ldots, J; \; k = 1, 2, \ldots, K \quad (1)$$

where $a_{in}$, $b_{jn}$, and $c_{kn}$ correspond to the underlying profile matrices $\mathbf{A}_{I\times N}$, $\mathbf{B}_{J\times N}$, and $\mathbf{C}_{K\times N}$ of $\boldsymbol{X}$ $(I \times J \times K)$, respectively. The term $e_{ijk}$ is the element of the three-way residual array $\boldsymbol{E}$ $(I \times J \times K)$. Then the modeled part of $x_{ijk}$ is trilinear in the parameter sets $a_{in}$, $b_{jn}$, and $c_{kn}$ [1,16,17,23,25,26,55,63,64].

The simulated EEMs trilinear model is illustrated in Figure 4. In this illustration, the three-way array $\boldsymbol{X}$ $(I \times J \times K)$ represents the simulated excitation-emission-sample (EX-EM-Sample) three-way data array; the profile matrix

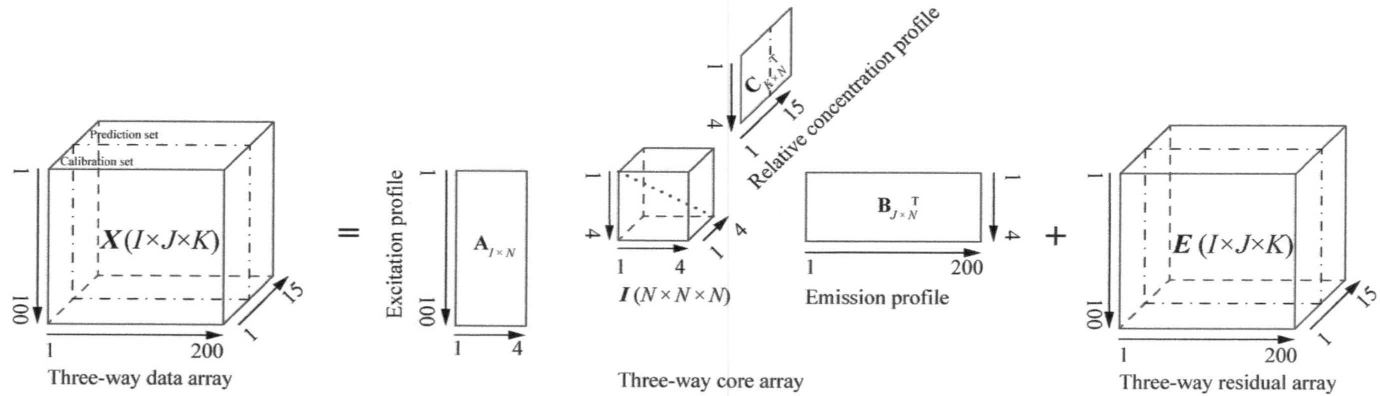


**FIGURE 4** Illustration of the simulated EEMs trilinear model in the three-way calibration, with the underlying number of components $N = 4$.

**A** is the excitation spectrum matrix, the profile matrix **B** is the emission spectrum matrix, and the profile matrix **C** is the relative concentration matrix; the three-way array $\boldsymbol{I}$ $(N \times N \times N)$ represents the core array in which only the main superdiagonal elements are nonzero elements.

### 2.2.2 Quadrilinear Model

Considering a model of the real-valued four-way data array $\boldsymbol{X}$ $(I \times J \times K \times L)$, in which each element $x_{ijkl}$ can be expressed as follows:

$$x_{ijkl} = \sum_{n=1}^{N} a_{in} b_{jn} c_{kn} d_{ln} + e_{ijkl}, \quad \text{for} \quad i = 1, 2, \ldots, I; \; j = 1, 2, \ldots, J; \quad k = 1, 2, \ldots, K; \; l = 1, 2, \ldots, L. \tag{2}$$

where $a_{in}$, $b_{jn}$, $c_{kn}$, and $d_{ln}$ correspond to the underlying profile matrices $\mathbf{A}_{I \times N}$, $\mathbf{B}_{J \times N}$, $\mathbf{C}_{K \times N}$, and $\mathbf{D}_{L \times N}$ of $\boldsymbol{X}$ $(I \times J \times K \times L)$, respectively. The term $e_{ijkl}$ is the element of the four-way residual array $\boldsymbol{E}$ $(I \times J \times K \times L)$. Then the modeled part of $x_{ijkl}$ is quadrilinear in the parameter sets $a_{in}$, $b_{jn}$, $c_{kn}$, and $d_{ln}$ [1,16,17,23,25,26,55,63,64].

### 2.2.3 Data Preprocessing

The correctness of decomposition of a multilinear model requires that the multilinear model holds multilinearity. However, there are some non-multilinear factors, which can cause a multilinear model to deviate the multilinearity [1,2,16,17,63,65]. For example, in the case of the EEMs type trilinear model, severe Rayleigh scattering and Raman scattering often exist, which will lead a trilinear model to deviate the trilinearity; or in the case of the HPLC-DAD type trilinear model, the time shift problem among different runs will also lead a trilinear model to deviate the trilinearity. If these problems occur, the data arrays in multiway calibrations must need appropriate data preprocessing before a multilinear decomposition.

## 2.3 Multilinear Decomposition

Regardless of scaling and permutation, the decomposition of a multilinear model will be a unique one [18,66]. The uniqueness of the decomposition is what the "second-order advantage" depends on, which makes decompositions of multilinear models very different with orthogonal tensor decompositions and curve resolution methods.

### 2.3.1 Trilinear Decomposition

In the trilinear decomposition in three-way calibration, the columns of the profile matrices $\mathbf{A}_{I \times N}$ and $\mathbf{B}_{J \times N}$ are normalized to length one with the values absorbed into the columns of the profile matrix $\mathbf{C}_{K\times N}$. Regardless of scaling and permutation, the decomposition of the trilinear model will be a unique one [18,66] given that $k_{\mathbf{A}}+k_{\mathbf{B}}+k_{\mathbf{C}}\geq 2N+2$, where $k_{\mathbf{A}}$, $k_{\mathbf{B}}$, and $k_{\mathbf{C}}$ are the $k$-ranks of the profile matrices **A**, **B**, and **C**, respectively. Algorithms that can achieve the trilinear decomposition [16,17,20–22,25,26,29–31,33,35,37, 38,55,57,67–72] are introduced in Chapter 3. Each algorithm was designed for situations more or less favorable for the application of the algorithm.

In the example of the analysis of the simulated three-way EEMs data array by using three-way calibration method based on the ATLD algorithm, we use the ATLD algorithm with the number of components $N=4$ to decompose this simulated EEMs trilinear model. For this simulated three-way EEMs data array, the actual relative profiles and the decomposed relative profiles provided by using the ATLD algorithm in three modes, respectively, are given in Figure 5. With the number of components $N=4$, the ATLD method accurately decomposed the excitation spectrum matrix $\mathbf{A}_{100\times 4}$, the emission spectrum matrix $\mathbf{B}_{200\times 4}$, and the relative concentration matrix $\mathbf{C}_{15\times 4}$. The computation time is only 0.104 s and the number of iterations is only 6. The decomposed excitation, emission, and relative concentration profiles of both the analytes of interest are very consistent with the actual ones (dash-dot lines). Interestingly, the uncalibrated interferent is also resolved. And the varying background signal is modeled by an additional component.

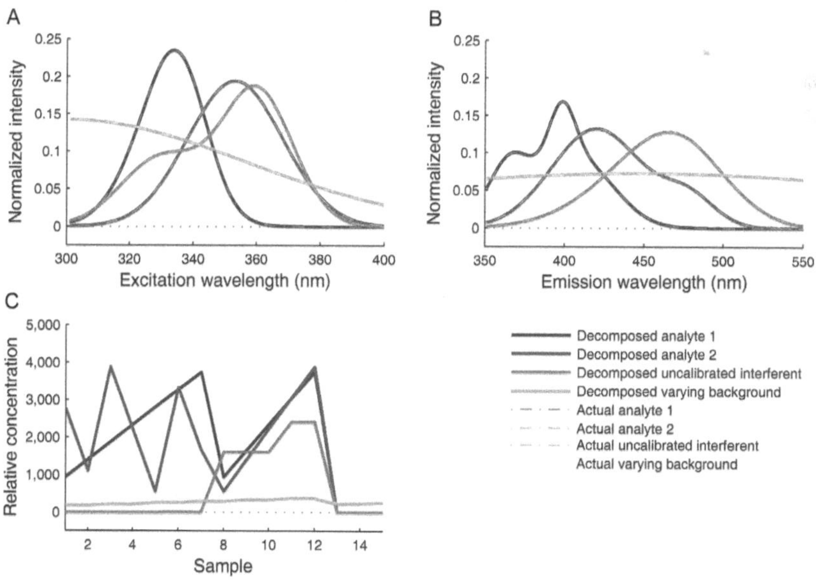


**FIGURE 5** Results of the decomposition in the three-way calibration method based on the ATLD algorithm with the number of components $N=4$, for the simulated EEMs three-way data array.

### 2.3.2 Quadrilinear Decomposition

In the quadrilinear decomposition in four-way calibration, the columns of the profile matrices $\mathbf{A}_{I\times N}$, $\mathbf{B}_{J\times N}$, and $\mathbf{C}_{K\times N}$ are normalized to length one with the values absorbed into the columns of the profiles matrix $\mathbf{D}_{L\times N}$. Regardless of scaling and permutation, the decomposition of the quadrilinear model will be a unique one [18,66], given that $k_{\mathbf{A}}+k_{\mathbf{B}}+k_{\mathbf{C}}+k_{\mathbf{D}}\geq 2N+3$, where $k_{\mathbf{A}}$, $k_{\mathbf{B}}$, $k_{\mathbf{C}}$, and $k_{\mathbf{D}}$ are the *k*-ranks of the profile matrices **A**, **B**, **C**, and **D**, respectively. Algorithms that can achieve the quadrilinear decomposition [1,3,23,32,35,37, 38,63,72–81] are also introduced in Chapter 3. Similarly, each algorithm was designed for situations more or less favorable for the application of the algorithm.

### 2.3.3 Number of Components

The number of components *N* for an investigated system should be estimated before the multilinear decomposition. Inclusion of too few components yields overly broad and featureless solutions. On the other hand, inclusion of too many factors often yields redundant solutions for most algorithms [2,16,17,37,38].

Initial estimate of the number of components *N* can be obtained by statistical, empirical, or *a priori* methods. Also, visual inspection of the decomposed profiles is beneficial. There already are many methods reported to estimate the number of components *N* for a multilinear model [18,26,37,38,82]. Considering the fact that estimating the number of components *N* is not always straightforward, especially in the presence of noise or complex matrix, we recommend to use different methods in evaluating the appropriate number of components.

It is worthy to notice that the ATLD method and some variants of ATLD are insensitive to the overestimated number of components. As long as the selected number of components is equal to or greater than the underlying real number of components, these algorithms or methods can always provide correct decomposed profiles and the excess components always appear as the shape of a noise which are easily recognized. For example, the ATLD algorithm with the number of components $N=4+1$ can still correctly decompose the simulated three-way EEMs data array. The actual relative profiles and the decomposed relative profiles provided by using the ATLD algorithm in three modes, respectively, are given in Figure 6. With the number of components $N=4+1$, the ATLD algorithm accurately decomposed the excitation spectrum matrix $\mathbf{A}_{100\times 5}$, the emission spectrum matrix $\mathbf{B}_{200\times 5}$, and the relative concentration matrix $\mathbf{C}_{15\times 5}$. The computation time is only 0.389 s, and the number of iterations is only 25.The decomposed excitation, emission, and relative concentration profiles of both the analytes of interest are very consistent with the actual ones (dash-dot lines). Interestingly, the uncalibrated interferent is also resolved. And the varying background signal is modeled by an additional component. Obviously, the fifth component (magenta solid line) is

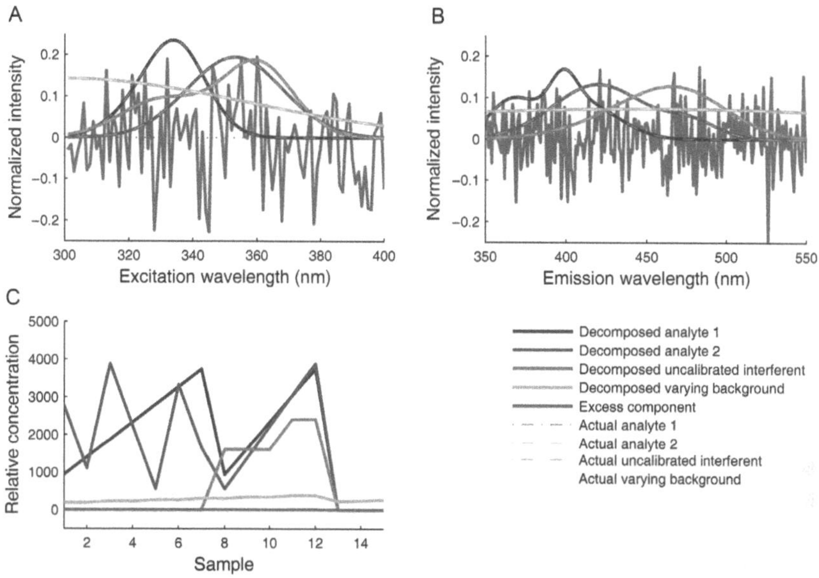


**FIGURE 6** Results of the decomposition in the three-way calibration method based on the ATLD algorithm with the number of components $N=4+1$, for the simulated EEMs three-way data array.

the excess component. Its profiles in the excitation and emission modes appear in the shape of a noise and its profile in the relative concentration mode is nearly the horizontal line of zero.

The ATLD algorithm even with the number of components $N=4+10$ can still correctly decompose the simulated three-way EEMs data array, as shown in Figure 7. The excitation and emission profiles of the excess components appear in the shape of a noise and their relative concentration profiles are nearly the horizontal line of zero. Theoretically, as long as the selected number of components is equal to or greater than the underlying real number of components, the ATLD algorithm and some variants of ATLD can provide correct decomposed profiles, and the excess components always appear in the shapes of noises, which are easily recognized. The excess components will not affect the convergence rate, the qualitative decomposition result, and the quantitative prediction result of the three-way calibration method based on these algorithms.

### *2.3.4 Solution Constraints*

Theoretically, the decomposition of a multilinear model will be a unique one [18,66], regardless of scaling and permutation. Therefore, there is no need to constrain the solution in the decomposition process of a multilinear model. However, slight non-multilinear factors presented in the system studied may

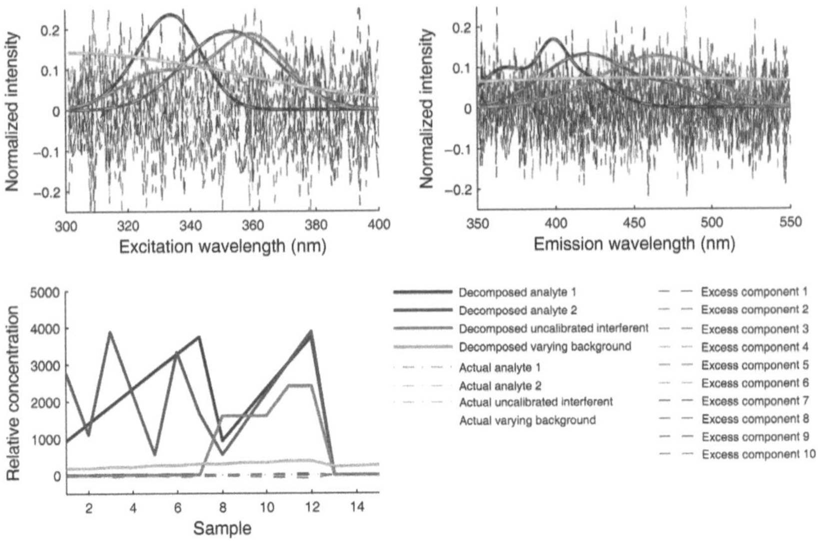


**FIGURE 7** Results of the decomposition in the three-way calibration method based on the ATLD algorithm even with the number of components $N=4+10$, for the simulated EEMs three-way data array.

result in decomposed profiles that are visually unsatisfactory and even containing significant quantitative errors. These effects can sometimes be minimized by using solution constraints that are based on *a priori* knowledge or assumptions of the data structure [2,3]. Potential constraints that could be used in the decomposition process of a multilinear model have nonnegativity constraint, unimodality constraint, known profiles constraint, and local-rank constraint, and so on.

Perhaps nonnegativity constraint is the most commonly used constraint in the decomposition process of a multilinear model. Negative values may occur sometimes in a decomposed profile of a multilinear model if the profile represents a spectrum, chromatogram, or relative concentration profile. The solution can be constrained to yield only nonnegative profile by the nonnegativity constraint [69,83], and one may see an improvement of the qualitative and quantitative results. But nonnegativity constraints should be used very carefully, since absorbance and quenching in a fluorescence spectrum or a chromatogram can often produce negative values in a profile, which can be measured and modeled actually as negative values in a profile.

## 2.4 Univariate Regression

After the multilinear decomposition in a multiway calibration, one can obtain the relative concentration profile matrix. The decomposition of the multilinear model joins the calibration set together with the prediction set, and so the concentration of an analyte of interest in a prediction sample can be predicted by the regression line obtained in a univariate regression of the decomposed relative concentration against the concentration based on the calibration set. The univariate regression in a multiway calibration is based on the linear model, which is very similar to the traditional univariate calibration (i.e., the standard curve, or one-way calibration in chemometric notation). But the signal used in the univariate regression in a multiway calibration is the decomposed relative concentration, not the same as the raw signal used in the traditional univariate calibration; therefore, some analysts prefer to call the univariate regression in a multiway calibration as a pseudo-univariate regression [64].

Let $y$ represents the decomposed relative concentration and $r$ represents the analyte concentration. Based on the linear model, there are two types of regression: (1) Taking the decomposed relative concentration as the dependent variable and the analyte concentration as the independent variable, i.e., $y=m_0+m_1r$, often known as classical univariate regression; (2) Taking the analyte concentration as the dependent variable and the decomposed relative concentration as the independent variable, i.e., $r=m_0+m_1y$, often known as inverse univariate regression.

### 2.4.1 Classical Univariate Regression

In classical univariate regression $y=m_0+m_1r$, the decomposed relative concentration $y$ of the $n$th analyte of interest is the dependent variable and the analyte concentration $r$ is the independent variable, as shown in Figure 8A. One can find the values of the slope $m_1$ and the vertical intercept $m_0$ by fitting the least squares line that minimizes the sum of the squares of the vertical distances from the points to the line [60,84]. Then the concentration of the $n$th analyte of interest in the $k$th sample can be predicted by $\bar{r}_k=(y_k-\bar{m}_0)/\bar{m}_1$.

This procedure assumes that the errors in the decomposed relative concentration $y$ are substantially greater than the errors in the analyte concentration $r$ [85] and uncertainties (standard deviations) in all the $y$ values are similar [60].

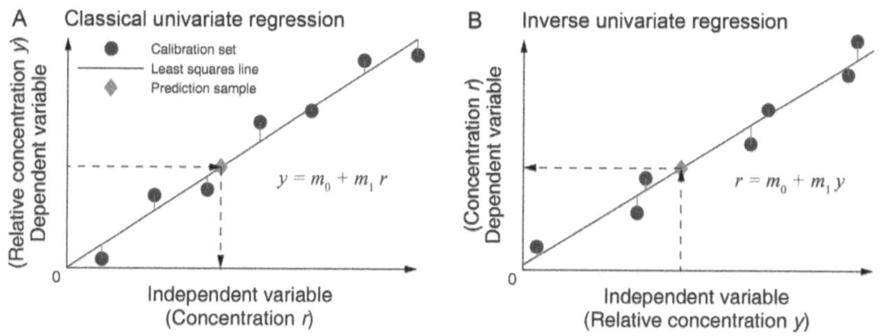


**FIGURE 8** Illustrations of classical univariate regression and inverse univariate regression in multiway calibration.

### 2.4.2 Inverse Univariate Regression

Several decades ago, many instruments were less reproducible than their modern counterparts, but the dramatic increase in instrumental performance has not been accompanied by an equally dramatic improvement in the quality of measurement burettes, pipettes, and balances. Nowadays, perhaps the greatest source of error is generally in sample preparation such as dilution, weighing, and extraction, rather than instrumental reproducibility. Therefore, some analysts now prefer to use the inverse univariate regression with the assumption that the errors in the analyte concentration $r$ are substantially greater than the errors in the decomposed relative concentration $y$.

In inverse univariate regression $r = m_0 + m_1 y$, the analyte concentration $r$ of the $n$th analyte of interest is the dependent variable and the decomposed relative concentration $y$ is the independent variable, shown in Figure 8B. One can find the values of the slope $m_1$ and the vertical intercept $m_0$ by fitting the least squares line that minimizes the sum of the squares of the vertical distances from the points to the line [60,84]. Then the concentration of the $n$th analyte of interest in the $k$th sample can be predicted by $\bar{r}_k = \bar{m}_0 + \bar{m}_1 y_k$.

For the simulated three-way EEMs data array, we also use the classical univariate regression for the regression in three-way calibration, since most chemists still use the classical univariate regression for quantitative analysis. The results are shown in Figure 9. For the analytes 1 and 2, the regression equations are $y = 46.7r - 0.1$ ($R^2 = 1.0000$) and $y = 55.4x + 0.3$ ($R^2 = 1.0000$), respectively. The values of $R^2$ are all above 0.9950, which deem a good linear fit for each analyte in its calibration concentration range.

For the simulated three-way EEMs data array, the quantitative results obtained by using three-way calibration method based on the ATLD algorithm are listed in Table 2. For the analytes 1 and 2, the root mean square errors of calibration (RMSECs) [2,3,64] are 0.9 and 1.7 pM, respectively, the average

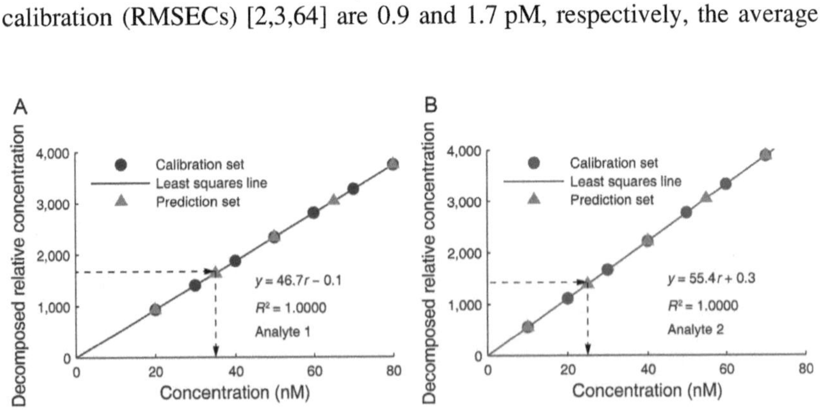


**FIGURE 9** Results of the regression in the three-way calibration method based on the ATLD algorithm with the number of components $N = 4$, for the simulated EEMs three-way data array.

**TABLE 2** Quantitative Results Obtained by the Three-Way Calibration Method Based on the ATLD Algorithm with the Number of Components $N=4$, for the Simulated EEMs Three-Way Data Array

| | | Actual concentration (nM) | | Predicted concentration (nM) [recovery (%)] | |
|---|---|---|---|---|---|
| No. | Sample | *Analyte 1* | *Analyte 2* | *Analyte 1* | *Analyte 2* |
| **Calibration set** | | | | | |
| 1 | cali01 | 20.0 | 50.0 | 20.0 [100] | 50.0 [100] |
| 2 | cali02 | 30.0 | 20.0 | 30.0 [100] | 20.0 [100] |
| 3 | cali03 | 40.0 | 70.0 | 40.0 [100] | 70.0 [100] |
| 4 | cali04 | 50.0 | 40.0 | 50.0 [100] | 40.0 [100] |
| 5 | cali05 | 60.0 | 10.0 | 60.0 [100] | 10.0 [100] |
| 6 | cali06 | 70.0 | 60.0 | 70.0 [100] | 60.0 [100] |
| 7 | cali07 | 80.0 | 30.0 | 80.0 [100] | 30.0 [100] |
| RMSEC (pM) | | | | 0.9 | 1.7 |
| Average recovery | | | | 100.0±0.0% | 100.0±0.0% |
| **Prediction set** | | | | | |
| 8 | pred01 | 20.0 | 10.0 | 20.0 [100] | 10.0 [100] |
| 9 | pred02 | 35.0 | 25.0 | 35.0 [100] | 25.0 [100] |
| 10 | pred03 | 50.0 | 40.0 | 50.0 [100] | 40.0 [100] |
| 11 | pred04 | 65.0 | 55.0 | 65.0 [100] | 55.0 [100] |
| 12 | pred05 | 80.0 | 70.0 | 80.0 [100] | 70.0 [100] |
| RMSEC (nM) | | | | 2.4 | 6.2 |
| Average recovery | | | | 100.0±0.0% | 100.0±0.0% |

recoveries (mean ± standard deviation) of the calibration set are 100.0±0.0% and 100.0±0.0%, respectively; the root mean square errors of prediction (RMSEPs) [2,3,64] are 2.4 and 6.2 pM, respectively, the average recoveries (mean ± standard deviation) of the prediction set are 100.0±0.0% and 100.0±0.0%, respectively. These satisfactory results show that, based on the EEMs trilinear model, the three-way calibration method based on the ATLD algorithm can successfully achieve the direct quantitative analysis of the two analytes of interest, even in the presence of an uncalibrated interferent and a varying background signal.

## 2.5 Figures of Merit

Figures of merit are analytical parameters used for evaluating the performance of a calibration method. Different approaches have been discussed in literatures for computing figures of merit for multiway calibration methods [86]. Here, we briefly describe one of approaches for computing the sensitivity, the limit of detection (LOD), and the limit of quantitation (LOQ), for three-way calibration and for four-way calibration.

In three-way calibration based on the trilinear model, the sensitivity (SEN) for the analyte $n$ can be computed by the following expression [86–88]:

$$\mathrm{SEN} = m_1 / \left\| n\text{th row of } \left( (\mathbf{I} - \mathbf{Z}_u \mathbf{Z}_u{}^{+}) \mathbf{Z}_c \right)^{+} \right\| \tag{3}$$

where the $(IJ) \times P$ $\mathbf{Z}_c = [\mathbf{b}_{c1} \odot \mathbf{a}_{c1} \ \cdots \ \mathbf{b}_{cP} \odot \mathbf{a}_{cP}]$, associated to the $P$ calibrated analytes, the $(IJ) \times [(I+J)Q]$ $\mathbf{Z}_u = [\mathbf{b}_{u1} \otimes \mathbf{I}_a \ \ \mathbf{I}_b \otimes \mathbf{a}_{u1} \ \cdots \ \mathbf{b}_{uQ} \otimes \mathbf{I}_a \ \ \mathbf{I}_b \otimes \mathbf{a}_{uQ}]$, containing information of the $Q$ unexpected interferents.

In four-way calibration based on the quadrilinear model, the SEN for the analyte $n$ is given by an expression [86] analogous to that of the three-way case:

$$\mathrm{SEN} = m_1 / \left\| n\text{th row of } \left( (\mathbf{I} - \mathbf{Z}_u \mathbf{Z}_u^{+}) \mathbf{Z}_c \right)^{+} \right\| \tag{4}$$

where the $(IJK) \times P$ $\mathbf{Z}_c = [\mathbf{c}_{c1} \odot \mathbf{b}_{c1} \odot \mathbf{a}_{c1} \ \cdots \ \mathbf{c}_{cP} \odot \mathbf{b}_{cP} \odot \mathbf{a}_{cP}]$, associated to the $P$ calibrated analytes, the $(IJK) \times [(I+J+K)Q]$ $\mathbf{Z}_u = [\mathbf{c}_{u1} \otimes \mathbf{b}_{u1} \otimes \mathbf{I}_a \ \ \mathbf{c}_{u1} \otimes \mathbf{I}_b \otimes \mathbf{a}_{u1} \ \ \mathbf{I}_c \otimes \mathbf{b}_{u1} \otimes \mathbf{a}_{u1} \ \cdots \ \mathbf{c}_{uQ} \otimes \mathbf{b}_{uQ} \otimes \mathbf{I}_a \ \ \mathbf{c}_{uQ} \otimes \mathbf{I}_b \otimes \mathbf{a}_{uQ} \ \ \mathbf{I}_c \otimes \mathbf{b}_{uQ} \otimes \mathbf{a}_{uQ}]$, containing information of the $Q$ unexpected interferents.

In a multiway calibration, the LOD and the LOQ can be estimated [86–88] as follows:

$$\mathrm{LOD} = 3.3 \sqrt{h_0 s_c^2 + h_0 \frac{s_x^2}{\mathrm{SEN}^2} + \frac{s_x^2}{\mathrm{SEN}^2}} \tag{5}$$

$$\mathrm{LOQ} = 10 \sqrt{h_0 s_c^2 + h_0 \frac{s_x^2}{\mathrm{SEN}^2} + \frac{s_x^2}{\mathrm{SEN}^2}} \tag{6}$$

where $h_0$ is the sample leverage, $s_c^2$ is the variance in calibration concentrations, $s_x^2$ is the variance in instrumental signals, and SEN is the analyte sensitivity.

For the simulated three-way EEMs data array, the figures of merit obtained by using three-way calibration method based on the ATLD algorithm are listed in Table 3. For the analytes 1 and 2, the sensitivities (SENs) are 24.3 and 25.4 $\mathrm{nM}^{-1}$, respectively, the LODs are 1.0 and 2.6 pM, respectively, and the LOQs are 2.9 and 7.8 pM, respectively. These results indicated the satisfactory sensitivity, LOD, and LOQ of the three-way calibration method based on the ATLD algorithm for the direct quantitative analysis of the two analytes of interest, even in the presence of an uncalibrated interferent and a varying background signal.

**TABLE 3** Figures of Merit Obtained by the Three-Way Calibration Method Based on the ATLD Algorithm with the Number of Components $N=4$, for the Simulated EEMs Three-Way Data Array

| Figures of merit | *Analyte* 1 | *Analyte* 2 |
|---|---|---|
| SEN (L $nmol^{-1}$) | 24.3 | 25.4 |
| LOD (pM) | 1.0 | 2.6 |
| LOQ (pM) | 2.9 | 7.8 |

## 3 APPLICATION FIELDS OF MULTIWAY CALIBRATION

Multiway calibration methods coupled with modern analytical instruments, such as EEMs and hyphenated chromatographic technology (HPLC-DAD, HPLC-MS), have been widely used for direct quantitative analysis of analytes of interest in complex systems [77,80,89–98]. Multiway calibration methods can fully extract and make the optimal use of relevant chemical information from analytical data. What is more, these methods greatly simplify sample preparation steps, saving time and labor. Multiway calibration methods have widespread applications in many fields such as biochemical, food, environmental, and pharmaceutical analysis.

In the analysis of biological samples, such as human plasma, urine and biological fluids, direct quantitative analysis of analytes of interest in complex systems is challenging due to the presence of unknown interferents. Numerous multiway calibration methods based on different instruments [79,99–105], such as EEM fluorospectroscopy [99,100], UV–vis spectroscopy [101], and $^{1}$H NMR [102], have been developed for biochemical analysis. Such applications of multiway calibration methods have been reviewed by Escandar and coworkers [90].

The increasing public interest in food quality in recent decades calls for higher standards in quality and process control of food production, which in turn requires more efficient analytical methods. Many multiway calibration methods have been developed for food analysis [106–109], and applications of multiway calibration methods combined with fluorescence spectroscopy in food analysis have been reviewed by Christensen and coworkers in 2006 [110].

Nowadays, there is a growing concern about environmental pollutants and their toxic effects on human health and wildlife. Various multiway calibration methods based on different instruments, such as GC×GC-MS, HPLC-DAD, and EEMs, have been developed for environmental analysis [59,78,111–115]. The applications of chemometric methods in environmental analysis have been reviewed by researchers from Spain and Argentina [92].

Pharmaceutical analysis focuses on the quality of natural or synthetic drugs. There are many related applications in pharmaceutical analysis by using multiway calibration methods [116–118] and several reviews on this aspect [91,119,120].

In recent years, many substandard cosmetic products have been exposed, which raises a concern about the toxicity of the contaminants in adulterated cosmetics. There are some related applications in cosmetic analysis by using three-way calibration methods [121–123].

For process analytical chemistry, multiway calibration methods also have excellent potential. For example, Liu and coworkers developed an approach based on online near-infrared spectroscopy and ATLD for industrial process analysis, and they concluded that ATLD was powerful to capture the intrinsic information hidden in the NIR spectra measured in a process and therefore very useful for industrial process analysis and quality insurance [124]. Malwade and coworkers demonstrated application of multiway calibration in chemometrics for extracting process information from analytical data of process streams, the multiway calibration method based on the trilinear model was used to extract chemical information such as number of impurities, their relative concentrations, and finally their identification from rather complex analytical chromatograms of flash column chromatography fractions during purification of Artemisinin from the crude extract of *Artemisia annua* [125].

## 4 PRACTICAL EXAMPLES OF THREE-WAY CALIBRATION

Examples of second-order instruments include HPLC-DAD, capillary electrophoresis-diode array detection and EEM fluorescence spectrophotometer. Other modern hyphenated instruments such as gas chromatography–mass spectrometry (GC-MS), liquid chromatography–mass spectrometry (LC-MS), capillary electrophoresis–mass spectrometry (CE-MS), and liquid chromatography-nuclear magnetic resonance spectrometer (LC-NMR) also belong to this second-order category of instruments.

In the following section, we illustrate three-way calibration methods based on the trilinear model by using some practical analytical examples.

### 4.1 Three-Way Calibration of EEM Fluorescence Type Data

#### *4.1.1 Case Study 1: Quantitative Analysis of Three Anticancer Drugs in Human Plasma*

There are many practical analytical applications of three-way calibration methods coupled with EEM fluorescence spectroscopy, based on the trilinear model [32,56,72,77,81,91,94,99,110,126–131], with the development of EEM fluorescence spectroscopy [56,132]. Here, we present a practical example of the quantitative analysis of three anticancer drugs in human plasma samples, by using the three-way calibration method based on the ATLD algorithm, coupled with EEM fluorescence spectroscopy [129].

Resveratrol, curcumin, and adriamycin are effective for curing many kinds of cancers, and combination of them can contribute to the prevention of cancer cells more effectively. Therefore, simultaneous quantification of resveratrol, curcumin, and adriamycin in human plasma is very important in clinical analysis. Up to now, routine quantitative analysis of resveratrol, curcumin, and adriamycin has been frequently performed by HPLC coupled with different detectors. Although tedious pretreatments have been performed before the HPLC analysis, chromatographic separations with acceptable resolution could not always be achieved, owing to either structural similarity among the analytes and coeluting interferents in systems studied, or the limited time range of single chromatographic run.

As resveratrol, curcumin, and adriamycin have strong intrinsic fluorescence, fluorescence spectroscopy seems to be a promising alternative to chromatographic methods for the quantitative analysis of them. Fluorescence measurement can provide high sensitivity [60,133], but fluorescence measurement cannot provide high selectivity. For the quantitative analysis of resveratrol, curcumin, and adriamycin in human plasma by classical fluorescence spectroscopy, using fluorescence signal at the maximum emission wavelength of one analyte, one can only determine this analyte at one time after separating it from the other two analytes and the plasma matrix, since the classical one-way (zero-order) calibration method [3,134] requires the analytical signal used must be fully selective for the analyte of interest. With two-way (first-order) multivariate calibration methods [2,7,9,35,38,134,135], for example, based on fluorescence emission spectra, the three analytes can be determined simultaneously. It makes sense to treat multivariate measurements simultaneously in data analysis, by which we could obtain more information in the form of correlation [2,5]. However, the spectrum for the analyte of interest must be partially different from the spectra of all other responding species. Additionally, calibration standards must be representative of the samples containing any interfering species in the system studied. With the rapid development of fluorescence instrumentation such as EEM fluorescence spectrophotometer [56,132], this prerequisite can be circumvented through three-way (second-order) calibration methods based on the trilinear model [32,56,72,77,81,91,94,99,110,126,128,129,131]. This is due to the second-order advantage [22,23,35,37,38,64,81], which means analytes can be quantitatively analyzed even in the presence of uncalibrated interferents.

#### 4.1.1.1 Experimental Design

In this example, there are three analytes of interest. They are resveratrol, curcumin, and adriamycin. The chemical structures of these analytes of interest are given in Figure 10. The calibration set consists of 10 samples, in which 10 concentration levels of the three analytes of interest are designed by a $U_{10}$ ($10^3$) uniform design [61]; since the uniform design could provide an appropriate experimental design for this 10-level, three-factor concentration

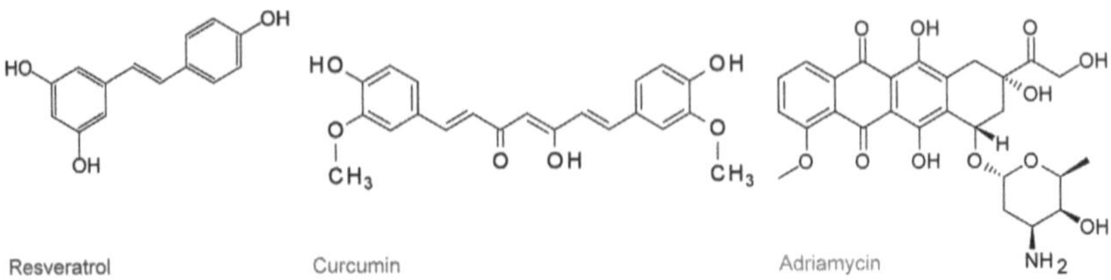


**FIGURE 10** Chemical structures of resveratrol, curcumin, and adriamycin.

design with only 10 experiments. The calibration samples are prepared in a simple solution and only contain the three analytes of interest. The prediction samples are prepared by spiking appropriate contents of all the analytes of interest with the appropriate volume of human plasma. The corresponding concentrations of the three anticancer drugs in the calibration and prediction samples are summarized in Table 4.

#### 4.1.1.2 EX-EM-Sample Trilinear Model

When a two-way EEM data array of a sample is measured at $I$ excitation wavelengths and $J$ emission wavelengths, the collection of EEMs of the calibration and prediction samples will form a three-way data array $\boldsymbol{X}$ dimensioned $I$ excitation wavelengths by $J$ emission wavelengths by $K$ samples, in which each element $x_{ijk}$ can be expressed as:

$$x_{ijk} = \sum_{n=1}^{N} a_{in} b_{jn} c_{kn} + e_{ijk}, \quad \text{for } i=1,2,\ldots,I;\ j=1,2,\ldots,J;\ k=1,2,\ldots,K. \quad (7)$$

where $x_{ijk}$ is the measured fluorescence intensity of the $k$th sample at the $i$th excitation wavelength and the $j$th emission wavelength; $N$ represents the number of components in the trilinear model. $a_{in}$, $b_{jn}$, and $c_{kn}$ correspond to the normalized excitation spectrum matrix $\mathbf{A}_{I\times N}$, the normalized emission spectrum matrix $\mathbf{B}_{J\times N}$,and the relative concentration matrix $\mathbf{C}_{K\times N}$ of the three-way EEMs data array $\boldsymbol{X}$ $(I\times J\times K)$, respectively. Specifically, $a_{in}$ and $b_{jn}$ represent the coefficients related to the spectral intensity of the $n$th component at the $i$th excitation wavelength and the $j$th emission wavelength; and $c_{kn}$ represents the relative concentration of the $n$th component in the $k$th sample. The EEMs trilinear model is illustrated in Figure 11. In this illustration, the three-way array $\boldsymbol{X}$ $(I\times J\times K)$ represents the measured three-way EX-EM-Sample data array; the profile matrix $\mathbf{A}$ is the normalized excitation spectrum matrix, the profile matrix $\mathbf{B}$ is the normalized emission spectrum matrix, and the profile matrix $\mathbf{C}$ is the relative concentration matrix; the three-way array $\boldsymbol{I}$ $(N\times N\times N)$ represents the core array in which only the main superdiagonal elements are nonzero elements.

In this example, the two-way EEM fluorescence data arrays are recorded at emission wavelengths between 320 and 617 nm (every 3 nm), at different

**TABLE 4** Concentration Design of Resveratrol, Curcumin, and Adriamycin

| No. | Sample | Added concentration ($\mu g\ mL^{-1}$) *Resveratrol* | *Curcumin* | *Adriamycin* | Human plasma ($\mu L\ mL^{-1}$) |
|---|---|---|---|---|---|
| **Calibration set** | | | | | |
| 1 | cali01 | 0.672 | 0.043 | 1.820 | |
| 2 | cali02 | 1.343 | 0.064 | 0.780 | |
| 3 | cali03 | 0.537 | 0.085 | 2.600 | |
| 4 | cali04 | 1.209 | 0.107 | 1.560 | |
| 5 | cali05 | 0.403 | 0.128 | 0.520 | |
| 6 | cali06 | 1.075 | 0.149 | 2.340 | |
| 7 | cali07 | 0.269 | 0.171 | 1.300 | |
| 8 | cali08 | 0.940 | 0.192 | 0.260 | |
| 9 | cali09 | 0.134 | 0.213 | 2.080 | |
| 10 | cali10 | 0.806 | 0.235 | 1.040 | |
| **Prediction set** | | | | | |
| 11 | pred01 | 0.672 | 0.043 | 1.820 | 10.0 |
| 12 | pred02 | 1.343 | 0.064 | 0.780 | 10.0 |
| 13 | pred03 | 0.537 | 0.085 | 2.600 | 10.0 |
| 14 | pred04 | 1.209 | 0.107 | 1.560 | 10.0 |
| 15 | pred05 | 0.403 | 0.128 | 0.520 | 10.0 |
| 16 | pred06 | 1.075 | 0.149 | 2.340 | 10.0 |
| 17 | pred07 | 0.269 | 0.171 | 1.300 | 10.0 |
| 18 | pred08 | 0.940 | 0.192 | 0.260 | 10.0 |
| 19 | pred09 | 0.134 | 0.213 | 2.080 | 10.0 |
| 20 | pred10 | 0.806 | 0.235 | 1.040 | 10.0 |
| **Human plasma samples** | | | | | |
| 21 | plasma01 | 0 | 0 | 0 | 10.0 |
| 22 | plasma02 | 0 | 0 | 0 | 10.0 |
| 23 | plasma03 | 0 | 0 | 0 | 10.0 |
| **Method blank samples** | | | | | |
| 24 | mblank01 | 0 | 0 | 0 | 0 |
| 25 | mblank02 | 0 | 0 | 0 | 0 |
| 26 | mblank03 | 0 | 0 | 0 | 0 |

Reproduced from Ref. [129], Table 1, p. 1229.

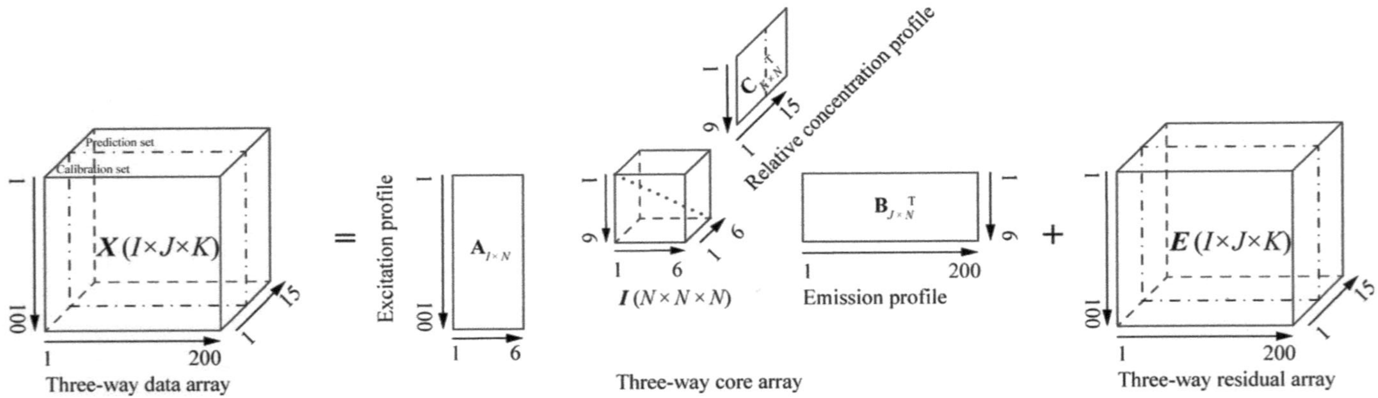


**FIGURE 11** Illustration of the EEMs trilinear model in three-way calibration, with the underlying number of components $N=6$.

excitation wavelengths between 250 and 547 nm (every 3 nm). For each sample, a two-way $100 \times 100$ EEM fluorescence data array is scanned. By arranging two-way 26 EEM fluorescence data arrays of 10 calibration samples, 10 spiked prediction samples, 3 human plasma samples, and 3 method blank samples, one can obtain a three-way $100 \times 100 \times 26$ EX-EM-Sample fluorescence data array. Here, we have added several additional samples such as the human plasma samples, which is some different from the data array analyzed in the original paper [129].

In Section 4.1.1.3, this data array will be analyzed by using three-way calibration method based on the ATLD algorithm.

#### 4.1.1.3 Trilinear Decomposition (with Correctly Estimated Number of Components)

The entire data analysis procedure can be summarized as four steps: first, non-trilinear factors in the data array are processed; second, the proper number of components is estimated; third, the ATLD algorithm is employed to decompose the trilinear model; Finally, the decomposed relative concentration for each analyte is regressed against the corresponding real concentration based on the calibration set, and thereafter the quantitative information of each analyte can be obtained. The ATLD algorithm developed by Wu *et al.* [26] is one of the most commonly used trilinear decomposition algorithms. The algorithm uses the Moore–Penrose generalized inverses strategy, which is based on the singular value decomposition, and thus has the advantages of converging fast, being insensitive to initialization and being insensitive to the overestimated number of components.

In all samples, severe Rayleigh scattering and Raman scattering are present. These non-trilinear factors can lead the EEMs array to deviate the trilinear model. Elimination of such interferences is a prerequisite for trilinear decomposition algorithms to decompose profile matrices correctly. Therefore, the elements of these non-trilinear scattering areas in the data array are simply set to be zero [129,136], prior to further analysis.

After the preprocessing of the data array, let us have a visual inspection of this system. The three-dimensional landscapes of EEM fluorescence for the calibration sample cali05, the human plasma sample plasma01, and the method blank sample mblank01 are given in Figure 12. From the landscape of cali05, one can see that the spectra of curcumin and adriamycin overlap seriously. From the landscape of plasma01, one can see that the spectra of the human plasma will seriously overlap with that of resveratrol. Obviously, there will be four spectral components at least in the spiked human plasma prediction samples. From the landscape of mblank01, one can see that there is a spectral background in the method blank samples, which may present a challenging interference to the trilinear decomposition of the three-way calibration method based on the ATLD algorithm; one can also see slight Raman

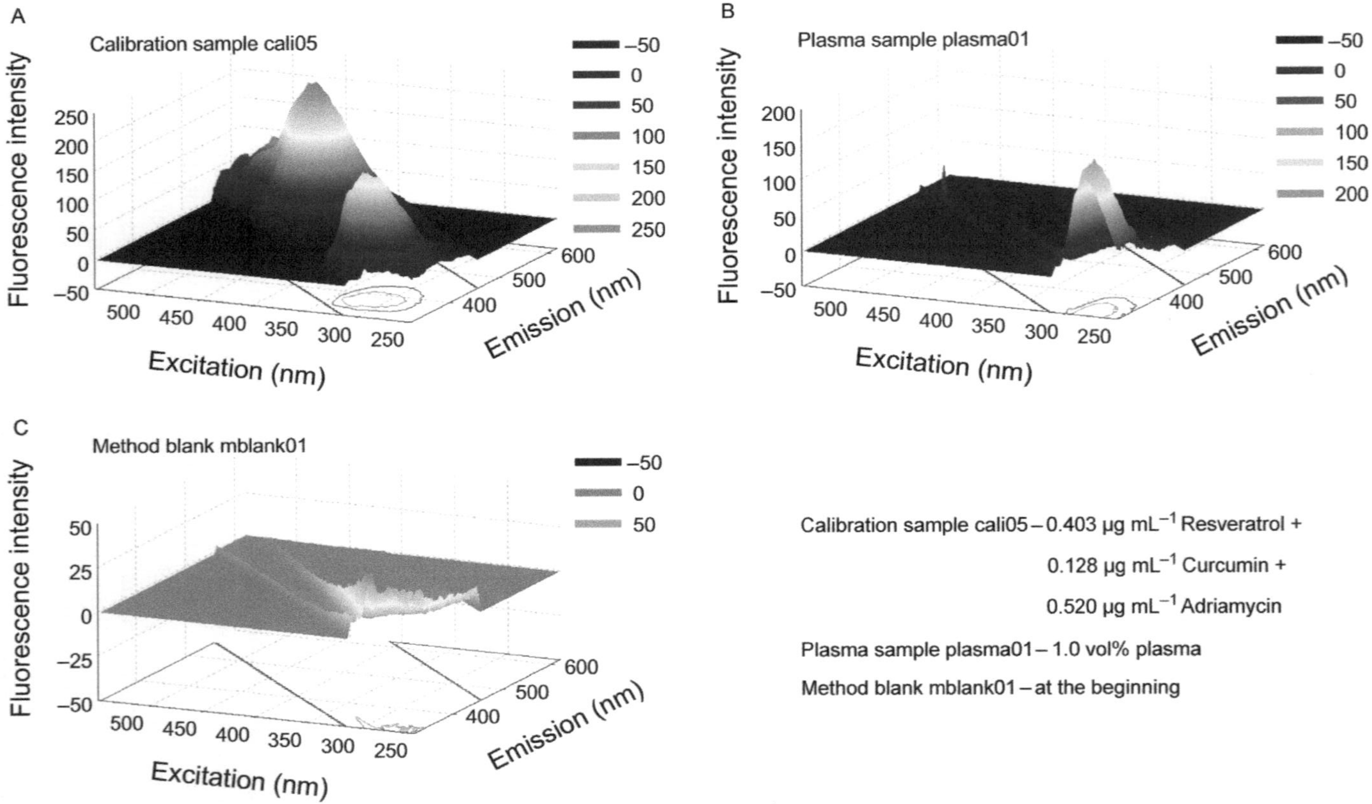


**FIGURE 12** Three-dimensional landscapes of EEM fluorescence for the calibration sample cali05, the human plasma sample plasma01, and the method blank sample mblank01.

scattering in this landscape, which is a weak fluorescence signal relative to the former five spectral components.

The number of components for the trilinear model should be estimated before the application of trilinear decomposition algorithms [16,17]. The estimated number of components in the original paper is $N=4$ [129] for a three-way EEMs data array, in which the method blank is subtracted. However, here we will directly decompose the three-way EEMs data array by using ATLD algorithm. In the three-way calibration method based on the trilinear model, the background can be modeled as additional components by using ATLD algorithm, which will be discussed later. For this three-way $100 \times 100 \times 26$ EX-EM-Sample fluorescence data array without subtracting the method blank, the appropriate number of components is estimated to be $N=6$, by the uses of add-one-up method [129,137] and visual inspection of the decomposed profiles.

Then, the ATLD algorithm with the number of components $N=6$ is used to decompose this trilinear model. With $N=6$ for this EEMs data array, the ATLD algorithm accurately decomposed the normalized excitation spectrum matrix $\mathbf{A}_{100\times6}$, the normalized emission spectrum matrix $\mathbf{B}_{100\times6}$, and the relative concentration matrix $\mathbf{C}_{26\times6}$. One of the main advantages of the ATLD algorithm is the fast convergence rate; the computation time is only 0.750 s and the number of iterations is only 61. Figure 13 shows the actual relative profiles (dash-dot lines) and the decomposed relative profiles (solid lines) in the three modes using the ATLD algorithm with $N=6$. From the decomposed profiles in the three modes, one can see that this is indeed a high collinear and complex system. The decomposed excitation and emission profiles of the three analytes of interest (blue (black in the print version), green (dark gray in the print version), and red (gray in the print version) solid lines) are in good agreement with the actual profiles based on pure analyte standards. The correlation coefficients between the decomposed and actual spectra are 0.9991 and 0.9990 in the excitation and emission modes, respectively, for resveratrol, 0.9985 and 0.9996 in the excitation and emission modes, respectively, for curcumin, and 0.9961 and 0.9955 in the excitation and emission modes, respectively, for adriamycin.

Interestingly, an uncalibrated interferent (pink (gray in the print version) solid line) which comes from human plasma is resolved by the fourth component, whose profiles in the excitation and emission modes seriously overlap with that of resveratrol. This indicates that the presence of uncalibrated interferents does not affect the trilinear decomposition of the three-way calibration method based on the ATLD algorithm, which is extremely important for direct quantitative chemical analysis, since traditional one-and two-way calibration methods cannot provide accurate prediction in the presence of uncalibrated interferents [2,3,23,26,56,60,63,64].

In addition, a varying background interferent (orange solid line (light gray in the print version)) is modeled by the fifth component. The excitation and

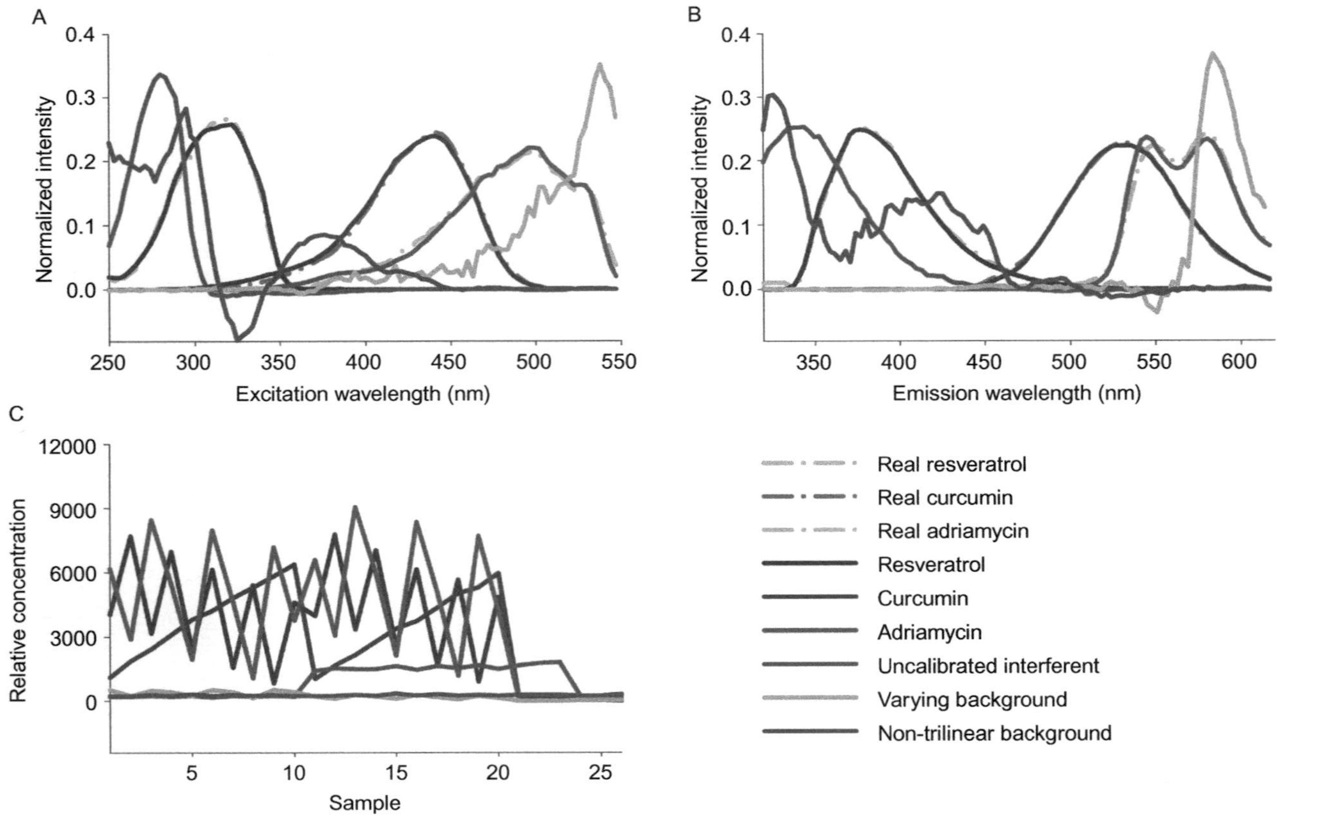


**FIGURE 13** Results of the decomposition in the three-way calibration method based on the ATLD algorithm with the number of components $N=6$, for the EEMs three-way data array.

emission profiles of this component are consistent with that of the landscape of mblank01 presented in Figure 12C, and the relative concentration profile of this component indicates that it is a varying background interferent. This is very interesting, since that varying background interferent (or called baseline drift) which sometimes occurs in spectral and chromatographic measurements is very hard to be handled by traditional methods [2,3,23,56,59,60,64]. This phenomenon indicates that the ATLD algorithm is a "soft" algorithm that can model weak spectral interferents (such as a weak varying background interferent) in mixtures.

Also, a slight non-trilinear background interferent (bluish-purple (dark gray in the print version) solid line) is resolved by the sixth component. One can see that some regions of its excitation profile are negative, which reminds us again that nonnegativity constraints should be used very carefully. Generally, non-trilinear factors cannot be modeled by the trilinear decomposition. However, this non-trilinear background is relatively weak, so that it could be modeled by using ATLD algorithm in practice. This phenomenon indicates that the ATLD algorithm is somewhat robust to non-trilinear factors. This feature of the ATLD algorithm benefits from its intelligence optimization strategy and its property of being insensitive to the overestimated number of components [26,37,63,64,138].

#### 4.1.1.4 Trilinear Decomposition (With Overestimated Number of Components)

Besides the advantages of converging fast and being insensitive to initialization, the ATLD algorithm has a very promising property of being insensitive to the excess number of components. Considering the fact that estimating the number of components $N$ is not always straightforward, especially in the presence of noise or complex matrix, this advantage of the ATLD algorithm is interesting and important.

With the overestimated number of components $N=6+1$ for the three-way EEMs data array studied, the ATLD algorithm accurately decomposed the normalized excitation spectrum matrix $\mathbf{A}_{100\times7}$, the normalized emission spectrum matrix $\mathbf{B}_{100\times7}$, and the relative concentration matrix $\mathbf{C}_{26\times7}$. The computation time is only 0.868 s and the number of iterations is only 66. Figure 14 shows the actual relative profiles (dash-dot lines) and the decomposed relative profiles (solid lines) in the three modes using the ATLD algorithm with the overestimated number of components $N=6+1$. The decomposed excitation and emission profiles of the three analytes of interest (blue (black in the print version), green (dark gray in the print version), and red (gray in the print version) solid lines), are in good agreement with the actual profiles based on pure analyte standards. The correlation coefficients between the decomposed and actual spectra are 0.9994 and 0.9992 in the excitation and emission modes, respectively, for resveratrol, 0.9985 and 0.9996 in the excitation and emission modes, respectively, for curcumin, and 0.9961 and 0.9956 in the excitation

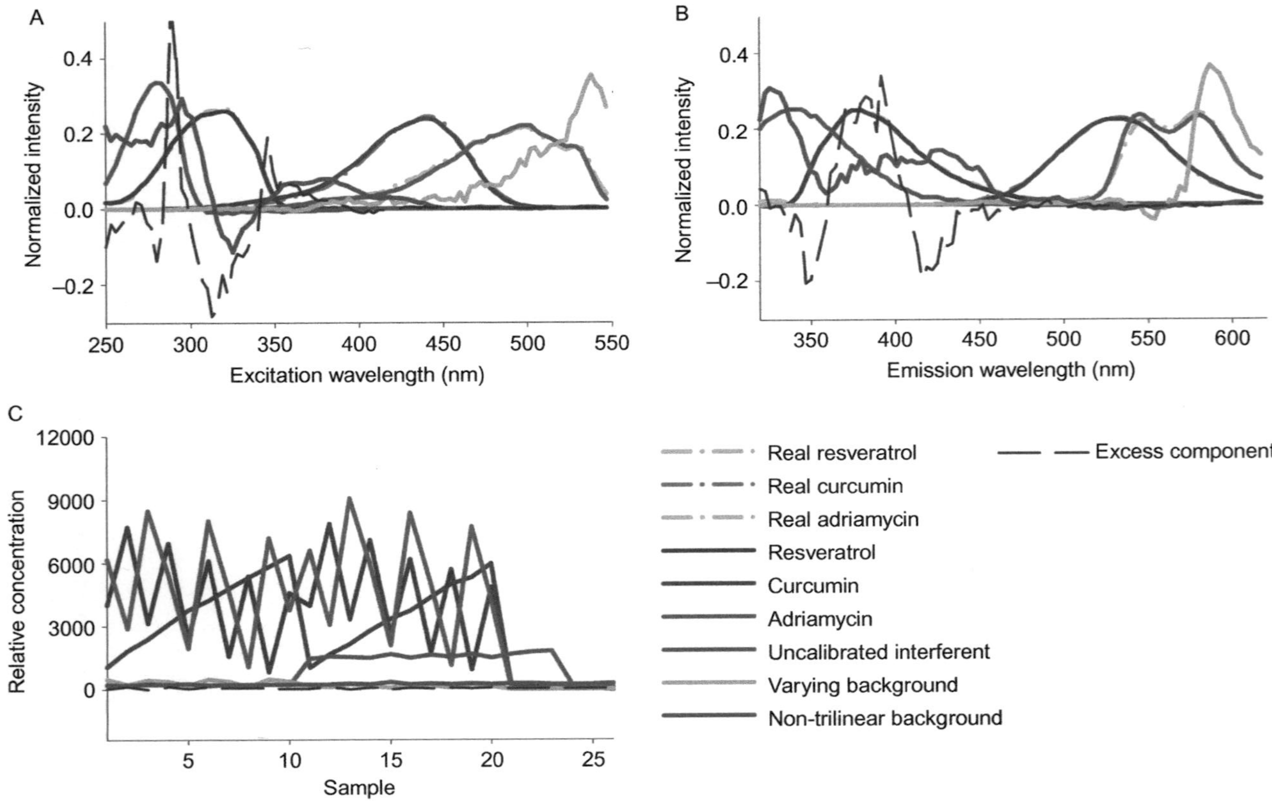


**FIGURE 14** Results of the decomposition in the three-way calibration method based on the ATLD algorithm with the overestimated number of components $N=6+1$, for the EEMs three-way data array.

and emission modes, respectively, for adriamycin. The uncalibrated interferent modeled by the fifth component (pink (light gray in the print version) solid line) and the weak non-trilinear background interferent modeled by the sixth component (bluish-purple (dark gray in the print version) solid line) are consistent with that decomposed with the appropriate number of components $N=6$. Very interestingly, the excess component is modeled by the seventh component (dark red (balck in the print version) long dash line), whose excitation and emission profiles appear in the shapes of a noise and can be easily recognized.

Even with the overestimated number of components $N=6+10$ for this three-way EEMs data array, the ATLD algorithm have still accurately decomposed the normalized excitation spectrum matrix $\mathbf{A}_{100\times16}$, the normalized emission spectrum matrix $\mathbf{B}_{100\times16}$, and the relative concentration matrix $\mathbf{C}_{26\times16}$. The computation time is only 2.857 s and the number of iterations is only 182. Figure 15 shows the actual relative profiles (dash-dot lines) and the decomposed relative profiles (solid lines) in the three modes using the ATLD algorithm even with the overestimated number of components $N=6+10$. The decomposed excitation and emission profiles of the three analytes of interest (blue (balck in the print version), green (dark gray in the print version), and red (gray in the print version) solid lines), are in good agreement with the actual profiles based on pure analyte standards. The correlation coefficients between the decomposed and actual spectra are 0.9994 and 0.9992 in the excitation and emission modes, respectively, for resveratrol, 0.9983 and 0.9997 in the excitation and emission modes, respectively, for curcumin, and 0.9936 and 0.9944 in the excitation and emission modes, respectively, for adriamycin. The uncalibrated interferent modeled by the fifth component (pink (light gray in the print version) solid line) and the slight non-trilinear background interferent modeled by the sixth component (bluish-purple (dark gray in the print version) solid line) are also consistent with that decomposed with the appropriate number of components $N=6$. Very interestingly, the excess components are modeled by the later 10 components (long dash lines), whose excitation and emission profiles appear in the shapes of noises and can be easily recognized.

The above two figures show the excess components do not affect the qualitative decomposition result of the ATLD algorithm. By comparing the convergence rates (the computation time and the convergence rate) of the ATLD algorithm with respect to different $N$, one can find that both computation time and the iteration number of the ATLD algorithm just increase slightly, with respective to $N$ equals to 6 up to 16. The excess components almost do not affect the convergence rate of the ATLD algorithm.

In the following section, we will compare the quantitative prediction results of the three-way calibration method based on the ATLD algorithm with respect to different $N$. Figure 16 gives the quantitative prediction results of the three-way calibration method based on the ATLD algorithm, with $N$

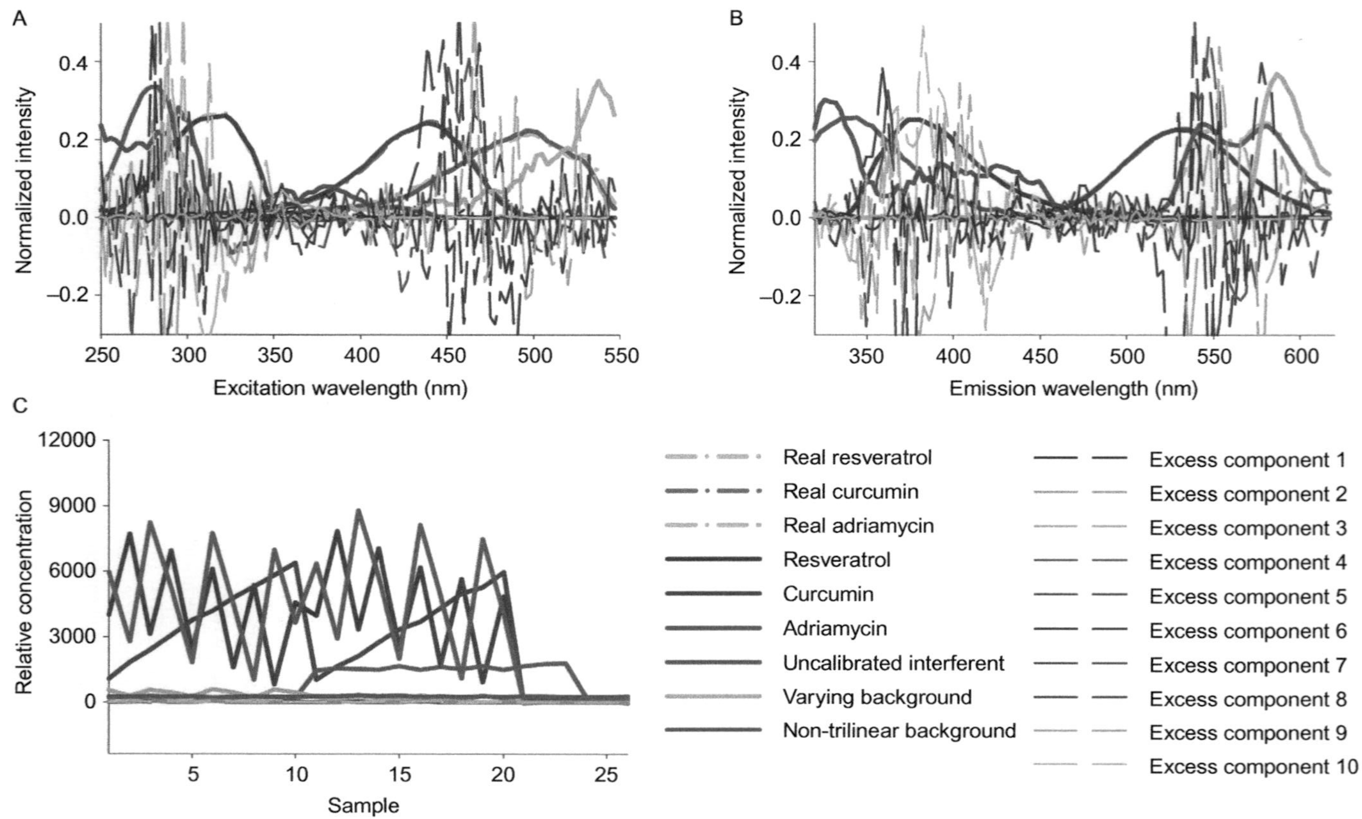


**FIGURE 15** Results of the decomposition in the three-way calibration method based on the ATLD algorithm even with the overestimated number of components $N=6+10$, for the EEMs three-way data array.

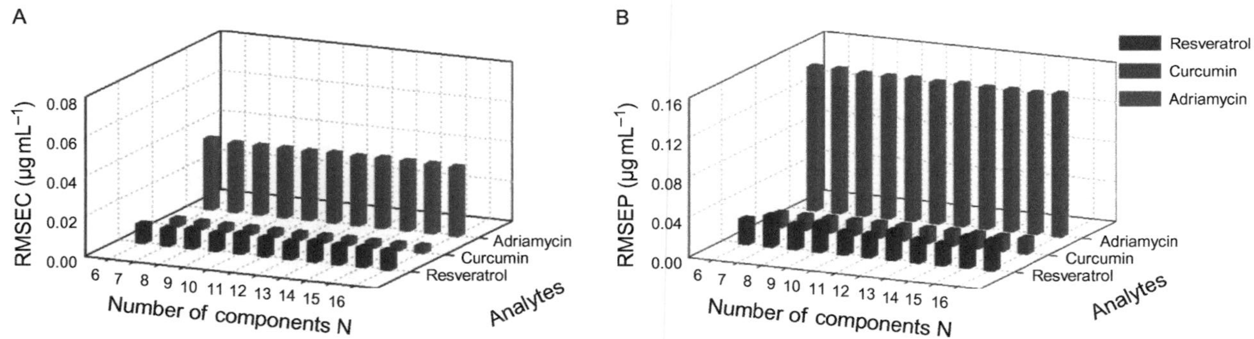


**FIGURE 16** Quantitative prediction results of the three-way calibration method based on the ATLD algorithm, with *N* from 6 up to 16.

equal to 6 up to 16. One can see that, for each analyte, the RMSECs and RMSEPs of the three-way calibration method based on the ATLD algorithm are always satisfactory, with $N$ equal to 6 up to 16. The excess components do not affect the quantitative prediction result of the three-way calibration method based on the ATLD algorithm.

These satisfactory results again indicate that the ATLD algorithm has the advantage of being insensitive to the overestimated number of components. Actually, some variants of ATLD [31,59,70,76] are also insensitive to the overestimated number of components. As long as the selected number of components is equal to or greater than the underlying real number of components, these algorithms can provide correct decomposed profiles, and the excess components always appear in the shapes of noises which are easily recognized.

#### 4.1.1.5 Varying Background (Baseline Drift)

From the landscape of mblank01 given in a former figure, one can see that there are spectral background fluorescence and the weak Raman scattering peak. These interferents, especially the Raman scattering peak, can often be reduced to some extent by subtracting the method blank sample. Here, we first subtract the method blank sample, then use the ATLD algorithm with the number of components $N=5$ to decompose the method blank subtracted trilinear model.

With $N=5$ for this method blank subtracted EEMs data array, the ATLD algorithm accurately decomposed the normalized excitation spectrum matrix $\mathbf{A}_{100\times5}$, the normalized emission spectrum matrix $\mathbf{B}_{100\times5}$, and the relative concentration matrix $\mathbf{C}_{26\times5}$. The computation time is only 0.13 s and the number of iterations is only 8. Figure 17 shows the actual relative profiles (dash-dot lines) and the decomposed relative profiles (solid lines) in the three modes using the ATLD algorithm with $N=5$. The decomposed excitation and emission profiles of the three analytes of interest (blue (black in the print version), green (dark gray in the print version), and red (gray in the print version)solid lines), are in good agreement with the actual profiles based on pure analyte standards. The correlation coefficients between the decomposed and actual spectra are 0.9991 and 0.9990 in the excitation and emission modes, respectively, for resveratrol, 0.9985 and 0.9997 in the excitation and emission modes, respectively, for curcumin, and 0.9960 and 0.9958 in the excitation and emission modes, respectively, for adriamycin. Also an uncalibrated interferent (pink (gray in the print version) solid line) which comes from human plasma is resolved by the fourth component.

In addition, a weak varying background interferent (orange (light gray in the print version) solid line) is still modeled and resolved by the fifth component. The excitation and emission profiles of this component are consistent with that of the fifth component (orange (light gray in the print version) solid line) presented in Figure 13. Although we have subtracted the method blank

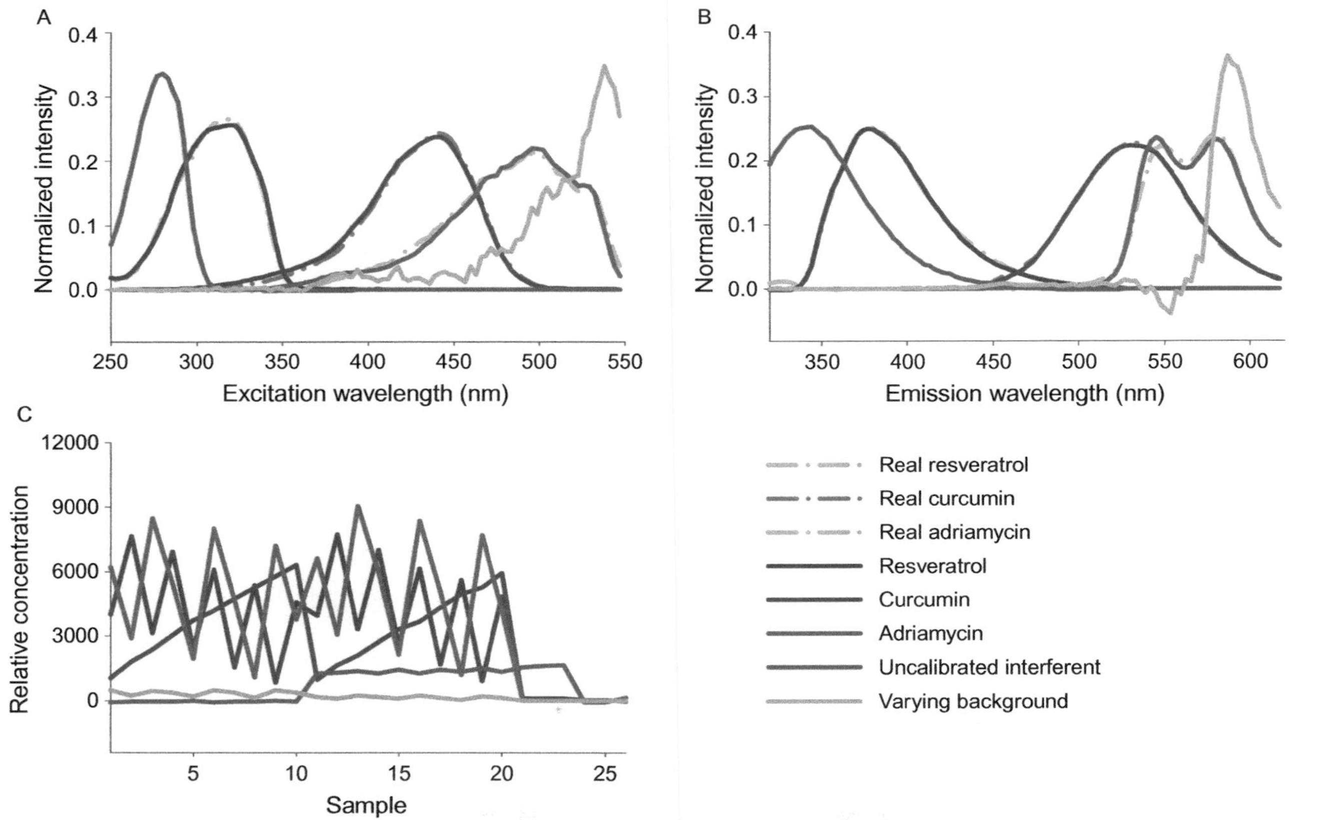


**FIGURE 17** Results of the decomposition in the three-way calibration method based on the ATLD algorithm with the number of components $N=5$, for the EEMs three-way data array with the method blank subtracted.

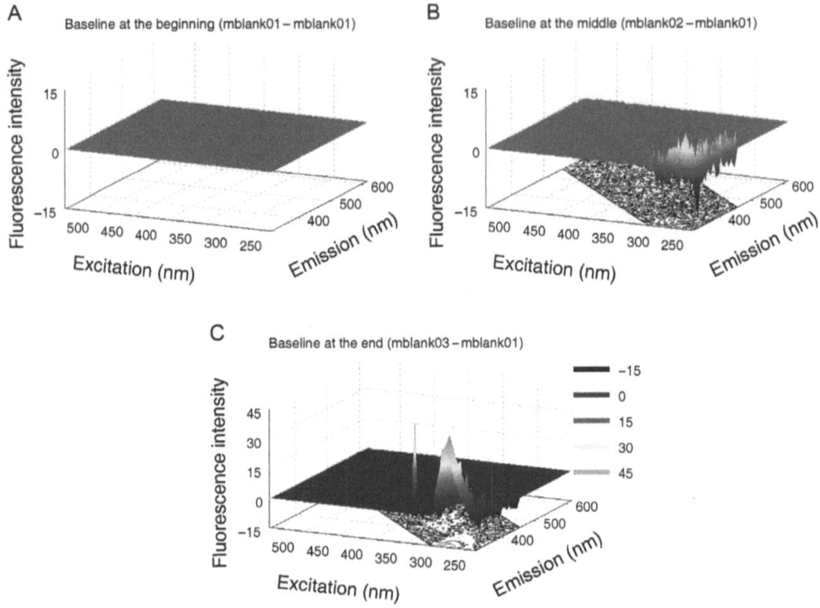


**FIGURE 18** Three-dimensional landscapes of EEM fluorescence for the method blank samples mblank01, mblank02, and mblank03, each minuses mblank01.

for this data array, the ATLD algorithm with the number of components $N=5$ have still resolved a weak varying background interferent. This should not be surprising, since a varying background interferent (or called baseline drift) may be reduced but cannot be completely removed by subtracting the corresponding method blank [2,3,23,56,59,60,64]. The experiment can validate this result. Figure 18 shows the three-dimensional landscapes of EEM fluorescence for the method blank samples mblank01, mblank02, and mblank03, each minus mblank01. One can see that, by taking mblank01 measured at the beginning of the experiment as the "baseline," the "baseline" measured at the middle of the experiment varies and is different from the one at the beginning, and the "baseline" measured at the end of the experiment further varies and deviates from the one at the beginning. In the three-way calibration based on the trilinear model, the varying background has been modeled by the ATLD algorithm with the additional fifth component (orange (light gray in the print version) solid line). Actually, the slight non-trilinear background interferent observed in the former trilinear decompositions (bluish-purple (black in the print version) solid line) is reduced by subtracting the method blank sample.

#### 4.1.1.6 Univariate Regression

After the decomposition in the three-way calibration method based on the ATLD algorithm with the appropriate number of components $N=6$, for this

EEMs three-way data array, the pure relative concentration of each analyte in each sample has been extracted. By the classical univariate regression in the three-way calibration method for each analyte, the concentration of each analyte in each sample can be predicted. For resveratrol, curcumin, and adriamycin, the regression equations are $y=5652.9r+93.3$ ($R^2=0.9993$), $y=26978.2r+131.0$ ($R^2=0.9963$), and $y=3225.8r+335.1$ ($R^2=0.9977$), respectively. The values of the squared correlation coefficient $R^2$ are all above 0.998, each $R^2$ deems a good linear fit for each analyte in its calibration range. Predicted concentrations for the calibration set and spiked prediction set are given in Table 5. For resveratrol, curcumin, and adriamycin, the average spike recoveries (mean $\pm$ standard deviation) of the spiked prediction set are $103.3 \pm 2.4\%$, $89.7 \pm 4.4\%$, and $107.6 \pm 4.8\%$, respectively.

#### 4.1.1.7 Figures of Merit

The SENs, LODs, and LOQs of the method are listed in Table 6. These satisfactory results indicate that the proposed analytical strategy can be used for simultaneously quantitative analysis of three anticancer drugs resveratrol, curcumin, and adriamycin in human plasma.

These satisfactory qualitative and quantitative results indicate that the three-way calibration method based on the ATLD algorithm, coupled with EEM fluorescence spectroscopy, can provide accurate and precise resolutions and predictions, with the "mathematical separation" replacing chemical or physical separation, for the simultaneous determination of the three anticancer drugs resveratrol, curcumin, and adriamycin in human plasma, even in the presence of an uncalibrated interferent, a slight non-trilinear background interferent, and a varying background interferent.

As has been shown, the three-way calibration method based on the ATLD algorithm, coupled with the EEM fluorescence spectroscopy, is a very fast and efficient method for the quantitative analysis of multiple anticancer drugs in complex systems containing uncalibrated spectral interferents and varying background interferents.

## 4.2 Three-Way Calibration of HPLC-DAD Type Data

### *4.2.1 Case Study 2: Quantitative Analysis of 11 Antihypertensives in Complex Systems*

There are already many practical applications of three-way calibration methods based on ATLDs, coupled with HPLC-DAD [26,80,108,112,114,117,139–146], since the first original application in 1998 [26]. Here, we present a practical example of the quantitative analysis of eleven antihypertensives in human plasma, health product and Chinese patent medicine samples, by using three-way calibration method based on the ATLD algorithm, coupled with HPLC-DAD [141].

**TABLE 5** Results of the Quantitative Analysis of the Three Anticancer Drugs in the Spiked Human Plasma Prediction Samples, by the Three-Way Calibration Method Based on the ATLD Algorithm

| | Spike concentration (μg mL$^{-1}$) | | | Predicted concentration (μg mL$^{-1}$) [spike recovery (%)] | | |
|---|---|---|---|---|---|---|
| Sample | *Resveratrol* | *Curcumin* | *Adriamycin* | *Resveratrol* | *Curcumin* | *Adriamycin* |
| **Calibration set** | | | | | | |
| cali01 | 0.672 | 0.043 | 1.820 | 0.696 [103.7] | 0.036 [83.7] | 1.806 [99.2] |
| cali02 | 1.343 | 0.064 | 0.780 | 1.344 [100.0] | 0.064 [100.2] | 0.789 [101.1] |
| cali03 | 0.537 | 0.085 | 2.600 | 0.543 [101.0] | 0.085 [99.6] | 2.521 [97.0] |
| cali04 | 1.209 | 0.107 | 1.560 | 1.212 [100.2] | 0.110 [102.8] | 1.580 [101.3] |
| cali05 | 0.403 | 0.128 | 0.520 | 0.409 [101.4] | 0.135 [105.5] | 0.498 [95.8] |
| cali06 | 1.075 | 0.149 | 2.340 | 1.067 [99.2] | 0.152 [101.5] | 2.368 [101.2] |
| cali07 | 0.269 | 0.171 | 1.300 | 0.257 [95.8] | 0.173 [101.2] | 1.335 [102.7] |
| cali08 | 0.940 | 0.192 | 0.260 | 0.936 [99.5] | 0.192 [99.8] | 0.225 [86.7] |
| cali09 | 0.134 | 0.213 | 2.080 | 0.131 [97.3] | 0.211 [98.7] | 2.121 [102.0] |
| cali10 | 0.806 | 0.235 | 1.040 | 0.795 [98.6] | 0.231 [98.4] | 1.058 [101.7] |
| RMSEC (μg mL$^{-1}$) | | | | 0.010 | 0.004 | 0.036 |
| Average recovery (mean ± standard deviation) | | | | 99.7 ± 2.2% | 99.1 ± 5.8% | 98.9 ± 4.8% |

| Prediction set | | | | | | |
|---|---|---|---|---|---|---|
| pred01 | 0.672 | 0.043 | 1.820 | 0.684 [101.8] | 0.034 [78.5] | 1.935 [106.3] |
| pred02 | 1.343 | 0.064 | 0.780 | 1.354 [100.8] | 0.057 [89.3] | 0.840 [107.7] |
| pred03 | 0.537 | 0.085 | 2.600 | 0.572 [106.5] | 0.075 [87.7] | 2.697 [103.7] |
| pred04 | 1.209 | 0.107 | 1.560 | 1.224 [101.3] | 0.098 [91.7] | 1.733 [111.1] |
| pred05 | 0.403 | 0.128 | 0.520 | 0.430 [106.6] | 0.120 [93.4] | 0.548 [105.4] |
| pred06 | 1.075 | 0.149 | 2.340 | 1.069 [99.5] | 0.133 [88.9] | 2.483 [106.1] |
| pred07 | 0.269 | 0.171 | 1.300 | 0.279 [103.9] | 0.156 [91.6] | 1.454 [111.8] |
| pred08 | 0.940 | 0.192 | 0.260 | 0.980 [104.3] | 0.180 [94.0] | 0.256 [98.3] |
| pred09 | 0.134 | 0.213 | 2.080 | 0.140 [104.1] | 0.191 [89.6] | 2.280 [109.6] |
| pred10 | 0.806 | 0.235 | 1.040 | 0.840 [104.2] | 0.216 [92.0] | 1.202 [115.6] |
| RMSEP (μg mL−1) | | | | 0.026 | 0.015 | 0.145 |
| Average spike recovery (mean ± standard deviation) | | | | 103.3 ± 2.4% | 89.7 ± 4.4% | 107.6 ± 4.8% |

Reproduced from Ref. [129], Table 2, p. 1231.

**TABLE 6** Figures of Merit Obtained by the Three-Way Calibration Method Based on the ATLD Algorithm

| | Figure of merit | | |
|---|---|---|---|
| Analytes | *Resveratrol* | *Curcumin* | *Adriamycin* |
| SEN (mL $\mu g^{-1}$) | 1.891 | 2.277 | 4.361 |
| LOD (ng $mL^{-1}$) | 0.107 | 2.300 | 1.900 |
| LOQ (ng $mL^{-1}$) | 0.323 | 7.000 | 5.800 |

Reproduced from Ref. [129], Table 3, p. 1231.

Hypertension is a cardiac chronic disease with a symptom of sustaining rise in systemic arterial blood pressure. It is evident that the elevated blood pressure plays a central role in the pathogenesis of both coronary heart disease and stroke which are the main contributors to the global burden of disease [147–149]. Therefore, the adequate control of hypertension is one of the biggest challenges to the public health worldwide. In practice, combination therapy is more and more popular [147], since it can not only help to meet the target blood pressure, but also decrease the morbidity and mortality of complications. Therefore, quantification of coadministrated antihypertensives in both biological matrix and pharmaceutical preparations is very important.

Up to now, routine quantitative analysis of antihypertensives have been frequently performed by high-performance liquid chromatography (HPLC) coupled with different detectors. Although tedious pretreatments have been performed before the HPLC analysis, chromatographic separations with acceptable resolution could not always be achieved, owing to either structural similarity among the analytes and coeluting interferents from systems studied, or the limited time range of single chromatographic run. Fortunately, with the aid of three-way calibration, the separation ability of chromatographic-based techniques can be further enhanced by using mathematical separations.

#### 4.2.1.1 Experimental Design

In this example, there are eleven analytes of interest. They are triamterene (TRI), captopril (CAP), bisoprolol (BIS), propranolol (PRO), furosemide (FUR), indapamide (IND), doxazosin (DOX), carvedilol (CAR), reserpine (RES), amlodipine (AML), and losartan (LOS). The chemical structures of these eleven analytes of interest are given in Figure 19. The calibration samples are prepared in the mobile phase and only contain the eleven analytes of interest. The corresponding concentrations of the eleven antihypertensives in the calibration samples are summarized in Table 7. And five validation samples also prepared in the mobile phase are designed in order to check the accuracy of the proposed strategy.

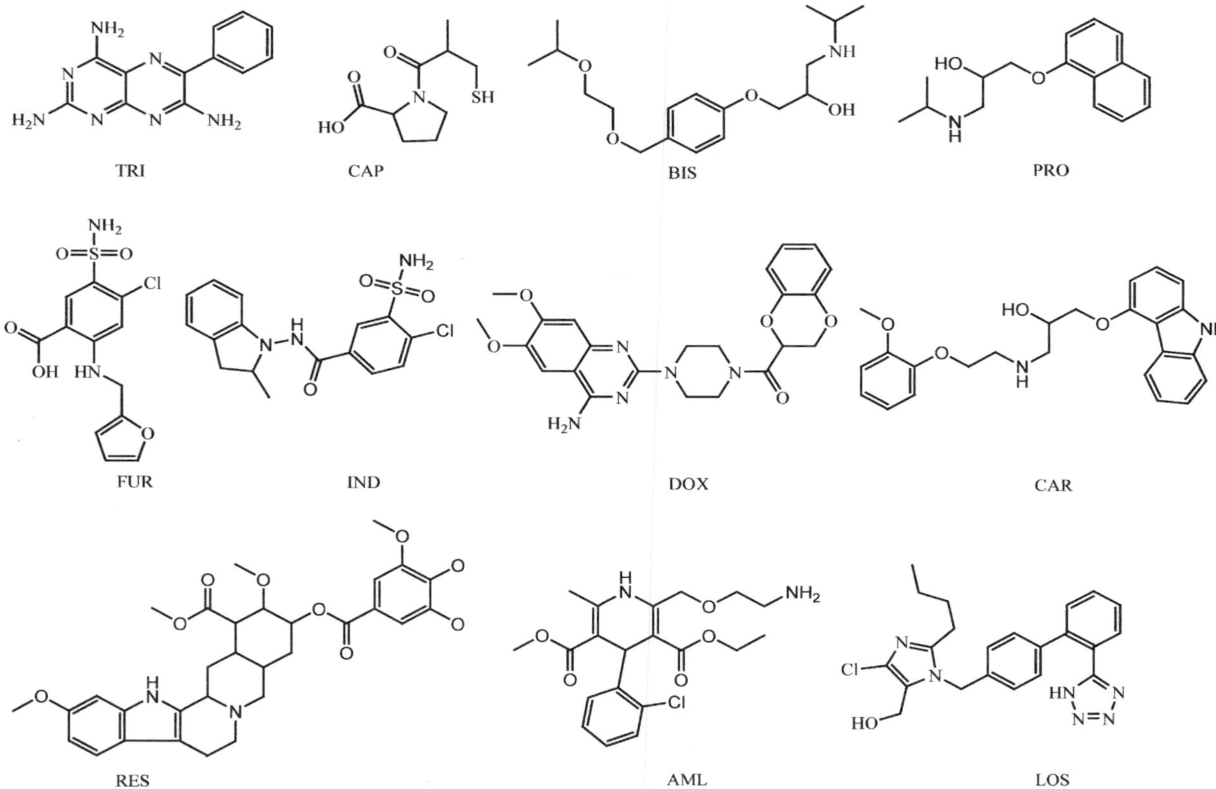


**FIGURE 19** Chemical structures of 11 analytes studied.

**TABLE 7** Retention Time $t_R$, Linear Range, and Corresponding Determination Coefficient $R^2$ for Each Analyte in the Calibration Samples

| Analytes | *TRI* | *CAP* | *BIS* | *PRO* | *FUR* | *IND* | *DOX* | *CAR* | *RES* | *AML* | *LOS* |
|---|---|---|---|---|---|---|---|---|---|---|---|
| $t_R$ (min) | 2.36 | 2.85 | 3.13 | 3.72 | 3.94 | 4.34 | 4.51 | 4.65 | 6.50 | 7.51 | 8.94 |
| Linearity ($\mu g\ mL^{-1}$) | 0.19–1.89 | 0.90–9.00 | 0.36–3.60 | 0.41–2.85 | 0.38–3.42 | 0.36–3.24 | 0.42–3.78 | 0.48–3.02 | 0.80–8.00 | 1.51–19.58 | 0.88–8.80 |
| $R^2$ | 0.9977 | 0.9991 | 0.9959 | 0.9997 | 0.9998 | 0.9987 | 0.9999 | 0.9996 | 1.0000 | 0.9996 | 0.9992 |

Three groups of prediction samples (Human plasma, AVL, and EUO samples) are designed. The concentrations of the eleven antihypertensives are within their corresponding concentration ranges in the calibration samples, respectively. For each group, five samples are prepared by spiking suitable volumes of working solutions to the corresponding system.

With the mobile phase consisting of methanol and 10 mM $KH_2PO_4$ (pH 2.60) at the ratio of 58:42 (v/v), all of the studied eleven antihypertensives are eluted within 10 min. The three-dimensional landscapes of the two-dimensional data arrays of the calibration sample cali08, the human plasma prediction sample HP05, the AVL prediction sample AVL05, and the EUO prediction sample EUO05 are given in Figure 20.

Based on the chromatogram of the calibration sample cali08, the entire chromatogram of the eleven drugs can be roughly divided into five regions according to their elution ranges: (I) an analyte TRI, eluted in the region of 2.20–2.50 min; (II) two analytes CAP and BIS, coeluted in the region of 2.60–3.40 min; (III) two analytes PRO and FUR, eluted in the region of 3.50–4.20 min with acceptable separation; (IV) three analytes IND, DOX and CAR, eluted together in the region of 4.18–4.95 min, their chromatographic profiles seriously overlap and only two elution peaks could be observed in this region; (V) the last three analytes RES, AML and LOS, eluted in the region of 5.60–9.60 min, which achieved successful separation. The three-way analysis of a data array based on each region is accomplished by trilinear decompositions in each one.

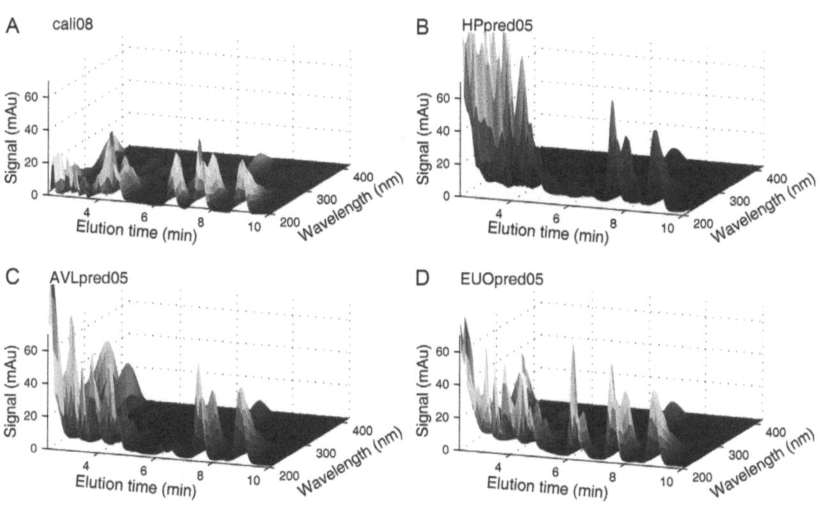


**FIGURE 20** Three-dimensional landscapes of the two-dimensional data arrays of the calibration sample cali08, the human plasma prediction sample HP05, the AVL prediction sample AVL05, and the EUO prediction sample EUO05.

Apart from the overlapping peaks among the analytes, it is worth noting that unexpected interferents which originate from the systems may occur in the prediction samples and overlap any of the peaks, shown in the three-dimensional landscapes of the two-dimensional data arrays of the human plasma prediction sample HP05, the AVL prediction sample AVL05, and the EUO prediction sample EUO05. If there are coeluted interferents, the concentrations of the eleven analytes might be determined wrongly by classical univariate calibration methods. Fortunately, with the aid of three-way calibration methods such as the three-way calibration method based on the ATLD algorithm, these problems can be solved mostly, if not completely. The underlying philosophy of this advantage lies in the fact that the trilinear decomposition of the trilinear model is always unique [18], which allows the relative chromatographic, spectral and concentration profiles of analytes of interest be resolved apart from those of uncalibrated interferents.

In addition, baseline drifts may bring trivial treatments for traditional univariate calibration methods. Compared with traditional strategies, three-way calibration methods have a significant advantage in handling with the baseline drift problem. In principle, baseline drifts can be easily modeled by additional components in a three-way calibration method based on the trilinear model.

### 4.2.1.2 HPLC-DAD-Sample Trilinear Model

In HPLC-DAD measurements, two-dimensional data array can be obtained for each sample in a single run and thus three-way data array can be obtained by arraying the two-dimensional data arrays of multiple samples along the third direction. In a HPLC-DAD-Sample three-way data array $\boldsymbol{X}$ $(I \times J \times K)$, each element $x_{ijk}$ can be expressed as:

$$x_{ijk} = \sum_{n=1}^{N} a_{in} b_{jn} c_{kn} + e_{ijk}, \quad \text{for} \quad i = 1, 2, \ldots, I; \quad j = 1, 2, \ldots, J; \quad k = 1, 2, \ldots, K. \tag{8}$$

where $x_{ijk}$ is the measured signal intensity of the $k$th sample at the $i$th elution time point and the $j$th wavelength point; $N$ represents the number of components in the trilinear model. $a_{in}$, $b_{jn}$, and $c_{kn}$, correspond to the normalized chromatogram matrix $\mathbf{A}_{I\times N}$, the normalized spectrum matrix $\mathbf{B}_{J\times N}$, and the relative concentration matrix $\mathbf{C}_{K\times N}$ of the HPLC-DAD-Sample three-way data array $\boldsymbol{X}$ $(I \times J \times K)$, respectively. Specifically, $a_{in}$ and $b_{jn}$ represent the coefficients related to the signal intensity of the $n$th component at the $i$th elution time point and the $j$th wavelength point; and $c_{kn}$ represents the relative concentration of the $n$th component in the $k$th sample. The term $e_{ijk}$ is the element of the three-way residual array $\boldsymbol{E}$ $(I \times J \times K)$. Figure 21 illustrates the general HPLC-DAD-Sample trilinear model.

There are four three-way data arrays. Data array I consists of the two-dimensional data arrays of the calibration samples and the five validation

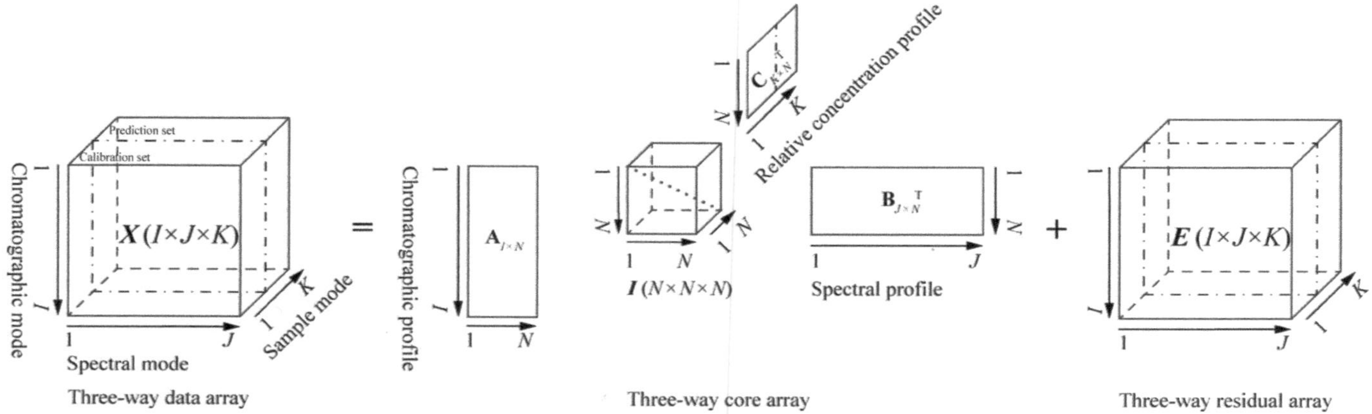


**FIGURE 21** Illustration of the HPLC-DAD-Sample trilinear model in three-way calibration.

samples; Data array II consists of the two-dimensional data arrays of the calibration samples and the five human plasma prediction samples; Data array III consists of the two-dimensional data arrays of the calibration samples and the five AVL tea prediction samples; Data array IV consists of the two-dimensional data arrays of the calibration samples and the five EUO tablet prediction samples.

#### 4.2.1.3 Trilinear Decomposition and Regression (Interferent-Free System)

The entire data analysis procedure can be summarized as four steps: First, characteristic information for each of the above-mentioned five ranges of retention is carefully selected according to the features of both chromatographic and spectral profiles; Second, the proper number of components is estimated for each elution range; Third, the ATLD algorithm is employed to decompose the trilinear models in specific elution ranges with the estimated proper number of components; At last, the decomposed relative concentration for each analyte is regressed against the corresponding real concentration, based on the calibration set, and thereafter the quantitative information of each analyte can be obtained.

The followings are the results of the trilinear decomposition and regression for these four data arrays. The ATLD algorithm developed by Wu *et al.* [26], is one of the most commonly used trilinear decomposition algorithms, and is particularly suitable for analyzing chromatographic-based data arrays. The ATLD algorithm uses the Moore–Penrose generalized inverses strategy which is based on the singular value decomposition, and thus has the advantages of converging fast, being insensitive to initialization, and being insensitive to the overestimated number of components.

Prior to further analysis, the data array I constructed by the two-dimensional data arrays of the calibration samples and the five interference-free validation samples, is analyzed by using ATLD algorithm. Along with the aforementioned data analysis steps, satisfactory results of the trilinear decomposition and regression are obtained. Figure 22 gives the decomposed chromatographic and spectral profiles of the data array I, by the ATLD algorithm.

By the use of the three-way calibration method based on the ATLD algorithm, the average recovery for each analyte of interest is $103.6 \pm 7.0\%$ (TRI), $97.3 \pm 4.2\%$ (CAP), $102.0 \pm 5.5\%$ (BIS), $103.0 \pm 2.4\%$ (PRO), $99.9 \pm 4.3\%$ (FUR), $104.8 \pm 7.1\%$ (IND), $99.8 \pm 7.4\%$ (DOX), $97.9 \pm 7.3\%$ (CAR), $98.8 \pm 6.0\%$ (RES), $101.1 \pm 4.0\%$ (AML), and $99.6 \pm 2.1\%$ (LOS), respectively.

For the analytes of interest in region V (RES, AML, and LOS) which reaches baseline separation in the calibration samples, it is advocated to quantify them by univariate calibration methods, for instance, by the regression of the chromatogram peak area versus the concentration. However, there are baseline drift problems occurring in the real samples, and the three-way

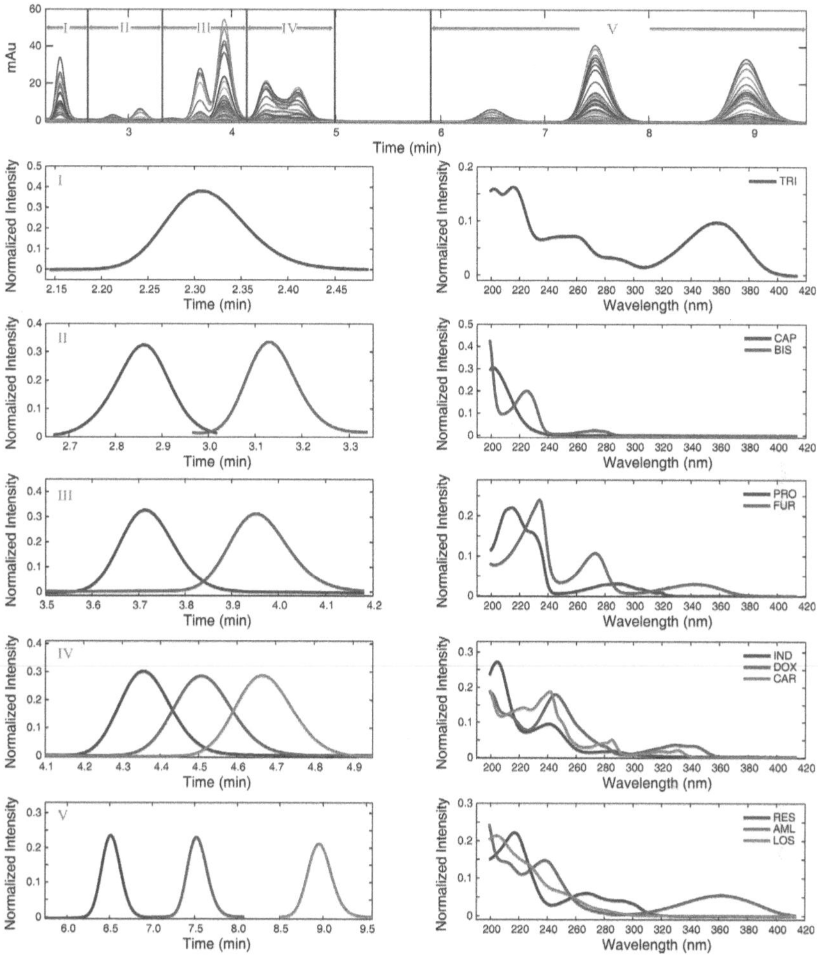


**FIGURE 22** Decomposed chromatographic and spectral profiles of the data array I (the interferent-free system) by the ATLD algorithm. The top subplot: the recorded chromatographic profiles of the calibration sample cali08 under multiwavelength channels.

calibration methods based on the trilinear model would be a more advisable choice. Here, the three-way data array corresponding to the last eluted analyte LOS, in data array I, is taken as an example to illustrate the benefit of introducing three-way calibration methods based on the trilinear model to solve the baseline drift problem. The baseline drift problem can be successfully eliminated by using an excess component, see Figure 23.

These results indicate that the three-way calibration method based on the ATLD algorithm, coupled with the HPLC-DAD second-order instrument, can achieve the quantitative analysis of the eleven antihypertensives even in the presence of the coelution problems and the varying background signals.

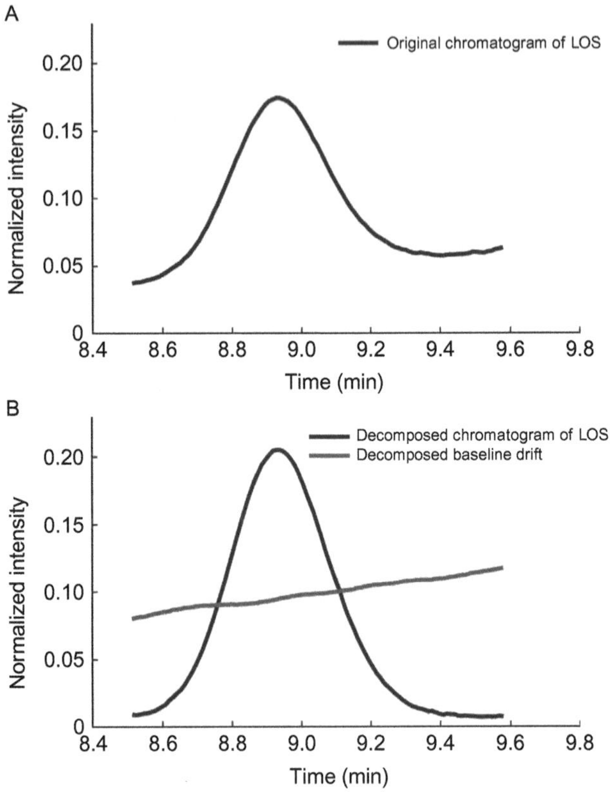


**FIGURE 23** Correction of the baseline drift problem by the trilinear decomposition based on the ATLD algorithm. (A) The original chromatographic profile of LOS, in which the upward baseline drift problem occurs; (B) the decomposed chromatographic profile, in which the upward baseline drift is modeled and separated by an additional component.

In the following parts, the performances of the method in three complex systems, i.e., human serum (HP) samples, health product (AVL) samples and Chinese patent medicine (EUO) samples are further investigated.

#### 4.2.1.4 Trilinear Decomposition and Regression (Human Plasma System)

Since combination therapy that joins several classes of antihypertensives has been proved to be more favorable than monotherapy [150], simultaneously monitoring the concentrations of multiclass antihypertensives inhuman plasma is very important.

The top subplot in Figure 24 demonstrates the chromatographic profiles of the spiked human plasma prediction sample pred05 under multiwavelength channels. A visual comparison of the chromatographic profiles of cali08 and

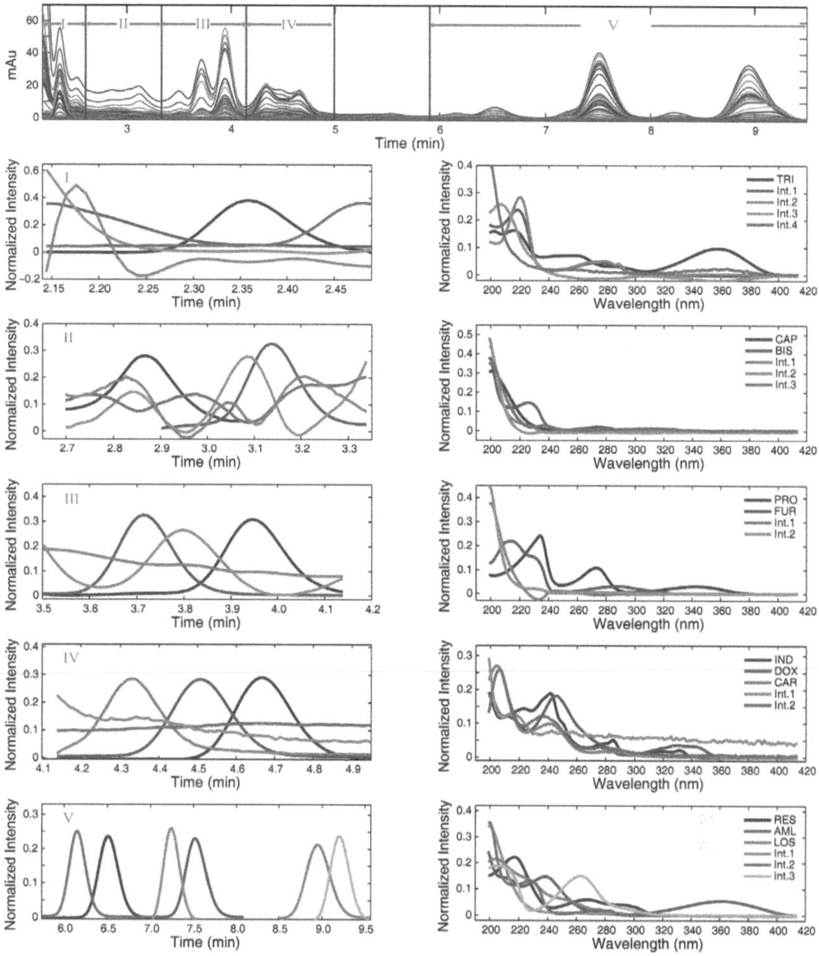


**FIGURE 24** Decomposed chromatographic and spectral profiles of the data array II (the human plasma system) by the ATLD algorithm. The top subplot: the recorded chromatographic profiles of the spiked human plasma prediction sample pred05 under multiwavelength channels.

that of the spiked human plasma prediction sample pred05, allows one to assess the high complexity of the human plasma system. Obviously, a large number of uncalibrated interferents coelutes with the analytes of interest all along the elution region, especially along TRI, CAP, and BIS, which are completely covered by uncalibrated interferents. These facts make big challenges for the successful implementation of univariate calibration methods.

The data array II constructed by the two-dimensional data arrays of the calibration samples and the five spiked human plasma prediction samples is analyzed by the ATLD algorithm. The above figure also gives the decomposed chromatographic and spectral profiles of the data array II (the human plasma system), by the ATLD algorithm. For the convenience of presentation, only the decomposed chromatographic and spectral profiles of the eleven antihypertensives and principal interferents are shown, while the profiles corresponding to the drifted baseline are omitted. By comparing the decomposed chromatographic and spectral profiles of the data array II (the human plasma system) with those of the data array I (the interferent-free system), one can see that the trilinear decompositions by the ATLD algorithm are satisfactory, even in the presence of serious uncalibrated interferents in the human plasma system and varying background signals.

By the regression in the three-way calibration method for each analyte, the quantitative concentration of each analyte in each spiked human plasma prediction sample can be obtained, as listed in Table 8. The satisfactory spike recoveries of the eleven analytes of interest in each spike level demonstrate that satisfactory quantitative results in the human plasma system are obtained.

#### 4.2.1.5 Trilinear Decomposition and Regression (AVL System)

*Apocynum venetum leaf* (AVL), which is known as Luobuma in China, consists of many bioactive constituents and has been receiving widespread applications as healthy beverage in the treatment of hypertension due to its efficient biological effects [147,151–153]. Consequently, monitoring the concentration levels of the eleven antihypertensives in health products is important. In this section, the eleven antihypertensives are simultaneously analyzed in the AVL tea.

The top subplot in Figure 25 demonstrates the chromatographic profiles of the spiked AVL prediction sample pred05 under multiwavelength channels. A visual comparison of the chromatographic profiles of cali08 and that of the spiked AVL prediction sample pred05, allows one to assess the high complexity of the AVL system. Obviously, a large number of uncalibrated interferents coelutes with the analytes of interest all along the elution region, especially for TRI, CAP, and BIS, which are completely covered by uncalibrated interferents. These facts make big challenges for the successful implementation of univariate calibration methods.

The data array III constructed by the two-dimensional data arrays of the calibration samples and the five spiked AVL prediction samples, is analyzed by the ATLD algorithm. The above figure also gives the decomposed chromatographic and spectral profiles of the data array III (the AVL system), by the ATLD algorithm. By comparing the decomposed chromatographic and spectral profiles of the data array III (the AVL system) with those of the data array I (the interferent-free system), one can see that the trilinear decompositions by the ATLD algorithm are satisfactory, even in the presence of serious uncalibrated interferents in the AVL system and varying background signals.

By the regression in the three-way calibration method for each analyte, the quantitative concentration of each analyte in each spiked AVL prediction

**TABLE 8** Quantitative Results of the 11 Antihypertensives in the Spiked Human Plasma Prediction Samples, by the Three-way Calibration Method Based on the ATLD Algorithm

| | Concentration (μg mL$^{-1}$) | | | | | | | | | | | | | | | | | | | | | |
|---|---|---|---|---|---|---|---|---|---|---|---|---|---|---|---|---|---|---|---|---|---|---|
| | *TRI* | | *CAP* | | *BIS* | | *PRO* | | *FUR* | | *IND* | | *DOX* | | *CAR* | | *RES* | | *AML* | | *LOS* | |
| Sample | *add* | *found* | *add* | *found* | *add* | *found* | *add* | *found* | *add* | *found* | *add* | *found* | *add* | *found* | *add* | *found* | *add* | *found* | *add* | *found* | *add* | *found* |
| vali01 | 0.28 | 0.28 | 9.00 | 9.59 | 3.60 | 3.06 | 1.43 | 1.36 | 0.76 | 0.72 | 0.72 | 0.76 | 3.78 | 3.80 | 1.91 | 2.03 | 7.60 | 7.39 | 2.01 | 1.94 | 1.32 | 1.38 |
| vali02 | 0.66 | 0.63 | 7.20 | 7.66 | 3.06 | 3.29 | 2.24 | 2.22 | 1.52 | 1.44 | 1.44 | 1.35 | 2.94 | 2.99 | 0.95 | 0.91 | 6.00 | 5.96 | 6.02 | 5.98 | 3.08 | 3.14 |
| vali03 | 1.04 | 1.04 | 5.40 | 5.27 | 2.52 | 2.74 | 2.65 | 2.62 | 1.90 | 1.94 | 1.80 | 1.66 | 2.10 | 2.26 | 2.86 | 2.98 | 4.40 | 4.35 | 10.04 | 10.03 | 4.84 | 4.72 |
| vali04 | 1.42 | 1.40 | 4.50 | 4.41 | 1.80 | 1.65 | 0.81 | 0.75 | 2.66 | 2.63 | 2.52 | 2.29 | 1.26 | 1.22 | 2.39 | 2.37 | 2.80 | 2.63 | 14.06 | 14.08 | 6.60 | 6.27 |
| vali05 | 1.79 | 1.90 | 2.70 | 2.47 | 1.26 | 1.16 | 1.83 | 1.85 | 3.42 | 3.55 | 3.24 | 2.97 | 0.42 | 0.42 | 1.43 | 1.48 | 1.20 | 1.13 | 18.07 | 18.05 | 8.36 | 8.75 |
| Average recovery | 99.8±3.9% | | 100.0±6.4% | | 96.9±10.5% | | 97.4±3.4% | | 98.9±4.0% | | 94.9±6.1% | | 101.3±4.1% | | 101.9±4.2% | | 96.7±2.6% | | 99.2±1.4% | | 100.7±4.3% | |

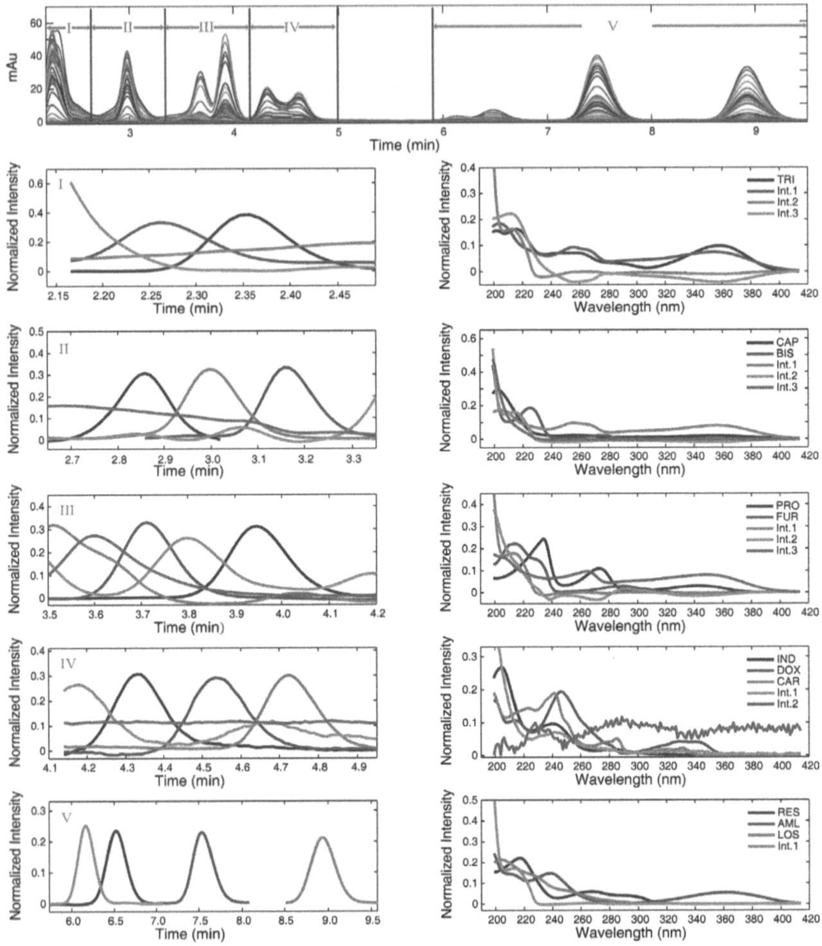


FIGURE 25 Decomposed chromatographic and spectral profiles of the data array III (the AVL system) by the ATLD algorithm. The top subplot: the recorded chromatographic profiles of the spiked AVL prediction sample pred05 under multiwavelength channels.

sample can be obtained, as listed in Table 9. The satisfactory spike recoveries of the eleven analytes of interest in each spike level demonstrate that satisfactory quantitative results in the AVL system are obtained.

#### 4.2.1.6 Trilinear Decomposition and Regression (EUO System)

*Eucommia ulmoides Oliv.* (EUO), widely called Du-zhong in China, is one of the commonly used herbs in Chinese patent medicine. Previous researchers have proved that EUO has many interesting features, such as antihypertensive and antioxidative, due to its bioactive compositions [154,155]. Du-zhong Pingya tablet is made from EUO leaves and often used as a substitute of

**TABLE 9** Quantitative Results of the Eleven Antihypertensives in the Spiked AVL Prediction Samples, by the Three-way Calibration Method Based on the ATLD Algorithm

| | Concentration ($\mu g\ mL^{-1}$) | | | | | | | | | | | | | | | | | | | | | |
|---|---|---|---|---|---|---|---|---|---|---|---|---|---|---|---|---|---|---|---|---|---|---|
| | *TRI* | | *CAP* | | *BIS* | | *PRO* | | *FUR* | | *IND* | | *DOX* | | *CAR* | | *RES* | | *AML* | | *LOS* | |
| Sample | *add* | *found* | *add* | *found* | *add* | *found* | *add* | *found* | *add* | *found* | *add* | *found* | *add* | *found* | *add* | *found* | *add* | *found* | *add* | *found* | *add* | *found* |
| vali01 | 0.28 | 0.29 | 9.00 | 8.63 | 3.60 | 3.49 | 1.43 | 1.44 | 0.76 | 0.76 | 0.72 | 0.73 | 3.78 | 3.71 | 1.91 | 1.78 | 7.60 | 6.89 | 2.01 | 2.10 | 1.32 | 1.40 |
| vali02 | 0.66 | 0.66 | 7.20 | 7.27 | 3.06 | 2.87 | 2.24 | 2.32 | 1.52 | 1.45 | 1.44 | 1.42 | 2.94 | 2.98 | 0.95 | 0.99 | 6.00 | 5.49 | 6.02 | 6.17 | 3.08 | 3.05 |
| vali03 | 1.04 | 1.13 | 5.40 | 5.09 | 2.52 | 2.33 | 2.65 | 2.80 | 1.90 | 2.01 | 1.80 | 1.85 | 2.10 | 2.17 | 2.86 | 2.91 | 4.40 | 4.04 | 10.04 | 10.90 | 4.84 | 5.10 |
| vali04 | 1.42 | 1.41 | 3.60 | 3.94 | 1.80 | 1.77 | 0.81 | 0.81 | 2.66 | 2.77 | 2.52 | 2.56 | 1.26 | 1.27 | 2.39 | 2.51 | 2.80 | 2.93 | 14.06 | 14.57 | 6.60 | 7.06 |
| vali05 | 1.79 | 1.91 | 1.80 | 1.70 | 1.26 | 1.17 | 1.83 | 1.90 | 3.42 | 3.48 | 3.24 | 3.48 | 0.84 | 0.82 | 1.43 | 1.46 | 1.20 | 1.20 | 18.07 | 17.74 | 8.36 | 8.56 |
| Average recovery | 103.3±4.2% | | 99.0±6.4% | | 94.8±2.8% | | 102.6±2.5% | | 101.5±3.9% | | 102.3±3.3% | | 100.2±2.4% | | 101.4±4.8% | | 95.7±6.3% | | 103.5±3.8% | | 104.0±3.3% | |

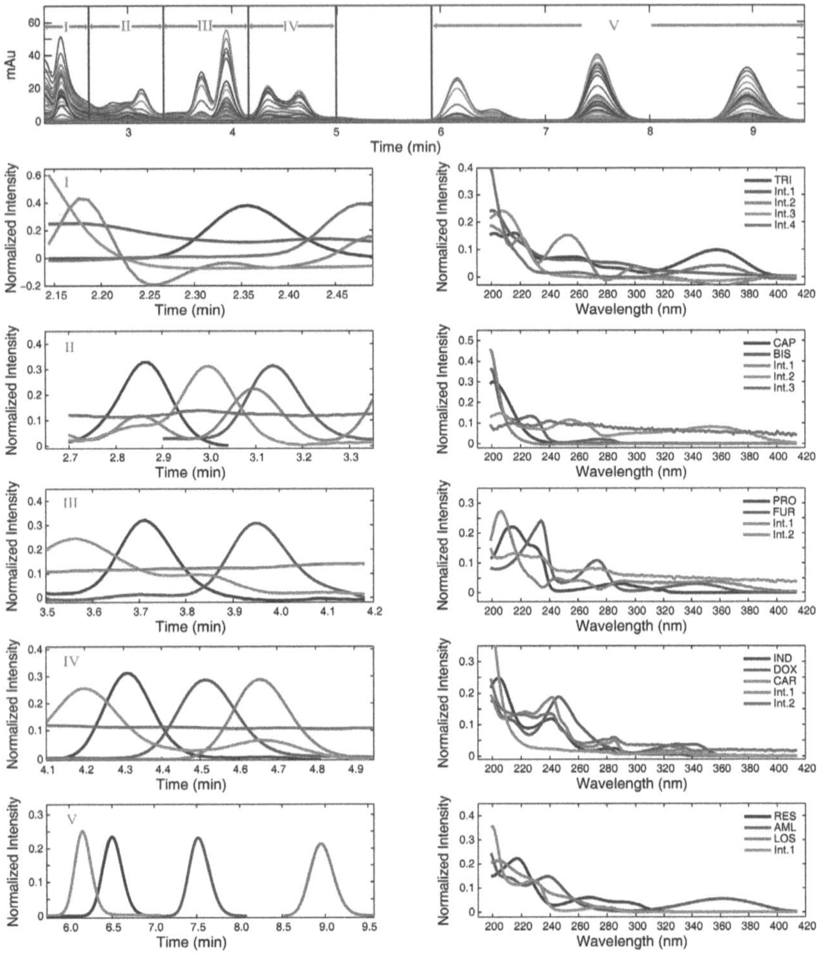


**FIGURE 26** Decomposed chromatographic and spectral profiles of the data array IV (the EUO system) by the ATLD algorithm. The top subplot: the recorded chromatographic profiles of the spiked EUO prediction sample pred05 under multiwavelength channels.

antihypertensive medicine. Therefore, the eleven antihypertensives which would potentially be adulterated are simultaneously determined in the EUO tablet.

The top subplot in Figure 26 demonstrates the chromatographic profiles of the spiked EUO prediction sample pred05 under multiwavelength channels. One can see that the coeluting interferents in the EUO system mainly occurred in the region 2.00–3.60 min and5.90–6.40 min.

The data array IV constructed by the two-dimensional data arrays of the calibration samples and the five spiked EUO prediction samples, is analyzed by the ATLD algorithm. The above figure also gives the decomposed chromatographic and spectral profiles of the data array IV (the EUO system), decomposed by the ATLD algorithm. By comparing the decomposed chromatographic and spectral profiles of the data array IV (the EUO system) with those of the data array I (the interferent-free system), one can see that the trilinear decompositions by the ATLD algorithm are satisfactory, even in the presence of serious uncalibrated interferents in the EUO system and varying background signals.

By the regression in the three-way calibration method for each analyte, the quantitative concentration of each analyte in each spiked EUO prediction sample can be obtained, as listed in Table 10. The satisfactory spike recoveries of the eleven analytes of interest in each spike level demonstrate that satisfactory quantitative results in the EUO system are obtained.

Although the coeluting interferents originated from the human plasma, AVL, and EUO systems are different, the decomposed chromatographic and spectral profiles of the eleven antihypertensives are in good agreement, which fully demonstrates the second-order advantage of the three-way calibration method based on the ATLD algorithm. In addition, the proposed strategy releases the traditional analytical strategy that explores specific separation condition for each system, suggesting that more labor force and toxic organic solvent would be saved. These features mean that the introduction of the three-way calibration method based on the ATLD algorithm to HPLC makes the analysis of multiple analytes in various complex systems a relatively easy analytical task which can be solved by quite simple ways.

#### 4.2.1.7 Figures of Merit

The RMSEPs, LODs, and LOQs of the eleven antihypertensives in the three types of systems are listed in Table 11. These satisfactory results indicate that the proposed analytical strategy can be used as a promising method for simultaneous quantitative analysis of multiple classes of antihypertensives in various complex systems.

As has been shown, the three-way calibration method based on the ATLD algorithm, coupled with the HPLC-DAD second-order instrument, is a general and reliable method for the quantitative analysis of multiple classes of antihypertensives in various complex systems, even in the presence of uncalibrated interferents, the coelution problems and the varying background signals, with a simple chromatographic condition.

## 5 PRACTICAL EXAMPLES OF FOUR-WAY CALIBRATION

Compared with three-way calibration methods (second-order tensor calibration methods), four-way calibration methods (third-order tensor calibration methods) can provide some additional advantages such as enhanced resolving power and higher sensitivity, apart from the second-order advantage. For complex systems with serious spectral interferents, one can step further to use four-way calibration methods.

**TABLE 10** Quantitative Results of the 11 Antihypertensives in the Spiked EUO Prediction Samples, by the Three-Way Calibration Method Based on the ATLD Algorithm

| | Concentration ($\mu g\ mL^{-1}$) | | | | | | | | | | | | | | | | | | | | | |
|---|---|---|---|---|---|---|---|---|---|---|---|---|---|---|---|---|---|---|---|---|---|---|
| | *TRI* | | *CAP* | | *BIS* | | *PRO* | | *FUR* | | *IND* | | *DOX* | | *CAR* | | *RES* | | *AML* | | *LOS* | |
| Sample | *add* | *found* | *add* | *found* | *add* | *found* | *add* | *found* | *add* | *found* | *add* | *found* | *add* | *found* | *add* | *found* | *add* | *found* | *add* | *found* | *add* | *found* |
| vali01 | 0.28 | 0.29 | 9.00 | 8.50 | 3.60 | 3.90 | 1.43 | 1.42 | 0.76 | 0.76 | 0.72 | 0.72 | 3.78 | 3.72 | 1.91 | 1.92 | 7.60 | 7.70 | 2.01 | 2.08 | 1.32 | 1.23 |
| vali02 | 0.66 | 0.63 | 7.20 | 6.89 | 3.06 | 3.30 | 2.24 | 2.21 | 1.52 | 1.46 | 1.44 | 1.42 | 2.94 | 2.96 | 0.95 | 1.01 | 6.00 | 6.08 | 6.02 | 6.33 | 3.08 | 3.03 |
| vali03 | 1.04 | 0.91 | 5.40 | 5.69 | 2.52 | 2.38 | 2.65 | 2.60 | 1.90 | 1.95 | 1.80 | 1.73 | 2.10 | 2.01 | 2.86 | 2.99 | 4.40 | 4.42 | 10.04 | 10.09 | 4.84 | 4.40 |
| vali04 | 1.42 | 1.40 | 3.60 | 4.44 | 1.80 | 1.92 | 0.81 | 0.76 | 2.66 | 2.66 | 2.52 | 2.38 | 1.26 | 1.24 | 2.39 | 2.46 | 2.80 | 2.89 | 14.06 | 13.47 | 6.60 | 6.41 |
| vali05 | 1.79 | 1.80 | 1.80 | 2.79 | 1.26 | 1.12 | 1.83 | 1.77 | 3.42 | 3.53 | 3.24 | 3.09 | 0.84 | 0.83 | 1.43 | 1.42 | 1.20 | 1.19 | 18.07 | 18.03 | 8.36 | 8.15 |
| Average recovery | 96.9±6.2% | | 99.5±4.7% | | 101.2±9.0% | | 97.3±2.1% | | 100.3±2.8% | | 97.1±2.5% | | 98.4±1.8% | | 102.8±2.7% | | 101.1±1.6% | | 100.9±3.6% | | 95.4±3.2% | |

**TABLE 11** Figures of Merit Obtained by the Three-Way Calibration Method Based on the ATLD Algorithm, for the Human Plasma, AVL, and EUO Systems

| Analytes | Figures of merit | | | | | | | | | | |
|---|---|---|---|---|---|---|---|---|---|---|---|
| | *TRI* | *CAP* | *BIS* | *PRO* | *FUR* | *IND* | *DOX* | *CAR* | *RES* | *AML* | *LOS* |
| **For the human plasma system ($\mu g\ mL^{-1}$)** | | | | | | | | | | | |
| RMSEP | 0.06 | 0.40 | 0.33 | 0.05 | 0.08 | 0.20 | 0.09 | 0.09 | 0.14 | 0.04 | 0.26 |
| LOD | 0.02 | 3.42 | 0.06 | 0.05 | 0.17 | 0.12 | 0.12 | 0.07 | 0.08 | 10.06 | 0.06 |
| LOQ | 0.07 | 10.36 | 0.17 | 0.16 | 0.50 | 0.38 | 0.36 | 0.20 | 0.22 | 3.20 | 0.18 |
| **For the AVL system ($\mu g\ g^{-1}$)** | | | | | | | | | | | |
| RMSEP | 9.38 | 37.50 | 20.00 | 11.38 | 11.25 | 15.00 | 6.75 | 12.13 | 60.00 | 66.25 | 36.25 |
| LOD | 10.00 | 33.75 | 10.00 | 2.25 | 2.75 | 1.75 | 8.25 | 2.88 | 21.25 | 6.75 | 2.25 |
| LOQ | 30.00 | 102.50 | 30.00 | 7.00 | 8.38 | 5.25 | 25.00 | 8.63 | 66.25 | 20.00 | 6.75 |
| **For the EUO system ($\mu g\ g^{-1}$)** | | | | | | | | | | | |
| RMSEP | 6.90 | 33.00 | 23.00 | 5.00 | 6.60 | 11.00 | 5.70 | 8.00 | 8.20 | 34.00 | 27.00 |
| LOD | 7.60 | 43.00 | 5.80 | 3.80 | 4.50 | 8.90 | 0.58 | 0.56 | 1.02 | 0.66 | 1.08 |
| LOQ | 23.00 | 130.00 | 18.00 | 12.00 | 13.00 | 27.00 | 1.80 | 1.69 | 3.10 | 2.01 | 3.26 |

As the modern instrumentation continues to develop toward a more advanced level, third-order data can be produced in a relatively easy way. Common strategies consist of recording EEM fluorescence (or phosphorescence) data as a function of reaction time [32,156,157], pH [59,78,158,159], volume of quencher [160], or decay time [161]. Another strategy uses hyphenated instruments, such as two-dimensional chromatography (LC × LC, GC × GC) coupled with mass spectrometry (MS) or diode array detection (DAD) [80].

The following section will illustrate the four-way calibration method based on the quadrilinear model using a practical analytical example based on the EEM phosphorescence spectroscopy.

## 5.1 Four-Way Calibration of EEM-Kinetics Phosphorescence Type Data

### *5.1.1 Case Study 3: Quantitative Kinetics of Carbaryl in Tap Water*

Here, we present a practical analytical example of the quantitative kinetics of carbaryl in tap water, using the four-way calibration method based on the alternating quadrilinear decomposition (AQLD) algorithm, coupled with EEM phosphorescence spectroscopy [156].

Carbaryl, a toxic pesticide for human systems, is widely used in agriculture to control pests on stored grain, grass lawns, fruits and vegetables [78]. The degradation product of carbaryl in alkaline medium is 1-naphthol, which presents weak phosphorescence. The EEM phosphorescence spectroscopy is used to continuously monitor the kinetics evolution of carbaryl, by recording EEPMs of the samples at different reaction times, based on the room-temperature phosphorescence (RTP) technique.

#### 5.1.1.1 Experimental Design

In this case, the analyte of interest is carbaryl. The degradation process of carbaryl in alkaline medium is given in Figure 27. The calibration set consists of five samples, in which the concentration levels of carbaryl are 0.61, 1.01,

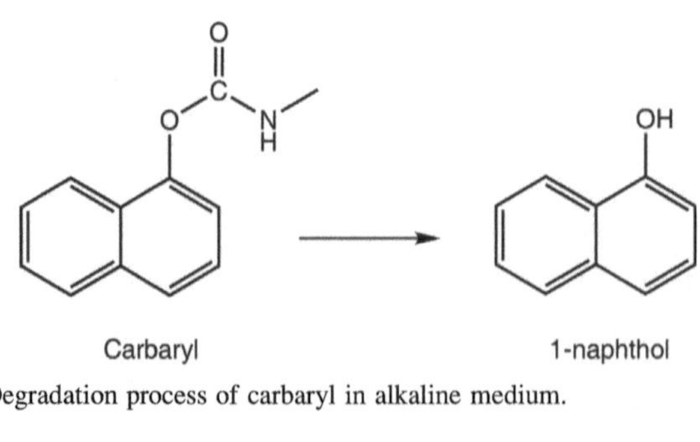


**FIGURE 27** Degradation process of carbaryl in alkaline medium.

2.02, 2.24, and 3.23 μg $mL^{-1}$, respectively. The calibration samples are prepared in a simple solution. The spiked prediction set consists of eight samples, prepared by spiking appropriate contents of the analyte of interest with the appropriate volume of tap water, in which the concentration levels of carbaryl are in the rang of the calibration set. In addition, three tap water samples are also prepared in the same manner.

When the buffer is mixed with the solutions containing carbaryl, the degradation reaction will start. The kinetic evolutions of EEM phosphorescence signals of these samples are then recorded. The EEM phosphorescence two-way data arrays are recorded at emission wavelengths between 250 and 350 nm (every 10 nm), at different excitation wavelengths between 430 nm and 580 nm (every 10 nm). For each sample, eight $11 \times 16$ EEM phosphorescence two-way data arrays at a series of specific moments (5.0, 14.0, 23.0, 32.0, 41.0, 50.0, 59.0, 68.0 min) are scanned, and therefore an $11 \times 16 \times 8$ EEM-Kinetics phosphorescence three-way data array is obtained per sample. By arranging the sixteen $11 \times 16 \times 8$ EEM-Kinetics phosphorescence three-way data arrays of the sixteen samples, one can obtain an $11 \times 16 \times 8 \times 16$ EEM-Kinetics-Sample phosphorescence four-way data array.

#### 5.1.1.2 EX-EM-Kinetics-Sample Quadrilinear Model

When a three-way EEM-Kinetics phosphorescence data array of a sample is measured at $I$ excitation wavelengths, $J$ emission wavelengths, and $K$ time points, the collection of three-way EEM-Kinetics phosphorescence data arrays of the calibration and prediction samples, forms a four-way EX-EM-Kinetics-Sample data array $\boldsymbol{X}$ $(I \times J \times K \times L)$, in which each element $x_{ijkl}$ can be expressed as follows:

$$x_{ijkl} = \sum_{n=1}^{N} a_{in} b_{jn} c_{kn} d_{ln} + e_{ijkl}, \quad \text{for} \quad i = 1, 2, \ldots, I; \; j = 1, 2, \ldots, J; \\ k = 1, 2, \ldots, K; \; l = 1, 2, \ldots, L \qquad (9)$$

where $x_{ijkl}$ is the measured phosphorescence intensity of the $l$th sample at the $i$th excitation wavelength, the $j$th emission wavelength, and the $k$th time point; $N$ represents the number of components in the quadrilinear model. $a_{in}$, $b_{jn}$, $c_{kn}$, and $d_{ln}$ correspond to the normalized excitation spectrum matrix $\mathbf{A}_{I \times N}$, the normalized emission spectrum matrix $\mathbf{B}_{J \times N}$, the normalized kinetic profile matrix $\mathbf{C}_{K \times N}$, and the relative concentration matrix $\mathbf{D}_{L \times N}$ of the EX-EM-Kinetics-Sample four-way data array $\boldsymbol{X}$ $(I \times J \times K \times L)$. Specifically, $a_{in}$, $b_{jn}$, and $c_{kn}$ represent the coefficients related to the spectral intensity of the $n$th component at the $i$th excitation wavelength, the $j$th emission wavelength, and the $k$th time point; and $d_{ln}$ represents the relative concentration of the $n$th component in the $l$th sample. And $e_{ijkl}$ is the element of the four-way residual array $\boldsymbol{E}$ $(I \times J \times K \times L)$.

In the following section, the measured four-way $11 \times 16 \times 8 \times 16$ EEM-Kinetics-Sample phosphorescence data array will be analyzed by using four-way calibration method based on the AQLD algorithm.

### 5.1.1.3 Quadrilinear Decomposition

The three-dimensional landscapes of EEM phosphorescence for the calibration sample cali02 and the tap water sample tapwater01 after 5.0 min of the reaction are given in Figure 28. From the landscape of cal02 at 5.0 min, one can see that carbaryl emits strong phosphorescence with the maximum excitation and emission intensities at 284 and 489/522 nm, respectively, which are consistent with the values reported in literature [162]. From the landscape of tapwater01 at 5.0 min, one can see that the tap water sample exhibits several unsmooth peaks

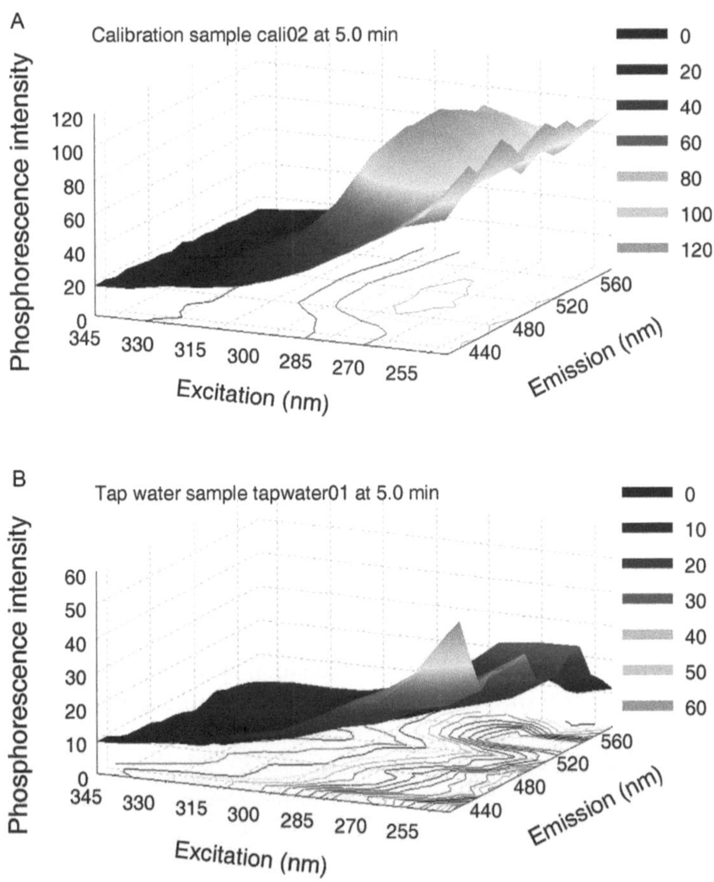


**FIGURE 28** Three-dimensional landscapes of EEM phosphorescence for the calibration sample cali02 and the tap water sample tapwater01, after 5.0 min of the reaction.

overlapped with that of carbaryl, which probably originate from inorganic or organic species in the tap water. Besides, it is clear that serious spectral baseline drift could be observed in the two samples, which may come from other interferents or the instrumental measurement itself.

These factors cause difficulties in directly studying the hydrolysis process of carbaryl in complex mixtures with ordinary RTP technique. Alternatively, one can turn to four-way calibration methods in the field of chemometrics, which might light up a new avenue by using the mathematical separation instead of physical or chemical separations. The AQLD algorithm uses the Moore–Penrose generalized inverses strategy which is based on the singular value decomposition, and thus it has the advantages of converging fast, being insensitive to initialization, and being insensitive to the overestimated number of components. Therefore, the quadrilinear decomposition is accomplished by the AQLD algorithm.

The number of components for the quadrilinear model should be estimated before the application of quadrilinear decomposition algorithms [16,17,37,38,64]. The number of components for this four-way data array is estimated as $N=3$ by the core consistency diagnostic test [82].

Then, the AQLD algorithm with the number of components $N=3$ is used to decompose this quadrilinear model. With $N=3$ for this EEMs data array, the AQLD algorithm accurately decomposed the normalized excitation spectrum matrix $\mathbf{A}_{11\times3}$, the normalized emission spectrum matrix $\mathbf{B}_{16\times3}$, the normalized kinetic profile matrix $\mathbf{C}_{8\times3}$, and the relative concentration matrix $\mathbf{D}_{16\times3}$. Figure 29 shows the actual relative profiles (dash-dot lines) and the decomposed relative profiles (solid lines) in the four modes using the AQLD algorithm with $N=3$. The decomposed excitation and emission profiles of carbaryl are in good agreement with the actual profiles based on pure analyte standards. The correlation coefficients between the decomposed and actual spectra are 0.9864 and 0.9965 in the excitation and emission modes, respectively, for carbaryl. Interestingly, two spectral interferents are resolved. The interferent 1 may come from the tap water, and the interferent 2 may represent the drifted baseline caused by the buffer or the instrumental measurement itself.

Considering that the AQLD is insensitive to the overestimated number of components, the quadrilinear decompositions by the AQLD algorithm with $N=4$ and 5 are also carried out, respectively. With $N=4$ for this EEMs data array, the AQLD algorithm accurately decomposed the normalized excitation spectrum matrix $\mathbf{A}_{11\times4}$, the normalized emission spectrum matrix $\mathbf{B}_{16\times4}$, the normalized kinetic profile matrix $\mathbf{C}_{8\times4}$, and the relative concentration matrix $\mathbf{D}_{16\times4}$. Figure 30 shows the actual relative profiles (dash-dot lines) and the decomposed relative profiles (solid lines) in the four modes using the AQLD with $N=4$. One can see that the decomposed excitation and emission profiles of carbaryl are in good agreement with the actual profiles based on pure analyte standards. The correlation coefficients between the decomposed and

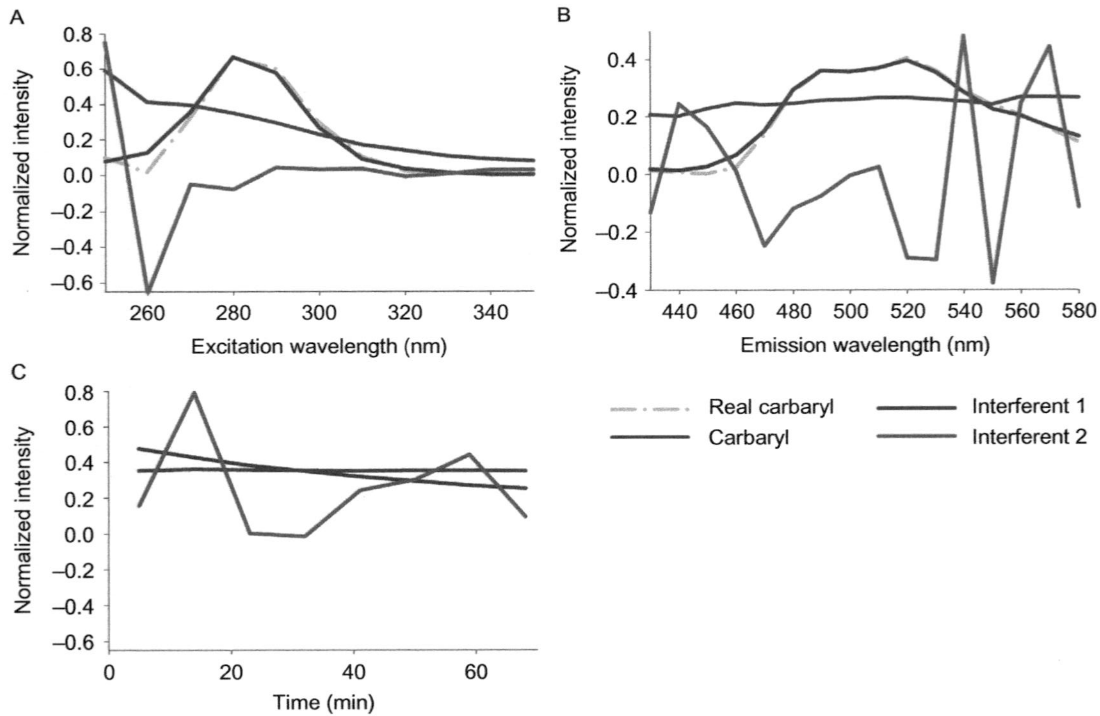


**FIGURE 29** Results of the decomposition in the four-way calibration method based on the AQLD algorithm with the number of components $N=3$, for the EEM-Kinetics-Sample four-way phosphorescence data array.

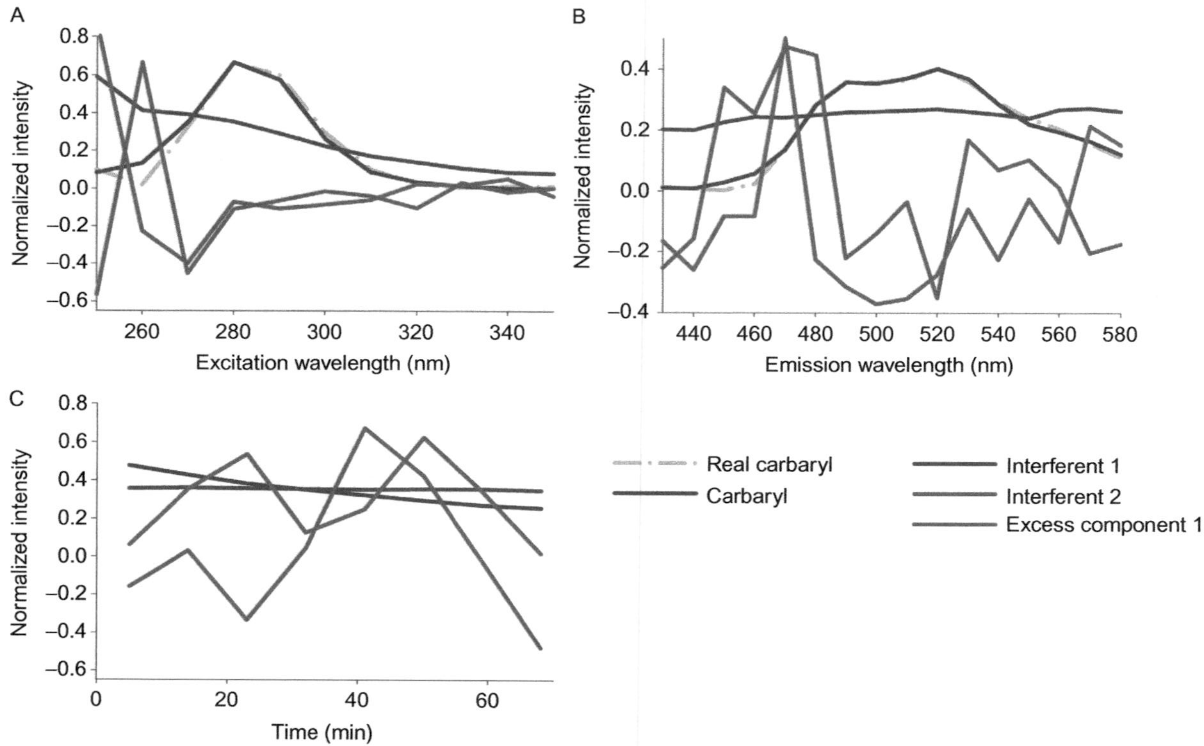


**FIGURE 30** Results of the decomposition in the four-way calibration method based on the AQLD algorithm with the number of components $N=4$, for the EEM-Kinetics-Sample four-way phosphorescence data array.

actual spectra are 0.9858 and 0.9965 in the excitation and emission modes, respectively, for carbaryl. The decomposed profiles of the analyte of interest are not influenced by the excess component.

With $N=5$ for this EEMs data array, the AQLD algorithm has accurately decomposed the normalized excitation spectrum matrix $\mathbf{A}_{11\times5}$, the normalized emission spectrum matrix $\mathbf{B}_{16\times5}$, the normalized kinetic profile matrix $\mathbf{C}_{8\times5}$, and the relative concentration matrix $\mathbf{D}_{16\times5}$. Figure 31 shows the actual relative profiles (dash-dot lines) and the decomposed relative profiles (solid lines) in the four modes using the AQLD with $N=5$. One can see that the decomposed excitation and emission profiles of carbaryl are in good agreement with the actual profiles based on pure analyte standards. The correlation coefficients between the decomposed and actual spectra are 0.9904 and 0.9965 in the excitation and emission modes, respectively, for carbaryl. The decomposed profiles of carbaryl were still not influenced by the excess components, and the quality of the decomposed excitation spectrum of carbaryl was even slightly improved.

#### 5.1.1.4 Univariate Regression

By the inverse univariate regression in the four-way calibration method for carbaryl, the concentration of carbaryl in each sample can be obtained. For the quadrilinear decompositions by the AQLD algorithm with $N=3$, 4, and 5, the regression equations for carbaryl are $r=0.0017y-0.16$, $r=0.0017y-0.16$, and $r=0.0020y-0.21$, respectively. Predicted concentrations for the spiked prediction samples are given in Table 12. For the quadrilinear decompositions by the AQLD algorithm with $N=3$, 4, and 5, the average spike recoveries (mean $\pm$ standard deviation) for carbaryl are $97.8\pm6.0\%$, $97.3\pm5.4\%$, and $98.2\pm4.5\%$, respectively.

#### 5.1.1.5 Figures of Merit

The SENs, LODs, and LOQs of the method are listed in Table 13. These satisfactory results indicate that the four-way calibration method based on the AQLD algorithm with $N=3$, 4, or 5, is sensitive enough for the quantitative analysis of carbaryl in tap water.

#### 5.1.1.6 Kinetic Analysis

Here, let us have a visual inspection of the degradation reaction of carbaryl in the buffer solution. Figure 32 shows the three-dimensional landscapes of EEM phosphorescence for the calibration sample cali02 as a function of the reaction time (5.0, 23.0, 50.0, and 68.0 min). One can see that, the phosphorescence intensity of carbaryl decreases as the reaction processes. This phenomenon coincides with the decomposed kinetic profile of carbaryl in the normalized kinetic profile matrix $\mathbf{C}_{K\times N}$.

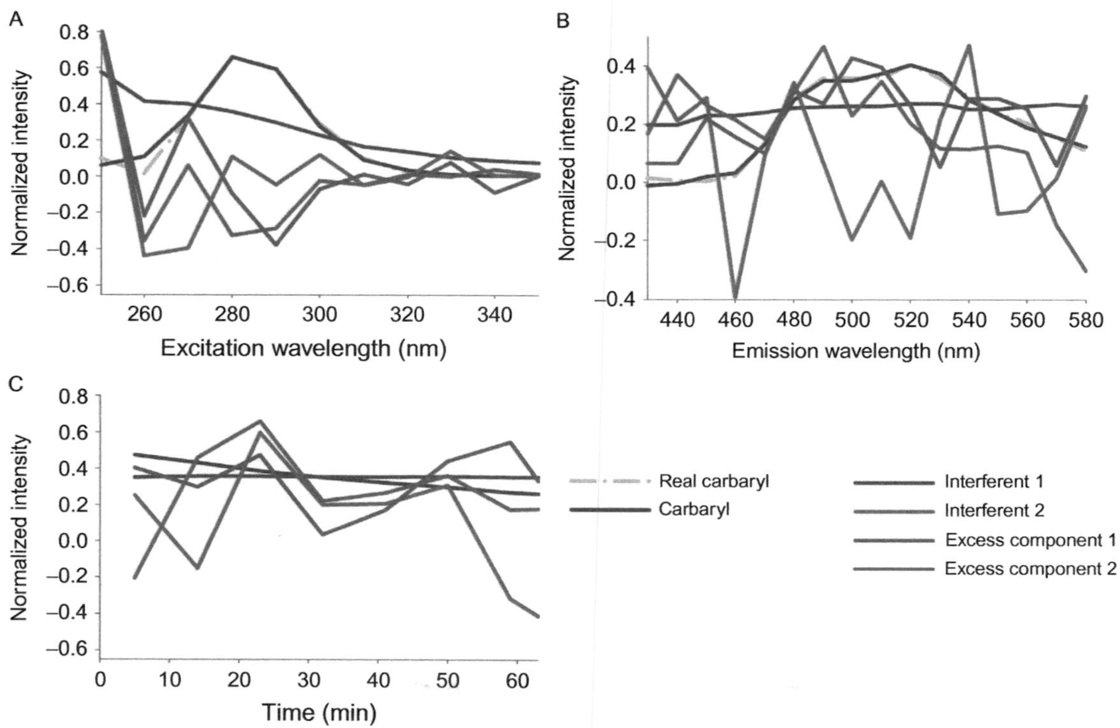


**FIGURE 31** Results of the decomposition in the four-way calibration method based on the AQLD algorithm with the number of components $N=5$, for the EEM-Kinetics-Sample four-way phosphorescence data array.

**TABLE 12** Results of the Quantitative Analysis of Carbaryl in the Spiked Prediction Samples, by the Four-Way Calibration Method Based on the AQLD Algorithm with Different Estimated Component Numbers

| Sample | Spike concentration ($\mu g\ mL^{-1}$) *Carbaryl* | Predicted concentration ($\mu g\ mL^{-1}$) [Spike recovery (%)] *N*=3 | *N*=4 | *N*=5 |
|---|---|---|---|---|
| **Spiked prediction set** | | | | |
| pred01 | 0.81 | 0.82[101.3] | 0.80[99.7] | 0.80[99.7] |
| pred02 | 0.91 | 0.91[100.6] | 0.90[99.1] | 0.90[99.1] |
| pred03 | 1.11 | 1.24[111.1] | 1.22[110.1] | 1.22[110.1] |
| pred04 | 1.21 | 1.03[85.0] | 1.06[87.1] | 1.07[88.2] |
| pred05 | 1.31 | 1.25[95.5] | 1.24[94.6] | 1.26[95.7] |
| pred06 | 1.52 | 1.54[102.0] | 1.54[101.7] | 1.56[103.3] |
| pred07 | 1.72 | 1.63[95.2] | 1.64[95.3] | 1.67[97.2] |
| pred08 | 1.82 | 1.67[91.7] | 1.65[90.6] | 1.70[93.6] |
| RMSEP ($\mu g\ mL^{-1}$) | | 0.11 | 0.10 | 0.09 |
| Average spike recovery (mean ± standard deviation) | | 97.8 ± 6.0% | 97.3 ± 5.4% | 98.2 ± 4.5% |

**TABLE 13** Figures of Merit Obtained by the Four-Way Calibration Method Based on the AQLD Algorithm with Different Estimated Component Numbers

| Analyte | Figures of merit *N*=3 | *N*=4 | *N*=5 |
|---|---|---|---|
| SEN ($mL\ \mu g^{-1}$) | 37.8 | 27.4 | 17.8 |
| LOD ($\mu g\ mL^{-1}$) | 0.03 | 0.07 | 0.10 |
| LOQ ($\mu g\ mL^{-1}$) | 0.10 | 0.21 | 0.31 |

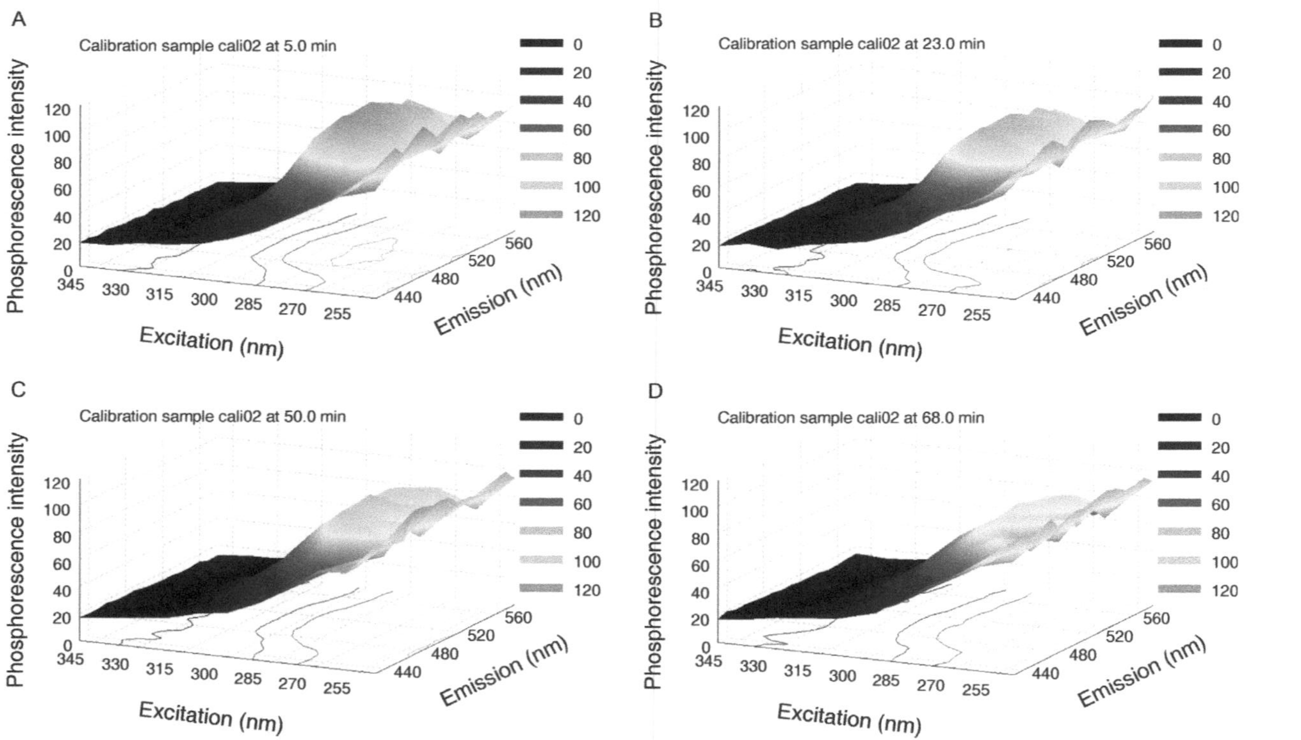


**FIGURE 32** Three-dimensional landscapes of EEM phosphorescence for the calibration sample cali02, as a function of the reaction time (5.0, 23.0, 50.0, and 68.0 min).

For the kinetic analysis of the degradation reaction of carbaryl in tap water, one can accomplish the analysis by related kinetic regressions, based on the decomposed pure kinetic profile in the normalized kinetic profile matrix $\mathbf{C}_{K \times N}$ and the reaction time. The plot of the natural logarithm of reciprocal of the relative kinetic profile intensity Ln($1/C_t$) versus the time $t-t_1$, is given in Figure 33. It is apparent that all the points fit nicely to the linear plot which shows the logarithm of the reciprocal of the relative kinetic profile intensity is linearly related to the reaction time.

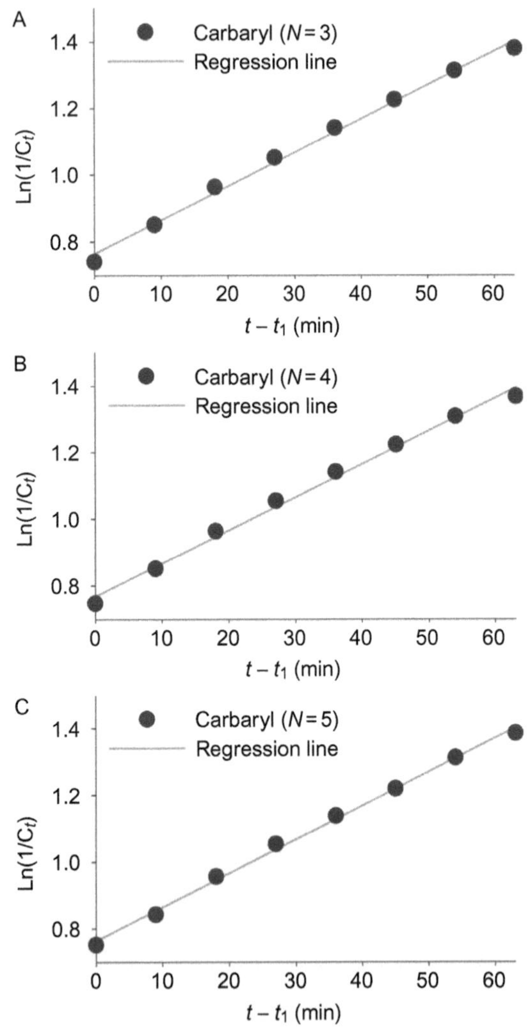


**FIGURE 33** Plot of the natural logarithm of reciprocal of the relative kinetic profile intensity Ln($1/C_t$) versus the time $t-t_1$, for different estimated component numbers.

**TABLE 14 Regression Equation, Correlation Coefficient, Rate Constant, and Half-Life of the Degradation Reaction of Carbaryl in Tap Water**

| The number of components | Kinetic parameters | | |
|---|---|---|---|
| | $N=3$ | $N=4$ | $N=5$ |
| Regression equation | $y=0.0101x+0.7646$ | $y=0.0099x+0.6488$ | $y=0.0101x+0.6488$ |
| Correlation coefficient | 0.9975 | 0.9969 | 0.9986 |
| Rate constant ($min^{-1}$) | 0.0101 | 0.0099 | 0.0101 |
| Half-life (min) | 62.6 | 63.6 | 62.6 |

The obtained kinetic parameters are shown in Table 14. Based on the pure kinetic profiles decomposed by the AQLD algorithm with $N=3$, 4, and 5, the rate constants are 0.0101, 0.0099, and 0.0101 $min^{-1}$, respectively; the values of half-life are 62.6, 63.6, and 62.6 min, respectively. Interestingly, one can see that the obtained kinetic parameters are consistent, by using the AQLD algorithm with $N=3$, 4, and 5, respectively.

These satisfactory qualitative and quantitative results indicate that the four-way calibration method based on the AQLD algorithm with $N=3$, 4, or 5, coupled with EEM phosphorescence spectroscopy, can successfully achieve the direct quantitative kinetics of carbaryl in tap water, even in the presence of unknown spectral interferents. These satisfactory results indicate that the use of the four-way calibration method based on the AQLD algorithm, coupled with the EEM fluorescence spectroscopy, can provide a promising tool for direct quantitative kinetics in complex systems containing unknown spectral interferents.

## 6 PERSPECTIVES

Multiway calibration methods based on multilinear models are gaining more and more attention in the field of analytical science. Based on the "mathematical separation," multiway calibration methods enable one to achieve direct quantitative analysis of analytes of interest in complex systems even in the presence of unknown, uncalibrated interferents, which is known as the "second-order advantage." Additionally, multiway calibration methods can provide some other advantages, such as autocorrecting varying background signals and providing a higher sensitivity. When combined with advanced analytical instruments capable of generating multidimensional arrays (e.g., EEMs, HPLC-DAD, and LC-MS), multiway calibration can be a smart and green quantitative analysis strategy for complex chemical systems, based on the "mathematical separation" partially or even completely replacing chemical and physical separations.

These multiway calibration methods have been used to direct quantitative analysis of multiple analytes of interest in complex systems in many fields such as biochemical, food, environmental, and pharmaceutical sciences. In the future, it can be expected that these multiway calibration methods based on multilinear models will become general calibration methods for quantitative analysis of multiple analytes of interest in complex chemical systems.

## ACKNOWLEDGMENTS

The authors gratefully acknowledge the Sasakawa Scientific Research Grant from the Japan Science Society, the National Natural Science Foundation of China (Grant Nos. 29735150, 29975007, 20475014, 20775025, 21175041, and 21221003), and the National Basic Research Program (Nos. 2007CB216404 and 2012CB910602) as well as the Grants from the Ministry of Education of China for financial support. The authors would also like to thank Mr. Zhao-Yang Ai and Ms. Ding Zeng for their English language review. Finally, the authors would like to thank our friends for their continuing support.

## APPENDIX MATLAB DEMO OF THE SIMULATION

% Matlab demo:

```
% State Key Laboratory of Chemo/Biosensing and Chemometrics,
% College of Chemistry and Chemical Engineering,
% Hunan University, Changsha 410082, China.
% ® Hai-Long Wu (hlwu@hnu.edu.cn)
% I. Simulation of EX-EM-Sample = 100*200*(7+5+3)
I = 100; J = 200; K = 15;
% The first mode (Excitation spectra, 301:1:400 nm)
A(:,1) = peak(1,16,30,I) + peak(1,12,36,I);
A(:,2) = peak(0.5,20,60,I) + peak(1,20,50,I);
A(:,3) = peak(1,16,60,I) + peak(0.5,18,30,I);
A(:,4) = peak(0.5,80,0,I);
% The second mode (Emission spectra, 351:1:550 nm)
B(:,1) = peak(0.5,12,48,J) + peak(0.5,20,16,J) + peak(0.5,31,60,J);
B(:,2) = peak(1,40,70,J)+ peak(0.3,24,128,J);
B(:,3) = peak(1,52,100,J) + peak(1,40,124,J);
B(:,4) = peak(0.8,300,100,J);
% The third mode
% 2 analytes + 1 uncalibrated interferent + 1 varying background
% 7 calibration samples + 5 prediction samples + 3 method blanks
```

```
C = [20.00 50.00  0     5.00
     30.00 20.00  0     5.00
     40.00 70.00  0     6.00
     50.00 40.00  0     6.00
     60.00 10.00  0     7.00
     70.00 60.00  0     7.00
     80.00 30.00  0     8.00
     20.00 10.00 20.00  8.00
     35.00 25.00 20.00  9.00
     50.00 40.00 20.00  9.00
     65.00 55.00 30.00 10.00
     80.00 70.00 30.00 10.00
      0     0     0     6.00
      0     0     0     6.50
      0     0     0     7.00];
% II. Construct the three-way data array
for k = 1:K
XIJKp(:,:,k) = A * diag(C(k,:)) * B';
End
% Add noise
for k = 1:K
  EIJK(:,:,k) = 0.01 * randn(I,J) * mean(mean(XIJKp(:,:,k)));
end
XIJK = XIJKp + EIJK;
% III. Three-way Calibration Method based on ATLD
N = 5;
epsilon = 1e-6;
tic
[A_ATLD,B_ATLD,C_ATLD,SSR_ATLD,iter_ATLD] = ATLD(XIJK,N,epsilon);
t_ATLD = toc
% - 3 Regression - omitted
```

## REFERENCES


[1] Smilde A, Bro R, Geladi P. Multi-way analysis with applications in the chemical sciences. Chichester: Wiley; 2005.

[2] Gemperline P. Practical guide to chemometrics. 2nd ed. Boca Raton: CRC press; 2006.

[3] Brown SD, Tauler R, Walczak B. Comprehensive chemometrics: chemical and biochemical data analysis. Amsterdam: Elsevier; 2009.

[4] Wold S, Ruhe A, Wold H, Dunn III WJ. The collinearity problem in linear regression. The partial least squares (PLS) approach to generalized inverses. SIAM J Sci Statist Comput 1984;5:735–43.

[5] Wold S, Esbensen K, Geladi P. Principal component analysis. Chemom Intell Lab Syst 1987;2:37–52.

[6] Kvalheim OM, Liang YZ. Heuristic evolving latent projections: resolving two-way multicomponent data. 1. Selectivity, latent-projective graph, datascope, local rank, and unique resolution. Anal Chem 1992;64:936–46.
[7] Liang YZ, Kvalheim OM, Keller HR, Massart DL, Kiechle P, Erni F. Heuristic evolving latent projections: resolving two-way multicomponent data. 2. Detection and resolution of minor constituents. Anal Chem 1992;64:946–53.
[8] Zeng YS, Hopke PK. A new receptor model: a direct trilinear decomposition followed by a matrix reconstruction. J Chemom 1992;6:65–83.
[9] Tauler R, Kowalski B, Fleming S. Multivariate curve resolution applied to spectral data from multiple runs of an industrial process. Anal Chem 1993;65:2040–7.
[10] Walczak B, Massart D. Robust principal components regression as a detection tool for outliers. Chemom Intell Lab Syst 1995;27:41–54.
[11] Jaumot J, Gemperline PJ, Stang A. Non-negativity constraints for elimination of multiple solutions in fitting of multivariate kinetic models to spectroscopic data. J Chemom 2005;19:97–106.
[12] Huang X, Xu QS, Liang YZ. PLS regression based on sure independence screening for multivariate calibration. Anal Methods 2012;4:2815–21.
[13] Luo JS, Yu LZ, Guo YZ, Li ML. Functional classification of secreted proteins by position specific scoring matrix and auto covariance. Chemom Intell Lab Syst 2012;110:163–7.
[14] Shao XG, Du GR, Jing M, Cai WS. Application of latent projective graph in variable selection for near infrared spectral analysis. Chemom Intell Lab Syst 2012;114:44–9.
[15] Wang Y, Ma X, Wen YD, Liu JJ, Cai WS, Shao XG. Discrimination of plant samples using near-infrared spectroscopy with a principal component accumulation method. Anal Methods 2012;4:2893–9.
[16] Carroll JD, Chang JJ. Analysis of individual differences in multidimensional scaling via an N-way generalization of "Eckart-Young" decomposition. Psychometrika 1970;35:283–319.
[17] Harshman RA. Foundations of the PARAFAC procedure: models and conditions for an "explanatory" multimodal factor analysis. UCLA working papers in phonetics, 16;1970. p. 1–84.
[18] Kruskal JB. Three-way arrays: rank and uniqueness of trilinear decompositions, with application to arithmetic complexity and statistics. Linear Algebra Appl 1977;18:95–138.
[19] Wilson BE, Lindberg W, Kowalski BR. Multicomponent quantitative analysis using second-order nonbilinear data: theory and simulations. J Am Chem Soc 1989;111:3797–804.
[20] Sanchez E, Kowalski BR. Tensorial resolution: a direct trilinear decomposition. J Chemom 1990;4:29–45.
[21] Kiers HA, Krijnen WP. An efficient algorithm for PARAFAC of three-way data with large numbers of observation units. Psychometrika 1991;56:147–52.
[22] Smilde AK. Three-way analyses problems and prospects. Chemom Intell Lab Syst 1992;15:143–57.
[23] Booksh KS, Kowalski BR. Theory of analytical chemistry. Anal Chem 1994;66:782–91.
[24] Tauler R. Multivariate curve resolution applied to second-order data. Chemom Intell Lab Syst 1995;30:133–46.
[25] Bro R. PARAFAC. Tutorial and applications. Chemom Intell Lab Syst 1997;38:149–71.
[26] Wu HL, Shibukawa M, Oguma K. An alternating trilinear decomposition algorithm with application to calibration of HPLC-DAD for simultaneous determination of overlapped chlorinated aromatic hydrocarbons. J Chemom 1998;12:1–26.
[27] de Juan A, Rutan SC, Tauler R, Massart DL. Comparison between the direct trilinear decomposition and the multivariate curve resolution-alternating least squares methods for the resolution of three-way data sets. Chemom Intell Lab Syst 1998;40:19–32.

[28] Hopke PK, Paatero P, Jia H, Ross RT, Harshman RA. Three-way (PARAFAC) factor analysis: examination and comparison of alternative computational methods as applied to ill-conditioned data. Chemom Intell Lab Syst 1998;43:25–42.
[29] Kiers HA. A three-step algorithm for CANDECOMP/PARAFAC analysis of large data sets with multicollinearity. J Chemom 1998;12:155–71.
[30] Linder M, Sundberg R. Second-order calibration: bilinear least squares regression and a simple alternative. Chemom Intell Lab Syst 1998;42:159–78.
[31] Chen ZP, Wu HL, Jiang JH, Li Y, Yu RQ. A novel trilinear decomposition algorithm for second-order linear calibration. Chemom Intell Lab Syst 2000;52:75–86.
[32] Nahorniak ML, Cooper GA, Kim YC, Booksh KS. Three- and four-way parallel factor (PARAFAC) analysis of photochemically induced excitation-emission kinetic fluorescence spectra. Analyst 2005;130:85–93.
[33] Olivieri AC. On a versatile second-order multivariate calibration method based on partial least-squares and residual bilinearization: second-order advantage and precision properties. J Chemom 2005;19:253–65.
[34] Lozano VA, Ibañez GA, Olivieri AC. Second-order analyte quantitation under identical profiles in one data dimension. A dependency-adapted partial least-squares/residual bilinearization method. Anal Chem 2010;82:4510–9.
[35] Olivieri AC. Recent advances in analytical calibration with multi-way data. Anal Methods 2012;4:1876–86.
[36] Ni YN, Gu Y, Kokot S. Multiway calibrations based on spectrofluorimetric data derived from kinetic analysis of the hydrolysis reaction of nitrofurans—a comparison of prediction performance. Chemom Intell Lab Syst 2012;112:55–64.
[37] Wu HL, Li Y, Yu RQ. Recent developments of chemical multiway calibration methodologies with second-order or higher-order advantages. J Chemom 2014;28:476–89.
[38] Escandar GM, Goicoechea HC, Muñoz de la Peña A, Olivieri AC. Second- and higher-order data generation and calibration: a tutorial. Anal Chim Acta 2014;806:8–26.
[39] Yu YJ, Wu HL, Kang C, Wang Y, Zhao J, Li YN, et al. Algorithm combination strategy to obtain the second-order advantage: simultaneous determination of target analytes in plasma using three-dimensional fluorescence spectroscopy. J Chemom 2012;26:197–208.
[40] Zhang XH, Wu H, Wang JY, Chen Y, Yu YJ, Nie C, et al. Second-order calibration applied to quantification of two active components of Schisandra chinensis in complex matrix. J Pharm Anal 2012;2:241–8.
[41] Zhang SR, Wu HL, Zhai M, Kang C, Yin XL, Yu RQ. Comparison for applicability of different trilinear decomposition algorithms to liquid chromatography-mass spectrometry data measured from multiple samples. Chin J Chromatogr 2013;31:550–5 [In Chinese].
[42] Xu H, Wu HL, Liu YJ, Su ZY, Zhang J, Nie CC, et al. Simultaneous determination of baicalein and wogonin in Chinese medicinal plant Scutellaria baicalensis using HPLC-DAD coupled with alternating trilinear decomposition. Fine Chem Intermed 2011;41:68–72 [In Chinese].
[43] Song JJ, Wu HL, Yu YJ, Huang L, Yu RQ. Determination of famciclovir and penciclovir in human plasma samples by using excitation-emission matrix fluorescent spectra coupled with second-order calibration. Sci Sinica Chim 2010;8:1136–41 [In Chinese].
[44] Wu HL, Liang YZ, Yu RQ. Analytical chemometrics. Anal Lab 1999;18:94–102 [In Chinese].
[45] Shao SZ, Wu HL, Li SF, Nie JF, Wang KY, Liang Y, et al. Determination of levodopa in both human urine and plasma samples using excitation-emission matrix fluorescence coupled with second-order calibration. J Anal Sci 2010;26:373–8 [In Chinese].

[46] Wu HL, Yu RQ. Aspects of recent developmemts of chemical multiway calibration methodologies. Chemistry 2011;74:771–82 [In Chinese].

[47] Wu HL, Zhang J, Wang JY, Shao SZ, Zhang XH, Yu RQ. Determination of alpinetin in cell culture medium samples by using three-dimensional fluorescence coupled with second-order calibration. J Hunan Univ (Nat Sci) 2011;38:50–4 [In Chinese].

[48] Zhang XH, Wu HL, Wang JY, Yu RQ. Determination of schizandrin A in body fluids and cell culture media by excitation-emission matrix fluorescence coupled with second-order calibration. Chem J Chin Univ 2011;32:1720–6 [In Chinese].

[49] Zhao J, Wu HL, Yu LL, Yu YJ, Niu JF, Wang Y, et al. Rapid quantitative analysis of ligustrazine and telmisartan in complex samples using excitation-emission matrix fluorescence coupled with second-order calibration method. J Anal Sci 2011;27:273–8 [In Chinese].

[50] Su ZY, Wu HL, Liu YJ, Xu H, Zhang J, Nie CC, et al. Simultaneous determination of main effective constituents in traditional chinese medicine Kudzuvine Root using HPLC-DAD coupled with second-order calibration based on alternating trilinear decomposition. Acta Chimi Sin 2012;70:459–64 [In Chinese].

[51] Sun YM, Wu HL, Wang JY, Yu RQ. Determination of melatonin in functional food by three-dimensional fluorescence spectroscopy combined with second-order calibration method. Chem J Chin Univ 2013;34:1084–91 [In Chinese].

[52] Zhang XH, Wu HL, Kang C, Wang Y, Yu RQ. Simultaneous and rapid determination of citalopram, fluoxetine and paroxetine contents in human serum samples using three-dimensional fluorescence technique coupled with chemometric second-order calibration. Fine Chem Intermed 2013;43:63–8 [In Chinese].

[53] Shao XG, Zhu WL, Wu HL, Li XX, Xu XJ. Progress of chemoinformatics (computational chemistry) during 2011–2013. Chemistry 2014;77:692–8 [In Chinese].

[54] Gu HW, Wu HL, Yin XL, Li Y, Liu YJ, Xia H, et al. Multi-targeted interference-free determination of ten β-blockers in human urine and plasma samples by alternating trilinear decomposition algorithm-assisted liquid chromatography–mass spectrometry in full scan mode: comparison with multiple reaction monitoring. Anal Chim Acta 2014;848:10–24.

[55] Appellof CJ, Davidson ER. Strategies for analyzing data from video fluorometric monitoring of liquid chromatographic effluents. Anal Chem 1981;53:2053–6.

[56] Booksh KS, Muroski AR, Myrick ML. Single-measurement excitation/emission matrix spectrofluorometer for determination of hydrocarbons in ocean water. 2. Calibration and quantitation of naphthalene and styrene. Anal Chem 1996;68:3539–44.

[57] Sanchez E, Kowalski BR. Generalized rank annihilation factor analysis. Anal Chem 1986;58:496–9.

[58] Liu YJ, Wu HL, Kang C, Gu HW, Nie JF, Li SS, et al. Four-way self-weighted alternating normalized residue fitting algorithm with application for the analysis of serotonin in human plasma. Anal Sci 2012;28:1097–104.

[59] Kang C, Wu HL, Yu YJ, Liu YJ, Zhang SR, Zhang XH, et al. An alternative quadrilinear decomposition algorithm for four-way calibration with application to analysis of four-way fluorescence excitation-emission-pH data array. Anal Chim Acta 2013;758:45–57.

[60] Harris DC. Quantitative chemical analysis. 7th ed. New York: W.H. Freeman and Company; 2007.

[61] Fang KT. Uniform design and uniform design tables. Beijing: Science Press; 1994.

[62] Fang KT, Lin DK, Winker P, Zhang Y. Uniform design: theory and application. Technometrics 2000;42:237–48.

[63] Kolda TG, Bader BW. Tensor decompositions and applications. SIAM Rev 2009;51:455–500.

[64] Olivieri A, Escandar GM. Practical three-way calibration. Amsterdam: Elsevier; 2014.

[65] Bahram M, Bro R, Stedmon C, Afkhami A. Handling of Rayleigh and Raman scatter for PARAFAC modeling of fluorescence data using interpolation. J Chemom 2006;20:99–105.
[66] Sidiropoulos ND, Bro R. On the uniqueness of multilinear decomposition of N-way arrays. J Chemom 2000;14:229–39.
[67] Geladi P. Analysis of multi-way (multi-mode) data. Chemom Intell Lab Syst 1989;7:11–30.
[68] Booksh K, Henshaw JM, Burgess LW, Kowalski BR. A second-order standard addition method with application to calibration of a kinetics-spectroscopic sensor for quantitation of trichloroethylene. J Chemom 1995;9:263–82.
[69] Paatero P. A weighted non-negative least squares algorithm for three-way 'PARAFAC' factor analysis. Chemom Intell Lab Syst 1997;38:223–42.
[70] Xia AL, Wu HL, Fang DM, Ding YJ, Hu LQ, Yu RQ. Alternating penalty trilinear decomposition algorithm for second-order calibration with application to interference-ree analysis of excitation-emission matrix fluorescence data. J Chemom 2005;19:65–76.
[71] Arancibia JA, Olivieri AC, Gil DB, Mansilla AE, Durán-Merás I, de la Peña AM. Trilinear least-squares and unfolded-PLS coupled to residual trilinearization: new chemometric tools for the analysis of four-way instrumental data. Chemom Intell Lab Syst 2006;80:77–86.
[72] Escandar GM, Olivieri AC, Faber NKM, Goicoechea HC, Muñoz de la Peña A, Poppi RJ. Second-and third-order multivariate calibration: data, algorithms and applications. TrAC Trends Anal Chem 2007;26:752–65.
[73] Paatero P. The multilinear engine-A table-driven, least squares program for solving multilinear problems, including the n-way parallel factor analysis model. J Comput Graph Stat 1999;8:854–88.
[74] Andersson CA, Bro R. The N-way toolbox for MATLAB. Chemom Intell Lab Syst 2000;52:1–4.
[75] de la Peña MA, Durán Merás I, Jiménez Girón A. Four-way calibration applied to the simultaneous determination of folic acid and methotrexate in urine samples. Anal Bioanal Chem 2006;385:1289–97.
[76] Xia AL, Wu HL, Li SF, Zhu SH, Hu LQ, Yu RQ. Alternating penalty quadrilinear decomposition algorithm for an analysis of four-way data arrays. J Chemom 2007;21:133–44.
[77] Wu HL, Nie JF, Yu YJ, Yu RQ. Multi-way chemometric methodologies and applications: a central summary of our research work. Anal Chim Acta 2009;650:131–42.
[78] Zhu SH, Wu HL, Xia AL, Nie JF, Bian YC, Cai CB, et al. Excitation-emission-kinetic fluorescence coupled with third-order calibration for quantifying carbaryl and investigating the hydrolysis in effluent water. Talanta 2009;77:1640–6.
[79] Fu HY, Wu HL, Yu YJ, Yu LL, Zhang SR, Nie JF, et al. A new third order calibration method with application for analysis of four way data arrays. J Chemom 2011;25:408–29.
[80] Arancibia JA, Damiani PC, Escandar GM, Ibañez GA, Olivieri AC. A review on second- and third-order multivariate calibration applied to chromatographic data. J Chromatogr B 2012;910:22–30.
[81] Murphy KR, Stedmon CA, Graeber D, Bro R. Fluorescence spectroscopy and multi-way techniques. PARAFAC Anal Methods 2013;5:6557–66.
[82] Bro R, Kiers HA. A new efficient method for determining the number of components in PARAFAC models. J Chemom 2003;17:274–86.
[83] Lawson CL, Hanson RJ. Solving least squares problems. Englewood Cliffs: Prentice Hall; 1974.
[84] Thomas GB, Weir MD, Hass J, Giordano FR. Thomas' calculus. Boston: Pearson; 2010.
[85] Centner V, Massart D, De Jong S. Inverse calibration predicts better than classical calibration. Fresenius J Anal Chem 1998;361:2–9.
[86] Olivieri AC. Analytical figures of merit: from univariate to multiway calibration. Chem Rev 2014;114:5358–78.

[87] Olivieri AC, Faber NKM. Standard error of prediction in parallel factor analysis of three-way data. Chemom Intell Lab Syst 2004;70:75–82.
[88] Olivieri AC, Faber K. New developments for the sensitivity estimation in four-way calibration with the quadrilinear parallel factor model. Anal Chem 2012;84:186–93.
[89] Hantao LW, Aleme HG, Pedroso MP, Sabin GP, Poppi RJ, Augusto F. Multivariate curve resolution combined with gas chromatography to enhance analytical separation in complex samples: a review. Anal Chim Acta 2012;731:11–23.
[90] Escandar GM, Damiani PC, Goicoechea HC, Olivieri AC. A review of multivariate calibration methods applied to biomedical analysis. Microchem J 2006;82:29–42.
[91] Gómez V, Callao MP. Analytical applications of second-order calibration methods. Anal Chim Acta 2008;627:169–83.
[92] Mas S, de Juan A, Tauler R, Olivieri AC, Escandar GM. Application of chemometric methods to environmental analysis of organic pollutants: a review. Talanta 2010;80:1052–67.
[93] Madsen R, Lundstedt T, Trygg J. Chemometrics in metabolomics—a review in human disease diagnosis. Anal Chim Acta 2010;659:23–33.
[94] SádeCká J, TóThoVá J. Fluorescence spectroscopy and chemometrics in the food classification—a review. Czech J Food Sci 2007;25:159–73.
[95] Forina M, Casolino M, De la Pezuela Martínez C. Multivariate calibration: applications to pharmaceutical analysis. J Pharm Biomed Anal 1998;18:21–33.
[96] Daszykowski M, Walczak B. Use and abuse of chemometrics in chromatography. TrAC Trends Anal Chem 2006;25:1081–96.
[97] Andrade-Eiroa Á, Canle M, Cerdá V. Environmental applications of excitation-emission spectrofluorimetry: an in-depth review I. Appl Spectrosc Rev 2013;48:1–49.
[98] Ni YN, Zhang QL, Kokot S. Analysis of the interactions of mixtures of two β-agonists steroids with bovine serum albumin: a fluorescence spectroscopy and chemometrics investigation. Analyst 2010;135:2059–68.
[99] Muñoz de la Peña A, Espinosa Mansilla A, González Gómez D, Olivieri AC, Goicoechea HC. Interference-free analysis using three-way fluorescence data and the parallel factor model. Determination of fluoroquinolone antibiotics in human serum. Anal Chem 2003;75:2640–6.
[100] Rodríguez Cáceres M, Guiberteau Cabanillas A, Bohoyo Gil D, Martinez Canas M. Quantification of danofloxacin and difloxacin in chicken tissues in the presence of sarafloxacin as interference. J Agric Food Chem 2009;57:7627–33.
[101] Zou HY, Wu HL, Zhang Y, Li SF, Nie JF, Fu HY, et al. Studying the interaction of pirarubicin with DNA and determining pirarubicin in human urine samples: combining excitation-emission fluorescence matrices with second-order calibration methods. J Fluoresc 2009;19:955–66.
[102] Montoliu I, Martin FPJ, Collino S, Rezzi S, Kochhar S. Multivariate modeling strategy for intercompartmental analysis of tissue and plasma 1H NMR spectrotypes. J Proteome Res 2009;8:2397–406.
[103] Olivieri AC, Arancibia JA, de la Peña AM, Duran-Meras I, Mansilla AE. Second-order advantage achieved with four-way fluorescence excitation-emission-kinetic data processed by parallel factor analysis and trilinear least-squares. Determination of methotrexate and leucovorin in human urine. Anal Chem 2004;76:5657–66.
[104] Damiani PC, Durán-Merás I, García-Reiriz A, Jiménez-Girón A, Muñoz de la Peña A, Olivieri AC. Multiway partial least-squares coupled to residual trilinearization: a genuine multidimensional tool for the study of third-order data. Simultaneous analysis of procaine and its metabolite p-aminobenzoic acid in equine serum. Anal Chem 2007;79:6949–58.

[105] Zhang SR, Wu HL, Chen Y, Zhang XH, Wang JY, Li Y, et al. An investigation on hydrogen bonding between 3-methylindole and ethanol using trilinear decomposition of fluorescence excitation–emission matrices. Chemom Intell Lab Syst 2013;121:9–14.
[106] Tan FY, Tan C, Zhao AP, Li ML. Simultaneous determination of free amino acid content in tea infusions by using high-performance liquid chromatography with fluorescence detection coupled with alternating penalty trilinear decomposition algorithm. J Agric Food Chem 2011;59:10839–47.
[107] Zhu SH, Wu HL, Li BR, Xia AL, Han QJ, Zhang Y, et al. Determination of pesticides in honey using excitation–emission matrix fluorescence coupled with second-order calibration and second-order standard addition methods. Anal Chim Acta 2008;619:165–72.
[108] Yu YJ, Wu HL, Shao SZ, Kang C, Zhao J, Wang Y, et al. Using second-order calibration method based on trilinear decomposition algorithms coupled with high performance liquid chromatography with diode array detector for determination of quinolones in honey samples. Talanta 2011;85:1549–59.
[109] Zhang XH, Wu HL, Yin XL, Li LH, Wang JY, Chen Y, et al. A combined theoretical and experimental study for the chiral discrimination of naproxen enantiomers by molecular modeling and second-order standard addition method. Anal Methods 2013;5:710–7.
[110] Christensen J, Nørgaard L, Bro R, Engelsen SB. Multivariate autofluorescence of intact food systems. Chem Rev 2006;106:1979–94.
[111] Parastar H, Radović JR, Jalali-Heravi M, Diez S, Bayona JM, Tauler R. Resolution and quantification of complex mixtures of polycyclic aromatic hydrocarbons in heavy fuel oil sample by means of GC × GC-TOFMS combined to multivariate curve resolution. Anal Chem 2011;83:9289–97.
[112] Li YN, Wu HL, Qing XD, Li Q, Li SF, Fu HY, et al. Quantitative analysis of triazine herbicides in environmental samples by using high performance liquid chromatography and diode array detection combined with second-order calibration based on an alternating penalty trilinear decomposition algorithm. Anal Chim Acta 2010;678:26–33.
[113] Li YN, Wu HL, Qing XD, Nie CC, Li SF, Yu YJ, et al. The maintenance of the second-order advantage: second-order calibration of excitation–emission matrix fluorescence for quantitative analysis of herbicide napropamide in various environmental samples. Talanta 2011;85:325–32.
[114] Wang JY, Wu HL, Chen Y, Sun YM, Yu YJ, Zhang XH, et al. Fast analysis of synthetic antioxidants in edible vegetable oil using trilinear component modeling of liquid chromatography–diode array detection data. J Chromatogr A 2012;1264:63–71.
[115] García-Reiriz A, Damiani PC, Olivieri AC, Cañada-Cañada F, Muñoz de la Peña A. Nonlinear four-way kinetic-excitation-emission fuorescence data processed by a variant of parallel factor analysis and by a neural network model achieving the second-order advantage: malonaldehyde determination in olive oil samples. Anal Chem 2008;80:7248–56.
[116] Ding YJ, Wu HL, Xia AL, Cui H, Yu RQ. Simultaneous determination of two antituberculosis drugs using alternating fitting residue algorithm combined with HPLC-DAD. J Anal Sci 2008;24:1.
[117] Ouyang LQ, Wu HL, Nie JF, Zhang Y, Zou HY, Fu HY, et al. Simultaneous determination of psoralen and isopsoralen in plasma and chinese medicine Xian Ling Gu Bao capsule by using HPLC-DAD coupled with alternating trilinear decomposition algorithm. Anal Chim Acta 2009;650:160–6.
[118] Zou HY, Wu HL, OuYang LQ, Zhang Y, Nie JF, Fu HY, et al. Fluorescent quantification of terazosin hydrochloride content in human plasma and tablets using second-order

calibration based on both parallel factor analysis and alternating penalty trilinear decomposition. Anal Chim Acta 2009;650:143–9.

[119] Gindy AE, Hadad GM. Chemometrics in pharmaceutical analysis: an introduction, review, and future perspectives. J AOAC Int 2012;95:609–23.

[120] Azzouz T, Tauler R. Application of multivariate curve resolution alternating least squares (MCR-ALS) to the quantitative analysis of pharmaceutical and agricultural samples. Talanta 2008;74:1201–10.

[121] Nie JF, Wu HL, Zhu SH, Han QJ, Fu HY, Li SF, et al. Simultaneous determination of 6-methylcoumarin and 7-methoxycoumarin in cosmetics using three-dimensional excitation–emission matrix fluorescence coupled with second-order calibration methods. Talanta 2008;75:1260–9.

[122] Nie JF, Wu HL, Wang XM, Zhang Y, Zhu SH, Yu RQ. Determination of testosterone propionate in cosmetics using excitation-emission matrix fluorescence based on oxidation derivatization with the aid of second-order calibration methods. Anal Chim Acta 2008;628:24–32.

[123] Tu DZ, Wu HL, Li YN, Zhang J, Li Y, Nie CC, et al. Measuring estriol and estrone simultaneously in liquid cosmetic samples using second-order calibration coupled with excitation–emission matrix fluorescence based on region selection. Anal Methods 2012;4:222–9.

[124] Liu JJ, Ma X, Wen YD, Wang Y, Cai WS, Shao XG. Online near-infrared spectroscopy combined with alternating trilinear decomposition for process analysis of industrial production and quality assurance. Ind Eng Chem Res 2011;50:7677–81.

[125] Malwade CR, Qu H, Rong B-G, Christensen LP. Chemometrics for analytical data mining in separation process design for recovery of Artemisinin from Artemisia annua. Ind Eng Chem Res 2014;53:5582–9.

[126] Tan YX, Jiang JH, Wu HL, Cui H, Yu RQ. Resolution of kinetic system of simultaneous degradations of chlorophyll *a* and *b* by PARAFAC. Anal Chim Acta 2000;412:195–202.

[127] Liu YJ, Wu HL, Nie JF, Wang JY, Ouyang LQ, Yu RQ. Simultaneous determination of dextromethorphan and its metabolite dextrorphan in plasma samples using second-order calibration coupled with excitation-emission matrix fluorescence. Anal Sci 2010;27:663–6.

[128] Gu HW, Wu HL, Liu YJ, Yin XL, Su ZY, Li SS, et al. Simultaneous determination of metoprolol and α-hydroxymetoprolol in human plasma using excitation-emission matrix fluorescence coupled with second-order calibration methods. Bioanalysis 2012;4:2781–93.

[129] Wang Y, Wu HL, Yu YJ, Wang JY, Zhao J, Chen Y, et al. Simultaneous determination of three anticancerogens in plasma samples using second-order calibration method based on alternating trilinear decomposition coupled with fluorescence spectroscopy. Sci Sinica Chim 2012;42:1226–33.

[130] Wang JY, Wu HL, Chen Y, Zhai M, Qing XD, Yu RQ. Quantitative determination of butylated hydroxyanisole and n-propyl gallate in cosmetics using three-dimensional fluorescence coupled with second-order calibration. Talanta 2013;116:347–53.

[131] Kang C, Wu HL, Xiang SX, Xie LX, Liu YJ, Yu YJ, et al. Simultaneous determination of aromatic amino acids in different systems using three-way calibration based on the PARAFAC-ALS algorithm coupled with EEM fluorescence: exploration of second-order advantages. Anal Methods 2014;6:6358–68.

[132] Muroski AR, Booksh KS, Myrick ML. Single-measurement excitation/emission matrix spectrofluorometer for determination of hydrocarbons in ocean water. 1. Instrumentation and background correction. Anal Chem 1996;68:3534–8.

[133] Lakowicz JR. Principles of fluorescence spectroscopy. 3rd ed. New York: Springer; 2006.

[134] Olivieri AC. Analytical advantages of multivariate data processing. One, two, three, infinity? Anal Chem 2008;80:5713–20.
[135] Wold H, Lyttkens E. Nonlinear iterative partial least squares (NIPALS) estimation procedures. Bulletin of the International Statistical Institute: proceedings of the 37th session, London, 43:1969. p. 29–51.
[136] Thygesen LG, Rinnan Å, Barsberg S, Møller JK. Stabilizing the PARAFAC decomposition of fluorescence spectra by insertion of zeros outside the data area. Chemom Intell Lab Syst 2004;71:97–106.
[137] Chen ZP, Liu Z, Cao YZ, Yu RQ. Efficient way to estimate the optimum number of factors for trilinear decomposition. Anal Chim Acta 2001;444:295–307.
[138] Meyers RA, Albani JR. Encyclopedia of analytical chemistry. New York: John Wiley & Sons; 2000.
[139] Zhang Y, Wu HL, Ding YJ, Xia AL, Cui H, Yu RQ. Simultaneous determination of cortisol and prednisolone in body fluids by using HPLC–DAD coupled with second-order calibration based on alternating trilinear decomposition. J Chromatogr B 2006;840:116–23.
[140] Zhang Y, Wu HL, Xia AL, Han QJ, Cui H, Yu RQ. Interference-free determination of Sudan dyes in chilli foods using second-order calibration algorithms coupled with HPLC-DAD. Talanta 2007;72:926–31.
[141] Zhao J, Wu HL, Niu JF, Yu YJ, Yu LL, Kang C, et al. Chemometric resolution of coeluting peaks of eleven antihypertensives from multiple classes in high performance liquid chromatography: a comprehensive research in human serum, health product and Chinese patent medicine samples. J Chromatogr B 2012;902:96–107.
[142] Liu YJ, Wu HL, Zhu SH, Kang C, Xu H, Su ZY, et al. Rapid determination of costunolide and dehydrocostuslactone in human plasma sample and chinese patent medicine Xiang Sha Yang Wei capsule using HPLC-DAD coupled with second-order calibration. Chin J Chem 2012;30:1137–43.
[143] Yu YJ, Wu HL, Niu JF, Zhao J, Li YN, Kang C, et al. A novel chromatographic peak alignment method coupled with trilinear decomposition for three dimensional chromatographic data analysis to obtain the second-order advantage. Analyst 2013;138:627–34.
[144] Zhang XH, Wu HL, Wang JY, Tu DZ, Kang C, Zhao J, et al. Fast HPLC-DAD quantification of nine polyphenols in honey by using second-order calibration method based on trilinear decomposition algorithm. Food Chem 2013;138:62–9.
[145] Wang JY, Wu HL, Sun YM, Gu HW, Liu Z, Liu YJ, et al. Simultaneous determination of phenolic antioxidants in edible vegetable oils by HPLC–FLD assisted with second-order calibration based on ATLD algorithm. J Chromatogr B 2014;947:32–40.
[146] Sun YM, Wu HL, Wang JY, Liu Z, Zhai M, Yu RQ. Simultaneous determination of eight flavonoids in propolis using chemometrics-assisted high performance liquid chromatography-diode array detection. J Chromatogr B 2014;962:59–67.
[147] Chobanian AV, Bakris GL, Black HR, Cushman WC, Green LA, Izzo JL, et al. Seventh report of the joint national committee on prevention, detection, evaluation, and treatment of high blood pressure. Hypertension 2003;42:1206–52.
[148] Asmar R, Pannier B, Brisac A-M, Tichet J, El Hasnaoui A. High blood pressure and associated cardiovascular risk factors in France. J Hypertens 2001;19:1727–32.
[149] Hill J, Bullock I, Alderson P. A summary of the methods that the national clinical guideline centre uses to produce clinical guidelines for the national institute for health and clinical excellence. Ann Intern Med 2011;154:752–7.
[150] Chalmers J, MacMahon S, Mancia G, WHO-ISH hypertension guidelines committee. 1999 world health organization-international society of hypertension guidelines for the management of hypertension. J Hypertens 1999;17:151–85.

[151] Xiong Q, Fan W, Tezuka Y, Adnyana IK, Stampoulis P, Hattori M, et al. Hepatoprotective effect of Apocynum venetum and its active constituents. Planta Med 2000;66:127–33.
[152] Kim DW, Yokozawa T, Hattori M, Kadota S, Namba T. Effects of aqueous extracts of Apocynum venetum leaves on spontaneously hypertensive, renal hypertensive and NaCl-fed-hypertensive rats. J Ethnopharmacol 2000;72:53–9.
[153] Elgawish MS, Mostafa SM, Elshanawane AA. Simple and rapid HPLC method for simultaneous determination of atenolol and chlorthalidone in spiked human plasma. Saudi Pharm J 2011;19:43–9.
[154] Takeshi D, Sansei N, Yoshihisa N. Constituents and pharmacological effects of Eucommia and Siberian ginseng. Acta Pharmacol Sin 2001;22:1057–70.
[155] Yen GC, Hsieh CL. Antioxidant activity of extracts from Du-zhong (Eucommia ulmoides) toward various lipid peroxidation models in vitro. J Agric Food Chem 1998;46:3952–7.
[156] Qing XD, Wu HL, Yan XF, Li Y, Ouyang LQ, Nie CC, et al. Development of a novel alternating quadrilinear decomposition algorithm for the kinetic analysis of four-way room-temperature phosphorescence data. Chemom Intell Lab Syst 2014;132:8–17.
[157] Goicoechea HC, Yu S, Olivieri AC, Campiglia AD. Four-way data coupled to parallel factor model applied to environmental analysis: determination of 2, 3, 7, 8-tetrachloro-dibenzo-para-dioxin in highly contaminated waters by solid–liquid extraction laser-excited time-resolved Shpol'skii spectroscopy. Anal Chem 2005;77:2608–16.
[158] Maggio RM, Muñoz de la Peña A, Olivieri AC. Unfolded partial least-squares with residual quadrilinearization: a new multivariate algorithm for processing five-way data achieving the second-order advantage. Application to fourth-order excitation-emission-kinetic-pH fluorescence analytical data. Chemom Intell Lab Syst 2011;109:178–85.
[159] Kang C, Wu HL, Xie LX, Xiang SX, Yu RQ. Direct quantitative analysis of aromatic amino acids in human plasma by four-way calibration using intrinsic fluorescence: exploration of third-order advantages. Talanta 2014;122:293–301.
[160] Rodríguez N, Ortiz MC, Sarabia LA. Fluorescence quantification of tetracycline in the presence of quenching matrix effect by means of a four-way model. Talanta 2009;77:1129–36.
[161] Goicoechea HC, Yu S, Moore AFT, Campiglia AD. Four-way modeling of 4.2 K time-resolved excitation emission fluorescence data for the quantitation of polycyclic aromatic hydrocarbons in soil samples. Talanta 2012;101:330–6.
[162] Machicote RG, Bruzzone L. Simultaneous determination of carbaryl and 1-naphthol by first-derivative synchronous non-protected room temperature phosphorescence. Anal Sci 2009;52:623–6.

Chapter 5

# Multivariate Curve Resolution for Quantitative Analysis

Romà Tauler*,[1] and Anna de Juan†,[1]
*Department of Environmental Chemistry, IDAEA-CSIC, Barcelona, Spain
†Department of Analytical Chemistry, Universitat de Barcelona, Barcelona, Spain
[1]Corresponding authors: e-mail: Roma.Tauler@idaea.csic.es; anna.dejuan@ub.edu

## Chapter Outline



## 1 MCR: BILINEAR MODEL AND AMBIGUITIES (PERMUTATION, SCALE, ROTATION)

Multivariate curve resolution (MCR) is the generic denomination for a family of methods meant to solve the mixture analysis problem by providing a chemically meaningful bilinear model of pure contributions from the sole information of an original data matrix, **D**, including a mixed measurement [1–6]. Such a bilinear model can be described as the sum of the products of dyads of profiles related to pure contributions (see Equation 1) or in the classical compact form of a bilinear model (see Equation 2)

Data Handling in Science and Technology, Vol. 29. http://dx.doi.org/10.1016/B978-0-444-63527-3.00005-9

$$\mathbf{D} = \sum\nolimits_i \mathbf{c}_i \mathbf{s}_i^{\mathrm{T}} + \mathbf{E} \qquad (1)$$

$$\mathbf{D} = \mathbf{C}\mathbf{S}^{\mathrm{T}} + \mathbf{E}. \qquad (2)$$

In these two equations, **D** represents the original data set, which often contains instrumental responses, e.g., spectra, collected as a function of a certain variable, e.g., elution time in chromatography. When defining the dyads in Equation (1), $\mathbf{s}_i^{\mathrm{T}}$ is usually the pure response profile of a component and $\mathbf{c}_i$ the related concentration profile. Equation (2) is the form most often used to express the bilinear MCR model. In many examples of MCR analysis, the row profiles in matrix $\mathbf{S}^{\mathrm{T}}$ refer to the qualitative information of the components in the system analyzed (pure instrumental responses or basic signatures of different kinds), whereas the column profiles in matrix **C** express the abundance or apportionment of the qualitative pure responses along the row direction of the data set.

Traditionally, MCR was initially conceived for evolutionary analytical data coming from a process or an analytical measurement [2–5]. The reason is that many analytical measurements are described by a natural bilinear model of pure signal contributions, e.g., the Beer–Lambert law in spectroscopy. Therefore, the first examples of MCR analysis were processes (reactions) monitored spectroscopically or chromatographic elutions with a multivariate spectroscopic detection [7,8] (see Figure 1a). MCR was also meant to take advantage of the evolutionary character in the concentration direction (related to a process, to a chromatographic elution). This is why MCR is traditionally applied to data sets where both directions in the bilinear model have a clear meaning.

Nowadays, the application of MCR has grown significantly and the fields of application have gained in complexity and diversity. Within the spectroscopic field, the structure in the concentration direction is no longer a requirement and this has allowed analyzing hyperspectral images, in which the image cube with two spatial directions ($x$ and $y$) and a spectroscopic direction can be unmixed previous unfolding of the cube into a data matrix with rows designing the pixel spectra and columns the spectral channels measured. After MCR, an operation of refolding of the concentration profiles is needed to recover the spatial structure of the distribution maps of the pure image components (see Figure 1b) [9–11].

Whether the pure contributions of a data set follow a natural bilinear model or not, MCR can always be used to furnish a simple and meaningful bilinear description of the variation in the data that may include some natural properties related to the problem of interest. The concept of component is then redefined according to the new kind of problem studied. This would be the case of environmental data, which can be described with bilinear models of contamination sources, where source composition profiles can be appropriately weighted by apportionment profiles linked to geographic/season

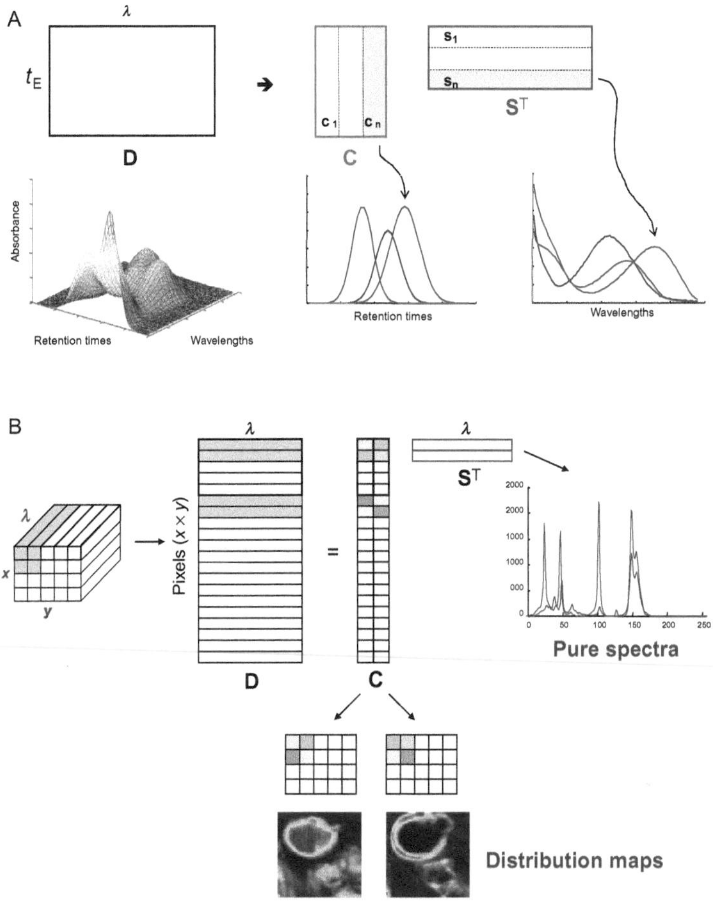


FIGURE 1 Bilinear model provided by MCR for: (a) an HPLC-DAD chromatographic data set and (b) a hyperspectral image.

information [12,13]. Within the -omics field, genomic data, such as those coming from DNA microarrays, could also be described by a bilinear model formed by dyads having basic gene expression signatures linked to sample profiles of cell lines [14,15].

Whereas the main asset of MCR models is the simplicity and ease of interpretation because of the chemically meaningfulness of the resolved profiles, the main limitation is the uncertainty present in this model, linked specially to the phenomena of ambiguity, in addition to the data modeling ubiquitously present noise propagation effects. The concept of ambiguity means that different combinations of sets of concentration profiles in **C** and response profiles in $\mathbf{S}^T$ can reproduce the original data set with the same fit quality [16]. Three different kinds of ambiguity can be observed:

- *Permutation ambiguity*. There is no sorting order on the MCR components. Therefore, they can be shuffled in the **C** and $\mathbf{S}^T$ matrix (always keeping the right correspondence of the dyads) and obtain identical results (see Figure 2a).
- *Intensity ambiguity*. Dyads of profiles having the same shape but different relative scales between concentration profile and related pure response profile reproduce equally well the original data set. This is the reason why concentration values and pure response intensities in $\mathbf{c}_i$ and $\mathbf{s}_i^T$ profiles are always in arbitrary units unless reference information on real intensities is available and actively used in the resolution process. This ambiguity can be understood through the two equivalent equations below:

$$\mathbf{D} = \sum\nolimits_i \mathbf{c}_i \mathbf{s}_i^T \tag{3}$$

$$\mathbf{D} = \sum\nolimits_i (\mathbf{c}_i k_i)\left(\mathbf{s}_i^T \frac{1}{k_i}\right), \tag{4}$$

where $k_i$ is a scaling factor that can be different for each pure component in the system. Normalization of concentration profiles or resolved spectra or use of reference concentration values within the optimization helps to suppress the intensity ambiguity (see Figure 2b).

- *Rotational ambiguity*. Sets of concentration profiles and spectra with different shapes can reproduce the original data set with the same fit quality. This is the most relevant kind of ambiguity and can be expressed as follows:

$$\mathbf{D} = \mathbf{C}\mathbf{S}^T + \mathbf{E} \tag{2}$$

$$\mathbf{D} = (\mathbf{C}\mathbf{T})\left(\mathbf{T}^1\mathbf{S}^T\right) + \mathbf{E}, \tag{5}$$

where **T** is any transformation matrix which keeps providing profiles that still obey the constraints set in the MCR optimization (see Figure 2c).

The main way to decrease/suppress ambiguities in MCR solutions is by means of the introduction of constraints and by the use of multiset data structures, as it will be described in next sections [17,18].

There are several approaches to assess the ambiguity in MCR solutions [16]. To find the full range of feasible solutions, only few approaches exist and they cannot go beyond systems with four components [19–21]. Other algorithms find the minimum and maximum boundaries of a certain objective function, i.e., amount of component signal over total signal with dyads of profiles obeying the constraints (also called signal (component) contribution function, SCF, of the considered component) [22,23]

$$f_{i,\min} = \min \frac{\left\|\mathbf{c}_i\mathbf{s}_i^T\right\|}{\left\|\mathbf{C}\mathbf{S}^T\right\|}, \quad f_{i,\max} = \max \frac{\left\|\mathbf{c}_i\mathbf{s}_i^T\right\|}{\left\|\mathbf{C}\mathbf{S}^T\right\|}. \tag{6}$$

The results provided are the two extreme MCR dyads related to the minimum and maximum of the objective function for each component in the

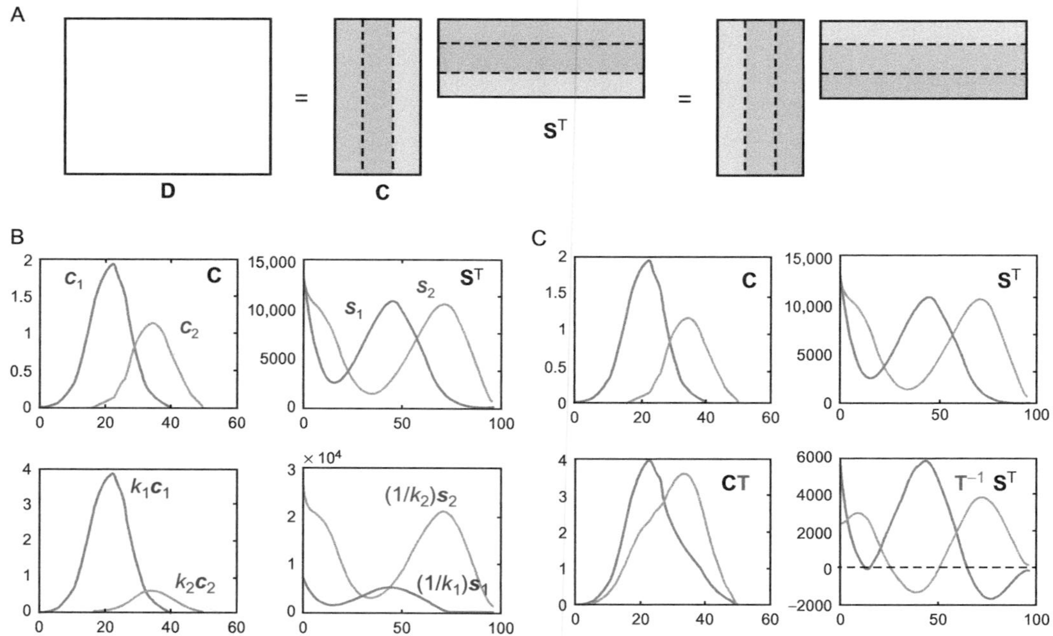


**FIGURE 2** Graphical representations of the ambiguity phenomenon in MCR. (a) Permutation ambiguity (dyads of profiles can be shuffled and the data set reproduction remains invariant), (b) intensity ambiguity (top plots: real solution; bottom plots: feasible solution with scale modifications), and (c) rotational ambiguity (top plots: real solution; bottom plots: feasible rotated solution).

system. These methods can work with systems with unlimited of number of components. When ambiguity is assessed, two different aspects should be examined, namely:

(1) *The extent of ambiguity (linked to each individual component)*, by calculating $f_{i,\max} - f_{i,\min}$. If the subtraction is zero, the component is resolved in a unique way. The higher the difference, the larger the ambiguity associated with the resolution of a particular component.
(2) *The location of ambiguity (linked separately to each of the profiles in the dyad of a pure component).* Plotting the dyads of profiles related to $f_{i,\max}$ and $f_{i,\min}$ for a particular component, it can be seen where the ambiguity affects are affecting concentration profiles and/or pure response profiles for a particular component. It is relevant to note that ambiguity is a concept related to each component in the system and to each profile within a dyad. Therefore, an ambiguous solution of MCR can be very satisfactory if the components of interest, e.g., analytes in a calibration problem, are uniquely resolved or affected by a very low degree of ambiguity, even if the rest of the system is not perfectly defined.

Other alternative approaches have been proposed to calculate the boundaries of the areas of MCR feasible solutions (AFS). These approaches have an undoubtedly importance from a theoretical point of view and to understand rotation ambiguity of MCR solutions. However, their application until now has been restricted to systems with a relatively small number of components (up to four) and under a limited number of constraints (nonnegativity and normalization). See Refs. [19–21] for a more detailed description of these other methods of evaluation of rotation ambiguities and MCR feasible band boundaries.

## 2 MCR-ALS: MODUS OPERANDI AND CONSTRAINTS

There are different MCR algorithms meant to provide the bilinear model of pure contributions from the sole mixed original data table. The main distinction is among noniterative methods and iterative methods [24,25]. Most noniterative methods were conceived for data sets with a structured/sequential structure in the concentration direction that allowed defining the windows of existence of the different components. Then, subspaces combining suitably windows with different conditions of component overlap (in terms of component identity and rank) were furnishing either the concentration profiles or the spectra, and the counterpart of the bilinear model was obtained by a single least-squares step. These algorithms are nowadays less in use because setting the concentration windows in systems with too complex or nonstructured concentration direction becomes too difficult and no extension of these algorithms for multiset analysis.

Iterative methods work by optimizing a set of initial estimates (guesses) of concentration profiles or spectra under the action of constraints until a convergence criterion is achieved [24]. The most known are ITTFA (iterative target transformation factor analysis) [26,27] and multivariate curve resolution-alternating least-squares (MCR-ALS) [4,5,17,18,25]. ITTFA works optimizing the concentration profile direction under constraints and recovering the spectra matrix in the bilinear model by least-squares, whereas MCR-ALS performs an alternating optimization of both concentration profiles and pure spectra in each iterative cycle.

Iterative methods are usually the preferred option nowadays because it is not needed to have a structured concentration direction for them to be applied and much chemical and/or mathematical information can be introduced in the optimization process under the form of constraints [4,5,17,24]. Besides, algorithms such as MCR-ALS have been adapted to work with multiset structures and have allowed for the introduction of many different constraints related to chemical information [4,5,18,24], to mathematical properties (rank conditions or model structure) [17,28–30], or to data analysis tasks usually associated with other kind of algorithms (e.g., deterministic hard-modeling [31] or calibration [32]). This is why MCR-ALS is the method that will be described more in detail and taken as the reference method in all that is described in next sections.

MCR-ALS works on a data set by optimizing initial estimates (concentration profiles or spectra) in an ALS way within each iterative cycle under the action of suitable constraints until a convergence criterion is fulfilled. The convergence criterion can be a preset number of iterations or a threshold value defining the difference in fit improvement between consecutive iterations. Once the optimization process has been finished, the MCR results are the set of concentration profiles and spectra and quality parameters related to the model fit, such as the variance explained or the lack of fit (LOF)

$$\%\text{LOF} = 100\sqrt{\frac{\sum_{i,j} e_{ij}^2}{\sum_{i,j} d_{ij}^2}}, \tag{7}$$

where $d_{ij}$ is an element of the data matrix **D** and $e_{ij}$ is the related residual.

To obtain satisfactory results, it is relevant to pay attention to the steps of selection of initial estimates and constraints. Initial estimates are the starting point of the optimization process and one has to try to begin with sensible guesses. On the other hand, constraints are the properties that the profiles should obey. Therefore, selecting the appropriate ones and knowing how to apply them is the most crucial point to ensure obtaining meaningful and reliable solutions, affected by as less uncertainty (ambiguity) as possible.

Initial estimates in MCR-ALS can be concentration profiles or pure responses (or, in general, pure column or row profiles). The golden rule is starting with sensible chemically meaningful initial guesses, which present the same general properties that will be sought in the final resolved profiles, i.e., initial random values are of no help if profiles with clearly defined characteristics are sought [4,5,21]. Initial estimates can come from previous knowledge if available or can be provided by methods devoted to this purpose.

The main distinction is between methods designed to obtain initial estimates for process-like data and general methods that can work when no structure in the concentration direction is available. Among methods designed for process-like data, evolving factor analysis (EFA) is the most known [33,34]. EFA is a local rank analysis method that detects the emergence and decay of the components in the data set and provides concentration profiles assuming a sequential order of emergence-decay for all components in the system. EFA provides a set of initial estimates and relevant additional information, such as the windows of existence of the components in the system (used to set selectivity or local rank constraints, see afterward).

Initial estimates can also be obtained by the so-called pure variable selection methods, which can work irrespectively of the presence or absence of a structured concentration direction in the data set [35]. Some of the most known are simple-to-use self-modeling analysis (SIMPLISMA) [36], orthogonal projection approach [37], and key set factor analysis [38]. All of them aim at selecting the most dissimilar rows or columns in a data set providing initial estimates of spectra or of concentration profiles, respectively. A general advice in using these methods is applying them in the direction of least overlap among components.

After having selected appropriate initial estimates, the MCR-ALS optimization of concentration profiles and pure responses under constraints can start. Constraints are the corner stone of iterative methods and can be defined as systematic properties that are used to give a meaningful shape to the row and column profiles in the bilinear model of the data set. Constraints are implemented as mathematical conditions and drive the MCR optimization process to the final solutions. There are constraints linked to natural properties of the measurement of the data set and others that express mathematical conditions or models [16,17,24] (see Figure 3).

Among the most known natural constraints are:

- *Nonnegativity*. Forces the profiles to be formed by null or positive values. Can be implemented replacing negative values by zeros or with softer algorithms, such as nonnegative least-squares or fast nonnegative least-squares. It applies to all concentration profiles and to many instrumental responses.
- *Unimodality*. The presence of a single maximum per profile is allowed. It is the condition fulfilled by peak-shaped signals, e.g., elution profiles and monotonic reaction profiles (always increasing or decreasing).
- *Closure*. It is the expression of the mass balance condition. Applies to some concentration profiles of reaction systems.
- *Known pure spectra/concentration profiles*. This is a kind of equality constraint, which makes the concentration profile and/or spectrum of a component to be equal to a certain known predefined shape.

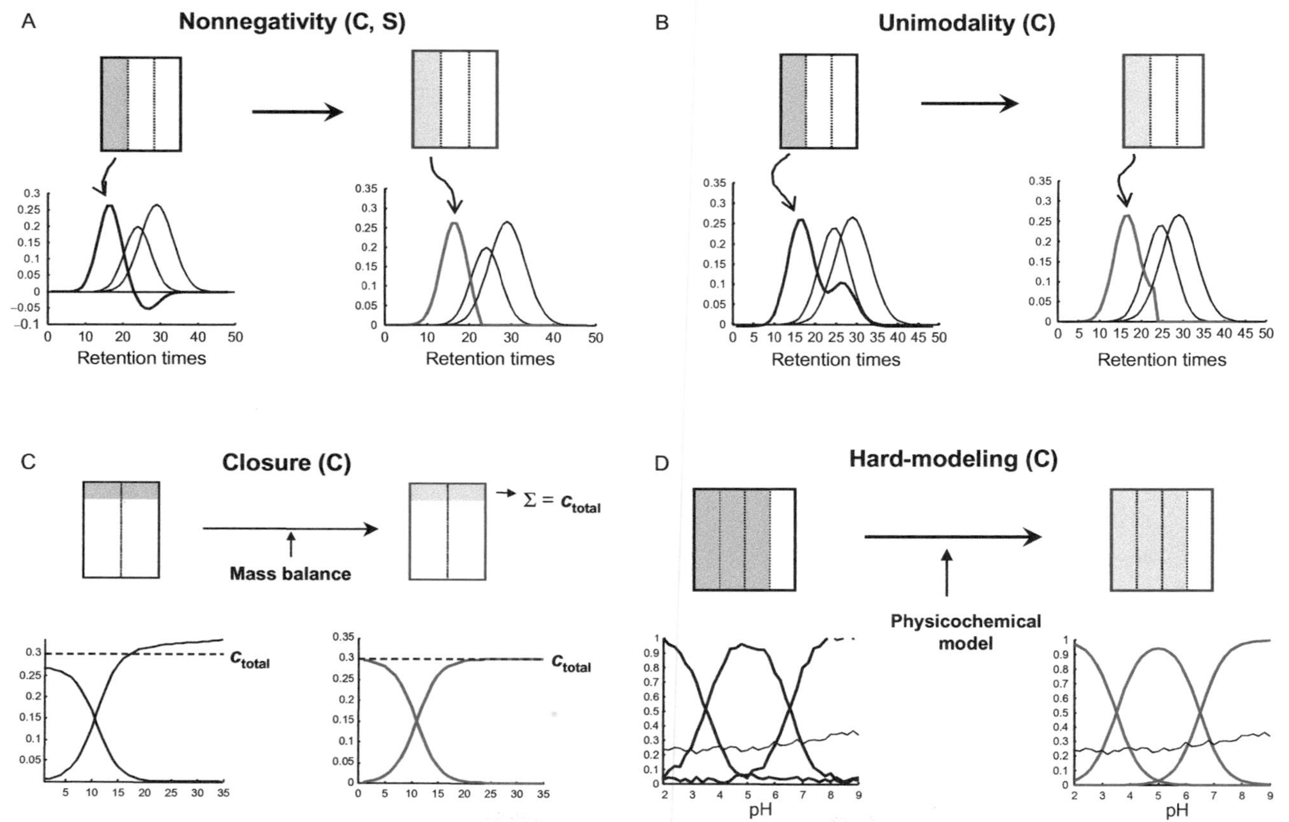


**FIGURE 3** Examples of common constraints used in MCR. (a) nonnegativity, (b) unimodality, (c) closure, and (d) hard-modeling. Bold black profiles on the left plots are unconstrained. Red (gray in the print version) profiles on the right profiles have been constrained.

Constraints related to mathematical conditions are:

- *Local rank/selectivity*. It relates to the zones of absence of components in some regions (windows) in the profiles. Usually, it applies to concentration profiles. The zero-concentration regions in the different profiles are determined by local rank analysis methods, such as EFA [33,34]. When only a component is present in a certain concentration window, we have selectivity. When only some components are absent, we talk about local rank information. Suitable local rank conditions suppress the ambiguity in the profiles retrieved by MCR [39].
- *Correspondence of species*. This constraint is only applied when several data matrices from different samples are treated together (see Section 3) and expressed the correspondence and presence/absence of components in the analyzed samples [5,17,18,40]. This constraint is very relevant to suppress ambiguity, similarly to that of local rank. It will also be of very much use in calibration problems because the composition of standard samples (presence/absence of analytes and interferences) is very well known and should be actively used.
- *Model constraints*. These are constraints applicable to multiset and only to the augmented matrices that may be present in the decomposition model. Although MCR gives by default a bilinear model, the conditions of trilinearity [18,28], multilinearity [30], or factor interaction (like in Tucker models) [29] can be implemented in a component-wise way. Therefore, completely bilinear, completely trilinear, or hybrid models can be set to be obtained in the final MCR results (see Section 5.4 for more detail).

Finally, a last group of constraints is linked to some tasks that can be done integrated in the MCR process.

- *Hard-modeling*. Forces the concentration profiles to be fitted by a parametric physicochemical model and the parameters of the model are obtained as an additional output. It can also be applied to pure analytical responses when the shape of the pure signal can be defined by a parametric equation. This constraint implies to perform a model fitting task during the iterative optimization process [18,31]. When this constraint is applied, the related concentration profiles do not present ambiguity. It is relevant to note that the hard-modeling constraint and, particularly, some of the fitted parameters, e.g., total concentration in equilibrium or enzymatic models [41,42], have been used for quantitative analysis purposes.
- *Correlation constraint*. In this case, internal univariate calibration models can be applied to particular concentration profiles of components in the system. The model is established between concentration values obtained with MCR (in arbitrary units) and real concentrations in calibration samples. The model is used to predict real concentrations in unknown samples [5,32,43,44]. Much more detail on this constraint will be given in Section 5.4.

All constraints are optionally applied and can affect differently the row and column profiles of the data set, the different components in the data set, and the different subsets in a multiset structure. The application of constraints can also be modulated according to tolerance criteria. The flexibility in the application of the constraints explains the versatility of MCR algorithms, which can adapt to very diverse scenarios through the proper selection of these restricting conditions.

## 3 MCR APPLIED TO MULTISET DATA: MATRIX AUGMENTATION

MCR analysis can be empowered significantly when multiple data sets or multiset data are simultaneously analyzed using the extension of this method [17,40,45]. The usual way to perform such an analysis is applying MCR to augmented data matrices.

A single data matrix can be row-wise, column-wise, or row- and column-wise augmented to form a new multiset data structure appending in the appropriate direction other data matrices either with the same number of rows, columns, or both simultaneously (see Figure 4). To have a new meaningful data structure, all individual data matrices (data slices) in this type of data set should share some information with the other appended matrices; otherwise, it does not make sense to produce such a new data arrangement.

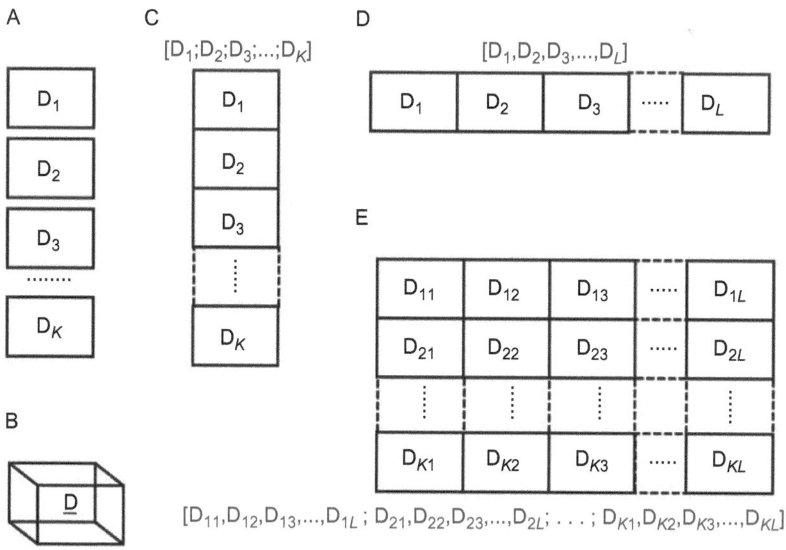


**FIGURE 4** Multiset data arrangements. (a) Single data matrices, (b) multiway data set (data cube), and (c–e) multiset data arrangements. (c) Column-wise augmented data matrix, (d) row-wise augmented data matrix, and (e) row- and column-wise augmented data matrix.

The first outstanding advantage of this new augmented data matrix is that the good features for resolution presented by one or some of the included data matrices will have always a positive effect on the resolution of the most complex ones. Introducing selectivity or favorable local rank conditions as constraints during MCR-ALS optimization into this new type of augmented data sets is generally easy and possible [39,46]. This can be done, for instance, by appending individual data matrices coming from simpler systems or processes or by appending data matrices from the pure response of one or few of the analytes (standards) present in a much more complex mixture data matrix [4,5,18].

As shown in Figure 4, three data arrangements are possible and may be written in a more concise way using MATLAB notation programming language [47]. Column-wise augmented matrices, $\mathbf{D}_{\text{aug}}$, are written as $[\mathbf{D}_1;\mathbf{D}_2;\mathbf{D}_3;\ldots;\mathbf{D}_K]$ where the semicolon ";" notation is used to indicate that the different data matrices $\mathbf{D}_k$, $k=1,\ldots,K$, are column-wise appended, one on top of each other, keeping the same number of columns. In this data arrangement, the different data matrices are supposed to share their column vector space. For instance, in the typical data arrangement where $\mathbf{D}_k$ matrices have wavelengths as column indices, appended data matrices are assumed to share some or all species/component pure spectra. Conversely, the row vector space of each data matrix is left unshared, to be independent of the other matrices. For this type of data matrix augmentation, the MCR bilinear model can be written using Equation (8)

$$\begin{pmatrix} \mathbf{D}_1 \\ \mathbf{D}_2 \\ \mathbf{D}_3 \\ \cdots \\ \mathbf{D}_K \end{pmatrix} = \begin{pmatrix} \mathbf{C}_1 \\ \mathbf{C}_2 \\ \mathbf{C}_3 \\ \cdots \\ \mathbf{C}_K \end{pmatrix} \mathbf{S}^{\mathrm{T}} + \begin{pmatrix} \mathbf{E}_1 \\ \mathbf{E}_2 \\ \mathbf{E}_3 \\ \cdots \\ \mathbf{E}_K \end{pmatrix} = \mathbf{C}_{\text{aug}}\mathbf{S}^{\mathrm{T}} + \mathbf{E}_{\text{aug}} \tag{8}$$

or, in the more compact (MATLAB notation) form:

$$\begin{aligned}[\mathbf{D}_1;\mathbf{D}_2;\mathbf{D}_3;\ldots;\mathbf{D}_K] &= [\mathbf{C}_1;\mathbf{C}_2;\mathbf{C}_3;\ldots;\mathbf{C}_K]\ \mathbf{S}^{\mathrm{T}} + [\mathbf{E}_1;\mathbf{E}_2;\mathbf{E}_3;\ldots;\mathbf{E}_K] \\ &= \mathbf{C}_{\text{aug}}\mathbf{S}^{\mathrm{T}} + \mathbf{E}_{\text{aug}} \end{aligned} \tag{9}$$

The bilinear model described by Equations (8) and (9) is formulated using a single spectra matrix $\mathbf{S}^{\mathrm{T}}$ with the pure spectra of the different components present in all the considered $\mathbf{D}_k$ data matrices, i.e., this matrix describes the changes in the column spectral vector space. In contrast, the augmented concentration matrix $[\mathbf{C}_1;\mathbf{C}_2;\mathbf{C}_3;\ldots;\mathbf{C}_K]$ describes freely the concentration changes of the resolved components in each related $\mathbf{D}_k$ data matrices. Therefore, the bilinear model of Equations (8) and (9) assumes that the components in the $\mathbf{D}_k$ data matrices included in the column-wise augmented data matrix have or share the same pure spectra, whereas they can have different concentration profiles. This augmentation strategy would be typically applied to a chemical systems monitored by the same spectroscopy and studied under different initial conditions. The chemical constituents in the different data sets (matrices) are the same (at least some of them) and they have the same spectra, but their concentration (reaction, elution, evolution) profiles can be different (in scale and in shape) depending on the conditions of the system. This is more the rule than the exception in chemical reaction systems.

Figure 5a displays the bilinear model of a data set arranged in a column-wise augmented data matrix. As it can be seen, this extended bilinear model is equally valid for data matrices formed with equal or unequal number of rows. Note that there is complete freedom in the shape of the concentration profiles (matrices $\mathbf{C}_1$, $\mathbf{C}_2$, and $\mathbf{C}_3$ in $\mathbf{C}_{\text{aug}}$) allows for appending matrices with different row sizes. This advantage is especially useful when the same chemical system is submitted to different types of stimuli or conditions, such as changes in the temperature, light, pH, and time scales monitored with the same kind of spectroscopic measurement, in which it is clear that the nature of row indices in the different data matrices will be completely different.

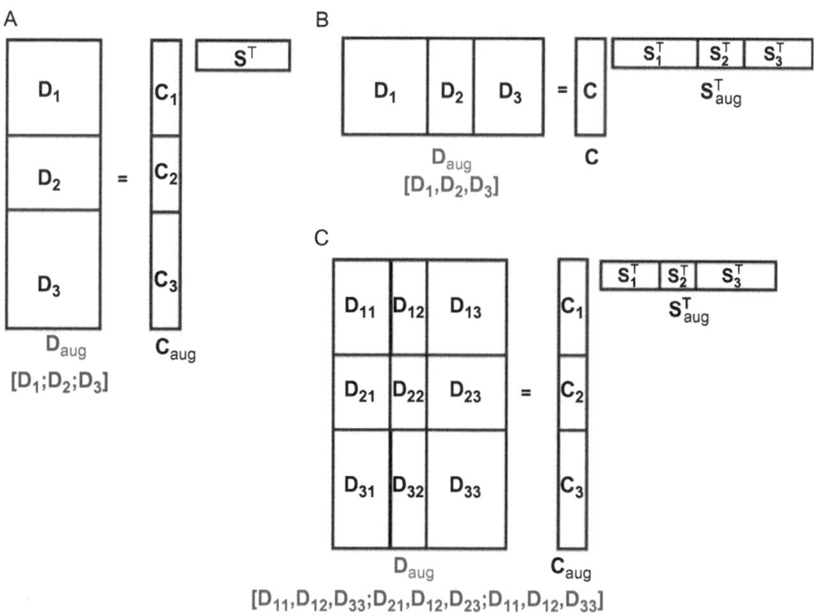


**FIGURE 5** Graphical representation of the bilinear model applied to different types of augmented data matrices. (a) Column-wise augmented data matrix, (b) row-wise augmented data matrix, and (c) column- and row-wise augmented data matrix.

Row-wise augmented data matrices are written as $[\mathbf{D}_1,\mathbf{D}_2,\mathbf{D}_3,\ldots,\mathbf{D}_l]$, where the comma "," notation is used now to indicate that the different data matrices $\mathbf{D}_l$, $l=1,\ldots,L$, are set one besides the other. In this data arrangement (Figure 5b), the different data matrices are supposed to share their row vector (concentration) space. For instance, this is the typical data arrangement formed by $\mathbf{D}_l$ matrices with the same samples as row indices. This would be the case in the investigation of the same chemical system using different spectroscopic methods or even mixing completely different types of instrumental analysis methods, such as voltammetry and spectroscopy [4,5]. In this case, the column vector space of each $\mathbf{D}_l$ data matrix would be unshared (the different spectroscopies or instrumental methods produce different type of spectra or signals), but the row (sample) space would be common. The extension of the MCR bilinear model can also be written for row-wise augmented data matrices (Equations 10 and 11)

$$\begin{aligned}[\mathbf{D}_1\ \ \mathbf{D}_2\ \ \mathbf{D}_3\ \ \ldots\ \ \mathbf{D}_L] &= \mathbf{C}\left[\mathbf{S}_1^{\mathrm{T}}\ \ \mathbf{S}_2^{\mathrm{T}}\ \ \mathbf{S}_3^{\mathrm{T}}\ \ \ldots\ \ \mathbf{S}_L^{\mathrm{T}}\right]+[\mathbf{E}_1\ \ \mathbf{E}_2\ \ \mathbf{E}_3\ \ \ldots\ \ \mathbf{E}_L] \\ &= \mathbf{C}\ \mathbf{S}_{\mathrm{aug}}^{\mathrm{T}}+\mathbf{E}_{\mathrm{aug}}\end{aligned} \tag{10}$$

or, in compact (MATLAB notation) form:

$$\begin{aligned}[\mathbf{D}_1,\mathbf{D}_2,\mathbf{D}_3,\ldots,\mathbf{D}_L] &= \mathbf{C}\left[\mathbf{S}_1^{\mathrm{T}},\mathbf{S}_2^{\mathrm{T}},\mathbf{S}_3^{\mathrm{T}},\ldots,\mathbf{S}_{\mathrm{L}}^{\mathrm{T}}\right]+[\mathbf{E}_1,\mathbf{E}_2,\mathbf{E}_3,\ldots,\mathbf{E}_L] \\ &= \mathbf{C}\ \mathbf{S}_{\mathrm{aug}}^{\mathrm{T}}+\mathbf{E}_{\mathrm{aug}}.\end{aligned} \tag{11}$$

The bilinear model described by Equations (10) and (11) shows a single concentration matrix $\mathbf{C}$ describing the changes in the concentration of the same components in all $\mathbf{D}_l$ data matrices, but an augmented spectra matrix $[\mathbf{S}_1^{\mathrm{T}},\mathbf{S}_2^{\mathrm{T}},\mathbf{S}_3^{\mathrm{T}},\ldots,\mathbf{S}_L^{\mathrm{T}}]$ is needed to describe the different spectra (instrumental responses) in the different individual data matrices. This data arrangement matches the expression "data fusion," where one particular system is investigated using different spectroscopic or different instrumental analysis techniques. Figure 5b displays that in this case, it is not necessary that the different appended data matrices have the same number of columns, only the same number of rows (samples) is needed to allow for the row-wise matrix augmentation strategy.

Going one step further, column- and row-wise data matrix augmentation can be employed simultaneously. In this case, the semicolon ";" and comma "," notations are used to indicate the two types of matrix augmentation, column- and row-wise; i.e., some of the matrices are set on top of each other and others are set besides the others. This means that some of the matrices are supposed to share the same column (spectra) vector space and some other their row vector (concentration) space. The bilinear models of Equations (8) and (10) can be also extended to data matrices obtained by simultaneous column- and row- augmentation, as shown in Equations (12) and (13) and displayed in Figure 5c.

$$\begin{bmatrix} \mathbf{D}_{11} & \mathbf{D}_{12} & \mathbf{D}_{13} & \dots & \mathbf{D}_{1L} \\ \mathbf{D}_{21} & \mathbf{D}_{22} & \mathbf{D}_{23} & \dots & \mathbf{D}_{2L} \\ \mathbf{D}_{31} & \mathbf{D}_{32} & \mathbf{D}_{33} & \dots & \mathbf{D}_{3L} \\ \dots & \dots & \dots & \dots & \dots \\ \mathbf{D}_{K1} & \mathbf{D}_{K2} & \mathbf{D}_{K3} & \dots & \mathbf{D}_{Kl} \end{bmatrix} = \begin{pmatrix} \mathbf{C}_1 \\ \mathbf{C}_2 \\ \mathbf{C}_3 \\ \dots \\ \mathbf{C}_K \end{pmatrix} \begin{bmatrix} \mathbf{S}_1^{\mathrm{T}} & \mathbf{S}_2^{\mathrm{T}} & \mathbf{S}_3^{\mathrm{T}} & \dots & \mathbf{S}_L^{\mathrm{T}} \end{bmatrix} + \begin{pmatrix} \mathbf{E}_{11} & \mathbf{E}_{12} & \mathbf{E}_{13} & \dots & \mathbf{E}_{1L} \\ \mathbf{E}_{21} & \mathbf{E}_{22} & \mathbf{E}_{23} & \dots & \mathbf{E}_{2L} \\ \mathbf{E}_{31} & \mathbf{E}_{32} & \mathbf{E}_{33} & \dots & \mathbf{E}_{3L} \\ \dots & \dots & \dots & \dots & \dots \\ \mathbf{E}_{K1} & \mathbf{E}_{K2} & \mathbf{E}_{K3} & \dots & \mathbf{E}_{Kl} \end{pmatrix} = \mathbf{C}_{\mathrm{aug}}\mathbf{S}_{\mathrm{aug}}^{\mathrm{T}} + \mathbf{E}_{\mathrm{aug}} \tag{12}$$

or, in a more compact form:

$$\begin{aligned} [\mathbf{D}_{11}, \mathbf{D}_{12}, \dots, \mathbf{D}_{1L}; \dots; \mathbf{D}_{K1}, \mathbf{D}_{K2}, \dots, \mathbf{D}_{KL}] &= [\mathbf{C}_1; \mathbf{C}_2; \dots; \mathbf{C}_K] \\ [\mathbf{S}_1^{\mathrm{T}}, \mathbf{S}_2^{\mathrm{T}}, \dots, \mathbf{S}_L^{\mathrm{T}}] &+ [\mathbf{E}_{11}, \mathbf{E}_{12}, \dots, \mathbf{E}_{1L}; \dots; \mathbf{E}_{K1}, \mathbf{E}_{K2}, \dots, \mathbf{E}_{KL}] \\ &= \mathbf{C}_{\mathrm{aug}}\mathbf{S}_{\mathrm{aug}}^{\mathrm{T}} + \mathbf{E}_{\mathrm{aug}}. \end{aligned} \tag{13}$$

In the bilinear model described by Equations (12) and (13), a row-wise augmented spectra matrix $[\mathbf{S}_1^{\mathrm{T}},\mathbf{S}_2^{\mathrm{T}},\mathbf{S}_3^{\mathrm{T}},\dots,\mathbf{S}_L^{\mathrm{T}}]$ is obtained to describe the different spectra in the different individual data matrices (i.e., different spectroscopies), and a column-wise augmented concentration matrix $[\mathbf{C}_1;\mathbf{C}_2;\mathbf{C}_3;\dots;\mathbf{C}_K]$ is obtained to describe the concentration changes in the different individual data matrices. The $\mathbf{D}_{kl}$ matrices in this data arrangement are assumed to share one of the two vector spaces, column- or row- space, but not both simultaneously. Thus, $\mathbf{D}_k$ matrices, i.e., matrices with identical index $k$, share the row (concentration) space, whereas $\mathbf{D}_l$ matrices, i.e., matrices with identical index $l$, share the column (spectral) space (see below for a deeper discussion of this issue in the implementation of trilinear and multilinear constraints in MCR methods below).

The ALS algorithm used by MCR-ALS when applied to multiset data sets is similar to the one applied to the analysis of a single data matrix. However, the extension of MCR-ALS to these more complex data sets should not be considered equivalent to the decomposition of a single data matrix, because of the submatrix-by-submatrix possible application of constraints. A first clear advantage of the simultaneous analysis of multiple data matrices using Equations (8)–(13) is that there is a net increase in the overdetermination of the linear system of equations to be solved and that this produces more stable and precise least-squares estimations in the general case. On the other hand, calculation of $\mathbf{C}$ and $\mathbf{S}^{\mathrm{T}}$ pseudoinverse matrices may be improved by appropriate selection and design of more orthogonal individual data sets to be included in the multiset arrangement. More interesting and relevant to multiset analysis is the possible application of new constraints. One of these constraints is the correspondence among the components in the different simultaneously analyzed matrices. This constraint fixes the sequence and the presence or absence of components in a particular $\mathbf{C}_k$ and/or $\mathbf{S}_l^{\mathrm{T}}$ matrices when Equations (8)–(13) are solved. This presence/absence information is coded in binary form and introduced into the MCR algorithm. As a consequence, when a particular component does not exist in a particular $\mathbf{C}_k$ and/or $\mathbf{S}_l^{\mathrm{T}}$, the elements in the related profile are set to zero. This type of constraint contributes significantly to the elimination of rotation ambiguities, because of the presence of zero values in factor $\mathbf{C}_k$ and $\mathbf{S}_l^{\mathrm{T}}$ matrices, which facilitates the achievement of unique resolution conditions [17,39,40].

## 4 MULTILINEAR CONSTRAINTS IN MCR ANALYSIS OF MULTIWAY DATA

One of the most popular multiset data arrangements is the multiway data structure. In this type of data structure, multiple two-way data sets (data matrices) equally sized, i.e., having the same number of rows and columns, are arranged in a higher ordered data structure. This structure can be a data cube or a parallelepiped, $\underline{\mathbf{D}}$, also called three-way data set (see Figures 4 and 6) or, when the number of common data directions (ways, modes) is higher, a hypercube data structure for four-, five- or higher way data structures. The MCR model previously described for two-way multiset data structures can be identically applied to multiway data structures by means of properly data matrix augmentation strategies. However, in the case of multiway data (also called multimode data), the bilinear model previously described for data matrices can also be extended to the so-called multilinear models, like the PARAFAC trilinear model [48,49] or the Tucker models [48,50]. In the case of the MCR-ALS method, these new type of multilinear models are implemented as constraints during the

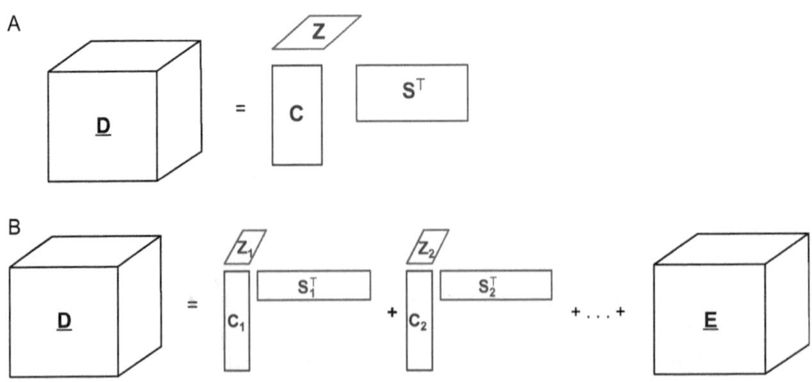


**FIGURE 6** Decomposition of a three-way data set according to a trilinear model. (a) Expressed as decomposition in loading matrices and (b) expressed as the sum of pure component triads.

ALS optimization. Next, the implementation of these constraints is briefly described for the case of the trilinear and quadrilinear models, and for the case of other more complex multilinear model extensions which take into account possible interaction between components in the different data modes (ways), usually referred in the literature as Tucker3 models [48,50].

## 4.1 The Trilinearity Constraint in MCR-ALS

The expression of the trilinear model (also called parallel factor analysis or PARAFAC model) [48,49], to describe the decomposition of a three-way data set, is given element-wise by Equation (14) as:

$$d_{ijk} = \sum_{n=1}^{N} c_{in} s_{jn} z_{kn} + e_{ijk} \tag{14}$$

and the reproduction of each slice data matrix (slice) is carried out as follows:

$$\mathrm{D}_k = \mathbf{C}\ \mathbf{Z}_k \mathbf{S}^{\mathrm{T}} + \mathbf{E}_k \tag{15}$$

where $d_{ijk}$ represents the $ijk^{\text{th}}$ element in the three-way data set ($I = 1,\ldots,I$, $j = 1,\ldots,\mathrm{J}$, and $k = 1,\ldots,K$), $n$ is the number of components (chemical rank) common to the three modes ($n = 1,\ldots,N$), $c_{in}$, $s_{jn}$, and $z_{kn}$ are the elements of $\mathbf{C}$, $\mathbf{S}^{\mathrm{T}}$, and $\mathbf{Z}$ factor matrices (component profiles in the three modes) and $e_{ijk}$ is the residual term (part of the data not explained by the model). This trilinear decomposition of a three-way array $\underline{\mathbf{D}}$ is represented graphically in Figure 6.

Analysis of a three-way data set $\underline{\mathbf{D}}$ using a trilinear model generates directly three sets of component profiles (loading) matrices with equal number of components in the three modes. **C** (concentration matrix in many chemistry examples) is the matrix of loadings in the first mode (also called scores in PCA of two-way data sets) and $\mathbf{S}^{\mathrm{T}}$ (spectra matrix in many chemistry examples) is the matrix of loadings in the second mode. The chemical meaning is the same as for the analogous matrices considered in the description of the bilinear model decomposition in MCR analysis. The new factor matrix $\mathbf{Z}_k$ of loadings in the third direction or mode is a diagonal matrix giving the relative amounts of every component in each considered data matrix $\mathbf{D}_k$ (when this happens, column profiles in **C** and row profiles in $\mathbf{S}^{\mathrm{T}}$ are normalized and $\mathbf{Z}_k$ describes all scaling information about components). As the matrix notation in Equation (15) describes, $\mathbf{Z}_k$ will be the only factor matrix in the trilinear model that changes in the trilinear decomposition for the different matrices $\mathbf{D}_k$. Both **C** and $\mathbf{S}^{\mathrm{T}}$ matrices have the same (unique, invariant) concentration profile and spectrum for each individual component in all simultaneously analyzed matrices $\mathbf{D}_k$. Trilinear model enforces decompositions to give unique solutions for the three factor matrices (apart from scale and trivial permutation rotation ambiguities), and avoids the presence of rotational ambiguities associated with lower structured bilinear models [22].

Though MCR-ALS is especially able to cope with nontrilinear three-way data sets formed by matrix augmentation taking individual data matrices of varying sizes, it can also work with three-way data sets formed by data matrices with the same row and column sizes (data cubes) and take advantage of their structure if they follow the premises of the trilinear model. Because of the inherent freedom in the modeling of the profiles of the augmented $\mathbf{C}_{\text{aug}}$ matrix, the so-called trilinear structure is incorporated as an optional constraint during the ALS optimization of the $\mathbf{C}_{\text{aug}}$ profiles (see Figure 7) [28].

An outstanding aspect of the trilinearity constraint in ALS is that the application of this constraint is performed separately and optionally on the concentration profiles of each component. As shown in Figure 7, when the trilinear constraint is applied during each iteration of the ALS optimization, the concentration profiles of the same component in the different matrices, $\mathbf{c}_k^n$ ($k=1,2,3$, $n=1,2$ components in this case), are forced to be invariant. To implement this, they are first placed one besides the other to form a one-component concentration profiles matrix, $\mathbf{C}^n$ with $I$ rows (number of rows of each data matrix) and $K=3$ columns (number of slices/matrices simultaneously analyzed in the figure), and $n$ indicating which component is considered ($n=1,\ldots,N$). For

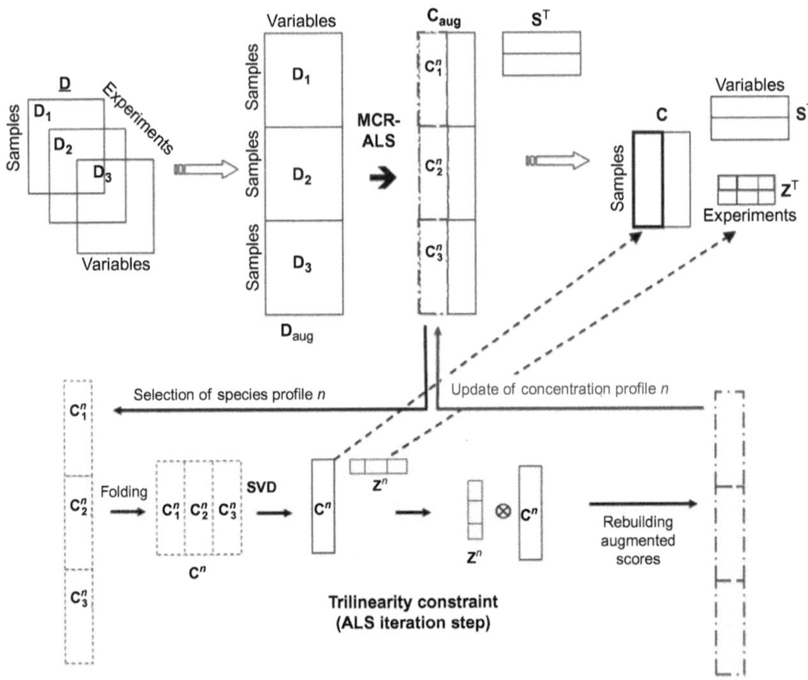


FIGURE 7 Implementation of the trilinearity constraint in the MCR-ALS algorithm (see text for details of the procedure).

a full trilinear model, every component matrix, $\mathbf{C}^n$, is approximated by its one-component bilinear decomposition (using for instance PCA or SVD) as follows:

$$\mathbf{C}^n \approx \mathbf{c}^n \mathbf{z}^n \tag{16}$$

where $\mathbf{c}^n$ is one column vector ($I$ rows) which contains the common (average) concentration profile of this $n$ component in the different $K$ matrices and $\mathbf{z}^n$ is a row vector ($N$ columns) with the relative amounts of this concentration profile in the different $K$ matrices. The appropriate Kronecker product of these two vectors rebuilds the augmented concentration vector of this component [51]. $\mathbf{C}_{\text{aug}}$ augmented matrix is finally obtained and updated by appropriate rearrangement of each of its columns corresponding to the concentration of different individual components in the different matrices. $\mathbf{c}^n$ and $\mathbf{z}^n$ vector profiles give the current estimation of the first and third mode loading profiles of the considered component. This is precisely what trilinear models described by Equations (14) and (15) are considering explicitly. Therefore, results obtained by MCR-ALS with the trilinearity constraint applied to all the components of the system will give practically the same results as the application of other trilinear model-based methods (like PARAFAC) when applied to the same system [28,52]. However, an advantage of the trilinearity constraint as it is implemented in MCR-ALS is that it is applied independently and optionally to each component of the data set, giving more flexibility to the whole data analysis and allowing for full trilinear and for partial trilinear models. This flexibility allows a more representative modeling of some real situations, such as those of systems where profiles of some components may behave in a trilinear fashion (e.g., analytes) and profiles of some others do not (e.g., strong background and baseline contributions, interferences in natural samples) [8].

When MCR-ALS optimization finishes, only the recovered spectral information in $\mathbf{S}^{\mathrm{T}}$ matches with the PARAFAC profiles in three modes. However, the matrix $\mathbf{C}_{\text{aug}}$ contains implicitly the information related to matrices $\mathbf{C}$ and $\mathbf{Z}$, which can be easily recovered using a similar procedure as in the application of the trilinearity constraint in Figure 7. This postresolution strategy can also be used to check the departure of the trilinearity structure for any set of profiles obtained from MCR-ALS analysis of a three-way data set using only the assumption of the bilinear model. The adequacy of the trilinear model in the analysis of a three-way data set can be also checked by other methods like applying and comparing SVD analysis of the different augmented matrices (row- and column-wise) and by comparison of data fits obtained under the assumption of both type of models, bilinear and trilinear models [17,52].

## 4.2 The Quadrilinearity Constraint in MCR-ALS

The procedure previously described for the implementation of the trilinearity constraint for three-way data can be extended to the implementation of the quadrilinearity constraint for four-way data. The expression of the quadrilinear model to describe the decomposition of a four-way data set is given element-wise by Equation (17) as:

$$d_{ijkl} = \sum_{n=1}^{N} c_{in} s_{jn} z_{kn} y_{\text{ln}} + e_{ijkl} \tag{17}$$

A four-way data set "$\mathbf{D}_{IJK,L}$", of dimensions $I$,$J$,$K$, and $L$ in the first, second, third, and fourth mode, respectively, can be arranged in an augmented column-wise manner ($\mathbf{D}_{\text{aug}}$) where three of the modes ($I$, $K$, and $L$) are concatenated and intermixed in the direction of the columns of the data matrix, and the other mode $J$ is left invariant in the row direction (Figure 8). MCR-ALS can also be applied to this augmented data matrix as previously given in Equation (9), $\mathbf{D}_{\text{aug}} = \mathbf{C}_{\text{aug}}\mathbf{S}^{\text{T}} + \mathbf{E}_{\text{aug}}$, where $\mathbf{C}_{\text{aug}}$ is the augmented concentration matrix containing the first, second, and third mode profiles and $\mathbf{S}^{\text{T}}$ is the spectra (loadings) matrix for the second mode, and $\mathbf{E}_{\text{aug}}$ is the error term.

Schematic representation of the quadrilinearity constraint implemented in MCR-ALS model (MCR-ALS$_{\text{Q}}$) to analyze a four-way data set, $\underline{\mathbf{D}}_{I,J,K,L}$, is also provided in Figure 8. Observe that in this figure, notation has been extended to cope with different modeling strategies and to give and make more clear the dimensions of vectors (lowercase in bold) and matrices (uppercase in bold), and their algebraic operations. In the same figure, the relation with bilinear (MCR-ALS$_{\text{B}}$) and trilinear modeling (MCR-ALS$_{\text{T}}$) and the application of corresponding modeling constraints are also displayed. In brief, the quadrilinear constraint is applied during the ALS optimization of the augmented concentration matrix ($\mathbf{C}_{\text{aug}}$). This quadrilinear constraint can be applied independently and optionally to each component of the data set, giving more flexibility to the whole data analysis and allowing testing full and partial quadrilinear models. Further details can be found elsewhere [30,53].

As shown in Figure 8, the augmented concentration profile of the selected component $n'$, $\mathbf{c}^{n}_{IKL,1}$, of dimensions ($IKL$,1) is initially folded to give the one-component profile matrix, ($\mathbf{C}^{n}_{IK,L}$) with $IK$ rows and $L$ columns, where ($n = 1,\ldots,N$). Matrix ($\mathbf{C}^{n}_{IK,L}$) is then approximated by its first-component bilinear SVD decomposition

$$\mathbf{C}^{n}_{IK,L} \approx \mathbf{c}^{n}_{IK,1}\ \mathbf{y}^{n}_{1,L} \tag{18}$$

giving a loadings vector ($\mathbf{c}^{n}_{IK,1}$) of dimensions ($IK$,1) having loadings for the $n^{\text{th}}$ component in the first and third modes (intermixed), and vector ($\mathbf{y}^{n}_{1,L}$) of dimensions (1,L) with loadings for the component $n^{\text{th}}$ in the fourth mode. This process is repeated for every component, giving the loadings matrix ($\mathbf{Y}^{\text{T}}$) in the fourth mode, of dimensions ($N$,$L$), where $N$ is the number of desired or selected components ($n = 1,\ldots,N$) and $L$ ($l = 1,\ldots,L$) refers to the total number of conditions in the fourth mode.

To recover the loading profiles in first and third modes, the previously obtained one-component profile vector ($\mathbf{c}^{n}_{IK,1}$) of dimensions ($IK$,1) is folded

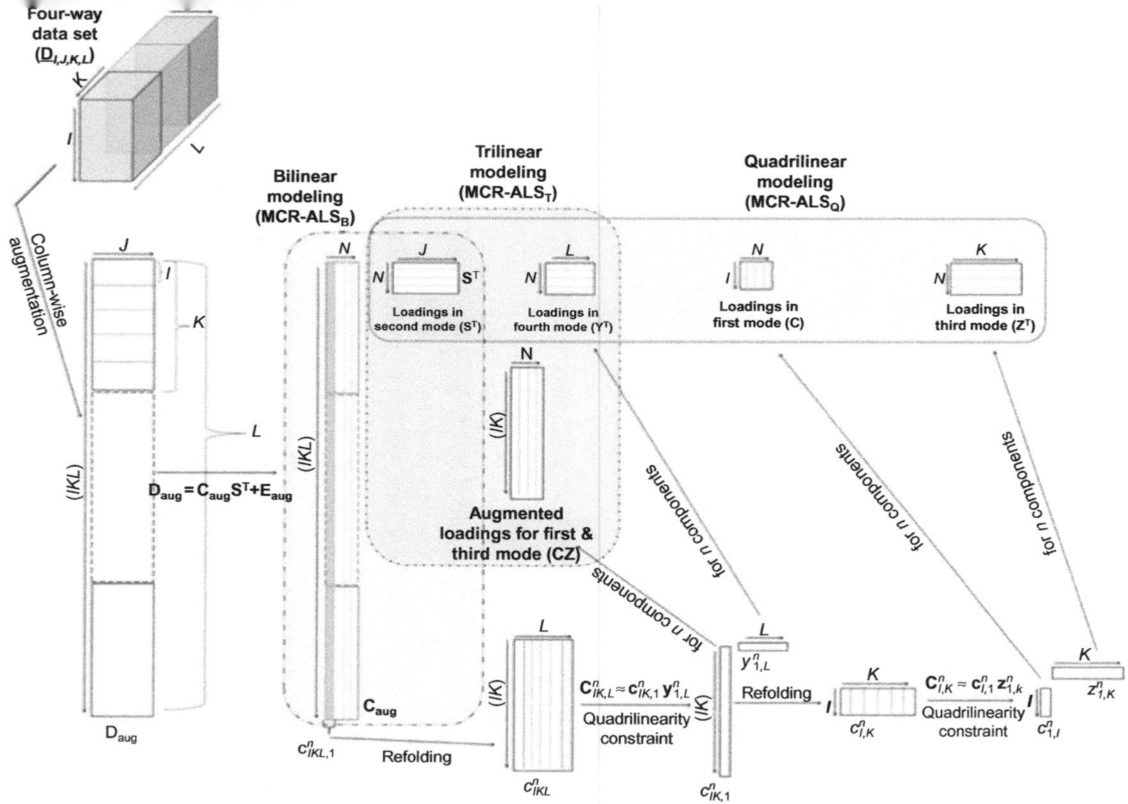


**FIGURE 8** Implementation of the quadrilinearity constraint in the MCR-ALS algorithm (see text for details of the procedure). *Adapted from Chemometrics and Intelligent Laboratory Systems 135, 223–234, Amrita Malik, Roma Tauler, Performance and validation of MCR-ALS with quadrilinear constraint in the analysis of noisy datasets, with permission from Elsevier.*

again to give the single component profile matrix ($\mathbf{C}^n_{I,K}$) with $I$ rows and $K$ columns (dimensions ($I$,$K$)). Matrix ($\mathbf{C}^n_{I,K}$) is then approximated by its first-component bilinear SVD decomposition

$$\mathbf{C}^n_{I,K} \approx \mathbf{c}^n_{I,1}\ \mathbf{z}^n_{1,k} \tag{19}$$

where, vector $\mathbf{c}^n_{L,1}$ of dimensions ($I$,1) has the loadings of the $n^{\text{th}}$ component in the first mode, and $\mathbf{z}^n_{1,k}$ of dimensions (1,$K$) has the loadings of $n^{\text{th}}$ component in the third mode. This process is repeated for every component targeted to be quadrilinear, which is then rearranged to give the corresponding loadings matrices in the first and third modes, $\mathbf{C}$ of dimensions ($N$,$I$), and $\mathbf{Z}^{\text{T}}$ of dimensions ($N$,$K$). These steps are repeated during ALS optimization, giving the final loading matrices $\mathbf{C}$, $\mathbf{Z}^{\text{T}}$, and $\mathbf{Y}^{\text{T}}$, corresponding to the quadrilinear decomposition of the four-way data set $\underline{\mathbf{D}}_{I,J,K,L}$, first, third, and fourth modes, respectively. The loading matrix in second mode, $\mathbf{S}^{\text{T}}$, is the only one directly obtained from the bilinear decomposition of $\mathbf{D}_{\text{aug}}$.

Reconstruction of the full augmented scores matrix "$\mathbf{C}_{\text{aug}}$" of dimensions ($IKL$,$N$) is achieved using the Kronecker tensor product [48,51] of the column vectors of $\mathbf{C}$, and of row vectors of $\mathbf{Z}^{\text{T}}$ and $\mathbf{Y}^{\text{T}}$ loadings matrices with elements $\mathbf{c}^n_{I,1}$ $\mathbf{z}^n_{1,k}$, and $\mathbf{y}^n_{1,L}$, to give $\mathbf{C}^n_{IKL,1}$ according to the following equations:

$$\mathbf{c}^n_{I,1} \otimes \mathbf{z}^n_{1,k} = \mathbf{c}^n_{IK} \tag{20}$$

$$\mathbf{c}^n_{IK,1} \otimes \mathbf{y}^n_{I,L} = \mathbf{C}^n_{IKL,1} \tag{21}$$

where $\otimes$ stands for the Kronecker product which performs element-to-element multiplication of the factor matrices, and $n$ represents the selected component of the matrix. The approach described by previous Equations (19)–(21) can be used similarly to recover the loading profiles of the different components in different modes of any four-way data set from the augmented scores ($\mathbf{C}_{\text{aug}}$), irrespective of the application of the multilinearity constraint, i.e., by appropriate refolding of each of the augmented profiles in $\mathbf{C}_{\text{aug}}$ into a one-component matrix and subsequent SVD analysis of them to recover the loading profiles in the different modes. This approach can be extended algorithmically to any multilinear model. See Refs. [30] and [53] for preliminary applications of this constraint to environmental data sets.

## 4.3 Interaction Between Profiles of Different Components

The general expression to describe the decomposition of three-way data sets where interaction between components is possible is given in Equation (22):

$$d_{ijk} = \sum_{p=1}^{Np} \sum_{q=1}^{Nq} \sum_{r=1}^{Nr} g_{pqr} c_{ip} s_{jq} z_{kr} + e_{ijk} \tag{22}$$

where $d_{ijk}$ represents the $ijk^{\text{th}}$ element in the three-way data set, $c_{ip}$, $s_{jq}$, and $z_{kr}$ are the elements in $\mathbf{C}(I,N_{\text{p}})$, $\mathbf{S}(J,N_{\text{q}})$, and $\mathbf{Z}(K,N_{\text{r}})$ factor matrices (loadings in

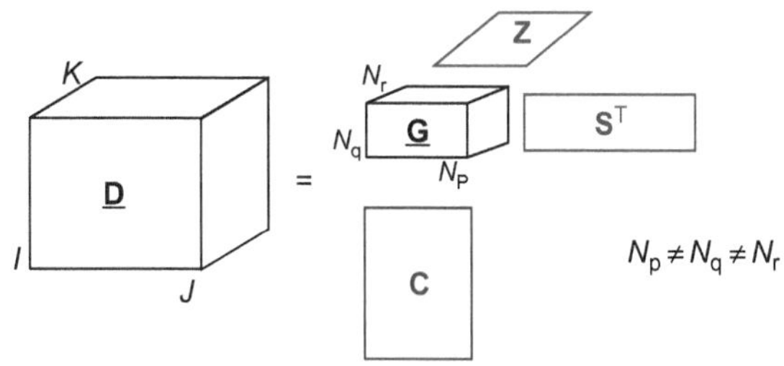


**FIGURE 9** Decomposition of a three-way data set according to a Tucker3 model.

the three modes, see Figure 9) used to reconstruct the $d_{ijk}$ element of $\underline{\mathbf{D}}(I,J,K)$, and $e_{ijk}$ is the residual term in $\underline{\mathbf{E}}(I,J,K)$. $N_p$, $N_q$, and $N_r$ are the number of components considered in each of the three modes, not necessarily equal as in the trilinear model, in which $N_p=N_q=N_r=N$ (Equation 17). $g_{pqr}$ is the pqr$^{th}$ element of the core array $\underline{\mathbf{G}}(N_p,N_q,N_r)$, where the nonnull elements are spread out in different manners depending on each particular data set. The magnitude of the pqr$^{th}$ element in the core matrix is proportional to the contribution of the triad formed by the p profile in **C**, the q profile in $\mathbf{S}^T$, and the r profile in **Z,** in the reproduction of the original three-way array, $\underline{\mathbf{D}}$. The decomposition of a three-way array $\underline{\mathbf{D}}$ according to Equation (22) is called in the multiway literature the Tucker3 model [48,50] and is represented graphically in Figure 9. Apart from the profiles in the three modes (as in PARAFAC), a three-way core array $\underline{\mathbf{G}}$ is additionally obtained, which has the dimensions of the number of profiles in each of the three modes. Tucker3 model has complete freedom in the correspondence of profiles among different components and modes. It may happen that two components have the same profile in one of the modes, e.g., the concentration profile or spectrum of two different components is the same or that one component needs more than one profile in one of the three modes to be properly described. Tucker3 is a complex model and usually difficult to interpret, partly because its flexibility implies nonunique solutions and the presence (like in bilinear MCR models) of rotational and intensity ambiguities [22]. Restricted Tucker3 models are a simpler type of Tucker3 models, where only a small number of possible selected interactions (triads) are allowed. To do so, the elements of the core matrix $\underline{\mathbf{G}}$ unrelated to the selected triads are set equal to zero. The restricted Tucker3 models are much more easily interpretable and have been proposed in different cases to solve chemical problems [48,50].

MCR-ALS can be adapted to implement situations where the components interact, with different number of profiles (loadings) in the different modes. Figure 10 shows graphically how this can be achieved for an example where two components have the same shape for their concentration profiles. In the example shown in this figure, the two suitable column profiles on the augmented $\mathbf{C}_{aug}$ matrix are first grouped to give the row-wise augmented concentration matrix $[\mathbf{C}^1,\mathbf{C}^2]$. This $[\mathbf{C}^1,\mathbf{C}^2]$ matrix has a number of rows equal to the

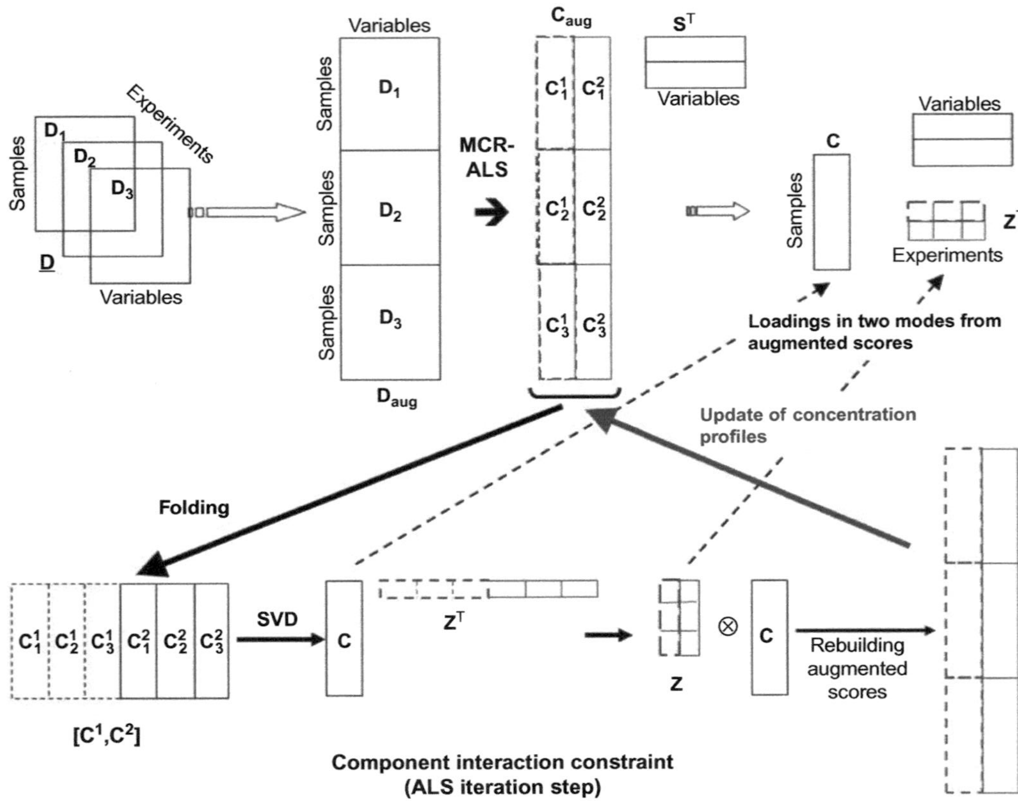


**FIGURE 10** Implementation of the Tucker3 model constraint in the MCR-ALS algorithm (see text for details of the procedure).

number of rows in the individual matrices and a number of columns equal to twice the number of matrices simultaneously analyzed because two components have the same profile shape ($2 \times K$). This folded matrix containing the profiles of the two components with common shape for the concentration profile is then approximated by their bilinear decomposition (using for instance the first component of PCA or SVD) as in

$$\left[\mathbf{C}^1, \mathbf{C}^2\right] \approx \mathbf{c}\,\mathbf{z}^{\mathrm{T}} \tag{23}$$

This first-component bilinear decomposition gives directly the loadings (the shape of the concentration profile) in the first mode $\mathbf{c}$, and the loadings (scaling factor) in the third mode $\mathbf{z}^{\mathrm{T}}$, which after adequate rearrangement component-wisely gives matrix $\mathbf{Z}$ (see Figure 10). The appropriate Kronecker product [50,51] of them gives the two new augmented profile vectors which substitute (update) the corresponding two columns of the $\mathbf{C}_{\mathrm{aug}}$ matrix. Observe that in this case only one profile (one loading vector) will be recovered in the first mode for the matrix $\mathbf{C}$. On the contrary, two profiles (two loading vectors) are recovered for the third mode matrix $\mathbf{Z}^{\mathrm{T}}$, as well as for the second mode, in the $\mathbf{S}^{\mathrm{T}}$ matrix. When these component interaction constraints are inserted during each step of the ALS iterative optimization procedure, the results obtained are analogous to those obtained for the restricted Tucker3 model with one component in the first mode and two components in the two other modes. This procedure can be generalized to other cases and it has been tested in some data examples [29,54].

Summarizing, three-way data set decompositions using trilinear and Tucker3 models yield explicitly loadings in the three modes ($\mathbf{C}$, $\mathbf{S}^{\mathrm{T}}$, and $\mathbf{Z}^{\mathrm{T}}$). Instead, column-wise (and also similarly for row-wise) augmented data matrix decompositions using bilinear model MCR-ALS give directly only the loadings in the second mode ($\mathbf{S}^{\mathrm{T}}$ matrix), whereas the loadings in the other two modes ($\mathbf{C}$ and $\mathbf{Z}^{\mathrm{T}}$ in PARAFAC and Tucker3) are confounded in the augmented mode ($\mathbf{C}_{\mathrm{aug}}$ matrix). Using a proper profile rearrangement and SVD analysis on the suitable augmented concentration profiles, as proposed in Figures 7–9, the loadings in the different modes can be recovered and the results compared with those obtained by trilinear, quadrilinear, or restricted Tucker3 model-based methods.

## 5 USE OF MCR OUTPUT FOR QUANTITATIVE ANALYSIS

### 5.1 General Approach

When the main purpose of MCR is quantitative analysis, some general aspects are worth to be mentioned. Quantitative analysis is generally associated with MCR analysis of multiset structures, i.e., which implies the use of a second-order data set or a full data matrix per sample. The full multiset structure contains, therefore, as many submatrices as samples subject to quantitative analysis. Because of the kind of data used to define a sample, the second-order advantage is fulfilled and this means that compounds of interest (analytes) can be quantified in the presence of other unknown compounds (interferences), even if those interferences are not present in the samples of the calibration set [55–57]. Being MCR a multivariate resolution method, every dyad of concentration profile and related pure response refers to a particular compound in the samples. This allows separating the information of the analytes from that of interferences. Besides, since it is a multivariate method, several analytes can be determined together in a single MCR analysis.

To explain how quantitative information is extracted from MCR results, we should go back to the meaning of the bilinear model associated with a multiset structure. Multisets for quantitative analysis always present a column-wise augmentation, which means that the matrices from the different samples are structured one under the other. Each sample can be represented by raw measurements from a single instrumental technique, which would give a column-wise augmented raw data set, or from several techniques and then a row- and column-wise augmented raw data set would be obtained (see Section 4). These data arrangements yield a bilinear model with a single (or augmented) $\mathbf{S}^{\mathrm{T}}$ matrix and always an augmented $\mathbf{C}$ matrix, with as many submatrices $\mathbf{C}_i$ as samples in the original data set. When multiset MCR analysis is performed, the profiles in the pure response matrix $\mathbf{S}^{\mathrm{T}}$ are normalized. As a consequence, and in order to adequately describe the difference in intensity of the raw measurements from different samples, the submatrices of concentration profiles ($\mathbf{C}_i$) enclose both the information related to the shape (e.g., elution peak in chromatography) and the scale/intensity (quantitative information) of the concentration profiles. Thus, the intensity (expressed in different ways) of the concentration profiles is used appropriately to perform quantitative analysis.

In an MCR model, the information from the different analytes in the set of samples analyzed is collected in separate augmented concentration profiles. This resolved (selective) information is used to build as many separate pseudounivariate calibration models as analytes in the data set. To do so, it is sufficient to regress the intensity values of the concentration profiles for a particular analyte in the calibration set samples against the reference concentration values (see Figure 11). Later on, the calibration line built can be used to retrieve the real concentration of unknown samples from the intensity of their related concentration profiles (see Figure 11). In the absence of reference concentration values, e.g., when only a set of real samples is available or when interferences are the compounds studied, relative quantitative information can still be obtained by performing the ratio between the intensity values of the concentration profiles of the different samples and the intensity of the concentration profile of a sample taken as the reference. The intensity of concentration profiles should never be straightforwardly interpreted as a concentration value because of the presence of intensity ambiguity in the resolution results [16,18] (see Section 1) unless explicit information on proper area scaling is provided within the MCR analysis (see Section 5.4.2).

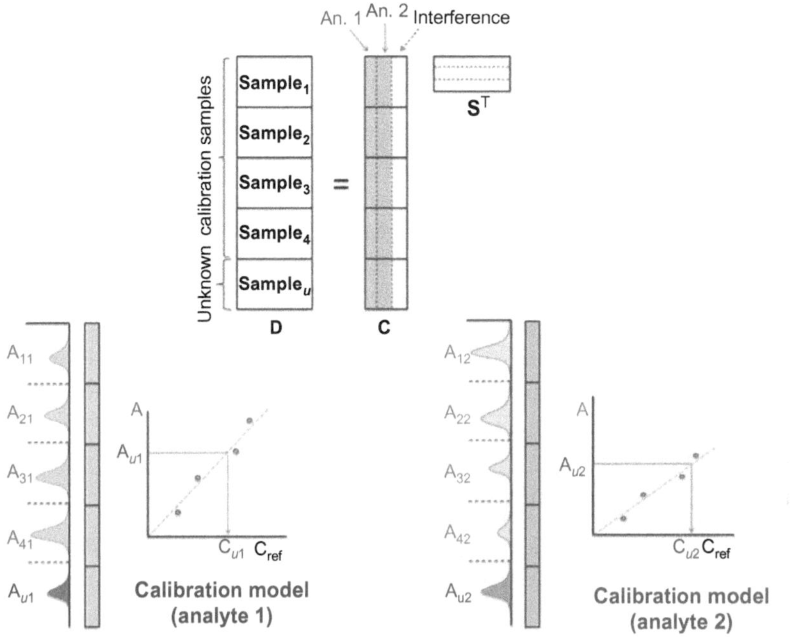


**FIGURE 11** Establishment of pseudounivariate calibration models from MCR results. The example refers to a three-component system with two analytes (and, hence, two calibration models) and an interference.

The intensity/scale of a concentration profile can be expressed by different parameters depending on the kind of concentration profile (or analytical measurement) used for quantitative analysis. A very common scaling measure is the integrated area under the concentration profile. This is the usual option in chromatography (peak areas) and in other analytical techniques, such as fluorescence measurements, i.e., excitation-emission matrices, where the area under the nonnormalized augmented direction (excitation or emission spectra) is taken to build calibration models. The advantage of taking the area under profiles instead of other parameters, such as peak height, is that concentration profiles are not restricted to have this unimodal signal shape and that the quantitative model is not affected by the fact that the shape of concentration profiles may vary from sample to sample [8,58,59]. Therefore, the use of peak height for calibration models should be restricted to analytical signals with a very reproducible peak shape, e.g., some electrochemical signals [60–62], or to peak profiles where some constraint devoted to fix the shape of the concentration profile in all samples, e.g., trilinearity, has been applied [18,63].

Signal areas or peak heights are scaling parameters related to analytical continuous profiles, such as spectra or elution profiles in chromatography. However, other kinds of measurements do not provide this kind of smooth concentration profiles, such as hyperspectral images (see Figure 1b). The concentration profile of an imaged sample is a stretched array of relative concentration values related to the different pixels in the image. This kind of profile does not present a smooth or continuous shape because of the often high heterogeneity of the sample surface and the fact that the 2D pixel neighborhood in the sample surface is not preserved in the related concentration profile. In this case, instead of taking integrated areas, the scaling parameter used tends to be the mean (or the median) of all the elements (pixels) in the concentration profile of the imaged sample. Such an approach has a double advantage: on the one hand, it allows working with images of different sizes because the different number of pixels does not affect the average scaling parameter obtained and, on the other hand, the calibration lines built in this way are useful to predict both global and local quantitative information on unknown images when the mean value of the concentration profile of a new image or when each pixel element in the new image concentration profile are used, respectively (see Figure 12) [64,65].

## 5.2 MCR Quantitative Analysis for Multilinear Data

Multiset analysis by MCR provides bilinear decompositions by default. However, as mentioned in Section 4, multilinear models can be accommodated

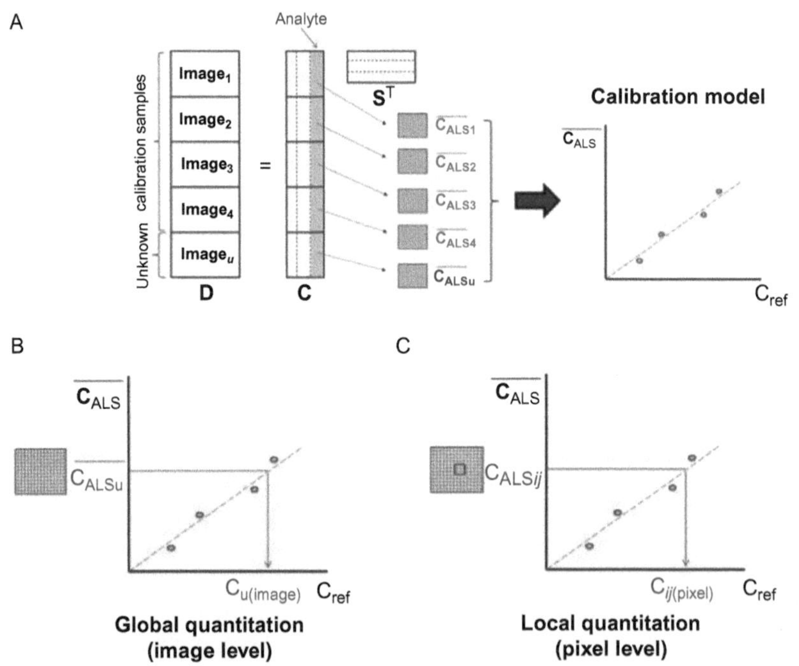


**FIGURE 12** (a) Establishment of a calibration model for quantitative analysis of hyperspectral images, (b) use of the model for global quantitative analysis (at image level), and (c) use of the model for local quantitative analysis (at a pixel level).

into the MCR analysis as model constraints. This means that MCR provides a framework adaptable to the linearity degree of the different data sets [18,28–30]. In difference to other algorithms, the choice of the decomposition model to be applied is not restricted by the nature of the algorithm used, but it is decided based on the natural behavior of the data sets. Derivation of quantitative information when coming from a multilinear MCR decomposition would be done as explained in Section 5.1.

When full multilinear decompositions are performed by MCR, the solutions are unique, as with any other intrinsically multilinear algorithms, such as PARAFAC, GRAM, or DTD [49,52,66,67]. As it has been seen in recent works, MCR, as PARAFAC, can deal with trilinear structures and also apply models with a higher multilinear order (quadrilinear, etc.) [30]. In the case of completely trilinear systems, such as those associated with many of the excitation-emission fluorescence measurements, MCR would provide analogous results to DTD, PARAFAC, or GRAM. MCR can also work with models presenting factor interaction, such as those represented by TUCKER models [29].

However, a relevant algorithmic difference between MCR and other multilinear algorithms is the way the multilinear structure is imposed. In MCR, the trilinear or multilinear condition is applied profile-wise. This offers the possibility to work with hybrid models, where some contributions are modeled following a trilinear or multilinear behavior, whereas some others are modeled following a bilinear structure. Such a feature is unique of the MCR method and avoids forcing the full chemical system to behave in a multilinear way, which can be a too rigid framework for many real data sets [18,52].

Hybrid models contemplate a situation that can be sometimes encountered in analytical quantitative analysis, i.e., the analyte has a quite reproducible concentration profile among the samples analyzed, but other contributions to the signal (instrumental or from interference compounds) not. This situation has been encountered in EEM examples (when minor compounds want to be modeled and some residual scattering contributions are present [68]) or in some chromatographic cases (in complex natural samples, where the elution pattern of the analyte can be quite reproducible, but many other chemical interferences or background contributions that could not be adequately removed contribute to the measured signal [63]). In these cases, the analyte(s) are constrained to follow a trilinear model, whereas the rest of contributions, much more variable in behavior, are modeled in a bilinear way. The consequence is that the analyte profiles are resolved in a unique way and this fact increases the quality of the quantitative information derived from the system. Although the final MCR solution may have some ambiguity, this phenomenon is restricted to interferences or background profiles, which are not the main targets in quantitative analysis.

## 5.3 Adaptation of Analytical Calibration Strategies to MCR

As mentioned before, MCR in multiset mode presents the second-order advantage, which means that we can quantify analytes in the presence of interferences by using standards that do not contain these unknown compounds [55,56]. Although this formulation is mathematically correct, analytical problems that prevent working with pure standards or taking scaling information in a straightforward way should not be disregarded. Similar strategies to the ones applied in classical univariate calibration can be used in these instances to build suitable data set structures and to handle appropriately the results provided by second-order calibration methods.

One of the most relevant analytical problems in calibration is the presence of the so-called sample matrix effect. This phenomenon makes that the relationship between instrumental response and concentration varies when the analyte is in an aqueous solution (pure standard) or when it is present in the sample to be analyzed because of the presence of other compounds (sample matrix). In this situation, a calibration line built with pure standards would be useless to predict concentration values in real samples because the response/concentration ratio (slope in the calibration line) would not be correct. To avoid the influence of this effect, the standard addition strategy is used, i.e., small known amounts of analyte are added to the original sample and the suitable analytical measurement is performed. In this case, each standard addition provides a second-order data set. After resolution of the multiset containing the second-order matrices of all the standard additions and the original sample, the pseudounivariate calibration lines are built by regressing the area under the analyte profile against the amount of analyte added. The crossing point between the calibration line and the abscissa axis corresponds to the negative concentration value of the analyte in the sample. This being said, nothing prevents appending pure standards in the multiset analyzed. These pure standards would help to obtain better defined concentration profiles and spectra, but would never be used to build the calibration line (see Figure 13a) [69–72].

Another difficulty linked to calibration is the intrinsic lack of reproducibility of the intensity of certain instrumental responses, such as chromatographic peak areas coming from gas chromatography (GC). This phenomenon increases the error in the calibration models when the intensity of the instrumental responses is used as such. This problem is analytically solved adding a known amount of internal standards to all samples. The idea behind is that a single response intensity (represented by area, height, etc) can have reproducibility problems, but the ratio between areas of pairs of compounds obtained in the same analysis remains constant. In such a case, the multiset structure can be built with samples that contain the analytes and the internal standard(s). The calibration plot is built regressing the ratio of the areas of analyte/internal standard versus concentration of analyte (see Figure 13b) [69].

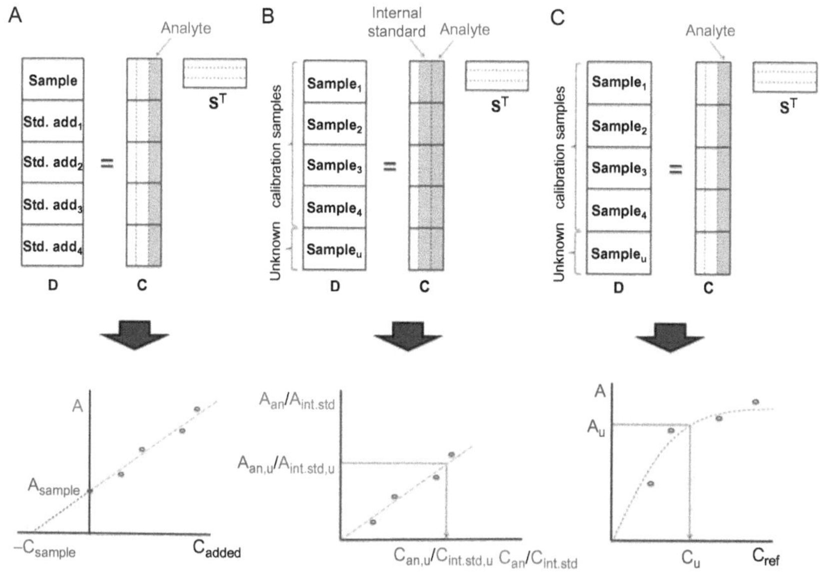


**FIGURE 13** Use of analytical calibration strategies in quantitative MCR analysis. Multiset structures and related calibration models. (a) Standard addition method, (b) internal standard method, and (c) model for inner matrix effect.

A last problem is the presence of the so-called inner filter effect, specifically linked to fluorescence measurements. In this case, above a certain concentration range, the linearity between the fluorescence intensity and the concentration is lost due to the absorption of light by the sample that causes a decrease in the fluorescence intensity expected. In such a case, the bilinear decomposition provided by MCR is not compromised, since the spectral shape of the excitation-emission spectra dyad remains invariant. However, when the calibration line is built, one should take into account that mathematical expressions more complex than a straight line will be needed to properly reflect the varying relationship among response and analyte concentration (see Figure 13c) [73]. In cases where the inner filter effect could modify the shape of the fluorescence spectra, a standard addition strategy would be used [45].

## 5.4 Introducing the Calibration Concept in MCR as a Constraint

Calibration concepts previously exposed can be used as constraints to improve MCR solutions (see Section 2). Two different situations can be encountered in practice. The first one occurs when the data analyzed are first-order data, meaning that one spectrum (one data vector) is obtained for every analyzed sample. This is the common situation in multivariate calibration, for instance, when food or agricultural samples are analyzed by NIR spectroscopy [74] and some individual constituents or parameters are determined. In this scenario, a set of samples covering all possible sources of variance is used for calibration and the concentration of the sought analyte or the value of a particular parameter is estimated independently. Using this set of NIR spectra, a calibration model is built to predict optimally the concentration of the constituent or any other parameter in the calibration samples. This calibration model is then validated and the parameters of the model adjusted using an external set of samples not used for the model construction. Once the calibration model is validated, the concentrations or parameters in new unknown samples can be predicted with this model. This strategy has been used successfully in many applied areas and there are a great number of applications already described, especially for the partial least-squares [75], but also for other methods [76]. One of the advantages of the multivariate calibration approach for first-order data is that calibration and prediction are possible in the presence of unknown interferences and that they allow multicomponent analysis. The drawback of this approach is, however, that calibration samples should contain the same variance sources present in the new samples to be predicted. Therefore, a large number of calibration samples are usually needed to describe the data variance. A second possible situation is when the analysis of a single sample already gives a second-order response data set, i.e., a data matrix. This is the case, for instance, when samples are analyzed by excitation-emission spectroscopy (fluorescence) or by any typical hyphenated chromatographic technique coupled to multiwavelength UV or to full scan MS spectroscopic detection methods. Under these circumstances, when multiple calibration samples are simultaneously analyzed, the data produced can be analyzed by the previously mentioned multilinear calibration methods (like PARAFAC [49], GRAM [66], or others), which confer unique solutions and the so-called second-order advantage. The second-order advantage implies that calibration is possible in the presence of unknown (nonmodeled) interferences and that calibration is possible using pure analyte responses without the need of having a large number of calibration samples where the interferences should be present and modeled implicitly during the calibration step. If multilinear models are not adequate, there is still the possibility to build up calibration models for the analytes (see Section 5.1). In this context, the application of MCR to multiset data described in this chapter fits well in this scenario. In these two situations, a way to improve calibration results and MCR concentration predictions is to implement the calibration concept within the algorithm through a correlation constraint as it is described below.

### 5.4.1 Correlation Constraint Applied to Individual Concentration Values (First-Order Calibration)

In this case, the proposed correlation constraint builds internal univariate calibration models between the concentration values calculated by the MCR models and the reference values from calibration samples [66,77,78]. These models are afterward used to predict new concentration values in validation and test samples. As a consequence, the concentration profiles are expressed in real concentration units. In each iteration, the concentration values $c_{\text{cal}}^{\text{ALS}}$ in calibration samples obtained from the selected MCR concentration profile (in **C** matrix) are regressed against the respective reference known concentration values $c_{\mathrm{cal}}^{\mathrm{ref}}$ of the analyte in these samples (see Figure 14). The slope $b$ and offset $b_0$ are obtained by fitting a linear least-squares regression model between $c_{\mathrm{cal}}^{\mathrm{ref}}$ and $c_{\mathrm{cal}}^{\mathrm{ALS}}$ values

$$c_{\mathrm{cal}}^{\mathrm{ALS}} = b c_{\mathrm{cal}}^{\mathrm{ref}} + b_0 \tag{24}$$

$$\hat{c}_{\mathrm{test}} = \frac{c_{\mathrm{test}}^{\mathrm{ALS}} - b_0}{b}. \tag{25}$$

Once the parameters $b$ and $b_0$ in Equation (24) are obtained, a vector $\hat{c}_{\mathrm{test}}$ with the predicted concentrations is obtained by Equation (25) using the

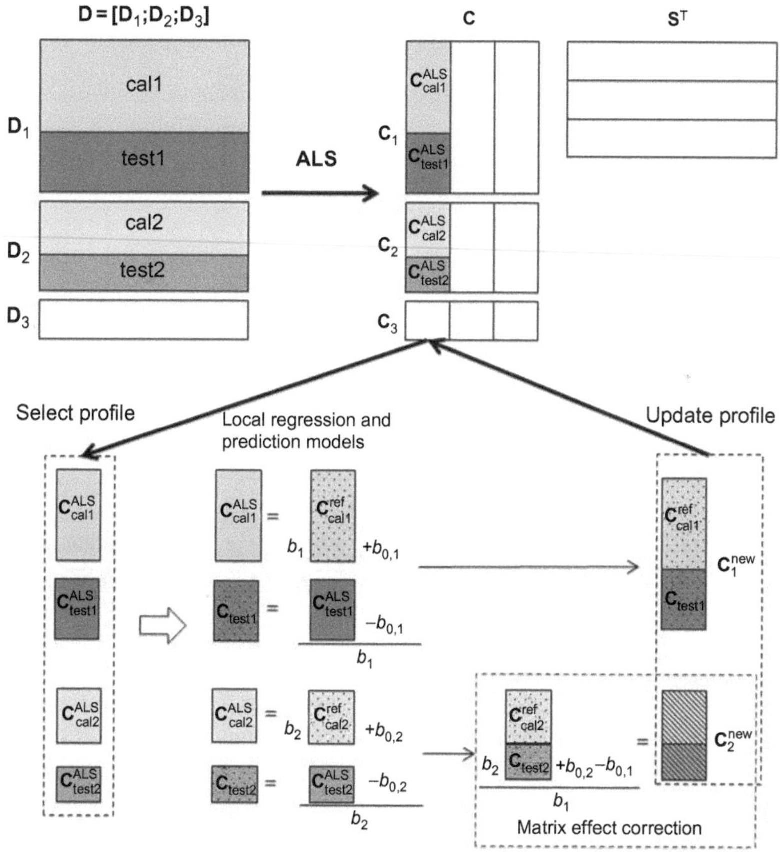


**FIGURE 14** Implementation of the MCR-ALS correlation constraint in MRC-ALS algorithm for first-order multivariate calibration (see text for details of the procedure). *Adapted from Talanta 125, 233–241, Rodrigo Rocha de Oliveira, Kássio Michell Gomes de Lima, Romà Tauler, Anna de Juan, Application of correlation constrained multivariate curve resolution alternating least-squares methods for determination of compounds of interest in biodiesel blends using NIR and UV–visible spectroscopic data, with permission from Elsevier.*

concentration values in the **C** profile for the test samples, $c_{\text{test}}^{\text{ALS}}$. The unconstrained $c^{\text{ALS}}$ vector is updated by original values for calibration samples, $c_{\text{cal}}^{\text{ref}}$, and for the new predicted values in test and/or unknown samples $\hat{c}_{\text{test}}$. The same procedure is repeated during ALS iterations until convergence. As any other constraint, the correlation constraint can be applied to one or more analyte concentration profiles, and a different calibration model (Equation 24) is obtained for every component to be calibrated.

Figure 14 describes the correlation constraint in different scenarios for multiple sample subsets and when matrix effects can be present [79,80]. The correlation constraint can be applied in a flexible way as described below:

**(a)** *Correlation constraint with a single regression model for all sample subsets.* This is the conventional way to apply the correlation constraint, when all the calibration samples in the data set are used to build a single calibration model per analyte;

**(b)** *Correlation constraint with local models per individual sample subset or group of sample subsets.* Three different cases are possible:

*Case (1) A single regression model using selected sample subsets.* Just one regression model is obtained per each analyte correlating the real analyte concentration to calibration samples of the selected subsets, as in case 0, and freely modeling the other sample subsets not considered in the correlation constraint. This happens when a pure analyte spectrum is included in the data set to enable the application of the selectivity constraint, but this selective sample is not used to build the regression model.

*Case (2) Local models.* In this case, separate regression models are calculated for each individual sample subset or group of sample subsets per analyte. It is also possible not to consider a certain data subset in the correlation constraint, as in case (1).

*Case (3) Local models with matrix effect correction.* Separate models for different sample subsets can be calculated to overcome sample matrix effect problems among samples of different subsets. Matrix effect means that there are different linear relationships between concentration values $\mathbf{C}_i$ and signal responses $\mathbf{D}_i$ of the analytes for each $i$ sample subset affected by a different matrix effect. In such a case, the pure spectrum of the analyte in $\mathbf{S}^{\text{T}}$ would be different in scale for each sample subset. To overcome this modeling problem, one of the sample subsets should be taken as a reference and a rescaling procedure must be applied to the concentration values of the other sample subsets affected by a different matrix effect before updating its corresponding concentration profile during ALS iterations. Real unscaled concentration values predicted by each local regression model for calibration and test samples are stored in a separate output and can be recovered at the end of the

MCR-ALS optimization. Once the concentration profile for a certain analyte is selected for the correlation constraint, different local regression models (one for every sample subset) are built as described above. For every subset a regression model is built relating $c_{\text{cal}}^{\text{ALS}}$ to $c_{\text{cal}}^{\text{ref}}$, and with their corresponding slopes and intercepts $b_1$ and $b_{0,1}$. These models are then used to predict real concentration values in test samples, $\hat{c}_{\text{test}}$. Since sample matrix effects exist, direct update in **C** matrix by the actual concentrations predicted is not possible in this case. Therefore, $\hat{c}_{\text{test}}$ values predicted by every submodel are stored in a separate output and they are used to obtain the constrained concentration profile to be introduced in the ALS optimization by means of a rescaling step. For instance in the case of having two data subsets, the model coming from first subset (with $b_1$ and $b_{0,1}$ parameters for the first subset and $b_2$ and $b_{0,2}$ parameters) is adopted as reference, and calibration and test samples for the second subset are rescaled according the following equation,

$$\mathbf{C}_2^{\text{new}} = \frac{b_2 \begin{bmatrix} c_{\text{cal}_2}^{\text{ref}} \\ \hat{c}_{\text{test}_2} \end{bmatrix} + b_{0,2} - b_{0,1}}{b_1} \tag{26}$$

Rescaled values for calibration and test samples, $\mathbf{C}_2^{\text{new}}$, are then updated in the corresponding concentration profile in **C** matrix, and ALS optimization continues.

Examples of application of the previously described correlation constraint can be found in previous works [32,43,44,77–80].

### 5.4.2 Correlation Constraint Applied to Component Profiles Properties (Area, Height)

This case is different to the previous one of correlation constraint for first-order data. In this case, the correlation model is built between known concentrations and one of the profile global properties of the resolved profiles for the components in matrix **C**. In this case, the property or concentration value does not refer to the individual values of the resolved component profile but to one global property of it, like its area or its height, or any other profile property. For instance, this can be the case of LC-DAD analysis of multiple samples using the previously explained multiset data arrangement (column-wise data matrix augmentation, see Figures 4 and 11) and MCR-ALS analysis. During the ALS optimization, the area or height of one of the (concentration) component profiles in calibration samples can be correlated to a known concentration value of this profile in these samples. A local correlation model is built and used to adjust the intensity of the profile of this component in the calibration samples to the known values. The model parameters are then used to predict and adjust the intensity of the profiles of the selected components in unknown samples.

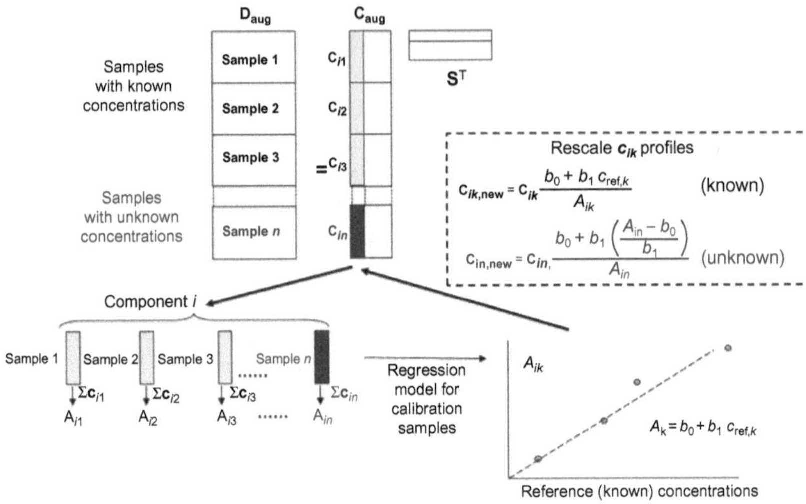


**FIGURE 15** Implementation of the MCR-ALS correlation constraint in MCR-ALS algorithm for second-order data multivariate calibration (see text for details of the procedure).

In Figure 15, the correlation constraint is described graphically for the case where the area of the ALS concentration profile (i.e., sum of individual values), $A_{ik}$, for the $i^{th}$ component in the $k^{th}$ calibration sample is correlated with a known external concentration reference values ($c_{\text{ref},k}$) of the same calibration sample. Whereas values from the calibration samples are updated according to the known values, the concentration of unknown samples, $c_{\text{in}}$, is predicted submitting their ALS profile areas to the calibration model. Examples of application of this strategy are being developed at present.

## 5.5 Validation and Error in Quantitative MCR Analysis

Validation is a necessary step in all quantitative analyses based on calibration procedures and it aims at testing the quality and predictive ability of a particular calibration model when used on new independent samples. Validation is carried out by using a set of known samples (validation set), similar in nature to those used in the calibration set and with available information about the concentration of the analyte(s). The validation samples are submitted to the calibration model and the retrieved quantitative information extracted is compared with the real reference concentration values. If the error in prediction has a satisfactory value, the model is validated and ready to be used for prediction in unknown samples.

Validation can and should also be done when MCR is used for quantitative analysis [8,64,81–83]. Thinking about the concentration values predicted from MCR-based calibration models, we can see that the quality of prediction depends on several factors, namely: (a) the quality of the pseudounivariate calibration lines, which are the regression models between the intensity of concentration profiles (so-called sometimes MCR score values) and the reference concentration values, and (b) the quality of the resolution step, because the intensity of the concentration profiles is linked to the resolved concentration profiles retrieved during the MCR analysis.

Validation can, therefore, be envisioned in two different ways:

**(a)** by building a multiset structure containing both the calibration and validation samples (in this case, the resolution step, $\mathbf{D}=\mathbf{C}\mathbf{S}^{\mathrm{T}}$, is common to all samples and only the quality in prediction of the pseudounivariate calibration line is tested) or

**(b)** by performing a resolution model on the calibration set, $\mathbf{D}_{\mathrm{cal}}=\mathbf{C}_{\mathrm{cal}}\mathbf{S}_{\mathrm{cal}}^{\mathrm{T}}$, which is used afterward on the validation set to predict concentration profiles by a single least-squares step, $\mathbf{C}_{\mathrm{val}}=\mathbf{D}_{\mathrm{val}}\left(\mathbf{S}_{\mathrm{cal}}^{\mathrm{T}}\right)^{+}$ and from the calculated concentration profiles to derive areas or peak heights that will be used in the calibration model to predict concentrations (in this case, both the resolution step and the quality of the pseudounivariate calibration line are tested at the same time).

From studies carried out on this subject, similar results are obtained with the two different kinds of validation as long as calibration samples span correctly the compound space of validation samples [64]. It should be noted, however, that out of the validation context, calibration and unknown samples should always be resolved together when interferences present in the real samples are absent in the calibration set.

Figures of merit are necessary to provide an accurate description of the properties and quality of calibration models. Parameters, such as sensitivity, limits of detection, and quantification, or error in prediction, should be also furnished when a second-order calibration model is going to be used.

Given the difficulty to come up with analytical expressions for the figures of merit in second-order calibration, empirical approaches based on resampling techniques are often used and provide very good approximations of the uncertainty linked to the propagation of the experimental error into the predicted concentrations by the calibration model [84]. Thus, the uncertainty linked to the prediction of a particular analyte concentration in a sample can be estimated using the results obtained from a large number of MCR analyses performed on multisets containing the set of calibration samples and a real test sample, the latter modified according to a suitable resampling strategy. The multisets used in resampling procedures should always include reproduced noise-free calibration samples to assess adequately the effect of propagation of experimental error only in the test sample [85–87]. Multisets are modified by adding to the real test sample a certain amount of Gaussian noise of the same level of the experimental noise (noise addition method) or by adding on a reproduced noise-free test sample a certain amount of noise (Monte Carlo method) [84]. After all the MCR analyses, a large number of values for the predicted analyte concentration in the test sample is obtained and the error in the predicted concentration can be obtained by calculating the standard deviation ($s_c$)

$$s_c = \sqrt{\frac{\sum_{i=1}^{n} (c_i - \bar{c})^2}{n-1}}, \tag{27}$$

where $c_i$ is the predicted concentration in the $i^{\text{th}}$ MCR analysis, $n$ is the number of MCR analyses carried out in the resampling procedure, and $\bar{c}$ is the mean value of the predicted concentration in all MCR analyses. From this estimated uncertainty, other interesting figures of merit for a particular sample can be derived, such as the sensitivity, which would be expressed as:

$$\text{SEN} = \frac{s_{\text{test}}}{s_c} \tag{28}$$

where $s_{\text{test}}$ is the standard deviation of the Gaussian noise introduced in the test sample signal when the Monte Carlo method has been used for resampling [88].

The first approach to provide analytical expressions for figures of merit obtained by MCR-based calibration models was inspired by the fact that quantitative analysis in MCR was performed by using pseudounivariate calibration lines displaying intensity values of concentration profiles versus reference concentration values [83,89,90]. Since each one of these pseudounivariate models referred to a single analyte, this scenario clearly reminded classical univariate calibration models where a selective univariate instrumental response is regressed against the analyte concentration. As a consequence, figures of merit identical to those used in univariate calibration models, recommended by the IUPAC [83,91,93,94], were adopted to describe the quality of pseudounivariate calibration models. This same strategy had also been used to provide figures of merit for multilinear resolution methods, such as PARAFAC [83].

MCR-based pseudounivariate calibration models can be interpreted as classical univariate calibration models when figures of merit, such as limit of detection (LOD) or limit of quantification (LOQ), are going to be retrieved. In this case, these parameters would be described as follows:

$$\text{LOD} = 3.3\frac{s_R}{b}\sqrt{\frac{1}{m} + \frac{1}{n} + \frac{\overline{r^2}}{b^2 \sum_i (c_i - \bar{c})^2}} \tag{29}$$

$$\text{LOQ} = 10\frac{s_R}{b}\sqrt{\frac{1}{m} + \frac{1}{n} + \frac{\overline{r^2}}{b^2 \sum_i (c_i - \bar{c})^2}} \tag{30}$$

where $c$ refers to real concentration values of calibration samples, $m$ is the number of replicates of a sample, $n$ the number of samples, $b$ the slope of the calibration line, and $s_R$ is the standard deviation of the residuals in the pseudounivariate calibration line, calculated as:

$$s_R = \sqrt{\frac{\sum_{i=1}^{n} (\hat{r}_i - r_i)}{n-1}} \quad (31)$$

where $r_i$ represents the response value (i.e., the intensity of the concentration profile provided by MCR) for the $i^{th}$ sample and $\hat{r}_i$ the reproduced value using the pseudounivariate calibration line.

However, this simple approach has several limitations when other figures of merit have to be estimated, such as sensitivity and other parameters that derive from the knowledge of this one. Indeed, sensitivity relates variations of the original instrumental response to variations in the analyte concentration and is largely affected by the overlap of the analyte with other compounds in the sample. Therefore, it is a sample-dependent magnitude. Knowing these facts, it is easy to understand that pseudounivariate calibration plots are insufficient to properly estimate this figure of merit. The reasons are: (a) first, the response used in these plots is the intensity of resolved concentration profiles, which is expressed in arbitrary units and does not represent appropriately the intensity and variations of the original instrumental signal and (b) sensitivity is a constant magnitude in univariate calibration, equal to the slope of the calibration line, because the response of all test samples is selective and, therefore, unaffected by any interference. Conversely, the responses of test samples in the pseudounivariate calibration models do not come from a common original selective scenario and, therefore, each one of them should be treated according to the nature and composition of the sample of origin.

In an attempt to take into account the different nature of samples in the determination of sensitivity for second-order MCR calibration models, a recent work proposes an analytical way to estimate this figure ($SEN_{MCR}$) [88]. Based on the results of a resolved multiset formed by the calibration samples and the test sample, the sensitivity can be estimated as:

$$SEN_{MCR} = m_n \left[ J \left( S^T S \right)_{nn}^{-1} \right]^{-1/2} \quad (32)$$

where $n$ is the index of the analyte to be determined, $m_n$ the slope of the related pseudounivariate calibration plot, $J$ is the number of elements in the concentration profile of the test sample (note that the related MCR score is derived as the sum of all elements in the profile), $\mathbf{S}^T$ is the matrix containing the resolved spectra of all components in the multiset and the expression $(S^T S)_{nn}^{-1}$ is a way to design the $(n,n)$ element of the $(S^T S)^{-1}$ matrix (for more detail in the derivation of this expression, see Ref. [81]). Some important comments to make on this expression are: (a) the fact that the calculated sensitivity ($SEN_{MCR}$) is always lower than the slope of the calibration line ($m_n$), which is logical in the presence of interferences and (b) that the sensitivity gets modified as a function of the term expressing the overlap among compounds in the sample analyzed $(S^{T}S)_{nn}^{-1}$.

Once this figure of merit has been derived, other related parameters, such as the selectivity ($SEL_{MCR}$), LOD, and limit of quantitation (LOQ), can be easily calculated as follows:

$$SEL = \left[ (S^{T}S)_{nn}^{-1} \right]^{1/2} \tag{33}$$

$$LOD = \frac{3.3 s_{dtest}}{SEN_{MCR}} \tag{34}$$

$$LOQ = \frac{10 s_{dtest}}{SEN_{MCR}} \tag{35}$$

For the case of LOD and LOQ, the values agree with those provided by direct derivation of these parameters from the pseudounivariate calibration line in Equations (29) and (30) [89], which supports the validity of the proposed estimation of sensitivity.

All what has been described above about the error linked to MCR-based calibration models holds as long as the analyte concentration profile resolved in the MCR analysis do not present ambiguity [84,88]. In such a case, the only source of error in the calibration models comes from the propagation of the experimental error contained in the original raw measurements. When rotational ambiguity is present in the analyte concentration profile, this means that there is a range of feasible concentration profiles and, hence, of related profile intensities that would equally describe the variation of the data set. The fact that these intensity values can move within a certain range contributes to the uncertainty in the quantitative information retrieved by MCR. There are recent works devoted to estimate how the uncertainty in the prediction of concentrations would change as a function of the rotational ambiguity left in the MCR solutions [92]. In this study, again, the importance of the proper use of constraints during the resolution of the data set is revealed to be crucial.

Although the potential presence and effect of rotational ambiguity should be checked whenever a quantitative analysis is carried out by MCR, we should take into account that this problem can be very often suppressed or extremely alleviated in a calibration context. A very important fact to be taken into account is that, in calibration problems, the main goal is resolving in a unique way the concentration profiles of analyte(s) because only the intensity of these profiles is used for quantitative purposes. Modeling in a unique way the rest of contributions present in the samples is only a secondary interest.

Contributing to the unique resolution of analyte contributions in MCR can be done by applying the trilinearity constraint to the analyte profiles [8,18,63,68]. As mentioned before, this can be done in some kind of measurements, such as excitation-emission fluorescence measurements, and in other kinds of analytical measurements by using hybrid models that include trilinearity only for the analyte profiles. An additional way to suppress or largely decrease ambiguity in nontrilinear chemical systems is by using adequately local rank constraints. Indeed, as mentioned in Section 2, local rank constraints can ensure unique recovery of concentration profiles and spectra if the suitable conditions of compound overlap are fulfilled by the data set (in a calibration context, the multiset structure containing calibration and test samples). In this sense, the correct spectrum of the analyte would be straightforwardly obtained if pure standards are used. As for the recovery of concentration profiles, the related resolution theorem states that: "The concentration profile of a compound can be correctly and uniquely recovered if all compounds inside its concentration window are also present outside" [39]. In this sense, we should remind that we refer to the augmented concentration profiles in the multiset structure, formed by the submatrices from the calibration samples and the test samples. This means that, many times, we can achieve this uniqueness condition by doing a proper design of the calibration samples to be included in the multiset structure and using explicitly the information of presence/absence of compounds through the constraints of correspondence of species (when referred to full concentration profiles of particular samples) or local rank (when referred to a particular region of the concentration direction within a sample). Actually, the concentration profile of analytes in test samples could be correctly recovered even if completely embedded under interference profiles. Only the case where an interference was completely embedded under the analyte profile in a test sample would prevent the unique recovery of this concentration profile and, in this case, that would be a minor compound and would not greatly affect the quantitative analysis.

## REFERENCES

[1] Rutan SC, de Juan A, Tauler R. Introduction to multivariate curve resolution. In: Brown S, Tauler R, Walczak R, editors. Comprehensive chemometrics, vol. 2. Oxford: Elsevier; 2009. p. 249–59.

[2] Lawton WH, Sylvestre EA. Self modeling curve resolution. Technometrics 1971;13:617.

[3] Craig Hamilton J, Gemperline PJ. Mixture analysis using factor analysis. II: self-modeling curve resolution. J Chemom 1990;4:1.

[4] de Juan A, Tauler R. Chemometrics applied to unravel multicomponent processes and mixtures. Revisiting latest trends in multivariate resolution. Anal Chim Acta 2003;500:195.

[5] de Juan A, Tauler R. MCR from 2000: progress in concepts and applications. Crit Rev Anal Chem 2006;36:163.

[6] Ruckebusch C, Blanchet L. Multivariate curve resolution: a review of advanced and tailored applications and challenges. Anal Chim Acta 2013;765:28–36.

[7] Jaumot J, Eritja R, Gargallo R. Chemical equilibria studies using multivariate analysis methods. Anal Bioanal Chem 2011;399:1983.

[8] de Juan A, Tauler R. Factor analysis of hyphenated chromatographic data. Exploration, resolution and quantification of multicomponents systems. J Chromatogr A 2007;1158:184.

[9] de Juan A, Piqueras S, Maeder M, Hancewicz T, Duponchel L, Tauler R. Chemometric tools for image analysis. In: Salzer R, Siesler HW, editors. Infrared and Raman spectroscopic imaging. 2nd ed.: Weinheim: Wiley-VCH; 2014. p. 57–110.

[10] de Juan A, Tauler R, Dyson R, Marcolli C, Rault M, Maeder M. Spectroscopic imaging and chemometrics: a powerful combination for global and local sample analysis. TrAC Trends Anal Chem 2004;23:70.

[11] Amigo JM, Cruz J, Bautista M, Maspoch S, Coello J, Blanco M. Study of pharmaceutical samples by NIR chemical-image and multivariate analysis. Trends Anal Chem 2008;27:696.

[12] Tauler R. Interpretation of environmental data using chemometrics. In: Barceló D, editor. Sample handling and trace analysis of pollutants: techniques, applications and quality assurance. Amsterdam: Elsevier; 2000. p. 689 [chapter 16].

[13] Terrado M, Barceló D, Tauler R. Quality assessment of the multivariate curve resolution alternating least squares method for the investigation of environmental pollution patterns in surface water. Environ Sci Technol 2009;43:5321.

[14] Jaumot J, Tauler R, Gargallo R. Exploratory data analysis of DNA microarrays by multivariate curve resolution. Anal Biochem 2006;358:76.

[15] Wentzell PD, Karakach TK, Roy S, Martínez MJ, Allen CP, Werner-Washburne M. Multivariate curve resolution of time course microarray data. BMC Bioinf 2006;7:343 art.

[16] Tauler R, Maeder M. Two-way data analysis: multivariate curve resolution—error in curve resolution. In: Brown S, Tauler R, Walczak R, editors. Comprehensive chemometrics, vol. 2. Oxford: Elsevier; 2009. p. 345–63.

[17] Tauler R, Smilde AK, Kowalski BR. Selectivity, local rank, three-way data analysis and ambiguity in multivariate curve resolution. J Chemom 1995;9:3.

[18] Tauler R, Maeder M, de Juan A. Two-way extended curve resolution. In: Brown S, Tauler R, Walczak R, editors. Comprehensive chemometrics, vol. 2. Oxford: Elsevier; 2009. p. 473–505.

[19] Abdollahi H, Maeder M, Tauler R. Calculation and meaning of feasible band boundaries in multivariate curve resolution of a two-component system. Anal Chem 2009;81:2115.

[20] Rajkó R, István K. Analytical solution for determining feasible regions of self-modeling curve resolution (SMCR) method based on computational geometry. J Chemom 2005;19:448.

[21] Sawall M, Kubis C, Selent D, Börner A, Neymeyr K. A fast polygon inflation algorithm to compute the area of feasible solutions for three-component systems. I: concepts and applications. J Chemom 2013;27:106.

[22] Tauler R. Calculation of maximum and minimum band boundaries of feasible solutions for species profiles obtained by multivariate curve resolution. J Chemom 2001;15:627.

[23] Gemperline PJ. Computation of the range of feasible solutions in self-modeling curve resolution algorithms. Anal Chem 1999;71:5398.

[24] de Juan A, Rutan SC, Tauler R. Two-way data analysis: multivariate curve resolution–iterative resolution methods. In: Brown S, Tauler R, Walczak R, editors. Comprehensive chemometrics, vol. 2. Oxford: Elsevier; 2009. p. 325–44.

[25] Liang Y. Two-way data analysis: multivariate curve resolution- non-iterative methods. In: Brown S, Tauler R, Walczak R, editors. Comprehensive chemometrics, vol. 2. Oxford: Elsevier; 2009. p. 309–24.

[26] Vandeginste BGM, Derks W, Kateman G. Multicomponent self-modeling curve resolution in high performance liquid chromatography by iterative target transformation factor analysis. Anal Chim Acta 1985;173:253.

[27] Gemperline PJ. A priori estimates of the elution profiles of the pure components in overlapped liquid chromatography peaks using target factor analysis. J Chem Inf Comput Sci 1984;24:206.
[28] Tauler R, Marqués I, Casassas E. Multivariate curve resolution applied to three-way trilinear data: study of a spectrofluorimetric acid–base titration of salicylic acid at three excitation wavelengths. J Chemom 1998;12:55.
[29] Alier M, Felipe M, Hernández I, Tauler R. Trilinearity and component interaction constraints in the multivariate curve resolution investigation of NO and O-3 pollution in Barcelona. Anal Bioanal Chem 2011;399:2015.
[30] Malik A, Tauler R. Extension and application of multivariate curve resolution-alternating least squares to four-way quadrilinear data-obtained in the investigation of pollution patterns on Yamuna River, India—a case study. Anal Chim Acta 2013;794:20.
[31] de Juan A, Maeder M, Martínez M, Tauler R. Combining hard- and soft-modelling to solve kinetic problems. Chemom Intell Lab Syst 2000;54:123.
[32] Antunes MC, Simão JEJ, Duarte AC, Tauler R. Multivariate curve resolution of overlapping voltammetric peaks: quantitative analysis of binary and quaternary metal mixtures. Analyst 2002;127:809.
[33] Maeder M, Zuberbühler AD. The resolution of overlapping chromatographic peaks by evolving factor-analysis. Anal Chim Acta 1986;181:287.
[34] Maeder M, de Juan A. Two-way data analysis: evolving factor analysis. In: Brown S, Tauler R, Walczak R, editors. Comprehensive chemometrics, vol. 2. Oxford: Elsevier; 2009. p. 261–74.
[35] Windig W. Two-way data analysis: detection of purest variables. In: Brown S, Tauler R, Walczak R, editors. Comprehensive chemometrics, vol. 2. Oxford: Elsevier; 2009. p. 275–307.
[36] Windig W, Guilment J. Interactive self-modeling mixture analysis. Anal Chem 1991;63:1425.
[37] Cuesta Sánchez F, van den Bogaert B, Rutan SC, Massart DL. Multivariate peak purity approaches. Chemom Intell Lab Syst 1996;34:139.
[38] Malinowski ER. Obtaining the key set of typical vectors by factor analysis and subsequent isolation of component spectra. Anal Chim Acta 1982;134:129.
[39] Manne R. On the resolution problem in hyphenated chromatography. Chemom Intell Lab Syst 1995;27:89.
[40] Tauler R. Multivariate curve resolution applied to second order data. Chemom Intell Lab Syst 1995;30:133.
[41] Amigo JM, de Juan A, Coello J, Maspoch S. A mixed hard- and soft-modelling approach to study and monitor enzymatic systems in biological fluids. Anal Chim Acta 2006;567:236.
[42] Diewok J, de Juan A, Maeder M, Tauler R, Lendl B. Application of a combination of hard and soft modeling for equilibrium systems to the quantitative analysis of pH-modulated mixture samples. Anal Chem 2003;75:641.
[43] Mas S, Miguel C, João Melo M, Lopes JA, de Juan A. Screening and quantification of proteinaceous binders in medieval paints based on μ-Fourier transform infrared spectroscopy and multivariate curve resolution alternating least squares. Chemom Intell Lab Syst 2014;134:148.
[44] Jaumot J, Igne B, Anderson CA, Drennen JK, de Juan A. Blending process modelling and control by multivariate curve resolution. Talanta 2013;117:492.
[45] Schenone AV, Culzoni MJ, Galera MM, Goicoechea HC. Second-order advantage achieved by modeling excitation-emission fluorescence matrices affected by inner filter effects using a

strategy which combines standardization and calibration: reducing experimental and increasing analytical sensitivity. Talanta 2013;109:107.
[46] Geladi P, Wold S. Local principal component models, rank maps and contextuality for curve resolution and multi-way calibration inference. Chemom Intell Lab Syst 1987;2:273.
[47] MATLAB computer and visualization environment, Release 13, The MathWorks Inc.: Natick, MA, USA; 2004.
[48] Smilde A, Bro R, Geladi P. Multi-way analysis. Applications in the chemical sciences. West Sussex, England: John Wiley & Sons, Ltd; 2004.
[49] Bro R. PARAFAC. Tutorial and applications. Chemom Intell Lab Syst 1997;38:149.
[50] Kroonenberg P. Three mode principal component analysis. Leiden: DSWO Press; 1983.
[51] Burdick DS. An introduction to tensor-products with applications to multiway data-analysis. Chemom Intell Lab Syst 1995;28:229.
[52] de Juan A, Tauler R. Comparison of 3-way resolution methods for non-trilinear data sets. J Chemom 2001;15:749.
[53] Malik A, Tauler R. Performance and validation of MCR-ALS with quadrilinear constraint in the analysis of noisy datasets. Chemom Intell Lab Syst 2014;135:223–34.
[54] Peré-Trepat E, Ginebreda A, Tauler R. Comparison of different multiway methods for the analysis of geographical metal distributions in fish, sediments and river waters in Catalonia. Chemom Intell Lab Syst 2007;88:69.
[55] Booksh K, Kowalski BK. Theory of analytical chemistry. Anal Chem 1994;66:782A–791A.
[56] Escandar GM, Faber NM, Goicoechea HC, Muñoz de la Peña A, Olivieri AC. Second- and third-order multivariate calibration: data, algorithms and applications. TrAC Trends Anal Chem 2007;26:752.
[57] Smilde AK, Tauler R, Saurina J, Bro R. Calibration methods for complex second-order data. Anal Chim Acta 1999;398:237.
[58] Mas S, de Juan A, Tauler R, Olivieri AC, Escandar GM. Application of chemometric methods to environmental analysis of pollutants: a review. Talanta 2010;80:1052.
[59] Saurina J, Leal C, Companyó R, Granados M, Tauler R, Prat MD. Determination of triphenyltin in sea-water by excitation-emission matrix fluorescence and multivariate curve resolution. Anal Chim Acta 2000;409:237.
[60] Esteban M, Ariño C, Díaz-Cruz JM. Chemometrics in electrochemistry. In: Brown S, Tauler R, Walczak R, editors. Comprehensive chemometrics, vol. 4. Oxford: Elsevier; 2009. p. 425.
[61] Esteban M, Ariño C, Díaz-Cruz JM. Chemometrics for the analysis of voltammetric data. TrAC Trends Anal Chem 2006;25:86.
[62] Brown SD, Bear RS. Chemometric techniques in electrochemistry—a critical-review. Crit Rev Anal Chem 1993;24:99.
[63] Tauler R, Barceló D. Multivariate curve resolution applied to liquid-chromatography diode-array detection. TrAC Trends Anal Chem 1993;12:319.
[64] Piqueras S, Burger J, Tauler R, de Juan A. Relevant aspects of quantification and sample heterogeneity in hyperspectral image resolution. Chemom Intell Lab Syst 2012;117:169.
[65] Amigo JM, Cruz J, Bautista M, Maspoch S, Coello J, Blanco M. Study of pharmaceutical samples by NIR chemical-image and multivariate analysis. TrAC Trends Anal Chem 2008;27:696.
[66] Sánchez E, Kowalski BR. Generalized rank annihilation factor-analysis. Anal Chem 1986;58:496.
[67] Sánchez E, Kowalski BR. Tensorial resolution: a direct trilinear decomposition. J Chemom 1990;4:29.

[68] Elcoroaristizábal S, de Juan A, García JA, Durana N, Alonso L. Comparison of second-order multivariate methods for screening and determination of PAHs by total fluorescence spectroscopy. Chemom Intell Lab Syst 2014;132:63.
[69] Peré-Trepat E, Lacorte S, Tauler R. Alternative calibration approaches for LC-MS quantitative determination of coeluted compounds in complex environmental mixtures using multivariate curve resolution. Anal Chim Acta 2007;595:228.
[70] Saurina J, Tauler R. Strategies for solving matrix effects in the analysis of triphenyltin in sea-water samples by three-way multivariate curve resolution. Analyst 2000;125:2038.
[71] Lozano VA, Tauler R, Ibáñez GA, Olivieri AC. Standard addition analysis of fluoroquinolones in human serum in the presence of the interferent salicylate using lanthanide-sensitized excitation-time decay luminescence data and multivariate curve resolution. Talanta 2009;77:1715.
[72] Mamian-Lopez MB, Poppi RJ. Standard addition method applied to the urinary quantification of nicotine in the presence of cotinine and anabasine using surface enhanced Raman spectroscopy and multivariate curve resolution. Anal Chim Acta 2013;760:53.
[73] Antunes MCG, da Silva JCGE. Multivariate curve resolution analysis excitation-emission matrices of fluorescence of humic substances. Anal Chim Acta 2005;546:52.
[74] Martens H, Naes T. Multivariate calibration. Chichester: John Wiley; 1989.
[75] Wold S, Sjöström M, Eriksson L. PLS-regression: a basic tool of Chemometrics. Chemom Intell Lab Syst 2001;58:109.
[76] Ni W, Nørgaard L, Mørup M. Non-linear calibration models for near infrared spectroscopy. Anal Chim Acta 2014;813:1.
[77] Azzouz T, Tauler R. Application of multivariate curve resolution alternating least squares (MCR-ALS) to the quantitative analysis of pharmaceutical and agricultural samples. Talanta 2008;74:1201.
[78] Goicoechea HC, Olivieri AC, Tauler R. Application of the correlation constrained multivariate curve resolution alternating least-squares method for analyte quantitation in the presence of unexpected interferences using first-order instrumental data. Analyst 2010; 135:636.
[79] Rocha de Oliveira R, de Lima KMG, Tauler R, de Juan A. Application of correlation constrained multivariate curve resolution alternating least-squares methods for determination of compounds of interest in biodiesel blends using NIR and UV-visible spectroscopic data. Talanta 2014;125:233.
[80] Lyndgaard LB, van den Berg F, de Juan A. Quantification of paracetamol through tablet blister packages by Raman spectroscopy and multivariate curve resolution-alternating least squares. Chemom Intell Lab Syst 2013;125:58.
[81] Gargallo R, Tauler R, Cuesta-Sánchez F, Massart DL. Validation of alternating least-squares multivariate curve resolution for chromatographic resolution and quantitation. TrAC Trends Anal Chem 1996;15:279.
[82] Tauler R, Lacorte S, Barceló D. Application of multivariate self-modeling curve resolution to the quantitation of trace levels of organophosphorus pesticides in natural waters from interlaboratory studies. J Chromatogr A 1996;730:177.
[83] Ortiz MC, Sarabia L. Quantitative determination in chromatographic analysis based on n-way calibration strategies. J Chromatogr A 2007;1158:94.
[84] Jaumot J, Gargallo R, Tauler R. Estimation of error propagation and prediction intervals in MCR-ALS using resampling methods. J Chemom 2004;18:327.
[85] Olivieri AC, Faber NM. A closed-form expression for computing the sensitivity in second-order bilinear calibration. J Chemom 2005;19:583.

[86] Olivieri AC. Computing sensitivity and selectivity in parallel factor analysis and related multiway techniques: the need for further developments in net analyte signal theory. Anal Chem 2005;77:4936.

[87] Olivieri AC, Faber NM. New developments for the sensitivity estimation in four-way calibration with the quadrilinear parallel factor model. Anal Chem 2012;84:186.

[88] Bauza MC, Ibáñez GA, Tauler R, Olivieri AC. Sensitivity equation for quantitative analysis with multivariate curve resolution-alternating least-squares: theoretical and experimental approach. Anal Chem 2012;84:8697.

[89] Saurina J, Leal C, Compañó R, Granados M, Prat MD, Tauler R. Estimation of figures of merit using univariate statistics for quantitative second-order multivariate curve resolution. Anal Chim Acta 2001;432:241.

[90] Rodríguez-Cuesta MJ, Boqué R, Rius FX, Vidal JLM, Frenich AG. Development and validation of a method for determining pesticides in groundwater from complex overlapped HPLC signals and multivariate curve resolution. Chemom Intell Lab Syst 2005;77:251.

[91] Rodríguez-Cuesta MJ, Boqué R, Rius FX, Zamora DP, Galera MM, Frenich AG. Determination of carbendazim, fuberidazole and thiabendazole by three-dimensional excitation-emission matrix fluorescence and parallel factor analysis. Anal Chim Acta 2003;491:47.

[92] Ahmadi G, Abdollahi H. A systematic study on the accuracy of chemical quantitative analysis using soft modeling methods. Chemom Intell Lab Syst 2013;120:59.

[93] Danzer K, Currie LA. Guidelines for calibration in analytical chemistry - Part 1. Fundamentals and single component calibration (IUPAC recommendations 1998). Pure Appl Chem 1998;70:993.

[94] Olivieri AC, Faber NM, Ferré J, Boqué R, Kalivas JH, Mark H. Uncertainty estimation and figures of merit for multivariate calibration. Pure Appl Chem 2006;78:633–61.

Chapter 6

# Multivariate Curve Resolution Methods for Qualitative and Quantitative Analysis in Analytical Chemistry

Hadi Parastar[1]
*Department of Chemistry, Sharif University of Technology, Tehran, Iran*
[1]*Corresponding author: e-mail: h.parastar@sharif.edu; h.parastar@gmail.com*

**Chapter Outline**



## 1 INTRODUCTION

Getting precise and accurate analytical results (i.e., qualitative and quantitative) from different chemical systems is one of the keys to make right decisions about the systems under study. In analytical chemistry, investigations concerning complex sample matrices, such as environmental, petrochemical,

**Data Handling in Science and Technology, Vol. 29. http://dx.doi.org/10.1016/B978-0-444-63527-3.00006-0**

and biological samples, have been widely reported [1]. These samples are particularly challenging because of their high chemical diversity along with an ever increasing number of chemical constituents. Consequently, the assessment of target analytes or screening of these samples may be affected by different problems during chemical analysis. The analytical procedure is an especially designed method to obtain qualitative and quantitative information about the composition and amount of target analyte(s) [2].

The increasing use of chemometrics and especially multivariate resolution techniques for qualitative and quantitative analysis in analytical chemistry during the past two decades confirm the potential of chemometric techniques in analytical measurements [3–5]. However, difficulties on the precise qualitative and quantitative chemical analysis of target analyte(s) in analytical samples are related to the complexity of their matrices, the low analyte levels to be analyzed and lack of selectivity for most of the analytical methods especially in complex samples. Therefore, multivariate curve resolution (MCR) techniques are emerged as powerful methods in order to overcome lack of selectivity limitation and to complement these analytical methodologies [4,6]. On the other side, with the developments of higher-order analytical instrumentations and data collection, which produces two- or higher-order data array for a single sample, MCR techniques with second-order calibration purposes are gaining special attention in analytical chemistry community. The attractive merit is their second-order advantage, which means accurately extract quantitative information of analyte(s) of interest even in the presence of unknown interferences.

The MCR of a multicomponent analytical data is based on the description of the variation of the measurements as a linear model of the contributions of their pure component profiles [7]. Therefore, relevant and sufficiently informative experimental data, such as those obtained from different analytical systems, are required. MCR methods are powerful approaches that do not require a lot of prior information about the analytical systems, such as number and nature of the pure components in the system. The bilinear structure of data is the main requirement for the MCR approaches. Fortunately, MCR techniques have shown potential to use for qualitative and quantitative analysis of analytical data. In this regard, proper arrangement of analytical data in data arrays is of prime importance. Figure 1 shows the ever-increasing trend of the papers in literature for the use of MCR methods in chemistry in recent years (2004–2014, "Multivariate curve resolution" keyword in Scopus).

Owing to the significant role of MCR in analytical chemistry, in this chapter, potential of MCR techniques for the analysis of different types of analytical data for qualitative and quantitative purposes is discussed. In this regard, first some theoretical backgrounds on MCR methods including data structures, data arrangement, bilinear model, and artifacts are presented. Then, MCR scenarios for qualitative and quantitative purposes in analytical chemistry are discussed. Finally, some of the recent applications of MCR methods are shown in details.

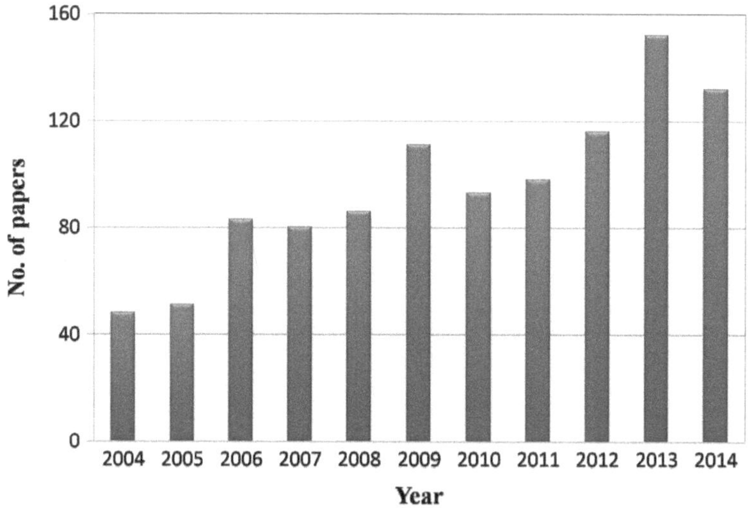


**FIGURE 1** Number of published papers regarding the use of MCR methods in chemistry in the past 10 years (2004–2014).

# 2 THEORETICAL CONCEPTS BEHIND MCR METHODS

## 2.1 Data Structures of Analytical Data for MCR

Figure 2 shows different types of analytical data with emphasis on their progression from the simplest zeroth order to multivariate data [8]. Zeroth-order data correspond to instruments producing a single response per sample, such as the absorbance at a single wavelength or the response of an ion-selective electrode (Figure 2A). First-order data for a given sample are arranged as a vector (Figure 2B), such as spectra obtained using UV–visible spectrophotometry (UV–Vis), fluorescence (FL), Fourier-transform infrared spectroscopy (FTIR), near-infrared spectroscopy (NIR), and nuclear magnetic resonance (NMR). Also, electrochemical signals (e.g., voltammogram) produce first-order analytical data. Similarly, obtained chromatograms using univariate detector response, such as gas chromatography coupled with flame ionization detector (GC-FID), electron capture detector (GC-ECD), and thermal conductivity detector (GC-TIC) and high-performance liquid chromatography coupled with single-wavelength UV–Vis, fluorescence, and refractive index detectors can produce first-order analytical data. Matrix data for a single sample are considered to be of second-order (Figure 2C). They can be recorded in three ways: (1) using a single instrument, such as fluorescence excitation–emission matrices (EEMs), diode-array spectrophotometer tracking the kinetics of a chemical reaction, or spectrophotometric titrations, (2) coupling two first-order instruments to make hyphenated instruments, such gas chromatography-mass spectrometry (GC-MS), liquid chromatography-diode

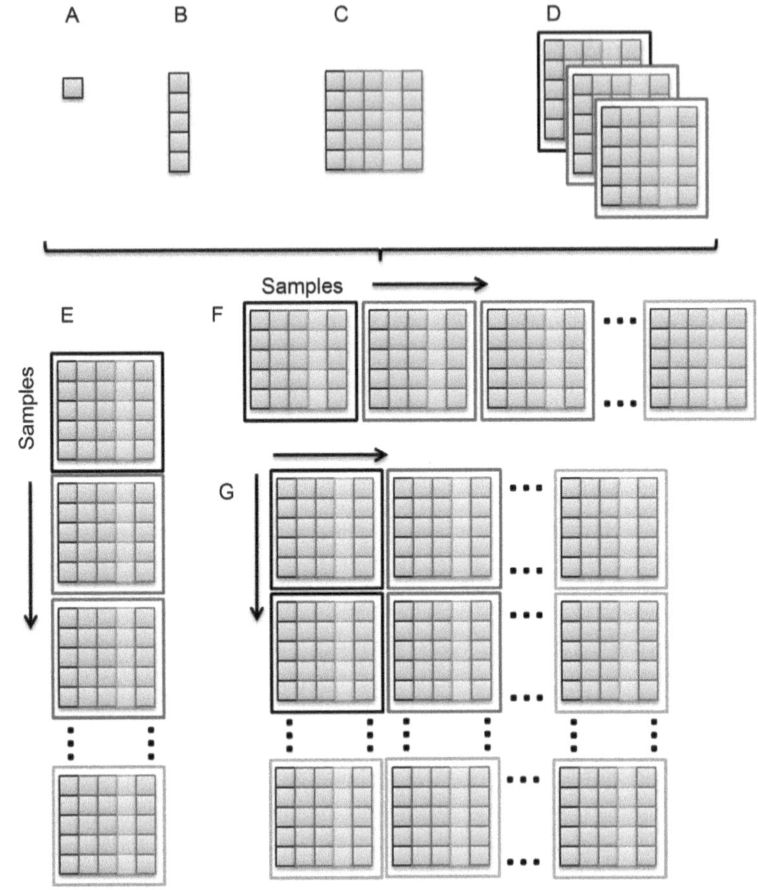


**FIGURE 2** Different types of data structures and data arrangement for analytical data: (A) zeroth-, (B) first-, (C) second-, and (D) third-order data. (E) Column-wise data arrangement, (F) row-wise data arrangement, and (G) simultaneous row- and column-wise data arrangement.

array detector (LC-DAD), liquid chromatography-mass spectrometry (LC-MS), and capillary electrophoresis-mass spectrometry, and (3) coupling two first-order instruments to make hypenated instruments, such as comprehensive two-dimensional gas chromatography (GC × GC) or comprehensive two-dimensional liquid chromatography (LC × LC). Furthermore, coupling two spectroscopic instruments can also produce second-order data, such as tandem mass spectrometry ($MS^2$) or two-dimensional nuclear magnetic resonance (2D-NMR). However, these data sets are not necessarily bilinear. Introducing an extra dimension in the data provides higher-order data including third-, fourth- and fifth-order data (Figure 2D). In this regard, folding second-order data for several samples into the third dimension cause producing a three-way array. Examples of three-way arrays are those obtained by EEM fluorescence spectroscopy, and GC-MS/LC-MS for several samples. In addition, advanced chromatographic instruments such as GC × GC-MS, LC × LC-DAD, LC × LC-MS, and GC × GC × GC-FID can produce three-way data for a single sample. In a similar way, grouping third-order data for several samples into the fourth dimension cause producing a four-way array. As an instance, data arrays obtained from the analysis of several samples using GC × GC-MS, LC × LC-DAD, LC × LC-MS, and GC × GC × GC-FID can be arranged in fourth-order (four-way) arrays [9].

Extensive complexity of analytical data demands for more involved data arrangements. Figure 2 shows different types of data arrangements for analytical data to be analyzed by bilinear MCR models. In this regard, three types of data arrangement are possible, including row-wise (Figure 2E), column-wise (Figure 2F), and simultaneous row- and column-wise augmentations (Figure 2G). The types of data arrangement depend on the data structure and aim of MCR modeling. For example, in the case of simultaneous analysis of multiple hyphenated chromatographic analyses (multiset data), column-wise data arrangement is recommended. Moreover, simultaneous row- and column-wise augmentation is also possible for multiset data fusion, which means for instance when combination of different spectroscopic data like for the same chromatography (i.e., LC-DAD-MS) is possible. In the particular case of multidimensional chromatographic techniques coupled to multivariate spectroscopic detectors, such as GC × GC-MS and LC × LC-MS, relatively more complicated data structures are obtained. In these cases, data amounts significantly increase and they should be arranged in more involved data structures. Column-wise superaugmentation approaches are the regular way of chromatographic data arrangement in the case of multiple samples analysis using MCR methods [9].

## 2.2 MCR Bilinear Model

MCR bilinear-based models are potent techniques for decomposition of mixed signals into the contributions of pure component profiles [4,10,11]. The bilinear structure of data set is the main prerequisite for the MCR approaches, which means they do not require prior information about the number and the nature of the pure components in the analytical system. An MCR bilinear model decomposes data matrix $\mathbf{X}$ into the product of two data matrices of $\mathbf{C}$ and $\mathbf{S}^{\mathrm{T}}$ (where T means the transpose of the matrix). Each of $\mathbf{C}$ and $\mathbf{S}^{\mathrm{T}}$ matrices contains the pure response profiles of the $n$ mixture components associated with the row and the column directions of the initial data matrix, respectively. In matrix notation, the general MCR bilinear model is as follows:

$$\mathbf{X} = \mathbf{C}\mathbf{S}^{\mathrm{T}} + \mathbf{E} \quad (1)$$

where $\mathbf{X}(I \times J)$ is the original data matrix, $\mathbf{C}(I \times n)$ and $\mathbf{S}^{\mathrm{T}}(n \times J)$ are the matrices containing the pure component profiles. $\mathbf{E}(I \times J)$ is the error matrix contains the unmodeled part of data matrix, **X**. The variables $I$ and $J$ represent the number of rows and columns of the original data matrix **X**, respectively. Also, $n$ is the number of chemical constituents in the matrix (i.e., chemical rank).

In case of simultaneous analysis of multiple runs (i.e., multiset data) of the same sample or of multiple samples having some components in common (e.g., different unknown mixtures of them containing possible interferences, with or without additional known analyte mixture samples used for calibration), the general MCR bilinear model given in Equation (1) can be easily extended to their simultaneous analysis as follows [9]:

$$\begin{pmatrix} \mathbf{X}_1 \\ \mathbf{X}_2 \\ \vdots \\ \mathbf{X}_\mathbf{K} \end{pmatrix} = \begin{pmatrix} \mathbf{C}_1 \\ \mathbf{C}_2 \\ \vdots \\ \mathbf{C}_\mathbf{K} \end{pmatrix} \mathbf{S}^\mathrm{T} + \begin{pmatrix} \mathbf{E}_1 \\ \mathbf{E}_2 \\ \vdots \\ \mathbf{E}_\mathbf{K} \end{pmatrix} \tag{2}$$

where now $\mathbf{X}_1, \mathbf{X}_2, \ldots, \mathbf{X}_\mathbf{K}$ are, respectively, the data matrices corresponding to the $k=1,\ldots,K$ runs having components in common to be analyzed simultaneously, $\mathbf{C}_1, \mathbf{C}_2, \ldots, \mathbf{C}_\mathbf{K}$ are the concentration matrices with the profiles of the resolved components in the $k=1,\ldots,K$ runs, $\mathbf{S}^\mathrm{T}$ is the matrix of spectral profiles of these eluted components, and $\mathbf{E}_1, \mathbf{E}_2, \ldots, \mathbf{E}_\mathbf{K}$ are the corresponding error matrices containing the part of the measured data unexplained by the proposed bilinear model. In a more compact form, this equation can be written as:

$$\begin{aligned} \mathbf{X}_\mathrm{aug} &= [\mathbf{X}_1; \mathbf{X}_2; \ldots; \mathbf{X}_\mathbf{K}] = [\mathbf{C}_1; \mathbf{C}_2; \ldots; \mathbf{C}_\mathbf{K}]\,\mathbf{S}^\mathrm{T} + [\mathbf{E}_1; \mathbf{E}_2; \ldots; \mathbf{E}_\mathbf{K}] \\ &= \mathbf{C}_\mathrm{aug}\mathbf{S}^\mathrm{T} + \mathbf{E}_\mathrm{aug} \end{aligned} \tag{3}$$

in which $\mathbf{X}_\mathrm{aug}$, $\mathbf{C}_\mathrm{aug}$, and $\mathbf{E}_\mathrm{aug}$ stand for the column-wise augmented matrices containing, respectively, the analytical data of the different runs, the concentration profiles of the resolved components in the different runs, and the error part of data in the different runs. In these equations, the notation ";" is used to indicate column-wise augmentation. Figure 3 shows decomposition of a single data matrix (Figure 3A) and a column-wise augmented data matrix containing multiple runs (Figure 3B). The similar model can be used for decomposing the row-wise and row- and column-wise data matrices. This is one of the interesting aspects of MCR models that they can analyze analytical data sets, which were arranged in various modes.

The reliability of most MCR solutions depends on the types of the constraints applied during the resolution of pure component profiles. Constraints can be considered as chemical or mathematical properties that concentration (**C**) and/or spectral ($\mathbf{S}^\mathrm{T}$) profiles should fulfill [3,5]. Therefore, when a particular profile is constrained, its shape is modified accordingly to fulfill a preselected property. Nonnegativity constraints are probably the most frequently used constraint in analytical data [12]. Apart from the application of

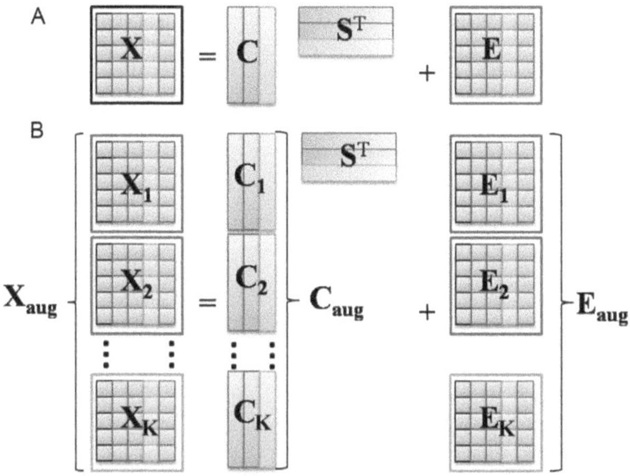


**FIGURE 3** General MCR bilinear model for decomposition of (A) second-order data matrix and (B) column-wise augmented data matrix into the pure component profiles.

nonnegativity to both pure concentration and spectral profiles, as imposed also in other chemical data sets, the particular properties of the concentration direction allow for the introduction of constraints related to the peak shape or to the sequential elution pattern of compounds. Thus, unimodality can be selected to preserve the presence of only one maximum in each pure profile [13]. On the other hand, the knowledge of the time windows where the components elute can provide very useful selective information. As an instance, time channels where a particular component elutes and where it does not can be used to introduce local rank information (selective and/or nonselective information). Also, when multiple analytical runs are simultaneously analyzed, correspondence among the components in the different runs fixes the sequence and presence, absence, and correspondence of each component in the different simultaneously analyzed data submatrices [14]. Apart from all of the mentioned constraints, in the case of simultaneous data analysis (multiset data) by MCR methods, there is possibility to force data to follow trilinear model assumption like multiway methods (e.g., parallel factor analysis, PARAFAC [15]). Although MCR methods are designed to handle nontrilinear three-way data sets formed by matrix augmentation taking individual data matrices of varying sizes, but it can be also used for the analysis of three-way data sets formed by data matrices with the same row and column sizes (data cubes) and take advantage of their structure if they follow the premises of the trilinear model [14]. All these constraints are extremely helpful to guarantee the proper resolution of the mixed components in the analyzed samples and increase the reliability of the quantitative information derived from the

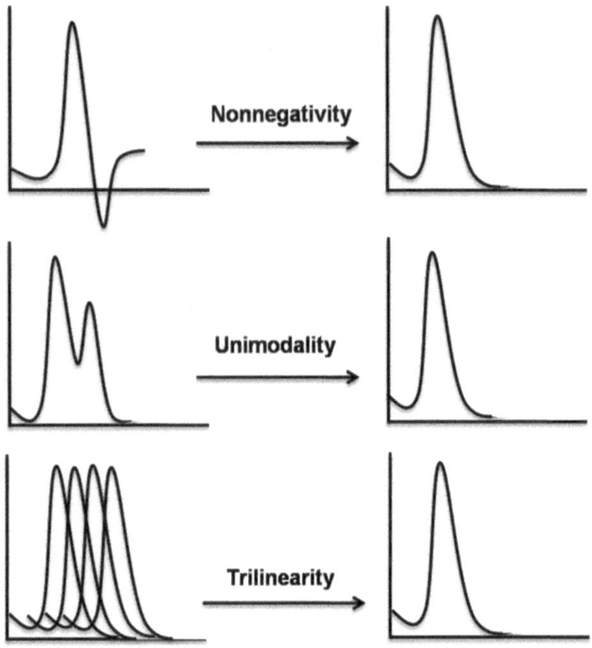


**FIGURE 4** Performance of nonnegativity, unimodality, and trilinearity constraints in MCR models.

corresponding quantitative analysis. These constraints can be applied in one step during resolution procedure (e.g., in noniterative MCR (NIMCR) methods) or in each step of an iterative resolution procedure (e.g., in non-iteative MCR (NIMCR) methods).

Figure 4 shows performance of three constraints of nonnegativity, unimodality, and trilinearity for a chromatographic profile as example. As it can be seen, by applying these constraints to the component profiles, a particular profile is constrained and its shape is modified to fulfill a preselected property.

## 2.3 Classification of MCR Algorithms

In general, MCR techniques can be classified into two classes of NIMCR and IMCR methods [6]. Indeed, after the first developed MCR algorithm by Lawton and Sylvestre for spectrophotometric analysis of binary mixtures [16], many other works with different mathematical backgrounds have been proposed to solve different types of chemical problems. Table 1 summarizes the classification of the MCR methods and their historical evolution.

Briefly, noniterative techniques aimed at finding a unique solution, in which the pure variables are uniquely defined according to the mathematical

**TABLE 1 Classification of Iterative and Noniterative MCR Methods**

| Method | Abbreviation | Year | References |
|---|---|---|---|
| **NIMCR** | | | |
| Self-modeling curve resolution | SMCR | 1971 | [16] |
| Evolving factor analysis | EFA | 1987 | [17] |
| Window factor analysis | WFA | 1992 | [18] |
| Heuristic evolving latent projection | HELP | 1992 | [19] |
| Orthogonal projection regression | OPR | 1994 | [20] |
| Subwindow factor analysis | SFA | 1999 | [21] |
| Evolving window orthogonal projection | EWOP | 1999 | [22] |
| Parallel vector analysis | PVA | 2002 | [23] |
| Alternative moving window factor analysis | AMWFA | 2006 | [24] |
| **IMCR** | | | |
| Iterative key set factor analysis | IKSFA | 1982 | [25] |
| Iterative target testing factor analysis | ITTFA | 1985 | [26] |
| Alternating regression | AR | 1989 | [27] |
| Simple-to-use interactive self-modeling mixture analysis | SIMPLISMA | 1991 | [28] |
| Multivariate curve resolution-alternating least square | MCR-ALS | 1993 | [29] |
| Orthogonal projection approach | OPA | 1996 | [30] |
| Positive matrix factorization | PMF | 1997 | [31] |
| Nonnegative matrix factorization | NMF | 1999 | [32] |
| Elementary matrix transformation | GENTLE | 2000 | [33] |
| Modified window target testing factor analysis | MWTTFA | 2000 | [34] |
| Iterative orthogonal projection | IOP | 2001 | [35] |
| Resolving factor analysis | RFA | 2001 | [36] |
| Simplex-based resolution method | SIMPLEX | 2003 | [37] |
| Penalty function alternating least squares | PALS | 2003 | [38] |
| Adaptive immune algorithm | AIA | 2004 | [39] |
| Mean-field independent component analysis | MFICA | 2006 | [40] |

*Continued*

**TABLE 1 Classification of Iterative and Noniterative MCR Methods—Cont'd**

| Method | Abbreviation | Year | References |
|---|---|---|---|
| Stochastic nonnegative independent component analysis | SNICA | 2006 | [41] |
| Multivariate curve resolution-weighted alternating least squares | MCR-WALS | 2006 | [42] |
| Multivariate curve resolution-objective function minimization | MCR-FMIN | 2007 | [43] |
| Window independent component analysis | WICA | 2008 | [44] |
| Nonnegative immune algorithm | NNIA | 2009 | [45] |
| Multivariate curve resolution-particle swarm optimization | MCR-PSO | 2013 | [46] |

principles involved. NIMCR methods are fast techniques that are based on one-step calculation algorithm. Additionally, they are applicable to systems with partial information that allows recovery of pure component profiles in one-at-a-time manner. In spite of the mentioned advantages, these methods suffer from the following limitations: (1) application of only nonnegativity constraint, (2) inability to extend to higher-order data, and (3) dependency of their solutions to the correct definition of the concentration window. In other words, the solutions of NIMCR methods are directly depend on the degree of overlap of component signals in the mixture and possibility of correct definition of their selective windows.

In contrast, IMCR techniques attempt to find pure solutions, in which the pure variables do not violate the chemical constraints. They can recover the pure component profiles one-at-a-time or simultaneously at each cycle of the optimization process. In addition, they are applicable to the systems with partial selective information or even no selective ones. Furthermore, these methods can be easily extended to the analysis of higher-order analytical data. However, IMCR solutions depend on initial estimates of concentration or spectral profiles. Additionally, their solutions are not unique and this can bring some additional limitations for these methods (i.e., ambiguities). Ambiguities in IMCR solutions imply production of a set of feasible solutions with the same fit that do not violate the constraints incorporated in the iterations; thus, it can cause serious problems in qualitative and quantitative analysis. Nevertheless, application of different types of constraints, such as nonnegativity, unimodality, normalization, selectivity, component correspondence, and trilinearity, can reduce the range of feasible solutions or even totally eliminate it [47].

On the other hand, various types of objective functions can be used for obtaining optimized **C** and $\mathbf{S}^T$ matrices by fulfilling the constraints. For example, alternating least squares (ALS) optimization [10], where constraints are applied to **C** and $\mathbf{S}^T$ matrices at each iteration of the two separated linear regression steps is one of these types. Equation (4) shows the ALS objective function.

$$\begin{aligned} \mathbf{E} &= \left\| \mathbf{X} - \mathbf{C}\mathbf{S}^T \right\| \\ \text{Objective function} &= \min(\text{SSE}) = \min\left(\sum_{i=1}^{I}\sum_{j=1}^{J} e_{ij}^2\right) \end{aligned} \tag{4}$$

where **E** is the error matrix and SSE stands for sum of squares of error. According to this equation, minimization of SSE is the objective function and is equal to the fulfillment of constraints on pure component profiles, **C** and $\mathbf{S}^T$.

Also, nonlinear optimization techniques of Quasi-Newton, Simplex [43], and particle swarm optimization (PSO) [46] can be used for optimization where the goal is minimization of an objective function defined directly from the nonfulfillment of constraints (Equation 5).

$$f(\mathbf{T}) = c_{\text{norm}}(\mathbf{T}) + c_{\text{non-neg}}(\mathbf{T}) + c_{\text{unimodal}}(\mathbf{T}) + c_{\text{clos}}(\mathbf{T}) + c_{\text{equal}}(\mathbf{T}) + \cdots \tag{5}$$

where $f$ (**T**) is the scalar objective function to be minimized and $c_{\text{norm}}(\mathbf{T})$, $c_{\text{non-neg}}(\mathbf{T})$, $c_{\text{unimodal}}(\mathbf{T})$, $c_{\text{clos}}(\mathbf{T})$, and $c_{\text{equal}}(\mathbf{T})$ are also scalar functions for normalization, nonnegativity, unimodality, closure, and equality constraints, respectively. Also, **T** is rotation or transformation matrix.

The final optimum values for objective functions can be used to test for the constraints fulfillment. The solution obtained for concentration and spectral profiles using the new rotation matrix are then assessed, and the lack of fit (LOF) and coefficient of multiple determinations ($R^2$) parameters were examined. LOF and $R^2$ can be calculated using Equations (6) and (7), respectively.

$$\text{LOF} = \sqrt{\frac{\sum_{i=1}^{I}\sum_{j=1}^{J}\left(x_{i,j} - \hat{x}_{i,j}\right)^2}{\sum_{i=1}^{I}\sum_{j=1}^{J} x_{i,j}^2}} \times 100 \tag{6}$$

$$R^2 = \left(1 - \frac{\sum_{i=1}^{I}\sum_{j=1}^{J}\left(x_{i,j} - \hat{x}_{i,j}\right)^2}{\sum_{i=1}^{I}\sum_{j=1}^{J} x_{i,j}^2}\right) \times 100 \tag{7}$$

where $x_{i,j}$ is each element in the experimental data matrix and $\hat{x}_{i,j}$ is each element in the MCR-modeled data matrix ($\hat{\mathbf{X}} = \mathbf{C}\mathbf{S}^T$).

Figure 5 depicts, for example, the general procedure undertaken by IMCR algorithms. In this regard, first, the desired experimental data are preprocessed

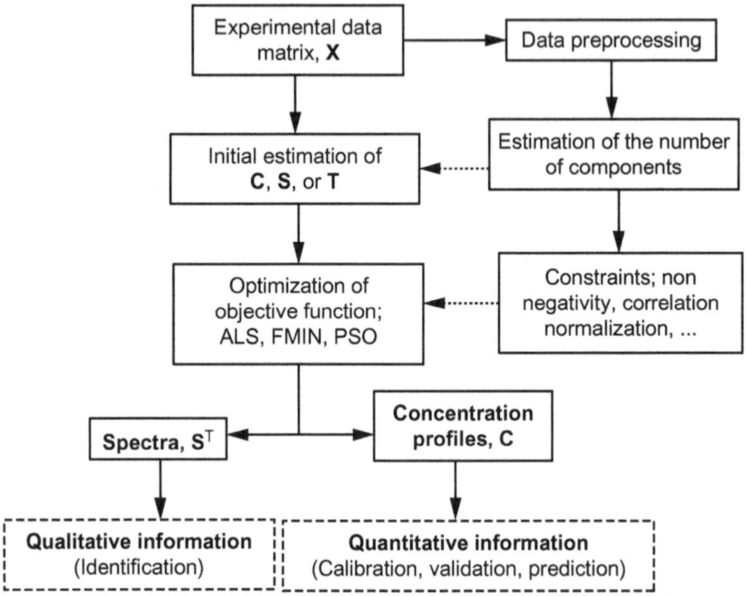


**FIGURE 5** General framework of iterative MCR models.

by proper methods to reduce the effects of artifacts, such as noise, baseline, and background. Then, the number of chemical components in the desired data matrix (single and/or multiple runs) is determined. Estimation of the initial values of concentration, spectral, or rotation matrix is the next step in the resolution procedure to start iterative optimization procedure. In the mean time, application of proper constraints during optimization procedure can be preformed to obtain chemically meaningful solutions. Finally, the optimized **C** and **S** matrices are used for qualitative (identification) and quantitative (calibration, validation, and prediction) chemical analysis.

## 2.4 Uncertainties in MCR Solutions

Even though MCR techniques give proper results regarding the pure component profiles using only the raw analytical data, but they suffer from uncertainties in their solutions. Generally, rotational and intensity (scale) ambiguities may present in the MCR solutions, which means instead of a single (unique) solution, a range of feasible solutions that fit the data equally well may be obtained [48]. These two kinds of uncertainties in MCR solutions can be shown as follows:

$$\mathbf{X} = \mathbf{C}\left(\mathbf{T}\mathbf{T}^{-1}\right)\mathbf{S}^{\mathrm{T}} = (\mathbf{C}\mathbf{T})\left(\mathbf{T}^{-1}\mathbf{S}^{\mathrm{T}}\right) = \mathbf{C}_{\mathrm{new}}\mathbf{S}^{\mathrm{T}}{}_{\mathrm{new}} \tag{8}$$

$$\mathbf{X} = \sum_{i=1}^{n}\left(\frac{1}{k_i}\mathbf{c}_i\right)\left(k_i\mathbf{s}^{\mathrm{T}}\right) \tag{9}$$

where **T** is rotation (transformation) matrix and $k$ is a scalar. The extent of ambiguities can be reduced or even eliminated by the use of appropriate constraints. As ambiguities produce a set of feasible solutions that do not violate the constraints incorporated in the iterations, thus, the estimation of the range of feasible solutions is of prime importance for the evaluation of the MCR results. Indeed, Lawton and Sylvestre have proposed first method for the estimation of the extent of rotational ambiguity in MCR solutions [16]. After this work, several methods have been proposed to address the problem of finding the range of feasible solutions. Simplex rotation algorithm, linear programming, and unmix multivariate receptor model, minimization and maximization of relative component contribution (RCC) function, Borgen plots, grid search and simplex optimization for finding the area of feasible solutions are the main attempts in this direction [48–55]. Also, a user-friendly software entitled MCR-BANDS has been developed by Jaumot and Tauler for the calculation of the extent of rotational ambiguity in an easy way based on maximization and minimization of RCC function [56]. The latest method for calculation of the extent of rotational ambiguity has been presented by Sawall and Neymeyr, which is based on a fast polygon inflation algorithm to compute the area of feasible solutions for three component systems [57]. FACPACK is developed MATLAB toolbox by this group for estimating the extent of rotational ambiguities in MCR solutions.

# 3 EXPERIMENTAL ARTIFACTS IN MCR MODELS

## 3.1 Deviations to Bilinearity in MCR Models

Bilinearity is an assumed property by MCR algorithms which means a data matrix can be written as a sum of single component data matrices (rank one), where each of them decomposed into the product of two vectors containing the concentration and spectral profiles [58]. However, in many practical analytical situations, slight or considerable deviations from this assumption may be observed, such as in the presence of interactions between different individual components in spectroscopic measurements or changing the second dimensions elution times of different modulations taken from first chromatographic column in GC × GC. When there is deviation from bilinear model assumption, then MCR models can still decompose the mixed experimental data into the pure component profiles. However, more components are needed to explain whole variance in the data compared to the ideal bilinear case. Therefore, the information related to one component is divided between more components and for sure these extra components do not really exist and cause wrong qualitative and quantitative analytical estimations. This makes data modeling and interpretation much more complicated, especially for complex analytical data sets. In such cases, this deviation must be corrected before MCR analysis. Deviations from bilinearity can be considered as one of the main consequences of analytical complexity, which can cause more complicated data structures for subsequent arrangement and MCR analysis. It is important to note that for the analytical data with at least one spectral dimension, such as GC-MS, EEMs, LC-MS, GC × GC-MS, LC × LC-DAD, violation from bilinear model assumption cannot be observed because of the presence of reproducible detector variables. In such cases, when several samples are analyzed, row- or column-wise augmentation can bring the required bilinearity prerequisite for MCR modeling.

Figure 6 demonstrates one of these types of deviation for GC × GC data [59]. In this figure, two different cases of bilinear data structure (ideal) and non-bilinear are depicted. The problem is appearing when elution profiles of the same component in the second chromatographic dimension of the same GC × GC run (within run shift) are compared. Figure 6A shows the GC × GC landscape from the first dimension. As it can be seen, five modulations are taken and analyzed in the second GC column. Figure 6B demonstrates the ideal situation where all the modulations taken from the first column have same shapes and positions (i.e., elution time) and modulations differ only in intensity. Therefore, the data have a bilinear structure. In other words, in

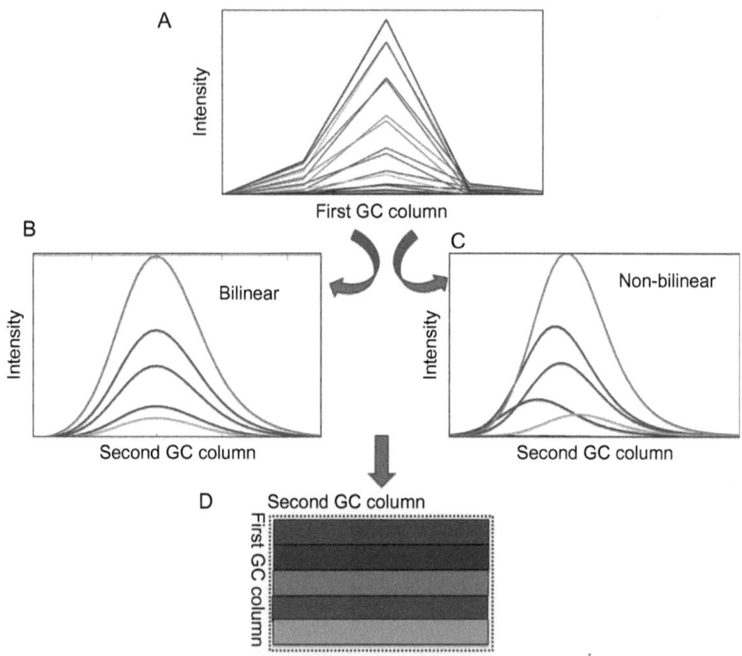


**FIGURE 6** Chemical shift in GC × GC data and deviation from bilinearity. (A) First chromatographic dimension landscape for a single-component system, (B) second chromatographic dimension modulations in ideal situation (bilinear), (C) second chromatographic dimension modulations in the presence of chemical shift (non-bilinear), and (D) GC × GC data as a second-order data matrix.

practical situations due to the increase in the temperature within the same chromatographic run, there is a significant and progressive shift of the second-dimension elution peaks to lower retention times in different modulations (Figure 6C). When these within-run retention time shifts are accompanied to coelution problems in GC × GC analysis of the same sample, discrimination of shift from coelution effects becomes troublesome and they can severely affect the proper resolution and quantification of the mixture components. In such situations, the data structure are non-bilinear and peak profiles are systematically shifted in the same direction due to the programmed gradient temperature within chromatographic run.

## 3.2 Noise

In analytical chemistry, noise can be considered as undesired part of data which can be defined as a high-frequency signal that arises from uncontrollable variables of the chemical processes and instruments. In general, noise can be classified in three different classes of homoscedastic, heteroscedastic, and correlated [6]. The presence of noise makes data structures more complex and demands for more appropriate tools to reduce it. Different denoising and smoothing methods have been developed to reduce the effect of noise in analytical signals. In this regard, denoising methods removes the small amplitude components of signals regardless of their frequency. In most practical situations, denoising methods have been used to reduce the effect of homoscedastic noise in data. On the other hand, the high-frequency components of signals can be removed using smoothing methods regardless of the amplitude. In contrast to denoising methods, smoothing methods have been used to convert heteroscedastic noise in data to homoscedastic one. Different methods have been proposed for denoising and smoothing including Fourier transform [6], Savitzky-Golay [60], wavelet transform, roughness penalty approach [6], and some methods based on the frequency difference between signal and noise to distinguish the signal and noise patterns in two-way data. On the other hand, principal component analysis (PCA) [61] is one of the best methods for noise reduction in analytical data. Since MCR methods has similar decomposition pathway like PCA, hence, they can be also used for noise reduction in analytical data.

## 3.3 Baseline/Background Contribution

The baseline/background contribution is associated with the systematic response from the analytical systems not related to the analyte. Baseline/background contribution can originate from various sources, such as matrix components signal in spectroscopy, formation of electrical double layer (non-faradic current) in electrochemistry, changes in mobile phase composition which influence detector signals in LC or chemical bleed signatures from stationary phases as they degrade in GC. Moreover, variations of pressure and temperature and fluctuations caused by impurities can bring additional baseline/background contributions in analytical measurements. On the other side, noise contributions are associated with unknown random variations and they are usually related to the sensitivity of the detector. Baseline/background contributions, particularly associated with noisy signals, may cause problems, especially for quantification. When the S/N is low, discrimination between noise and baseline/background contribution becomes difficult in practice, but in order to achieve a clear analytical signal, it is necessary to identify and eliminate the interferences caused by noise and baseline/background contributions. Existence of baseline/background contribution in analytical data causes difficulty in determination of the pure contribution of target analyte(s) particularly in low concentration levels where the S/N is low.

## 3.4 Shift

Shift is ubiquitous in analytical data and it can cause severe issues in qualitative and qualitative analyses. Shift in analytical data can be categorized into the random and progressive ones. Both types of shifts can be observed in spectroscopic, electrochemical, and chromatographic data. For example, when variations in the spectral information occur, the same chemical component identified in two analyses can conclude in different matrix columns. Therefore, the peak positions and peak shapes are accompanied by variations in peak positions and peak shapes not directly related to the sample. On the other hand, spectral shifts are always random, and thus, prediction of its amount in an analysis is impossible. Possible reasons for spectral shifts are differences in temperature, pressure, viscosity, pH, matrix effects, and imperfections in the recording procedure.

Elution time shifts in chromatography is another type of shift frequently observed in chromatographic separation [9]. Possible reasons for elution time shifts in 1D and 2D chromatography are associated with uncontrollable fluctuations in temperature, flow rate and pressure, matrix effects, stationary phase degradation, change in mobile phase composition in LC, and errors in the timing of manual injections. Elution time shifts in chromatography are almost random. However, there is another type of elution time shift in multidimensional chromatography (e.g., GC $\times$ GC) when temperature programming is applied to second chromatographic column to improve resolution and to avoid general peak overlap problems in the multicomponent analysis of complex samples. In these situations, this type of elution time shift is progressive and it can make the data structure more complex. Figure 7 shows elution time shift in GC $\times$ GC in two different situations. The first situation is for a single-component system where the progressive shift between different modulations is observed (Figure 7A). The second situation is for a two-component system where progressive elution time shift and peak overlap are simultaneously occurred (Figure 7B).

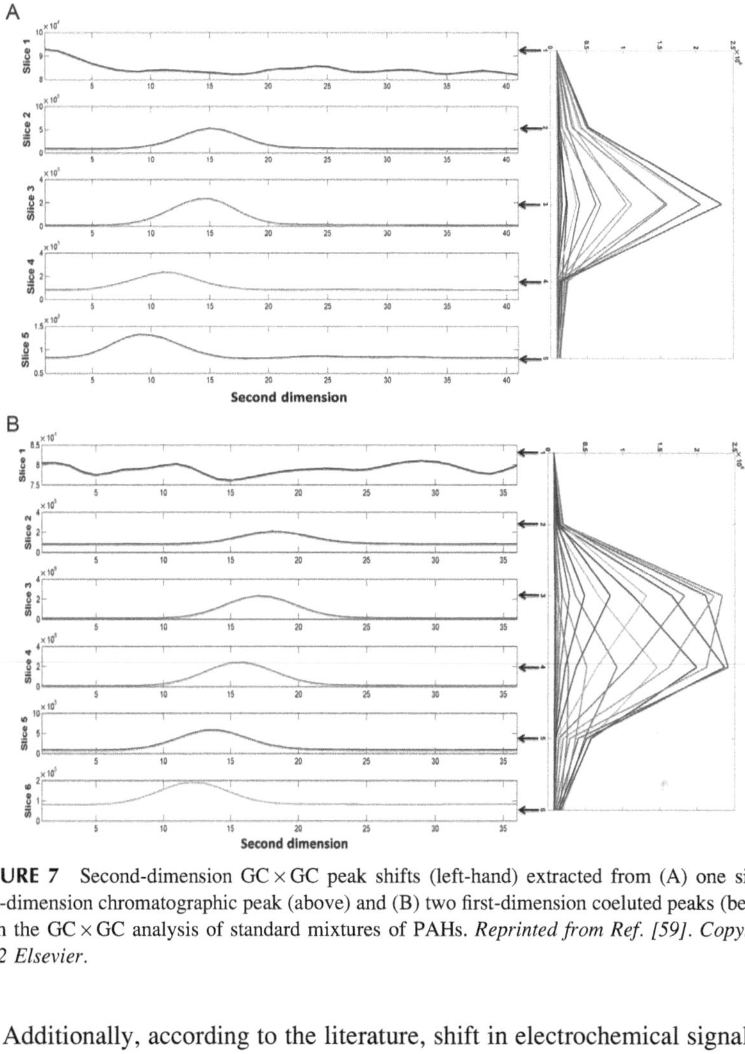


**FIGURE 7** Second-dimension GC × GC peak shifts (left-hand) extracted from (A) one single first-dimension chromatographic peak (above) and (B) two first-dimension coeluted peaks (below) from the GC × GC analysis of standard mixtures of PAHs. *Reprinted from Ref. [59]. Copyright 2012 Elsevier.*

Additionally, according to the literature, shift in electrochemical signals is common and it can be originated from adsorptive phenomena on the electrode surface, pH variations in the cell, or fluctuations in the composition of cell solution. Therefore, random and progressive shifts can be seen in electroanalytical data and it can cause potential shift in such data. This potential shift can cause deviation from bilinear model assumption discussed in Section 3.1.

## 4 MCR SCENARIOS FOR QUALITATIVE AND QUANTITATIVE ANALYSIS

The results from bilinear MCR methods when applied to different types of analytical data sets are the pure concentration and spectral profiles of the sought constituents of the analyzed samples. Figure 8 shows the general

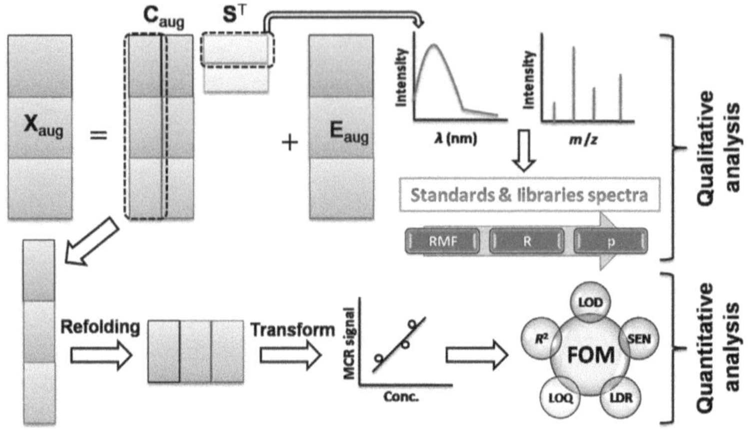


**FIGURE 8** General MCR scenarios for qualitative and quantitative analysis in analytical chemistry.

procedure by MCR for qualitative and quantitative chemical analysis in analytical chemistry. As it can be seen, first the data sets are arranged in a proper data matrix (e.g., column-wise augmentation). Then, MCR algorithms (i.e., iterative and/or noniterative ones) are applied to this augmented data matrix under application of proper constraints. The output of MCR analysis is a matrix that contains the pure concentration profiles in an augmented form and a matrix containing the pure spectral profiles.

Resolved MCR spectral profiles can be used to identify these components by comparing their resolved spectra with those of the authentic standards or standard spectra in available libraries. Different criteria can be used to evaluate the purity of the resolved spectral profiles, such as spectral correlation coefficient ($R$), match factor (MF), reverse match factor (RMF), and statistical probability ($p$) [6]. All of these criteria calculate the angle between the MCR-resolved profile (each profile forms a vector) with those of the standard spectral profile in different ways. For example, MF and RMF are the square-root of the normalized dot product between the MCR-resolved spectrum and the possible standard mass spectrum candidates from MS libraries (e.g., NIST or Wiley libraries). Also, there are some methods, which are based on the calculation of the statistical probability ($p$), similarity, identity and so forth.

On the other hand, resolved concentration profiles can be used for quantitative analysis. For example, in hyphenated chromatography, first the resolved elution profiles in augmented matrix are folded to form a new data matrix with elution times as rows and number of calibration samples as columns of this data matrix. The peak area/height of the target analyte in different concentrations can be used to build a calibration graph. Thus, relative quantitative information for one target compound can be then directly derived from the comparison of MCR-resolved elution profiles for different samples under the assumption of linear relation between relative peak areas/heights of the resolved elution profiles and their relative concentrations. Also, this linear relation can be conducted between the ratios of the peak areas of the target analyte to the peak area of internal standard (I.S.). Furthermore, various calibration strategies of external calibration, standard addition, and I.S. can be used in this way. After building calibration graphs for sought analytes, analytical figures of merits (AFOMs) including calibration sensitivity (i.e., slope of calibration curve), analytical sensitivity, limit of detection (LOD), limit of quantitation (LOQ), linear dynamic range, regression coefficient ($R^2$), and relative error (RE) in calibration concentrations can be calculated. These parameters are mainly used to evaluate the capability and versatility of the proposed analytical method. Owing to the fact that MCR methods have the second-order advantage, therefore, they can easily adopt to get pure analytical information gathered from analytical system under study. However, utilizing multivariate chemometric methods (i.e., MCR methods) demands new ways to develop multivariate AFOMs instead of the univariate ones [62]. Fortunately, new ways have been proposed to calculate multivariate AFOMs using MCR solutions instead of pseudounivariate calibration parameters. These multivariate parameters are sensitivity (SEN), analytical sensitivity ($\sigma$), selectivity (SEL), LOD, LOQ, and RE. Sensitivity is the most important AFOM to define other figures of merit. More details about multivariate AFOMs for MCR will be discussed in Section 5 and can be found in the literature [62].

## 5 MULTIVARIATE AFOMs

Estimation of the multivariate AFOMs for MCR with second-order calibration purposes has been the object of important theoretical and experimental efforts in the last decade [62]. Multivariate AFOMs are numerical parameters for evaluation of the performance of an analytical instrument or method. Since, MCR methods bring the second-order advantage for the analysis of different types of analytical data, therefore, conventional FOMs obtained using pseudounivariate calibration curve cannot be used for evaluation of the performance of MCR methods. Thus, development of new ways for calculation of AFOMs for multivariate methods (e.g., MCR methods) is of prime importance. SEN and SEL are important AFOMs in MCR methods, regularly employed for comparison of the analytical performance of MCR methods. SEN is defined as the change in the response of the instrument divided by the corresponding change in the concentration of the sought analyte. In other words, SEL is the extent to which a method can be used to determine particular analyte(s) in mixtures without interferences from other constituents. Different attempts have been paid to define SEN and SEL for MCR models and especially MCR-ALS algorithm. Among different studies, using resampling techniques or Monte Carlo noise addition approaches, considering the pseudounivariate scores-concentration calibration graph generated by MCR-ALS and utilizing error propagation analysis as the inverse of the ratio of predicted concentration uncertainty to signal uncertainty are the most important studies in this field [62,63]. The multivariate AFOM expressions suitable for MCR-ALS by the most recent method in analytical studies is derived and supported by Monte Carlo noise addition simulations and experimental data [63]. This new method allows for precise definition of multivariate AFOMs based on MCR pure component profiles. Apart from SEN and SEL, other parameters like LOD as the minimum analyte concentration that is detectable with a certain degree of confidence, LOQ as the concentration level for which the relative prediction error is 10%, and RE as the measure of the difference between true and predicted concentration values can be used for evaluation of analytical methods. Multivariate AFOMs for MCR are as follows:

$$\mathrm{SEN_{MCR}} = m_n \left[ J \left( \mathbf{S}^{\mathrm{T}} \mathbf{S} \right)^{-1}_{nn} \right]^{-1/2} \tag{10}$$

$$\mathrm{SEL_{MCR}} = \left[ \left( \mathbf{S}^{\mathrm{T}} \mathbf{S} \right)^{-1}_{nn} \right]^{-1/2} \tag{11}$$

$$\mathrm{LOD_{MCR}} = \frac{3.3 s_{\mathrm{dtest}}}{\mathrm{SEN_{MCR}}} \tag{12}$$

$$\mathrm{LOQ_{MCR}} = \frac{10.0 s_{\mathrm{dtest}}}{\mathrm{SEN_{MCR}}} \tag{13}$$

$$\mathrm{RE}(\%) = \frac{\sum_i (c_i - \hat{c}_i)^2}{\sum_i c_i^2} \times 100 \tag{14}$$

In the above equations, $J$ in Equation (10) is the number of data points in each submatrix in the augmented mode, $m$ is the slope of the MCR-ALS pseudounivariate graph (built by plotting analyte scores vs. nominal calibration concentrations), and $n$ is the index for the sought analyte. Also, $\mathbf{S}$ in Equations (10) and (11) is the MCR-resolved profiles in nonaugmented mode (e.g., spectral profiles). In addition, $s_{\mathrm{dtest}}$ in Equations (12) and (13) is the standard deviation for test sample. In Equation (14), $c_i$ is the known concentration of target analyte in calibration standard $i$ and $\hat{c}_i$ is its concentration calculated using MCR method. See Refs. [62] and [63] for more details about multivariate AFOM.

Owing to the fact that multivariate FOMs for MCR are very new, thus, they are rarely applied for experimental data. One of the recent examples of the calculation of multivariate AFOMs is for simultaneous determination of five selected pharmaceuticals in water samples using solid-phase extraction as preconcentration step combined to HPLC-DAD [64]. In this regard, the HPLC-DAD data sets for different concentrations are column-wise augmented with elution times as rows and detector wavelengths as columns. The MCR-ALS was used to recover pure elution and spectral profiles. The resolved profiles were have been used to calculate the multivariate FOMs according to Equations (10)–(14). In general, there is an improvement in using multivariate FOMs instead of univariate FOMs. Inspection of the results showed that the values of SEN are significantly changing from one analyte to another, and also in some cases, the univariate SEN values are better than the multivariate ones. In other words, the values of LOD and LOQ are in general better than the univariate ones.

## 6 MCR APPLICATIONS FOR QUALITATIVE PURPOSES

Qualitative analysis means identification of target analyte(s) according to their chemical characteristics, such as spectrum. Nowadays, there is an increasing interest in identification of chemical composition of complex samples by their chemical signals that is called chemical fingerprints. A chemical fingerprint can be defined as a characteristic profile reflecting the complex chemical composition of the analyzed sample and can be obtained by spectroscopic, chromatographic, or electrochemical techniques. This type of identification is always based on pattern recognition methods. Applications of pattern recognition methodologies for qualitative analysis in analytical chemistry are important. Among different fingerprinting techniques, chromatographic fingerprinting has been proposed as a promising tool in qualitative analysis of complex natural products. Furthermore, chromatographic fingerprinting has been widely used for identification, authentication, characterization and classification of complex natural samples, such as herbal medicines in the last decade.

For example, Figure 9 shows the different steps of MCR analysis combined with classification techniques for chemical fingerprinting [65]. It should be pointed out that this strategy can be easily extended to other types of analytical data (i.e, spectroscopic and electrochemical data) for qualitative analysis. In

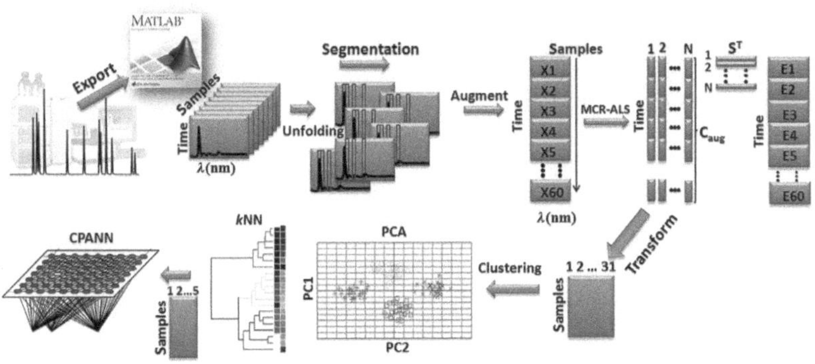


**FIGURE 9** General framework of combination of MCR and classification techniques for chromatographic fingerprint analysis (HPLC-DAD data in this case). *Reprinted from Ref. [65]. Copyright 2014 Elsevier.*

this regard, the chemical fingerprint data for each sample was segmented to desired number of submatrices. For electrochemical or spectroscopic data, this step can be eliminated and all data are prepared for data arrangement and analysis. Then, each segment or all data for all samples are arranged in an appropriate format, for example, column-wise augmentation of hyphenated chromatographic data with elution times as rows and wavelengths as columns. The augmented data matrix for each segment is then analyzed using one of the MCR techniques under application of proper constraints. The relative concentrations of the resolved components can be obtained using simple peak integration of resolved elution profiles for each component. These concentrations are used to build a new matrix containing relative concentrations of all resolved components for all different samples. Then, this new data matrix is analyzed by multivariate clustering methods (unsupervised classification techniques) of PCA, *k*-nearest neighbors (*k*NN), and *k*-means to cluster data to desired number of clear-cut groups and show the similarity and dissimilarity of samples according to the relative concentration of different components. Finally, supervised classification techniques, such as counter-propagation artificial neural network (CP-ANN) can be used to confirm the results of PCA, *k*NN, and *k*-means for classification of chemical fingerprints and to find the most effective variables in discrimination of clear-cut groups.

Table 2 shows some of the recent applications of MCR techniques combined to classification techniques for qualitative analysis in analytical chemistry.

Also, as it is already mentioned, in the simplest way, the resolved spectral profiles using MCR techniques can be used for identification of the identity of the unknown constituents by comparing the resolved spectral profile with those of standards present in the laboratory or stored in available libraries. Some of these types of MCR applications are also included in Table 2.

## 7 MCR APPLICATIONS FOR QUANTITATIVE PURPOSES

### 7.1 MCR in Chromatography

Development and evolution of new chromatographic systems for separation, identification, and quantification of chemical compounds in complex samples matrices has become one of the most important developments in analytical chemistry in the recent years. Nowadays, most of the chromatographic studies are focused on the analysis of trace and ultratrace amount of constituents in complex samples with minimum sample preparation and on reducing its timescales and sample size requirements [9]. Analysis of complex samples has been a great challenge in analytical chemistry. Despite the fact that many analytical methods have been developed, most of them are not feasible to be used for analysis of real samples without tedious separation and concentration operations. Indeed, chromatographic techniques have been revolutionized the

**TABLE 2 Examples of Application of MCR Methods for Qualitative Analysis in Analytical Chemistry**

| Sample | Technique | Chemometric technique | References |
|---|---|---|---|
| Salvia R. | HPLC-DAD | MCR-ALS, PCA, *k*NN, CPANN | [65] |
| Citrus fruits | GC-MS | MCR-ALS, PCA, *k*-means, CPANN | [66] |
| Wine | HPLC-DAD | MCR-ALS, PCA | [67] |
| Cell | UPLC-MS | MCR-ALS | [68] |
| Plant essential oils | GC-MS | Various MCR methods | [6] |
| Photodegradation products | LC-DAD-MS | MCR-ALS | [2] |
| Caviar | FL | MCR-ALS | [69] |
| Protein-nanomaterial | UV–Vis | MCR-ALS | [70] |
| Food | HNMR | ICA, MCR | [71] |
| Azo dyes | UV–Vis, HNMR | MCR-ALS | [72] |
| Oils | FT-IR | MCR | [73] |
| Serum | LSV, CV, DPV | MCR-ALS | [74] |
| Small-moleculebiopolymer | Voltammetry, FL | MCR-ALS | [75] |
| Metal binding | DPP, CD | MCR-ALS | [76] |
| Heavy metals-phytochelatins | Voltammetry | MCR-ALS | [77] |

routines in analytical chemistry for the detection, quantification, and identification of chemical constituents. On the other side, enormous efforts have been conducted on coupling chromatographic methods with different spectroscopic detection methods (hyphenation), such as gas and liquid chromatography-mass spectrometry (GC-MS and LC-MS) in order to provide different hyphenated chromatographic methods. Although the ability to isolate components in complex mixtures has extremely increased, separation systems with the capacity to isolate thousands of species simultaneously are still inadequate when applied to commonplace natural mixtures such as fuel, environmental, or biological samples. Furthermore, with the hyphenated instruments, a huge amount of data can be produced. For most of the cases, to obtain a satisfactory separation, optimized experimental conditions should be employed, which is usually laborious and time-consuming. Therefore, in spite of advanced hyphenated chromatographic systems, the incomplete separation issues still exist in the analysis of complex samples. This incomplete separation directly reflects the lack of selectivity in the chromatographic systems. Therefore, exploiting pure chromatographic and spectroscopic information of target compounds in the presence of interferences is very difficult. As a consequence, reliable qualitative and quantitative information cannot be obtained [9].

In other words, multidimensional chromatographic techniques (i.e., hyphenation) have emerged as powerful techniques suitable for the separation of very complex mixtures because of their higher resolution and higher peak capacity in recent decade, which can be considered as an important development in chromatographic history [78]. However, one of the main challenges in multidimensional chromatography is related to the difficulty on the analysis and interpretation of the enormous amount of data obtained in these cases. This is due to the existence of different chromatographic challenges in their chromatographic or spectroscopic dimensions, which are much more severe than in 1D chromatography [9]. Therefore, it is clear that both types of hyphenated and multidimensional chromatographic analyses suffer from artifacts that directly affect their qualitative and quantitative analytical results. These challenges are not different from those derived from any other instrument; being based on both chemical and nonchemical aspects of the analysis. In general, these challenges can be divided into three different classes of (1) baseline/background contributions and noise, (2) elution time shifts, and (3) coelution or peak overlap (overlapped and embedded peaks) [79,80].

Chemometricians have made great efforts to extract pure analytical information from hyphenated and multidimensional chromatographic data. Among different chemometric methods, MCR techniques have shown great potential to compensate the lack of selectivity problem in chromatographic systems and to obtain pure chromatographic and spectroscopic profiles of target compounds in the presence of interferences (which is called second-order advantage).

In general, the chromatographic quantitative pathways using MCR can be classified in three classes of (1) semiquantitative analysis, (2) pseudounivariate calibration, and (3) multivariate calibration.

The first approach is based on semiquantitative analysis. By using this technique, only some information about the relative percent (composition) of each component in the whole chromatogram can be obtained. This strategy is common when the pure standards of the target analyte(s) are not accessible. In this regard, the overall volume integration is usually used for calculating the amount of each component after resolving the elution and spectral profiles of pure components. The advantage of this quantitative method over the general peak area integration is that all spectral intensities are taken into account.

As an example of such situations, the total ion current (TIC) chromatogram for a plant metabolites data (grapefruit essential oil) is shown in

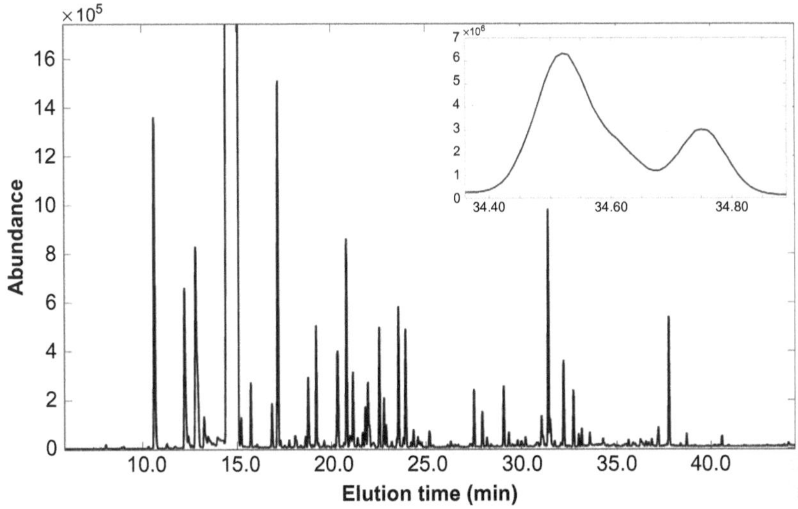


**FIGURE 10** Total ion current (TIC) chromatogram of grapefruit essential oil. Inset shows the selected chromatographic segment to analyze by MCR-PSO. *Reprinted from Ref. [19]. Copyright 2013 Elsevier.*

Figure 10 [46]. As it can be seen, the sample matrix is very complex having several components along with different chromatographic problems. Inset in Figure 10 demonstrates the chromatographic segment to be analyzed by MCR (e.g., MCR-PSO algorithm in this case) between elution times 34.3 and 34.9 min. Figure 11A shows the 3D plot of the selected chromatographic region from the GC-MS-TIC data of grapefruit essential oil. The resolved MCR-PSO elution profiles for three components are shown in Figure 11B. In addition, resolved mass spectra for three components are demonstrated in Figure 11C. The LOF value for this chromatographic region was 4.5%, which was acceptable according to the noise level in data. Matching the resolved mass spectra with the standard mass spectra in NIST MS library confirmed the presence of decanal ($C_{10}H_{20}O$), *cis*-carveol ($C_{10}H_{16}O$), and octylacetate ($C_{10}H_{20}O_2$), respectively. The values of RMF for the resolved mass spectra were respectively 984, 890, and 979. As it can be seen, these spectral matches are quite high and establish the possible identity of the resolved components in this chromatographic window. The percent of these components in the analyzed essential oil can be obtained by dividing the peak area of each component to the total peak area of TIC.

As can be judged from this example, the MCR-resolved elution profiles are used for obtaining semiquantitative information from target compounds. This type of information can be good for some purposes, such as –omics science.

The second quantitative approach is based on the pseudounivariate calibration curves. The general procedure for qualitative chromatographic analyses

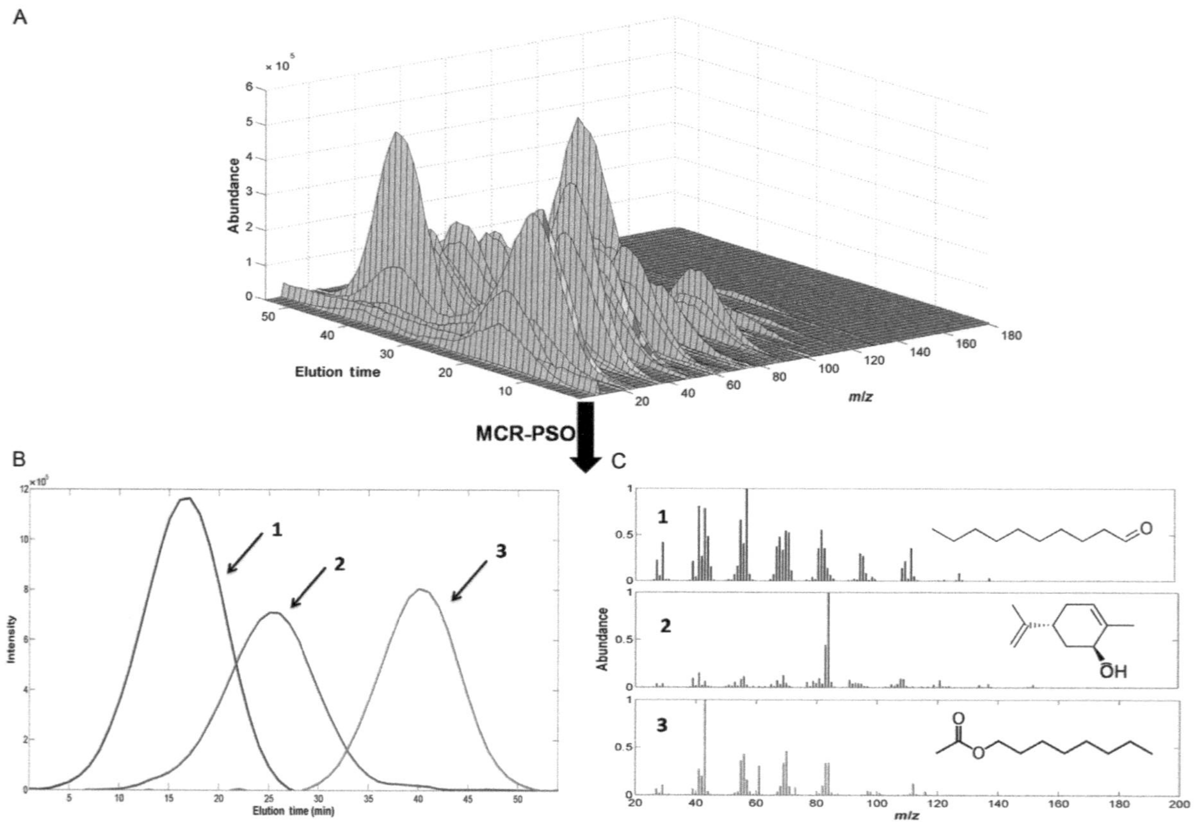


**FIGURE 11** Resolved MCR-PSO profiles for selected chromatographic segment in Figure 10. (A) 3D visualization of the data, (B) resolved elution profiles, and (C) resolved mass spectra of decanal (1), *cis*-carveol (2), and octylacetate (3) using similarity search in NIST MS database. *Reprinted from Ref. [19]. Copyright*

in this approach is similar to the procedure shown in Figure 8. Three types of calibration strategies of external calibration, standard addition, and I.S. can be used. As it has already discussed, relative quantitative information for one target compound can be then directly derived from the comparison of MCR-resolved elution profiles for different samples under the assumption of linear relation between relative peak areas or heights of the resolved elution profiles and their relative concentrations (pseudounivariate calibration graphs). This is one of the interesting aspects of MCR methods, which can reduce or even totally eliminate the matrix effect by calculating the pure contribution of each component in the mixture. This strategy is commonly used MCR approach for quantitative analysis using hyphenated chromatographic techniques in the literature. However, for obtaining similar quantitative chemical information from the analysis of complex mixtures using advanced chromatographic instruments, more involved MCR procedure is required. For example, Figure 12 shows the general procedure for the global proposed strategy for GC × GC-TOFMS data arrangement and their MCR analysis [79,81].

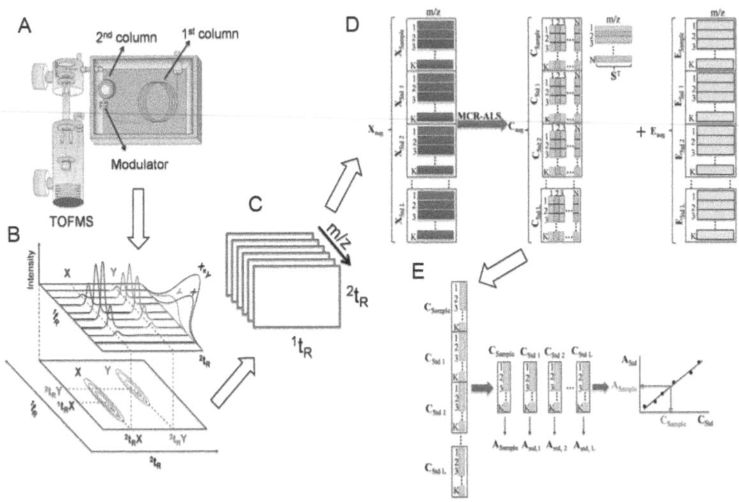


**FIGURE 12** (A) GC × GC-TOFMS instrumentation design, (B) signal acquisition of two overlapped components X and Y eluting from the first column and injected to the second column and corresponding 2D contour plot of the second column collected chromatograms, (C) three-way data arrangement of the whole GC × GC-TOFMS signal acquired during the analysis of a single sample, (D) $\mathbf{X}_{aug}$ is the column-wise super-augmented matrix built using all augmented data matrices corresponding to different samples. $\mathbf{X}_{std}$ are the individual augmented data matrices corresponding to the GC × GC-TOFMS analysis of every standard mixture samples and $\mathbf{X}_{sample}$ is the individual augmented data matrix corresponding to the unknown HFO sample. $\mathbf{C}_{aug}$ is the matrix of elution profiles of every component in the standard mixture samples, $\mathbf{C}_{std}$, and in the unknown sample, $\mathbf{C}_{sample}$, and $\mathbf{S}^T$ are the pure mass spectra of every component resolved by MCR-ALS. $\mathbf{E}_{aug}$ is the residual matrix, and (E) external calibration strategy to obtain quantitative information from the peak areas of the MCR-ALS-resolved elution profiles. *Reprinted from Ref. [79]. Copyright 2011 American Chemical Society.*

The GC × GC-TOFMS instrumentation design and signal acquisition where the entire first column chromatogram is modulated into a series of high-speed short secondary chromatograms of a length equal to $P_M$ are shown in Figure 12A and B. Each modulation produces a data matrix, $\mathbf{X}(I,J)$ where $I$ is the number of data points collected, which could be addressed by the second column retention time and $J$ is the number of $m/z$. The modulations obtained at the different retention times in the first dimension can be then combined for every $m/z$ to describe the elution pattern by means of 2D-contour plots $(I,K)$ as shown in Figure 12B where $K$ is the number of slices taken from the first column of the total run. This contour plot will be obtained for every $m/z$ in the GC × GC-TOFMS analysis. When all the $m/z$ values ($J$) are simultaneously considered, the whole set of acquired data in one GC × GC run (in the analysis of a single sample) gives a three-way data cube with dimensions equal to $(I,J,K)$ where $I$, $J$, and $K$ are like before (Figure 12C). When several standard and unknown samples are simultaneously analyzed, a four-way hypercube data set will be obtained with dimensions equal to $(I,J,K,L)$ where $I$, $J$, and $K$ are the same as before, and $L$ is the number of analyzed samples. This complex data arrangement can be simplified joining the augmented data matrices obtained in the analysis of every individual sample with dimensions $(I \times K, J)$ to give the column-wise superaugmented data matrix with dimensions $(I \times K \times L, J)$ for the simultaneous analysis of the different $L$ samples (Figure 12D). Since the number of $m/z$ is equal for all slices and for all chromatographic runs, the column-wise data arrangement shown in Figure 12D results to be very flexible and adequate to bilinear modeling requirements. For instance, the number of chromatographic ranges analyzed can be different in each slice and the presence of time shifts among different slices and among different samples do not destroy the bilinear model assumption associated to the column-wise data augmentation strategy. It is important to note that each modulation in the superaugmented data matrix can have different number of rows (retention times in the first and second dimensions). This is a very flexible property of matrix augmentation, which adapts very well to GC × GC-TOFMS data because of ubiquitous retention time peak shifts of the eluted components from slice-to-slice (in second column) and from sample-to-sample (in first and second columns). Furthermore, with this superaugmented data arrangement, there is no limitation in the number of included submatrices (number of slices and samples). For quantitative analysis for instance, it is possible to include the second dimension slices taken from the first column for the different standard mixture and unknown samples, and their replicates in the same column-wise superaugmented data matrix and perform their simultaneous analysis in one shot.

An illustrative example of this situation is given in Figure 13. In this case, MCR-ALS is applied to the resolution and quantification of a mixture of polyaromatic hydrocarbons (PAHs) in standards and in a North Sea oil sample

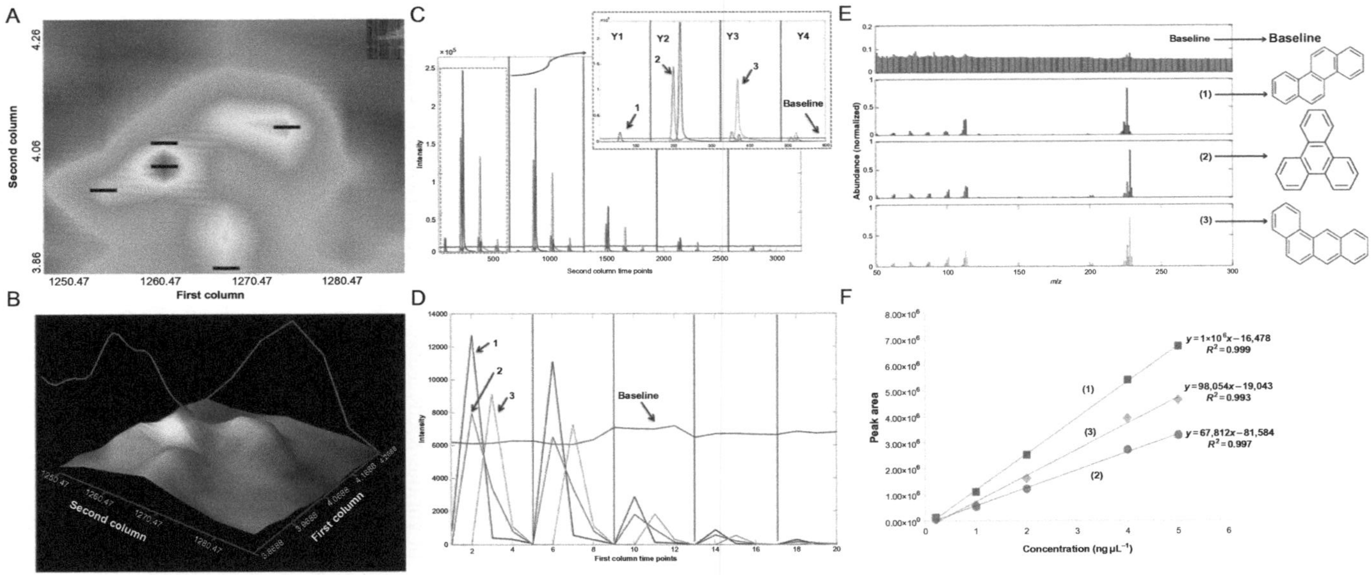


**FIGURE 13** Resolved MCR-ALS pure component profiles for a chromatographic region taken from the GC × GC-TOFMS analysis of North Sea oil sample: (A) 2D contour plot and (B) 3D chromatogram of this chromatographic region. (C) Resolved second-dimension elution profiles in five different standard samples. Inset shows the resolved profiles for three components (blue (dark gray in the print version) (1), green (gray in the print version) (2), and light green (light gray in the print version) (3)) as well as baseline/background contribution (red (gray in the print version)) for four different modulations. (D) Reconstructed first-dimension elution profiles of corresponding components in (C). (E) Resolved MCR-ALS mass spectra identified using NIST MS database as chrysene (blue (dark gray in the print version), 1), triphenylene (green (gray in the print version), 2), and benz[*a*]anthracene (light green (light gray in the print version), 3). (F) External calibration curves built up using five different standards for chrysene (blue (dark gray in the print version), 1), triphenylene (green (gray in the print version), 2), and benz[*a*]anthracene (light green (light gray in the print version), 3). Calibration equations and regression coefficients for each component are shown next to the calibration curve. *Reprinted from Ref. [81]. Copyright 2013 Springer.*

analyzed by GC × GC-TOFMS [81]. Figure 13A and B shows the contour and 3D plots for the selected chromatographic region. Figure 13C shows the resolved elution profiles in the second chromatographic dimension for the four different modulations from first dimension and for five different standard samples. MCR-ALS resolved first-dimension elution profiles are shown in Figure 13D. Resolved mass spectra for the corresponding components are shown in Figure 13E. In addition, areas of the resolved MCR-ALS elution profiles in the second chromatographic dimension can be used to build calibration curves for three resolved chemical components as shown in Figure 13F. All of these results confirm the power and reliability of MCR-ALS analysis to address the fundamental chromatographic challenges occurring during GC × GC-TOFMS analysis, with proper qualitative and quantitative analytical results.

The univariate or multivariate AFOMs can be used to evaluate the quality of calibration procedure. As it has already mentioned these values, SEN, SEL, LOD, LOQ, RE, and so forth can incorporate the pure information obtained by MCR resolution procedure in the calibration procedure.

The third MCR quantitative approach is based on the multivariate calibration pathway. In this regard, the data are divided into three calibration, validation, and prediction sets. Then, a multivariate calibration model is build using the MCR by considering all the chemical components present in the calibration set. Afterward, this model is evaluated using the validation set and some statistical parameters, such as root mean square error, $R^2$, and RE. Finally, the conducted MCR model is used to predict the concentration of target analyte(s) in unknown samples. As it can be seen, this approach is the same as other multivariate calibration models, such as partial least squares (PLS) [82].

Table 3 shows some of the recent applications of MCR methods to solve chromatographic challenges and to obtain pure quantitative analytical information. In this table, combination of different chromatographic techniques (i.e., hyphenated and multidimensional) with various MCR techniques are shown.

In recent decade, great efforts have been made by different chemometric groups for developing software for implementation of different MCR methods for analyzing multicomponent systems. Multivariate curve resolution of chromatographic data (MCRC software) is a dedicated software for MCR analysis of different types of chromatographic data [98]. Also, it can also be used for other types of multivariate data (i.e., electrochemical and spectroscopic data). It consists of five groups of techniques for preprocessing, chemical rank determination, local rank analysis, multivariate resolution, and peak integration. This software has the ability of the analysis of complex multicomponent chromatographic signals of GC-MS and HPLC-DAD. The software allows a user to apply the implemented methods in an easy way and it gives a straightforward possibility to visualize the obtained results. The main features of the presented software are providing a number of preprocessing techniques, implementation

**TABLE 3** Application of MCR Methods in Chromatographic Analysis of Different Sample Matrices

| MCR technique | Chromatography | Analyte(s) | Sample | References |
|---|---|---|---|---|
| HELP | GC-MS | FAs | Fish oil | [83] |
| MCR-ALS | GC × GC-MS | PAHs | Heavy fuel oil | [79,81] |
| MCR-COW-MCR | HPLC-DAD | Phamaceutics | Water | [64] |
| IKSA-MCR-ALS | LC × LC-DAD | Metabolites | Human urine | [84] |
| MCR-FMIN | GC-MS | Essential oil | Saffron | [85] |
| MCR-PSO | GC-MS | Essential oil | Grape fruit | [46] |
| HELP | GC-MS | Essential oil | Rose | [86] |
| MCR-ALS | HPLC-DAD | Sulfonamide | Wastewater | [87] |
| MCR-ALS | GC × GC-FID | PAHs | Crude oil | [59] |
| MCR-ALS | GC × GC-FID | Monoterpens | Essential oil | [88] |
| MCR-ALS | LC-MS | Pesticides | Sediment, Water | [89] |
| MCR-ALS | HPLC-DAD | Phenobarbital, carbamazepine | Human serum albumin | [90] |
| MF-ICA | GC-MS | Monoterpens | Citrus fruit | [91] |
| MCR-ALS | LC-DAD-MS | Biocide mixture | Sediment and wastewater | [92] |
| MCR-ALS | GC × GC/qMS | Terpenes, cannabinoids | Cannabis sativa | [93] |
| MCR-ALS | HPLC-DAD | Phenolic acids | Strawberry | [94] |
| MCR-ALS | LC-FTIR | Carbohydrate, nitrophenol | Mixtuers | [95] |
| EFA-ALS | GC-MAD | Vapors | Mixtures | [96] |
| MPCA, MCR-ALS | GC × GC-FID | Biodiesel | Biodiesel blends | [97] |

of different chemical rank determination methods, usage of IMCR and NIMCR resolution techniques, and a user-friendly environment with variety of graphical outputs. There is also other software for this purpose, such as PLS-Toolbox, MCR-ALS toolbox, GUIPRO, MVC2, and MVC3 [6,62].

## 7.2 MCR in Electrochemistry

The application of MCR techniques to electroanalytical data with the aim of resolving dynamic multicomponent systems, such as metal complexation equilibria and chemically modified electrode, is rather limited, especially compared to their use in chromatography and/or spectroscopy. However, application of MCR in electrochemistry is growing steadily in recent years [99]. The main application of MCR to electroanalytical determinations has been in the analytical resolution of complex systems and consequently quantitative analysis of target analyte(s) in the presence of interferences. In general, these applications can be classified as (1) multivariate resolution of mixtures of analytes that cannot be physically isolated, giving highly overlapping signals, (2) multivariate calibration of one or more analytes in complex mixtures in the presence of interferences, and (3) chemical model identification in complex dynamic systems. On the other hand, for quantitative purposes in complex electroanalytical systems, in general, two different strategies are considered. In the first strategy, the simplest way is to resolve the electroanalytical signal (usually a voltammogram) contributed by the different analytes and then perform a simple calibration on the resolved signals using peak heights or peak areas. In this approach, the second- or third-order electroanalytical data (e.g., voltammograms) are analyzed individually, one by one, or in an augmented form. In this regard, only the region containing the target analyte(s) are considered for data analysis. This strategy is similar to the one which is presented in Figure 8. In the second strategy, which can be denoted as a multivariate calibration approach, the whole data are collected and then analyzed, so a data matrix is thoroughly analyzed to get the proper information. In this regard, the data in calibration set are used to build in a calibration model and then, the developed model is used to predict the concentration of target analyte(s) in unknown samples. In both strategies, three types of calibration strategies (i.e., external, standard addition, and I.S.) can be used.

The MCR methods are very powerful tools for interpretation of electroanalytical systems, especially for multianalyte calibration and modeling in multicomponent mixtures. Also, modern electrochemical instrumentation can provide reliable and reproducible data, which are the basis of the analyses with very low LOD. Therefore, the combined use of modern electroanalytical instruments with MCR techniques is considered as significant improvements in the capabilities of electroanalytical measurements.

Among different MCR techniques, MCR-ALS has been attracted more attentions in electrochemical studies due to its flexibility in handling various types of data structures. The main advantage of MCR-ALS in the study of electrochemical multicomponent systems is that MCR-ALS allows one to decompose the experimental current data matrix into a matrix containing the information of the concentration profiles and another matrix containing the pure voltammograms. From the concentration matrix, it is possible to figure out information of the evolution of the different species during the experiment and the stoichiometries. From the electroanalytical signal matrix (e.g., voltamograms), it is possible to infer information about the characteristics of the involved electrochemical processes.

Figure 14 depicts the general procedure proposed using MCR-ALS to get pure concentration profiles and pure electrochemical signals. Before starting MCR procedure, the raw experimental data can be processed (e.g., smoothed and/or baseline corrected) to reduce the effects of artifacts in data. Afterward, the number of electroactive chemical components is estimated and then, the ALS optimization is started using the proper initial estimates.

Table 4 gives examples of application of MCR techniques (especially MCR-ALS) to diverse electroanalytical problems in which quantitative analysis of the target compounds was required.

### 7.3 MCR in Spectroscopy

Decomposing linear mixtures into their pure component profiles is a problem occurring in different field of analytical chemistry, especially in spectroanalytical measurements. In spectroscopy, it is desired to decompose spectra of chemical mixtures into the contribution of individual components in order to identify the emitting or absorbing mixture constituents and quantify their abundances. Analytical mixtures often contain chemical constituents with similar or identical structural groups (e.g., isomers) and in some special cases, analysis is complicated by extensive spectral details, such as vibrational structure in UV spectra of PAHs [114]. Chemical analysis of such mixtures and obtaining pure quantitative chemical information is a nontrivial task and MCR can be a way out of the situation when highly overlapping regions present in the spectra of multicomponent mixtures.

Figure 15 shows an example of the application of MCR techniques (MCR-ALS in this case) on the integrated EPR data **D**, to obtain the spectral contribution $\mathbf{S}^T$ and the concentration profiles **C** of the three paramagnetic ions, $Cu^{2+}$ (green (light gray in the print version) line), $VO^{2+}$ (red (gray in the print version) line), and $Mn^{2+}$ (blue (dark gray in the print version) line) [115]. From the extracted spectral profiles $\mathbf{S}^T$, one can notice the very high spectral overlapping between chemical compounds. It is clear that using the intensity at a particular magnetic field as a way to estimate concentration for a given spectrum can be risky because of this lack of spectral specificity. From this point of view, the multivariate MCR-ALS approach is a much better alternative to analyze EPR data sets. From the extracted concentration profiles **C**, one can also notice that

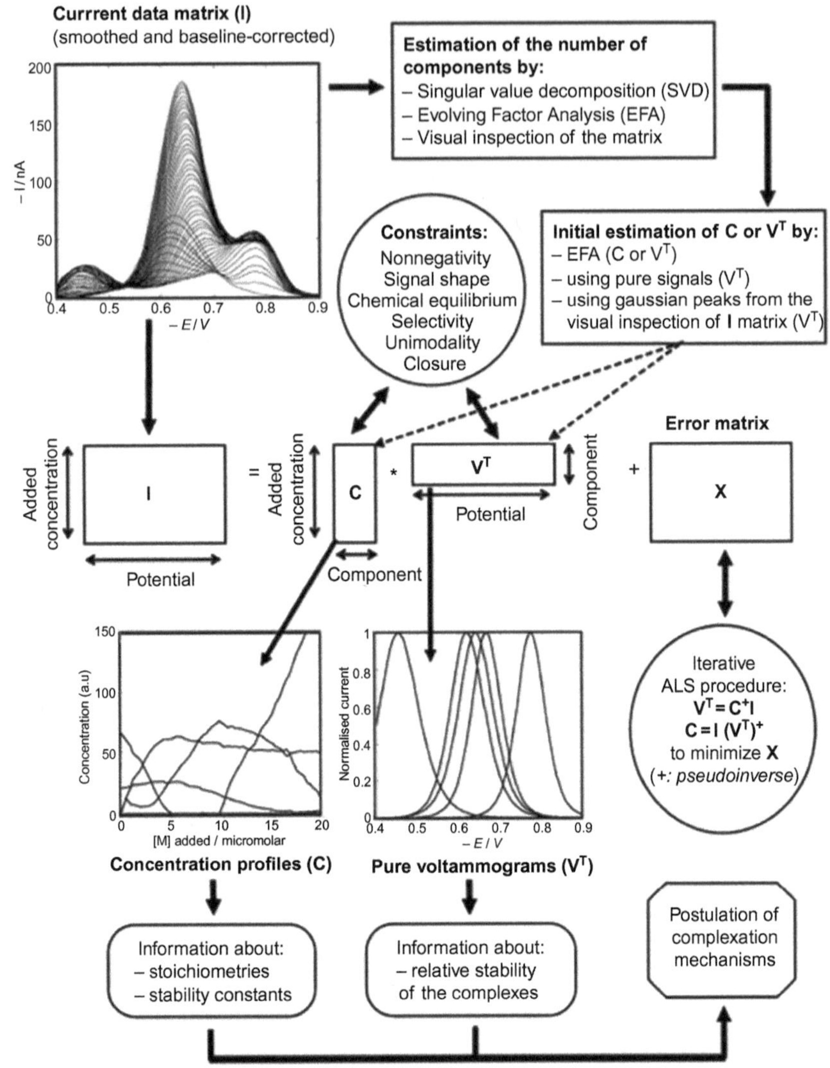


**FIGURE 14** Schematic of MCR-ALS model for analysis of electroanalytical data. *Reprinted from Ref. [99]. Copyright 2009 Elsevier.*

it is possible to detect and extract low concentration compounds, as it is the vanadium ion (red (gray in the pint version) line). This contribution would have been nearly undetectable for most spectra in the data set without the use of chemometric tools.

One of the most challenging problems in mixture decomposition using MCR is determination of the right number of chemical components in the

**TABLE 4** Application of MCR Methods for Obtaining Quantitative Information from Electroanalytical Data

| MCR technique | Electroanalytical method | Analyte(s) | Sample | Ref. |
|---|---|---|---|---|
| MCR-ALS | DPV | Vitamin B7 | Serum albumin | [74] |
| MCR-ALS | DPV | Catechin | Green tea | [100] |
| MCR-ALS | CV, LSV, SWV | Carbidopa | Human serum | [101] |
| MCR-ALS | Staircase CV | Simulated | Synthetic mixture | [102] |
| MCR-ALS | SWV | Tocopherols | Vegetable oil | [103] |
| MCR-ALS | DPASV | Pb, Cd | River water | [104] |
| MCR-ALS | EIS, CV | Betaxolol,atenolol | Human plasma | [105] |
| MCR-ALS | EIS, CV | Dopamine | Human plasma | [106] |
| MCR-ALS | DPV | Tryptophan | Synthetic sample | [107] |
| MCR-ALS | DPV | $Hg^{2+}$ -lipoic acid | Standards | [108] |
| MCR-ALS | DPV | Pb | River water | [109] |
| MCR-ALS | DPV | Quercetin-BSA | Bovin serum | [110] |
| MCR-ALS | NPV, RPV | Heavy metals | Standards | [111] |
| MCR-ALS | DPP | Pb | Standards | [112] |
| MCR-ALS | DPP | Pb-phytochelatin | Standards | [113] |

mixture and therefore, identification of the chemical nature of the components and estimation of their concentrations.

Quantitative spectroscopic methods open a window into exploring complex mixtures by combining some unique features, including a unified approach (many of them treat various spectral signals, such as IR, UV–Vis, EPR, NMR, FL, MS, and X-ray), nondestructive measurements (especially important for *in vivo* analysis of natural samples) and an opportunity to separate measurements and data processing. Accordingly, for quantitative purposes in complex spectroscopic systems, two different strategies are usually used. In the first strategy, the spectrum of different analytes is resolved using MCR techniques and then a simple calibration model on the peak heights or peak areas of resolved spectra and relative concentration is built. In this approach, the second- or third-order spectral data are analyzed individually or in an augmented form. In this regard, only the region containing the target analyte(s) are considered for data analysis. In the second strategy, which can be denoted as a multivariate

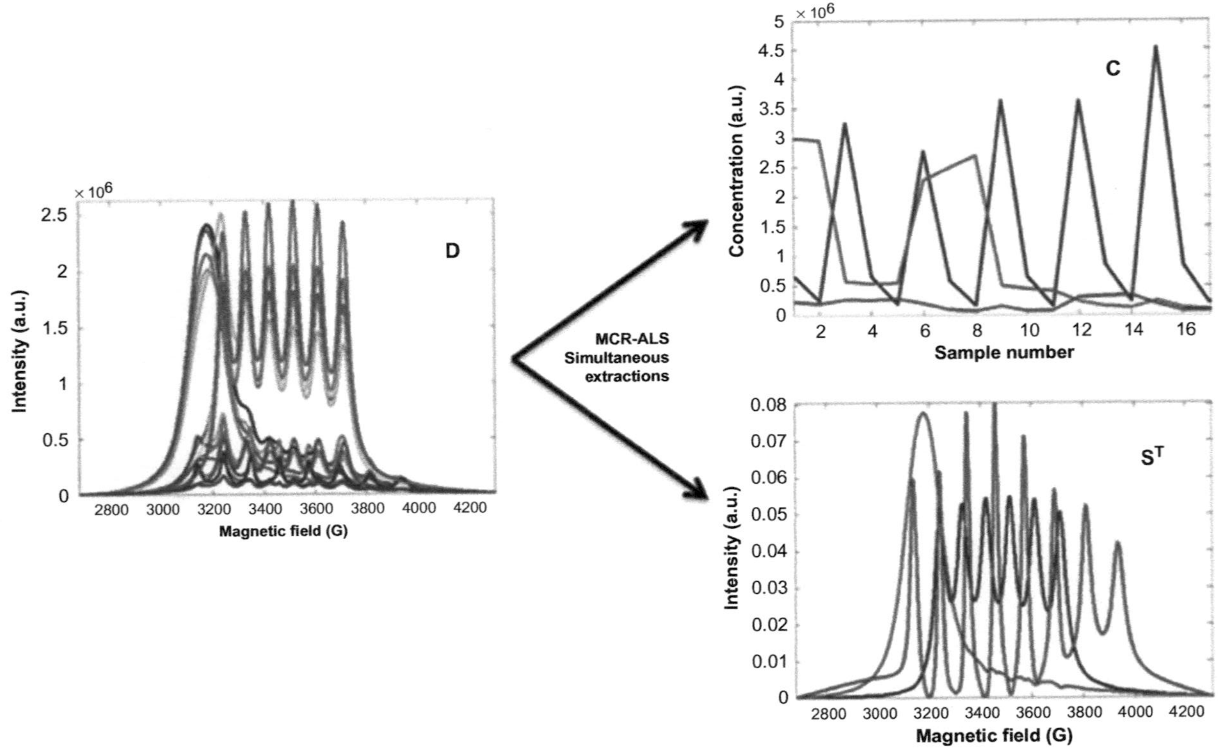


**FIGURE 15** Decomposition of EPR data using MCR-ALS model into the contributions of pure concentration and spectral profiles. *Reprinted from Ref. [115]. Copyright 2014 Elsevier.*

calibration approach, the whole data are collected and then analyzed, so a data matrix is thoroughly analyzed to get the proper information. In this regard, the data in calibration set are used to build in a calibration model and then, the developed model is used to predict the concentration of target analyte(s) in unknown samples. Again, in both strategies, three types of calibration strategies (i.e., external, standard addition, and I.S.) can be used.

Figure 16 shows the general framework of using MCR methods for quantitative purposes using first approach. In this regard, the experimental data matrix containing the spectra at different calibration concentrations is decomposed and pure concentration and spectral profiles for target analyte(s) and interferences are obtained [116]. The concentration profile for target analyte is used to build up calibration graph (e.g., standard addition in this case) and obtaining AFOMs.

The second approach for quantitative purposes in spectroscopic measurements is discussed using a new potent type of constraint. One of the most important developments of MCR techniques for multivariate calibration purposes is the introduction and development of a new constraint namely "correlation constraint" [117]. The MCR-ALS method has been shown to provide an improved resolution compared to other MCR methods and to allow quantitative determinations in the analysis of complex mixtures using spectroscopic means. The MCR-ALS with correlation constraint can be used as a proper tool for quantitative spectroscopic measurements. The first application of this constraint has been introduced for the simultaneous quantitative analysis of mixtures of metal ions using voltammetric analysis and it implies the establishment of calibration models for MCR-ALS, to be used for the quantitative determination of the analytes in the presence of unknown interferences [118]. This constraint is extended to quantitative analysis of spectral data in the simultaneous analysis of different analytes in samples of increasing complexity, including forage samples. This correlation constraint is explained in detail in Figure 17, and it consists of a series of steps performed during each iteration of the ALS optimization. Concentrations of a particular analyte in calibration samples, $\mathbf{c}_{\mathrm{ALS}}^{\mathrm{cal}}$, obtained by ALS at each iteration are correlated with previously known reference concentration values of the analyte $\mathbf{c}_{\mathrm{ref}}$ in these samples. A local linear model between the values is then built up. Using this local model, the concentration of calibration samples is calculated and then, in order to predict the concentration of target analytes in prediction samples, the parameters of calibration model are used. Each ALS iteration is then completed after updating the obtained values of prediction.

Until now, the correlation constraint has always been applied in a global context to a single data matrix, where a calibration set and a test set are included. The modified implementation of correlation constraint allows to work with multiset structures. The major advantage of using a multiset structure, compared to a single data matrix, is the possibility of expanding the amount of data and thereby improving the robustness of the regression results.

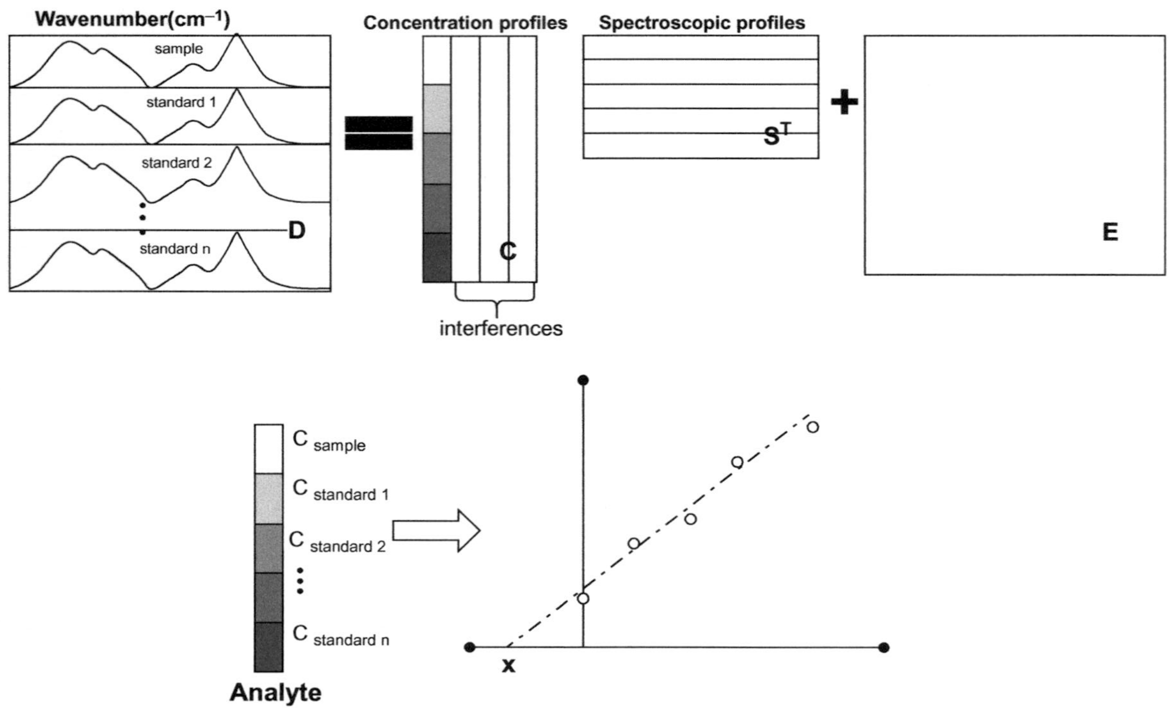


**FIGURE 16** General procedure of MCR method for quantitative analysis in spectroscopy. *Reprinted from Ref. [116]. Copyright 2013 Elsevier.*

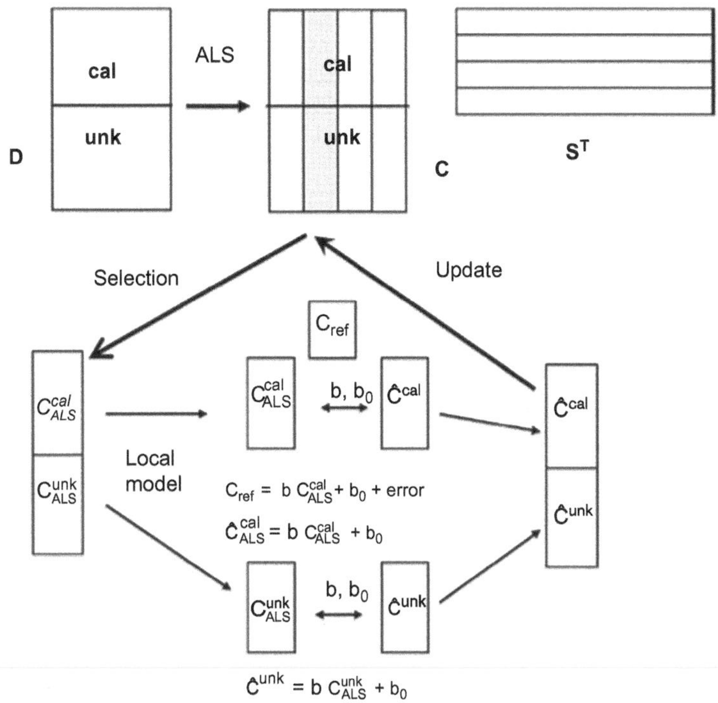


**FIGURE 17** Detailed description of MCR-ALS with correlation constraint. *Reprinted from Ref. [117]. Copyright 2008 Elsevier.*

A multiset can be the combination of subsets of data from different experiments, techniques, or data collected under different conditions.

Table 5 gives some of the recent examples in which different MCR techniques have been coupled to various spectroscopic techniques to get pure quantitative information of the target compounds using two discussed approaches.

## 8 MCR FOR PREPROCESSING OF ANALYTICAL DATA

Apart from outstanding performance of MCR techniques in obtaining pure qualitative and quantitative analytical information, these techniques can be used to reduce the effects of artifacts on analytical data. As it is already pointed out, the most important artifacts in analytical data are noise, baseline/background contribution, and chemical shifts. MCR methods like many other chemometric techniques extract the contribution of significant variables in the system from those related to the noise. Therefore, noise reduction is the primary mission of most of chemometric methods, such as MCR. On the other hand, MCR methods have clear advantages to model baseline/background

**TABLE 5** MCR Methods Applied to Quantitative Analysis of Different Sample Matrices Using Various Spectroscopic Methods

| MCR technique | Spectroscopic method | Analyte(s) | Sample | References |
|---|---|---|---|---|
| MCR-ALS | FT-IR | Sugars, ethanol | Beer | [119] |
| ICA | HNMR | Organic compounds | Food | [71] |
| ICR | NIR | Glucose | Synthetic mixture | [120] |
| MCR-ALS | FT-IR | Proteinaceous binders | Medieval paints | [121] |
| MCR-ALS | EEM | PAHs | Aerosol | [122] |
| MILCA | UV–Vis | Vitamins | Multivitamin drug | [123] |
| MCR | $MS^2$ | Isomers | Binary mixture | [124] |
| MCR-ALS | Time-resolved spectroscopy | Simulated | Mixtures | [125] |
| MCR-ALS | CD | Pb, Cd | Complexes | [76] |
| MCR-ALS | Raman | Paracetamol | Tablet | [126] |
| MCR-ALS | X-ray | Catalyst | Bulk samples | [127] |
| PCA | IMS | Peptides | Synthetic | [128] |
| MCR-ALS | UV–Vis | Paracetamol | Biological | [129] |
| MCR-ALS | FT-Raman, FT-IR | Six oils | Vegetable oils | [130] |
| MCR-ALS | EPR | Paramagnetic species | Synthetic mixtures | [115] |
| MCR-ALS-CC | NIR, UV–Vis | Antioxidants | Biodisels | [131] |
| MCR-ALS | SERS | Nicotine | Urine | [116] |

contributions because of the bilinear model assumption properties associated to these models. For example, in simultaneous analysis of multiple chromatographic runs by MCR-ALS, no limitation exists on baseline shape changes among all samples. In other words, when, for example, hyphenated or multidimensional chromatographic data is column-wise augmented (retention times in rows and spectral channels in columns), baseline chromatographic profile can change from sample to sample, but it keeps the same spectral characteristics among all samples. This is a very important advantage of MCR-ALS bilinear modeling in multirun chromatographic measurements, since baseline changes from run to run (sample to sample) are common place, either due to changes in sample matrix or due to changes in instrumental parameters (see Figure 13 as an instance). It should be pointed out that multiple background components may be needed to properly correct for background/baseline issues.

On the other side, bilinear model assumption used in MCR models can be used as a criterion to correct chemical shift in different types of analytical data. For example, this idea can be used to correct elution time shift in multidimensional chromatography (e.g., GC × GC coupled with univariate detector) [59]. In this regard, the progressive shifts in different second chromatographic dimension chromatograms taken from the first chromatographic column are removed by forcing data to follow this prerequisite. As an instance, Figure 18 depicts overlaid second-dimension chromatograms for different modulations taken from the first GC column. As can be seen, there is a progressive shift in the chromatograms due to the temperature program in GC run. Each modulation in GC × GC is defined as the second-dimension chromatogram(s) obtained by injection of a small sample fraction from first-dimension chromatogram, sampled by the modulator. For example, for a single chromatographic peak in first dimension, there are at least three sampling times in the modulator and therefore, three second-dimension

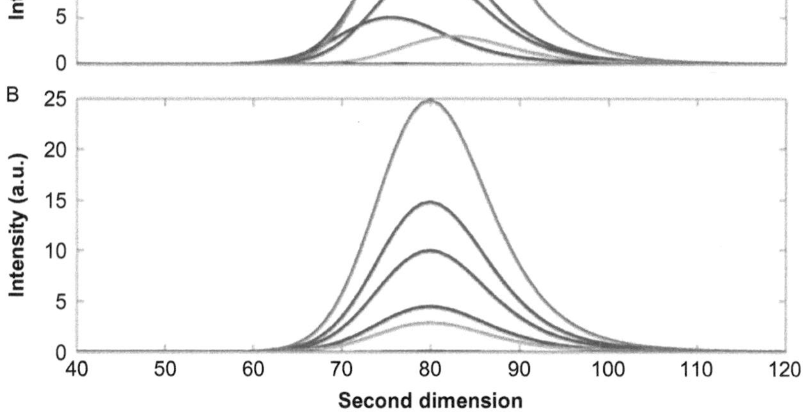


**FIGURE 18** Performance of the BPA method to correct progressive retention time shifts within the same GC × GC run: (A) one-component peak-simulated data and (B) recovered data using the BPA method. *Reprinted from Ref. [59]. Copyright 2012 Elsevier.*

modulations. In GC × GC experiments, due to the within-run retention time shifts between different modulations taken from the first dimension, the data matrix is not bilinear and the number of significant singular values will be larger than the number of eluted components. Therefore, transformation of the original measured data to follow the bilinear model using MCR techniques and correction for retention time shifts would be a good idea. This new method has been called "bilinear peak alignment, BPA" to handle retention time shifts within GC × GC runs. The most important parameter to determine in this method is the number of components. Similar to other MCR techniques, LOF values and relative errors ($E_r$) in recovered peak areas before and after bilinear peak alignment can be used for the correct determination of the number of components in the data matrix. It is important to note that bilinear peak alignment method can be used for GC × GC only if the peak shape does not change a lot between different slices. If the change in peak shape from one modulation to another for a particular component is very large, then this method cannot be used.

Figure 18 shows the performance of MCR bilinear peak alignment method for elution time shift correction for one-component simulated data. Figure 18A depicts the uncorrected data, and Figure 18B shows the corrected data using MCR method. It is clear from this figure that the progressive shift in retention times was suitably corrected by assuming the presence of one single component and applying the MCR BPA method.

Another interesting application of MCR for correction of chemical shifts has been proposed by Tistaert and Van der Heyden as a novel alignment procedure for chromatographic signals [132]. In this method to overcome the misalignment problem of correlation optimized warping (COW) [133] in the case of complex chromatographic signals, at first the complexity of the chromatographic signals is reduced by proper arrangement of chromatographic data sets in a bilinear way and then, performing MCR bilinear decomposition on an augmented data matrix to obtain pure elution and spectral profiles (Equation 2). It should be pointed out that different constraints can be applied during MCR analysis prior to COW, such as nonnegativity constraints in both elution and spectral profiles and unimodality constraint in elution profiles. After MCR bilinear decomposition, the COW alignment is performed on the elution profiles of samples individually, which minimizes the risk of aligning noncorresponding information. It is important to note that MCR-based methods are based on bilinear model assumption, which means that changes in peak shape and peak position cannot affect the quality of the resolved elution and spectral profiles because of the presence of reproducible spectral dimension. However, existence of absurd components, such as baseline/background contribution, can be removed from resolved data to make subsequent analysis more reliable. Then, the augmented matrix is reshaped to individual data matrices. To do this, they are first placed one beside the other to form a one-component elution profile matrix. Afterward, the unfolded data matrices are individually aligned by the COW algorithm. In most of the cases, the median of the elution profiles is used as "reference." After this, the full augmented elution matrix is rebuilt and updated by replacing their misaligned elution profiles by the aligned profiles obtained from the MCR-COW model. Figure 19 shows the above mentioned procedure for HPLC-DAD data sets as an example [134].

Figure 20 shows an example of the application of MCR-COW method to correct elution time shifts in the HPLC-DAD analysis of a complex mixture containing five pharmaceutics [64]. In this case, after ALS optimization, the resolved elution and spectral profiles of pure components as well as baseline/background contribution were obtained. Figure 20A shows the resolved elution profiles for target analytes in different concentration levels used to build external calibration curves. The presence of elution time shifts between different chromatographic runs is evident in this figure. For example, inset in Figure 20A shows the chromatographic region containing mefnamic acid and gemfibrozil where elution time shift and peak overlap occurred simultaneously. Then, the resolved elution profiles were aligned using the COW method. The median of the elution profiles in five different concentrations was used as reference for the COW algorithm. Figure 20B depicts the aligned elution profiles using the MCR-COW method. As can be seen, this complex chromatographic data is correctly aligned using MCR-COW. In addition, MCR-COW can discriminate elution time shift from peak overlap (see inset in Figure 20B). It is important to note that due to the presence of heavy overlap in elution and spectral profiles, it was very difficult to correct elution time shifts using conventional alignment algorithms, such as COW.

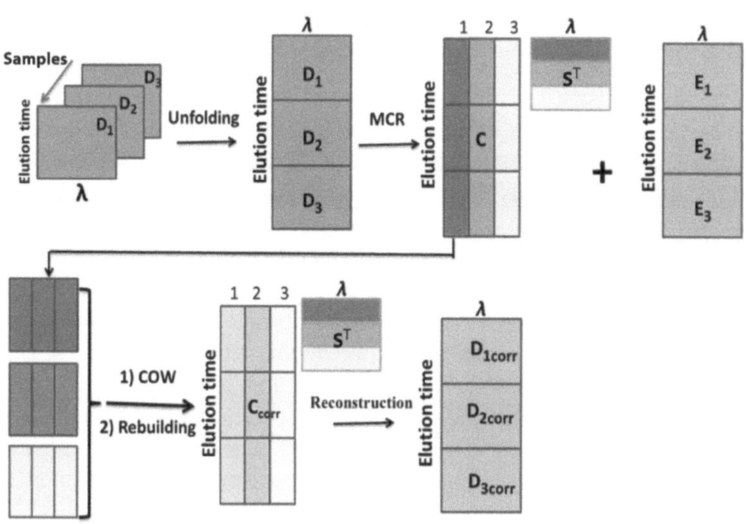


**FIGURE 19** Scheme of MCR-COW algorithm for chemical shift correction in chromatographic data. *Reprinted from Ref. [134]. Copyright 2014 Elsevier.*

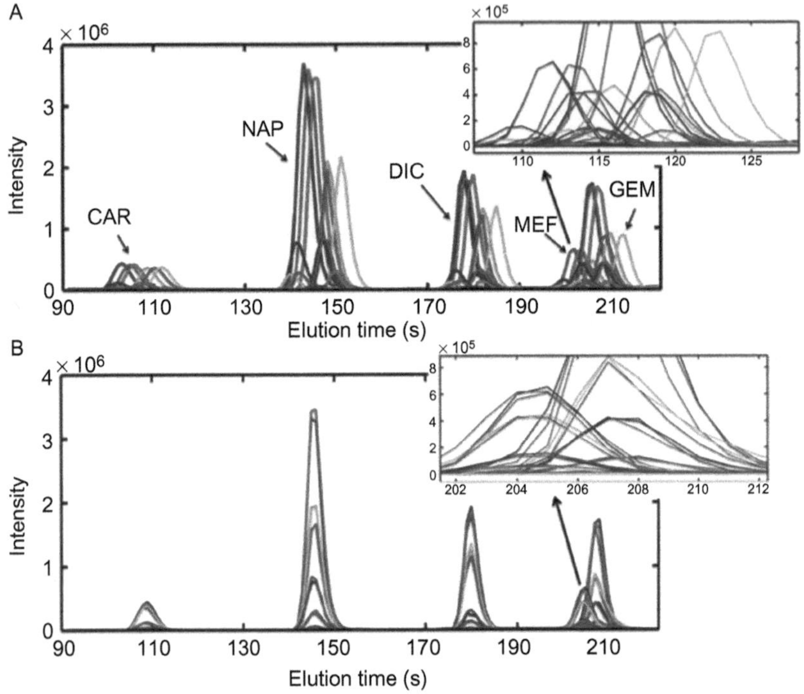


**FIGURE 20** Resolved MCR-ALS elution profiles of HPLC-DAD chromatograms for calibration samples (A) before elution time shift correction and (B) after alignment using COW. *Reprinted from Ref. [64]. Copyright 2014 Elsevier.*

## 9 CONCLUSIONS AND FUTURE OUTLOOK

Analytical chemistry has been significantly improved in recent years from both sides of instrumentations and complexity of the analyzed samples. Therefore, enormous amount of data can be obtained, which need more sophisticated data analysis tools to get pure qualitative and quantitative analytical information. MCR techniques are proved to be very potent tools for this purpose. This chapter is focused on the applications of the MCR techniques in improving the quality of analysis of the very complicated analytical systems. This chapter shows the strength of the MCR methods to those researchers interested in analysis of different analytical systems using various instrumental techniques. MCR techniques are also flexible to handle responses of varying complexity, from single spectra to data matrices coming from chromatographic, spectroscopic, and electrochemical measurements and more complex data arrays (augmented or superaugmented data matrices coming from multidimensional chromatographic measurements).

The key advantages of MCR methods are (1) compensating the lack of selectivity of analytical methods, (2) their increasing resolution power at the time that they decrease expenses on time, chemicals, solvents, and money, (3) allowing at the same time a better understanding of the global analytical process, (4) more flexibility and possible automation of tedious individual steps, and (5) getting high-quality qualitative and quantitative analytical information even in the presence of different artifacts, such as noise, baseline, chemical shifts, and especially peak overlap. Undoubtedly, the combination of analytical systems together with MCR methods provides an extremely powerful approach which complements and, in many cases, increases the analytical potential of this new approach. In such a manner, that once demonstrated, it should attract the attention of the analytical chemistry community. It is concluded that by introducing these techniques to different analytical systems and getting familiar with their strength, the rate of growth in the application of these techniques would increase in analytical chemistry in the near future.

## ACKNOWLEDGMENTS

H.P. would like to thank Research Council of Sharif University of Technology (SUT) for financial support of this work. H.P. would like to thank Prof. Mehdi Jalali-Heravi from Department of Chemistry at SUT as "Father of Chemometrics in Iran" for introduction and development of chemometrics in Iran and getting him familiar with this interesting field. Also, H.P. appreciates Prof. Roma Tauler from Department of Environmental Chemistry at Institute of Environmental and Water Research (IDAEA-CSIC), Barcelona, Spain, who taught him the basic concepts in chemometrics and specially MCR. Finally, H.P. thanks Dr. Behrooz Zekavat from Department of Chemistry at University of Maine, Maine, USA, for his favors and discussions.

## REFERENCES

[1] Wei GL, Zeng EY. Gas chromatography-mass spectrometry and high-performance liquid chromatography-tandem mass spectrometry in quantifying fatty acids. Trends Anal Chem 2011;30:1429–36.

[2] Mas S, Tauler R, de Juan A. Chromatographic and spectroscopic data fusion analysis for interpretation of photodegradation processes. J Chromatogr A 2011;1218:9260–8.

[3] de Juan A, Tauler R. Factor analysis of hyphenated chromatographic data. Exploration, resolution and quantification of multicomponent systems. J Chromatogr A 2007;1158:184–95.

[4] de Juan A, Tauler R. Multivariate Curve Resolution (MCR) from 2000: progress in concepts and applications. Crit Rev Anal Chem 2006;36:163–76.

[5] De Juan A, Tauler R. Chemometrics applied to unravel multicomponent processes and mixtures: revisiting latest trends in multivariate resolution. Anal Chim Acta 2003;500: 195–210.

[6] Jalali-Heravi M, Parastar H. Recent trends in application of multivariate curve resolution approaches for improving gas chromatography-mass spectrometry analysis of essential oils. Talanta 2011;85:835–49.

[7] de Juan A, Rutan SC, Tauler R. Introduction to multivariate curve resolution. In: Tauler R, Brown S, Walczak R, editors. Comprehensive chemometrics: chemical and biochemical data analysis. Oxford: Elsevier; 2009. p. 249–59.

[8] Olivieri AC, Escandar GM, Peña AMDL. Second-order and higher-order multivariate calibration methods applied to non-multilinear data using different algorithms. Trends Anal Chem 2011;30:607–17.

[9] Parastar H, Tauler R. Multivariate curve resolution of hyphenated and multidimensional chromatographic measurements: a new insight to address current chromatographic challenges. Anal Chem 2014;86:286–97.

[10] Tauler R. Multivariate curve resolution applied to second order data. Chemom Intell Lab Syst 1995;30:133–46.

[11] Parastar H, Jalali-Heravi M, Tauler R. Is independent component analysis appropriate for multivariate resolution in analytical chemistry? Trends Anal Chem 2012;31:134–43.

[12] Bro R, De Jong S. A fast non-negativity-constrained least squares algorithm. J Chemom 1997;11:393–401.

[13] De Juan A, Vander Heyden Y, Tauler R, Massart DL. Assessment of new constraints applied to the alternating least squares method. Anal Chim Acta 1997;346:307–18.

[14] Tauler R, Maeder M, De Juan A. Multiset data analysis: extended multivariate curve resolution. In: Tauler R, Brown S, Walczak R, editors. Comprehensive chemometrics: chemical and biochemical data analysis. Oxford: Elsevier; 2009. p. 473–503.

[15] Bro R. PARAFAC. Tutorial and applications. Chemom Intell Lab Syst 1997;38:149–71.

[16] Lawton WH, Sylvestre EA. Self modeling curve resolution. Technometrics 1971;13:617–33.

[17] Maeder M. Evolving factor analysis for the resolution of overlapping chromatographic peaks. Anal Chem 1987;59:527–30.

[18] Malinowski ER. Window factor analysis: theoretical derivation and application to flow injection analysis data. J Chemom 1992;6:29–40.

[19] Kvalheim OM, Liang YZ. Heuristic evolving latent projections: resolving two-way multicomponent data. 1. Selectivity, latent-projective graph, datascope, local rank, and unique resolution. Anal Chem 1992;64:936–46.

[20] Liang YZ, Kvalheim OM. Diagnosis and resolution of multiwavelength chromatograms by rank map, orthogonal projections and sequential rank analysis. Anal Chim Acta 1994;292:5–15.

[21] Manne R, Shen H, Liang Y. Subwindow factor analysis. Chemom Intell Lab Syst 1999;45:171–6.

[22] Xu CJ, Jiang JH, Liang YZ. Evolving window orthogonal projections method for two-way data resolution. Analyst 1999;124:1471–6.

[23] Jiang JH, Šaašić S, Yu RQ, Ozaki Y. Resolution of two-way data from spectroscopic monitoring of reaction or process systems by parallel vector analysis (PVA) and window factor analysis (WFA): inspection of the effect of mass balance, methods and simulations. J Chemom 2003;17:186–97.

[24] Zeng ZD, Liang YZ, Wang YL, Li XR, Liang LM, Xu QS, et al. Alternative moving window factor analysis for comparison analysis between complex chromatographic data. J Chromatogr A 2006;1107:273–85.

[25] Malinowski ER. Obtaining the key set of typical vectors by factor analysis and subsequent isolation of component spectra. Anal Chim Acta 1982;134:129–37.

[26] Vandeginste BGM, Derks W, Kateman G. Multicomponent self-modelling curve resolution in high-performance liquid chromatography by iterative target transformation analysis. Anal Chim Acta 1985;173:253–64.

[27] Karjalainen EJ. The spectrum reconstruction problem. Use of alternating regression for unexpected spectral components in two-dimensional spectroscopies. Chemom Intell Lab Syst 1989;7:31–8.

[28] Windig W, Guilment J. Interactive self-modeling mixture analysis. Anal Chem 1991;63: 1425–32.
[29] Tauler R, Barcelo D. Multivariate curve resolution applied to liquid chromatography-diode array detection. Trends Anal Chem 1993;12:319–27.
[30] Cuesta Sánchez F, Toft J, Van den Bogaert B, Massart DL. Orthogonal projection approach applied to peak purity assessment. Anal Chem 1996;68:79–85.
[31] Xie YL, Hopke PK, Paatero P. Positive matrix factorization applied to a curve resolution problem. J Chemom 1998;12:357–64.
[32] Lee DD, Seung HS. Learning the parts of objects by non-negative matrix factorization. Nature 1999;401:788–91.
[33] Manne R, Grande BV. Resolution of two-way data from hyphenated chromatography by means of elementary matrix transformations. Chemom Intell Lab Syst 2000;50:35–46.
[34] Lohnes MT, Guy RD, Wentzell PD. Window target-testing factor analysis: theory and application to the chromatographic analysis of complex mixtures with multiwavelength fluorescence detection. Anal Chim Acta 1999;389:95–113.
[35] Gong F, Liang YZ, Cui H, Chau FT, Chan BTP. Determination of volatile components in peptic powder by gas chromatography-mass spectrometry and chemometric resolution. J Chromatogr A 2001;909:237–47.
[36] Mason C, Maeder M, Whitson A. Resolving factor analysis. Anal Chem 2001;73:1587–94.
[37] Jiang J-H, Liang Y-Z, Ozaki Y. On simplex-based method for self-modeling curve resolution of two-way data. Chemom Intell Lab Syst 2003;65:51–65.
[38] Gemperline PJ, Cash E. Advantages of soft versus hard constraints in self-modeling curve resolution problems. Alternating least squares with penalty functions. Anal Chem 2003;75:4236–43.
[39] Shao X, Wang G, Wang S, Su Q. Extraction of mass spectra and chromatographic profiles from overlapping GC/MS signal with background. Anal Chem 2004;76:5143–8.
[40] Wang G, Cai W, Shao X. A primary study on resolution of overlapping GC-MS signal using mean-field approach independent component analysis. Chemom Intell Lab Syst 2006;82:137–44.
[41] Astakhov SA, Stögbauer H, Kraskov A, Grassberger P. Monte Carlo algorithm for least dependent non-negative mixture decomposition. Anal Chem 2006;78:1620–7.
[42] Jaumot J, Tauler R, Gargallo R. Exploratory data analysis of DNA microarrays by multivariate curve resolution. Anal Biochem 2006;358:76–89.
[43] Tauler R. Application of non-linear optimization methods to the estimation of multivariate curve resolution solutions and of their feasible band boundaries in the investigation of two chemical and environmental simulated data sets. Anal Chim Acta 2007;595:289–98.
[44] Liu Z, Cai W, Shao X. Sequential extraction of mass spectra and chromatographic profiles from overlapping gas chromatography-mass spectroscopy signals. J Chromatogr A 2008;1190:358–64.
[45] Liu Z, Cai W, Shao X. High-throughput approach for analysis of multicomponent gas chromatographic-mass spectrometric signals. J Chromatogr A 2009;1216:1469–75.
[46] Parastar H, Ebrahimi-Najafabadi H, Jalali-Heravi M. Multivariate curve resolution-particle swarm optimization: a high-throughput approach to exploit pure information from multicomponent hyphenated chromatographic signals. Anal Chim Acta 2013;772:16–25.
[47] de Juan A, Rutan SC, Tauler R. Two-way data analysis: multivariate curve resolution - iterative resolution methods. In: Brown DS, Tauler R, Walczak B, editors. Comprehensive chemometrics: chemical and biochemical data analysis. Oxford: Elsevier; 2009. p. 325–44.

[48] Tauler R. Calculation of maximum and minimum band boundaries of feasible solutions for species profiles obtained by multivariate curve resolution. J Chemom 2001;15:627–46.
[49] Golshan A, Abdollahi H, Maeder M. Resolution of rotational ambiguity for three-component systems. Anal Chem 2011;83:836–41.
[50] Vosough M, Mason C, Tauler R, Jalali-Heravi M, Maeder M. On rotational ambiguity in model-free analyses of multivariate data. J Chemom 2006;20:302–10.
[51] Rajkó R, István K. Analytical solution for determining feasible regions of self-modeling curve resolution (SMCR) method based on computational geometry. J Chemom 2005;19:448–63.
[52] Kim BM, Henry RC. Extension of self-modeling curve resolution to mixtures of more than three components. Part 2. Finding the complete solution. Chemom Intell Lab Syst 1999;49:67–77.
[53] Gemperline PJ. Computation of the range of feasible solutions in self-modeling curve resolution algorithms. Anal Chem 1999;71:5398–404.
[54] Henry RC, Kim BM. Extension of self-modeling curve resolution to mixtures of more than three components. Part 1. Finding the basic feasible region. Chemom Intell Lab Syst 1990;8:205–16.
[55] Borgen OS, Kowalski BR. An extension of the multivariate component-resolution method to three components. Anal Chim Acta 1985;174:1–26.
[56] Jaumot J, Tauler R. MCR-BANDS: a user friendly MATLAB program for the evaluation of rotation ambiguities in Multivariate Curve Resolution. Chemom Intell Lab Syst 2010;103:96–107.
[57] Sawall M, Neymeyr K. On the area of feasible solutions and its reduction by the complementarity theorem. Anal Chim Acta 2014;828:17–26.
[58] Ruckebusch C, Blanchet L. Multivariate curve resolution: a review of advanced and tailored applications and challenges. Anal Chim Acta 2013;765:28–36.
[59] Parastar H, Jalali-Heravi M, Tauler R. Comprehensive two-dimensional gas chromatography (GC × GC) retention time shift correction and modeling using bilinear peak alignment, correlation optimized shifting and multivariate curve resolution. Chemom Intell Lab Syst 2012;117:80–91.
[60] Savitzky A, Golay MJE. Smoothing and differentiation of data by simplified least squares procedures. Anal Chem 1964;36:1627–39.
[61] Wold S, Esbensen K, Geladi P. Principal component analysis. Chemom Intell Lab Syst 1987;2:37–52.
[62] Olivieri AC. Analytical figures of merit: from univariate to multiway calibration. Chem Rev 2014;114:5358–78.
[63] Bauza MC, Ibañez GA, Tauler R, Olivieri AC. Sensitivity equation for quantitative analysis with multivariate curve resolution-alternating least-squares: theoretical and experimental approach. Anal Chem 2012;84:8697–706.
[64] Akvan N, Parastar H. Second-order calibration for simultaneous determination of pharmaceuticals in water samples by solid-phase extraction and fast high-performance liquid chromatography with diode array detector. Chemom Intell Lab Syst 2014;137:146–54.
[65] Hakimzadeh N, Parastar H, Fattahi M. Combination of multivariate curve resolution and multivariate classification techniques for comprehensive high-performance liquid chromatography-diode array absorbance detection fingerprints analysis of Salvia reuterana extracts. J Chromatogr A 2014;1326:63–72.
[66] Parastar H, Jalali-Heravi M, Sereshti H, Mani-Varnosfaderani A. Chromatographic fingerprint analysis of secondary metabolites in citrus fruits peels using gas chromatography-mass

spectrometry combined with advanced chemometric methods. J Chromatogr A 2012;1251:176–87.

[67] Pisano PL, Silva MF, Olivieri AC. Exploration of liquid chromatographic-diode array data for Argentinean wines by extended multivariate curve resolution. Chemom Intell Lab Syst 2014;132:1–7.

[68] Lima KMG, Bedia C, Tauler R. A non-target chemometric strategy applied to UPLC-MS sphingolipid analysis of a cell line exposed to chlorpyrifos pesticide: a feasibility study. Microchem J 2014;117:255–61.

[69] Airado-Rodríguez D, Skaret J, Wold JP. Assessment of the quality attributes of cod caviar paste by means of front-face fluorescence spectroscopy. J Agric Food Chem 2010;58:5276–85.

[70] Wang Y, Ni Y. Combination of UV-vis spectroscopy and chemometrics to understand protein-nanomaterial conjugate: a case study on human serum albumin and gold nanoparticles. Talanta 2014;119:320–30.

[71] Monakhova YB, Tsikin AM, Kuballa T, Lachenmeier DW, Mushtakova SP. Independent component analysis (ICA) algorithms for improved spectral deconvolution of overlapped signals in 1H NMR analysis: application to foods and related products. Magn Reson Chem 2014;52:231–40.

[72] Fernández C, Pilar Callao M, Soledad Larrechi M. UV-visible-DAD and 1H-NMR spectroscopy data fusion for studying the photodegradation process of azo-dyes using MCR-ALS. Talanta 2013;117:75–80.

[73] Gracia N, Thomas S, Thibault-Starzyk F, Lerasle O, Duponchel L. Combination of mid-infrared spectroscopy and curve resolution method to follow the antioxidant action of alkylated diphenylamines. Chemom Intell Lab Syst 2011;106:210–5.

[74] Gholivand MB, Jalalvand AR, Goicoechea HC, Gargallo R, Skov T. Chemometrics: an important tool for monitoring interactions of vitamin B7 with bovine serum albumin with the aim of developing an efficient biosensing system for the analysis of protein. Talanta 2015;132:354–65.

[75] Zhang Q, Ni Y, Kokot S. Combined voltammetric and spectroscopic analysis of small molecule-biopolymer interactions: the levodopa and serum albumin system. Talanta 2012;88:524–32.

[76] Gusmão R, Cavanillas S, Ariño C, Díaz-Cruz JM, Esteban M. Circular dichroism and voltammetry, assisted by multivariate curve resolution, and mass spectrometry of the competitive metal binding by phytochelatin PC5. Anal Chem 2010;82:9006–13.

[77] Cruz BH, Díaz-Cruz JM, Ariño C, Esteban M. Complexation of heavy metals by phytochelatins: voltammetric study of the binding of Cd2+ and Zn2+ ions by the phytochelatin (γ-Glu-Cys)3Gly assisted by multivariate curve resolution. Environ Sci Technol 2005;39:778–86.

[78] Semard G, Adahchour M, Focant JF. Basic instrumentation for GCxGC. In: Ramos L, editor. Comprehensive two dimensional gas chromatography. Amsterdam: Elsevier; 2009. p. 15–48.

[79] Parastar H, Radović JR, Jalali-Heravi M, Diez S, Bayona JM, Tauler R. Resolution and quantification of complex mixtures of polycyclic aromatic hydrocarbons in heavy fuel oil sample by means of GC × GC-TOFMS combined to multivariate curve resolution. Anal Chem 2011;83:9289–97.

[80] Amigo JM, Skov T, Bro R. ChroMATHography: solving chromatographic issues with mathematical models and intuitive graphics. Chem Rev 2010;110:4582–605.

[81] Parastar H, Radović JR, Bayona JM, Tauler R. Solving chromatographic challenges in comprehensive two-dimensional gas chromatography-time-of-flight mass spectrometry using multivariate curve resolution-alternating least squares. Anal Bioanal Chem 2013;405:6235–49.

[82] Wold S, Sjöström M, Eriksson L. PLS-regression: a basic tool of chemometrics. Chemom Intell Lab Syst 2001;58:109–30.
[83] Jalali-Heravi M, Vosough M. Characterization and determination of fatty acids in fish oil using gas chromatography-mass spectrometry coupled with chemometric resolution techniques. J Chromatogr A 2004;1024:165–76.
[84] Bailey HP, Rutan SC. Chemometric resolution and quantification of four-way data arising from comprehensive 2D-LC-DAD analysis of human urine. Chemom Intell Lab Syst 2011;106:131–41.
[85] Jalali-Heravi M, Parastar H, Ebrahimi-Najafabadi H. Self-modeling curve resolution techniques applied to comparative analysis of volatile components of Iranian saffron from different regions. Anal Chim Acta 2010;662:143–54.
[86] Jalali-Heravi M, Parastar H, Sereshti H. Development of a method for analysis of Iranian damask rose oil: combination of gas chromatography-mass spectrometry with Chemometric techniques. Anal Chim Acta 2008;623:11–21.
[87] Vosough M, Mashhadiabbas Esfahani H. Fast HPLC-DAD quantification procedure for selected sulfonamids, metronidazole and chloramphenicol in wastewaters using second-order calibration based on MCR-ALS. Talanta 2013;113:68–75.
[88] de Godoy LAF, Hantao LW, Pedroso MP, Poppi RJ, Augusto F. Quantitative analysis of essential oils in perfume using multivariate curve resolution combined with comprehensive two-dimensional gas chromatography. Anal Chim Acta 2011;699:120–5.
[89] Peré-Trepat E, Lacorte S, Tauler R. Solving liquid chromatography mass spectrometry coelution problems in the analysis of environmental samples by multivariate curve resolution. J Chromatogr A 2005;1096:111–22.
[90] Vosough M, Ghafghazi S, Sabetkasaei M. Chemometrics enhanced HPLC-DAD performance for rapid quantification of carbamazepine and phenobarbital in human serum samples. Talanta 2014;119:17–23.
[91] Jalali-Heravi M, Parastar H, Sereshti H. Towards obtaining more information from gas chromatography-mass spectrometric data of essential oils: an overview of mean field independent component analysis. J Chromatogr A 2010;1217:4850–61.
[92] Peré-Trepat E, Tauler R. Analysis of environmental samples by application of multivariate curve resolution on fused high-performance liquid chromatography-diode array detection mass spectrometry data. J Chromatogr A 2006;1131:85–96.
[93] Omar J, Olivares M, Amigo JM, Etxebarria N. Resolution of co-eluting compounds of Cannabis Sativa in comprehensive two-dimensional gas chromatography/mass spectrometry detection with Multivariate Curve Resolution-Alternating Least Squares. Talanta 2014;121:273–80.
[94] Mas S, Fonrodona G, Tauler R, Barbosa J. Determination of phenolic acids in strawberry samples by means of fast liquid chromatography and multivariate curve resolution methods. Talanta 2007;71:1455–63.
[95] Kuligowski J, Quintás G, Tauler R, Lendl B, de la Guardia M. Background correction and multivariate curve resolution of online liquid chromatography with infrared spectrometric detection. Anal Chem 2011;83:4855–62.
[96] Bryant-Genevier J, Scholten K, Kim SK, Zellers ET. Multivariate curve resolution of co-eluting vapors from a gas chromatograph with microsensor array detector. Sens Actuators B 2014;202:167–76.
[97] Mogollon NGS, Ribeiro FADL, Lopez MM, Hanto LW, Poppi RJ, Augusto F. Quantitative analysis of biodiesel in blends of biodiesel and conventional diesel by comprehensive two-dimensional gas chromatography and multivariate curve resolution. Anal Chim Acta 2013;796:130–6.

[98] Jalali-Heravi M, Parastar H, Kamalzadeh M, Tauler R, Jaumot J. MCRC software: a tool for chemometric analysis of two-way chromatographic data. Chemom Intell Lab Syst 2010;104:155–71.

[99] Esteban M, Ariño C, Díaz-Cruz JM. Chemometrics in electrochemistry. In: Brown S, Tauler R, Walczak B, editors. Comprehensive chemometrics. Oxford: Elsevier; 2010. p. 425–58.

[100] Masoum S, Behpour M, Azimi F, Motaghedifard MH. Potentiality of chemometric approaches for the determination of (+)-catechin in green tea leaves at the surface of multi-walled carbon nanotube paste electrode. Sens Actuators B 2014;193:582–91.

[101] Gholivand MB, Jalalvand AR, Goicoechea HC. Multivariate analysis for resolving interactions of carbidopa with dsDNA at a fullerene-C60/GCE. Int J Biol Macromol 2014;69:369–81.

[102] Safavi A, Hemmateenejad B, Honarasa F. Chemometrics assisted resolving of net faradaic current contribution from total current in potential step and staircase cyclic voltammetry. Anal Chim Acta 2013;766:34–46.

[103] Robledo SN, Zachetti VGL, Zon MA, Fernández H. Quantitative determination of tocopherols in edible vegetable oils using electrochemical ultra-microsensors combined with chemometric tools. Talanta 2013;116:964–71.

[104] Alves GMS, Magalhães JMCS, Tauler R, Soares HMVM. Simultaneous anodic stripping voltammetric determination of Pb and Cd, using a vibrating gold microwire electrode, assisted by chemometric techniques. Electroanalysis 2013;25:1895–906.

[105] Khoobi A, Ghoreishi SM, Masoum S, Behpour M. Multivariate curve resolution-alternating least squares assisted by voltammetry for simultaneous determination of betaxolol and atenolol using carbon nanotube paste electrode. Bioelectrochemistry 2013;94:100–7.

[106] Khoobi A, Ghoreishi SM, Behpour M, Masoum S. Three-dimensional voltammetry: a chemometrical analysis of electrochemical data for determination of dopamine in the presence of unexpected interference by a biosensor based on gold nanoparticles. Anal Chem 2014;86:8967–73.

[107] Kooshki M, Abdollahi H, Bozorgzadeh S, Haghighi B. Second-order data obtained from differential pulse voltammetry: determination of tryptophan at a gold nanoparticles decorated multiwalled carbon nanotube modified glassy carbon electrode. Electrochim Acta 2011;56:8618–24.

[108] Chekmeneva E, Díaz-Cruz JM, Ariño C, Esteban M. Complexation of Hg2+ with α-Lipoic and dihydrolipoic acids: study by differential pulse voltammetry on rotating Au-Disk electrode and ESI-MS. Electroanalysis 2010;22:177–84.

[109] Abdollahi H, Kooshki M. Second-order data obtained from differential pulse voltammetry: determination of lead in river water using multivariate curve resolution-alternating least-squares (MCR-ALS). Electroanalysis 2010;22:2245–53.

[110] Ni Y, Zhang X, Kokot S. Spectrometric and voltammetric studies of the interaction between quercetin and bovine serum albumin using warfarin as site marker with the aid of chemometrics. Spectrochim Acta A 2009;71:1865–72.

[111] Garrigosa AM, Díaz-Cruz JM, Ariño C, Esteban M. Multivariate curve resolution as a tool to minimize the effects of electrodic adsorption in normal pulse voltammetry. Electrochim Acta 2008;53:5579–86.

[112] Alberich A, Díaz-Cruz JM, Ariño C, Esteban M. Combined use of the potential shift correction and the simultaneous treatment of spectroscopic and electrochemical data by multivariate curve resolution: analysis of a Pb(ii)-phytochelatin system. Analyst 2008;133:470–7.

[113] Alberich A, Ariño C, Díaz-Cruz JM, Esteban M. Soft modelling for the resolution of highly overlapped voltammetric peaks: application to some Pb-phytochelatin systems. Talanta 2007;71:344–52.

[114] Monakhova YB, Astakhov SA, Kraskov A, Mushtakova SP. Independent components in spectroscopic analysis of complex mixtures. Chemom Intell Lab Syst 2010;103:108–15.

[115] Fadel MA, De Juan A, Touati N, Vezin H, Duponchel L. New chemometric approach MCR-ALS to unmix EPR spectroscopic data from complex mixtures. J Magn Reson 2014;248:27–35.

[116] Mamián-López MB, Poppi RJ. Standard addition method applied to the urinary quantification of nicotine in the presence of cotinine and anabasine using surface enhanced Raman spectroscopy and multivariate curve resolution. Anal Chim Acta 2013;760:53–9.

[117] Azzouz T, Tauler R. Application of multivariate curve resolution alternating least squares (MCR-ALS) to the quantitative analysis of pharmaceutical and agricultural samples. Talanta 2008;74:1201–10.

[118] Antunes MC, Simão JEJ, Duarte AC, Tauler R. Multivariate curve resolution of overlapping voltammetric peaks: quantitative analysis of binary and quaternary metal mixtures. Analyst 2002;127:809–17.

[119] Grassi S, Amigo JM, Lyndgaard CB, Foschino R, Casiraghi E. Assessment of the sugars and ethanol development in beer fermentation with FT-IR and multivariate curve resolution models. Food Res Int 2014;62:602–8.

[120] Al-Mbaideen A, Benaissa M. Determination of glucose concentration from NIR spectra using independent component regression. Chemom Intell Lab Syst 2011;105:131–5.

[121] Mas S, Miguel C, Melo MJ, Lopes JA, de Juan A. Screening and quantification of proteinaceous binders in medieval paints based on μ-Fourier transform infrared spectroscopy and multivariate curve resolution alternating least squares. Chemom Intell Lab Syst 2014;134:148–57.

[122] Elcoroaristizabal S, De Juan A, García JA, Elorduy I, Durana N, Alonso L. Chemometric determination of PAHs in aerosol samples by fluorescence spectroscopy and second-order data analysis algorithms. J Chemom 2014;28:260–71.

[123] Monakhova YB, Mushtakova SP, Kolesnikova SS, Astakhov SA. Chemometrics-assisted spectrophotometric method for simultaneous determination of vitamins in complex mixtures. Anal Bioanal Chem 2010;397:1297–306.

[124] Zampronio CG, Moraes LAB, Eberlin MN, Poppi RJ. Multivariate curve resolution applied to MS/MS data obtained from isomeric mixtures. Anal Chim Acta 2001;446:495–502.

[125] Debus B, Sliwa M, Miyasaka H, Abe J, Ruckebusch C. Multivariate curve resolution—alternating least squares to cope with deviations from data bilinearity in ultrafast time-resolved spectroscopy. Chemom Intell Lab Syst 2013;128:101–10.

[126] Lyndgaard LB, Van den Berg F, De Juan A. Quantification of paracetamol through tablet blister packages by Raman spectroscopy and multivariate curve resolution-alternating least squares. Chemom Intell Lab Syst 2013;125:58–66.

[127] Voronov A, Urakawa A, Beek WV, Tsakoumis NE, Emerich H, Rønning M. Multivariate curve resolution applied to in situ X-ray absorption spectroscopy data: an efficient tool for data processing and analysis. Anal Chim Acta 2014;840:20–7.

[128] Zekavat B, Solouki T. Chemometric data analysis for deconvolution of overlapped ion mobility profiles. J Am Soc Mass Spectrom 2012;23:1873–84.

[129] Feng X, Zhang Q, Cong P, Zhu Z. Determination of the paracetamol degradation process with online UV spectroscopic and multivariate curve resolution-alternating least squares methods: comparative validation by HPLC. Anal Methods 2013;5:5286–93.

[130] Muik B, Lendl B, Molina-Diaz A, Valcarcel M, Ayora-Cañada MJ. Two-dimensional correlation spectroscopy and multivariate curve resolution for the study of lipid oxidation in edible oils monitored by FTIR and FT-Raman spectroscopy. Anal Chim Acta 2007;593:54–67.
[131] De Oliveira RR, De Lima KMG, De Juan A, Tauler R. Application of correlation constrained multivariate curve resolution alternating least-squares methods for determination of compounds of interest in biodiesel blends using NIR and UV-visible spectroscopic data. Talanta 2014;125:233–41.
[132] Tistaert C, Vander Heyden Y. Bilinear decomposition based alignment of chromatographic profiles. Anal Chem 2012;84:5653–60.
[133] Skov T, Van Den Berg F, Tomasi G, Bro R. Automated alignment of chromatographic data. J Chemom 2006;20:484–97.
[134] Parastar H, Akvan N. Multivariate curve resolution based chromatographic peak alignment combined with parallel factor analysis to exploit second-order advantage in complex chromatographic measurements. Anal Chim Acta 2014;816:18–27.

Chapter 7

# Unfolded and Multiway Partial Least-Squares with Residual Multilinearization: Fundamentals


Alejandro C. Olivieri*,[1], Graciela M. Escandar*, Héctor C. Goicoechea[†] and Arsenio Muñoz de la Peña[‡]

*Instituto de Química Rosario (CONICET-UNR), Facultad de Ciencias Bioquímicas y Farmacéuticas, Universidad Nacional de Rosario, Rosario, Argentina

[†]Laboratorio de Desarrollo Analítico y Quimiometría (LADAQ), Cátedra de Química Analítica I, Facultad de Bioquímica y Ciencias Biológicas, Universidad Nacional del Litoral, Ciudad Universitaria, Santa Fe, Argentina

[‡]Department of Analytical Chemistry, University of Extremadura, Badajoz, Spain

[1]Corresponding author: e-mail: aolivier@fbioyf.unr.edu.ar; olivieri@iquir-conicet.gov.ar


## Chapter Outline



## 1 INTRODUCTION

In other chapters of this book, algorithms are described that permit the processing of multiway data when the latter can be assumed to follow some physicochemical structural model. One particularly appealing model is the multilinear one because it often leads to uniqueness in the decomposition of a multiway


Data Handling in Science and Technology, Vol. 29. http://dx.doi.org/10.1016/B978-0-444-63527-3.00007-2

data array, leading to the retrieval of pure component profiles in each data mode. When the data are not multilinear, they can be frequently arranged into a data matrix following a bilinear model, which is another option for multiway data processing that eventually leads to physically interpretable results in terms of pure component properties [1].

There are additional nonmultilinear data, however, which require algorithms with increased flexibility, to be applied when the data cannot be modeled as either a multilinear array or a bilinear matrix. In a general case, the most flexible option is to unfold the original data structure for a given sample, transforming the raw data arrays into a vector. This leads to some loss in chemical interpretability because information on pure component profiles is lost during the unfolding operation.

A powerful algorithm for processing vectorial signals per sample is partial least-squares (PLS) regression analysis [2,3]; when applied to unfolded multiway data, it gives rise to the so-called unfolded PLS (U-PLS) methodology [4]. A complementary technique, which maintains the original data structure but models them with the PLS philosophy, has also been developed: multiway PLS (N-PLS) [5]. Both U-PLS and N-PLS intend to provide multiway data processing with enough flexibility to face calibration protocols based on complex data, mainly because they work under the latent variable philosophy. In this chapter, these approaches are discussed in the versions in which each analyte is modeled at a time (corresponding to the so-called PLS-1 version in first-order calibration).

It is important to notice that neither U-PLS nor N-PLS allows one to model the presence of unexpected constituents in a test sample, as required by multiway studies of samples containing interferents which have not been considered in the calibration phase. In other words, they do not achieve the important second-order advantage [6]. For the analysis of second-order data (which give rise to three-way arrays when data for a group of samples are joined), a complementary technique was developed and called residual bilinearization (RBL) [7], a name which derives from its intention to model the residues of U-PLS or N-PLS for the test sample as a sum of bilinear contributions from the unexpected components. Since RBL is also a latent-variable tool, the resulting hybrid methodologies, U-PLS/RBL and N-PLS/RBL, should be intrinsically very flexible and adequate for tackling some nonmultilinear data which cannot be covered by other algorithms based on more rigid models. Regrettably, the increased flexibility comes at the price of some disadvantages; the most conspicuous one is the inability of U-PLS/RBL and N-PLS/RBL to render approximations to pure constituent profiles, thereby loosing chemical interpretability for the sake of accurate analyte quantitation. It is interesting that U-PLS/RBL and N-PLS/RBL were developed in 1990 [7], later abandoned and recently rediscovered [8], and successfully applied to a variety of analytical systems [9,10].

For processing multiway data with additional data modes, RBL has been further extended by developing the companion procedures residual tri- and quadrilinearization (RTL and RQL, respectively, see Refs. [11,12] for RTL and Ref. [13] for RQL), which are able to recognize the presence of potential interferents when studying samples for which third- and fourth-order data have been measured. The latter data, when grouped for a set of samples, give rise to four- and five-way data, respectively. As the instrumental technology for measuring complex data progresses toward data with even larger number of modes, there might be corresponding developments in this sequel of residual multilinearization (RML) procedures.

The organization of this chapter begins with the simplest methodology, i.e., unfolded partial least-squares with residual bilinearization or U-PLS/RBL, for the analysis of second-order/three-way data. This is, by far, the most applied one, and at the same time, the easiest to explain. We then develop the N-PLS/RBL variant and proceed to increase the number of modes to the coupling of U- and N-PLS to higher-order RML counterparts. Finally, we review the classical version of U-PLS/RML based on modeling the calibration data with classical least-squares instead of with PLS.

## 2 UNFOLDED PLS

The first step in U-PLS calibration is to convert the calibration data arrays into vectors. This would produce from a $J \times K$ data matrix, a $JK \times 1$ vector, from a $J \times K \times L$ three-way data array, a $JKL \times 1$ vector, from a $J \times K \times L \times M$ four-way data array, a $JKLM \times 1$ vector, and so on [4]. With all the unfolded calibration data, a new calibration matrix $\mathbf{X}_{cal}$, suitable for the application of PLS regression, is built by placing all the column vectors adjacent to each other:

$$\mathbf{X}_{cal} = \left[\mathbf{x}_{cal,1} \middle| \mathbf{x}_{cal,2} \middle| \cdots \middle| \mathbf{x}_{cal,I}\right] \tag{1}$$

where $\mathbf{x}_{cal,i}$ is the calibration vector obtained for the $i$th sample after unfolding the corresponding original array. The latter $\mathbf{X}_{cal}$ matrix can therefore be of size $JK \times I$ ($I$ = number of calibration samples) for second-order data, $JKL \times I$ for third-order data, etc., and is subjected to the classical and well-known PLS regression analysis. This involves decomposition of $\mathbf{X}_{cal}$ into the product of two matrices:

$$\mathbf{X}_{cal} = \mathbf{P}\mathbf{T}^{T} + \mathbf{E}_{cal} \tag{2}$$

where $\mathbf{P}$ is called the loading PLS matrix and $\mathbf{T}$ is the score PLS matrix, while $\mathbf{E}_{cal}$ collects the residuals. The loadings $\mathbf{P}$ and the scores $\mathbf{T}$ matrices are usually estimated via an iterative algorithm, whose aim is to explain as much variance in the $\mathbf{X}_{cal}$ matrix, as well as its covariance with the nominal analyte concentrations in the calibration samples [2]. PLS has the advantage of producing sets of loadings and scores which are analyte specific, and hence, PLS calibration is conducted each analyte at a time. In any case, the loadings and scores do not bear physically recognizable information, as they are linear combinations of real profiles and concentrations, as in other latent-structured methodologies.

The scores **T** in Equation (2) are regressed against nominal analyte concentrations contained in the vector $\mathbf{y}_{cal}$ as follows:

$$\mathbf{y}_{cal} = \mathbf{T}\mathbf{v} + \mathbf{e}_y \tag{3}$$

where $\mathbf{e}_y$ collects the regression errors and **v** is the vector of regression coefficients. The latter is provided by least-squares minimization of the residuals $\mathbf{e}_y$.

One important issue in connection with the PLS decomposition of Equation (2) concerns the sizes of the matrices **P** and **T**. The former one has a size which depends on the original data, i.e., $JK \times A$ for second-order data, $JKL \times A$ for third-order, etc., and the latter has a size $I \times A$, where $A$ is the optimum number of latent variables, necessary to explain the variance in $\mathbf{X}_{cal}$, and leaving errors in $\mathbf{E}_{cal}$ of the order of the instrumental uncertainty. There is a reliable way of estimating $A$, which involves the prediction of concentrations and its comparison with their nominal values. It is important to notice that the value of $A$ should not be blindly accepted as the number provided by a black box without questions. The optimum number of latent variables should be close to the known or suspected number of chemically responsive constituent in the calibration set, i.e., those producing measurable signals. There are several methodologies for reliably estimating the optimum value of $A$, among which probably the most popular one is leave-one-out cross-validation, to be described below in detail. Additional phenomena causing instrumental variations from sample to sample, such as baseline drifts, nonlinearities, interactions, etc., may contribute with additional latent variables to the overall value of $A$. The optimum value should always be amenable to some chemically reasonable interpretation.

Once the number of latent variables $A$ adequate for modeling the $\mathbf{X}_{cal}$ matrix in connection with the prediction of the concentration of a particular analyte is found (see below), the calibration model renders the calibration loadings and scores. It also furnishes the so-called vector of regression coefficients, which is useful for predicting the analyte concentration in future samples, according to the following predictive expression:

$$y = \mathbf{v}^T \mathbf{t} \tag{4}$$

where **t** is the test sample score vector, obtained as the projection of the unfolded data matrix for the test sample in the space defined by the calibration PLS loadings. Notice that for mean-centered data, the mean calibration concentration should be added to the right-hand side of Equation (4).

## 3 MULTIWAY PLS

One of the first versions of PLS applicable to three-way data was developed in 1989 [14] and optimized and generalized to multiway PLS (N-PLS) in

1996 [5]. The three-way version of N-PLS decomposes the three-way array of instrumental data for a group of samples into a trilinear model that is similar to the PARAFAC model, but seeking, in accordance with the PLS philosophy, to describe the covariance of the analyte concentrations and the instrumental variables. This is achieved by fitting a multilinear model of the instrumental signals and a regression model relating the latter model to the analyte concentrations.

Specifically, the three-way array of signals for the calibration samples $\boldsymbol{X}_{\mathrm{cal}}$ is decomposed using the following trilinear model:

$$x_{ijk} = \sum_{a=1}^{A} \sum_{a=1}^{A} w_{ja}^{J} w_{ka}^{K} t_{ia} + e_{ijk} \tag{5}$$

where $x_{ijk}$ is a generic element of $\boldsymbol{X}_{\mathrm{cal}}$, $A$ is the number of N-PLS components, $t_{ia}$ is the score for sample $i$ and component $a$, $w_{ja}^{J}$ and $w_{ka}^{K}$ are elements of two loading matrices $\mathbf{W}^{J}$ (size $J \times A$) and $\mathbf{W}^{K}$ (size $K \times A$) describing the $a$th component behavior along the instrumental data modes, and $e_{ijk}$ is a model residual. N-PLS is implemented by finding $\mathbf{W}^{J}$ and $\mathbf{W}^{K}$ such that the score matrix $\mathbf{T}$ (whose elements are the $t_{ia}$ values in Equation 5) and the calibration concentration vector $\mathbf{y}_{\mathrm{cal}}$ have maximal covariance. The prediction model is found using a regression expression analogous to Equation (4) for U-PLS.

Although it is not apparent, Equation (5) is the three-way analogous of Equation (2), a fact which can be made transparent if the latter is written in the following way:

$$x_{ij} = \sum_{a=1}^{A} p_{ja} t_{ia} + e_{ij} \tag{6}$$

where $x_{ij}$ and $e_{ij}$ are elements of the $\mathbf{X}_{\mathrm{cal}}$ and $\mathbf{E}_{\mathrm{cal}}$ matrices, respectively, while $p_{ja}$ and $t_{ia}$ are the corresponding elements of $\mathbf{P}$ and $\mathbf{T}$.

Generalization to multiway N-PLS requires the introduction of additional loading matrices to account for the extra instrumental data modes, i.e., $\mathbf{W}^{L}$ for four-way N-PLS, $\mathbf{W}^{M}$ for five-way N-PLS, etc., Equation (5), in turn, adopts a general multilinear form which includes the corresponding elements of these loading matrices ($w_{la}^{L}$, $w_{ma}^{M}$, etc.). The score matrix $\mathbf{T}$, in turn, is still of size $I \times A$ and participates in the regression to concentration values in an analogous fashion to Equation (3).

## 4 ESTIMATING THE NUMBER OF CALIBRATION LATENT VARIABLES

One of the most popular techniques to estimate the number of calibration PLS latent variables $A$ is called leave-one-out cross-validation [15]. Each sample is left out from the calibration set, and its concentration is predicted using a model built with the data for the remaining samples. The squared error for the prediction of the left-out sample is summed into a parameter called PRESS (predicted error sum of squares):

$$\text{PRESS} = \sum_{i=1}^{I} \left(y_{i,\text{nom}} - y_{i,\text{pred}}\right)^2 \tag{7}$$

where $I$ is the number of calibration samples, and "nom" and "pred" stand for nominal and predicted, respectively. The PRESS is estimated for values of $A$ ranging from 1 to a certain maximum, larger than the suspected optimal value. It initially decreases with increasing values of $A$, but then tends to increase, because the last latent variables represent noise rather than true chemical effects. Intuitively, one would select the optimum $A$ as that corresponding to the minimum PRESS, but some uncertainty is always involved in the CV process and thus in the PRESS values. The principle of parsimony suggests that the optimum $A$ is the minimum value whose PRESS is not statistically larger than the minimum PRESS. It can be obtained by first computing the following $F(A)$ ratios:

$$F(A) = \text{PRESS}(A < A^*)/\text{PRESS}(A^*) \tag{8}$$

where $A^*$ leads to the minimum PRESS. Then a probability $p(A)$ is assigned to each $F(A)$, corresponding to the Snedecor's $F$ value with $I$ degrees of freedom for both the numerator and denominator [15]. The optimum value of $A$ is then selected as the one leading to a probability of less than 75% [i.e., $p(A) < 0.75$]. Detailed examples on the operation of cross-validation will be given below.

## 5 PROCESSING SECOND-ORDER DATA WITH U-PLS/RBL

In this section, we describe the processing of second-order data with U-PLS/RBL as example. Proper extension of these concepts would allow one to introduce N-PLS and the complementary techniques of RBL for higher number of data modes.

If no unexpected components occurred in the test sample, i.e., when all test sample constituents are adequately modeled in the calibration phase, Equation (4) would be confidently employed as the predictive PLS expression. On the other hand, when unexpected constituents occur in the test data matrix $\mathbf{X}_{\text{test}}$, the sample scores $\mathbf{t}$ obtained by projecting the unfolded $\mathbf{X}_{\text{test}}$ onto the PLS loadings are unsuitable for analyte prediction through Equation (4). One indication that this is indeed the case comes from the inspection of the U-PLS residuals of the specific model for a test sample:

$$s_{\text{U-PLS}} = \|\text{vec}(\mathbf{X}_{\text{test}}) - \mathbf{Pt}\|/(JK - A)^{1/2} \tag{9}$$

where the product ($\mathbf{P}\,\mathbf{t}$) represents the approximation of the PLS model to the signals of the test sample, and vec() implies the unfolding or vectorization operator. In the presence of unexpected sample components, the $s_{U\text{-}PLS}$ residuals of Equation (9) will be significantly larger than the typical instrumental noise level, because the product ($\mathbf{P}\,\mathbf{t}$) cannot successfully reconstruct the test sample vector vec($\mathbf{X}_{test}$). The mission of RBL is to free the raw test sample scores $\mathbf{t}$ from the effect of the potential interferents, leaving filtered scores which do only contain information on calibrated constituents to be employed in Equation (4) for analyte prediction.

As the name implies, RBL intends to model the residuals (the part of the test sample unexplained by PLS) assuming that they can be arranged into a bilinear matrix [7,8]. One convenient and flexible bilinear analysis which is useful in this regard is principal component analysis (PCA). Specifically, the RBL procedure fits the sample data to the sum of two contributions: (1) the portion of the test data, which can be explained by the calibration PLS loadings, and (2) the contribution from the potential interferents modeled by a certain number of principal components ($N_{RBL}$). The complete U-PLS/RBL modeling equation involves a residual error term to be minimized by least-squares:

$$\mathbf{X}_{test} = \text{reshape}(\mathbf{P}\mathbf{t}_{RBL}) + \mathbf{B}_{RBL}\,\mathbf{T}_{RBL}^{T} + \mathbf{E}_{RBL} \tag{10}$$

In the latter expression, the product ($\mathbf{B}_{RBL}\,\mathbf{T}_{RBL}^{T}$) is the PCA model for the residual matrix [$\mathbf{X}_{test} - \text{reshape}(\mathbf{P}\,\mathbf{t}_{RBL})$] with $N_{RBL}$ principal components, with "reshape" meaning transformation of $JK \times 1$ vector into a $J \times K$ data matrix, i.e., the inverse of the unfolding operation. Minimization of $\mathbf{E}_{RBL}$ in Equation (10) allows one to retrieve the final score vector $\mathbf{t}_{RBL}$. Initially, the residual matrix [$\mathbf{X}_{test} - \text{reshape}(\mathbf{P}\,\mathbf{t}_{RBL})$] contains contributions from both the calibrated analytes and the potential interferents. Modeling this latter matrix with PCA extracts $N_{RBL}$ bilinear components; the more these bilinear components resemble the unexpected contributions, the better the product ($\mathbf{P}\,\mathbf{t}_{RBL}$) is able to model the behavior of the analytes in the test sample, leading to a continuous decrease in the $\mathbf{E}_{RBL}$ residuals.

There are two ways in which RBL can be carried out in order to fit the test signal to Equation (10): (1) a Gauss–Newton minimization and (2) an iterative algorithm. The first alternative is implemented as the default method in the well-known toolbox MVC2 [16] written in the MATLAB environment [17]. It has been shown that in certain special cases, i.e., in the presence of several potential interferents whose signals are significantly larger than those from the analytes, the RBL procedure may fall in a false minimum with no chemical significance [18]. In order to avoid this undesirable outcome, the RBL minimization has been proposed to be preceded by a search of the score space, in such a way that the initial value to start the RBL is closer to the global minimum [18]. This is possible thanks to a metaheuristic procedure known as particle swarm optimization (PSO) [19], an algorithm based on the so-called swarm intelligence strategy, which includes a partially stochastic methodology able to surpass local minima in a multidimensional surface [20]. PSO has been proposed to mimic the social behavior, which is typical of bird flocking and schooling [21].

Once the residuals $\mathbf{E}_{RBL}$ are minimized in Equation (10), the output is a final $\mathbf{t}_{RBL}$ vector which represents the true contribution of the calibrated analytes to the test sample. Analyte concentrations are then obtained by introducing the final $\mathbf{t}_{RBL}$ vector instead of $\mathbf{t}$ in Equation (4). As can be seen, while calibration data are unfolded to be processed by U-PLS with greater flexibility, test data should be maintained in matrix form in order to be able to apply PCA within the RBL procedure.

Information on the profiles for the potential interferents is contained in the PCA loading matrices $\mathbf{B}_{RBL}$ and $\mathbf{T}_{RBL}$ (sizes $J \times N_{RBL}$ and $K \times N_{RBL}$, respectively; the latter is usually column-wise normalized to unit length and named $\mathbf{C}_{RBL}$, for consistency with other methodologies). When $N_{RBL} = 1$, the $\mathbf{B}_{RBL}$ and $\mathbf{C}_{RBL}$ profiles should be excellent approximations to those for the potential interferent in each data mode. Unfortunately, for additional unexpected components, these matrices contain linear combinations of profiles, because they are the result of a latent-structured PCA study, and these profiles are no longer recognizable as true unexpected profiles.

Figure 1 illustrates this property, showing typical RBL profiles retrieved in the presence of either one or two potential interferents when analyzing a two-analyte system with U-PLS/RBL. As can be seen in this figure, for spectral overlapping of Gaussian bands in both data modes used for simulation (Figure 1A and B), when there is a single potential interferent, the U-PLS/RBL profiles for the latter agrees well with the true pure constituent profile (Figure 1C and D, to be compared with the black solid lines in the top plots). However, for two potential interferents, the retrieved profiles no longer resemble true spectra (Figure 1E and F, compare with the black solid and green (grey in the print version) dash-dotted lines in the top plots). It is apparent that in the latter case, RBL retrieves linear combinations of true profiles, as expected from the PCA model underlying the RBL procedure.

We notice that the aim which guides the RBL procedure is the minimization of the residual error $s_{RBL}$ to a level compatible with the degree of noise present in the measured signals, with $s_{RBL}$ given by [22]:

$$s_{RBL} = \|\mathrm{vec}(\mathbf{E}_{RBL})\| / [(J - N_{RBL})(K - N_{RBL}) - A]^{1/2} \quad (11)$$

Therefore, if more than one unexpected component is considered, RBL should select the simplest model giving a residual value which is not statistically different than the minimum one. These considerations are the basis for the estimation of the correct number of RBL components, as described below for selected examples. Another recently introduced criterion is to study the change in the so-called generalized cross-validation error, which is a penalized residual fit, given by [22]:

$$s_{GCV} = \|\mathrm{vec}(\mathbf{E}_{RBL})\| (JK)^{1/2} / [(J - N_{RBL})(K - N_{RBL}) - A] \quad (12)$$

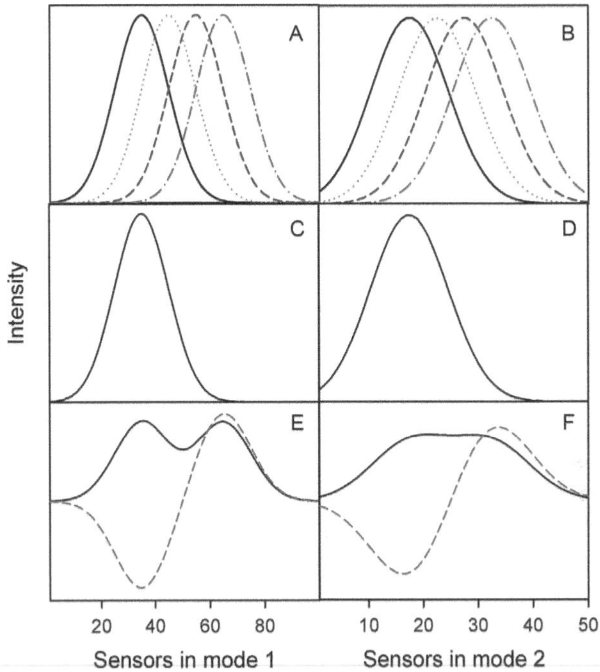


**FIGURE 1** Results from U-PLS/RBL data processing of second-order data for a two-analyte system in the presence of potential interferents. (A, B) Pure component profiles in both data modes (red (light gray in the print version)-dotted and blue (dark gray in the print version)-dashed lines, calibrated analytes; black solid and green (gray in the print version) dash-dotted lines, potential interferents). (C) and (D) Profiles retrieved by U-PLS/RBL in both data modes when a single potential interferent is present in a test sample. (E) and (F) Profiles retrieved by U-PLS/RBL when two interferents are present in a test samples (black solid line, first RBL component, green (gray in the print version) dashed line, second RBL component). For these simulated systems, a central composite, nine-sample calibration design was employed, with random component concentrations in the test samples.

The advantage of the GCV parameter is that it clearly stabilizes or even increases when the correct number of RBL components has been reached.

Figure 2 summarizes the way in which the complete U-PLS/RBL methodology works.

## 6 PROCESSING SECOND-ORDER DATA WITH N-PLS/RBL

In the case of N-PLS/RBL, the discussion above is also pertinent, with the exception that calibration is based on N-PLS instead of on U-PLS. This only requires to modify Equation (10) for the N-PLS/RBL model as follows:

$$\mathbf{X}_{\text{test}} = \text{reshape}\left[\left(\mathbf{W}^{J}|\otimes|\mathbf{W}^{K}\right)\mathbf{t}_{\text{RBL}}\right] + \mathbf{B}_{\text{RBL}}\,\mathbf{T}_{\text{RBL}}^{\text{T}} + \mathbf{E}_{\text{RBL}} \tag{13}$$

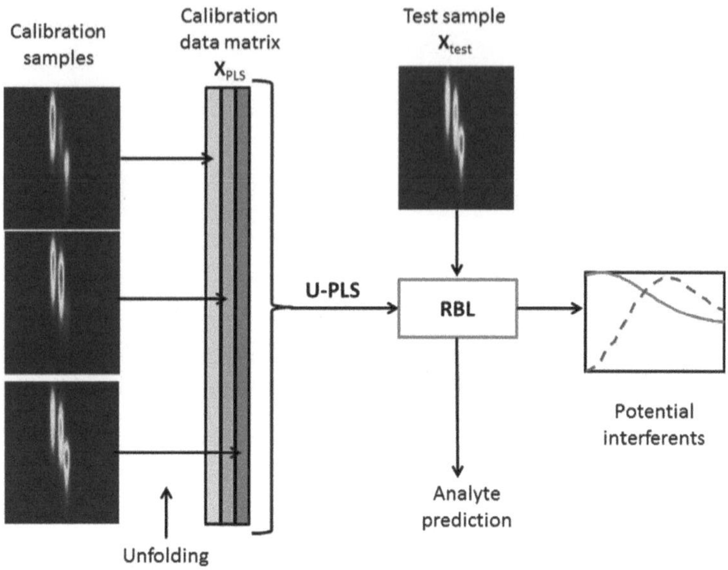


**FIGURE 2** Scheme illustrating how U-PLS/RBL works for second-order data. Three typical calibration samples (of a possibly much larger calibration set) are shown at the left, from which the unfolding operation leads to the $\mathbf{X}_{\mathrm{PLS}}$ calibration data matrix. The latter is processed with U-PLS and submitted, together with the test sample data $\mathbf{X}_{\mathrm{test}}$, to RBL for separating the contribution of the potential interferents, yielding accurate analyte prediction.

where $|\otimes|$ is the Khatri–Rao operator [23] and $\mathbf{W}^J$ and $\mathbf{W}^K$ are the N-PLS loading matrices, which are kept constant at the calibration values during the RBL procedure, while $\mathbf{t}_{\mathrm{RBL}}$ is varied until the final RBL residual error is minimized. An additional change, of course, is that in Equation (10), the product $(\mathbf{B}_{\mathrm{RBL}}\,\mathbf{T}_{\mathrm{RBL}}^{\mathrm{T}})$ is the PCA model for the residual matrix $\left(\mathbf{X}_{\mathrm{test}} - \left[\left(\mathbf{W}^J|\otimes|\mathbf{W}^K\right)\mathbf{t}_{\mathrm{RBL}}\right]\right)$.

The N-PLS/RBL approach is thus very similar to the U-PLS/RBL tool described in detail in the Section 5. Notice that the matrix $\left(\mathbf{W}^J|\otimes|\mathbf{W}^K\right)$ is of size $JK \times A$, so that the product $\left[\left(\mathbf{W}^J|\otimes|\mathbf{W}^K\right)\mathbf{t}_{\mathrm{RBL}}\right]$ is a vector of size $JK \times 1$, and the "reshape" operation in Equation (13) transforms the latter vector into a $J \times K$ matrix, as required by the sizes of the remaining mathematical objects appearing in Equation (13). This is equivalent to transforming the vector $(\mathbf{P}\,\mathbf{t}_{\mathrm{RBL}})$ in Equation (10) into a $J \times K$ matrix. In both cases, the resulting matrices represent the portion of the experimental data matrix $\mathbf{X}_{\mathrm{test}}$ registered for a test sample, which can be modeled by the PLS loadings from the calibration phase (either from U-PLS or N-PLS).

The remainder of the discussion on the application of RBL for obtaining the second-order advantage after N-PLS calibration is formally identical to that of the Section 5 concerning U-PLS/RBL.

## 7 PROCESSING THIRD-ORDER DATA WITH U-PLS/RTL AND N-PLS/RTL

When data with more than two modes are collected, i.e., third- or fourth-order data, the calibration information can be processed using U-PLS as described before, because once the data are vectorized, the original data structure is lost, and the $\mathbf{X}_{\text{cal}}$ matrix is formally identical to that described for second-order data above.

For third-order data, the companion of RBL is called RTL. Inspection of Equation (10) suggests that RTL can be applied to third-order data simply by replacing the PCA modeling of the residuals by an appropriate Tucker3 model [24], which is the natural sequel of PCA for three data modes. The Tucker3 model is an extension of ordinary PCA to a three-way array $\boldsymbol{X}$; for $N1$, $N2$, and $N3$ components in the 1st, 2nd, and 3rd mode, respectively, it can be formulated as:

$$x_{jkl} = \sum_{n1=1}^{N1} \sum_{n2=1}^{N2} \sum_{n3=1}^{N3} b_{jn1} c_{kn2} d_{ln3} g_{n1n2n3} + e_{jkl} \qquad (14)$$

where $(g_{n1n2n3})$ is an element of the three-way array $\boldsymbol{G}$ $(N1 \times N2 \times N3)$ reflecting the importance of the interaction between factors. For orthonormal components, the squared element $(g_{n1n2n3})^2$ reflects the explained variation by the combination of factor $n_1$ in the first mode, factor $n_2$ in the second, and factor $n_3$ in the third. Equation (14) is usually written in compact form as:

$$\boldsymbol{X} = \text{Tucker3}(\boldsymbol{X}) + \boldsymbol{E} \qquad (15)$$

where $\boldsymbol{E}$ is a three-way array of the same size as $\boldsymbol{X}$, whose elements are the $e_{jkl}$ values in Equation (14). The RTL modeling expression, equivalent to the RBL Equation (10), is thus:

$$\boldsymbol{X}_{\text{test}} = \text{reshape}(\mathbf{P}\mathbf{t}_{\text{RTL}}) + \text{Tucker3}[\boldsymbol{X}_{\text{test}} - \text{reshape}(\mathbf{P}\mathbf{t}_{\text{RTL}})] + \boldsymbol{E}_{\text{RTL}} \qquad (16)$$

where $\boldsymbol{X}_{\text{test}}$ represents the third-order data array for the test sample and $\boldsymbol{E}_{\text{RTL}}$ the corresponding RTL residuals. In Equation (16), "reshape" implies transforming a $JKL \times 1$ vector back into a $J \times K \times L$ array, while the Tucker3 model proceeds using up to $N_{\text{RTL}}$ components in each mode. This means that for $N_{\text{RTL}} > 1$, several different Tucker3 models could in principle be constructed, because the number of loadings may be different in each dimension (i.e., $N1$, $N2$, and $N3$ in Equation 14). The aim which guides the RTL procedure is the minimization of the residual error term represented by $\boldsymbol{E}_{\text{RTL}}$ in Equation (16). Therefore, if two unexpected components are considered, for example, one should explore the possible Tucker3 models having one or two loadings in each dimension and select the simplest model, giving a residual value which is not statistically different than the minimum one. For more unexpected components, a similar procedure is recommended. The final Tucker3 model selected to model the unexpected effects is the simplest one that provides a residual value which is not statistically different than the minimum.

Recently, it has been shown that for multiway instrumental signals of interest in analytical chemistry, the Tucker3 model can be conveniently replaced by a parallel factor (PARAFAC) model [25]. The use of a Tucker3 model in RTL is the most general form of describing the contribution of the potential interferents to the overall multiway signal. However, in practical cases where the signals arise from known chemical phenomena, simpler models may be employed. For example, PARAFAC analysis may be used to model these contributions instead of Tucker models, if interferent signals are suspected to be adequately modeled by PARAFAC. The latter is a special case of a Tucker model, with the restriction that: (1) the number of loadings in each instrumental mode are the same in PARAFAC but can be different in Tucker models, and (2) no interactions are allowed among profiles in PARAFAC, in contrast to Tucker models. The most general form of RTL using Tucker models will require more fitting parameters, because interactions among the profiles in the different data modes are allowed.

The remainder of the discussion in the latter RBL section applies here, with the following exceptions: RTL renders three unit length normalized profiles for each of the $N_{\text{RTL}}$ potential interferents, which could be called $\mathbf{B}_{\text{RTL}}$, $\mathbf{C}_{\text{RTL}}$, and $\mathbf{D}_{\text{RTL}}$, each of them of size $J \times N_{\text{RTL}}$, $K \times N_{\text{RTL}}$, and $L \times N_{\text{RTL}}$, respectively. They contain the normalized profiles for the $N_{\text{RTL}}$ potential interferents that have been considered in the analysis of a specific test sample. A final score vector $\mathbf{t}_{\text{RTL}}$ is also provided, which is useful for analyte prediction through Equation (4).

In the case of calibration with N-PLS, the corresponding four-way expression for applying RTL in combination with the former calibration approach is as follows:

$$\begin{aligned} \boldsymbol{X}_{\text{test}} = & \operatorname{reshape}\left[\left(\mathbf{W}^J|\otimes|\mathbf{W}^K|\otimes|\mathbf{W}^L\right)\mathbf{t}_{\text{RTL}}\right] \\ & + \operatorname{Tucker3}\left[\boldsymbol{X}_{\text{test}} - \operatorname{reshape}\left[\left(\mathbf{W}^J|\otimes|\mathbf{W}^K|\otimes|\mathbf{W}^L\right)\mathbf{t}_{\text{RTL}}\right] + \boldsymbol{E}_{\text{RTL}}\right. \end{aligned} \tag{17}$$

where all symbols have analogous meanings to above.

## 8 PROCESSING FOURTH-ORDER DATA WITH U-PLS/RQL AND N-PLS/RQL

The above discussion can be analogously extended to fourth-order data. In the latter case, the relevant U-PLS/RQL modeling expression is as follows:

$$\boldsymbol{X}_{\text{test}} = \operatorname{reshape}(\mathbf{P}\mathbf{t}_{\text{RQL}}) + \operatorname{Tucker4}[\boldsymbol{X}_{\text{test}} - \operatorname{reshape}(\mathbf{P}\mathbf{t}_{\text{RQL}})] + \boldsymbol{E}_{\text{RQL}} \tag{18}$$

where now RQL stands for residual quadrilinearization, and the Tucker4 model of the residuals is carried out using the first $N_{\text{RQL}}$ components (or a suitable PARAFAC model, unless interactions among phenomena take place in the different modes). As before, this analysis gives rise to the normalized profiles in the fourth data modes for the potential interferents, in the form of $J \times N_{\text{RTL}}$, $K \times N_{\text{RTL}}$, $L \times N_{\text{RTL}}$, and $M \times N_{\text{RTL}}$ matrices called $\mathbf{B}_{\text{RQL}}$, $\mathbf{C}_{\text{RQL}}$, $\mathbf{D}_{\text{RQL}}$, and $\mathbf{E}_{\text{RQL}}$, and a $\mathbf{t}_{\text{RQL}}$ score vector adequate for analyte prediction.

In the case of calibration with N-PLS, the corresponding five-way expression for applying RQL in combination with the former calibration approach is as follows:

$$\begin{aligned} \boldsymbol{X}_{\text{test}} = & \text{reshape}\left[\left(\mathbf{W}^J|\otimes|\mathbf{W}^K|\otimes|\mathbf{W}^L|\otimes|\mathbf{W}^M\right)\mathbf{t}_{\text{RQL}}\right] \\ & + \text{Tucker4}\left[\boldsymbol{X}_{\text{test}} - \text{reshape}\left[\left(\mathbf{W}^J|\otimes|\mathbf{W}^K|\otimes|\mathbf{W}^L|\otimes|\mathbf{W}^M\right)\mathbf{t}_{\text{RQL}}\right] + \boldsymbol{E}_{\text{RQL}}\right. \end{aligned} \quad (19)$$

where all symbols have analogous meanings to above.

Notice that RBL, RTL, and RQL all belong to a general technique known as RML, and therefore, U-PLS/RML or N-PLS/RML are algorithms which can be in principle applied to data of any number of modes.

## 9 THE CLASSICAL VARIANT MULTILINEAR LEAST-SQUARES WITH RML

Another alternative to work with unfolded data which should be mentioned in this chapter is bilinear least-squares (BLLS) for second-order data [26,27], and its trilinear least-squares (TLLS) and quadrilinear least-squares (QLLS) counterparts, which have been developed and employed for some time to process third- and fourth-order data, respectively [11,12]. Generally speaking, all of them operate in a manner which resembles U-PLS, except that calibration models are constructed on the basis of classical least-squares instead of using a PLS philosophy. All require RML (i.e., RBL, RTL, or RQL, depending on the number of data modes) for achieving the second-order advantage.

The BLLS model has been originally developed as a truly multiway decomposition algorithm [26], although it has been later shown to be partially analogous to an unfolded classical least-squares methodology [28]. Since the latter is simpler to follow than the original formulation, we will describe it here, in the so-called SVD-LS (singular value decomposition-least-squares) version. BLLS operates by first introducing concentration information into the calibration step (without including data for the unknown sample), in order to obtain approximations to pure-analyte matrices $\mathbf{S}_n$ at unit concentration. To estimate the latter ones, the calibration data matrices are vectorized and grouped into a $JK \times I$ matrix, identical to $\mathbf{X}_{\text{cal}}$ above for U-PLS, and classical least-squares is applied [28]:

$$\mathbf{V}_S = \mathbf{X}_{\text{cal}}\mathbf{Y}_{\text{cal}}\mathrm{T}^{+} \quad (20)$$

where $\mathbf{Y}_{\text{cal}}$ is an $I_{\text{cal}} \times N_{\text{cal}}$ matrix collecting the nominal calibration concentrations, $N_{\text{cal}}$ is the number of calibrated analytes, and $\mathbf{V}_S$ (size $JK \times N_{\text{cal}}$) contains the vectorized $\mathbf{S}_n$ matrices:

$$\mathbf{V}_S = [\text{vec}(\mathbf{S}_1)|\,\text{vec}(\mathbf{S}_2)|\cdots|\text{vec}(\mathbf{S}_{N_{\text{cal}}})] \quad (21)$$

To obtain the profiles in both modes which are present in $\mathbf{S}_n$, the most reliable procedure is the so-called singular value decomposition (SVD) profile estimator. Component profiles are obtained by single-component singular value decomposition ($\text{SVD}_1$) of each of the $J \times K$ $\mathbf{S}_n$ matrices:

$$(g_n, \mathbf{b}_n, \mathbf{c}_n) = \text{SVD}_1(\mathbf{S}_n) \quad (22)$$

where $g_n$ is the first singular value, and $\mathbf{b}_n$ and $\mathbf{c}_n$ are the $J \times 1$ and $K \times 1$ left and right eigenvectors of $\mathbf{S}_n$, respectively. It should be noticed that the identification of calibrated components is not required in SVD-LS, since this is automatically performed by the algorithm. This completes the calibration process.

After calibration, the preferred concentration predictor is the least-squares one. If there were no potential interferents in the test samples, $\mathbf{S}_{\text{cal}}$ and $\mathbf{Y}_{\text{cal}}$ could be employed to predict the analyte concentrations [28]:

$$\mathbf{y} = \mathbf{S}_{\text{cal}}^{+}\,\text{vec}(\mathbf{X}_{\text{test}}) \quad (23)$$

where $\mathbf{y}$ is an $N_{\text{cal}} \times 1$ vector holding the concentrations of the $N_{\text{cal}}$ analytes in the sample, and $\mathbf{S}_{\text{cal}}$ is a $JK \times N_{\text{cal}}$ matrix given by:

$$\mathbf{S}_{\text{cal}} = [g_1(\mathbf{c}_1 \otimes \mathbf{b}_1)|g_2(\mathbf{c}_2 \otimes \mathbf{b}_2)|\cdots|g_{N_c}(\mathbf{c}_{N_c} \otimes \mathbf{b}_{N_{\text{cal}}})] \quad (24)$$

where $\otimes$ denotes the Kronecker product [29].

When unexpected constituents occur in $\mathbf{X}_{\text{test}}$, then the calibration $\mathbf{S}_{\text{cal}}$ matrix given by Equation (24) is unsuitable for the analyte prediction through Equation (23). In this case, the residuals of the BLLS prediction step ($s_{\text{BLLS}}$, see Equation 25) will be abnormally large in comparison with the typical instrumental noise level:

$$s_{\text{BLLS}} = \|\text{vec}(\mathbf{X}_{\text{test}}) - \mathbf{S}_{\text{cal}}\mathbf{y}\| / (JK - N_{\text{cal}})^{1/2} \quad (25)$$

This situation can also be handled by RBL, minimizing the norm of the residual vector $\mathbf{E}_{\text{RBL}}$, computed while fitting the sample data to the sum of the relevant contributions [26,27]:

$$\mathbf{X}_{\text{test}} = \text{reshape}(\mathbf{S}_{\text{cal}}\mathbf{y}) + \mathbf{B}_{\text{RBL}}\,\mathbf{T}_{\text{RBL}}^{\text{T}} + \mathbf{E}_{\text{RBL}} \quad (26)$$

where $\mathbf{B}_{\text{RBL}}$, $\mathbf{T}_{\text{RBL}}^{\text{T}}$, and $\mathbf{E}_{\text{RBL}}$ have the same meaning as for U-PLS above. During the RBL procedure, $\mathbf{S}_{\text{cal}}$ is kept constant, and $\mathbf{y}$ is varied until $\|\mathbf{E}_{\text{RBL}}\|$ is minimized in Equation (26). Analyte concentrations are provided by the final $\mathbf{y}$ vector [30].

The complementary TLLS and QLLS methods work similarly to BLLS, except that Tucker models of appropriate dimensionality are required in Equation (26) to retrieve the profiles for the calibrated components. The application of these Tucker models follows the same premises discussed in connection with U-PLS/RML above.

**TABLE 1 Residual Multilinearization Methods for Processing Multiway Data**

| Data order/ways[a] | Calibration phase | Second-order advantage | Algorithm |
|---|---|---|---|
| Second-order/three-way | U-PLS | RBL | U-PLS/RBL |
| | N-PLS | RBL | N-PLS/RBL |
| | BLLS | RBL | BLLS/RBL |
| Third-order/four-way | U-PLS | RTL | U-PLS/RTL |
| | N-PLS | RTL | N-PLS/RTL |
| | TLLS | RTL | TLLS/RTL |
| Fourth-order/five-way | U-PLS | RQL | U-PLS/RQL |
| | N-PLS | RQL | N-PLS/RQL |
| | QLLS | RQL | QLLS/RQL |

*[a]The order corresponds to data for a single sample; the number of ways to data for a group of samples.*

These classical counterparts of U-PLS show some disadvantages: (1) they are less flexible than the latent-variable models, and (2) they require special *ad hoc* adjustments for coping with linearly dependent phenomena, e. g., when the calibrated analytes participate in acid–base or kinetic phenomena and have various equilibrating species [31].

Table 1 shows a summary of all the methods described in this chapter.

## 10 CONCLUSIONS

When multiway signals do not follow simple structural models such as the multilinear model or the bilinear matrix model, analyst require to have available more flexible possibilities. In this regard, useful alternatives are based on the processing of the calibration data using latent variables, either by first unfolding the multiway calibration arrays into vectors, as in the U-PLS algorithm, or by modeling the data with a multiway N-PLS algorithm, which keeps the original data arrays in their raw formats. In any of these cases, the contribution of the potential interferents to each new sample needs to be taken into account by the powerful RML procedure, which separates the portion of the test signal which can be modeled by the calibration loadings, permitting accurate analyte quantitation and achieving the second-order advantage.

The presently discussed models not only allow one to calibrate with data for any number of modes but also to analyze data deviating from multilinearity in a general and flexible way. This implies that they are also useful to study data following the multilinear or bilinear matrix models. A future chapter will show experimental examples where U-PLS/RML and N-PLS/RML have been exploited to obtain relevant analytical results in complex analytical systems.

## ACKNOWLEDGMENTS

Universidad Nacional de Rosario, Universidad Nacional del Litoral, CONICET (Consejo Nacional de Investigaciones Científicas y Técnicas), ANPCyT (Agencia Nacional de Promoción Científica y Tecnológica, Projects No. PICT-2013-0136 and PICT-2011-0005), and Ministerio de Economía y Competitividad of Spain (Project CTQ2014-52309) and Gobierno de Extremadura (Project GR15-Research Group FQM003), both cofinanced by the European FEDER funds, are gratefully acknowledged for financial support.

## REFERENCES

[1] Escandar GM, Olivieri AC. Practical three-way calibration. 1st ed. Amsterdam: Elsevier; 2014.

[2] Martens H, Næs T. Multivariate calibration. 1st ed. Chichester: John Wiley; 1989 [chapter 3].

[3] Wold S, Sjostrom M, Eriksson L. PLS-regression: a basic tool of chemometrics. Chemometr Intell Lab Syst 2001;58:109–30.

[4] Wold S, Geladi P, Esbensen K, Øhman J. Multiway principal components and PLS analysis. J Chemometr 1987;1:41–56.

[5] Bro R. Multiway calibration. Multilinear PLS. J Chemometr 1996;10:47–61.

[6] Sanchez E, Kowalski BR. Tensorial calibration: II. Second-order calibration. J Chemometr 1988;2:265–80.

[7] Öhman J, Geladi P, Wold S. Residual bilinearization. Part I. Theory and algorithms. J Chemometr 1990;4:79–90.

[8] Olivieri AC. On a versatile second-order multivariate calibration method based on partial least-squares and residual bilinearization. Second-order advantage and precision properties. J Chemometr 2005;19:253–65.

[9] Escandar GM, Faber NM, Goicoechea HC, Muñoz de la Peña A, Olivieri AC, Poppi RJ. Second and third-order multivariate calibration: data, algorithms and applications. Trends Anal Chem 2007;26:752–65.

[10] Olivieri AC. Recent advances in analytical calibration with multi-way data. Anal Methods 2012;4:1876–86.

[11] Arancibia JA, Olivieri AC, Bohoyo Gil D, Muñoz de la Peña A, Durán-Merás I, Espinosa Mansilla A. Trilinear least-squares and unfolded-PLS coupled to residual trilinearization: new chemometric tools for the analysis of four-way instrumental data. Chemometr Intell Lab Syst 2006;80:77–86.

[12] Damiani PC, Durán-Merás I, García-Reiriz A, Jiménez-Girón A, Muñoz de la Peña A, Olivieri AC. Multi-way partial least-squares coupled to residual trilinearization: a genuine multi-dimensional tool for the study of third-order data. Simultaneous analysis of procaine and its metabolite p-amino benzoic acid in equine serum. Anal Chem 2007;79:6949–58.

[13] Maggio RM, Muñoz de la Peña A, Olivieri AC. Unfolded partial least-squares with residual quadrilinearization: a new multivariate algorithm for processing five-way data achieving the

second-order advantage. Application to fourth-order excitation-emission-kinetic-pH fluorescence analytical data. Chemometr Intell Lab Syst 2011;109:178–85.

[14] Ståhle L. Aspects of analysis of three-way data. Chemometr Intell Lab Syst 1989;7:95–100.

[15] Haaland DM, Thomas EV. Partial least-squares methods for spectral analyses. 1. Relation to other quantitative calibration methods and the extraction of qualitative information. Anal Chem 1988;60:1193–202.

[16] Olivieri AC, Wu HL, Yu RQ. MVC2: a MATLAB graphical interface toolbox for second-order multivariate calibration. Chemometr Intell Lab Syst 2009;96:246–51.

[17] MATLAB, The Mathworks Inc., Natick, MA.

[18] Bortolato SA, Arancibia JA, Escandar GM, Olivieri AC. Improvement of residual bilinearization by particle swarm optimization for achieving the second-order advantage with unfolded partial least-squares. J Chemometr 2007;21:557–66.

[19] Parsopoulos KE, Vrahatis MN. Recent approaches to global optimization problems through particle swarm optimization. Nat Comput 2002;1:235–306.

[20] Beni G, Wang J. Swarm intelligence in cellular robotic systems. In: Proceedings of the NATO advanced workshop on robots and biological systems, Tuscany, Italy; 1989.

[21] Kennedy J, Eberhart RC. Particle swarm optimization. In: Proceedings of the IEEE International Conference on Neural Networks, vol. IV. Piscataway, NJ: IEEE Service Center; 1995. p. 1942–8.

[22] Bortolato S, Arancibia JA, Escandar GM. Chemometrics-assisted excitation-emission fluorescence spectroscopy on nylon membranes. Simultaneous determination of benzo[*a*]pyrene and dibenz[*a*,*h*]anthracene at parts-per-trillion levels in the presence of the remaining EPA-PAH priority pollutants. Anal Chem 2008;80:8276–86.

[23] Rao CR, Mitra S. Generalized inverse of matrices and its applications. 1st ed. New York: Wiley-Interscience; 1971.

[24] Tucker LR. Some mathematical notes on three-mode factor analysis. Psychometrika 1966;31:279–311.

[25] Allegrini FA, Olivieri AC. Analytical figures of merit for partial least-squares coupled to residual multi-linearization. Anal Chem 2012;84:10823–30.

[26] Linder M, Sundberg R. Second-order calibration: bilinear least squares regression and a simple alternative. Chemometr Intell Lab Syst 1998;42:159–78.

[27] Linder M, Sundberg R. Precision of prediction in second-order calibration, with focus on bilinear regression methods. J Chemometr 2002;16:12–27.

[28] Faber NM, Ferré J, Boqué R, Kalivas JH. Second-order bilinear calibration: the effects of vectorising the data matrices of the calibration set. Chemometr Intell Lab Syst 2002;63:107–16.

[29] Horn RA, Johnson CR. Topics in matrix analysis. 1st ed. Cambridge, UK: Cambridge University Press; 1991.

[30] Haimovich A, Orselli R, Escandar GM, Olivieri AC. Sensitivity and prediction error for spectroscopic bilinear least-squares exploiting the second-order advantage. Theoretical and experimental study. Chemometr Intell Lab Syst 2006;80:99–108.

[31] Goicoechea HC, Olivieri AC. A new robust bilinear least-squares method for the analysis of spectral-pH matrix data. Appl Spectrosc 2005;59:926–33.

Chapter 8

# Unfolded and Multiway Partial Least-Squares with Residual Multilinearization: Applications

Alejandro C. Olivieri*,[1], Graciela M. Escandar*, Héctor C. Goicoechea[†] and Arsenio Muñoz de la Peña[‡]

*Instituto de Química Rosario (CONICET-UNR), Facultad de Ciencias Bioquímicas y Farmacéuticas, Universidad Nacional de Rosario, Rosario, Argentina

[†]Laboratorio de Desarrollo Analítico y Quimiometría (LADAQ), Cátedra de Química Analítica I, Facultad de Bioquímica y Ciencias Biológicas, Universidad Nacional del Litoral, Ciudad Universitaria, Santa Fe, Argentina

[‡]Department of Analytical Chemistry, University of Extremadura, Badajoz, Spain

[1]Corresponding author: e-mail: olivieri@iquir-conicet.gov.ar; aolivier@fbioyf.unr.edu.ar

## Chapter Outline



## 1 INTRODUCTION

In Chapter 7, the fundamentals of unfolded and multiway partial least-squares with residual multilinearization have been summarized, and a discussion about the advantages and limitations of these algorithms for the processing of multiway data and multiway calibration has been addressed.

**Data Handling in Science and Technology, Vol. 29. http://dx.doi.org/10.1016/B978-0-444-63527-3.00008-4**

In this chapter, several practical applications of the outlined methodologies are presented in detail. The applications include examples using different kinds of second-order data, as excitation-emission fluorescence, kinetic-spectroscopic, pH-spectral, HPLC/CE spectral detection, electrochemical, total synchronous spectroscopic, and tandem MS/MS matrices. Also, examples of third-order data, as kinetic-excitation-emission-fluorescence, decay time-excitation-emission-fluorescence, and liquid chromatography-excitation-emission cubes, and fourth-order data, as kinetic-pH-excitation-emission-fluorescence hypercubes, are covered.

The selected applications include the resolution of practical analytical problems in the environmental, bioanalytical, and food fields.

## 2 SECOND-ORDER DATA

### 2.1 Excitation-Emission Matrices

Excitation-emission fluorescence matrix (EEFM) data can be obtained by registering emission spectra at various excitation wavelengths. In order to generate an EEFM, a proper emission spectral range is usually selected, including the wavelengths where the sample constituents are known to emit. Generally, emission data are registered every 0.5–2 nm, and the main idea is to obtain spectra covering the whole band for the analyte (which might be ca. 50–100 nm wide). On the other hand, the number of excitation wavelengths to be used should be selected gathering at least 10–20 data points. The steps are manually selected; thus, if 20 data points are needed for defining a spectrum 100 nm wide, then the step should be of at most 5 nm. It should be considered that increasing the number of data points in the excitation mode by reducing the step will unavoidably lead to more experimental time in collecting an entire data matrix, and a reasonable trade-off between working time and excitation resolution is desirable [1].

Other signals, different from fluorescence phenomena, may be detected by the instrument when recording an EEFM. They are (1) the Rayleigh dispersion, (2) the Raman dispersion, and (3) the second-order harmonic of the Rayleigh dispersion. The first one appears as a narrow and intense signal centered at the excitation wavelength. In this case, the exciting light is dispersed by the solvent and small solid particles suspended in the solution and recorded by the emission detector. On the other hand, the solvent Raman signal corresponds to inelastic dispersion and usually appears at longer wavelengths with respect to the excitation value. Finally, the second-order harmonics of the diffraction grating for the incident radiation can be also recorded as a narrow peak centered at twice the excitation wavelength.

Several procedures were developed to cope with the above-mentioned dispersion phenomena, which are not trilinear and therefore they make the data analysis considerably less simple. Among these procedures, replacing the Rayleigh dispersion values by zeros or by the so-called missing values is one of the most used ones [2]. Restricting the excitation and spectral ranges so that the dispersion signals do not contribute to the overall signal either during experimental data collection or by subsequent digital manipulation is another choice.

As will be shown in further examples, these phenomena and also additional problems like poor sensitivity or matrix effects originated by compounds present in complex samples call for improved flexibility in data modeling, a subject in which the algorithm based on unfolded partial least-squares with residual bilinearization (U-PLS/RBL) has shown an excellent performance [3–6].

*Example 1*

Let us describe an experiment to analyze complex biological samples in the presence of analyte–background interactions by using EEFM data modeled with U-PLS/RBL [4].

- Step 1, describing the analytical problem and the procedure

  The samples are composed of different human sera which were spiked with the antibiotic tetracycline. This substance is known for its strong interaction with serum proteins, and thus, it is important to notice that calibration should be done in the presence of different samples of the biological matrix. Hence, analyte–background interactions (and the associated spectral changes) may differ from sample to sample. This may be a challenge to trilinear data analysis based on, e.g., parallel factor analysis (PARAFAC), which needs reproducible analyte profiles in all samples, leading the data to be trilinear. This is not necessarily true for U-PLS/RBL, which requires calibration data to be modeled by a suitable number of latent loadings, and the presence of unexpected substances by RBL based on singular value decomposition. Hence, it is natural to expect U-PLS/RBL to show a more flexible structure.

  In the presence of analyte–background interactions, on the other hand, an alternative is to process the second-order data in the standard addition mode. However, when analyzing each test sample, standard addition uses a reduced number of samples in comparison with external calibration. Additionally, it is experimentally more demanding because each test sample should be spiked with various increasing amounts of pure analyte. Thus, the benefits of standard addition come at the price of a significant increment in sample size, which in the case of blood analysis is strongly unadvisable. In addition, this analytical system presents the problem of poor sensitivity for the analyte, i.e., the serum signal is ca. 10 times larger than the analyte signal.

  Calibration sets were built by adding pure standards of the analyte to normal sera. Five standards were built for tetracycline calibration, with the following concentrations: 0.5, 1.0, 2.0, 3.0, and 4.0 mg $mL^{-1}$.

The selected concentrations cover the therapeutic range of tetracycline. After suitable dilution (1:3), the latter concentrations render values which have been previously verified to lie in the known linear fluorescence–concentration range, i.e., 0.167–1.333 mg mL$^{-1}$. Several test sets of samples were prepared (the composition of each set is shown in Table 1). The test samples are composed of normal sera and tetracycline, with the addition, in certain cases, of four different drugs which are occasionally coadministered with tetracycline, i.e., salicylate, paracetamol, ibuprofen, and doxorubicin (DXR).

Appropriate amounts of the stock solution of tetracycline were added to 1.00 mL of serum in order to obtain the desired concentrations. The samples were then homogenized, and 2.00 mL of magnesium acetate ($5 \times 10^{-3}$ mol L$^{-1}$) in ammonium chloride (3 mol L$^{-1}$) was added. After homogenization and within a period of 30 min from the sample preparation (in order to avoid signal intensity changes), the EEFM was recorded in the following spectral ranges: $\lambda_{em}$ = 400–600 nm every 2 nm and $\lambda_{exc}$ = 290–440 nm every 5 nm ($101 \times 31 = 3131$ data points per sample matrix). The excitation and emission slit widths were 5 nm. Finally, the excitation was restricted to 290–400 nm in order to eliminate the Rayleigh dispersion, originating a matrix of $101 \times 23$ (see the rectangle marked in Figure 1).

- Step 2, modeling

The first step in U-PLS/RBL modeling is to estimate the number of calibration components. This is usually made by the leave-one-out cross-validation tool, as explained in Chapter 7. In the present study system, two latent variables were chosen (one for the analyte and another one corresponding to the serum signal).

Let us consider the analysis of one of the samples containing a single unexpected component, in this case sample C-2, which contains salicylate (at 12 μg mL$^{-1}$) and the analyte (at 2.20 μg mL$^{-1}$). If the prediction of tetracycline concentration is performed without RBL components, a large prediction error can be expected because the test sample contains potential interferents not included in the calibration set, and hence, RBL would be required to obtain the second-order advantage.

When no RBL components are considered, two important modeling figures should be considered: (a) the average calibration residue (the average residual fit to the calibration sample data is 1.9 units, close to the known value of the instrumental noise of 2 units) and (b) the residual fit for the sample under study (for this sample, it is equal to 30, i.e., considerably larger than the noise level). These are strong indications that this sample contains potential interferents, demanding RBL for accurate analyte quantitation. The predicted value is equal to 2.67 μg g mL$^{-1}$, which represents a recovery of 121.4%.

The next step is to introduce one RBL component in the modeling process. In this case, the residual fit for the sample C-2 is 2.9, a value that is

**TABLE 1** Validation and Test Samples for Example 1 and Results Using the U-PLS/RBL Model in the Prediction of Tetracycline Concentration

| Sample[a] | Unexpected component[b] ($\mu g\ mL^{-1}$) | Tetracycline concentration ($\mu g\ mL^{-1}$) | | |
|---|---|---|---|---|
| | | | *Predicted* | |
| | | *Nominal* | *U-PLS/RBL*[c] | *Recovery (%)* |
| A-1 | | 1.30 | 1.36 (1) | 104.6 |
| A-2 | – | 1.30 | 1.29 (1) | 99.2 |
| A-3 | – | 1.30 | 1.20 (1) | 92.3 |
| B-1 | – | 1.80 | 1.78 (2) | 98.9 |
| B-2 | – | 1.80 | 1.76 (2) | 97.8 |
| B-3 | – | 1.80 | 1.82 (2) | 101.1 |
| B-4 | – | 1.80 | 1.78 (2) | 98.9 |
| B-5 | – | 1.80 | 1.83 (2) | 101.7 |
| C-1 | – | 2.20 | 1.78 (1) | 80.9 |
| C-2 | Salicylate (12) | 2.20 | 2.31 (1) | 105.0 |
| C-3 | Paracetamol (12) | 2.20 | 1.86 (2) | 84.5 |
| C-4 | Ibuprofen (20) | 2.20 | 1.92 (2) | 87.3 |
| C-5 | Doxorubicin (12) | 2.20 | 2.30 (2) | 104.5 |
| D-1 | – | 2.80 | 3.23 (1) | 115.4 |
| D-2 | Salicylate (12) | 2.80 | 3.24 (2) | 115.7 |
| D-3 | Paracetamol (12) | 2.80 | 3.23 (2) | 115.4 |
| D-4 | Ibuprofen (20) | 2.80 | 3.28 (2) | 117.1 |
| D-5 | Doxorubicin (12) | 2.80 | 3.02 (2) | 107.9 |
| RMSEP | | | 0.26 | |
| REP% | | | 12.6 | |

[a]*The capital letter corresponds to the type of serum used. All samples were prepared using different sera than those for calibration samples.*
[b]*Concentration of the unexpected component in parenthesis.*
[c]*Number of unexpected components ($N_{unx}$, see text) in parenthesis.*

closer to the instrumental noise value than when no RBL was implemented. The predicted value in this opportunity is equal to 2.31 $\mu g\ mL^{-1}$, which represents a recovery of 105.5%.

Furthermore, the plot showing the RBL profiles in both modes for the potential interferent can be examined in Figure 2. Interestingly, although

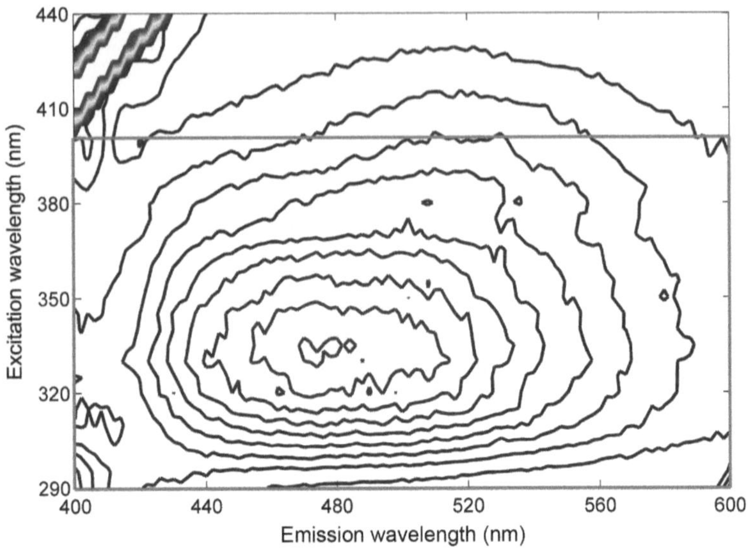


**FIGURE 1** Contour plot corresponding to an EEFM for a plasma solution spiked with tetracycline.

these profiles have been obtained through a latent-structured principal component analysis of the sample residuals, they still have some physically recognizable characteristic (they are excellent approximations to the pure profiles for salicylate).

It should be commented that the same samples were analyzed with PARAFAC, obtaining worst results than those obtained with U-PLS/RBL. Two possibilities are envisaged to account for this fact: (a) the predictive ability of the latter algorithm is better under conditions of poor sensitivity toward the analyte and/or (b) U-PLS takes better account of the analyte–background interactions due to its latent variable properties.

## 2.2 Kinetic-Spectroscopic Matrices

When a chemical reaction takes place in a sample in which the analyte changes its chemical composition with time, recording UV–visible absorption or emission fluorescence spectra as a function of time constitutes a useful way for second-order data generation. The kinetics should develop in a suitable time window, i.e., on the order of minutes, which can be conveniently followed using a diode array spectrophotometer or a fast-scanning fluorimeter. A fast kinetics can originate problems associated with lack of reproducibility (which can be minimized by using a stopped-flow reaction cell), while a very slow kinetics may be impractical as it requires long experimental times. It should be noticed that if various analytes react and follow very similar kinetics, then the selectivity in the time mode will be severely diminished [1].

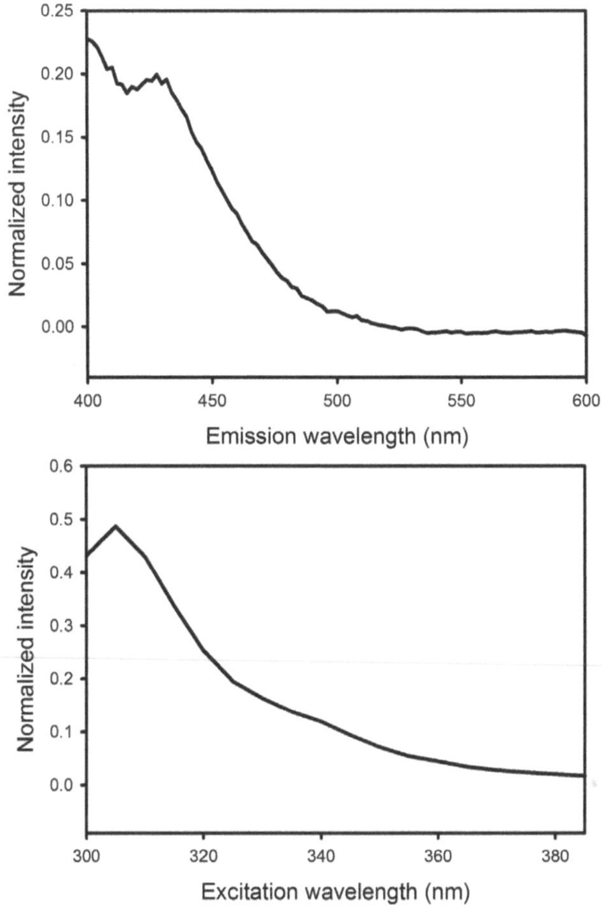


**FIGURE 2** Profiles in both modes for the potential interferent extracted by the RBL procedure.

These kinds of second-order data have been traditionally processed by the application of multivariate curve resolution-alternating least-squares (MCR-ALS). However, several applications can be found in the literature, in which the modeling was successfully performed with U-PLS and multiway PLS (N-PLS) with RBL [7,8].

Figure 3 shows a typical kinetic evolution of UV–visible spectra as a function of reaction time throughout the reaction between two dyes (tartrazine and brilliant blue) and potassium bromate in the presence of Fe(II) [9].

During the oxidation with potassium bromate, the absorbance of tartrazine (peak maximum at 430 nm) decreases, leading to a colorless component. On the other hand, brilliant blue does also react with bromate, suffering a kinetic degradation and yielding a product whose spectrum with an absorption maximum similar to tartrazine. An additional dye, sunset yellow, also interferes in

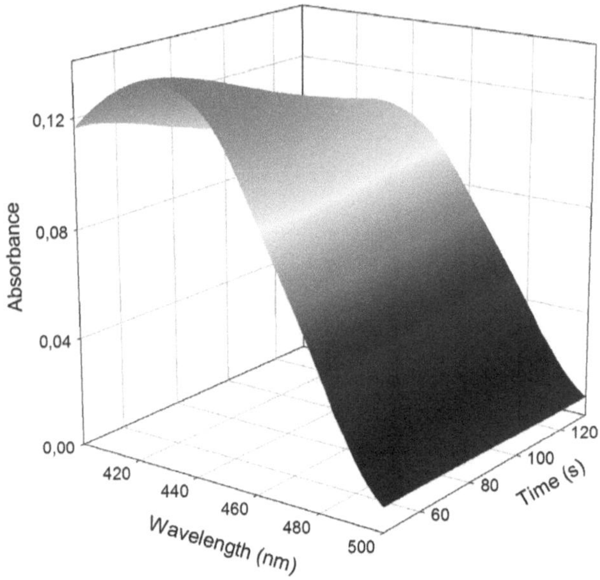


**FIGURE 3** Typical kinetic evolution of UV–visible spectra as a function of reaction time.

the determination, owing to the fact that it also presents a small peak at 420 nm, whose absorbance also decreases in the presence of the oxidant, generating a colorless product. The normalized spectra of the three dyes can be appreciated in Figure 4A, as well as the one from the reaction product of brilliant blue with potassium bromate. Figure 4B shows the kinetic curves of these compounds at their absorption maxima, i.e., 430, 430, 630, and 565 nm for tartrazine and the brilliant blue product, brilliant blue and sunset yellow, respectively.

As can be appreciated, there is a small difference between the shape and peak positions of the spectra corresponding to the analyte and the main interference (brilliant blue). On the other hand, the kinetic profiles are different enough to warrant a good data modeling and an accurate prediction of tartrazine. Data processing with U-PLS/RBL in a similar way as described for EEFM in Example 1 leads to acceptable results.

## 2.3 pH-Spectral Matrices

These data can be generated producing a double pH gradient in a flow injection system after injecting an alkaline solution in an acid carrier (or vice versa) and letting the flow reach a DAD or a fast-scanning fluorescence detector. A typical landscape, measured for determination of DXR in human plasma when acid is injected into an alkaline mixture containing the analyte as carrier and the flow reaches a DAD, is shown in Figure 5 [10]. As was commented for spectral-time signals, the ideal situation is when high

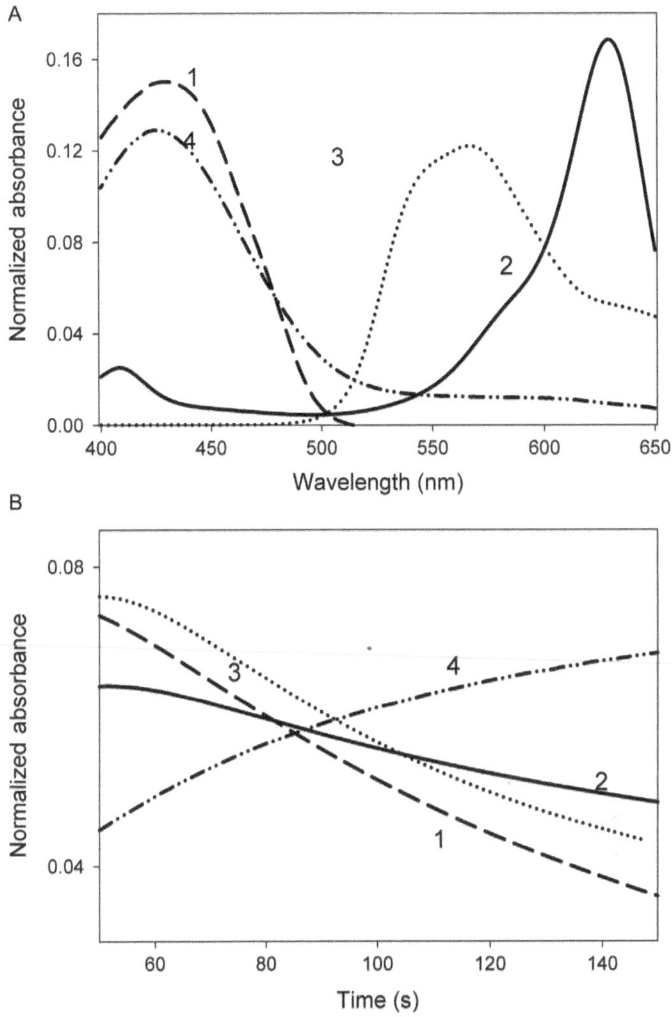


**FIGURE 4** (A) Normalized spectra of tartrazine (1), brilliant blue (2), sunset yellow (3), and a degradation product of brilliant blue (4). (B) Kinetic curves of these compounds at their absorption maxima: 430, 430, 630, and 565 nm for tartrazine, brilliant blue product, brilliant blue, and sunset yellow, respectively.

selectivity exists both in the spectral and pH modes, which demands the sample constituents to be weak acids with sufficiently different p*K*a values.

These data (as the ones discussed above) are in principle nontrilinear with one trilinearity-breaking mode. If the variation of pH along the pH mode is not reproducible among samples, because the pH gradient mode breaks the trilinearity, the data are nontrilinear. Conversely, if gradients can be generated in a reproducible way from sample to sample, then spectral-pH data are

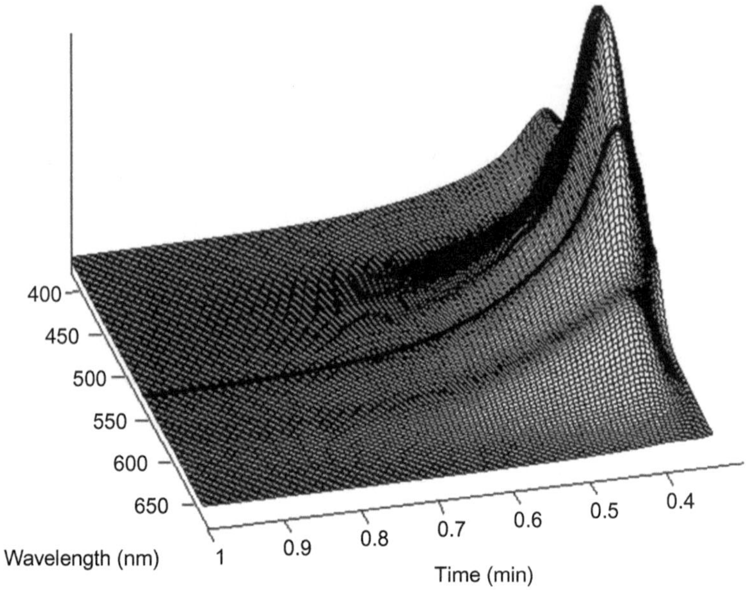


**FIGURE 5** Landscape of second-order UV–visible absorption-pH data. In the pH mode, the units are time data points, because the pH gradient was generated in a flow system, injecting an acid solution into an alkaline carrier stream, and a different pH corresponds to a different time along the flow gradient.

trilinear. On the other hand, in this example three-component profiles occur along the pH mode, corresponding to the different acid–base species of the analyte, which are mutually correlated. In this particular scheme, the sum of the three pH profiles is constant, since a mass balance should be obeyed. Figure 6A shows the three species of DXR, while Figure 6B displays the pH profiles. This fact causes dependency among profiles, which is part of a more general manifestation of proportionality or linear relationship among constituent profiles. This phenomenon should be taken into account, because it might be potentially troublesome when it comes to interpreting the results of data processing or to determining an analyte suffering these effects [11,12].

Ghasemi and Niazi reported the quantitation of lorazepam in pharmaceutical formulations and biological fluids by using spectrophotometric data recorded at different pHs (in the pH range 1.0–12.0). The study was performed using PARAFAC and PLS (without exploiting the second-order advantage with the RBL procedure). In this particular case, better performance was obtained with PARAFAC [13]. Other examples can be cited in which this kind of data were employed to quantitate analytes in the presence of unexpected compounds, but most of them resorted to MCR-ALS or bivariate least-squares modeling [14]. Olivieri demonstrated that U-PLS/RBL is able to manage pH-DAD data, obtaining good analytical results [15].

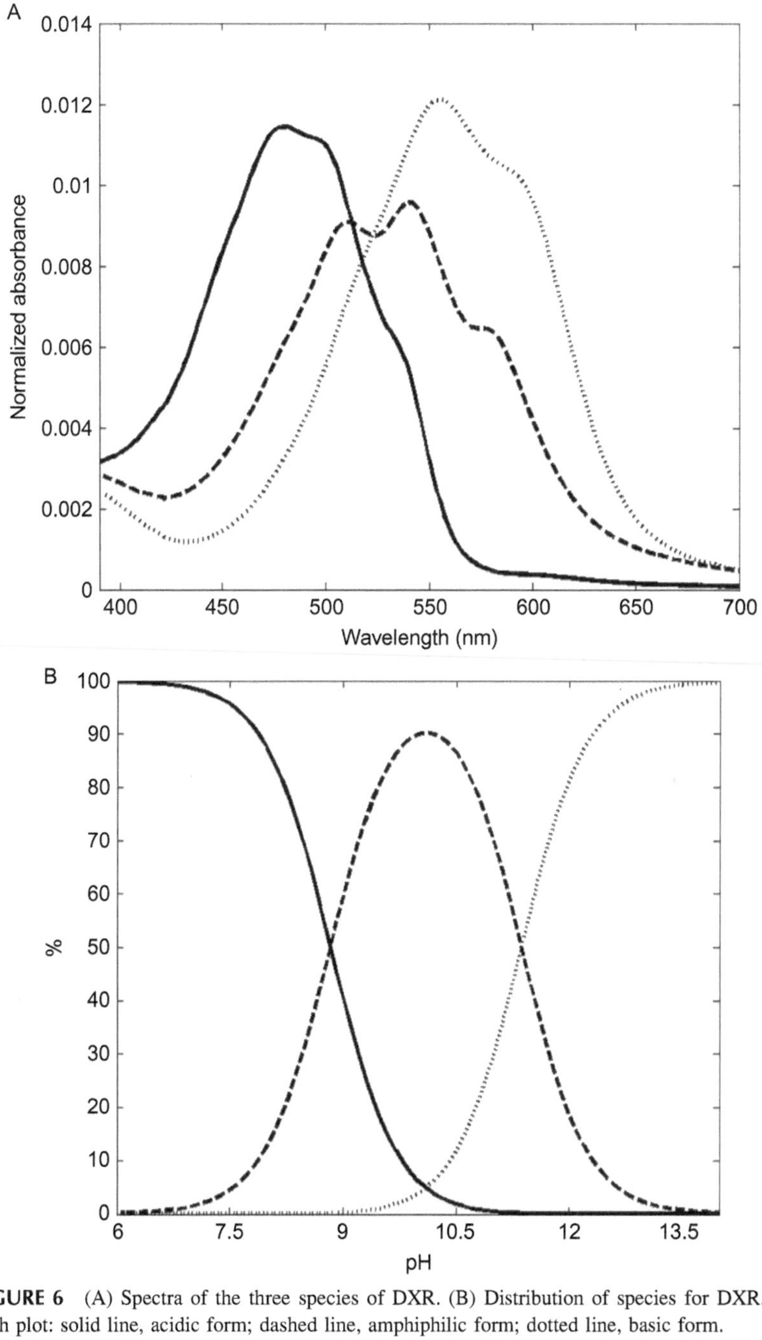


**FIGURE 6** (A) Spectra of the three species of DXR. (B) Distribution of species for DXR. In both plot: solid line, acidic form; dashed line, amphiphilic form; dotted line, basic form.

## 2.4 HPLC/CE Spectral Detection Matrices

In chromatography, the retention factor ($k$), which is the degree of retention of the sample constituent in the column, plays an important role in most analyses: in practice, analytes elute with retention factors between 1 and 20, with a peak having $k$ equal to 0 indicating a constituent which does not interact with the stationary phase [16]. Chromatographic separations can become a difficult task when complex samples have to be analyzed because of the presence of constituents with similar retention factors. Nevertheless, the use of multiway calibration may provide a useful resource for accurate analyte quantitation when complete separation is not accomplished, or new constituents present in the sample are being analyzed [17].

The hyphenation of liquid (LC) or gas (GC) chromatography (or capillary electrophoresis, CE) with spectroscopic techniques can yield second-order data which combine instrumental signals built from both spectral and time domains. Popular techniques are LC-DAD (UV–visible diode array detection), LC-FSFD (fast-scanning fluorescence detection), or LC-MS (mass spectrometric detection). The responses are arranged as a data matrix, where each column corresponds to a wavelength (or $m/z$ ratio) and each row corresponds to a different elution time. Figure 7 shows a typical example of a second-order data matrix generated by CE-DAD corresponding to the fast determination of six quinolones in the presence of several unexpected components in water samples [18]. As can be appreciated, the incomplete separation of six analytes (plus interferents) was accomplished in ca. 1.5 min. Implementation of a second-order calibration by using pure standard samples matrices furnished highly accurate results when analyzing real water samples. The terms "chroMATHography" [19] and "chromametrics" [20] have been coined to describe the combination of chromatography and chemometrics, and special issues have been devoted to chromametrics by specialized journals [21,22].

An important number of reports have been presented focusing on the resolution of really complex samples by using liquid chromatography and exploiting the second-order advantage [18–23]. In this context, extremely important issues such as reduction in analysis time and consequently in costs and amount of contaminant solvents should be considered.

Several algorithms can be used to model this kind of data, but MCR-ALS has become the choice in most of the published works. This may be due to the fact that differential migration of the constituents originates dissimilarities in both retention times and peak shapes (this is even more significant in CE data), leading to data without the property of trilinearity (see above). MCR-ALS can efficiently solve this type of problems by resorting to the mathematical resource of matrix augmentation. PARAFAC2 is a variant of PARAFAC attempting to model such changes, but apparently is less powerful than MCR-ALS in this regard, especially when time shifts or band shapes are significant and when potential interferents are present in test samples. However, when

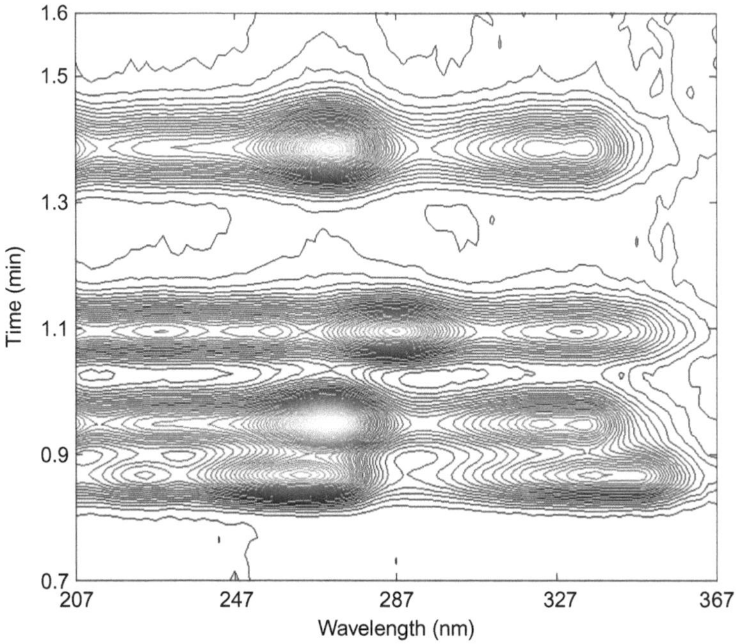


**FIGURE 7** Second-order data matrix generated by CE-DAD.

data are conveniently pretreated in order to alleviate the above-mentioned problems, using adequate alignment strategies of elution profiles, good results can be obtained by using PARAFAC or RBL-based algorithms [17,23]. For a detailed description of the different pretreatment approaches, see Ref. [17] and a recent tutorial [24].

*Example 2*

Let us describe an experiment in which U-PLS/RBL was applied for first time on second-order LC-DAD data and compared with MCR-ALS for the simultaneous determination of eight tetracyclines (tetracycline, oxytetracycline, meclocycline, minocycline, metacycline, chlortetracycline, demeclocycline, and doxycycline) in wastewaters.

- Step 1, describing the analytical problem and the procedure

  Tetracyclines were preconcentrated using Oasis Max C18 cartridges and then separated on a Thermo Aquasil C18 (150 mm × 4.6 mm, 5 μm) column. A baseline correction pretreatment was applied allowing to correct the effect of breakthrough and to reduce the total interferences retained after preconcentration of wastewaters. The results showed that the eight tetracycline antibiotics can be successfully determined in

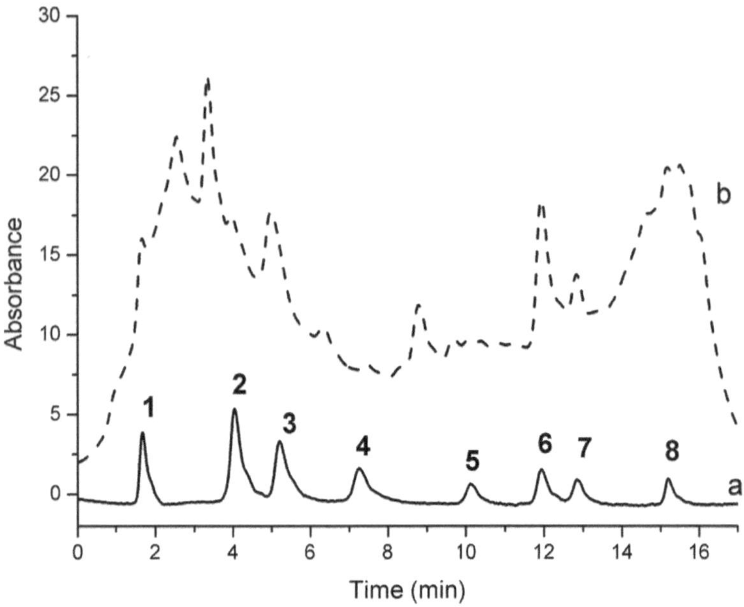


**FIGURE 8** Chromatograms (absorption $\lambda = 360$ nm) corresponding to (a) a standard mixture containing the eight tetracyclines at 0.50 mg $L^{-1}$ and (b) a real sample extract obtained after spiking it with the eight tetracyclines at 2.00 mg $L^{-1}$. (1) Minocycline, (2) oxytetracycline, (3) tetracycline, (4) demeclocycline, (5) chlortetracycline, (6) metacycline, (7) doxycycline, and (8) meclocycline.

wastewaters, the drawbacks due to matrix interferences being adequately handled and overcome by using U-PSL/RBL [25].

Figure 8 provides evidence for the complexity of the analytical problem at hand, i.e., eight compounds which could be successfully separated, immersed in such a complex matrix containing unexpected compounds which interfere in the analysis.

Working standard solutions of the analytes were prepared daily by dilution of the appropriate aliquot in 1% (v/v) formic acid and were filtered through Millipore membrane PTFE filters before injection into the chromatographic system. 250 mL of aqueous (Milli-Q or wastewater) samples were spiked with the appropriate volumes of standards and prepared for extraction by adding 1 mL of 5% $Na_2EDTA$ and 285 mg of citric acid and then adjusting the pH to 2.8 with HCl immediately prior to extraction. A set of seven samples, each of them in triplicate, containing 0.05, 0.10, 0.20, 0.40, 0.60, 0.80, and 1.00 mg $L^{-1}$ of the eight tetracyclines, was prepared in pure solvent and was used as the calibration set. A validation set, containing six samples and four replicates of another different sample, was also prepared by spiking different amounts of tetracyclines in a final effluent domestic wastewater.

As regards the data collection, photometric detection was performed in the 230–400 nm range with a spectral resolution of 1.5 nm and the period of integration was 0.4 s per spectrum. Thus, every data matrix had 1950 rows (elution times between 3 and 16 min) and 114 columns (number of wavelengths), i.e., each sample generates a matrix of size $1950 \times 114$. In addition, with the object of simplifying the analysis, the total chromatographic data registered for each sample were partitioned in eight regions, each of them corresponding to each analyte. Thus, a model for each tetracycline with only one analyte and the corresponding interferent was built.

- Step 2, modeling

The first step in U-PLS/RBL modeling is to estimate the number of calibration components. This is usually made by the leave-one-out cross-validation tool, as explained in Chapter 7. In the present study system, one latent variable was chosen because each region contains a single analyte (see above). Then, the presence of unexpected components should be considered, with the aim to decrease the prediction residuals for the test samples until they stabilized at a value compatible with the instrumental noise. The variation of the prediction residuals as a function of trial values of unexpected compounds ($N_{unx}$) allows one to select the appropriate number. In this case, the comparison was made with an instrumental noise of ca. 0.16 absorbance units in this system. $N_{unx}$ depends on both the number of interferents retained in the solid-phase extraction (SPE) procedure and the success of the background correction procedure. Supporting the latter assert, $N_{unx}$ was different for data matrices before and after data preprocessing (for example, $N_{unx}$ was equal to 3 and 1 before and after preprocessing, respectively) (for more details, see Ref. [25]).

A comparative analysis of the performance of MCR-ALS and U-PLS/RBL can be made by comparison of the retrieved profiles in both cases. It should be noted that U-PLS/RBL does not extract profile for pure analytes, but for the interferents. Figure 9 shows the chromatographic profiles corresponding to the analyte and the interference, extracted by MCR-ALS, and the interference profile extracted by U-PLS/RBL from an unknown sample for the second region (corresponding to the analyte oxytetracycline). This figure also shows the corresponding pure normalized instrumental signal. As can be seen, two components were considered in the MCR-ALS analysis (both of them plotted as solid lines). The matching between the solid line, which represents the chromatographic profile corresponding to the analyte, and the circles line, which represents the normalized pure component chromatogram at 360 nm, strongly suggests that the extracted profile corresponds to the analyst of interest. This analysis allows to compare the MCR-ALS extracted profile for the interference with those extracted by the RBL procedure (marked with the letters a and b in Figure 9, respectively).

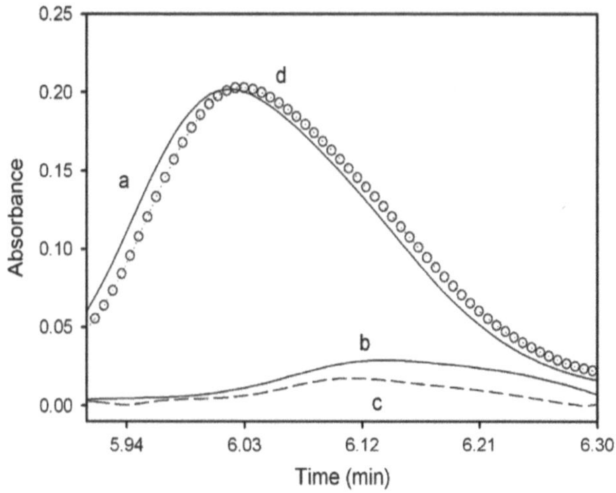


**FIGURE 9** MCR-ALS elution time profiles for peak 2: (a) concentration profile for oxytetracycline and (b) time profile for the interference. (c) U-PLS/RBL elution time profile for the interference present in peak 2. (d) Normalized pure component chromatogram for oxytetracycline. *Reprinted with permission from Ref. [25]. Copyright 2008 Elsevier.*

Predictions obtained when analyzing the real sample without and with spiked antibiotics were acceptable for most samples and analytes, despite the complexity of the samples. Notably poor recoveries were only obtained for minocycline by both methods. The fact that, for this analyte, both implemented methodologies furnished comparable values tells us about possible experimental errors in sample preparation, analysis, or analyte instability, but not about poor predictive ability of U-PLS/RBL.

## 2.5 Electrochemical Second-Order Data

Another kind of second-order data which have been scarcely exploited as regards the generation of second-order data are those obtained by electrochemistry techniques. First-order electrochemical data can be obtained through voltammetry carried out in different options, such as differential pulse voltammetry (DPV) or square wave voltammetry. A time window or the pulse height can be introduced to generate the second mode. For example, Khoobi *et al.* reported second-order data built from variable pulse heights of DPV and analyzed by MCR-ALS for the quantitation of betaxolol and atenolol using a carbon nanotube paste electrode [26]. Another example is the generation of second-order data recording voltammograms at different adsorption kinetic times for the determination of fenitrothion and its metabolites in river water samples [27]. In this latter report, PARAFAC, N-PLS/RBL, and U-PLS/RBL were the algorithms of choice, with the latter furnishing the best analytical results. Few reports can be found in the literature on this subject, most of them applying MCR-ALS [14]. It should be noticed that in general a limited number of data points can be collected in the second mode (time, pulse height, etc.), an issue which becomes a challenge for researchers.

## 2.6 Total Synchronous Fluorescence Spectroscopy Matrices

Synchronous fluorescence spectra (SFS) are a kind of first-order data, which can be used to resolve complicated multicomponent samples with severely overlapping emission and/or excitation spectra. Synchronous spectrometry essentially consists in simultaneously scanning both monochromators, while maintaining a constant wavelength interval or offset ($\Delta\lambda$) between them. The simplification of the spectral profiles, together with the reduction in bandwidths, constitutes its main advantageous characteristics. Furthermore, there is a reduction in Rayleigh scattering, and corrections or preprocessing methods are no longer needed, as may happen for excitation-emission spectra (see above) [28].

It is important to note that although SFS are narrower and sharper compared to conventional fluorescence spectra, it is not always possible to avoid spectral overlap in certain multicomponent mixtures. However, combining SFS with a multivariate method makes the analysis simple and fast [29]. First-order algorithms allow one to identify new samples containing unexpected components, a property known as the first-order advantage, and may compensate for potential interferences, usually by their inclusion in the calibration set. An alternative is to obtain total synchronous fluorescence spectroscopy (TSFS) matrices, which provide the total synchronous fluorescence characteristics of a multifluorophoric sample at various possible wavelength intervals ($\Delta\lambda$). Every TFSF datum consists in a two-way array of dimensions (excitation wavelengths $\times$ number of $\Delta\lambda$s) and can be successfully handled by second-order multivariate calibration algorithms.

Interestingly, TSFS behaves as a nonbilinear matrix, meaning that most of the fluorophores show variations in their spectral shapes along with their spectral intensities as long as $\Delta\lambda$ changes. Each fluorophore in TSFS does not have a single unique SF spectrum, due to the fact that the SF spectral shape of a fluorophore varies with the change in offset [30]. Kumar and Mishra were able to simultaneously quantify five polycyclic aromatic hydrocarbons (PAHs) using TSFS with multivariate methods such as N-PLS, U-PLS, and MCR-ALS, but no second-order advantage was achieved, since all the components were added in both calibration and test samples [31]. In addition, the same authors have addressed the issue of lack of trilinearity, converting the TSFS data into an excitation-emission fluorescence matrix, so that PARAFAC modeling could be applied [32].

As very recently commented by Olivieri and Escandar [1], it is likely that U-PLS/RBL would be able to manage nonbilinear data, although the expression "residual bilinearization of a nonbilinear matrix" may seem contradictory.

However, a nonbilinear matrix can be modeled with a sum of several bilinear terms (considerably more than the number of constituents), whereas a bilinear matrix is modeled with a few bilinear terms (ideally equal to the number of constituents). Hence, RBL would simply need more principal components for modeling the potential interferents. Based on the latter hypothesis, two works were recently reported demonstrating that this is indeed possible [33,34]. In the first one, DXR (an antibiotic used in the treatment of a wide variety of cancers) was quantitated in human plasma samples in the presence of an unexpected component (riboflavin) by using TSFS data matrices modeled with U-PLS/RBL. In the second one, five hydroxy derivative polycyclic aromatic hydrocarbons (OH-PAHs)—namely, 1-hydroxypyrene (1-OHPyr), 2-hydroxyfluorene (2-OHFlu), 3-hydroxybenzo[*a*]pyrene (3-OHB[*a*]P), 4-hydroxybenzo[*a*]pyrene (4-OHB[*a*]P), and 6-hydroxybenzo[*a*]pyrene (6-OHB[*a*]P)—were solid-phase extracted from urine samples on C18 silica cartridges and directly determined in the solvent extract (methanol) without previous chromatographic separation. The advantage of modeling TSFS data over the straightforward implementation of both synchronous fluorescence spectroscopy and first-derivative synchronous fluorescence spectroscopy was illustrated with the analysis of 6-hydroxy-chrysene (6-OHChry) and 4-OHB [*a*]P, in the presence of relatively large concentrations of 1-OHPyr and 3-OHB[*a*]P, respectively. The presence of creatinine, naproxen, ibuprofen, diclofenac, and amoxicillin in urine samples was investigated as potential interferences in the analysis of OH-PAHs. It was demonstrated that the combination of TSFS to either U-PLS/RBL or N-PLS/RBL provides robust screening approaches for the analysis of OH-PAHs in urine samples of unknown composition.

*Example 3*

Let us describe the experiment of Ref. [33], carried out to analyze DOX in human plasma samples in the presence of riboflavin (RF) as unexpected component, by using TSFS data modeled with U-PLS/RBL.

- Step 1, describing the analytical problem and the procedure

  The calibration set was built by adding pure standard of DOX to plasma obtained from healthy patients due to the presence of interactions between the analyte and the sample background. The set was prepared in two steps: first, five water standards of DOX with the following concentrations: 8.0, 24.1, 40.0, 56.9, and 72.8 μg $mL^{-1}$ were prepared in duplicate. Then, 20.0 μL of the latter solutions were placed into an Eppendorf vial together with 600.0 μL of plasma. After adding 20 μL of TCA 20% (w/v) for protein precipitation, each standard sample was vortex mixed for 30 s and ultracentrifuged at 14,000 rpm for 10 min. Then, 500.0 μL of the supernatant was transferred to a quartz cell. The final concentration values of the calibration set were 0.25, 0.75, 1.25, 1.78, and 2.28 μg $mL^{-1}$. The protein precipitation step was performed to eliminate a possible matrix effect and to enhance the sensitivity of the method, since plasma proteins provide a significant quenching.

Additionally, a validation set of nine samples was prepared, following a central composite design of two factors with one center point, with different concentrations of DOX (from 16.0 to 64.0 μg $mL^{-1}$) and RF (from 2.4 to 9.6 μg $mL^{-1}$) in water. The same procedure described for the preparation of the calibration samples was then applied. The final concentrations of DOX and RF in the validation set are given in Table 2. These concentrations are within the therapeutic values of the studied drug in human plasma.

All spectrofluorimetric measurements were performed using slits for the excitation and emission monochromators kept at 10 nm, and the detector voltage was 650 V. TSF matrices were acquired with excitation between 489.5 and 614.5 nm at a resolution of 0.5 nm, with offsets

**TABLE 2** Predictions and Statistical Analysis of DOX in Test Samples

| | Nominal | | Predicted DOX | |
|---|---|---|---|---|
| Sample | *DOX* (*μg* $mL^{-1}$) | *RF* (*μg* $mL^{-1}$) | *U-PLS/RBL*[a] (*μg* $mL^{-1}$) | *Rec. (%)* |
| 1 | 0.50 | 0.19 | 0.43(1) | 86.0 |
| 2 | 1.78 | 0.27 | 1.73(1) | 97.2 |
| 3 | 1.25 | 0.30 | 1.28(1) | 102.4 |
| 4 | 1.99 | 0.19 | 1.91(1) | 96.0 |
| 5 | 1.25 | 0.19 | 1.32(1) | 105.6 |
| 6 | 0.72 | 0.27 | 0.73(1) | 101.4 |
| 7 | 1.25 | 0.075 | 1.36(1) | 108.8 |
| 8 | 0.72 | 0.11 | 0.69(1) | 95.8 |
| 9 | 1.78 | 0.11 | 1.80(1) | 101.1 |
| Rec. (%)[a,b] | | | | 99(7) |
| RMSE (μg $mL^{-1}$)[c] | | | | 0.06 |
| REP (%)[d] | | | | 5 |

[a]*Between parenthesis the standard deviation.*
[b]*Mean recovery (%).*
[c]*Root mean square error,* $RMSE=\left[\frac{1}{I}\sum_{1}^{I}(c_{act}-c_{pred})^2\right]^{1/2}$, where $I=9$.
[d]*Relative error of prediction, REP= 100× RMSE/$\bar{c}$*, where $\bar{c}$ is the mean calibration concentration.

ranging from 30 to 160 nm (in steps of 10 nm) between excitation and emission monochromators. Hence, the size of each data matrix was 251 × 14. Figure 10A and B shows emission spectra obtained at different excitation wavelength (EEFM), and synchronous spectra recorded as was commented above (TSFSM) for a standard sample of DOX. In these figures, it can be clearly appreciated that the EEFM data matrix is bilinear, while the TSFSM data matrix is apparently nonbilinear.

- Step 2, modeling

  The number of calibration components estimated by the leave-one-out cross-validation was equal to two latent variables, as expected considering both the analyte and the plasma signal.

  Let us consider the analysis of sample number 3, which contains DOX at 1.25 μg $mL^{-1}$ and RF at 0.30 μg g $mL^{-1}$. If the prediction of DOX concentration is performed with no RBL components, a large prediction error can be expected, as commented above. Whereas the average calibration residue is close to the known value of the instrumental noise, the residual fit for the sample under study is considerably larger than the noise level. These are strong indications that this sample contains potential interferents, demanding RBL for accurate analyte quantitation.

  The next step is to introduce one RBL component in the modeling process. In this case, the residual fit for sample number 3 is still larger than

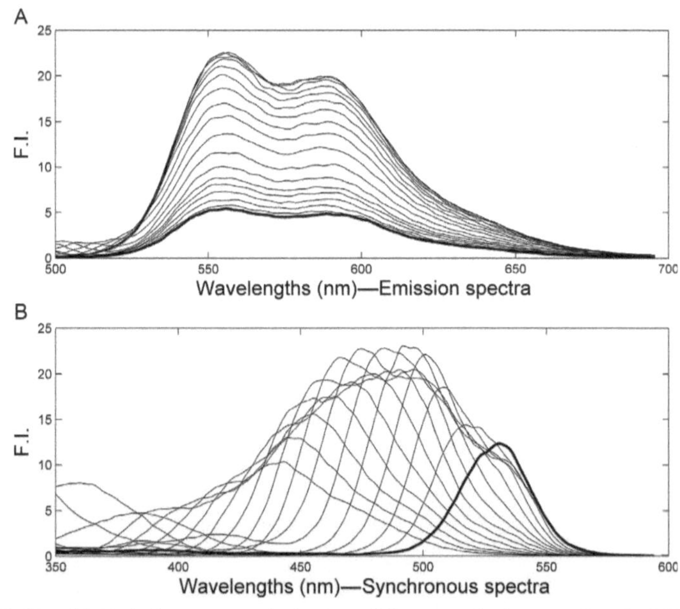


**FIGURE 10** (A) Emission spectra obtained at different excitation wavelength (EEFM) and (B) synchronous spectra recorded at different offsets (TSFSM), both for a standard sample of DOX.

the noise level. This is indicative that one component is not enough to fit the presence of one unexpected component (RF). If two more RBL components are introduced, the residual fit value results closer to the instrumental noise value, a fact that agrees well with the previous comment about the possibility of modeling the interference adding an excessive number of RBL components. The predicted value in this opportunity is equal to 1.28 μg $mL^{-1}$, which represents a recovery of 102.4%.

### 2.7 Tandem Mass Spectrometry (MS/MS)

The requirement of equal profiles in all modes and all samples for a given component is not fulfilled in the case of MS/MS data, because a given component displays various different profiles in one of the data modes. N-PLS with the aid of data preprocessing was found to give good results when this type of nontrilinear data was analyzed. However, all the components were also present in training and test samples, making unnecessary the use of RBL and the second-order advantage [35].

## 3 THIRD-ORDER DATA

Four-way data are obtained when joining third-order data for a set of samples into a four-dimensional mathematical object. It is interesting to note that four-way data not only retain the "second-order advantage" inherent to three-way/second-order calibration [36] but can also have additional advantages [37]. They would display the obvious advantage of providing richer analytical information, implying more stable methods toward interferences and matrix effects, and less prone to minor changes in reaction conditions, which should allow for an improvement in predictive ability. In addition, improvements in sensitivity and the resolution of collinearity problems have been reported [38]. However, it is a fact that only few experimental four-way data have been recorded and used to develop analytical methodologies to date, which can be attributed to the fact that the practical acquisition of these data arrays is still difficult to implement. In addition, a thorough understanding of their analytical advantages is still needed.

Among the procedure for obtaining four-way data with a single instrument, the most commonly used one is the recording of luminescence excitation-emission matrices as a function of some factors such as reaction time, decay time, or any additional variable affecting the analytical signal (pH, sample volume, quenching effects, etc.), introduced as independent analytical modes to construct higher-way data to exploit more inherent information in samples. Chromatography, in the conventional one-dimensional configuration and in the different two-dimensional approaches, also allows to obtain third-order data. Most of the reported applications have been combined with U-PLS and/or N-PLS multiway calibration procedures.

To work with third-order data, all calculations can be done using different routines and the graphical interface MVC3 (MultiVariate Calibration for third-order), an integrated MATLAB toolbox for third-order calibration developed by Olivieri *et al.* [39], which allows performing third-order calibration with different modeling methods, including U-PLS and N-PLS, both coupled to residual trilinearization (RTL). The software can be freely downloaded from the webpage www.iquir-conicet.gov.ar/descargas/mvc3.rar.

## 3.1 Kinetic-EEFM Third-Order Data

The first example of processing four-way time-evolving EEFMs was the modeling of the degradation of chlorophylls *a* and *b* extracted from spinach by Tan *et al.*, although the procedure was not used for analytical determination of the studied chlorophylls [40]. On the strictly analytical side, several successful third-order multivariate calibrations have been developed using an analogous strategy. Nonfluorescent adrenaline and noradrenaline were derivatized to the fluorescent 3,5,6-trihydroxyindole derivatives and analyzed by N-PLS. The method derives selectivity from the different rates at which the derivatives (lutines) of the catecholamines are formed and degraded [41]. For running the lutine reaction, a flow system was used, and to acquire the EEFMs as a function of time, a laboratory-constructed charge-coupling device fluorometer was used, as previously described by Muroski [42].

Apart from this pioneering work, a fast-scanning fluorimeter is usually employed to record the full EEFMs at different times, as reported for the first time for the determination of methotrexate and leucovorin in human urine samples [43]. By following this approach, and based on analyte oxidation by permanganate, these determinations were reported: methotrexate and leucovorin in human urine, by trilinear least-squares and unfolded-PLS (U-PLS), both coupled to RTL [43,44], folic acid and methotrexate in urine by N-PLS [45] and in serum by U-PLS/RTL [46], and folic acid in the presence of its two main metabolites, 5-methyltetrahydrofolic acid and tetrahydrofolic acid by U-PLS/RTL and N-PLS/RTL [47].

Similarly, the kinetic reaction of malonaldehyde with methylamine was used for four-way malonaldehyde quantitation in olive oils, using a nonlinear variant of PARAFAC and a neural network coupled to RTL [48].

Similarly, carbaryl and 1-naphthol in water samples, in the presence of uncalibrated interferents, such as other commonly employed agrochemicals, were determined by kinetic-EEFM four-way data in combination with U-PLS/RTL calibration [49].

*Example 4*

Let us describe the experiment in which the algorithm N-PLS/RTL was described for the first time and applied to the four-way analyses of procaine (PRO) and its metabolite *p*-amino benzoic acid (PABA) in equine serum, based on an alkaline hydrolysis reaction [50].

- Step 1, describing the analytical problem and the procedure

  PRO is used as a local anesthetic, and in human plasma, it is almost completely hydrolyzed by the enzyme pseudocholinesterase to PABA. It is also usually used in horses as an anesthetic and a stimulating drug, and horse plasma does not promote the complete hydrolysis of PRO into PABA, and therefore, the simultaneous determination of both compounds is important in monitoring fair horse sports.

  A calibration set was constructed composed of 11 samples, using a central composite design with the central point replicated three times. The concentrations were in the ranges 0–900 ng $mL^{-1}$ for PRO and 0–600 ng $mL^{-1}$ for PABA, selected according with the drug therapeutic levels. For preparing a given concentration sample, the analytes were mixed in the measuring cell, by taking the appropriate volumes of PRO and PABA solutions and diluting with 0.1 M NaOH to the desired concentration. The horse serum samples (0.50 mL) were treated with 1.5 mL of 2% (v/v) trichloroacetic acid and centrifuged at 3000 rpm during 5 min for deproteinization. Finally, the 0.40 mL of the supernatant was diluted to a final volume of 2.00 mL with 0.1 M NaOH.

  The experimental parameters for the hydrolysis reaction were studied and a pH 13 and a temperature of 40 °C were selected to obtain appropriate kinetics and fluorescence properties of the product originated. As can be seen in Figure 11, in which the kinetic has been followed during 18 min, PRO is weakly fluorescent, but its hydrolysis in basic medium generates PABA, which is a strongly fluorescent compound.
- Step 2, modeling

  When applying N-PLS to the serum samples, it is necessary to assess the number of unexpected components ($N_{unx}$) to be employed in the RTL procedure. This can be done by analyzing the residuals $s_u$ as a function of a trial number of unexpected components. A typical result for one of the analyzed samples was as follows: $s_u = 45$ and 6.2 for $N_{unx} = 0$ and 1, respectively. For $N_{unx} = 2$, all possibilities having different number of loadings in each mode have to be tried in order to construct the Tucker3 model for the potential interferents, giving $s_u = 4.8$ [2 2 2], 6.1 [1 1 2], 6.1 [1 2 1], 6.2 [2 1 1], 4.8 [2 2 1], 6.0 [2 1 2], and 6.1 [1 2 2] (the values are in arbitrary fluorescence units, with the number of loadings used in the emission, excitation, and time mode, respectively, between square brackets). Since blank replication allowed to estimate an instrumental noise level of ca. 5.0 arbitrary fluorescence units, for this particular sample $N_{unx} = 1$ is a reasonable choice. This is because the value of $(s_u)^2$, measuring the residual variance, is statistically indistinguishable from the squared noise level (25 arbitrary fluorescence units$^2$) using an appropriate Fischer $F$ test for the comparison of variances. Similar results were obtained for the remaining samples. The choice of $N_{unx} = 1$ can also be confirmed to be reasonable because, using this value, the predicted analyte concentrations

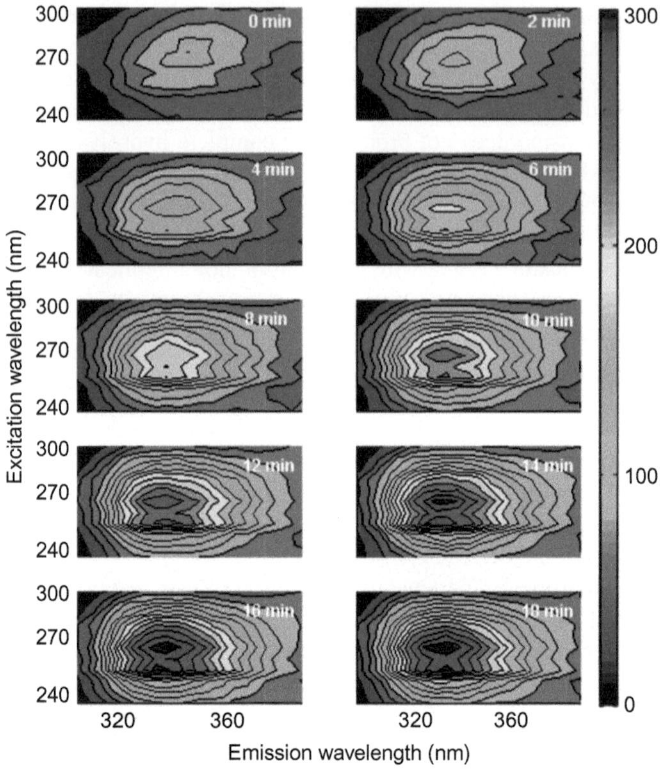


**FIGURE 11** Contour plots of the excitation-emission matrices as a function of time (as indicated) for a solution of PRO (700 ng $mL^{-1}$). The bar on the right shows the fluorescence intensities. *Reprinted with permission from Ref. [50]. Copyright 2007 American Chemical Society.*

did not significantly change with respect to the use of higher number of unexpected components. Application of the N-PLS/RTL methodology to this set led to the retrieval of unexpected serum spectral and time profiles, which can be compared with those expected for background serum. It is interesting to note that the presently applied method is based on a kinetic phenomenon, and therefore, the precise timing of the instrumental measurements may be crucial for the success of a given algorithm. The following causes may lead to multilinearity losses: (1) the time needed to record the complete excitation-emission matrix of a sample in comparison with the reaction progress and (2) the fact that the analytes are the reagent and product of the hydrolysis reaction. Concerning the specific prediction results, better figures of merit were obtained with N-PLS/RTL in comparison with those furnished with PARAFAC, because of the more flexible internal structure of N-PLS/RTL, which makes it a more appropriate approach for processing four-way data, which are not strictly quadrilinear.

### 3.2 Decay Time-EEFM Third-Order Data

In this way, for obtaining third-order data using luminescence spectroscopy, several different approaches can be found in the literature. The first one is based on EEFM-fluorescence lifetime data: the first publication dates back to as early as 1990 [51], by incorporating fluorescence lifetime as a third mode of information to the excitation-emission matrix (EEFM), using the phase modulation technique. The approach was applied to the resolution of a two-constituent mixture composed of benzo[*b*]fluoranthene and benzo[*k*]fluoranthene. A second approach involves laser-excited time-resolved EEFM-fluorescence data, as in the collection of fluorescence time-resolved excitation-emission data at 4.2 K via laser-excited time-resolved Shpol′skii spectroscopy, which has been recently proposed for the resolution of mixtures of 15 PAHs [52]. These data could be successfully processed using U-PLS/RTL and more information about the procedure can be found in Chapter 12.

### 3.3 HPLC-EEFM Third-Order Data

Bro was the first to explore the possibility of obtaining four-way data in unidimensional chromatography by recording full-fluorescence EEFMs [53]. In another report, the obtaining of four-way LC-DAD-kinetic data allowed to follow the hydrolysis of Aly pesticide [54]. Recently, two different approaches have been reported to obtain four-way data by recording the complete EEFMs of the mixture of interest at each elution time in a chromatographic run. In the first approach, the method was used for the analysis of chlorophylls and pheophytins in olive oil samples [55], and in the second approach, the quantitation of fluoroquinolones in water samples was proposed [56].

In the first approach [55], each sample was injected into the chromatograph eight times, each time exciting with a different excitation wavelength, and the emission spectra were recorded along the full chromatogram. The third-order data thus obtained were joined into a four-way data array, which was subsequently analyzed with PARAFAC, U-PLS/RTL, and N-PLS/RTL. The best algorithm to perform the multiway calibration was U-PLS/RTL.

In Figure 12, the excitation-emission matrices are shown for different elution times, as extracted from the third-order sample data. Contour plots are shown for the complete landscape of fluorescence intensity as a function of excitation and emission wavelengths, at 10 different elution times, for a sample containing chlorophylls (chl) and pheophytins (phe). It can be appreciated that a significant overlap occurs between them. The elution order is as follows: chl *b* elutes between 0.36 and 0.45 min, chl *a* at 0.54 min, phe *b* between 0.72 and 0.81 min, and phe *a* at 0.90 min. The profiles of emission spectra were very similar for the four constituents, and their maxima are very close, being 672 nm for chl *a*, 663 nm for chl *b*, 671 nm for phe *a*, and 658 nm for phe *b*. The largest differences correspond to the excitation spectra.

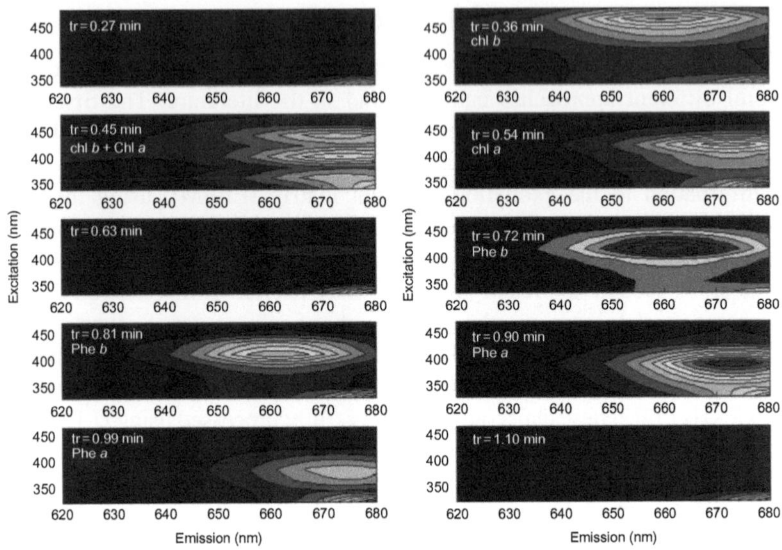


**FIGURE 12** Contour plots of the EEFMs obtained for different elution times (tr) to illustrate the chromatographic evolution of chl *a*, chl *b*, phe *a*, and phe *b* of a sample containing 1.00 μg $mL^{-1}$ of each constituent. Fluorescence intensity has been coded in colors, with deep blue (dark gray in the print version) indicating the lowest level and deep red (dark gray in the print version) the largest one. The 10 contour plots correspond to 10 different elution times selected through the chromatogram, from 0.27 to 1.1 min. *Reprinted with permission from Ref. [55]. Copyright 2013 Elsevier.*

The results with PARAFAC were not good, due to the fact that this algorithm requires that the data show the property of quadrilinearity, which can be lost if chromatographic elution times are not exactly reproducible. Another challenge in this system is that the spectral modes show a large degree of overlapping, as with the emission spectra of these four analytes.

The method was applied to the determination of chlorophylls *a* and *b* and pheophytins *a* and *b* in olive oil samples, in the presence of other interfering fluorescent constituents and without previous sample treatment. The experimental data can be easily obtained and in a short time with an ultrafast chromatographic system operating in gradient mode, and with an automatic injector. It was demonstrated that better results can be obtained by using the complete EEFM of the sample across the chromatogram instead of the emission spectra at a fixed excitation wavelength.

In the second approach [56], a method for the quantitation of fluoroquinolones in water samples, based on sample preconcentration with SPE, and generation of third-order chromatographic data was developed. For this purpose, a Zorbax Eclipse XDB-C18 column was used, and the column temperature was controlled by setting the oven temperature at 35 °C. The mobile phase consisted in a 10 mmol $L^{-1}$ acetic acid buffer (pH 4.0)–acetonitrile–methanol mixture (71:8:20), the flow rate was maintained at 2.00 mL min$^{-1}$, and the complete analysis was carried out in 2 min.

The procedure for generating the data was carried out collecting chromatographic fractions every 2 s in an ELISA plate of 96 wells. The collection started 47 s of chromatographic run, and it was performed during 34 s, thus 17 fractions were collected for each chromatogram. The ELISA plate was set at the end of the chromatograph and operated with a homemade device which automatically moves the plate, allowing the collection of 60 μL in every well. Then, using a fluorimeter equipped with a plate reader accessory coupled to an optical filter, the complete EEFMs were recorded for each fraction.

EEFMs were collected by varying the emission wavelength between 380 and 500 nm each 5 nm and registering the excitation spectra from 260 to 340 nm each 5 nm. Hence, the size of each data matrix was $17 \times 25$. In Figure 13, 12 EEFMs registered for chromatographic fractions collected between 47 and 70 s, for a sample containing 88.3 μg mL$^{-1}$ of ofloxacin and 132.4 μg mL$^{-1}$ of ciprofloxacin is presented. A visual inspection of this figure makes evident the large amount of information that can be obtained by using the described data generating procedure.

The data were modeled with three algorithms: PARAFAC, U-PLS/RTL, and MCR-ALS, the latter conveniently adapted to third-order data. MCR-ALS furnished the better results, when the quadrilinearity was lost due to the occurrence of temporal shifts. It is interesting to note that by obtaining the EEFM at every chromatographic time, more information will be available compared with the acquisition of an emission spectrum at one excitation wavelength (usually selected as a compromise value between the optimal wavelengths for each analyte). Thus, optimal signal for every analyte can be gathered, exciting at all possible different wavelengths across the excitation spectrum.

## 4 FOURTH-ORDER DATA

### 4.1 Kinetic-pH-EEFM Fourth-Order Data

Five-way data can be obtained by joining fourth-order data for a sample set. A procedure has been reported by following the kinetic evolution of EEFM-pH data [57]. The algorithm U-PLS/RQL (residual quadrilinearization) has been developed as a new latent-structured algorithm for the processing of these instrumental data. In order to check its analytical predictive ability, fluorescence excitation-emission-kinetic-pH data were measured and processed [57]. The concentration of the fluorescent pesticide carbaryl was determined in the presence of uncalibrated interferents, in the first reported example of fourth-order multivariate calibration. The hydrolysis of the analyte was followed at different pH values using a fast-scanning spectrofluorimeter and recording the EEFMs during its evolution to produce 1-naphthol, which does also emit fluorescence (Figure 14) [58].

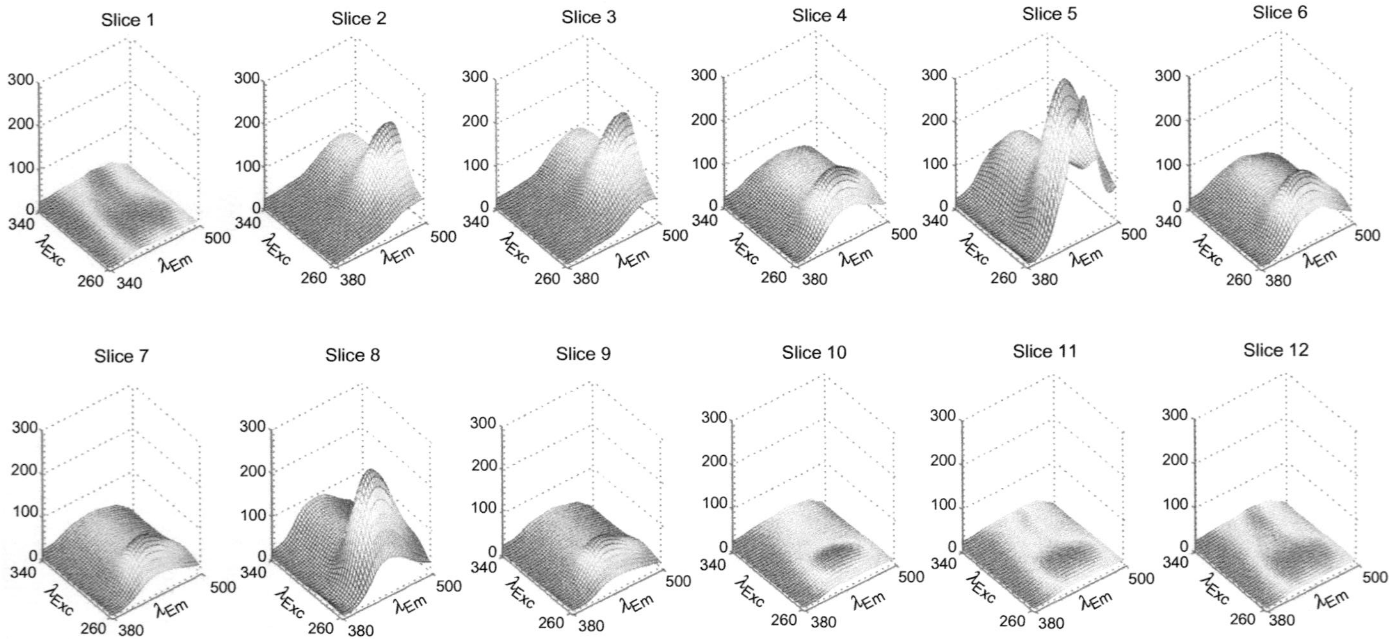


**FIGURE 13** Twelve EEFMs registered for chromatographic fractions between 47 and 70 s for a sample containing 84.3 μg mL$^{-1}$ ofloxacin and 132.4 μg mL$^{-1}$ ciprofloxacin. *Reprinted with permission from Ref. [56]. Copyright 2013 Elsevier.*

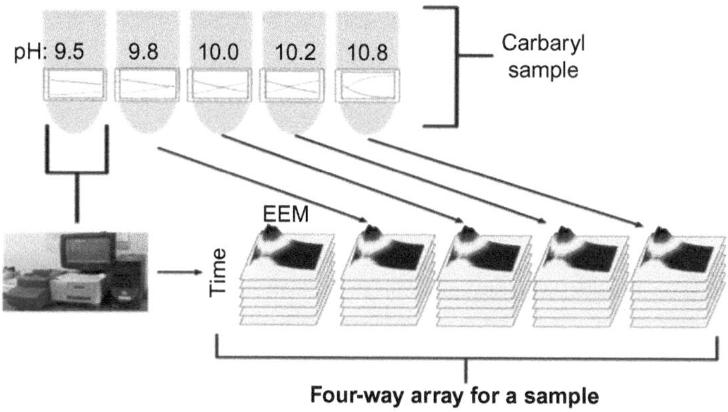


**FIGURE 14** Building of a four-way data array for a single sample of carbaryl. *Reprinted with permission from Ref. [58]. Copyright 2014 Elsevier.*

# 5 CONCLUSIONS

Multiway data, when combined with advanced multiway chemometric modeling methods, allow the successful determination of the analytes of interest in complex samples, as shown in the practical applications reported in this chapter, in the bioanalytical, food, and environmental fields. Particularly, unfolded and multiway partial least-squares algorithms, that are genuine higher-order latent variable methods, once combined with the residual multilinearization procedure, achieve the second-order advantage needed to successfully quantify the chosen analytes. The more flexible internal structure of these algorithms, in comparison with other more rigid methodologies, makes them more appropriate approaches for processing multiway data, which are not strictly multilinear. The use of multiway data, as the presently reported second-, third-, and fourth-order data, opens new strategies for resolving analytical problems in complex samples.

# ACKNOWLEDGMENTS

Universidad Nacional de Rosario, Universidad Nacional del Litoral, CONICET (Consejo Nacional de Investigaciones Científicas y Técnicas), ANPCyT (Agencia Nacional de Promoción Científica y Tecnológica, Projects No. PICT-2010-0084 and No. PICT-2011-0005) and Ministerio de Economía y Competitividad of Spain (Project CTQ2014-52309) and Gobierno de Extremadura (Project GR15-Research Group FQM003), both cofinanced by the European FEDER funds, are gratefully acknowledged for financial support.

# REFERENCES

[1] Olivieri AC, Escandar GM. Practical three-way calibration. 1st ed. Waltham, MA: Elsevier; 2014.

[2] Thygesen LG, Rinnan Å, Barsberg S, Møller JKS. Stabilizing the PARAFAC decomposition of fluorescence spectra by insertion of zeros outside the data area. Chemometr Intell Lab Syst 2004;71:97–106.

[3] Bohoyo Gil D, Muñoz de la Peña A, Arancibia JA, Escandar GM, Olivieri AC. Second-order advantage achieved by unfolded-partial least-squares/residual bilinearization modeling of excitation-emission fluorescence data presenting inner filter effects. Anal Chem 2006;78:8051–8.

[4] Culzoni MJ, Goicoechea HC, Pagani AP, Cabezón MA, Olivieri AC. Evaluation of partial least-squares with second-order advantage for the multi-way spectroscopic analysis of complex biological samples in the presence of analyte–background interactions. Analyst 2006;131:718–23.

[5] Vatsavai K, Goicoechea HC, Campiglia AD. Direct quantification of monohydroxy-polycyclic aromatic hydrocarbons in synthetic urine samples via solid-phase extraction–room-temperature fluorescence excitation–emission matrix spectroscopy. Anal Biochem 2008;376:213–20.

[6] Bortolato SA, Arancibia JA, Escandar GM. Chemometrics-assisted excitation-emission fluorescence spectroscopy on nylon membranes. Simultaneous determination of benzo[a]pyrene and dibenzo[a,h]anthracene at parts-per-trillion levels in the presence of the remaining EPA PAH priority pollutants as interferences. Anal Chem 2008;80:8276–86.

[7] Culzoni MJ, Damiani PC, García-Reiriz A, Goicoechea HC, Olivieri AC. Experimental study of non-linear second-order analytical data with focus on the second-order advantage. Analyst 2007;132:654–63.

[8] García-Reiriz A, Damiani PC, Olivieri AC. Different strategies for the direct determination of amoxicillin in human urine by second-order multivariate analysis of kinetic–spectrophotometric data. Talanta 2007;71:806–15.

[9] Schenone AV, Culzoni MJ, Marsili NR, Goicoechea HC. Determination of tartrazine in beverage samples by stopped-flow analysis and three-way multivariate calibration of non-linear kinetic-spectrophotometric data. Food Chem 2013;138:1928–35.

[10] Alcaráz MR, Schenone AV, Culzoni MJ, Goicoechea HC. Modeling of second-order spectrophotometric data generated by a pH-gradient flow injection technique for the determination of doxorubicin in human plasma. Microchem J 2014;112:25–33.

[11] Goicoechea HC, Olivieri AC. A new robust bilinear least-squares method for the analysis of spectral-pH matrix data. Appl Spectrosc 2005;59:926–33.

[12] Marsili NR, Lista A, Fernandez Band BS, Goicoechea HC, Olivieri AC. Evaluation of complex spectral-pH three-way arrays by modified bilinear least-squares: determination of four different dyes in interfering systems. Analyst 2005;130:1291–8.

[13] Ghasemi J, Niazi A. Two- and three-way chemometrics methods applied for spectrophotometric determination of lorazepam in pharmaceutical formulations and biological fluids. Anal Chim Acta 2005;533:169–77.

[14] Gómez V, Callao MP. Analytical applications of second-order calibration methods. Anal Chim Acta 2008;627:169–83.

[15] Olivieri AC. On a versatile second-order multivariate calibration method based on partial least-squares and residual bilinearization: second-order advantage and precision properties. J Chemom 2005;19:253–65.

[16] Dong MW. Modern HPLC for practicing scientists. Wareham, MA: Wiley, Synomics Pharmaceutical Services; 2006.

[17] Daszykowski M, Walczak B. Use and abuse of chemometrics in chromatography. Trends Anal Chem 2006;25:1081–96.

[18] Alcaráz MR, Vera-Candioti L, Culzoni MJ, Goicoechea HC. Ultrafast quantitation of six quinolones in water samples by second-order capillary electrophoresis data modeling with multivariate curve resolution–alternating least squares. Anal Bioanal Chem 2014;406:2571–80.

[19] Amigo JM, Skov T, Bro R. ChroMATHography: solving chromatographic issues with mathematical models and intuitive graphics. Chem Rev 2010;110:4582–605.
[20] van Mispelaar VG. Chromametrics. PhD Thesis, The Netherlands: University of Amsterdam; 2005.
[21] Synovec RE, editor. Enhancing chemical separations with chemometric data analysis. In: Talanta 2011;83:1067–316.
[22] Araujo P, Grung B, editors. Chemometrics in chromatography. In: J Chromatogr B 2012;910:1–172.
[23] Goicoechea HC, Culzoni MJ, Gil García MD, Martínez Galera M. Chemometric strategies for enhancing the chromatographic methodologies with second-order data analysis of compounds when peaks are overlapped. Talanta 2011;83:1098–107.
[24] Bloemberg TG, Gerretzen J, Lunshof A, Wehrens R, Buydens LMC. Warping methods for spectroscopic and chromatographic signal alignment: a tutorial. Anal Chim Acta 2013;781:14–32.
[25] Gil García MD, Culzoni MJ, De Zan MM, Valverde R, Goicoechea HC, Martinez Galera M. Solving matrix-effects exploiting the second order advantage in the resolution and determination of eight tetracycline antibiotics in effluent wastewater by modelling liquid chromatography data with multivariate curve resolution-alternating least squares and unfolded-partial least squares followed by residual bilinearization algorithms. II. Prediction and figures of merit. J Chromatogr A 2008;1179:115–24.
[26] Khoobi A, Ghoreishi SM, Masoum S, Behpour M. Multivariate curve resolution-alternating least squares assisted by voltammetry for simultaneous determination of betaxolol and atenolol using carbon nanotube paste electrode. Bioelectrochemistry 2013;94:100–7.
[27] Galeano-Díaz T, Guiberteau-Cabanillas A, Espinosa-Mansilla A, López-Soto MD. Adsorptive stripping square wave voltammetry (Ad-SSWV) accomplished with second-order multivariate calibration determination of fenitrothion and its metabolites in river water samples. Anal Chim Acta 2008;618:131–9.
[28] Wehry EL. In: Guilbault GG, editor. Practical fluorescence. New York: Marcel Dekker; 1990.
[29] Patra D, Mishra AK. Recent developments in multi-component synchronous fluorescence scan analysis. Trends Anal Chem 2002;21:787–98.
[30] Kumar K, Mishra AK. Application of 'multivariate curve resolution alternating least square (MCR-ALS)' analysis to extract pure component synchronous fluorescence spectra at various wavelength offsets from total synchronous fluorescence spectroscopy (TSFS) data set of dilute aqueous solutions of fluorophores. Chemometr Intell Lab Syst 2012;116:78–86.
[31] Kumar K, Mishra AK. Simultaneous quantification of dilute aqueous solutions of certain polycyclic aromatic hydrocarbons (PAHs) with significant fluorescent spectral overlap using total synchronous fluorescence spectroscopy (TSFS) and N-PLS, unfolded-PLS and MCR-ALS analysis. Anal Methods 2011;3:2616–24.
[32] Kumar K, Mishra AK. Application of parallel factor analysis to total synchronous fluorescence spectrum of dilute multifluorophoric solutions: addressing the issue of lack of trilinearity in total synchronous fluorescence data set. Anal Chim Acta 2012;755:37–45.
[33] Schenone AV, Culzoni MJ, Campiglia AD, Goicoechea HC. Total synchronous fluorescence spectroscopic data modeled with first- and second-order algorithms for the determination of doxorubicin in human plasma. Anal Bioanal Chem 2013;405:8515–23.
[34] Calimag-Williams K, Knobel G, Goicoechea HC, Campiglia AD. Achieving second order advantage with multi-way partial least squares and residual bi-linearization with total synchronous fluorescence data of monohydroxy- polycyclic aromatic hydrocarbons in urine samples. Anal Chim Acta 2014;811:60–9.

[35] Zampronio CG, Gurden SP, Moraes LAB, Eberlin MN, Smilde AK, Poppi RJ. Direct sampling tandem mass spectrometry (MS/MS) and multiway calibration for isomer quantitation. Analyst 2002;127:1054–60.
[36] Booksh KS, Kowalski BR. Theory of analytical chemistry. Anal Chem 1994;66:782A–791A.
[37] Olivieri AC. Analytical advantages of multivariate data processing. One, two, three, infinity? Anal Chem 2008;80:5713–20.
[38] Wu HL, Li Y, Yu RQ. Recent developments of chemical multiway calibration methodologies with second-order or higher-order advantages. J Chemom 2014;28:476–89.
[39] Olivieri AC, Wu HL, Yu RQ. MVC3: a MATLAB graphical interface toolbox for third-order multivariate calibration. Chemometr Intell Lab Syst 2012;116:9–16.
[40] Tan Y, Jiang JH, Wu HL, Cui H, Yu RQ. Resolution of kinetic system of simultaneous degradations of chlorophyll a and b by PARAFAC. Anal Chim Acta 2000;412:195–202.
[41] Nikolajsen RPH, Booksh KS, Hansen AM, Bro R. Quantifying catecholamines using multi-way kinetic modelling. Anal Chim Acta 2003;475:137–50.
[42] Muroski AR, Booksh KS, Myrick ML. Single-measurement excitation/emission matrix spectrofluorometer for determination of hydrocarbons in ocean water. 1. Instrumentation and background correction. Anal Chem 1996;68:3534–8.
[43] Olivieri AC, Arancibia JA, Muñoz de la Peña A, Durán Merás I, Espinosa Mansilla A. Second-order advantage achieved with four-way fluorescence excitation-emission-kinetic data processed by parallel factor analysis and trilinear least-squares. Determination of methotrexate and leucovorin in human urine. Anal Chem 2004;76:5657–66.
[44] Arancibia JA, Olivieri AC, Bohoyo Gil D, Espinosa Mansilla A, Durán Merás I, Muñoz de la Peña A. Trilinear least-squares and unfolded-PLS coupled to residual trilinearization: new chemometric tools for the data analysis of four-way instrumental data. Chemom Intell Lab Syst 2006;80:77–86.
[45] Muñoz de la Peña A, Durán Merás I, Jiménez Girón A. Four-way calibration applied to the simultaneous determination of folic acid and methotrexate in urine samples. Anal Bioanal Chem 2006;385:1289–97.
[46] Muñoz de la Peña A, Durán Merás I, Jimenez Girón A, Goicoechea HC. Evaluation of unfolded-partial least-squares coupled to residual trilinearization for four-way calibration of folic acid and methotrexate in human serum samples. Talanta 2007;72:1261–8.
[47] Jiménez Girón A, Durán Merás I, Espinosa Mansilla A, Muñoz de la Peña A, Cañada Cañada F, Olivieri AC. On line photochemically induced excitation-emission-kinetic four-way data. Analytical application for the determination of folic acid and its two main metabolites in serum by U-PLS and N-PLS/residual trilinearization (RTL) calibration. Anal Chim Acta 2008;622:94–103.
[48] García Reiriz AG, Damiani PC, Olivieri AC, Cañada Cañada F, Muñoz de la Peña A. Non-linear four-way kinetic-excitation-emission fluorescence data processed by a variant of parallel factor analysis and by a neural network model achieving the second-order advantage: malonaldehyde determination in olive oil samples. Anal Chem 2008;80:7248–56.
[49] Maggio RM, Damiani PC, Olivieri AC. Four-way kinetic-excitation-emission fluorescence data processed by multi-way algorithms. Determination of carbaryl and 1-naphthol in water samples in the presence of fluorescent interferents. Anal Chim Acta 2010;677:97–107.
[50] Damiani PC, Durán-Merás I, Garcia-Reiriz A, Jimenez Girón A, Muñoz de la Peña A, Olivieri AC. Multiway partial least-squares coupled to residual trilinearization: a genuine multidimensional tool for the study of third-order data. Simultaneous analysis of procaine and its metabolite p-aminobenzoic acid in equine serum. Anal Chem 2007;79:6949–58.

[51] Burdick DS, Tu XM, McGown LB, Millican DW. Resolution of multicomponent fluorescent mixtures by analysis of the excitation-emission-frequency array. J Chemom 1990;4:15–28.

[52] Goicoechea HC, Yu S, Moore AFT, Campligia AD. Four-way modeling of 4.2 time-resolved excitation-emission fluorescence data for the quantitation of polycyclic aromatic hydrocarbons in soil samples. Talanta 2012;101:330–6.

[53] Bro R. Multi-way analysis in the food industry. Models, algorithms and applications. Doctoral Thesis, The Netherlands: University of Amsterdam; 1998, p. 253.

[54] Bezemer E, Rutan SC. Analysis of three- and four-way data using multivariate curve resolution-alternating least squares with global multi-way kinetic fitting. Chemometr Intell Lab Syst 2006;81:82–93.

[55] Lozano VA, Muñoz de la Peña A, Durán-Merás I, Espinosa Mansilla A, Escandar GM. Four-way multivariate calibration using ultra-fast high-performance liquid chromatography with fluorescence excitation-emission detection. Application to the direct analysis of chlorophylls a and b and pheophytins a and b in olive oils. Chemometr Intell Lab Syst 2013;125:121–31.

[56] Alcaráz MR, Siano GG, Culzoni MJ, Muñoz de la Peña A, Goicoechea HC. Modelling four and three-way fast high-performance liquid chromatography with fluorescence detection data for quantitation of fluoroquinolones in water samples. Anal Chim Acta 2014;809(2014):37–46.

[57] Maggio RM, Muñoz de la Peña A, Olivieri AC. Unfolded partial least-squares with residual quadrilinearization: a new multivariate algorithm for processing five-way data achieving the second-order advantage. Application to fourth-order excitation-emission-kinetic-pH fluorescence analytical data. Chemometr Intell Lab Syst 2011;109:178–85.

[58] Escandar GM, Goicoechea HC, Muñoz de la Peña A, Olivieri AC. Second- and higher-order data generation and calibration: a tutorial. Anal Chim Acta 2014;806:8–26.

Chapter 9

# Simultaneous Determination of Organic Dyes Using Second-Order Data

M. Pilar Callao[1] and M. Soledad Larrechi
*Department of Analytical and Organic Chemistry, Rovira i Virgili University, Tarragona, Spain*
*[1]Corresponding author: e-mail: mariapilar.callao@urv.cat*

## Chapter Outline



## 1 INTRODUCTION

Organic dyes are used in a wide range of industrial applications in sectors such as textiles (the most significant), food products, cosmetics, pharmaceuticals, or paper printing. Several groups of dyes can be distinguished on the basis of their compound structure: for example, phtalocyanins, anthraquinones, quinone-imines, and xanthenes. The most commonly used dyes in industrial applications are azo dyes because they are cost effective to synthesize. Dyes are hazardous to the environment because many are toxic to living organisms, either directly or through the absorption and reflection of sunlight entering the water, which interferes with the growth of aquatic organisms.

Sometimes a particular color can only be obtained by mixing two or more dyes, so analytical methods need to be able to determine multiple components. Chromatographic techniques are the most commonly used for analyzing

**Data Handling in Science and Technology, Vol. 29. http://dx.doi.org/10.1016/B978-0-444-63527-3.00009-6**

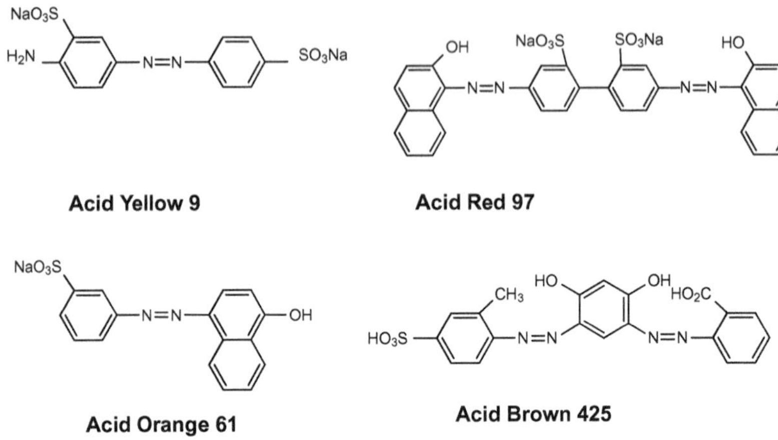


**FIGURE 1** The azo dyes studied.

multiple species. The main disadvantage is that they often require the sample to be pretreated so they are both costly and time consuming.

Another option is to use analytical methods based on spectroscopic measurements and chemometric data treatment. Spectroscopic techniques respond to numerous organic compounds and rarely require the sample to be pretreated. However, they are not selective. Organic dyes and most of their precursors and degradation products respond to UV–visible absorption, fluorescence, Raman and nuclear magnetic resonance (NMR) spectroscopies. However, the signals from the various dyes present in the sample, all probably of a similar chemical nature, and the degradation products that they may give rise to are expected to overlap to quite a considerable extent. In this context, multiway analysis or second-order data-treatment techniques can provide information about concentration and spectra profiles of the species of interest by recovering the pure analyte response from the analysis of a gross signal that contains other interference.

In this chapter, we shall discuss a set of case studies that are of environmental interest. All the cases presented aimed to illustrate the possibilities of multiway data analysis techniques in an area that is arousing considerable interest in the scientific community (more than 7000 scientific articles published since 2005, according to Scopus sources 11–2014) in the hope that the procedures and results can be extrapolated to the study of other analytes and/or similar areas. The analytes that have been used in the case studies are four azo dyes, the structures of which are presented in Figure 1.

## 2 STRATEGIES FOR OBTAINING SECOND-ORDER DATA

Experimental measurements generated by analytical instruments are classified as second-order data, when one data matrix is obtained per sample. In the applications discussed in this chapter, the two modes are: (a) mode 1:

UV–visible absorption spectra and proton nuclear magnetic resonance ($^1$H-NMR) and (b) mode 2: measurements over time of degradation processes or in a sequential injection analyzer.

## 2.1 UV–Visible (UV–vis) Absorption Spectroscopy

UV–vis spectra are obtained by measuring the absorbance of monochromatic radiation over a range of wavelengths passing through a solution in a cell. Following Lambert Beer's law, for a given system there is a linear relationship between the absorbance and the concentration of the species that are sensitive to the radiation [1].

Organic azo dyes are chemical compounds that contain the functional group RN=N—R′ where R and R′ are aryl groups. Because of electron delocalization through the N=N and the aryl groups, these compounds are naturally colored. So, the visible spectral region (400–700 nm) is highly appropriate for analysis. The visible range is not the only range of interest, however, because information about other species (e.g., intermediates in dye degradation processes) can also be obtained from the UV range (190–400 nm).

Other features of this technique that should also be taken into account are its reasonable instrumental and maintenance costs, its sensitivity (appropriate for the type of analyte), and its high reproducibility, robustness, and safety.

## 2.2 Proton-Nuclear Magnetic Resonance ($^1$H-NMR)

NMR is based on the behavior of the nuclei of atoms with spin (H in the case of $^1$H-NMR) under the action of an external magnetic field $B_0$. These nuclei behave like magnets, aligning in the same direction as $B_0$ (lower energy state) or in the opposite direction (higher energy state). The difference in energy between the two states ($\Delta E$) is the radiofrequency region. When the external energy applied coincides with the $\Delta E$ between the two states, they switch from one to the other (resonance), which generates a specific profile (spectrum) of the sample studied.

NMR spectroscopy has largely been used to obtain structural information about a compound from the number, position, intensity, and multiplicity of signals [2]. Qualitative and quantitative information can also be obtained by regarding the spectrum as a data vector containing a number of variables (each one of which corresponds to a shift) with a particular value (intensity).

Breakthroughs in NMR technology have led to measurements with increased sensitivity, resolution, and reproducibility; so, the resulting data are increasingly of higher quality. It is, therefore, an efficient method for treating multiway data. It is also a nondestructive technique; it does not require the sample to be treated; it is fast; and it is very suitable for identifying unstable compounds.

## 2.3 Photodegradation

Photodegradation is the process by which the absorption of photons—particularly those with wavelengths in the UV–visible spectrum—causes a molecule to degrade. During the process, the species evolve and turn into new ones. There are three types of photodegradation:

*Photolysis* is a process that uses radiation with ultraviolet (UV) light to generate such reactive species as radicals, ions, and excited molecules [3,4].

The effect of radiation on a compound depends on the nature of the compound and the amount of energy that is transferred by the radiation. This process is intensified in aqueous solutions by the primary products formed from the photolysis of water, which produces electrons ($e^-_{aq}$), hydroxyl radicals ($^{\cdot}OH$), and hydrogen radicals ($^{\cdot}H$) as intermediate species that decompose the solute.

*Photooxidation processes* combine oxidants such as $O_3$, $H_2O_2$, or Fenton reagent with UV light so that the degradation is more effective than UV alone. The hydroxyl radicals and perhydroxyl radicals ($^{\cdot}HO_2$) produced in subsequent reactions are the main oxidizing species.

*Photocatalysis* consists of a photoinduced reaction which is accelerated by the presence of a catalyst. The process starts when a photon with sufficient energy (equal to or higher than the band gap energy of the catalyst) is absorbed. This absorption leads to charge separation because an electron ($e^-$) is promoted from the valence band of the catalyst to the conduction band, thus generating a hole (h+). This electron–hole pair can migrate to the catalyst surface where it can enter into a redox reaction with other species present on the surface.

Several semiconductors (e.g., $TiO_2$, ZnO, $Fe_2O_3$, CdS, and ZnS) can act as catalysts due to the electronic structure of their metal atoms, which is characterized by a filled valence band and an empty conduction band separated by an energy gap or a band gap. The semiconductor that is most widely used as a catalyst is $TiO_2$ because it is chemically and biologically inert, photocatalytically stable, relatively easy to produce and use, and catalyzes reactions efficiently. It is also cheap and does not pose any risk to the environment or humans [5]. ZnO is also widely used.

The main problem of using heterogeneous catalysts is the undesired electron–hole recombination which, in the absence of a proper electron acceptor (or donor), is extremely efficient. One strategy for inhibiting electron–hole recombination is to add irreversible electron acceptors to the reaction system. In this context, the addition of $H_2O_2$ is beneficial, but it can also become a scavenger of valence band holes and $^{\cdot}OH$ when present at high concentrations.

Figure 2 shows a photoreactor like the one that has been used in some of the examples in this chapter.

## 2.4 Sequential Injection Analysis [6]

Figure 3 shows the scheme of a basic sequential injection analysis (SIA) system. During the analytical sequence, an automatic pump or a syringe is used to aspirate the carrier, reagents, and sample into the holding coil, and then push them into the reaction coil and toward the detector. A computer controls the functions of each component.

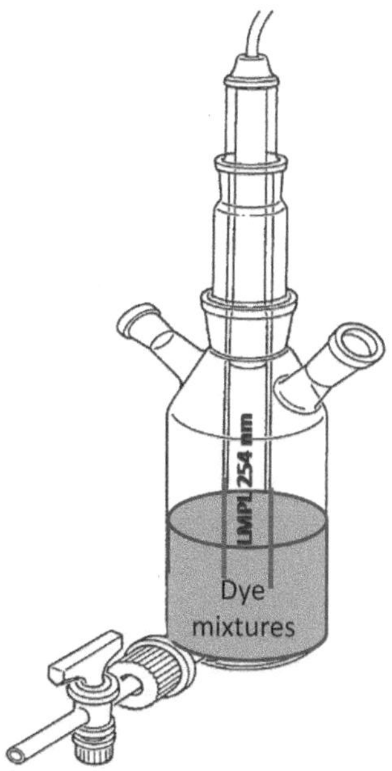


FIGURE 2 Cylindrical annular batch photoreactor.

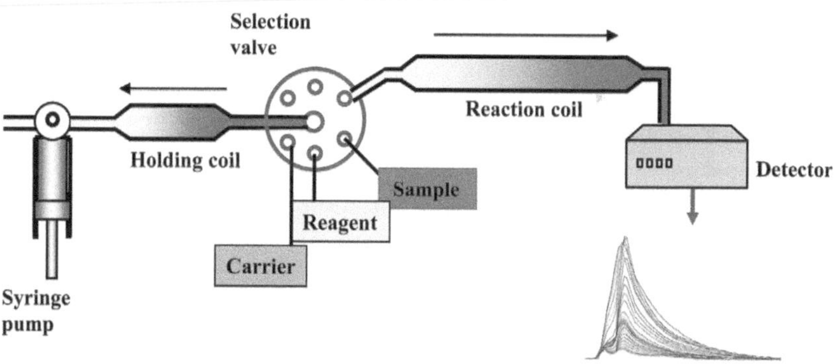


FIGURE 3 Scheme of a sequential injection analyzer.

During the process, reagents and sample are mixed by dispersion. The type of data provided by SIA (scalar, vector, or matrix) depends on the type of detector used and whether the reagents totally or partially interact with the analyte before passing through the detector. If the reagent only interacts partially, a diffusive mixture of reagent, analyte, and product will be observed as the solution passes though the detector. In these cases, if the detector can provide a data vector, as diode array detectors can, a data matrix is obtained for each sample. This means that various multiway tools can be used to provide chemical information about the system.

## 3 CHEMOMETRIC BACKGROUND

When a sample of dyes is analyzed using one of the instrumental configurations described in Section 2, the data obtained can be set out in the form of a matrix **X** ($J \times K$) where $J$ is the number of measurements made and $K$ is the number of variables associated with the instrumental measurement made. The two modes represent the variation in the composition of the sample and the variation in the instrumental response, respectively. The data in **X** are bilinear, so the total variability of the data collected can be expressed as an additive model of the individual contributions.

The data in matrix **X** can be subject to multivariate analysis to evaluate the components that contribute to the variability throughout the process studied. Curve resolution is an effective approach to this end because all curve resolution methods [7–11] mathematically decompose matrix **X** into the product of two matrices **C** and $\mathbf{S}^{\mathrm{T}}$ according to Equation (1)

$$\mathbf{X} = \mathbf{C}\mathbf{S}^{\mathrm{T}} + \mathbf{E} \tag{1}$$

where **C** ($J \times N$) is the matrix of the concentration profiles of the $N$ pure components sensitive to the instrumental signal and $\mathbf{S}^{\mathrm{T}}$ ($N \times K$) are the pure profiles of their instrumental responses. **E** ($J \times K$) is the error matrix that contains the variability of the data that is not related to a chemical contribution. Figure 4 provides an example of the degradation of a dye in which data matrix **X**, which contains the visible UV spectra obtained during the process, has been decomposed into two matrices: matrix **C**, which contains the information about the evolution of the dye's concentration and an intermediate product that is generated; and matrix $\mathbf{S}^{\mathrm{T}}$, which corresponds to the spectra of both species.

A wide variety of methods and algorithms can be used to decompose matrix **X** according to Equation (1). Their basic features have been discussed in a recently compiled bibliography [12]. The applications presented here have used the method of curve resolution based on alternating least squares (MCR-ALS).

MCR-ALS has been discussed in depth by Tauler's group [13–15] and its theoretical aspects have been dealt with in full in this book and in the reference provided [12]. This section will merely introduce the most basic aspects and the concepts that will be mentioned during the discussion of the practical cases.

MCR-ALS estimates the number of $N$ components in the data matrix **X**, selects an initial estimate either from the matrix of concentrations **C** or from the matrix of spectra $\mathbf{S}^{\mathrm{T}}$, and sets the restrictions that are to be applied in the iteration process until the optimal solution is found.

The number of components ($N$) that contribute to the variability of the data collected in the matrix **X** needs to be estimated before the resolution and optimization process can be initiated. However, during the resolution

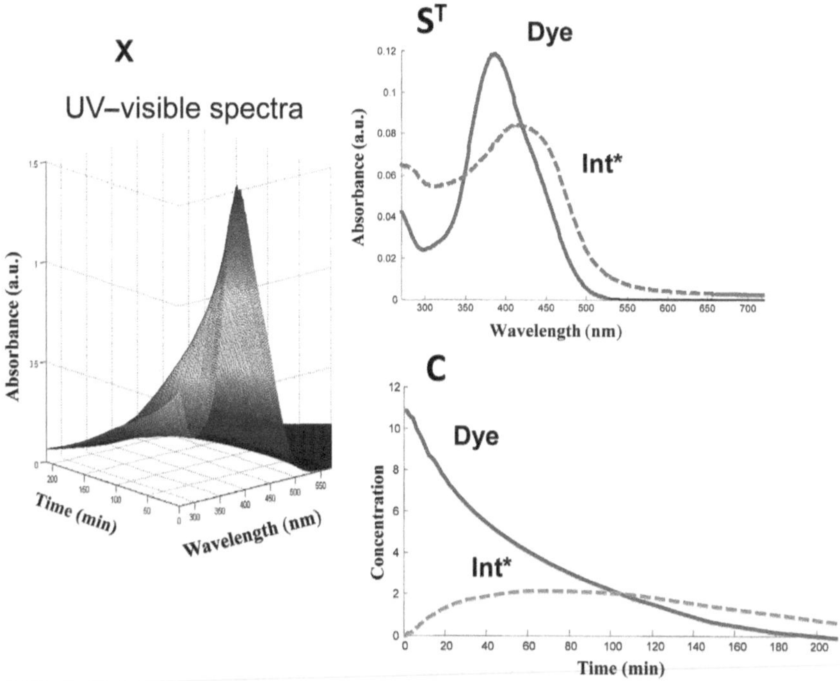


**FIGURE 4** Schematic representation of the MCR-ALS resolution of a spectral data matrix **X** according to equation $\mathbf{X}=\mathbf{C}\mathbf{S}^{\mathrm{T}}+\mathbf{E}$.

process this initial value may need to be reconsidered and the process restarted with a new value for $N$.

An initial estimate of $N$ can be made using principal component analysis (PCA) [9,16,17] of data matrix **X**. The variability in the response matrix **X** is extracted in a few factors or principal components. These factors are linear combinations of the original variables [18]. Mathematically, this analysis can be expressed as the decomposition of matrix **X** into the product of two matrices according to Equation (2)

$$\mathbf{X}=\mathbf{T}\mathbf{P}^{\mathrm{T}}+\mathbf{E} \tag{2}$$

where **T** ($J\times N$, with $N$ being equal to the number of significant components) is the matrix of the scores that have information about the rows in matrix **X**, **P** ($J\times N$) is the matrix of the loadings that contains the information from the columns in matrix **X**, and **E** is the matrix of residuals. Various numerical procedures can be used for PCA decomposition [12,19]. In the examples that we shall discuss, the PCA decomposition of data matrix **X** has been carried out using singular value decomposition (SVD) [19], as expressed in Equation (3)

$$\mathbf{X} = \mathbf{USV}^{\mathrm{T}} + \mathbf{E} \tag{3}$$

where **U** is a matrix with "normalized" scores of **X** as column ($J \times N$) subject to $\mathbf{U}^{\mathrm{T}}\mathbf{U} = \mathbf{I}$; **V** is a matrix with the normalized loadings of **X** as columns ($K \times N$) subject to $\mathbf{V}^{\mathrm{T}}\mathbf{V} = \mathbf{I}$; **S** is a matrix with "singular values" as "diagonal" elements ($N \times N$).

The singular values of matrix **S** are ordered from highest to lowest. They are the square roots of the eigenvalues ($\lambda$) of the decomposition of matrix **X**, and they can be used to extract information about the number of factors or components that contain useful information about the system under study. Those factors that have low associated singular values generally provide information about the noise associated with experimental measures. In this way, the chemical rank of the data matrix [19,20] can be determined: it is the same as the number of significant components of the data set and enables the data matrix to be reconstructed without losing any important information.

Determining how many principal components are required to ensure that there is only noise in matrix **E** is by no means a trivial task in PCA [18,21–24] although various statistical approaches can be used to assess whether the contribution of a new component can be regarded as significant. In some of the practical examples discussed in this chapter, the selection criterion is to retain a number of components for which the singular value is always greater than the second singular value found in the decomposition using the SVD of a matrix **B** ($J \times K$) obtained when monitoring, in the same conditions, a sample that does not contain the species under study. Table 1 displays the singular values obtained in the SVD analysis of a matrix **X** that contains the UV–visible spectra represented in Figure 4 and the values obtained from the SVD analysis of a matrix **B** obtained in the same conditions as **X** for a sample blank. The value associated with the second singular value

**TABLE 1** Singular Values Obtained from SVD of the Photodegradation Data Matrix (X) and Blank Data Matrix (B)

| Factors | B (Blank matrix) | X (Data matrix) |
|---|---|---|
| 1 | 85.488 | 972.621 |
| 2 | 22.051 | 135.322 |
| 3 | 0.3117 | 46.741 |
| 4 | 0.0675 | 18.442 |
| 5 | 0.0173 | 0.6436 |

Extracted from Ref. [43] with permission of Elsevier.

obtained from the SVD analysis of the **B** matrix is considered representative of the noise, so it is concluded that matrix **X** has three sources of variability.

There are a variety of strategies for estimating either matrix **C** or matrix $\mathbf{S}^T$ [12], ranging from applying chemometric techniques to randomly extracting a number of columns or rows from matrix **X** equal to the number of components desired.

In the cases presented, the data are obtained from spectroscopic measurements of a sample that is still evolving, so the concentration profiles of the components involved in the process are expected to decrease and emerge sequentially. Figure 4 displays the profiles of a photodegradation process, for which evolving factor analysis (EFA) [25,26] is particularly appropriate because it shows the factors that emerge or decrease. EFA performs subsequent PCA runs on windows that are gradually enlarged by adding rows in the process direction. EFA runs by moving the window in the direction of the concentrations from the beginning to the end (forward EFA) and then in the opposite direction (backward). EFA plots provide initial estimates of the concentration profiles of the compounds [27].

Another method that is widely used is a method that computes the purest variables [28,29]. This method sequentially selects the initial estimates of $\mathbf{S}^T$ or **C** by assessing the relative standard deviation of the columns (or rows, depending on the choice) in the data matrix:

$$p_i = s_i/m_i + f \tag{4}$$

where $p_i$ is the relative standard deviation, $s_i$ is the standard deviation, and $m_i$ the mean for the column (or row). In an attempt to prevent columns (or rows) from being overvalued with a small mean value (generally corresponding to the noise), a correction factor $f$ is introduced (normally 0.1 or 1 depending on the level of noise accepted).

The column (or row) with the highest relative standard deviation is determined and then normalized. The second column or row selected as well as having the largest relative standard deviation must have the smallest correlation with the first. The procedure then continues until the predetermined $N$ values are found.

The estimates obtained are of $\mathbf{S}^T$ or **C** depending on whether the selection has been made in the direction of the concentrations or in the direction of the responses.

The constraints can be defined as any mathematical or chemical property that is systematically satisfied by the whole system or by some of its pure contributions [13,14,30]. They are imposed on the **C** and/or $\mathbf{S}^T$ matrices during the MCR-ALS optimization step, in order to find an optimal solution with physicochemical meaning.

Typical constraints are:

*Nonnegativity:* this constraint forces all values of a profile to be greater than or equal to zero [31]. It can be applied to all concentration profiles and spectral profiles.

*Closure*: this constraint means that at all times the sum of the concentrations of the species present in the system under study (or some of them) must be equal to a constant [14]. It can be applied to a chemical reaction [30] if all the species involved in the process respond to the instrumental technique being used.

*Unimodality:* this constraint imposes the existence of a single maximum in one or both response profiles.

*Hard-modeling:* this constraint forces the concentration or response profiles to satisfy a mathematical function [32]. In the case of the **X** matrices illustrated in this chapter, it imposes physical–chemical models attributed to the degradation kinetics of dyes [33].

There are other mathematical constraints that are independent of the chemistry. The constraints of this sort are associated with the concept of local rank: for example, the constraint of *selectivity* implies the existence of regions, in one or both profiles, in which only one species is present [15,34].

MCR-ALS can also be used for the simultaneous analysis of more than one matrix [14,15,35]. For example, a matrix $\mathbf{X_1}$ ($J_1 \times K$) can be fused with another matrix $\mathbf{X_2}$ ($J_2 \times K$), if they have the same type of variable. In this case, the fusion would give an augmented matrix $\mathbf{X_{aug}}$ (($J_1 + J_2) \times K$). Using MCR-ALS, the augmented matrix can be decomposed according to the following expression:

$$\mathbf{X_{aug}} = \mathbf{C_{aug}}\mathbf{S}^{\mathrm{T}} + \mathbf{E_{aug}} = [\mathbf{C_1}; \mathbf{C_2}]\mathbf{S}^{\mathrm{T}} + \mathbf{E_{aug}} \tag{5}$$

Another possibility is to fuse $\mathbf{X_1}$ ($J \times K_1$) and $\mathbf{X_2}$ ($J \times K_2$), to give $\mathbf{X_{aug}}$ ($J \times (K_1 + K_2)$). The solution gives:

$$\mathbf{X_{aug}} = \mathbf{C}\mathbf{S}^{\mathrm{T}}_{\mathbf{aug}} + \mathbf{E_{aug}} = \mathbf{C}\left[\mathbf{S}^{\mathrm{T}}_{\mathbf{1}}, \mathbf{S}^{\mathrm{T}}_{\mathbf{2}}\right] + \mathbf{E_{aug}} \tag{6}$$

These strategies have significant advantages for studying complex processes because the constraints applied to the augmented matrix and all the individual matrices can be different.

# 4 PRACTICAL EXAMPLES

## 4.1 Second-Order Calibration

Calibration is fundamental to analytical chemistry. The purpose of calibration is to take measurements of samples that have known concentrations of analytes to establish a function that relates the signals to these concentrations. Once this function has been obtained, the ultimate objective is to determine the amount of analyte of interest in unknown samples that might contain (unknown) interferents. Second-order calibration is used for second-order data (one data matrix per sample).

MCR-ALS can be used to solve an augmented matrix $\mathbf{X_{aug}} = [\mathbf{X_1};\ldots;\mathbf{X_i};\mathbf{X_{unk}}]$ created by column-wise combining *i* data matrices obtained by analyzing standard samples of known concentration with data matrices obtained by

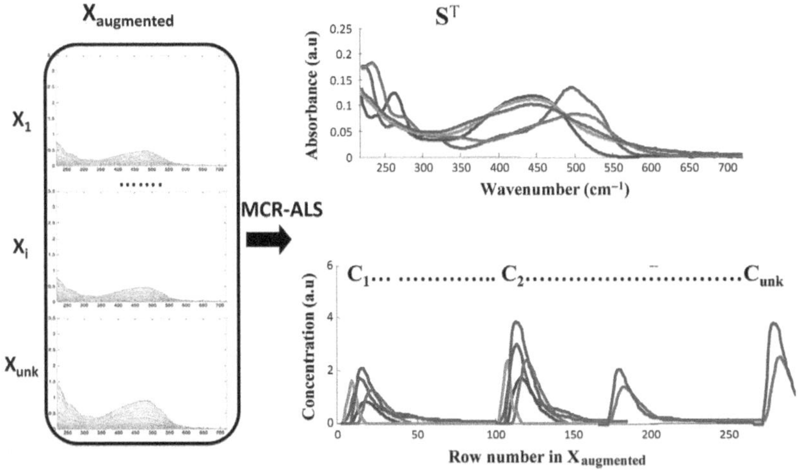


**FIGURE 5** Column-wise augmented data matrix used to perform a second-order calibration.

analyzing samples of unknown concentration to provide the concentration profiles $[\mathbf{C}_1;\ldots;\mathbf{C}_i;\ \mathbf{C}_{unk}]$ and the spectral profiles $\mathbf{S}^T$ of the species in the samples (see Figure 5). From the concentration profiles, the area ($A$) of each standard and the unknown sample can be obtained for each of the resolved species. Using the areas of the standards ($A$) and their concentration ($c$) a univariate linear regression is obtained of the type:

$$A = b_1 c + b_0 \tag{7}$$

where $b_0$ and $b_1$ are the parameters of the regression straight line. The concentration of the analyte, $c_{unk}$, is obtained from the corresponding value $A_{unk}$ according to the expression:

$$c_{unk} = \frac{A_{unk} - b_0}{b_1} \tag{8}$$

If numerous samples have to be analyzed, there is no need to work with an augmented matrix of high dimensionality if an independent calibration is carried out that is subsequently applied to the samples.

To this end, a sample standard of known concentration ($c_{st}$) is prepared and the corresponding matrix ($\mathbf{X}_{st}$) is obtained. This matrix is added to all the resolution processes as an augmented matrix to obtain the area ($A_{st}$) of the concentration profile.

The calibration and the evaluation of the concentration are carried out as relative measurements, with the matrix $\mathbf{X}_{st}$ as the internal standard, which leads to a greater reproducibility of the results.

$$\frac{A}{A_{st}} = b_1 \left(\frac{c}{c_{st}}\right) + b_0 \tag{9}$$

$$c_{unk} = c_{st} * \left( \frac{(a_{unk}/a_{st}) - b_0}{b_1} \right) \tag{10}$$

The simultaneous determination of dyes in samples from a coloring process in the leather industry was carried out by a second-order calibration [36]. The dyes to be determined are Acid Red 97, Acid Orange 61, and Acid Brown 425 (see Figure 1). The synthetic dyes are polyprotic species that have different spectra for the basic and acidic species, and the corresponding UV–visible spectra have an overlapped signal.

The analytical procedure is carried out in an SIA system with a diode array spectrophotometer like the one shown in Figure 3. If an analytical sequence is programmed into which the sample (which contains the three dyes) and a strong acid (e.g., $H_2SO_4$) or a base (NaOH) is introduced, the interdiffusion process of the sample and reactant lead to a pH gradient, and the acid/basic species arrive sequentially at the detector [37].

Figure 6 shows spectra of the various species according to the pH. For Acid Red, we obtained spectrum *a* up to pH 11, between pH 11 and 12 the dye evolved, and at higher pHs we obtained spectrum *b*. For Acid Orange, we obtained spectrum *d* up to pH 2, between pH 2 and 3 the dye evolved, and at higher pHs we obtained spectrum *c*. Finally, for Acid Brown, we

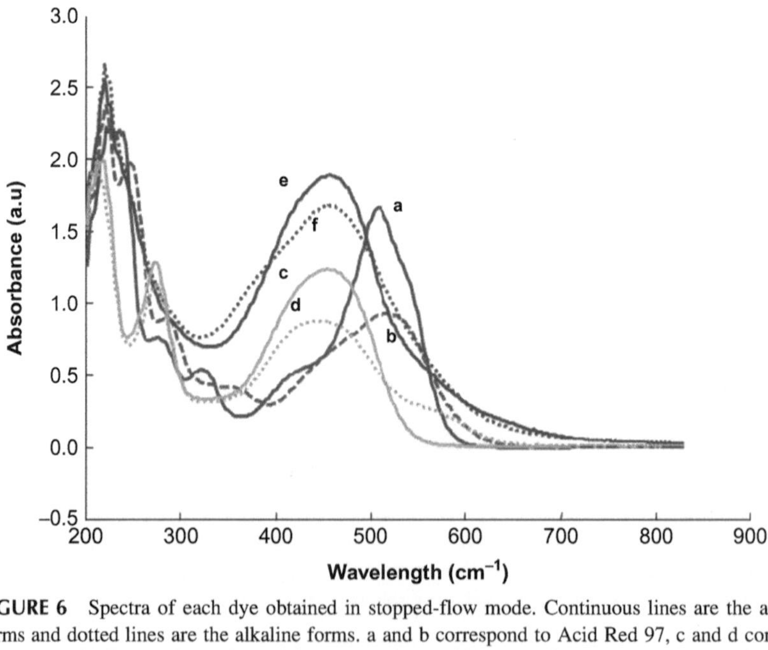


**FIGURE 6** Spectra of each dye obtained in stopped-flow mode. Continuous lines are the acid forms and dotted lines are the alkaline forms. a and b correspond to Acid Red 97, c and d correspond to Acid Orange 61, and e and f correspond to Acid Brown 425. *Extracted from Ref. [36] with permission of Elsevier.*

obtained spectrum *e* up to pH 8, between pH 8 and 9 the dye evolved, and at higher pHs we obtained spectrum *f*.

Because a wide range of pHs are required if the six species are to be observed, it was decided to work with an alkaline analytical sequence rather than a complete gradient, even though the concentration and spectra profiles of Acid Orange could not be obtained after the resolution. The system had five components, which arrived at the detector in the following order: Acid Red, Basic Red, Acid Brown, Basic Brown, and Basic Orange.

NaOH and sample volumes were selected taking into account the dimensions of the reactors of the system and the carrier volume necessary to prevent the reagents and the sample from entering the syringe. The optimal conditions for analyzing mixtures containing the three dyes were: a NaOH concentration of 0.025 M, a NaOH volume of 0.083 mL, and a sample volume of 0.25 mL.

Figures of merit for each calibration line are shown in Table 2. As calibration was modeled with relative values of both the signal (area quotient) and the concentration (concentration quotient), ideally the regression line should have a slope of 1 and an ordinate at the origin. The accuracy of the method is checked by calculating the concentration of dye in real-spiked samples to which known concentrations of standards had been added. The difference

**TABLE 2** Figures of Merit for Calibrations of the Three Dyes: Acid Red, Acid Brown, and Acid Orange

| | Dye | | |
|---|---|---|---|
| | *Red* | *Brown* | *Orange* |
| $b_1$ | 0.9744 | 1.0451 | 0.9675 |
| $b_0$ | 0.0197 | −0.0291 | 0.0474 |
| $R$ | 0.9991 | 0.9979 | 0.9994 |
| $s_{b_1}$ | 0.0146 | 0.0239 | 0.0118 |
| $s_{b_0}$ | 0.0088 | 0.0145 | 0.0072 |
| $s_e$ | 0.0132 | 0.0217 | 0.0107 |
| $n$ | 5 | 5 | 5 |
| $x_d$ (ppm) | 2.6 | 3.9 | 2.12 |
| $F_{cal}$ | 1.29 | 1.03 | 8.22 |
| $F_{critic}$ (0.05,2,3) = 16.04 | | | |

$b_1$ and $b_0$ are the slope and the intercept of the calibration curve; $s_{b_1}$ and $s_{b_0}$ are the standard deviation of the slope and the intercept: $R$ is the correlation coefficient; $s_e$ is the standard error; $n$ is the number of points; $x_d$ is the limit of detection; $F_{cal}$ is the calculated value; $F_{critic}$ is the stadistic value.

Extracted from Ref. [36] with permission of Elsevier.

between each pair of results (concentrations of the spiked samples and concentrations of the samples without spiking) should ideally be the concentration of the added standard. With these data (differences in concentration/concentration of added standards) a paired *t*-test is applied [38], using the null hypothesis in which the differences between both paired series of data correspond to a population with $\mu_d = 0$. After applying the corresponding *t*-test, it is concluded that there is no significant difference at 90% confidence.

## 4.2 Optimization of Dye Removal

The most commonly used degradation techniques are those that use a radiation treatment with UV light, either with a catalyst (photocatalysis) or without (photolysis). Another common type of process is adsorption, which is very effective with all dye types. The most used adsorbent is active carbon.

A great deal of interest has been taken in whether one new condition (the use of a new reagent or adsorbent, etc.) has a positive effect on the selected criteria (or response). The main criterion when choosing a dye removal process is that the removal rate should be high. However, other criteria such as removal time, cost, etc., should also be considered.

In dye removal processes, numerous parameters can affect the result. Generally, the effect that a factor has on the response is known: for example, in adsorption processes, in which the response is the % of dye removal, the amount of adsorbate, and the time of contact have a positive effect. Nevertheless, the real effect of each factor is difficult to predict because it depends on the range of factor values studied (domain) and on the values of other factors that could have synergistic effects on the factor considered. Table 3 shows the

**TABLE 3** Most Influential Factors in the Various Dye Removal Processes

| Elimination dye process | Studied factors |
|---|---|
| Adsorption or coagulation | Dye concentration<br>Amount of adsorbate<br>Concentration of coagulant<br>Kind of adsorbate or coagulant<br>Time contact |
| Advanced oxidation processes (AOP) | pH<br>Dye concentration<br>Kind of catalyst ($TiO_2$, ZnO,...)<br>Concentration of catalyst<br>Irradiation (if sample is irradiate)<br>Treated volume<br>Stirring speed<br>Concentration and kind of oxidant (e.g., $H_2O_2$) |

relationship between the removal techniques and the most important factors they are affected by [39].

The use of experimental designs to optimize processes for removing dyes is on the rise. These methodologies provide maximum information about the process in the minimum number of experiments.

The optimization of processes for removing three azo dyes (Acid Red 97, Acid Brown 425, and Acid Orange 61) has been studied using catalytic photodegradation [40,41] and adsorption on active carbon [42]. The photodegration studies were carried out in a cylindrical annular batch reactor (see Figure 2). The total volume of the solution was 500 mL. The pH was adjusted using $H_2SO_4$ or NaOH solutions. The reactions were carried out at room temperature with continuous stirring until the solution lost its color.

The factors considered are: the initial concentration of each dye, the amount of catalyst ($TiO_2$), and the pH. The responses studied are the half-life time of each dye, obtained by applying the MCR-ALS resolution of the degradation data matrices using the strategy of augmented matrices [15].

Each augmented matrix $\mathbf{X_{aug}}$ was constructed by appending in the column-wise direction a matrix $\mathbf{X_1}$ containing the spectra of the 13 standards of the three dyes; a matrix $\mathbf{X_2}$ containing the spectra recorded during the photodegradation process; and the vectors $\mathbf{v_i}$ containing the spectra of each species present in the solution.

By way of example, Figure 7 shows the graph to data matrix ($\mathbf{X_2}$) of one of the experiments designed. When SVD was applied to the data matrices augmented for each of the 16 degradation experiments (a $2^{5-1}$ design), four significant components were obtained in all cases. Figure 8 shows the spectral and concentration profiles for the three dyes and the fourth component obtained by MCR-ALS. The half-life time of each dye is obtained from the respective kinetic plot (see Figure 9).

The results show that: (a) the concentration of red dye has an important effect on the behavior of the three dyes and that it is not affected by the presence of the others. This dye has a considerable tendency to be adsorbed on the catalyst, which blocks the surface of the catalyst to some extent or another (depending on the concentration), thus affecting the behavior of other species. (b) The concentration of catalyst is not an important factor in the degradation of any dye. When the catalyst loading increases, the adsorption and degradation also increase. Therefore, one possible interpretation of the results is that the range of catalyst concentrations studied was too small for its influence to be detected and, as a result, small changes in its concentration do not affect the degradation process. (c) The pH has no effect on the degradation of the red dye but it does on the degradation of the other dyes.

On the basis of these results, further experimentation was carried out to increase the number of variables considered. Three of the five variables studied were qualitative: the type of catalyst ($TiO_2$ and ZnO); the presence of the oxidant $H_2O_2$ or not; and the presence of stirring or not. The other two

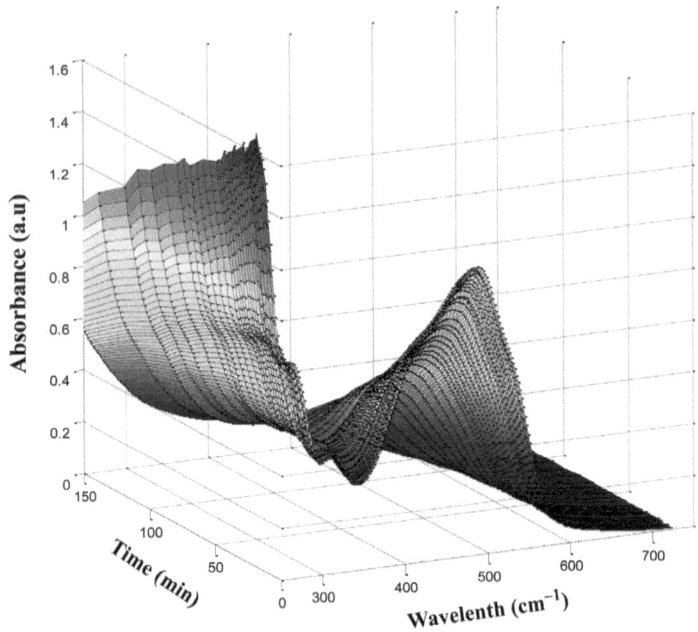


**FIGURE 7** UV–vis spectra recorded during the photodegradation process of a mixture of Acid Red 97, Acid Orange 61, and Acid Brown 425. *Extracted from Ref. [40] with permission of Elsevier.*

variables (pH and catalyst concentration) were quantitative. The catalyst concentration used in the previous study was increased so that its possible influence could be determined.

It has been confirmed that brown dye is always the most persistent in the solution and the presence of hydroxide peroxide notably reduces the half-life time of the degradation of the three dyes. Because of the extent of the influence of $H_2O_2$, the degradation rate in a homogeneous medium was evaluated without a catalyst. Using photolysis instead of photocatalysis could be interesting from a practical, economic, and environmental point of view.

Photolytic experiments have revealed that the degradation process is efficient. The efficiency value quadratically depends on the concentration of $H_2O_2$ and the other dyes.

The adsorption onto activated carbon [42] was studied by immersing the desired amount of carbon in the solution of dyes with agitation at ambient temperature. The solution and solid phase were separated by filtration with nylon syringe filters. An aliquot of the supernatant was removed and analyzed for dyes.

The concentrations of the various dyes have been evaluated by second-order calibration (MCR-ALS) using data obtained with the SIA system. A pH gradient and DAD spectroscopic detection are applied as reported in Section 4.1.

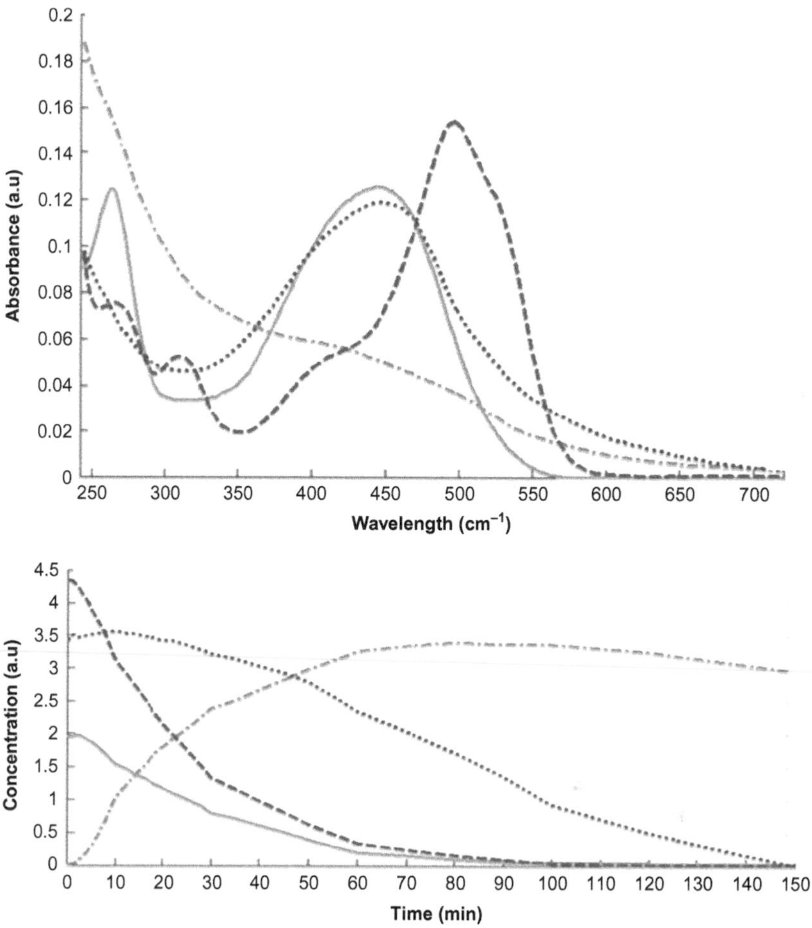


**FIGURE 8** Pure spectra and normalized concentration profiles obtained by multivariate curve resolution of the data matrix in Figure 7. (– –) corresponds to Acid Red 97, (–) to Acid Orange 61, (··) to Acid Brown 425 and (–··) to the new compound formed during the photodegradation. *Extracted from Ref. [40] with permission of Elsevier.*

Five factors are involved in the adsorption process: the initial concentration of each dye, mass of carbon, and contact time. The responses studied were the percentage of the relative concentration of each dye at the end of the process ($r_i$), the individual response, and the percentage of the relative concentration of the sum of the dyes ($r_t$), total response, calculated as:

$$r_i = \frac{c_{i,f}}{c_{i,0}} \cdot 100$$

$$r_t = \frac{\sum c_{i,f}}{\sum c_{i,0}} \cdot 100$$

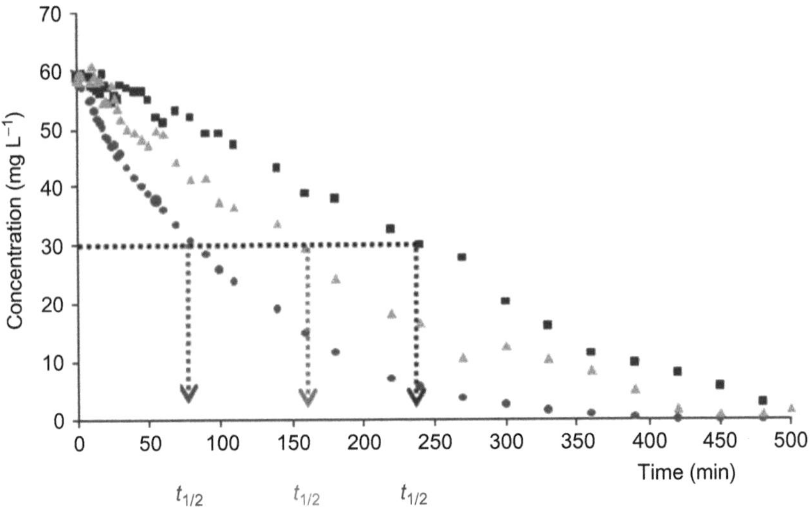


**FIGURE 9** Kinetic profiles obtained from the experiments corresponding to Figure 7. (●) is Acid Red 97, (▲) is Acid Orange 61, and (■) is Acid Brown 425. *Partially extracted and modified from Ref. [40] with permission of Elsevier.*

This study has demonstrated that the adsorption process is competitive and that the adsorption of the brown dye was clearly influenced by the presence of the other two dyes. The adsorption of the orange dye was less influenced by the red dye and the adsorption of the red dye was similar whether it was by itself in solution or in a mixture. For the total response, the most significant factor was the contact time, followed by the mass of carbon.

## 4.3 Detecting Intermediates

Photodegradation processes generate intermediates that are potentially dangerous, so it is essential that they be identified and, in particular, that at the end of the process it can be confirmed that they have degraded to innocuous molecules ($CO_2$, $H_2O$, etc).

By way of example it is studied the photooxidative degradation of the C.I. Acid Yellow 9 in the presence of $H_2O_2$ [43]. The process was monitored by UV–visible spectroscopy and a spectral data matrix $\mathbf{X}$ ($61 \times 451$) was obtained. Figure 10 shows the EFA plot of the $\mathbf{X}$ matrix. The horizontal line indicates the log (eigenvalue) associated with the noise, which was retrieved by SVD [19] analysis of the spectral data matrix obtained in the degradation of a sample of $H_2O_2$ under the same reaction conditions as the sample. The reaction times at which the factors attained significant values are indicated with an arrow.

In the forward direction of the EFA, two factors were significant in the first 5 min of the process and the other was significant after 15–20 min. These

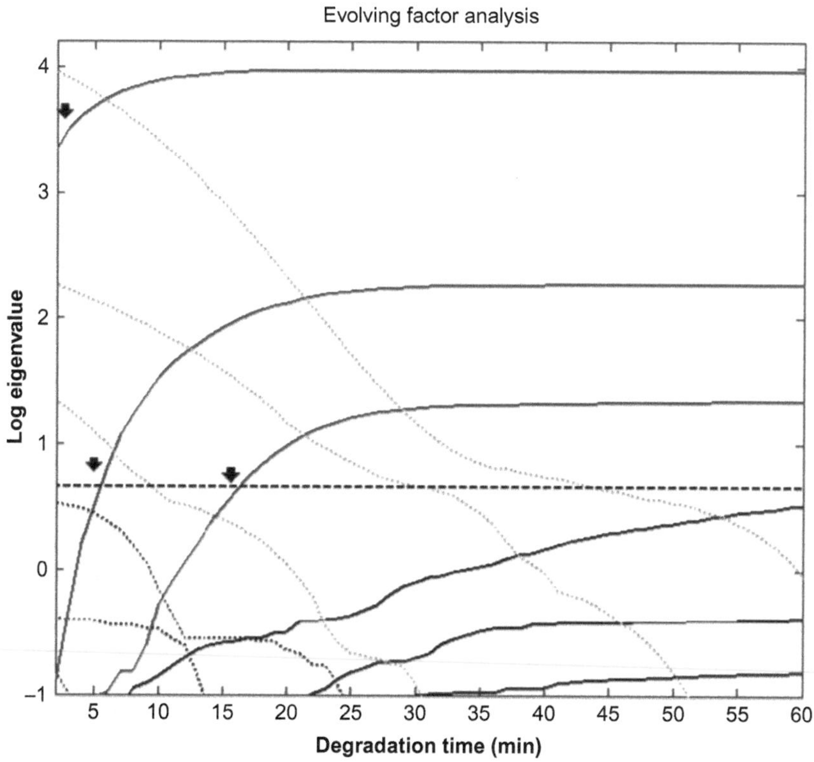


**FIGURE 10** EFA plot of the Acid Yellow 9 photodegradation matrix. Forward (continuous lines) and backward (dotted lines) directions. *Partially extracted and modified from Ref. [43] with permission of Elsevier.*

factors were associated with three different chemical species involved in the dye photooxidative process [43]. In the backward direction, the EFA plot shows no significant factors between minute 60 and 45, which is interpreted as the period in which the photooxidative degradation process of the dye finished.

Figure 11 contains the optimal spectrum recovered by MCR-ALS analysis of matrix **X** (nonnegativity constraints were used). The correlation coefficient between spectral profile 1 and the dye spectrum was higher than 0.99, so it was attributed to this compound. Spectral profile 2 suggests that it belongs to chemical species with an azo group in its chemical structure, because the spectrum shows absorbance at higher wavelengths than the initial dye. Spectral profile 3 is a typical UV spectrum, which is associated with aromatic molecules that are products of the breakdown of the azo linkage.

Mass spectrometry was used to analyze the intermediates formed. Considering that the degradation time during which there is a change in the number of factors detected by EFA is related to the chemical environment of the

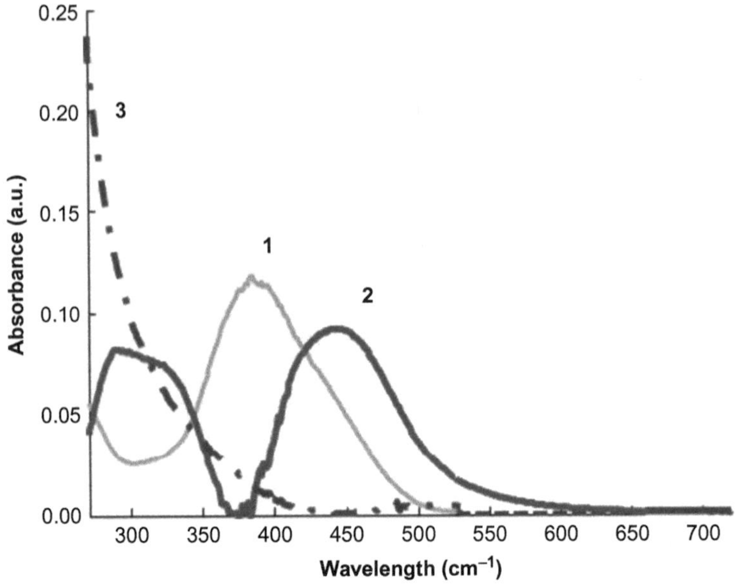


**FIGURE 11** UV–vis spectral profiles of the chemical species involved in Acid Yellow 9 photodegradation. *Extracted from Ref. [43] with permission of Elsevier.*

sample analyzed, the samples obtained after 0, 8, 18, 25, 30, and 60 min of reaction are used for the analysis.

The analysis of the MS spectra suggested the pathway shown in Figure 12. In this scheme, C.I. Acid Yellow 9 (a) is initially oxidized to (b) and then the azo linkage is broken to form (c) and (d). Subsequently, other reactions take place. At the end of the degradation time (60 min), $C_2H_4SO_3$ (e), $H_2SO_5$ (f), and $H_2SO_4$ (g) can be observed and the other compounds are nearly degraded. As can be seen in this figure, in the UV–vis range studied undetectable compounds can be formed in all steps of the main reaction. These results are in agreement with those obtained by MCR-ALS in Figure 11, where the dye (a) corresponds to compound **1**, the oxidized species (b) corresponds to **2**, and the species (c) and (d), which are aromatic compounds that absorb in the UV region, are assumed to be part of the shape of spectra **3**. The other compounds observed in MS (g, e, and f) are not active in the UV–visible region studied. This is why no compounds are observed at the end of degradation in Figure 11.

When the sample under study is more complex than the one discussed above—which may be the case when the three dyes are mixed together and the various compounds that can be generated throughout the degradation process need to be detected—further information may be required in the multivariate resolution. The case being studied is the one presented in Section 4.2: the photodegradation of the mixture of the three azo dyes—Acid

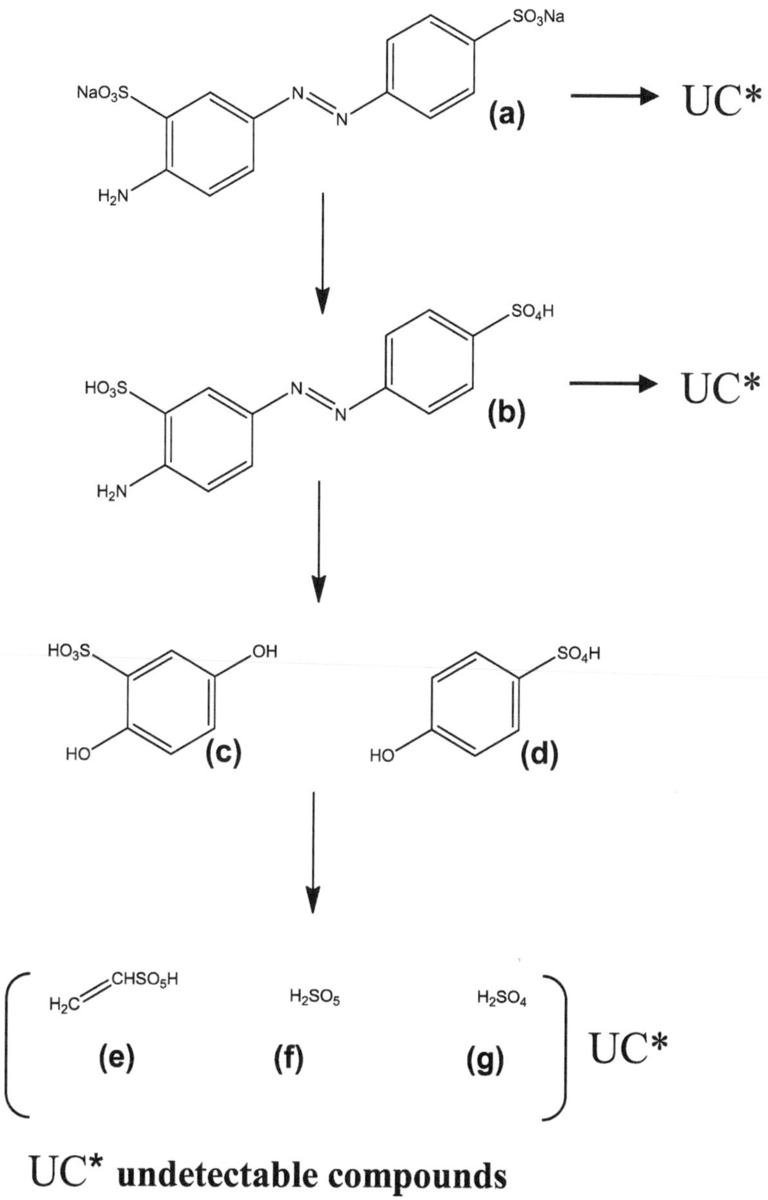


**FIGURE 12** Proposed reaction pathway for Acid Yellow 9 photodegradation. *Partially extracted and modified from Ref. [43] with permission of Elsevier.*

Orange61, Acid Red97, and Acid Brown425—which was monitored not only by UV–visible spectroscopy but also by nuclear magnetic resonance ($^1$H-NMR) [44].

The sets of spectra obtained by these two techniques led to two matrices $\mathbf{X}_{\mathbf{UV\text{-}visible}}$ (23 × 477) and $\mathbf{X}_{\mathbf{1H\text{-}NMR}}$ (23 × 1049) which contained the

UV–visible spectra (between 250 and 700 nm) and, for the same samples, the $^{1}$H-NMR spectra. These data were fused in an augmented matrix $\mathbf{X}_{\mathbf{aug}} = [\mathbf{X}_{\mathbf{UV\text{-}visible}}, \mathbf{X}_{\mathbf{1H\text{-}NMR}}]$ before they were simultaneously analyzed with MCR-ALS [44]. In this case, given the considerable difference in scale of the signals used, the data were normalized and all the variables of the $^{1}$H-NMR spectrum that provided no significant information were removed [45,46].

An initial estimation of the number of compounds involved in the process was made by individually analyzing each $\mathbf{X}_{\mathbf{UV\text{-}visible}}$ and $\mathbf{X}_{\mathbf{1H\text{-}NMR}}$ data matrix with SVD [19] and EFA [25,26]. As in the previous example, whereas the analysis of the matrix $\mathbf{X}_{\mathbf{UV\text{-}visible}}$ indicated the presence of an intermediate compound, the analysis of $\mathbf{X}_{\mathbf{1H\text{-}NMR}}$ revealed that up to seven factors could be regarded as significant, which suggests that four compounds were generated during the process. A chemical explanation for this difference may be that the intermediates are chemical species with naphthalene and benzenesulfonic groups that cannot be distinguished in the UV region [44].

This information was taken into account to initiate the process of simultaneously resolving matrix $\mathbf{X}_{\mathbf{aug}}$ with MCR-ALS. The fused data matrix was resolved by considering 4, 5, 6, and 7 factors, and the solutions were analyzed so that only those of greatest chemical significance were retained [44]. The results were best with six factors, which means that two more intermediate compounds were detected than Figure 8 shows.

## 4.4 Kinetic and Thermodynamic Studies

For any chemical process to be feasible, it must satisfy two basic conditions: it must be thermodynamically favorable (that is to say, the reaction must be total or, if equilibrium is reached, this equilibrium must be sufficiently shifted) and kinetically favorable (that is to say, it must take place at sufficient speed). This is why there is such interest in evaluating the constants of the reactions that take place in the processes of dye removal.

As has been mentioned in Section 2.3, photocatalysis, which generates strongly oxidizing hydroxyl radicals (•OH) that oxidize organic pollutants, is one of the most effective technologies for removing dye from the environment [39]. In heterogeneous photocatalysis, the reagents and the photocatalyst are present in two phases. The example described below aims to evaluate the adsorption and rate constant of Acid Yellow 9 [33].

All reactions involved in this process take place in one of four possible situations: (a) the reaction occurs while both the radical and the substrate molecule are adsorbed, (b) the reaction takes place between a radical in solution and an adsorbed substrate molecule, (c) a radical linked to the surface reacts with a substrate molecule in solution, or (d) the reaction takes place with a radical and a substrate molecule which are in solution. It has been proved [47–51] that all these mechanisms can give rise to the Langmuir–Hinshelwood model, commonly used to describe photocatalytic degradation kinetics. This model is expressed as

$$r = -\frac{dC}{dt} = \frac{k_r KC}{1 + KC} \tag{11}$$

where $r$ is the rate of dye degradation, $C$ is the concentration of the dye in mg $L^{-1}$ at any time, $k_r$ is the rate constant for photocatalysis, and $K$ is the adsorption constant of dye on the catalyst particles.

During the optimization step [32,33,43,52] of the resolution using MCR-ALS, this model can be used as a constraint on the dye concentration profile of a data matrix that contains the spectra recorded while the dye was degrading.

This strategy was used during the kinetic analysis of the photodegradation process using $TiO_2$ and ZnO as catalysts at different pHs [33]. One example of the UV–visible spectra obtained is shown in Figure 4 of Section 3. In order to impose a kinetic model as a constraint on the dye concentration profile, optimum values for the degradation rate ($k_r$) and adsorption constant ($K$) were obtained, which enabled the overall degradation rate ($r$) of the dye to be calculated according to Equation (11). The results show that the degradation process when $TiO_2$ was used as a catalyst was governed by the reaction involved in the adsorption process ($K$ between 1.3 and 0.04 L $mg^{-1}$ vs. $K$ between 0.013 and 0.05 L $mg^{-1}$ for ZnO), whereas when ZnO was used the solution reactions were of greater importance ($k_r$ between 1.5 and 0.3 mg $L^{-1}$ $min^{-1}$ vs. $k_r$ between 0.17 and 0.05 mg $L^{-1}$ $min^{-1}$ for $TiO_2$). The $r$ values obtained from the experiments carried out in the presence of ZnO ranged between 0.16 and 0.30 mg L $min^{-1}$, whereas when $TiO_2$ was used as catalyst they ranged between 0.04 and 0.07 mg L $min^{-1}$.

The example below describes the kinetics and thermodynamics of the adsorption of three azo dyes (Acid Red 97, Acid Brown 425, and Acid Orange 61) on active carbon [53]. Experiments were carried out on solutions that contained a single dye and mixtures of the three. Most of the studies deal with individual compounds and very few with binary or ternary mixtures, despite the fact that the samples used in technological processes often contain several dyes and other substances and competitive interaction with the solid surface and interaction with one another is to be expected.

The dye adsorption data were obtained from water solutions by the immersion method. The solution and solid phase were separated by filtration with nylon syringe filters. The adsorption capacity of dyes ($q$) was then calculated using the relation:

$$q = \frac{V\Delta C}{m} \tag{12}$$

where $V$ is the volume of the liquid phase, $m$ is the mass of the adsorbent, and $\Delta C$ is the difference between the initial and final concentration of dye in solution.

The concentrations of the various dyes were evaluated using a second-order calibration as has been described in Section 4.1. Although not strictly necessary, it was decided to use the same method to evaluate the concentrations in the solutions in which there was only one dye. For the kinetic experiments, the amounts of dye adsorption were determined by analyzing the solution at appropriate time intervals at room temperature. To study thermodynamic behavior, experiments were carried out at different initial concentrations of both single dyes and mixtures until equilibrium was reached.

Figure 13 shows the influence of contact time on solutions of individual dyes (dotted lines) and mixtures containing all three dyes (continuous line). It can be seen that the maximum amount adsorbed of each dye when they reach the equilibrium decreases when dyes are in a mixture because the total amount of dye in solution is higher. This reduction is not the same for all dyes: Acid Red 97 behaves almost the same when it is alone in solution or in a mixture; Acid Orange 61 behaves differently; and Acid Brown 425 behaves drastically differently. This behavior shows that dyes compete to occupy the adsorption sites of the activated carbon.

The adsorption of acid dyes from wastewater onto activated carbon obeys the pseudo-second-order kinetic model. The equation for this model is the following:

$$\frac{1}{q_e - q_t} = \frac{1}{q_e} + k_2 t \tag{13}$$

where $q_e$ and $q_t$ are the amount of dye adsorbed (mg $g^{-1}$) at equilibrium and at any time, $t$ (min), respectively, and $k_2$ is the equilibrium rate constant of the pseudo-second-order adsorption (g $mg^{-1}$ $min^{-1}$). The values of the rate constants are similar to those obtained when each dye is studied individually or in mixtures.

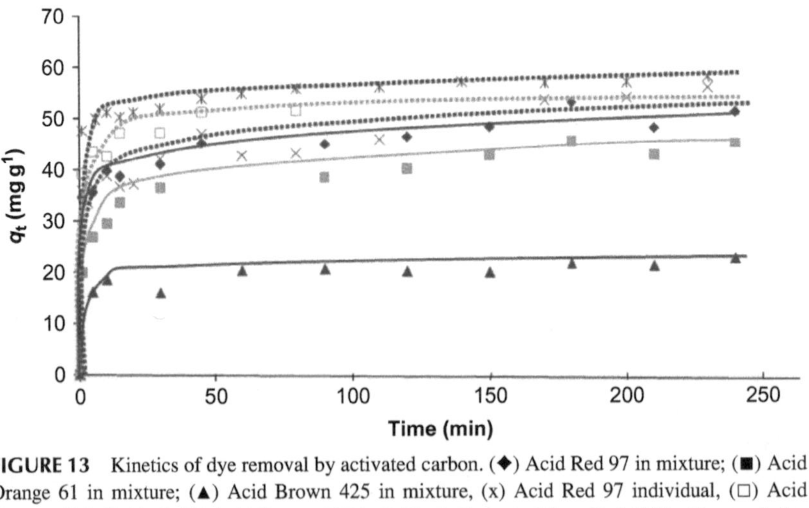


**FIGURE 13** Kinetics of dye removal by activated carbon. (◆) Acid Red 97 in mixture; (■) Acid Orange 61 in mixture; (▲) Acid Brown 425 in mixture, (x) Acid Red 97 individual, (□) Acid Orange 61 individual, (*) Acid Brown 425 individual. *Extracted from Ref. [53] with permission of Elsevier.*

Thermodynamic studies have shown that the behavior of each dye in individual solutions follows the Freundlich isotherm model, which is expressed by the following equation:

$$q_e = K_F C_e^{1/n} \tag{14}$$

where $q_e$ is the amount of dye adsorbed per gram of adsorbent (mg $g^{-1}$), $C_e$ is the concentration of the dye, $K_F$ is a rough indicator of the adsorption capacity, and $1/n$ is the adsorption intensity. In general, as the $K_F$ value increases, the adsorption capacity of an adsorbent for a given adsorbate increases. This type of model is for nonideal sorption that involves heterogeneous surface energy systems.

When dyes have been studied in mixtures, Freundlich models for multiconstituents have applied but the experimental data do not fit the model so it has not been possible to draw conclusions that enable the behavior of individual dyes and dyes in mixtures to be compared.

## 5 FINAL REMARKS

Finding methods of analysis that facilitate analysis in real time is a challenge for analytical chemists. Clearly, the main motivation is the speed with which analytical processes can be carried out but the recommendations made by "green chemistry" should also be taken into account (for example, low consumption of reagents, less need to handle chemical products, etc.). And the time they save also has economic repercussions.

A good strategy for developing and implementing real-time analytical methods is to use techniques and or procedures that provide a data matrix for each sample analyzed together with second-order chemometric techniques.

This chapter illustrates several problems that require the simultaneous determination of organic dyes. The quantitative determination of organic pollutants and their removal from the environment is one of the challenges of present-day society because of their presence in numerous products of daily life. The examples described are representative of a specific problem and can be generalized and applied to other pollutants.

Throughout the chapter, it has been shown that analytical methods based on second-order data can be used to tackle the issues in this area: the development of calibration models, the optimization of dye removal processes, the detection of intermediate species, and the determination of the kinetic and thermodynamic constants of the reactions involved in these processes.

The algorithm used in the examples is based on multivariate curve resolution with alternating least squares although many other algorithms could also be used for the same purpose. Evidence has been provided about how chemometric treatments can be improved by using a variety of strategies: augmented matrices, EFA graphs, hard and soft modeling, and data fusion techniques.

## REFERENCES

[1] Kellner R, Mermet JM, Otto M, Valcárcel M, Widmer HM. Analytical chemistry. A modern approach to analytical science. 2nd ed. Hardcover, Germany: Wiley-VCH; 2004.
[2] Wade Jr LG. Organic chemistry. 8th ed. Upper Saddle River, New Jersey: Prentice Hall; 2013.
[3] Rauf MA, Ashraf S. Radiation induced degradation of dyes—an overview. J Hazard Mater 2009;166:6–16.
[4] Wojnárovits L, Takács E. Irradiation treatment of azo dye containing wastewater: an overview. Radiat Phys Chem 2008;77:225–44.
[5] Akpan UG, Hameed H. Parameters affecting the photocatalytic degradation of dyes using TiO2 based photocatalyst: a review. J Hazard Mater 2009;170:520–9.
[6] Ruzicka J, Marshall GD. Sequential injection: a new concept for chemical sensors, process analysis and laboratory assays. Anal Chim Acta 1990;237:329–43.
[7] Borgen OS, Kowalski BR. An extension of the multivariate component-resolution method to three components. Anal Chim Acta 1985;174:1–26.
[8] Windig W. Self-modeling mixture analysis of spectral data with continuous concentration profiles. Chemom Intell Lab Syst 1992;16:1–16.
[9] de Juan A, Casassas E, Tauler R. Soft modeling of analytical data. Encyclopedia of analytical chemistry:instrumentation and applications. New York: Wiley; 2000. p. 9800.
[10] Liang Y, Kvalheim OM. Resolution of two-way data: theoretical background and practical problem-solving. Part 1: theoretical background and methodology. Fresenius J Anal Chem 2001;370:694–704.
[11] Jiang JH, Ozaki Y. Self-modelling curve resolution (SMCR): principles, techniques and applications. Appl Spectrosc Rev 2002;37:321–45.
[12] Brown SD, Tauler R, Edit WB. Comprehesive chemometrics. Chemical and biochemical data analysis. Amsterdam, The Netherland: Elsevier; 2009.
[13] Tauler R, de Juan A. Multivariate curve resolution. In: Gemperline P, editor. Practical guide to chemometrics. Boca Raton, FL: CRC Press; 2006. p. 417–74.
[14] Tauler R, Smilde AK, Kowalski BR. Selectivity, local rank, three-way data analysis and ambiguity in multivariate curve resolution. J Chemom 1995;9:31–58.
[15] Tauler R. Multivariate curve resolution applied to second order data. Chemom Intell Lab Syst 1995;30:133–46.
[16] Malinowski ER. Factor analysis in chemistry. 3rd ed. New York: Wiley-VCH; 2002.
[17] Massart DL, Vandeginste BGM, Buydens LMC, de Jong S, Lewi PJ, Smeyers-Verbeke J. Handbook of chemometrics and qualimetrics. Data handling in science and technology. vol. 20A. Amsterdam: Elsevier; 1997.
[18] Mardia KV, Kent JT, Bibby JM. Multivariate data analysis. London: Academic Press; 1979.
[19] Vandeginste BGM, Massart DL, Buydens LMC, de Jong S, Lewi PJ, Smeyers-Verbeke J. Handbook of chemometrics and qualimetrics. Data handling in science and technology. vol. 20B. Amsterdam: Elsevier; 1998.
[20] Brereton RG. Chemometrics: data analysis for the laboratory and chemical plant. Chichester: Wiley; 2003.
[21] Massart DL, Vandeginste BGM, Deming S, Michotte Y, Kaufmann L. Chemometrics: a textbook. Amsterdam, The Netherland: Elsevier; 1988.
[22] Johnson RA, Wichern DW. Applied multivariate statistical analysis. Englewood Cliffs, New Jersey: Prentice-Hall; 1988.
[23] Martens H, Næs T. Multivariate calibration. Chichester, England: Wiley; 1989.

[24] Brereton R. Chemometrics—applications of mathematics and statistics to laboratory systems. New York: Ellis Horwood; 1990.

[25] Gampp H, Maeder M, Meyer CJ, Zuberbühler AD. Evolving factor analysis of spectrophotometric titrations: forget about the law of mass action? Chimia 1985;39:315–7.

[26] Maeder M, Zuberbühler AD. The resolution of overlapping chromatographic peaks by evolving factor analysis. Anal Chim Acta 1986;181:287–91.

[27] Gampp H, Maeder M, Meyer CJ, Zuberbühler A. Calculation of equilibrium constants from multiwavelength spectroscopic data IV: model-free least-squares refinement by use of evolving factor analysis. Talanta 1986;33:943–51.

[28] Windig W, Guilment J. Interactive self-modeling mixture analysis. Anal Chem 1991;63:1425–32.

[29] Windig W, Gallagher NB, Shaver JM, Wise BM. A new approach for interactive self-modeling mixture analysis. Chemom Intell Lab Syst 2005;77:85–96.

[30] de Juan A, Vander Heyden Y, Tauler R, Massart DL. Assessment of new constraints applied to the alternating least squares method. Anal Chim Acta 1997;346:307–18.

[31] Bro R, Jong S. A fast non-negativity-constrained least squares algorithm. J Chemom 1997;11:393–401.

[32] de Juan A, Maeder M, Martınez M, Tauler R. Combining hard- and soft-modelling to solve kinetic problems. Chemom Intell Lab Syst 2000;54:123–41.

[33] Fernández C, de Juan A, Callao MP, Larrechi MS. Evaluation of the adsorption and rate constants of a photocatalytic degradation by means of HS-MCR-ALS. Study of process variables using experimental design. Chemom Intell Lab Syst 2012;114:64–71.

[34] Tauler R. Calculation of maximum and minimum band boundaries of feasible solutions for species profiles obtained by multivariate curve resolution. J Chemom 2001;15:627–46.

[35] de Juan A, Tauler R. Chemometrics applied to unravel multicomponent processes and mixtures. Revisiting latest trends in multivariate resolution. Anal Chim Acta 2003;500: 195–210.

[36] Gómez V, Font J, Callao MP. Sequential injection analysis with second order tretament for the determination of cyes in the exhaustion process of tanning effluents. Talanta 2007;71:1393–8.

[37] de Alba PL L, López-Martínez L, Wróbel-Kaczmarczyk K, Wróbel K, Amador-Hernández J. Anal Lett 1996;29:487.

[38] Miller JN, Miller JC. Statistics and chemometrics for analytical chemistry. 4th ed. Madrid: Prentice Hall; 2002.

[39] Fernandez C, Larrechi MS, Callao MP. An analytical overview of processes for removing organic dyes from wastewater effluents. Trends Anal Chem 2010;29:1202–11.

[40] Fernández C, Larrechi MS, Callao MP. Study of the influential factors in the simultaneous photocatalytic degradation process of three textile dyes. Talanta 2009;79:1292–7.

[41] Fernández C, Larrechi MS, Callao MP. Modelling of the simultaneous photodegradation of Acid Red 97, Acid Orange 61 and Acid Brown 425 using factor screening and response surface strategies. J Hazard Mater 2010;180:474–80.

[42] Gómez V, Callo MP. Modeling the adsorption of dyes onto activated carbon by using experimental designs. Talanta 2008;77:84–9.

[43] Fernández C, Callao MP, Larrechi MS. Kinetic analysis of C.I. Acid Yellow 9 photooxidative decolorization by UV-visible and chemometrics. J Hazard Mater 2011;190:986–92.

[44] Fernández C, Callao MP, Larrechi MS. UV–visible-DADand 1H-NMR spectroscopy data fusion for studying the photodegradation process of azo-dyes using MCR-ALS. Talanta 2013;117:5–80.

[45] Nørgaard L, Saudland A, Wagner J, Nielsen JP, Munck L, Engelsen SB. Interval partial least-squares regression (i-PLS): a comparative chemometric study with an example from near-infrared spectroscopy. Appl Spectrosc 2000;54:413–9.

[46] di Anibal C, Ruisánchez I, Callao MP. $^{1}$H-NMR variable selection approaches for classification. A case study: the determination of adulterated foodstuffs. Talanta 2011;86:316–23.

[47] Kumar KV, Porkodi K, Selvaganapathi A. Constrain in solving Langmuir– Hinshelwood kinetic expression for the photocatalytic degradation of Auramine O aqueous solutions by ZnO catalyst. Dyes Pigments 2007;75:246–9.

[48] Behnajady MA, Modirshahla N, Hamzavi R. Kinetic study on photocatalytic degradation of C.I. Acid Yellow 23 by ZnO photocatalyst. J Hazard Mater 2006;B133:226–32.

[49] Chakrabarti S, Dutta BK. Photocatalytic degradation of model textile dyes in wastewater using ZnO as semiconductor catalyst. J Hazard Mater 2004;B112:269–78.

[50] Sahel K, Perol N, Chermette H, Bordes C, Derriche Z, Guillard C. Photocatalytic decolorization of Remazol Black 5 (RB5) and Procion Red MX-5B — isotherm of adsorption, kinetic of decolorization and mineralization. Appl Catal B Environ 2007;77:100–9.

[51] Konstantinou K, Albanis TA. TiO2-assisted photocatalytic degradation of azo dyes in aqueous solution: kinetic and mechanistic investigations: a review. Appl Catal B Environ 2004;49:1–14.

[52] Mouton N, de Juan A, Sliwa M, Ruckebusch C. Hybrid hard- and soft-modeling approach for the resolution of convoluted femto second spectrokinetic data. Chemom Intell Lab Syst 2011;150:74–82.

[53] Gómez V, Larrechi MS, Callao MP. Kinetic and adsorption study of acid dye removal using activated carbon. Chemosphere 2007;69:1151–8.

Chapter 10

# Pixel-Level Data Analysis Methods for Comprehensive Two-Dimensional Chromatography

Karisa M. Pierce*, Brendon A. Parsons† and Robert E. Synovec[1,†]
*Department of Chemistry and Biochemistry, Seattle Pacific University, Seattle, Washington, USA
†Department of Chemistry, University of Washington, Seattle, Washington, USA
[1]Corresponding author: e-mail: synovec@chem.washington.edu

## Chapter Outline



## 1 INTRODUCTION

Comprehensive two-dimensional (2D) chromatography continues to evolve as an important instrumental platform to address challenges in the analysis of complex samples. The instrumentation primarily utilized is either comprehensive 2D gas or liquid chromatography (e.g., GC × GC or LC × LC) [1–10]. While other comprehensive 2D separation methods have been reported, this chapter focuses primarily on these two commonly applied instruments. Additionally, these instruments are often coupled to multichannel spectroscopic

Data Handling in Science and Technology, Vol. 29. http://dx.doi.org/10.1016/B978-0-444-63527-3.00010-2

detectors, namely time-of-flight mass spectrometry for GC × GC-TOFMS [11] and multiwavelength absorbance detection (i.e., diode array detection) for LC × LC-DAD [8]. In the analysis of complex samples, the resulting data can be very challenging to efficiently and accurately analyze. For this purpose, chemometric data analysis has emerged as the path forward to provide a successful, informative analysis. Many excellent chemometric text books and reviews are available [12–15]. Other chemometrics reviews focused primarily on comprehensive 2D separations data are also available in the literature [16–23]. Indeed, the data obtained from a single run from either GC × GC or LC × LC, when coupled with a univariate detector such as GC × GC-FID or LC × LC with single wavelength absorbance detection, produces second-order data, and third-order data when more than one sample run is simultaneously analyzed, so chemometric analysis is ideally suited to glean useful information from the rich data structure. The situation is even more urgent to apply chemometrics for the data that are inherently third order for a single sample run, i.e., using GC × GC-TOFMS or LC × LC-DAD. There are different approaches to "chemometrically" address data analysis, and in the context of this chapter, we refer to the dependence on the data-level. Data from comprehensive 2D chromatography separations can be analyzed, on the following three levels: pixel [24], peak table [25,26], or peak region [27,28] basis. While all of these three levels of data are utilized [29,30], the peak table-level and peak region-level approaches are firmly provided by the instrument manufacturers, and hence the analyst must apply such software in the commercially available state. Hence, there is generally little room for independent development of chemometric software at these two data levels, whereas at the pixel-level, there has been rapid growth in the development of chemometric tools for creative and informative analysis of comprehensive 2D chromatography separations data. Accordingly, the focus of this chapter is on the pixel-level chemometric analysis of comprehensive 2D chromatography separations data. For example, recently a novel tile-based data analysis approach with its fundamental basis at the pixel-level has been introduced for the comparative analysis of large numbers of GC × GC-TOFMS sample runs [31,32].

This chapter is organized into the following three sections: Sections 2–4. Section 2 provides an overview of the data preprocessing steps that are deemed essential for optimizing the use of pixel-level chemometric analyses of comprehensive 2D chromatography separations data. The following four generally essential preprocessing steps are covered: baseline correction, noise reduction, normalization, and retention time alignment. Additional preprocessing steps that are contingent on the particular chemometric method are discussed in due course. Since overlap of analyte peaks is commonplace in the analysis of complex samples, which are often studied using comprehensive 2D chromatography separation methods, Section 3 covers the basics of commonly applied chemometric deconvolution methods for pixel-level data. For this chapter, we due to consideration of space, we have limited the discussion primarily to generalized rank annihilation method (GRAM), multivariate curve resolution alternating least squares (MCR-ALS), and parallel factor analysis (PARAFAC), with some discussion of related methods. Finally, Section 4 provides an overview of commonly applied chemometric methods for fingerprinting and pattern recognition, in which the analyst is interested in initially comparing a series of samples on a broader scale (based upon the experimental design). In Section 4, we cover two commonly encountered experimental designs: unsupervised design using principal component analysis (PCA) and supervised design using either partial least squares (PLS) methods or Fisher ratio analysis. An overarching goal of this chapter is to provide the reader with the background and knowledge of the fundamentals of chemometric data analysis of comprehensive 2D chromatography separations data, in particular for GC × GC and LC × LC, so they can attack their own data sets at the pixel-level.

## 2 SIGNAL PREPROCESSING IN 2D CHROMATOGRAPHY WITH PIXEL-BASED DATA

Pixel-based analysis aims to discover the valuable chemical information in complex 2D chromatographic data with high sensitivity for real chemical signals and minimal interference from typical instrumental variations that manifest in the data. To achieve this objective, the chromatographic data must be carefully manipulated to reduce the impact of common variations, such as baseline drift and noise. These steps are performed prior to the analysis of the data, and are referred to as "preprocessing." Preprocessing of the raw chromatographic data reduces the impact of immaterial variations in the data with the goal of improving accuracy and precision of qualitative and quantitative analyses of the chemically relevant variations that are of interest to the analyst. Preprocessing is thus critically important, and is a step that, if performed properly, allows for maximal discovery of useful information while avoiding wasted experimental, instrumental, and analytical efforts. The most common preprocessing steps which we consider here are: baseline correction, noise reduction, normalization, and retention time alignment, which are presented here in the order by which they are performed.

### 2.1 Baseline Correction

The first preprocessing step in most chromatographic analysis work flows is baseline correction. Baseline drift is the low-frequency signal variation that occurs in the baseline due to column stationary phase bleed, background ionization, and low-frequency variations in the detector and/or instrument-controlled parameters (such as temperature or flow). Baseline rise is the steady increase in baseline observed in temperature-programmed or gradient elution separations. Baseline correction procedures aim to mitigate the effects of baseline drift and rise, improving the detection and quantification of analyte peaks that may otherwise be impacted by baseline interference. Baseline correction methods must be carefully implemented in order to preserve the relevant chemical variations, while avoiding overfitting the chromatogram or mistaking broadened features of coeluting compounds for baseline signals. Following baseline correction, the baseline noise should be centered on zero for the length of the chromatographic separation, and in the case of multichannel detectors, for all channels, as well.

The simplest baseline correction procedure is to collect a "blank" chromatogram and to subtract the blank chromatogram from the sample chromatogram being analyzed. This method is commonly employed in one-dimensional (1D) chromatography, but is less-commonly applied to 2D chromatography, as the presence of minor run-to-run misalignment may magnify the baseline when subtracting the blank chromatogram. The next simplest baseline correction algorithm simulates a blank chromatogram by polynomial least squares fitting and subtracts it from the sample chromatogram. Baseline fitting and subtraction may also be performed on manually selected chromatographic subregions [33] or individual peaks [34–36]. Penalized least squares regression or robust orthogonal regression may also be used to model and subtract baselines [37]. Delta2D image processing software uses a powerful "rolling ball" algorithm to perform baseline subtraction for 2D chromatograms [38]. Rolling minimum methods take advantage of the dead time regions on the second separation dimension in comprehensive 2D chromatography for rapid baseline correction.

## 2.2 Noise Reduction

Baseline correction procedures are designed to correct low-frequency noise and offsets, but not to alter higher frequency variations. Several techniques may be applied to reduce high-frequency variations and improve signal-to-noise (S/N) ratio in chromatograms. The most commonly implemented methods are smoothing and binning. The classic Savitzky–Golay method is a running smoothing function that fits a low-order polynomial to each data pixel and its neighbors, replacing the signal of each data pixel with the value provided by the polynomial fit [39]. Wavelet smoothing methods transform the chromatogram into the frequency domain, remove the high-frequency components that are assumed to be indeterminate noise, and then perform the reverse transform to the time domain yielding the smoothed chromatogram [40]. Similarly, the low-frequency components characteristic of slowly drifting baseline variations may also be removed with the wavelet method, assisting in baseline correction. Data reduction by simple interpolation and averaging, referred to as "binning," improves S/N and additionally reduces computational load when processing large volumes of 2D separations data [41]. COMForTS, a Fourier transformation method for comprehensive

2D chromatographic separations, improves S/N and additionally reduces the separation time and data density [42].

Noise reduction techniques are critical to successfully addressing a wide variety of analytical challenges. The improvement in S/N, as well as the concurrent data reduction with some methods, improves the ability of the following pixel-based analysis to discover meaningful chemical differences between samples, while reducing false discoveries due to noise. Regardless of the technique used, noise reduction is a delicate operation that requires careful choice of parameters to avoid compromising the real chemical signals that are present in the data. Careless choice of parameters for noise removal algorithms or excessive noise removal may artificially broaden peaks, reduce resolution, introduce artifacts, and/or complicate deconvolution.

## 2.3 Normalization

Chromatograms are often normalized to correct between-sample variations that are unavoidably introduced during sample collection, preparation, and injection. The most commonly applied normalization technique is the internal standard method [43–46]. However, it may be difficult to find an inert standard that is completely resolved from all native components for truly unknown samples. As an alternative, isotopically labeled internal standards may be of use, taking advantage of the isotope ratio method [47] if the standard chosen happens to be natively present in the sample. However, the use of isotopic standards is limited to separations coupled to mass spectrometry. When it is not practical to use the internal standard method, many users apply the sum-normalization method, which is briefly mentioned in chromatography textbooks, sometimes in terms of area-normalization [48,49]. The sum-normalization method uses the total sum of all baseline corrected signals in a chromatogram as the signal to be normalized. The ratio of the total signal for each sample to that of the mean of all samples is sometimes used as the normalization factor to subsequently adjust all chromatographic signals in the samples. The user must assume the samples in the data set are sufficiently similar, such that equal volumes of the samples should have sufficiently equal total signals at the detector. While this assumption is rarely strictly true, it is often assumed the principal source of variation is due to injection volume. Other normalization methods involve mathematically forcing the mean signal of each chromatogram to equal 1 [50] or mathematically forcing the maximum peak signal volume to equal 1 [51].

Normalization techniques should be carefully chosen to achieve normalization of the desired source of variation. If, for example, a user desires to correct for the efficiency of an extraction, and also the variation in injection volume, a normalization method must be chosen for each. An option may be to add internal standard(s) prior to extraction, as well as a different internal standard immediately prior injection. The normalization would thus adjust samples based first on injection and then on extraction. In some cases, the additional sample handling required for internal standard normalization may introduce larger variation than that which is being normalized. Analysts should observe typical variation in their samples and proceed accordingly.

Normalization is applied to samples (every data pixel in a chromatogram is normalized using the same normalization factor), while scaling is applied to variables (each data pixel in a chromatogram is scaled by a unique factor to meet some criteria). Usually, the purpose of scaling is to reduce the influence of large signals in comparative analyses. Common scaling methods are autoscaling (where the mean of each pixel across all samples is forced to be zero and each pixel is forced to have unit standard deviation across all samples) [52], or dividing each pixel by the mean value among samples [53], or log transformations, or applying power transformations.

## 2.4 Retention Time Alignment

Alignment is an important preprocessing step because even minor pressure, flow and temperature fluctuations cause retention time variations that may obscure chemical information, resulting in poor performance of most chemometric methods. Alignment algorithms are designed to shift the raw signal along the time axes of the 2D separations so the peak position of a given analyte matches from one sample run to the next. The accuracy of the 2D peak signal volumes should also be preserved during alignment. Alignment algorithms can vary from very simple approaches such as application of a simple scalar shift, to a targeted peak list approach, to more sophisticated approaches involving locally and globally optimized alignment algorithms. While all of these alignment algorithm approaches have merit, for the chemometric analysis of pixel-level data, the most commonly applied preprocessing approaches implemented are the latter. The locally and globally optimized alignment algorithms can handle severe and dynamic shifting in comprehensive 2D chromatography applications, because they are robust, powerful, and essentially automated in every regard after the initial parameter selections have been made by the analyst [24,50,51,54–59].

Alignment of a "sample" chromatogram to a "target" or "reference" chromatogram at the pixel-level requires the assumption that the signal due to analyte peaks in those chromatograms are truly matched. For data sets in which samples are relatively similar, this assumption is generally realized. However, for data sets in which samples vary greatly, it is more challenging for alignment to succeed. This challenge is more confidently addressed for comprehensive 2D chromatography instruments that implement multichannel spectroscopic detection, such as GC × GC-TOFMS and LC × LC-DAD, since the alignment algorithms are designed to utilize the spectral information to correctly match the signal due to analyte peaks across samples. It should also be noted that the use of local alignment algorithms can provide successful alignment when multichannel spectral information is not available.

# 3 PIXEL-LEVEL METHODS FOR DECONVOLUTION

## 3.1 Aims of Deconvolution

Analyte peaks often coelute despite the excellent peak capacity and resolving power provided by comprehensive 2D chromatography. The analyst may need to mathematically resolve analyte peaks that are not chromatographically resolved. Peak resolution algorithms can provide the needed resolution for pixel-level chromatographic data. GRAM, MCR-ALS, and PARAFAC are some of the well-known resolution methods which will be described herein. GRAM and MCR-ALS can decompose 2D data, while PARAFAC and sometimes MCR-ALS are primarily applied to 3D data. For all of these methods, the section of either 2D or 3D data must be of suitably low rank by containing a relatively small number of analyte components. To resolve peaks in an unknown sample, GRAM requires a standard sample and is noniterative, while MCR-ALS and PARAFAC are iterative and do not require a standard sample.

These pixel-level resolution methods rely heavily on linear algebra because chromatographic data, under ideal conditions, can be thought of as bilinear or trilinear matrices and arrays for 2D and 3D chromatography, respectively. This means that the pixel-level chromatographic data are ideally composed of unique, consistent, concentration-dependent signals from each independent and unique analyte source present, and those signals are additive, even across samples. Chromatographic instruments are designed to produce such results, though sources of uncontrollable and/or unavoidable variation do diminish the bilinearity and trilinearity of analyte signals in 2D and 3D chromatograms, e.g., various forms of retention time misalignment. Peak resolution algorithms do incorporate some methods of dealing with these sources of variation with varying degrees of success. In this text, the linear algebra notation is as follows: scalars are lower-case italics, vectors are bold lower-case text, 2D matrices are bold capitals, and 3D arrays are underlined bold capitals.

## 3.2 Generalized Rank Annihilation Method

GRAM can be used to mathematically resolve signals of bilinear and independent underlying factors (or components) such as overlapping analyte peaks in sections of 2D pixel-level chromatograms provided by comprehensive 2D separations [1], such as GC × GC, LC × LC, LC × CE, and so on [4,60–62]. To be bilinear, data must be composed of unique, consistent, and concentration-dependent signals from each independent and unique source present, and those signals must be additive. GRAM decomposition of such bilinear data provides the pure peak profiles as well as the concentration ratios of compounds in the sample relative to the standard. Several GRAM algorithms have been published [4,60,61]. Here, we describe the GRAM algorithm that is based on the standard eigenvalue method [61].

Consider a section of bilinear GC × GC-FID data of a standard sample represented by matrix **N** in which the column dimension represents the ${}^2t_R$ dimension and the row dimension represents the ${}^1t_R$ dimension. The standard might contain $K$ components that can be expressed according to Equation (1).

$$\mathbf{N} = \sum_{k=1}^{K} \left( \mathbf{x}_k c_{k,k} \mathbf{y}_k^T \right) = \mathbf{X}\mathbf{C}_N\mathbf{Y}^T \tag{1}$$

In Equation (1), **X** and **Y** are matrices whose columns $\mathbf{x}_k$ and $\mathbf{y}_k$ are vectors representing the pure component signals in the ${}^2t_R$ dimension and ${}^1t_R$ dimension of **N**. $\mathbf{C}_N$ is a diagonal matrix in which the elements $c_{k,k}$ represent the concentration of the $k$th component in the standard sample. The superscript T denotes a transpose.

Now consider extracting the same GC × GC-FID data section from an unknown sample chromatogram, represented by matrix **M**, again in which the column dimension represents ${}^2t_R$ and the row dimension represents ${}^1t_R$. **M** can be expressed according to Equation (2).

$$\mathbf{M} = \mathbf{X}\mathbf{C}_M\mathbf{Y}^T \tag{2}$$

In Equation (2), **X**, **Y**, and superscript T have the same meaning as in Equation (1) and $\mathbf{C}_M$ is a diagonal matrix in which the elements $c_{k,k}$ represent the concentration of the $k$th component in the unknown sample. Thus, given **N**, **M**, $\mathbf{C}_N$, and $K$, it is possible to calculate **X**, **Y**, and $\mathbf{C}_M$.

GRAM calculates the unique values for **X**, **Y**, and $\mathbf{C}_M$ relative to $\mathbf{C}_N$, given the following requirements. The unknown sample must contain relative analyte component concentrations that differ from the standard. The overlapped peaks must have some resolution on both ${}^1t_R$ and ${}^2t_R$. The analyte retention times and peak shapes must be equal between the unknown sample and the standard (i.e., the data must be bilinear). Figure 1 shows a standard (**N**) and sample (**M**) chromatographic data section that meets these requirements for GRAM. If retention time shifting is present, then it is necessary to align the data prior to application of GRAM [63,64].

Furthermore, the unknown sample can contain interference peaks, which are absent in the standard and vice versa. The key is that the user augments **M** and **N** into a single matrix so that all potentially present components can be modeled, even if some are absent in either **M** or **N**. For this step, GRAM requires that **M** is augmented row-wise with **N** to form a 2D bilinear augmented matrix, [**M**,**N**], where each row continues to represent a ${}^2t_R$ pixel and each column continues to represent a ${}^1t_R$ pixel. Next, **M** is also augmented column-wise with **N** to form another augmented matrix [**M**;**N**]. (The square brackets with comma or semi-colon between matrices denote row-wise and column-wise matrix augmentation, respectively). The GRAM algorithm then proceeds to perform singular value decomposition (SVD) on the augmented matrices, yielding Equations (3) and (4).

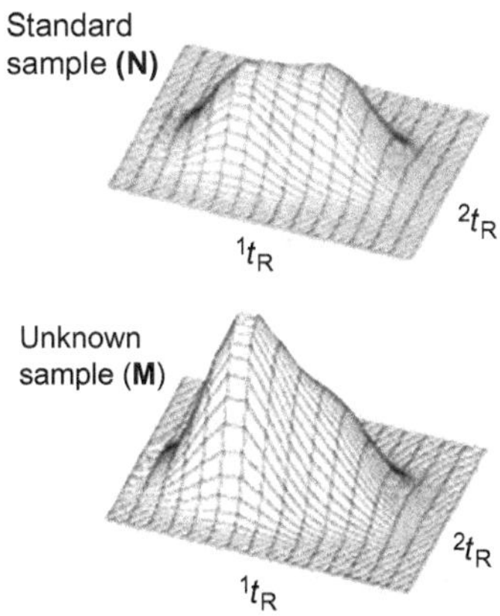


**FIGURE 1** Standard (N) and sample (M) chromatographic data sections that meet the requirements for GRAM to mathematically resolve the two overlapping analyte peaks.

$$[\mathbf{M}, \mathbf{N}] = \mathbf{U}_r \boldsymbol{\Theta}_r \mathbf{V}_r^T \tag{3}$$

$$[\mathbf{M}; \mathbf{N}] = \mathbf{U}_c \boldsymbol{\Theta}_c \mathbf{V}_c^T \tag{4}$$

In Equations (3) and (4), subscript r denotes the row-wise augmented matrix, subscript c denotes the column-wise augmented matrix, the decomposed components from SVD of the augmented matrices are $\boldsymbol{\Theta}$ (a diagonal matrix of singular values) and **U** and **V** (which are orthogonal matrices of eigenvectors). The estimated pseudorank of the augmented matrices should ideally equal the expected number of analyte plus interferent components present in the subregion ($K$) as indicated by estimating the number of significant factors in $\boldsymbol{\Theta}_r$ and $\boldsymbol{\Theta}_c$. **M** and **N** are transformed to square matrices $\mathbf{M}_{\mathbf{U}_r\mathbf{V}_c}$ and $\mathbf{N}_{\mathbf{U}_r\mathbf{V}_c}$ while preserving the pseudorank of **M** and **N**. Next, the generalized eigenvalue problem is solved for the square matrices using the Moler and Stewart algorithm [65] as in Equations (5)–(7), wherein $\boldsymbol{\lambda}$ is a matrix of eigenvalues [61]. Please see the cited references for more details about the linear algebra described herein [61,65].

$$\mathbf{M}_{\mathbf{U}_r\mathbf{V}_c} = \mathbf{U}_r^T \mathbf{M} \mathbf{V}_c = \left(\mathbf{U}_r^T \mathbf{X}\right) \mathbf{C}_M \left(\mathbf{V}_c^T \mathbf{Y}\right)^T \tag{5}$$

$$\mathbf{N}_{\mathbf{U}_r\mathbf{V}_c} = \mathbf{U}_r^T \mathbf{N} \mathbf{V}_c = \left(\mathbf{U}_r^T \mathbf{X}\right) \mathbf{C}_N \left(\mathbf{V}_c^T \mathbf{Y}\right)^T \tag{6}$$

$$\mathbf{M}_{\mathbf{U}_r\mathbf{V}_c} \mathbf{V}_r = \mathbf{N}_{\mathbf{U}_r\mathbf{V}_c} \mathbf{V}_r \boldsymbol{\lambda} \tag{7}$$

The GRAM algorithm next sorts the matrices **V** and $\boldsymbol{\lambda}$ according to magnitude and applies similarity transforms ($\mathbf{T}^\dagger$ and $\mathbf{T}^\ddagger$) to **V** and $\boldsymbol{\lambda}$ to ensure only noncomplex solutions are produced as in Equations (8)–(11) wherein superscript −1 denotes the inverse matrix [61].

$$\mathbf{V}^\dagger = \mathbf{V}\left(\mathbf{T}^\dagger\right)^{-1} \tag{8}$$

$$\boldsymbol{\lambda}^{\dagger} = \mathbf{T}^{\dagger}\boldsymbol{\lambda}\left(\mathbf{T}^{\dagger}\right)^{-1} \quad (9)$$

$$\mathbf{V}^{\ddagger} = \mathbf{V}^{\dagger}\left(\mathbf{T}^{\ddagger}\right)^{-1} \quad (10)$$

$$\boldsymbol{\lambda}^{\ddagger} = \mathbf{T}^{\ddagger}\boldsymbol{\lambda}^{\dagger}\left(\mathbf{T}^{\ddagger}\right)^{-1} \quad (11)$$

The pure component profiles in the chromatographic dimensions (**X** and **Y**) are then calculated according to Equations (12) and (13) wherein the superscript + denotes the pseudoinverse.

$$\mathbf{X} = \mathbf{U}_{\mathrm{r}}\mathbf{N}_{\mathbf{U}_{\mathrm{r}}\mathbf{V}_{\mathrm{c}}}\mathbf{V}^{\ddagger} \quad (12)$$

$$\mathbf{Y} = \mathbf{V}_{\mathrm{c}}\left(\left(\mathbf{V}^{\ddagger}\right)^{+}\right)^{\mathrm{T}} \quad (13)$$

The pure analyte peak profiles provided by GRAM resolution of the overlapping peaks from Figure 1 are shown in Figure 2 using $K=2$.

Furthermore, quantitative information can be acquired from the ratio of concentrations of the pure components (**R**) in **M** and **N** when calculated according to Equations (14)–(16).

$$\mathbf{C}_{\mathrm{M}} = \left(\mathbf{U}_{\mathrm{r}}^{\mathrm{T}}\mathbf{X}\right)^{-1}\mathbf{M}_{\mathbf{U}_{\mathrm{r}}\mathbf{V}_{\mathrm{c}}}\left(\left(\mathbf{V}_{\mathrm{c}}^{\mathrm{T}}\mathbf{Y}\right)^{\mathrm{T}}\right)^{-1} \quad (14)$$

$$\mathbf{C}_{\mathrm{N}} = \left(\mathbf{U}_{\mathrm{r}}^{\mathrm{T}}\mathbf{X}\right)^{-1}\mathbf{N}_{\mathbf{U}_{\mathrm{r}}\mathbf{V}_{\mathrm{c}}}\left(\left(\mathbf{V}_{\mathrm{c}}^{\mathrm{T}}\mathbf{Y}\right)^{\mathrm{T}}\right)^{-1} \quad (15)$$

$$\mathbf{R} = \mathbf{C}_{\mathrm{M}}\mathbf{C}_{\mathbf{N}}^{-1} \quad (16)$$

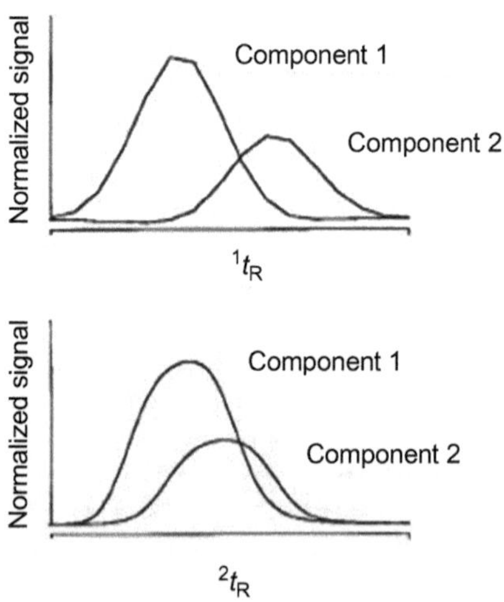


**FIGURE 2** Pure peak profiles provided by the GRAM resolution of the overlapping peaks in Figure 1.

Other GRAM algorithms have been developed and implemented [4,60,62]. For example, another algorithm [4] uses an addition matrix (**M**+**N**) instead of an augmented matrix to model all of the components common to both the standard and the unknown sample; this addition ensures common components will be modeled even if the standard has more components than the sample and vice versa. In this particular algorithm, the addition matrix is submitted to SVD decomposition yielding the diagonal matrix $\boldsymbol{\Theta}$ and the orthogonal matrices **U** and **V** as in Equation (17).

$$(\mathbf{M}+\mathbf{N}) = \mathbf{U}\boldsymbol{\Theta}\mathbf{V}^{\mathrm{T}} \tag{17}$$

Next, the SVD decomposition components are truncated according to $K$ (the expected number of components or estimated rank of the addition matrix). Then the eigenvalue problem is solved for the truncated matrices (which are "overlined" as in Equation 18) producing a matrix of eigenvectors (**T**) and a diagonal matrix of eigenvalues ($\boldsymbol{\Pi}$).

$$\left(\bar{\boldsymbol{\Theta}}^{-1}\bar{\mathbf{U}}^{\mathrm{T}}\mathbf{N}\bar{\mathbf{V}}\right)\mathbf{T} = \mathbf{T}\boldsymbol{\Pi} \tag{18}$$

The matrices **T** and $\boldsymbol{\Pi}$ are then submitted to a similarity transform to eliminate any complex solutions that arise [4,62]. The pure component profiles (**X** and **Y**) that are common to both **M** and **N** are determined using the eigenvectors and decomposed components from SVD following Equations (19) and (20).

$$\mathbf{X} = \bar{\mathbf{U}}\bar{\boldsymbol{\Theta}}\mathbf{T} \tag{19}$$

$$\mathbf{Y} = \bar{\mathbf{V}}\left(\mathbf{T}^{-1}\right)^{\mathrm{T}} \tag{20}$$

Finally, the relative concentration information for the unknown sample ($\mathbf{C}_{\mathrm{M}}$) is determined from $\boldsymbol{\Pi}$ and $\mathbf{C}_{\mathrm{N}}$ as in Equation (21).

$$\mathrm{diagonal}(\boldsymbol{\Pi}) = \frac{\mathbf{C}_{\mathrm{N}}}{\mathbf{C}_{\mathrm{M}} + \mathbf{C}_{\mathrm{N}}} \tag{21}$$

Ultimately, no matter which algorithm is applied, GRAM is designed to provide mathematically resolved chromatographic profiles of components that are present in both the standard and the unknown samples. Given the proper conditions, GRAM algorithms are designed to output results that are unambiguous and physically meaningful.

## 3.3 Multivariate Curve Resolution-Alternating Least Squares

MCR-ALS is used to mathematically resolve signals of bilinear and independent underlying factors (components) such as overlapping peaks in sections of 2D pixel-level chromatograms. It can also be applied to unfolded pixel-level LC × LC-DAD or GC × GC-TOFMS chromatograms wherein ${}^{1}t_{\mathrm{R}}$ and ${}^{2}t_{\mathrm{R}}$ are properly augmented in the same dimension or multiple sample chromatograms have been properly augmented to yield a bilinear 2D data matrix [66–70].

Consider an $(A \times B \times J)$ GC × GC-TOFMS data section containing $A$ pixels in the ${}^2t_R$ dimension, $B$ pixels in the ${}^1t_R$ dimension, and $J$ pixels in the mass spectral dimension. If significant retention time shifting occurs from run-to-run along the second GC dimension within a single sample run on the instrument, then this 3D data array may not be sufficiently trilinear [61]. Similarly, for LC × LC-DAD, under some circumstances the data may not be sufficiently trilinear, for example, when the data are collected over a relatively long period of time [70]. However, Tauler, Rutan, and others have shown that the data can be readily analyzed if the data array is properly unfolded so that the ${}^1t_R$ and ${}^2t_R$ indices are augmented, producing a bilinear 2D $(A{*}B \times J)$ matrix [66–70]. For the remaining discussion of the matrix dimensions and MCR-ALS, $A{*}B$ will be replaced with $I$, the number of pixels in the unfolded dimension. MCR-ALS can then be readily applied to resolve peaks, and the resulting pure component profiles can be refolded into the original 3D format. For example, Figure 3 shows a section of an unfolded GC × GC-TOFMS chromatogram of a blend of biodiesel and conventional diesel wherein the mass spectral dimension ($m/z$ 76–250) is perpendicular to the plane of the page. At least two components overlap in this region.

Figure 4 illustrates the MCR-ALS decomposition of the bilinear components in the unfolded GC × GC-TOFMS chromatographic data

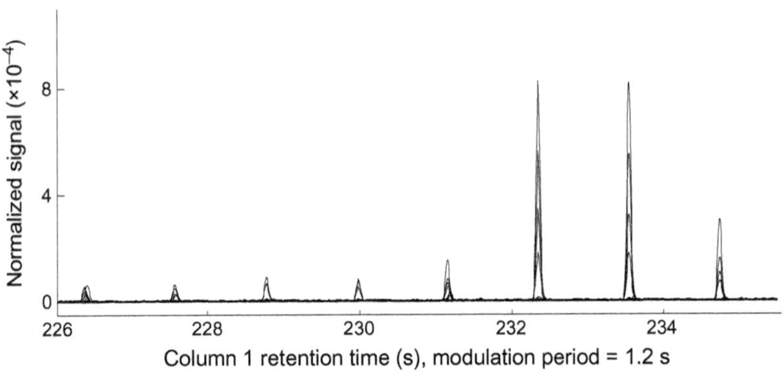


**FIGURE 3** Section of an unfolded GC × GC-TOFMS chromatogram of a blend of biodiesel and conventional diesel shown as overlaid chromatograms for ions of $m/z$ 76–250 with modulation period $(P_M) = 1.2$ s. The mass spectral dimension is perpendicular to the plane of the page.

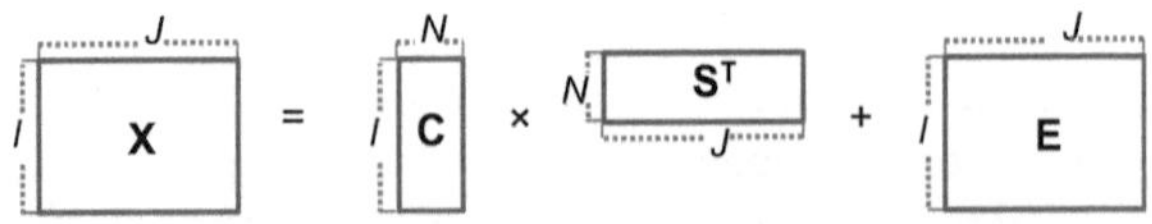


**FIGURE 4** An $(A \times B \times J)$ section of a GC × GC-TOFMS chromatogram was unfolded into a matrix (**X**) of dimensions $(I \times J)$, wherein $A{*}B = I$, and decomposed by MCR into its $N$ bilinear factors (pure components described by their $N$ pure chromatographic profiles in **C** and $N$ pure spectra in **S**) plus residual signals (**E**).

section. To be bilinear, data must be composed of unique, consistent, and concentration-dependent signals from each independent and unique source present, and those signals must be additive.

Equation (22) mathematically describes MCR-ALS applied to an unfolded GC × GC-TOFMS chromatographic data section of interest:

$$\mathbf{X} = \sum_{n=1}^{N} \mathbf{c}_n \mathbf{s}_n^{\mathrm{T}} + \mathbf{E} \tag{22}$$

In Equation (22), **X** represents the given unfolded ($I \times J$) chromatographic data section, which is equal to the sum (for $n=1$ to $N$ factors) of the product of the $n$th column ($\mathbf{c}_n$) of the ($I \times N$) loadings matrix **C** and the $n$th row ($\mathbf{s}_n$) of the ($N \times J$) transposed loadings matrix **S**, plus the ($I \times J$) error matrix **E**. Thus, each of the $N$ factors present is expressed by its pure mass spectrum ($\mathbf{s}_n$) multiplied by its resolved chromatographic profile ($\mathbf{c}_n$). Ideally, each of the $N$ columns and rows in **C** and **S**, respectively, fully describe a single analyte signal source. **E** contains all signals not accounted for in the $N$ factors and can be used as a metric for the fit between the model and the original data, **X**.

Given initial estimates of the pure concentration profiles and pure spectra of components in a matrix such as **X**, MCR-ALS implements an alternating least-squares (ALS) method to iteratively optimize the estimates of those pure analyte peak profiles and pure spectra while using appropriate constraints to achieve convergence. The alternating least-squares initialization values for **C** and **S** can be user-provided pure spectra or pure concentration profiles if the analysis is targeted and the information is available. However, for nontarget analysis, initialization values can be provided by evolving factor analysis, orthogonal projection approach, or simple-to-use self-modelling analysis [66]. These pure variable selection methods strive to detect the most dissimilar rows or columns in a matrix and use those as initial estimates of **C** and **S** for MCR-ALS. MCR-ALS goes on to calculate **C** and **S** while obeying appropriate constraints, which may include unimodality in the unfolded $^2t_R$ dimension and nonnegativity in both dimensions. The unimodality constraint is appropriate for chromatographic data which nominally produces Gaussian-like peaks, but if the user expects to observe a factor for noise or background contributions, then the unimodality constraint may not be appropriate. An MCR-ALS algorithm may be capable of accepting local rank/selectivity constraints wherein the user identifies regions in **X** where components are known to be present or absent [66]. The convergence criterion can be a preset number of iterations or a threshold value defining the difference in fit improvement between consecutive iterations. Metrics for fit between the model and original data may be the relative magnitude of **E**, the variance explained, or the lack-of-fit (LOF). LOF is expressed in Equation (23), where $x_{ij}$ is an element of the data matrix **X** and $e_{ij}$ is the corresponding element from matrix **E**.

$$\%\text{LOF} = 100\% \times \sqrt{\left(\sum_{i,j} e_{ij}^2\right) / \left(\sum_{i,j} x_{ij}^2\right)} \tag{23}$$

Ambiguous predictions may pose a challenge for MCR-ALS if and when various combinations of **C** and **S** reproduce **X** with equally good fit metrics. An example of intensity ambiguity is when $\mathbf{c}_n$ and $\mathbf{s}_n$ have signal intensity shifted from one dimension to the other such that **X** and **E** in Equation (1) are equal to **X** and **E** in Equation (24):

$$\mathbf{X} = \sum_{n=1}^{N} (\mathbf{c}_n k_i) \left( \mathbf{s}_n^{\mathrm{T}} \frac{1}{k_i} \right) + \mathbf{E} \tag{24}$$

In Equation (24), $k_i$ is a scalar that simultaneously upscales one dimension while downscaling the other. Thus, Equations (22) and (24) result in chromatographic and spectral profiles that have the *same shape* but differ in scale, so this is known as intensity ambiguity. To treat intensity ambiguity, the chromatographic dimension is often normalized so that the relative signal intensity information is stored in the spectral dimension. Another source of ambiguity in MCR-ALS can predict sets of pure chromatographic and spectral profiles that yield equally good fit metrics, yet have *different shapes*. This is known as rotational ambiguity. Any transform matrix (**T**) that can be multiplied by **C** while its inverse ($\mathbf{T}^{-1}$) is multiplied by $\mathbf{S}^{\mathrm{T}}$ will cause this type of ambiguity in MCR-ALS results. Constraints are the main method of suppressing rotational ambiguity. In addition, some degree of chemical selectivity in both separation dimensions is required for all factors present in the 2D data section to provide diversity that reduces the number of possible $k_i$ values and/or **T** matrices that adequately model **X** while fulfilling constraints. Quantitative information for an analyte can be obtained by summing all the elements of the outer product of the $n$th loadings vectors $\mathbf{c}_n$ and $\mathbf{s}_n$ to yield a scalar pure peak volume that should be proportional to relative concentration. When this quantification procedure is applied to a matrix composed of augmented sample chromatograms, the result yields concentration ratios between the common analyte components in the different samples.

The number of factors (value of $N$) is critical to the accuracy of the resolved profiles in **C** and **S**. If $N$ is too large (over fitting), the signal from a single source is described by two or more factors in the loadings $\mathbf{c}_n$ and $\mathbf{s}_n$. If $N$ is too small (under fitting), the signal from a single source might be absent in the loadings or it might be inappropriately combined with the loadings of another source. SVD may be used to estimate $N$. It is common practice to model a variety of $N$ values and then the user selects the MCR-ALS model that captures the highest degree of variance and predicts pure profiles with reasonable shapes. At this point, it may be necessary to note that MCR-ALS models are not nested. This means the $N$-resolved components in a chromatographic data section are represented by $N$ unique factors, which are all different than the factors produced by an MCR-ALS model for the same data section but built using $N-x$ factors, where $x$ is any positive integer from 1 to $N-1$.

An example of implementing MCR-ALS to decompose the GC × GC-TOFMS data from Figure 3 is provided in Figure 5. MCR-ALS yielded three pure chromatographic profiles using $N=3$, nonnegativity constraints in both dimensions, and 99.02% variance captured. The pure mass spectra (inset) reveal background contributions in component 3 were filtered out simultaneously as diesel components 1 and 2 were resolved.

## 3.4 Parallel Factor Analysis

PARAFAC [71–73] can be used to mathematically resolve signals of independent underlying factors (components) for third-order data, such as overlapping analyte peaks, background contributions, and/or noise in sections of 3D pixel-level GC × GC-TOFMS chromatograms where the first dimension is ${}^1t_R$, the second dimension is ${}^2t_R$, and the third dimension is the mass spectral dimension [74]. Figure 6 (top) is an example of a section of GC × GC-TOFMS data containing three diesel components where the mass spectral dimension has been temporarily summed for viewing purposes to produce a total ion current (TIC) chromatogram. Also shown (at the bottom of Figure 6) is the top-down view of the same section which shows the diesel components are unresolved, to some extent.

The PARAFAC decomposition of the trilinear components in a GC × GC-TOFMS chromatographic section is illustrated in Figure 7. To be trilinear, data must be composed of unique, consistent, concentration-dependent signals from each independent and unique analyte source present, and those signals must be additive.

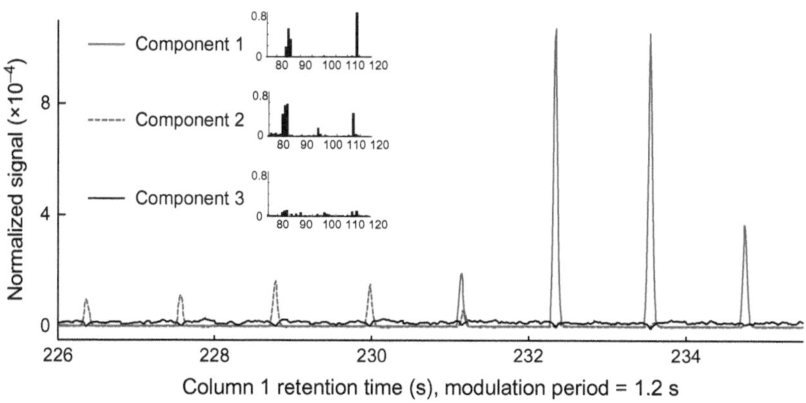


**FIGURE 5** Pure chromatographic profiles and pure mass spectra (inset) provided by MCR-ALS decomposition of three components in the GC × GC-TOFMS chromatographic data section wherein $P_M=1.2$ s.

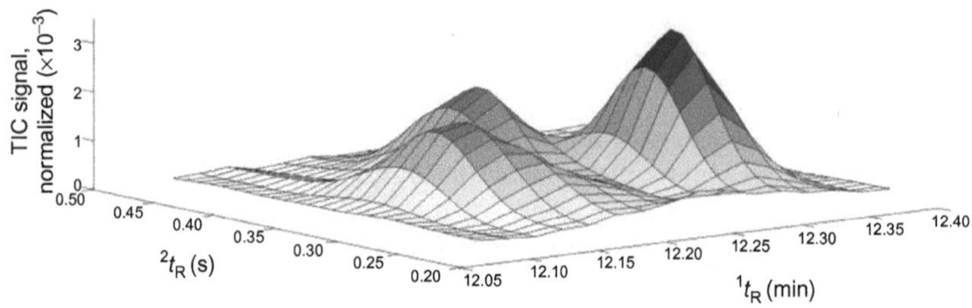


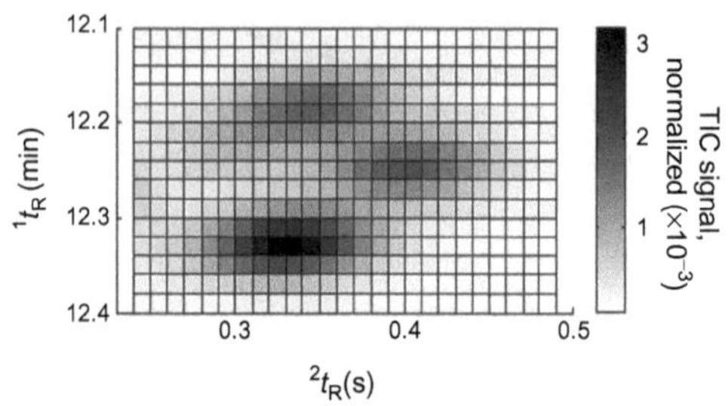


**FIGURE 6** Section of a GC × GC-TOFMS chromatogram of diesel: (top) The mass spectral dimension of the section is temporarily summed and shows unresolved peaks are present. (bottom) Top-down view of the same TIC section.

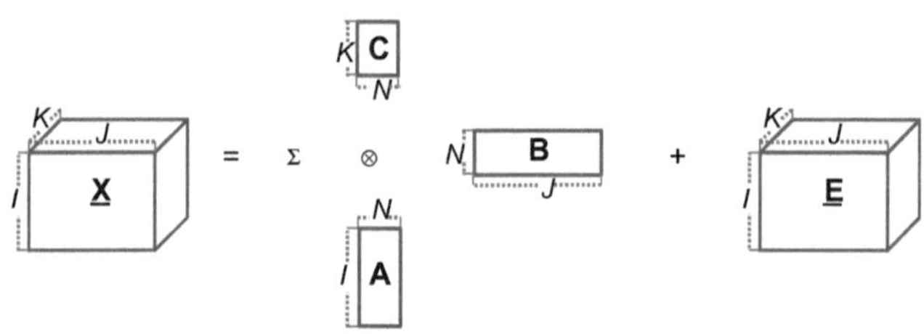


**FIGURE 7** PARAFAC decomposition of an ($I \times J \times K$) section of a GC × GC-TOFMS chromatogram ($\underline{\mathbf{X}}$) into its $N$ trilinear factors (pure components described by $N$ pure chromatographic profiles in **A** and **B**, as well as $N$ pure spectra in **C**) plus residual signals ($\underline{\mathbf{E}}$).

Equation (25) mathematically describes PARAFAC applied to a GC × GC-TOFMS data section:

$$\underline{\mathbf{X}} = \sum_{n=1}^{N} \mathbf{a}_n \otimes \mathbf{b}_n \otimes \mathbf{c}_n + \underline{\mathbf{E}} \tag{25}$$

In Equation (25), $\underline{\mathbf{X}}$ represents the given ($I \times J \times K$) chromatographic data section of interest. According to Equation (25), $\underline{\mathbf{X}}$ is equal to the sum of the outer product of each of the $n$th columns ($\mathbf{a}_n$, $\mathbf{b}_n$, and $\mathbf{c}_n$) of the respective ($I \times N$), ($J \times N$), and ($K \times N$) loadings matrices **A**, **B**, and **C**, for $n = 1$ to $N$, where $N$ is the number of factors, plus the $(I \times J \times K)$ error array $\underline{\mathbf{E}}$. Thus, $\mathbf{a}_n$, $\mathbf{b}_n$, and $\mathbf{c}_n$ essentially contain the resolved $^2t_R$ profile, the resolved $^1t_R$ profile, and the resolved mass spectral profile of each of the $N$ factors present in the chromatographic subregion. Ideally, each of the $N$ columns in **A**, **B**, and **C** describe a single signal source (e.g., factors such as analytes, noise, background contribution, interferents, etc). $\underline{\mathbf{E}}$ contains all signals not accounted for in the $N$ factors and it is a metric for the fit between the model and the original chromatographic data section.

PARAFAC can determine appropriate values for **A**, **B**, and **C** using the ALS method. Traditionally, ALS iteratively seeks values of **A**, **B**, and **C** that minimize $\underline{\mathbf{E}}$ while optionally applying constraints. The ALS algorithm can fail to locate the global minimum (true solution) when chromatographic sections contain a large number of factors, or if there is insufficient selectivity in every dimension, or if there are sufficiently large deviations from trilinearity. In such cases, repeatedly building PARAFAC models of a single given chromatographic data section may produce a variety of solutions. This instability can often be improved by applying various constraints such as unimodality in the $^1t_R$ and $^2t_R$ dimensions and nonnegativity in all three dimensions. The ALS initialization values for **A**, **B**, and **C** can be provided by trilinear decomposition (TLD) based on the generalized eigenvalue decomposition problem, which does not require any initial values but also does not allow constraints to be applied. Another way to initialize ALS may be to use random values and start from several different starting points of random values until the same solution is repeatedly reached. Other possible initialization values may be random orthogonalized values, or singular values. The stopping condition for ALS can be minimizing $\underline{\mathbf{E}}$, meeting a certain threshold value for sum of squares of the residuals ($\underline{\mathbf{E}}$), reaching a threshold *relative* change in $\underline{\mathbf{E}}$ between two iterations, allowing a maximum elapsed time for iterations, executing a maximum number of iterations, or meeting some other goodness-of-fit metric or convergence criterion.

The number of factors (value of $N$) is critical to the accuracy of the resolved peak profiles in **A**, **B**, and **C**. If $N$ is too big (over fitting), the signal from a single source is described by two or more factors in the loadings $\mathbf{a}_n$, $\mathbf{b}_n$, and $\mathbf{c}_n$. If $N$ is too small (under fitting), the signal from a single source might be absent in the loadings or it might be inappropriately combined with the loadings of another source. Some of the following methods can be used to determine the appropriate number of factors in a parsimonious manner. Some users seek the minimal value of $N$ that still provides a PARAFAC model with a relatively high core consistency value and explains a high percentage of the variance in $\underline{\mathbf{X}}$. Another method of determining $N$ is to examine the "noisiness" in $\underline{\mathbf{E}}$ and choose the lowest $N$ factor model, which appears to contain a high degree of randomness and where systematic variation is not visible. Another method is known as split-half analysis wherein the appropriate number of factors is revealed by that which produces identical PARAFAC models built for independent subsamples of the given data, though this is more amenable to data wherein one of the dimensions is the sample dimension, which is not the case with the GC × GC-TOFMS chromatogram considered herein. Another method of determining the appropriate value of $N$ for the PARAFAC model of a GC × GC-TOFMS data section is detection of "splitting" in the mass spectral loadings as $N$ is incrementally increased. Splitting is detected in an $N$-factor PARAFAC model when two or more decomposed mass spectra both match a common reference spectrum above a certain match value threshold because signal from a single source was spread across two or more factors. Thus, a PARAFAC model provided by $N-1$ or fewer factors is considered appropriate. At this point, it may be necessary to note that PARAFAC models are not nested. Also of note, PARAFAC can provide a unique solution for the final values of **A**, **B**, and **C** because there is no rotational problem when the appropriate number of factors is modeled and the data are sufficiently trilinear [73]. However, the final values of **A**, **B**, and **C** may have signal intensity shifted from one dimension to another when comparing two nominally equal PARAFAC models. To treat this ambiguity, both chromatographic dimensions are often normalized so that the signal intensity information is stored in the mass spectral dimension. Quantitative information for an analyte can be obtained by summing all the elements of the outer product of the $n$th loadings vectors $\mathbf{a}_n$, $\mathbf{b}_n$, and $\mathbf{c}_n$ to yield a scalar pure peak volume that is proportional to relative concentration.

Figure 8A shows the pure component profiles in the chromatographic dimensions provided by PARAFAC decomposition of the chromatographic data section shown in Figure 6. In this case, the PARAFAC software that was used was from the PLS Toolbox for MATLAB (Eigenvector Research, Inc) with ALS initiated by TLD using nonnegativity constraints in all three dimensions and three factors based on core consistency and prior knowledge of the sample). Figure 8B shows the pure component mass spectra that were also provided by PARAFAC for component 1 (top), component 2 (middle), and component 3 (bottom). According to a mass spectral library search, these components are alkylated benzenes. Figure 8C shows the product of the ${}^1t_R$ and ${}^2t_R$ pure component chromatographic profiles multiplied by (scaled by) the sum of the full pure component mass spectrum for component 1 (top), component 2 (middle), and component 3 (bottom). Manual comparison of Figure 8C to Figure 6 reveals these mathematically resolved pure component models provided by PARAFAC can indeed be superimposed or added to closely fit the original chromatographic data section of interest where the three peaks overlapped, to some extent.

Herein, PARAFAC was described for a GC × GC-TOFMS application, but PARAFAC is also suitable for LC × LC or any comprehensive 2D separation coupled with a selective multivariate detector [8–10]. PARAFAC decomposition requires the data to have sufficient selectivity in each of the three dimensions. For example, if two overlapping components have identical spectra,

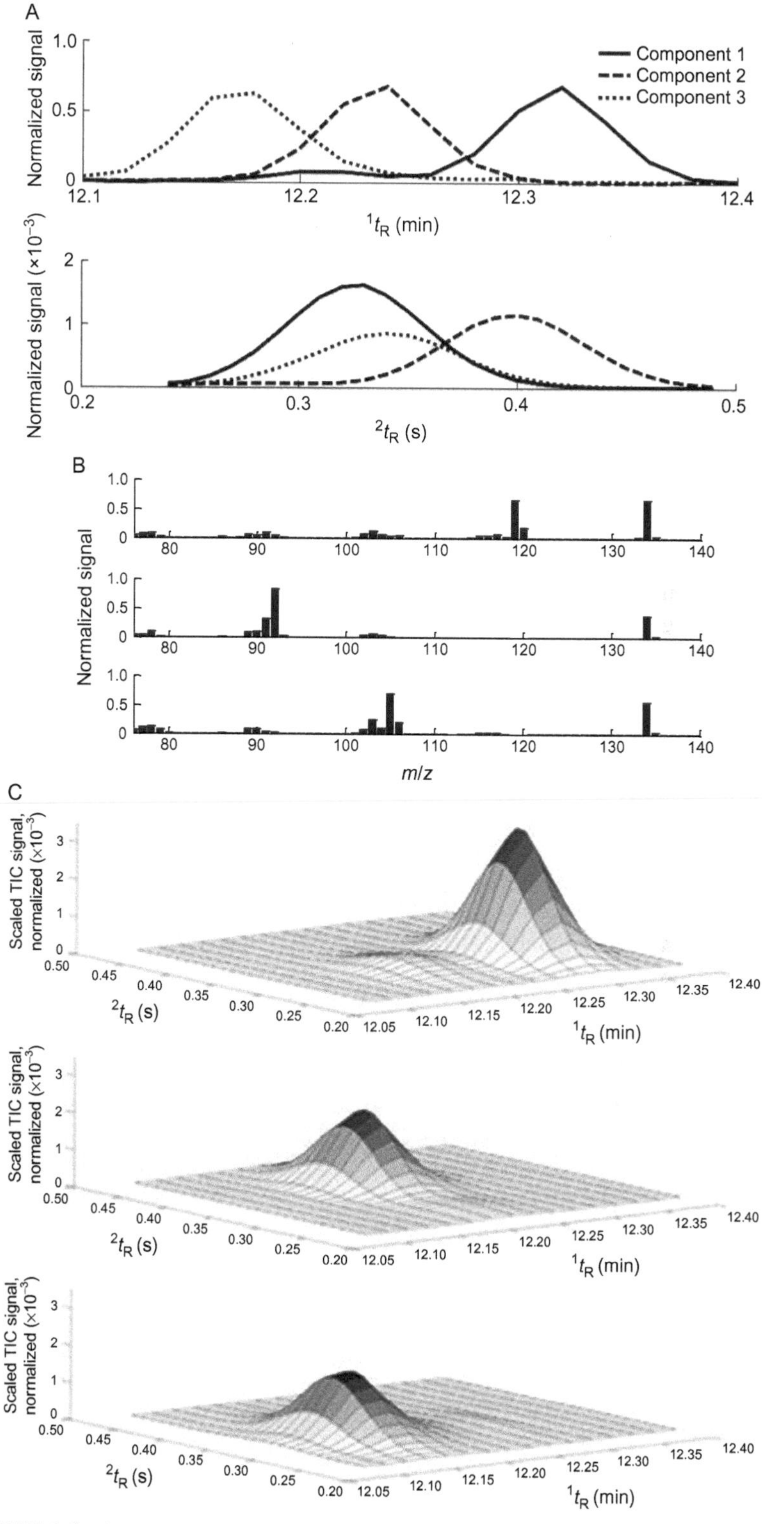


**FIGURE 8** See legend on next page.

then PARAFAC will not be able to resolve them. Prudent selection of complementary columns and detector parameters for GC × GC and LC × LC separations usually help provide sufficient selectivity in all three dimensions. PARAFAC decomposition also requires the data to be additive, which is often a valid assumption for stable sample concentrations that are within the linear range of the detectors on GC × GC and LC × LC instruments. PARAFAC decomposition also requires the data to be sufficiently trilinear. Experience has shown that GC × GC-TOFMS data do not need to be completely trilinear to achieve suitable quantitative accuracy, but sufficient trilinearity should be confirmed prior to application of PARAFAC [36,53,74,75]. Indeed, GC × GC and LC × LC instruments may be prone to retention time shifting in the second separation dimension within a single sample injection onto the instrument, so the rank may vary in the ${}^2t_R$ dimension, but not in the spectral dimension, causing deviations from trilinearity. If shifting is severe, pixel-level retention time alignment may be a necessary preprocessing step for PARAFAC. In general, while achieving trilinear data in LC × LC separations is now readily achieved for short-term studies [8–10], for long-term studies the trilinearity may be compromised so MCR-ALS is preferred for LC × LC-DAD [70]. For GC × GC-TOFMS, if within-run retention time shifting in the second GC dimension is severe, implementation of MCR-ALS for such cases should be explored, as discussed earlier in this chapter [69].

Creative and extremely useful variations and automations of PARAFAC have been described by experts. For example, software for automated non-targeted PARAFAC decomposition of *entire* GC × GC-TOFMS chromatograms was developed. The software automatically determines the optimal *N*-factor PARAFAC model for each component in a chromatographic subregion and then it continues to comprehensively analyze the entire chromatogram [75]. In another advancement, PARAFAC2 was shown to provide accurate solutions for GC × GC-TOFMS data exhibiting deviations from trilinearity caused by within-run retention time shifting in ${}^2t_R$ [76,77]. PARAFAC2 relaxes the strict trilinearity requirement by allowing the estimated loading matrix for the ${}^2t_R$ dimension to vary for each factor in the ${}^1t_R$ loadings matrix. As is the case for PARAFAC, the PARAFAC2 method provides a unique solution that cannot be rotated without a loss of fit and the solution provides pure peak profiles and pure mass spectra when the proper number of factors is used. Another advancement of note was PARAFAC applied to quantify analytes in comprehensive three-dimensional gas chromatography ($GC^3$) chromatograms [78].

**FIGURE 8** (A) Pure component profiles in the ${}^1t_R$ dimension (top) and the ${}^2t_R$ dimension (bottom) as provided by PARAFAC decomposition of the chromatographic subregion shown in Figure 1. (B) Pure component mass spectra provided by PARAFAC for component 1 (top), component 2 (middle), and component 3 (bottom). (C) Product of the ${}^1t_R$ and ${}^2t_R$ pure component profiles multiplied by (scaled by) the sum of the full pure component mass spectrum for component 1 (top), component 2 (middle), and component 3 (bottom).

# 4 FINGERPRINTING AND PATTERN RECOGNITION

## 4.1 Aim of Fingerprinting and Pattern Recognition

Chromatograms are composed of hundreds, if not thousands, of variables (retention time units) for which signals proportional to chemical concentration are recorded. Consider comparing many chromatograms of various samples of interest. Some of the chromatographic variables will reveal important chemical similarities (correlations) and important chemical differences (variations) among the samples. Manually examining chromatograms to discover interesting patterns is subjective, and large volumes of data may make manual observations unfeasible. Instead, an unsupervised pattern recognition algorithm can automatically explore a given data set and objectively reveal interesting patterns, variations, correlations, covariations, similarities, and differences among complex samples. Additionally, supervised pattern recognition algorithms use given quantitative and/or qualitative information to model a given data set and meet various goals such as discovering statistically significant similarities and differences among complex samples, or building a multivariate calibration model to predict a quantitative or qualitative property of an unknown sample. A major purpose of unsupervised and supervised pattern recognition algorithms is to reduce the data set dimensionality while preserving the original information, which allows more efficient manipulation of the data and perhaps a better understanding of underlying phenomena related to the samples.

For pattern recognition algorithms to be successful, sources of correlation and variation in the data must be true chemical correlations, manifested as relevant variations in the detector response. But chemical correlations and variations can be obscured by other sources of variation such as retention time shifting, uncontrollable instrumental fluctuations, injection volume discrepancies, and scaling variations among series of measurements with wide ranging magnitudes. Preprocessing techniques such as baseline correction, normalization, alignment or peak matching, mean centering, and scaling are often applied to data prior to pattern recognition analysis to diminish variations in the data that may obscure true chemical variations among samples.

Theoretically, if peak-level data are comprehensive, accurate, and precise, and if pixel-level data are bilinear and of suitable dimensions, then pattern recognition algorithms should return results that are equally accurate for both pixel-level and peak-level data. However, these requirements are often challenging to achieve for these two data types. The following sections explore the implementation of pattern recognition studies with pixel-level 2D separations data.

## 4.2 Unsupervised Pattern Recognition—PCA

PCA is used to find linear combinations of factors (latent variables (LVs) or principal components, PCs) in a bilinear 2D data matrix that model the reproducible variations and correlations (e.g., covariations, patterns) in that data matrix more succinctly. To be bilinear, the 2D data matrix must be composed of unique, consistent, concentration-dependent signals from each independent and unique source present, and those signals must be additive. The data matrix can be composed of many kinds of data including pixel-level data or peak-level data. PCA is certainly applicable to any kind of chromatography, but this text will use examples from pixel-level GC × GC-TOFMS data to explain PCA.

The first step in preparing pixel-level GC × GC-TOFMS chromatograms for PCA is to properly unfold and augment chromatograms from various samples of interest into a 2D matrix such that the first dimension is the sample dimension. Assume $I$ samples are included so the first dimension contains $I$ elements. The second dimension of the 2D matrix is the properly unfolded retention time dimension. For pixel-level analysis of GC × GC with a univariate detector (e.g., FID or TIC), the second dimension will contain $J$ elements, where $J = (\text{pixels in } {}^2t_R) \times (\text{pixels in } {}^1t_R)$. For pixel-level analysis of GC × GC with a multivariate detector, the second dimension will contain $J = (\text{pixels in } {}^2t_R) \times (\text{pixels in } {}^1t_R) \times (\text{pixels in detector dimension})$ elements, so in this case, $J$ is often extremely large. Without specialized computer equipment, analysts will encounter limits to the maximum size of the 2D matrix that can be successfully submitted to PCA algorithms. Thus, analysts may need to reduce data density prior to PCA using data binning techniques that were discussed earlier in this chapter. Assume the properly unfolded and bilinear chromatograms from many samples are augmented into matrix **X** of size $(I \times J)$ where the rows represent $I$ samples and the columns represent $J$ variables.

The goal of PCA is to produce a decomposition model of **X** which can be written in matrix notation as $\mathbf{X} = \mathbf{TP}^{\mathrm{T}} + \mathbf{E}$, wherein **T** is an $(I \times F)$ scores matrix, $\mathbf{P}^{\mathrm{T}}$ is a transposed $(F \times J)$ loadings matrix, **E** is an $(I \times J)$ residuals matrix, and $F$ is the number of PCs necessary to model 100% of the variation in **X**. The PCA decomposition model of **X** is illustrated in Figure 9. The scores contain information about how the samples (rows in **X**) relate to each other. The loadings contain information about how the variables (columns in **X**) relate to each other. Conceptually, if two samples have perfectly matching chromatograms, they should ideally have equal scores.

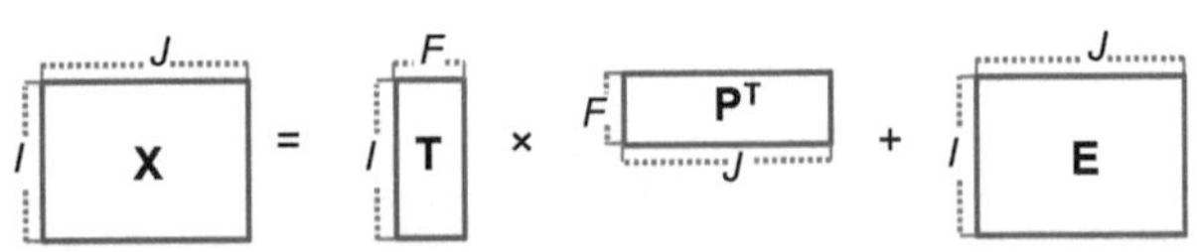


**FIGURE 9** Chromatograms composed of $J$ pixels from comprehensive 2D chromatography of $I$ samples were properly unfolded and augmented into a data matrix (**X**) of dimensions $(I \times J)$. PCA decomposes **X** into an $(I \times F)$ scores matrix (**T**), a transposed $(F \times J)$ loadings matrix ($\mathbf{P}^{\mathrm{T}}$), and an $(I \times J)$ residuals matrix (**E**). $F$ is less than or equal to the minimum of $I$ and $J$ and $F$ is the number of PCs necessary to model 100% of the variation in **X**.

The PCA model of $\mathbf{X}$ can also be written as the sum of the $F$ outer products of a score vector ($\mathbf{t}_f$) and its corresponding loadings vector ($\mathbf{p}_f$) plus residuals $\mathbf{E}$ as in Equation (26) [12,14,79,80].

$$\mathbf{X} = \sum_{f=1}^{F} \mathbf{t}_f \mathbf{p}_f^{\mathrm{T}} + \mathbf{E} \tag{26}$$

In Equation (26), $F$ is less than or equal to the minimum of $I$ and $J$, and $\mathbf{p}$ and $\mathbf{t}$ ideally preserve the important variations in $\mathbf{X}$. To preserve and summarize the important variations in $\mathbf{X}$ while finding $\mathbf{t}$ and $\mathbf{p}$, the PCA algorithm converts this model into a standard eigenvalue decomposition problem as in Equation (27).

$$\mathrm{cov}(\mathbf{X})\mathbf{p}_f = \boldsymbol{\lambda}_f \mathbf{p}_f \quad \text{wherein } \mathrm{cov}(\mathbf{X}) = \frac{(\mathbf{X}^{\mathrm{T}}\mathbf{X})}{(I-1)} \text{ for } I > J$$

or

$$\mathrm{cov}(\mathbf{X}) = \frac{\mathbf{X}\mathbf{X}^{\mathrm{T}}}{(I-1)} \text{ for } I \leq J \tag{27}$$

In Equation (27), cov($\mathbf{X}$) is defined as the covariance matrix of $\mathbf{X}$ when the columns of $\mathbf{X}$ have been mean centered. Mean centering is a preprocessing technique wherein the original mean value of each column of $\mathbf{X}$ is subtracted from each variable in the column to force the average of each column to be zero. If the columns of $\mathbf{X}$ have been autoscaled, then Equation (27) substitutes the correlation matrix of $\mathbf{X}$ (corr($\mathbf{X}$)) for cov($\mathbf{X}$). Autoscaling is a preprocessing technique wherein the columns of $\mathbf{X}$ are mean centered and then the variance in each column is normalized to unity by dividing each column by its original standard deviation. PCA loadings are a function of both signal variation and signal magnitude so autoscaling helps reduce the effect of signal magnitude on the loadings.

Equation (27) has a standard linear algebra solution that uses SVD to decompose cov($\mathbf{X}$) into the product of three terms: an orthogonal matrix $\mathbf{U}$, a diagonal singular values matrix $\mathbf{S}$, and a transposed orthogonal matrix $\mathbf{V}$. The $f$th squared singular value ($s_f$) is the same as the $f$th eigenvalue ($\boldsymbol{\lambda}_f$) in Equation (27) and so the unit normalized $\mathbf{p}_f$ (standardized eigenvectors) can be determined. The $\mathbf{p}_f$ loadings vectors are orthonormal to each other $\left(\mathbf{p}_i^{\mathrm{T}}\mathbf{p}_{\mathrm{j}} = 0 \,\text{for}\, i \neq j,\ \mathbf{p}_i^{\mathrm{T}}\mathbf{p}_j = 1 \text{ for } i = j\right)$. The $\boldsymbol{\lambda}_f$ reveal the percent variation captured in each factor. Finally, each score vector $\mathbf{t}_f$ is supposed to be the linear combination of the original $\mathbf{X}$ variables defined by $\mathbf{p}_f$, so using the $\mathbf{X}\mathbf{p}_f = \mathbf{t}_f$ projection reveals the $\mathbf{t}_f$ scores vectors which are orthogonal to each other ($\mathbf{t}_i^{\mathrm{T}}\mathbf{t}_j = 0$ for $i \neq j$). Ambiguity can be a problem for PCA because any scalar value can simultaneously upscale the $\mathbf{t}$ or $\mathbf{p}$ vector while downscaling the other vector and still solve Equations (26) and (27). Thus $\mathbf{p}$ is normalized to unit length in Equation (27) to treat this scaling ambiguity.

The $\mathbf{t}_f$, $\mathbf{p}_f$ pairs are sorted in descending order of variation based on $\boldsymbol{\lambda}_f$ values. The linear combinations of the pairs ($\mathbf{t}_f\,\mathbf{p}_f^{\mathrm{T}}$) are known as PCs (specifically PC 1 through PC $F$ for $f=1$ to $F$). PC 1 most closely models $\mathbf{X}$ in Equation (26) with residuals as minimal as possible, considering PC 1 is a linear factor trying to capture maximum variation in $\mathbf{X}$. Each successive PC models diminishing amounts of variation in $\mathbf{X}$. The PCs are nested, meaning a model retaining $F+1$ PCs has PC 1 through PC $F$ which are identical to a model retaining $F$ PCs, for any integer value of $F$. Therefore, the user can truncate $\mathbf{T}$ and $\mathbf{P}^{\mathrm{T}}$ to $F$ columns and rows, respectively, to keep only the initial PCs that capture sufficient variation to adequately model $\mathbf{X}$. Choosing the number of PCs to retain is a subjective matter and there is not a single best method for making this decision. Some of the methods that exist involve inspecting the ordered $\boldsymbol{\lambda}_f$ and retaining only the corresponding PCs which captured a relatively high percentage of variation compared to that of the final PCs which are assumed to represent noise.

The analyst can plot and inspect the model residuals to glean information about variables that were not modeled. This may reveal weaknesses or strengths in assuming the data set is suitable for PCA modeling. The analyst can examine their data set using various metrics related to PCA modeling, particularly outlier detection metrics. Hotelling's $\mathrm{T}^2$ measures the Mahalanobis distance of a score from the PCA model origin and accounts for different variation in different PCs. Samples with large Hotelling's $\mathrm{T}^2$ might be outliers, assuming the samples are supposed to represent a multivariate normal distribution. The metric sum square error ($Q$) measures the squared distance of a score from a plane of scores in a PCA model, and samples with large $Q$ might be considered outliers. When using such metrics the analyst should be conservative and always carefully review all assumptions about normal distributions and variations in the data set when deciding whether the metrics are appropriate for labeling a sample as an outlier.

In addition to the above description of PCA, the following geometric conceptualization of PCA might be helpful. Consider a data set of 10 sample chromatograms, each composed of 20,000 pixels. The first conceptual step in applying PCA to this data set is to imagine each preprocessed and unfolded chromatogram as a 1D data vector plotted in independent variable-space. This means each chromatogram is geometrically represented by a single vector endpoint located in the independent variable space that was defined by 20,000 orthogonal axes ($J=20{,}000$). Then, among those 10 vector endpoints (representing the $I=10$ sample chromatograms), PCA calculates a vector (PC 1) which is fit so that it captures the greatest variation in the locations of those 10 vector endpoints. A second principal component (PC 2) that is orthogonal to PC 1 is calculated and fit such that it captures the next greatest variation in those 10 vector endpoints. This is repeated until up to 10 orthogonal PCs are fit and ordered based on captured percent variation. The 10 PCs will cumulatively have modeled 100% of the variation in the locations of those original 10 vector endpoints. The 10 vector endpoints can now be thought of as being in PC space relative to (at most) 10 PC axes, rather than 10 points relative to the original 20,000 independent variable-axes. Since the PCs are ordered and nested, the analyst can truncate the later PCs and project the 10 vector endpoints onto only the first few primary PCs. The distance in PC space between each vector endpoint and each PC axis is the score for each sample. Samples that are similar to each other will have similar score values because the vector endpoints of similar samples cluster together in PC space. Samples that are different from each other will have different score values because the vector endpoints of different samples may be farther apart in PC space. When the 10 sample scores on PC 1 are plotted, this "scores plot" reveals chemical similarities (or differences) among the 10 samples (similar samples may have similar scores and different samples may have different scores).

As for a conceptual description of the loadings in PCA, imagine the 10 PCs in PC space are 10 axes that are projected onto the original 20,000 independent variable-axes. Each PC axis has a calculable angle between it and each of the original 20,000 independent variable-axes. If an original variable had the largest variations across samples, then it would have a large influence on the calculation of PC 1 and the angle at the intersection of the two axes is small. If an original variable had little variation across samples, then it would have a small influence on PC 1 and the angle at the intersection of the two axes is large. As that angle of intersection decreases from $\Pi/2$ to 0°, the cosine of the angle goes from 0 to 1. Summarily, the loadings values are the cosines of the angles between each PC axis and each original independent variable axis. For mean centered chromatograms, the loadings values range from $-1$ to 1. Original independent variables (retention times) that have the most influence on the clustering of vector endpoints in PC space will have the greatest absolute loadings. Retention times with the most substantial amounts of signal variation across all the samples will be the most highly loaded variables in the primary PCs. When loadings on PC 1 are plotted as a function of the original variables, the variables with greatest absolute magnitude in this "loadings plot" had the most influence on any clustering observed in the PC 1 scores plot.

Consider an example of PCA applied to a data set of normalized, properly unfolded, and mean centered GC × GC TIC chromatograms of seven samples (two pure conventional diesels and five blends of various biodiesels with various conventional diesels). In this case, the PCA software that was used was from the PLS Toolbox for MATLAB (Eigenvector Research, Inc.). A chromatogram of one of the blends is shown in Figure 10A. PCA is not given this information, but it is useful to know that the peaks eluting before 47 min are primarily conventional diesel components and the peaks eluting after 47 min are primarily biodiesel components, specifically fatty acid methyl esters (FAMEs). Figure 10B shows some results of PCA: the scores on PC 1 captured 60.78% of the variation in the seven chromatograms. It is apparent that samples 1 and 2 have negative scores, samples 3, 4, and 5 have moderately positive scores, and samples 6 and 7 have the highest positive scores in this data set. Conceptually, this means that samples 1 and 2 may be similar to each other, samples 3, 4, and 5 may be similar to each other, and samples 6 and 7 may be similar to each other, if some subjectivity is acceptable in deciding that three classes are present rather than two classes. Figure 10C

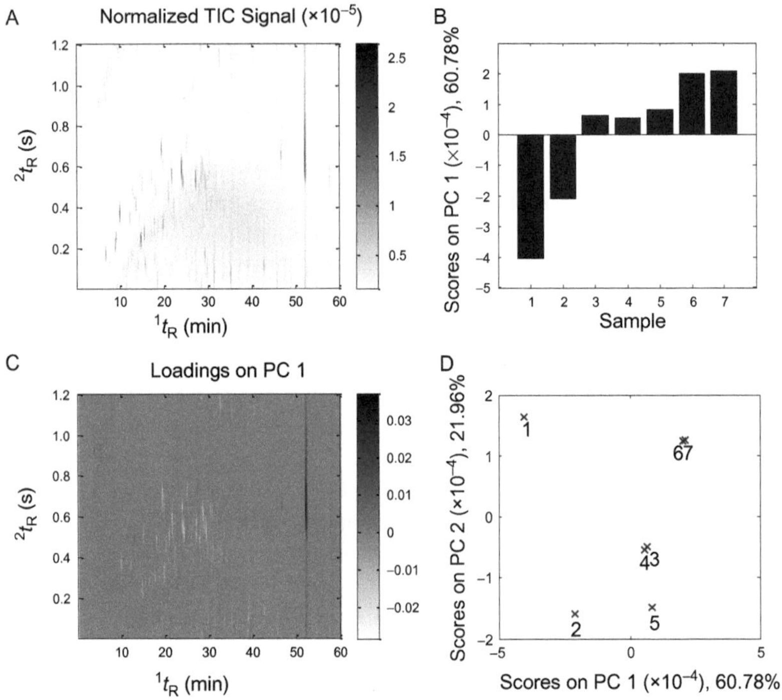


**FIGURE 10** GC × GC TIC chromatograms of seven samples* composed of various biodiesels and various conventional diesels were submitted to PCA. (A) The chromatogram of Sample 3. Peaks eluting before 47 min are primarily conventional diesel components. Peaks eluting after 47 min are primarily biodiesel components. (B) Scores on PC 1 which captured 60.78% of the variation. (C) Loadings plot for PC 1. The most positively loaded variables are black, while the most negatively loaded variables are white, and variables with moderate-to-zero absolute loadings values are gray. (D) Scores plot for PC 1 versus PC 2. PC 2 captured 21.96% variation.
*Sample 1 = 100% (v/v) Conventional Diesel K;
*Sample 2 = 100% (v/v) Conventional Diesel L;
*Sample 3 = 12.5% (v/v) Biodiesel P with 87.5% (v/v) Conventional Diesel L;
*Sample 4 = 12.5% (v/v) Biodiesel Q with 87.5% (v/v) Conventional Diesel L;
*Sample 5 = 12.5% (v/v) Biodiesel R with 87.5% (v/v) Conventional Diesel L;
*Sample 6 = 20% (v/v) Biodiesel P with 80% (v/v) Conventional Diesel M;
*Sample 7 = 20% (v/v) Biodiesel Q with 80% (v/v) Conventional Diesel N.

shows the loadings plot for PC 1. The most positively loaded chromatographic variables are black and it appears that these are primarily the FAMEs. The most negatively loaded variables are white and it appears that these are primarily the conventional diesel components. The variables with little to zero loadings (both positive and negative) are gray. Combining the basic conclusions from the scores plot with the basic conclusions from the loadings plot could lead an analyst to surmise that PCA reveals samples 1 and 2 have similar conventional diesel components, while samples 3, 4, 5, 6, and 7 have relatively similar FAME components. The analyst might further conclude that PCA reveals samples 3, 4, and 5 have biodiesel components similar to each other, while samples 6 and 7 are in a different category of biodiesel composition. In fact, the samples can be loosely classified into three categories: (category 1) samples 1 and 2 are pure conventional diesels, (category 2) samples 3, 4, and 5, are composed of 12.5% (v/v) biodiesel blended with conventional diesel, and (category 3) samples 6 and 7 are composed of 20% (v/v) biodiesel blended with conventional diesel. Figure 10D shows a plot of scores on PC 1 versus scores on PC 2. PC 2 captured 21.96% of the variation in the seven chromatograms and it is plotted on a slightly different scale than the PC 1 axis which captured the 60.78% variation mentioned earlier. In this scores plot, samples 6 and 7 cluster together, while samples 3 and 4 cluster together; this matches the tentative categorizations gleaned from inspecting the clustering in Figure 10B. However, contrary to what might be expected, samples 1 and 2 do not cluster together and instead sample 2 is closest to sample 5. This is because PC 2 captures variations that are orthogonal to PC 1. PC 1 primarily focused on the presence or absence of biodiesel FAMEs, but the variables highly loaded in PC 2 must be indicative of other variations, perhaps only variations in the conventional diesels; the analyst would have to look at the loadings on PC 2 to find out. Since the sample 1 score on PC 2 is positive while the sample 2 score on PC 2 is negative, an analyst might conclude there are chemical differences between these two samples of pure conventional diesel, so perhaps there really are more than three categories of samples present. Finally, sample compositions are revealed in the caption of Figure 10, and the reader can observe the similarities and differences among the samples. Perhaps, as Figure 10D indicated, there is some similarity between sample 2 and sample 5, after all. The reader can evaluate how well PCA, an unsupervised pattern recognition method, preserved and modeled important variations in the data set.

## 4.3 Supervised fingerprinting and pattern recognition

### *4.3.1 PLS Analysis and Multilinear Partial Least Squares for* N*-way analysis*

PLS analysis is a multivariate calibration method for modeling a relationship of maximum covariance between given 1D sample data (independent variables) and corresponding quantitative data (dependent variables) [81–83]. Once the relationship between the independent variables and the dependent variables is modeled, the values of the dependent variable for unknown samples can be predicted through regression.

Multilinear partial least squares for $N$-way analysis (NPLS) generalizes PLS from 1D data to higher order ND data (where $N$ is a small positive integer) [84–86]. NPLS can be applied to data sets of many 2D or 3D chromatograms that have been augmented into a single matrix where one dimension is the sample dimension. The augmented data matrix must be composed of trilinear data. To be trilinear, data must be composed of unique, consistent, concentration-dependent signals and those signals must be independently additive. NPLS is applicable to both pixel-level and peak-level data; theoretically, both levels of data may return equally accurate results as long as both are comprehensive and peaks are properly aligned across samples.

Consider a 3D data array ($\underline{\mathbf{X}}$) composed of augmented samples of pixel-level GC × GC-FID chromatograms (or chromatograms from any comprehensive 2D chromatography with a univariate detector). The dimensions of $\underline{\mathbf{X}}$ are ($I$ = number of samples × $J$ = data points on ${}^1t_R$ × $K$ = data points on ${}^2t_R$). Also consider a single column vector ($\mathbf{y}$) of given quantitative information for each sample. NPLS decomposes $\underline{\mathbf{X}}$ into $F$ factors (e.g., LVs or linear combinations of the original variables) defined by $F$ score vectors ($\mathbf{t}$ ($I \times 1$)), $F$ weight vectors in the ${}^1t_R$ dimension (${}^J\mathbf{w}$ ($J \times 1$)), and $F$ weight vectors in the ${}^2t_R$ dimension (${}^K\mathbf{w}$ ($K \times 1$)). These produce a model of $\underline{\mathbf{X}}$ that can be represented by Equation (28) and a model of $\mathbf{y}$ with regression coefficients ($\mathbf{b}$) that can be represented by Equation (29):

$$x_{ijk} = t_i {}^J w_j {}^K w_k \tag{28}$$

$$y_i = t_i b_i \tag{29}$$

NPLS finds $\mathbf{t}_i$, ${}^J\mathbf{w}$, and ${}^K\mathbf{w}$ that fit the model in Equation (28) while maximizing the covariance of $\underline{\mathbf{X}}$ with $\mathbf{y}$.

The following steps describe an algorithm to build an NPLS model of $\underline{\mathbf{X}}$ [84].

Step 1. Mean center the given $\underline{\mathbf{X}}$ and $\mathbf{y}$ and save the mean data so it can be later used during model regression and prediction. Do the same for scaling the data, if scaling is appropriate [87]. Initially, use $f = 1$ to calculate the first LV.

Step 2. Calculate the product of the cube $\underline{\mathbf{X}}$ and the vector $\mathbf{y}$ to yield the matrix $\mathbf{Z}$ ($J \times K$).

$$z_{jk} = \sum_{i=1}^{I} x_{ijk} y_i$$

Step 3. Calculate the weight vectors ${}^J\mathbf{w}$ ($J \times 1$) and ${}^K\mathbf{w}$ ($K \times 1$), each normalized to unit length, which yield a solution to the model in Equation (28) by using the first set of normalized vectors from SVD of $\mathbf{Z}$. Some texts interchange the terms "weights" and "loadings."

$$\max_{{}^J\mathbf{w}{}^K\mathbf{w}}\left[\left({}^J\mathbf{w}\right)^{\mathrm{T}}\mathbf{Z}^K\mathbf{w}\right]\approx\left({}^J\mathbf{w}^K\mathbf{w}\right)=\mathrm{SVD}(\mathbf{Z})$$

Step 4. Calculate the **t** that has maximal covariance with **y**. When the **w** vectors are known and of length one, the solution to the problem of finding **t** is found by projection of $\underline{\mathbf{X}}$ onto the **w** vectors. This can be described in summation notation and combined with the prior expression for $z_{jk}$ to yield [84]:

$$\max_{{}^J\mathbf{w}{}^K\mathbf{w}}\left[\sum_{i=1}^{I}t_i y_i | t_i=\sum_{j=1}^{J}\sum_{k=1}^{K}x_{ijk}{}^J w_j{}^K w_k\right]=\max_{{}^J\mathbf{w}{}^K\mathbf{w}}\left[\sum_{j=1}^{J}\sum_{k=1}^{K}z_{jk}{}^J w_j{}^K w_k\right]$$

An alternative format for this calculation of each element of **t** ($t_i$) under the constraint of maximizing covariance is [88,89]:

$$t_i=\sum_{j=1}^{J}\sum_{k=1}^{K}x_{ijk}{}^J w_j{}^K w_k$$

(As an aside, at this point the *f*th LV is definable by **t**, ${}^J\mathbf{w}$, and ${}^K\mathbf{w}$, and $\hat{x}_{ijk}=\mathbf{t}_i{}^J\mathbf{w}_j\left({}^K\mathbf{w}_k\right)^{\mathrm{T}}$ is the prediction of $x_{ijk}$ that represents the best least squares model of $\underline{\mathbf{X}}$. An alternative format of expressing the model of $\underline{\mathbf{X}}$ is $\hat{x}_{ijk}=\sum_{f=1}^{F}t_i={}^J w_j{}^K w_k$).

Step 5. Use **t** to calculate regression coefficients: $\mathbf{b}=(\mathbf{t}^{\mathrm{T}}\mathbf{t})^{-1}\mathbf{t}^{\mathrm{T}}\mathbf{y}$
(Note that at this point the prediction of **y** is $\hat{\mathbf{y}}=\mathbf{tb}$.)
Step 6. Calculate the residuals for the chromatograms: $\mathbf{X}_{i,\mathrm{res}}=\mathbf{X}_i-t_i{}^J\mathbf{w}\,({}^K\mathbf{w})^{\mathrm{T}}$.
Step 7. Calculate the residuals for the dependent variable values: $\mathbf{y}_{\mathrm{res}}=\mathbf{y}-\mathbf{tb}$.
Step 8. Iterate to the next factor ($f=f+1$) and return to Step 2 with the following deviations: (1) use the residuals from Steps 6 and 7 to replace $\mathbf{X}_i$ and **y**; (2) during Steps 5 and 7 replace **t** with the ($I\times F$) scores matrix (**T**) whose columns are the augmented **t** vectors for $f=1$ to $f$ [80,84]. Continue reiterating Steps 2–8 until convergence when a proper prediction of **y** is achieved.

To predict the dependent variable for unknown samples of interest using the NPLS model, use the saved mean data from Step 1 to center the unknown sample chromatogram. Do the same to scale the data, if appropriate. Then regress the preprocessed unknown sample chromatogram ($\mathbf{X}_{\mathrm{unk}}$) onto the model ($\mathbf{y}=\mathbf{bX}$) and predict the value of the dependent variable for the unknown sample of interest.

Generally, the model is built using an appropriate training set and the optimal number of LV ($F$) may be chosen by a cross-validation procedure that allows the user to appropriately minimize the root mean square error of cross-validation of predicted $y_i$ values to avoid over fitting or under fitting the model. Then the model is evaluated using an independent test set.

The NPLS algorithm described here is limited to only one dependent variable (single column vector **y**), but another version of this algorithm exists for calculating the NPLS model using several dependent variables for each sample (when **Y** is a matrix) [84]. The algorithm described here is also limited to only a univariate detector, but for comprehensive 2D chromatography with a multivariate detector the ND data array where *N* equals any small positive integer can be modeled by NPLS using more detailed versions of the algorithm [84,86]. If it is appropriate, 2D chromatograms could theoretically be unfolded into 1D and analyzed by PLS, but expect the results to differ because PLS calculates an additional set of weight vectors to maintain orthogonal scores while NPLS does *not* provide orthogonal scores and weights vectors [89].

Conventional figures-of-merit that are commonly reported with univariate calibration methods need multivariate propagated error techniques to be completely defined for multivariate calibration methods. In fact, instead of quantifying a single limit of detection (LOD) for a model, it is more appropriate to report an interval of LODs for PLS calibration models [90]. The standard error of prediction metric for multiway PLS regression is also available [91].

Examples of NPLS applied to comprehensive 2D chromatography include modeling and predicting the aromatic and naphthene content in naphtha samples and modeling and predicting the percent composition of blends of biodiesel and conventional diesel [92,93].

### 4.3.2 Supervised Pattern Recognition—Feature Selection

Supervised variable selection techniques are used to discover statistically significant chemical differences or similarities among known classes of complex samples. Since class membership must be known, experimental design is particularly important. Furthermore, eliminating uninformative variables and reducing voluminous data down to selective variables can improve predictive models. Two common techniques are uninformative variable elimination (UVE) [94] and Fisher ratios (*F*-ratio) feature selection [11,31,32] methods.

UVE is designed to inspect a given data set and eliminate the independent variables that contain no more information than that of random variables [94]. The original UVE algorithm is designed to analyze the regression coefficients ($\mathbf{b}$ $(1 \times K)$) of the PLS multivariate calibration model for a data set of independent variables ($\mathbf{X}$ $(J \times K)$) related to dependent variables ($\mathbf{y}$ $(J \times 1)$) with residuals ($\mathbf{e}$). The PLS model is represented by $\mathbf{y} = \mathbf{b} \times \mathbf{X} + \mathbf{e}$. UVE looks at the "reliability" ($\mathbf{c}$ $(1 \times K)$) of the regression coefficients for mean centered data, where each element of **c** is defined as each element of **b** divided by the standard deviation of **b** coefficients obtained by leave-one-out modeling. Independent variables that have the smallest **c** values are considered uninformative and are eliminated from the data set. The UVE algorithm defines the threshold for **c** as the value of **c** computed for random variables that are temporarily added to the data set. Variations of the algorithm exist to make it robust against outliers, to reduce influence of the maximum random **c** value on the final threshold value, and to calculate **c** using the absolute value of **b** coefficients from autoscaled data.

*F*-ratio feature selection methods are designed to inspect a data set and select variables that contain statistically significant differences among given sample classes [11,31,32,95]. The *F*-ratio of each independent variable across a given variety of sample classes is defined as the class-to-class variation of that independent variable divided by the sum of its within-class variations, which may be defined as follows when applied to chromatography [95].

Class-to-class variation $= \sigma_{\mathrm{cl}}^2 = \sum \frac{(\overline{x_i} - \bar{x})^2 n_i}{(k-1)}$ (where $n_i$ is the number of chromatograms in the $i$th class, $\bar{x}_i$ is the mean of the $i$th class, $\bar{x}$ is the overall mean, and $k$ is the number of classes).

Within-class variation $= \sigma_{\mathrm{err}}^2 = \frac{\sum\left(\sum(\overline{x_{ij}} - \bar{x})^2\right) - \left(\sum(\overline{x_i} - \bar{x})^2 n_i\right)}{(N-k)}$ (where $x_{ij}$ is the $i$th measurement of the $j$th class, and $N$ is the total number of chromatograms).

Finally, the $F$-ratio $= \frac{\sigma_{\mathrm{cl}}^2}{\sigma_{\mathrm{err}}^2}$.

Independent variables with large *F*-ratios are likely to be features of interest that differentiate sample classes, whereas independent variables with small *F*-ratios are likely to be representative of noise or features that do not differentiate sample classes. A recently introduced *F*-ratio algorithm called the tile-based Fisher-ratio software [31,32] is designed to comprehensively analyze 4D data sets of pixel-level GC × GC-TOFMS chromatograms to discover chemical features that differentiate given sample classes. The tile-based *F*-ratio software uses a novel indexing scheme that benefits from the advantages offered by pixel-level data analysis (namely, comprehensive nontargeted information is quickly gleaned) as well as the advantages offered by peak-level data analysis (namely, the algorithm is robust against retention time variations). The independent variables are sorted in descending order of *F*-ratio value which corresponds with the analyst's level of interest in each independent variable. *F*-ratio analysis is a robust, comprehensive, nontargeted analysis method that is useful for a variety of studies, particularly forensics experiments in which cause and effect chemical signatures can be cleverly evaluated by the experimental design. It is also useful for metabolomics experiments wherein the analyst often seeks to discover interesting patterns in the chemical fingerprints that characterize cellular processes.

## REFERENCES

[1] Giddings JC. Two-dimensional separations: concept and promise. Anal Chem 1984;56:1258–70. http://dx.doi.org/10.1021/ac00276a003.

[2] Liu Z, Phillips JB. Comprehensive two-dimensional gas chromatography using an on-column thermal modulator interface. J Chromatogr Sci 1991;29:227–31. http://dx.doi.org/10.1093/chromsci/29.6.227.

[3] Bushey MM, Jorgenson JW. Automated instrumentation for comprehensive two-dimensional high-performance liquid chromatography of proteins. Anal Chem 1990;62:161–7. http://dx.doi.org/10.1021/ac00201a015.

[4] Bruckner CA, Prazen BJ, Synovec RE. Comprehensive two-dimensional high-speed gas chromatography with chemometric analysis. Anal Chem 1998;70:2796–804. http://dx.doi.org/10.1021/ac980164m.

[5] Harynuk J, Marriott PJ. Fast GC×GC with short primary columns. Anal Chem 2006;78:2028–34. http://dx.doi.org/10.1021/ac0519413.

[6] Seeley JV, Micyus NJ, Bandurski SV, Seeley SK, McCurry JD. Microfluidic deans switch for comprehensive two-dimensional gas chromatography. Anal Chem 2007;79:1840–7. http://dx.doi.org/10.1021/ac061881g.

[7] Siegler WC, Fitz BD, Hoggard JC, Synovec RE. Experimental study of the quantitative precision for valve-based comprehensive two-dimensional gas chromatography. Anal Chem 2011;83:5190–6. http://dx.doi.org/10.1021/ac200302b.

[8] Porter SEG, Stoll DR, Rutan SC, Carr PW, Cohen JD. Analysis of four-way two-dimensional liquid chromatography-diode array data: application to metabolomics. Anal Chem 2006;78:5559–69. http://dx.doi.org/10.1021/ac0606195.

[9] Stoll DR, Cohen JD, Carr PW. Fast, comprehensive online two-dimensional high performance liquid chromatography through the use of high temperature ultra-fast gradient elution reversed-phase liquid chromatography. J Chromatogr A 2006;1122:123–37. http://dx.doi.org/10.1016/j.chroma.2006.04.058.

[10] Stoll DR, Li X, Wang X, Carr PW, Porter SEG, Rutan SC. Fast, comprehensive two-dimensional liquid chromatography. J Chromatogr A 2007;1168:3–43. http://dx.doi.org/10.1016/j.chroma.2007.08.054.

[11] Pierce KM, Hoggard JC, Hope JL, Rainey PM, Hoofnagle AN, Jack RM, et al. Fisher ratio method applied to third-order separation data to identify significant chemical components of metabolite extracts. Anal Chem 2006;78:5068–75. http://dx.doi.org/10.1021/ac0602625.

[12] Brereton RG. Chemometrics: data analysis for the laboratory and chemical plant. New York: Wiley; 2003.

[13] Massart DL. Chemometrics: a textbook. New York: Elsevier Sciences Ltd.; 1988

[14] Beebe KR, Pell RJ, Seasholtz MB. Chemometrics: a practical guide. New York: Wiley-Interscience; 1998.

[15] Amigo JM, Skov T, Bro R. ChroMATHography: solving chromatographic issues with mathematical models and intuitive graphics. Chem Rev 2010;110:4582–605. http://dx.doi.org/10.1021/cr900394n.

[16] Cortes HJ, Winniford B, Luong J, Pursch M. Comprehensive two dimensional gas chromatography review. J Sep Sci 2009;32:883–904. http://dx.doi.org/10.1002/jssc.200800654.

[17] Mondello L, Tranchida PQ, Dugo P, Dugo G. Comprehensive two-dimensional gas chromatography-mass spectrometry: a review. Mass Spectrom Rev 2008;27:101–24. http://dx.doi.org/10.1002/mas.20158.

[18] François I, Sandra K, Sandra P. Comprehensive liquid chromatography: fundamental aspects and practical considerations—a review. Anal Chim Acta 2009;641:14–31. http://dx.doi.org/10.1016/j.aca.2009.03.041.

[19] Reichenbach SE, Tian X, Tao Q, Stoll DR, Carr PW. Comprehensive feature analysis for sample classification with comprehensive two-dimensional LC. J Sep Sci 2010;33:1365–74. http://dx.doi.org/10.1002/jssc.200900859.

[20] Pierce KM, Kehimkar B, Marney LC, Hoggard JC, Synovec RE. Review of chemometric analysis techniques for comprehensive two dimensional separations data. J Chromatogr A 2012;1255:3–11. http://dx.doi.org/10.1016/j.chroma.2012.05.050.

[21] Cook DW, Rutan SC. Chemometrics for the analysis of chromatographic data in metabolomics investigations. J Chemom 2014;28:681–7. http://dx.doi.org/10.1002/cem.2624.

[22] Ruckebusch C, Blanchet L. Multivariate curve resolution: a review of advanced and tailored applications and challenges. Anal Chim Acta 2013;765:28–36. http://dx.doi.org/10.1016/j.aca.2012.12.028.

[23] Furbo S, Hansen AB, Skov T, Christensen JH. Pixel-based analysis of comprehensive two-dimensional gas chromatograms (color plots) of petroleum: a tutorial. Anal Chem 2014;86:7160–70. http://dx.doi.org/10.1021/ac403650d.

[24] Gröger T, Schäffer M, Pütz M, Ahrens B, Drew K, Eschner M, et al. Application of two-dimensional gas chromatography combined with pixel-based chemometric processing for the chemical profiling of illicit drug samples. J Chromatogr A 2008;1200:8–16. http://dx.doi.org/10.1016/j.chroma.2008.05.028.

[25] Almstetter MF, Appel IJ, Dettmer K, Gruber MA, Oefner PJ. Comparison of two algorithmic data processing strategies for metabolic fingerprinting by comprehensive two-dimensional gas chromatography–time-of-flight mass spectrometry. J Chromatogr A 2011;1218:7031–8. http://dx.doi.org/10.1016/j.chroma.2011.08.006.

[26] Welke JE, Manfroi V, Zanus M, Lazzarotto M, Alcaraz Zini C. Differentiation of wines according to grape variety using multivariate analysis of comprehensive two-dimensional gas chromatography with time-of-flight mass spectrometric detection data. Food Chem 2013;141:3897–905. http://dx.doi.org/10.1016/j.foodchem.2013.06.100.

[27] Harvey PM, Shellie RA. Data reduction in comprehensive two-dimensional gas chromatography for rapid and repeatable automated data analysis. Anal Chem 2012;84:6501–7. http://dx.doi.org/10.1021/ac300664h.

[28] Reichenbach SE, Tian X, Boateng AA, Mullen CA, Cordero C, Tao Q. Reliable peak selection for multisample analysis with comprehensive two-dimensional chromatography. Anal Chem 2013;85:4974–81. http://dx.doi.org/10.1021/ac303773v.

[29] Reichenbach SE, Tian X, Cordero C, Tao Q. Features for non-targeted cross-sample analysis with comprehensive two-dimensional chromatography. J Chromatogr A 2012;1226:140–8. http://dx.doi.org/10.1016/j.chroma.2011.07.046.

[30] Brokl M, Bishop L, Wright CG, Liu C, McAdam K, Focant J-F. Multivariate analysis of mainstream tobacco smoke particulate phase by headspace solid-phase micro extraction coupled with comprehensive two-dimensional gas chromatography–time-of-flight mass spectrometry. J Chromatogr A 2014;1370:216–29. http://dx.doi.org/10.1016/j.chroma.2014.10.057.

[31] Marney LC, Siegler WC, Parsons BA, Hoggard JC, Wright BW, Synovec RE. Tile-based Fisher-ratio software for improved feature selection analysis of comprehensive two-dimensional gas chromatography–time-of-flight mass spectrometry data. Talanta 2013;115:887–95. http://dx.doi.org/10.1016/j.talanta.2013.06.038.

[32] Parsons BA, Marney LC, Siegler WC, Hoggard JC, Wright BW, Synovec RE. Tile-based Fisher ratio analysis of comprehensive two-dimensional gas chromatography time-of-flight mass spectrometry (GC $\times$ GC–TOFMS) data using a null distribution approach. Anal Chem 2015;87:3812–9. http://dx.doi.org/10.1021/ac504472s.

[33] Bailey HP, Rutan SC, Carr PW. Factors that affect quantification of diode array data in comprehensive two-dimensional liquid chromatography using chemometric data analysis. J Chromatogr A 2011;1218:8411–22. http://dx.doi.org/10.1016/j.chroma.2011.09.057.

[34] Mondello L, Herrero M, Kumm T, Dugo P, Cortes H, Dugo G. Quantification in comprehensive two-dimensional liquid chromatography. Anal Chem 2008;80:5418–24. http://dx.doi.org/10.1021/ac800484y.

[35] Hoggard JC, Synovec RE. Parallel factor analysis (PARAFAC) of target analytes in GC $\times$ GC–TOFMS data: automated selection of a model with an appropriate number of factors. Anal Chem 2007;79:1611–9. http://dx.doi.org/10.1021/ac061710b.

[36] Tobias HJ, Sacks GL, Zhang Y, Brenna JT. Comprehensive two-dimensional gas chromatography combustion isotope ratio mass spectrometry. Anal Chem 2008;80:8613–21. http://dx.doi.org/10.1021/ac801511d.

[37] Daszykowski M, Wróbel MS, Bierczynska-Krzysik A, Silberring J, Lubec G, Walczak B. Automatic preprocessing of electrophoretic images. Chemom Intell Lab Syst 2009;97:132–40. http://dx.doi.org/10.1016/j.chemolab.2009.03.002.

[38] Schmarr H-G, Bernhardt J. Profiling analysis of volatile compounds from fruits using comprehensive two-dimensional gas chromatography and image processing techniques. J Chromatogr A 2010;1217:565–74. http://dx.doi.org/10.1016/j.chroma.2009.11.063.

[39] Stevenson PG, Mnatsakanyan M, Guiochon G, Shalliker RA. Peak picking and the assessment of separation performance in two-dimensional high performance liquid chromatography. Analyst 2010;135:1541–50. http://dx.doi.org/10.1039/B922759H.

[40] Soggiu A, Marullo O, Roncada P, Capobianco E. Empowering spot detection in 2DE images by wavelet denoising. In Silico Biol 2009;9:125–33. http://dx.doi.org/10.3233/ISB-2009-0393.

[41] Nadeau JS, Wilson RB, Hoggard JC, Wright BW, Synovec RE. Study of the interdependency of the data sampling ratio with retention time alignment and principal component analysis for gas chromatography. J Chromatogr A 2011;1218:9091–101. http://dx.doi.org/10.1016/j.chroma.2011.10.031.

[42] Trudgett MJE, Guiochon G, Shalliker RA. Theoretical description of a new analytical technique: comprehensive online multidimensional fast Fourier transform separations. J Chromatogr A 2011;1218:3545–54. http://dx.doi.org/10.1016/j.chroma.2011.03.061.

[43] van der Klift EJC, Vivó-Truyols G, Claassen FW, van Holthoon FL, van Beek TA. Comprehensive two-dimensional liquid chromatography with ultraviolet, evaporative light scattering and mass spectrometric detection of triacylglycerols in corn oil. J Chromatogr A 2008;1178:43–55. http://dx.doi.org/10.1016/j.chroma.2007.11.039.

[44] Castillo S, Mattila I, Miettinen J, Orešič M, Hyötyläinen T. Data analysis tool for comprehensive two-dimensional gas chromatography/time-of-flight mass spectrometry. Anal Chem 2011;83:3058–67. http://dx.doi.org/10.1021/ac103308x.

[45] Amador-Muñoz O, Marriott PJ. Quantification in comprehensive two-dimensional gas chromatography and a model of quantification based on selected summed modulated peaks. J Chromatogr A 2008;1184:323–40. http://dx.doi.org/10.1016/j.chroma.2007.10.041.

[46] Kallio M, Kivilompolo M, Varjo S, Jussila M, Hyötyläinen T. Data analysis programs for comprehensive two-dimensional chromatography. J Chromatogr A 2009;1216:2923–7. http://dx.doi.org/10.1016/j.chroma.2008.11.037.

[47] Humston EM, Hoggard JC, Synovec RE. Utilizing the third order advantage with isotope dilution mass spectrometry. Anal Chem 2010;82:41–3. http://dx.doi.org/10.1021/ac902184b.

[48] Robards K, Hadad PR, Jackson PE. Principles and practice of modern chromatographic methods. New York: Academic Press; 1994.

[49] Lee ML, Yang FJ, Bartle KD. Open tubular column gas chromatography. New York: John Wiley & Sons; 1984.

[50] Vial J, Pezous B, Thiébaut D, Sassiat P, Teillet B, Cahours X, et al. The discriminant pixel approach: a new tool for the rational interpretation of GC×GC-MS chromatograms. Talanta 2011;83:1295–301. http://dx.doi.org/10.1016/j.talanta.2010.07.059.

[51] Mohler RE, Tu BP, Dombek KM, Hoggard JC, Young ET, Synovec RE. Identification and evaluation of cycling yeast metabolites in two-dimensional comprehensive gas chromatography–time-of-flight-mass spectrometry data. J Chromatogr A 2008;1186:401–11. http://dx.doi.org/10.1016/j.chroma.2007.10.063.

[52] Carvalho PC, Hewel J, Barbosa VC, Yates III JR. Identifying differences in protein expression levels by spectral counting and feature selection. Genet Mol Res 2008;7:342–56. http://dx.doi.org/10.4238/vol7-2gmr426.

[53] Hoggard JC, Wahl JH, Synovec RE, Mong GM, Fraga CG. Impurity profiling of a chemical weapon precursor for possible forensic signatures by comprehensive two-dimensional gas chromatography/mass spectrometry and chemometrics. Anal Chem 2010;82:689–98. http://dx.doi.org/10.1021/ac902247x.

[54] Allen RC, Rutan SC. Investigation of interpolation techniques for the reconstruction of the first dimension of comprehensive two-dimensional liquid chromatography–diode array detector data. Anal Chim Acta 2011;705:253–60. http://dx.doi.org/10.1016/j.aca.2011.06.022.

[55] Johnson KJ, Wright BW, Jarman KH, Synovec RE. High-speed peak matching algorithm for retention time alignment of gas chromatographic data for chemometric analysis. J Chromatogr A 2003;996:141–55. http://dx.doi.org/10.1016/S0021-9673(03)00616-2.

[56] Tomasi G, van den Berg F, Andersson C. Correlation optimized warping and dynamic time warping as preprocessing methods for chromatographic data. J Chemom 2004;18:231–41. http://dx.doi.org/10.1002/cem.859.

[57] Gröger T, Zimmermann R. Application of parallel computing to speed up chemometrics for GC × GC–TOFMS based metabolic fingerprinting. Talanta 2011;83:1289–94. http://dx.doi.org/10.1016/j.talanta.2010.09.015.

[58] Zhang D, Huang X, Regnier FE, Zhang M. Two-dimensional correlation optimized warping algorithm for aligning GC×GC−MS data. Anal Chem 2008;80:2664–71. http://dx.doi.org/10.1021/ac7024317.

[59] Vial J, Noçairi H, Sassiat P, Mallipatu S, Cognon G, Thiébaut D, et al. Combination of dynamic time warping and multivariate analysis for the comparison of comprehensive two-dimensional gas chromatograms: application to plant extracts. J Chromatogr A 2009;1216:2866–72. http://dx.doi.org/10.1016/j.chroma.2008.09.027.

[60] Faber NM, Buydens LMC, Kateman G. Generalized rank annihilation method. I. Derivation of eigenvalue problems. J Chemom 1994;8:147–54. http://dx.doi.org/10.1002/cem.1180080206.

[61] Li S, Hamilton JC, Gemperline PJ. Generalized rank annihilation method using similarity transformations. Anal Chem 1992;64:599–607. http://dx.doi.org/10.1021/ac00030a007.

[62] Sanchez E, Kowalski BR. Generalized rank annihilation factor analysis. Anal Chem 1986;58:496–9. http://dx.doi.org/10.1021/ac00293a054.

[63] Prazen BJ, Synovec RE, Kowalski BR. Standardization of second-order chromatographic/spectroscopic data for optimum chemical analysis. Anal Chem 1998;70:218–25. http://dx.doi.org/10.1021/ac9706335.

[64] Fraga CG, Prazen BJ, Synovec RE. Objective data alignment and chemometric analysis of comprehensive two-dimensional separations with run-to-run peak shifting on both dimensions. Anal Chem 2001;73:5833–40. http://dx.doi.org/10.1021/ac010656q.

[65] Moler C, Stewart G. An algorithm for generalized matrix eigenvalue problems. SIAM J Numer Anal 1973;10:241–56. http://dx.doi.org/10.1137/0710024.

[66] de Juan A, Jaumot J, Tauler R. Multivariate curve resolution (MCR). Solving the mixture analysis problem. Anal Methods 2014;6:4964–76. http://dx.doi.org/10.1039/C4AY00571F.

[67] Parastar H, Radović JR, Jalali-Heravi M, Diez S, Bayona JM, Tauler R. Resolution and quantification of complex mixtures of polycyclic aromatic hydrocarbons in heavy fuel oil sample by means of GC × GC-TOFMS combined to multivariate curve resolution. Anal Chem 2011;83:9289–97. http://dx.doi.org/10.1021/ac201799r.

[68] Kuligowski J, Quintás G, Tauler R, Lendl B, de la Guardia M. Background correction and multivariate curve resolution of online liquid chromatography with infrared spectrometric detection. Anal Chem 2011;83:4855–62. http://dx.doi.org/10.1021/ac2004407.

[69] Parastar H, Tauler R. Multivariate curve resolution of hyphenated and multidimensional chromatographic measurements: a new insight to address current chromatographic challenges. Anal Chem 2014;86:286–97. http://dx.doi.org/10.1021/ac402377d.

[70] Bailey HP, Rutan SC. Chemometric resolution and quantification of four-way data arising from comprehensive 2D-LC-DAD analysis of human urine. Chemom Intell Lab Syst 2011;106:131–41. http://dx.doi.org/10.1016/j.chemolab.2010.07.008.

[71] Harshman RA. Foundations of the PARAFAC procedure: models and conditions for an explanatory multimodal factor analysis. UCLA working papers in phonetics, 16: Ann Arbor, Michigan: University Microfilms; 1970. p. 1–84, No. 10,085.

[72] Carroll JD, Chang J-J. Analysis of individual differences in multidimensional scaling via an n-way generalization of "Eckart-Young" decomposition. Psychometrika 1970;35:283–319. http://dx.doi.org/10.1007/BF02310791.

[73] Bro R. PARAFAC. Tutorial and applications. Chemom Intell Lab Syst 1997;38:149–71. http://dx.doi.org/10.1016/S0169-7439(97)00032-4.

[74] Hoggard JC, Synovec RE. Automated resolution of nontarget analyte signals in GC × GC-TOFMS data using parallel factor analysis. Anal Chem 2008;80:6677–88. http://dx.doi.org/10.1021/ac800624e.

[75] Hoggard JC, Siegler WC, Synovec RE. Toward automated peak resolution in complete GC × GC–TOFMS chromatograms by PARAFAC. J Chemom 2009;23:421–31. http://dx.doi.org/10.1002/cem.1239.

[76] Bro R, Andersson CA, Kiers HAL. PARAFAC2—Part II. Modeling chromatographic data with retention time shifts. J Chemom 1999;13:295–309. http://dx.doi.org/10.1002/(SICI)1099-128X(199905/08)13:3/4<295::AID-CEM547>3.0.CO;2-Y.

[77] Skov T, Hoggard JC, Bro R, Synovec RE. Handling within run retention time shifts in two-dimensional chromatography data using shift correction and modeling. J Chromatogr A 2009;1216:4020–9. http://dx.doi.org/10.1016/j.chroma.2009.02.049.

[78] Watson NE, Siegler WC, Hoggard JC, Synovec RE. Comprehensive three-dimensional gas chromatography with parallel factor analysis. Anal Chem 2007;79:8270–80. http://dx.doi.org/10.1021/ac070829x.

[79] Wise BM, Gallagher NB, Bro R, Shaver JM, Windig W, Koch SR. PLS Toolbox 3.5 for use with Matlab™. Eigenvector Research, Inc. ISBN: 0-97611840-8.

[80] Bro R, Smilde AK. Principal component analysis. Anal Methods 2014;6:2812–31. http://dx.doi.org/10.1039/C3AY41907J.

[81] Geladi P, Kowalski BR. Partial least-squares regression: a tutorial. Anal Chim Acta 1986;185:1–17. http://dx.doi.org/10.1016/0003-2670(86)80028-9.

[82] Haaland DM, Thomas EV. Partial least-squares methods for spectral analyses. 1. Relation to other quantitative calibration methods and the extraction of qualitative information. Anal Chem 1988;60:1193–202. http://dx.doi.org/10.1021/ac00162a020.

[83] de Jong S. SIMPLS: an alternative approach to partial least squares regression. Chemom Intell Lab Syst 1993;18:251–63. http://dx.doi.org/10.1016/0169-7439(93)85002-X.

[84] Bro R. Multiway calibration. Multilinear PLS. J Chemom 1996;10:47–61. http://dx.doi.org/10.1002/(SICI)1099-128X(199601)10:1<47::AID-CEM400>3.0.CO;2-C.

[85] Andersson CA, Bro R. The N-way Toolbox for MATLAB. Chemom Intell Lab Syst 2000;52:1–4. http://dx.doi.org/10.1016/S0169-7439(00)00071-X.

[86] Olivieri AC, Wu H-L, Yu R-Q. MVC3: a MATLAB graphical interface toolbox for third-order multivariate calibration. Chemom Intell Lab Syst 2012;116:9–16. http://dx.doi.org/10.1016/j.chemolab.2012.03.018.

[87] Gurden SP, Westerhuis JA, Bro R, Smilde AK. A comparison of multiway regression and scaling methods. Chemom Intell Lab Syst 2001;59:121–36. http://dx.doi.org/10.1016/S0169-7439(01)00168-X.

[88] Bro R, Heimdal H. Enzymatic browning of vegetables. Calibration and analysis of variance by multiway methods. Chemom Intell Lab Syst 1996;34:85–102. http://dx.doi.org/10.1016/0169-7439(96)00019-6.

[89] Smilde AK. Comments on multilinear PLS. J Chemom 1997;11:367–77. http://dx.doi.org/10.1002/(SICI)1099-128X(199709/10)11:5<367::AID-CEM481>3.0.CO;2-I.

[90] Allegrini F, Olivieri AC. IUPAC-consistent approach to the limit of detection in partial least-squares calibration. Anal Chem 2014;86:7858–66. http://dx.doi.org/10.1021/ac501786u.

[91] Faber N (Klaas) M, Bro R. Standard error of prediction for multiway PLS: 1. Background and a simulation study. Chemom Intell Lab Syst 2002;61:133–49. http://dx.doi.org/10.1016/S0169-7439(01)00204-0.

[92] Prazen BJ, Johnson KJ, Weber A, Synovec RE. Two-dimensional gas chromatography and trilinear partial least squares for the quantitative analysis of aromatic and naphthene content in naphtha. Anal Chem 2001;73:5677–82. http://dx.doi.org/10.1021/ac010637g.

[93] Pierce KM, Hoggard JC. Chromatographic data analysis. Part 3.3.4: handling hyphenated data in chromatography. Anal Methods 2014;6:645–53. http://dx.doi.org/10.1039/C3AY40965A.

[94] Centner V, Massart D-L, de Noord OE, de Jong S, Vandeginste BM, Sterna C. Elimination of uninformative variables for multivariate calibration. Anal Chem 1996;68:3851–8. http://dx.doi.org/10.1021/ac960321m.

[95] Johnson KJ, Synovec RE. Pattern recognition of jet fuels: comprehensive GC×GC with ANOVA-based feature selection and principal component analysis. Chemom Intell Lab Syst 2002;60:225–37. http://dx.doi.org/10.1016/S0169-7439(01)00198-8.

Chapter 11

# Application of Multiway Calibration in Comprehensive Two-Dimensional Gas Chromatography

Luiz A.F. de Godoy*, Marcio P. Pedroso[†], Fabio Augusto[‡] and Ronei J. Poppi[‡,1]

*Cristália Produtos Químicos Farmacêuticos Ltda, Itapira, São Paulo, Brazil

[†]Department of Chemistry, Federal University of Lavras – UFLA, Lavras, Minas Gerais, Brazil

[‡]Institute of Chemistry, University of Campinas – UNICAMP, Campinas, São Paulo, Brazil

[1]Corresponding author: e-mail: ronei@iqm.unicamp.br

**Chapter Outline**



## 1 COMPREHENSIVE TWO-DIMENSIONAL GAS CHROMATOGRAPHY (GC×GC)

Gas chromatography (GC) is a separation technique suitable for the analysis of samples whose constituents are volatiles and semivolatile compounds. However, samples such as petrochemical derivatives, flavors, and aromas, which are made up by hundreds or even thousands of components, have always represented a challenge for the complete separation. The conventional GC fails in the separation of all individual constituents from complex samples, and coelutions of two or more compounds take place several times along a chromatographic run. Historically, the first approach that increased dramatically the separation power of GC was the advent of the capillary columns in the 1970s. As the efficiency of the capillary columns had achieved its

Data Handling in Science and Technology, Vol. 29. http://dx.doi.org/10.1016/B978-0-444-63527-3.00011-4

maximum performance, the main attempts to increase the peak capacity in GC were related to the multidimensional separation. In multidimensional GC, the entire sample is sequentially submitted to different elution processes, as commonly performed in planar chromatography. The first successful attempt to increase the dimensionality of the GC system was the heartcut two-dimensional (2D) gas chromatography (GC-GC) [1]. In this technique, two capillary columns were connected in tandem through a Deans switching, and one or few fractions of the sample, where severe coelution had been apparent, were submitted to the separation on the second dimension. In addition, previous information is required to select the fraction (which contains the target analytes) to be submitted to the second column so that GC-GC peak capacity is slightly larger than conventional GC. Therefore, GC-GC is not a truly multidimensional system since the whole sample is not separated on two different columns, and consequently, this technique is not applied to screening-type analysis.

In 1991, Liu and Phillips developed the first prototype of the comprehensive 2D gas chromatography (GC × GC) [2]. The main result was the development of a device, called modulator, used to collect all effluent from the first column and to transfer to the second one, without increasing the run time or losing the separation on the first column. GC × GC can be considered a truly comprehensive separation since (i) the entire sample is eluted on two different columns, (ii) compounds separated on one-dimensional (1D) remain separated on 2D, and (iii) the elution profile from both dimensions is preserved. As a result, the peak capacity of GC × GC enhances geometrically when compared to conventional GC. Since the development of the GC × GC, remarkable results have been achieved in terms of resolution and number of detected peaks. There are excellent reviews about GC × GC covering the technique development and discussion of each part of the equipment and applications [3–8]. Thereupon, just a brief introduction about GC × GC will be presented in this chapter.

A GC × GC instrument is based on a conventional GC approach, but two columns are connected sequentially with a modulator between them. The first column or dimension ($^1$D) is a conventional column (usually 30 m × 0.25 mm I.D. × 0.25 μm), while the second column or dimension ($^2$D) is a shorter and efficient column as those commonly used in fast GC (generally a 1–2 m × 0.1 mm I.D. × 0.1 μm). A general scheme of a GC × GC system is presented in Figure **1**A: The $^1$D column is connected to the injector and the $^2$D column is fitted into the detector. Both columns can be placed in the same oven or each one separated. Between both columns (actually, at the beginning of the $^2$D), a modulator is placed. The modulator is the main part of the GC × GC instrument and its function can be summarized in three steps: (i) Collect or sample continuously small fractions of the effluent $^1$D, ensuring that the separation in that dimension is maintained; (ii) focus or refocus the effluent—not achieved with valves modulators—and (iii) quickly transfer

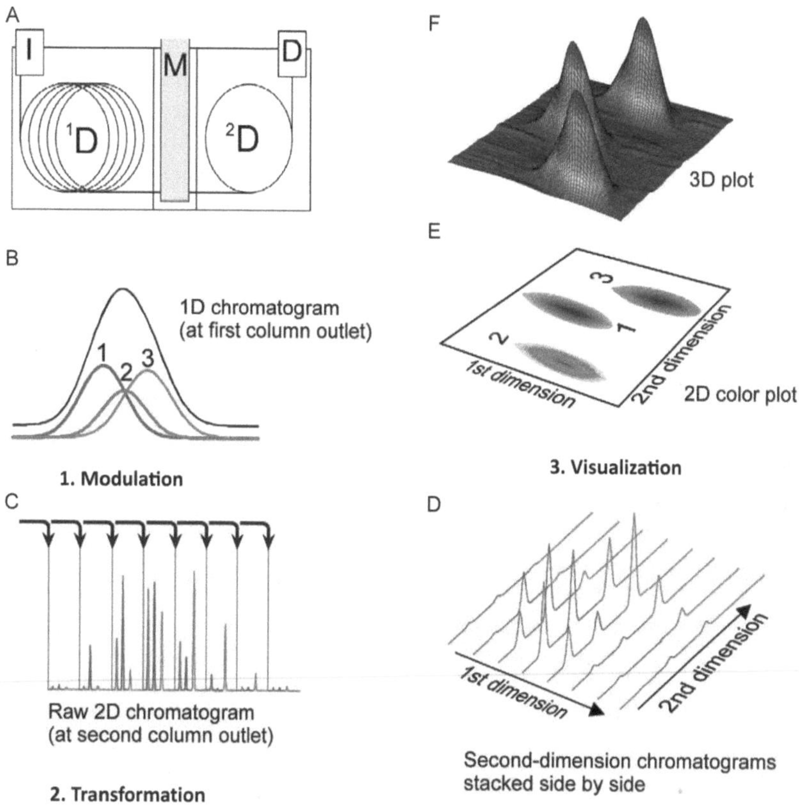


**FIGURE 1** Generation and visualization of a GC × GC chromatogram. *Adapted from Ref. [4], copyright 2003, with permission from Elsevier.*

the $^2$D collected and focused fraction as a narrow pulse. The combination of these three stages is called modulation cycle, which is repeated throughout the chromatographic run. The time required to perform a cycle is called the modulation period (MP), which is typically 2–10 s, and it depends on the time required for the compounds to be eluted in $^2$D. The MP should be as small as possible in order to avoid that separation obtained in $^1$D be lost. There are two different concepts of modulator being commercialized nowadays: the cryogenic modulators and the valve ones. The majority of the GC × GC instruments are equipped with cryogenic modulators that need high amounts of liquid nitrogen or carbon dioxide [5].

Another difference between GC × GC and conventional GC arises from the modulation step. A peak of only one compound eluted from $^1$D is converted, after modulation, in two or more high and narrow $^2$D peaks, resulting in a more complex chromatogram. For instance, in Figure **1**B, a wide chromatographic peak consisted of three nonseparated analytes on $^1$D is fractionated and eluted on $^2$D, generating a raw chromatogram (Figure **1**C). The detector signal registered along the time in a GC × GC system is a continuous and chained sequence of short chromatograms for each eluted fraction on $^2$D. In the $^2$D raw chromatogram, each compound (depicted by color) is represented by more than one peak due to the modulation process. Between two peaks of the same compound (same color), there are peaks from other compounds making the visualization of the raw chromatogram really confusing. Whenever these $^2$D short chromatograms are arranged side by side, it results in a three-dimensional (3D) structure (Figure **1**D) constituted by retention time on $^1$D ($^1t_R$) × retention time on $^2$D ($^2t_R$) × detector signal. Dedicated softwares generate a 3D landscape view chromatogram called 3D plot (Figure **1**F), which can be visualized from above resulting in a 2D chromatogram, a $^1t_R \times {}^2t_R$ contour plot, by means of colors or contour lines representing the signal intensity (Figure **1**E). As some huge peaks on 3D plot tend to hide the visualization of the several small peaks, the contour plot is the preferred presentation of a GC × GC chromatogram.

Regarding the stationary phase of the GC × GC columns, an orthogonal column set is desirable. In other words, the mechanism of retention on $^1$D is independent from the mechanism of retention on $^2$D [9]. The most orthogonal column set is the combination of a nonpolar $^1$D coupled with a polar $^2$D, which provides a better use of the separation space on the GC × GC chromatogram. Using the nonpolar/polar column set, the separation on $^1$D is, typically, based on vapor pressure of the compounds; thus, the retention on $^1$D is governed by volatility. Each fraction collected at the end of the $^1$D is composed of a group of analytes with closely similar volatilities. On a polar $^2$D, the separation mechanism is based on specific interaction of the analytes with the stationary phase (polarity) and on volatility. Since similar volatility compounds in each fraction are transferred to the $^2$D, volatility plays a small role on differentiation of these species on the $^2$D and, thus, analytes are separated due to the differential interaction with the stationary phase (polarity). As a result, the whole sample is submitted to independent mechanisms of separation on both columns (volatility and polarity).

Besides the increase in separation compared to conventional GC, the orthogonality on the column set provides a unique and powerful feature in the qualitative aspect: the chromatographic structure. Whenever a complex sample rich in isomers is analyzed by GC × GC, peaks of related substances (homologous series, chain, or position isomers) elute clustered in specific parts of the GC × GC chromatogram. The chromatographic structure is easily visualized in the weathered deepwater oil GC × GC chromatogram, presented in Figure **2**. Peaks of hydrocarbons having the same number of carbon atoms or belonging to the same class are grouped on $^1t_R \times {}^2t_R$ surface, resulting in the so-called roof-tile effect [8], which provides valuable information especially on compound identification and group type studies. Although the nonpolar/polar column set is preferred, the polar/nonpolar approach also gives successful results for a variety of samples [8].

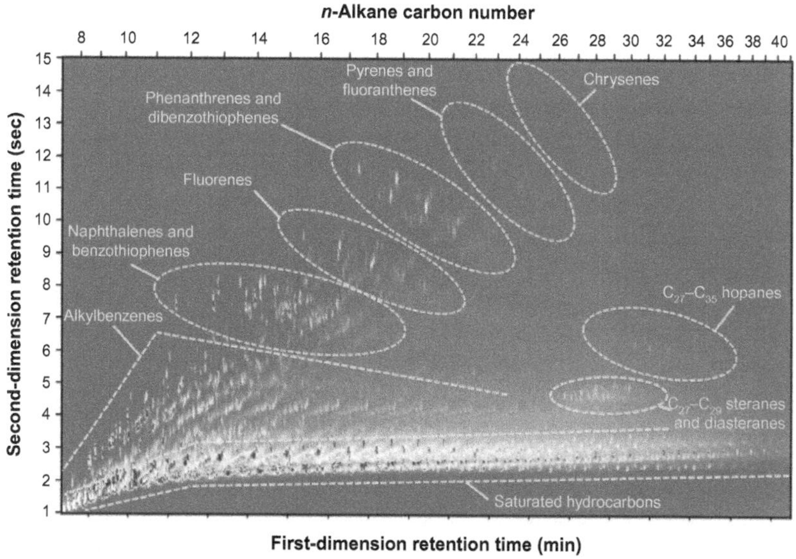


**FIGURE 2** GC × GC chromatogram (contour plot) of weathered deepwater oil. In dotted lines, the regions of various compound classes are given. *Adapted from Ref. [10], copyright 2013, with permission from Elsevier.*

The number of detected peaks in GC × GC tends to be larger than conventional GC, as well as the chromatogram complexity, due to the modulation process. In this case, the chromatographic band is compressed or focused (only on cryogenic modulators) and, in consequence, detectability increases dramatically. In opposite as a short and wide peak, each compound elutes as group of narrow and high intense peaks (elevated signal-to-noise ratio), as presented in Figure **1**C. Due to the focusing effect, the baseline peak wide on $^2$D ($^2w_b$) decreases roughly 10–50 times, resulting in modulated $^2$D peaks ranging from 50 to 500 ms. Only fast GC detectors (acquisition frequency higher than 100 Hz and low dead volume) are able to register such peaks without peak skewing. Considering detectors equipped in conventional GC systems, the flame ionization detector (FID) meets all requirements and, therefore, was the first choice to equip GC × GC systems. However, FID provides no structural information for identification analysis. On the other hand, mass spectrometry (MS) detection is indispensable to allow the identification of the resolved compounds in GC. Therefore, it has been carefully investigated for GC × GC due to the intrinsic differences between quadrupole (qMS) and time-of-flight (TOFMS) mass analyzers [6]. It must be emphasized that the mass spectra obtained through a GC × GC-MS are purer than those from GC-MS, because analyte peaks and column bleeding are frequently resolved on $^2$D, improving the identification step. In summary, mass spectrometer firstly subjects the column-eluted molecules to fragmentation through impact electron ionization. Secondly, MS detectors operate registering a wide range of *m/z* fragments produced in the ionization source (called scan mode) or monitoring only selected ions among those produced (SIM mode). Finally, the mass spectrum is displayed as a plot of the relative intensity of these fragment ions versus their mass-to-charge ratio (*m/z*). As result, every point in the chromatogram became a mass spectrum. Therefore, whenever GC × GC and MS are coupled, a new dimension is added to the GC × GC data and spectral information is present on GC × GC-MS (scan or SIM) chromatograms. For better visualization of the GC × GC chromatogram, a whole spectrum can be reduced to only one signal. It is obtained by summing the intensity of all ionic fragments that constitute the spectrum. A chromatogram built through such operation is named total ion chromatogram (TIC), and in this case, a GC × GC-MS chromatogram is similar to a GC × GC-FID: Quantitative aspects of the sample can be extracted from both, but spectral data are lost.

As a conclusion, GC × GC provides the impressive peak capacity resulting in (i) elevated separation among sample components; (ii) best qualitative and quantitative results because of reduced coelution, which lead to well-defined peak profile and purest mass spectra; and (iii) chromatography structure to group type or classification studies. The massive information from GC × GC chromatograms virtually makes impossible manual interpretation of such amount of data, requiring chemometric tools.

## 2 MULTIWAY CALIBRATION

Before the discussion about multiway calibration, the topic of this chapter, it is important to introduce a data classification according to their order. Data can be classified by different orders (Figure **3**), depending on the analytic equipment used for the analysis [12].

Zeroth-order data correspond to instruments producing a single response per sample, such as the absorbance at a single wavelength or the reading of an ion-selective electrode. Univariate calibration models are built using this kind of data, being often employed in routine analysis due to their simplicity. They have validation protocols well established and described in literature. However, the use of this kind of model requires that the instrumental signal of the target analyte must be free of interferents. First-order data for a given sample are arranged as a vector, such as infrared (IR) spectroscopy and GC. The sample set constituted of first-order data is composed of a matrix (Figure **3**). Multivariate calibration models are built using first-order data, and they present several advantages over univariate models, for example, analysis even in the presence of calibrated interferents, simultaneous determination of more than one target component, and instrumental signals of the analyte that do not need to be totally resolved. There are some chemometric algorithms to generate multivariate calibration models using first-order data, such as multiple linear regression, principal components regression, and

| Data order | Array | | Calibration |
|---|---|---|---|
| | One sample | A sample set | |
| Zero | Scalar | One-way | Univariate |
| First | Vector | Two-way | Multivariate |
| Second | Matrix | Three-way | Multiway |
| Third | Three-way | Four-way | Multiway |

**FIGURE 3** Different arrays that can be obtained for a single sample and for a set of samples. *Adapted from Ref. [11], copyright 2014, with permission from Elsevier.*

partial least squares (PLS). The main drawback of the calibration model of first-order data is the need of calibration set composed of many samples when calibrating complex samples [11–13].

Matrix data recorded for a single sample are considered to be of second order. A sample set will be composed of a 3D array and it is known as three-way data (Figure 3). They can be recorded in two ways: (1) using a single instrument, such as a spectrofluorometer registering excitation–emission matrices (EEM), and (2) coupling two hyphenated first-order instruments, for instance, tandem gas chromatography–MS and comprehensive 2D gas chromatography with flame ionization detection (GC × GC-FID). Besides all of the advantages of first-order calibration model described above, second-order calibration models are able to perform quantitative determination of one or more properties of interest in unknown samples even in the presence of uncalibrated interferences. In this way, a calibration step can be built with few samples only having the target component or property in an appropriate solvent instead of a large calibration set containing the target component/property and all possible interferences that would be found in the unknown samples. Certainly, it represents the main advantage of this kind of model and it is called as second-order advantage [11,12].

Introducing an extra dimension in second-order data leads to high-order data, in which the mathematical object obtained by grouping third-order data for several samples into the fourth dimension is known as a four-way array (Figure 3). Examples of four-way arrays are those obtained by following the kinetics of EEM fluorescence spectroscopy or by hyphenating three instruments, for example, GC × GC-MS (scan mode). The advantages of third-order data are the same of second-order one, except for the amount of information obtained for each sample that is higher in third-order data due to the additional dimensional, but it leads to the need of more sophisticated computers to generate calibration models [11,12].

Specifically for GC × GC data, sample data order depends on detector employed (e.g., FID or MS), detector operation mode, and matrix transformation. For instance, a GC × GC-FID contour plot is a ${}^{1}t_R \times {}^{2}t_R$ second-order data since FID is a single-channel detector. Thus, it can be classified as a second-order data (see Figure **1**D), or it can be unfolded to a first-order vector (as Figure **1**C). The information in both data sets is the same, despite different algorithms used in each case. Similarly, GC × GC-MS data are originally a ${}^{1}t_R \times {}^{2}t_R \times m/z$ third-order data since detector is set to scan mode or SIM mode (more than one $m/z$ monitoring). For example, if MS detector is set to a scan mode range of $m/z$ 50–250, the $m/z$ dimension in a GC × GC-MS (scan) data will be 200 points. In the same way, if MS detector is operated at SIM mode registering the 73, 111, and 215 $m/z$ channels, the $m/z$ dimension from such GC × GC-MS (SIM analysis) will be only three points. In contrast, whenever GC × GC-MS TIC data are provided, the $m/z$ dimension disappears and the data are reduced to second order, similar to a GC × GC-FID. Moreover, the GC × GC-MS TIC data can be unfolded and transformed into first-order data.

As mentioned earlier, compared to conventional GC-FID (or even GC-MS), the amount of information contained on a GC × GC chromatogram is considerably larger, which causes manual interpretation of a data set into a very difficult task (or, in many cases, virtually impossible), especially for complex samples. Therefore, the adoption of chemometric strategies for processing and interpretation of GC × GC data is desirable. Three basic categories of chemometrics have been used in data treatment from 2D separations: peak deconvolution, multivariate calibration, and pattern recognition [14]. Only the two first approaches can be employed to build multiway calibration models and they will be discussed in the following items. Examples of chemometric algorithms for peak deconvolution are parallel factor analysis (PARAFAC) [15], PARAFAC2 [16], generalized rank annihilation method (GRAM) [17], and multivariate curve resolution coupled with alternating least squares (MCR-ALS) [18]. Examples of multivariate calibration models are unfolded partial least squares (U-PLS) [19] and multiway partial least squares (N-PLS) [20]. A summary of the characteristics of algorithms is shown in Table **1**, where it can be seen that U-PLS and N-PLS do not obtain the second-order advantage, GRAM and PARAFAC are not able to deal with trilinearity deviations, and MCR-ALS cannot be applied to third-order data. Even though PARAFAC2 seems to be the ideal algorithm, it has rarely been used with GC × GC data in order to get multiway calibration models, and when it was used, the prediction results were worse than PARAFAC or N-PLS models, as it will be seen in the application examples. In conclusion, there is no perfect algorithm and it must be chosen taking into account some

**TABLE 1** Some Characteristics of Second-Order Algorithms

| | Algorithm | | | | | |
|---|---|---|---|---|---|---|
| **Characteristic** | *PARAFAC* | *PARAFAC2* | *GRAM* | *MCR-ALS* | *U-PLS* | *N-PLS* |
| Achievement of the second-order advantage | Yes | Yes | Yes | Yes | No | No |
| Presence of linear dependence | I, C | I, C | No | C | Yes | Yes |
| Handling of trilinearity deviations | No | Yes[a] | No | Yes[a] | Yes | Yes |
| Use of incomplete calibration | Yes | Yes | Yes | Yes | Yes | Yes |
| Extension to higher-order | Yes | Yes | No | No | Yes | Yes |
| Provision of physical information | Yes | Yes | Yes | Yes | No | No |

I, initialization; C, constraint.
[a]*Only if the deviations are produced by sample-to-sample changes in profiles in one of the data modes.*
Adapted from Ref. [12], copyright 2007, with permission from Elsevier

factors, such as aims of the study and sample characteristics. Since all the algorithms mentioned are well presented and discussed in the literature and the focus of this chapter is on the application of multiway calibration with GC × GC data, all the mathematics behind each algorithm will not be discussed here.

## 2.1 Multiway Calibration I: Two-Step Approach

Deconvolution of unresolved chromatographic peaks largely extends experimental capability by using chemometric tools, which has been widely employed to cope with overlapped signals from data acquired by comprehensive 2D gas chromatography using flame ionization or mass detector. In order to generate calibration models in GC × GC data, a two-step approach is accomplished: First, the deconvolution of unresolved chromatographic peaks is performed and then a univariate calibration model is developed.

The application of chemometric deconvolution techniques enables the analyst to carry out a quantitative research work without performing a broad chromatography optimization study to resolve the target(s) peak(s), which could mean many experiments. Therefore, the utilization of chemometric deconvolution techniques to build multiway calibration models using GC × GC chromatograms saves time, labor, and money. Since the chromatogram of pure components can be independently extracted from overlapped signals, peak area or peak volume can be used for relative quantification. After this, absolute quantification can be achieved with the help of standard calibration strategies, including the following methods: normalization, internal standard, external standard, and standard addition [21,22].

Among the algorithms presented in Table **1**, GRAM, PARAFAC, PARAFAC2, and MCR-ALS are the ones that perform deconvolution of overlapped signals, and they have been employed to generated calibration models with data set acquired by GC × GC; examples applying these algorithms to GC × GC chromatograms will be highlighted in the next item [11].

GRAM is used for analyzing two bilinear data samples at a time, forming a matrix with only rows, being one row the standard and the other one an unknown sample. It is a chemometric model that uses the bilinear structure of 2D separations both to deconvolute peaks and to compare a peak's magnitude to that of the standard for quantitation. Furthermore, it may offer an enhancement in signal-to-noise ratio [23]. However, the analytic data must follow some requirements for the successful application of GRAM. First, detector response must be linear with concentration. Second, peak profile shape for the target analyte must not change. For GC × GC chromatograms, this means that the retention times of each slice of a compound taken from the effluent of the first column must have a consistent second-column retention time, and all the retention times in both dimensions must be the same in the sample being studied and in the standard. Consequently, the more precise the run-to-run retention times, the more powerful the GRAM analysis becomes. The final requirement for appropriate use of GRAM is that no two compounds within the window of data analyzed can perfectly covary in concentration from the standard to the sample. This covariance possibility is minimized by performing GRAM on subsections of the entire 2D separation data set, which reduces the number of peaks analyzed at a time. Because of all the requirements appointed above, the fact that only one sample is analyzed per model and its impossibility to expand to higher-order data, GRAM had been applied to GC × GC data in the first half of the 2000s, being practically forgotten after that. Analytic chemists have mainly used PARAFAC or MCR-ALS to deconvolute GC × GC peaks since then [24,25].

In the beginning of the 1970s, PARAFAC was proposed independently by Carrol and Chang [26] under the name canonical decomposition and by Harshman [27] under the name PARAFAC. Initially, it was used in psychometrics field. In 1997, Bro [15] published a tutorial paper describing PARAFAC and its application to build models using chemical analytic data, and he made the algorithm available for free download in his website. Since then, it has been gaining more and more interest in chemometrics due to the analytic instrument improvements and computer development. PARAFAC is a decomposition method that can be considered as a possible generalization of principal component analysis (PCA) to higher-order arrays. In PARAFAC algorithm, a three-way array, such as a sample set of GC × GC-FID, is decomposed into three matrices, one containing the concentration profiles and the other two presenting the pure instrumental profiles. Then, PARAFAC is able to deconvolute from initially partially resolved data the pure component profile in all dimensions of the analytic experimental data. Talking specifically about PARAFAC model of GC × GC data, the deconvoluted 3D peak can be reconstructed by taking the outer product between the first- and second-dimension chromatographic profiles. This deconvoluted analytic peak can then be integrated to give a total peak volume that is proportional to the analyte concentration. Definitely, the main advantages of PARAFAC are the achievement of second-order advantage, its possibility of extension to higher-order data, and the availability of imposition of some constraints to the models. However, some requirements must be met by the experimental data for a successful application of PARAFAC: linear detector response with concentration variation and no shifts in the retention time of GC × GC peaks [22]. Only 2 years after the paper describing the application of PARAFAC in chemistry, Bro et al. [16] proposed PARAFAC2, a multiway chemometric method based on PARAFAC able to deal with trilinearity deviations in one mode of the experimental data, which seemed at that time to be an incredible improvement for the use of multiway chemometric analysis to build calibration models from GC × GC chromatograms. However, only few papers [28,29] described multiway calibration model combining PARAFAC2 to GC × GC data. Moreover, these papers compared PARAFAC2 prediction results to the ones obtained by PARAFAC and N-PLS, and this comparison showed that PARAFAC or N-PLS generated better models than PARAFAC2 in each case.

An important multivariate technique that has not been widely used with GC × GC data is the algorithm proposed by Tauler et al. in 1995 called multivariate curve resolution (MCR) [18]. This method has been employed in the analysis of complex mixtures through different analytic techniques [30–35], but only few papers describe its application to generate multiway calibration models using data set acquired by GC × GC, being the first one dated from 2010 [36]. Differently from PARAFAC, MCR is a bilinear method; in other words, the experimental data must be organized as a 2D matrix. There are different ways to transform the three and four-way sample set arrays of GC × GC-FID and GC × GC-MS, respectively, into a 2D matrix, which will be highlighted hereafter. In the MCR algorithm, the data set is decomposed into two matrices, one related to concentration profiles and another related to instrumental ones. These two matrices are iteratively adjusted to the data set through an alternating least squares (ALS) procedure, which starts with an initial estimate of either concentrations or instrumental profiles. During the ALS optimization, several constraints, such as nonnegativity, unimodality, closure, and selectivity (local rank), can be applied to obtain chemically meaningful solutions. The main advantages of MCR are the achievement of second-order advantage, its ability to deal with trilinearity deviations in one mode of the data, and the availability of imposition of some constraints to the models. Nevertheless, there is a drawback called rotational ambiguity, which is the intrinsic difficulty to encounter a unique solution of concentration and instrumental profiles for the measured experimental data. The rotational ambiguity can be overcome by the application of the right constraints to the experimental data being analyzed. Other drawbacks are that the target peaks must not be completely overlapped and a collinearity condition must be satisfied. In other words, there must be variations in the relative concentrations of the coeluting analytes between samples [37].

### *2.1.1 Examples of Application*

In 1998, Synovec and coworkers [38] described the application of the GRAM algorithm to resolve overlapped peaks of GC × GC-FID chromatograms. This paper probably represents the first use of a multiway chemometric strategy to perform a quantitative study through GC × GC-FID analysis. In this article, the chemometric algorithm was employed to quantify two test analytes, ethylbenzene and *m*-xylene, in white gasoline samples. The samples were prepared by adding between 0% and 3% (w/w) of the test analytes to white gasoline. *o*-Xylene at 1% (w/w) was added to all samples as an internal standard. The chromatographic peaks obtained for ethylbenzene and *m*-xylene were severely overlapped on both chromatographic dimensions. In this way, the feasibility of GRAM in quantifying severely overlapped peaks was demonstrated.

For the first study, six different samples varying the proportion of ethylbenzene/*m*-xylene in white gasoline were analyzed. As mentioned before, each GRAM model performs the analysis of only one sample using a standard. Thus, one sample was taken as the standard, and GRAM models were built to predict the amount of ethylbenzene and *m*-xylene in the other samples. It is important to emphasize that GRAM quantification results for a given sample/standard pair were the combination of 16 models, since each sample was run four times, which highlight one of the disadvantages of GRAM algorithm. In order to show the second-order advantage of this method, in a second study, a sample of white gasoline containing only *m*-xylene was chosen as standard. Actually, the standard had ethylbenzene in its composition, but not in an appreciable level. The chemometric predictive results were comparable to the ones obtained by a reference GC method.

Two years later, the same research group published a paper describing another use of GRAM algorithm for a quantification purpose. In this case, a high-speed quantitative analysis of three aromatic isomers (isopropylbenzene, propylbenzene, and 1,3,5-trimethylbenzene) in a jet fuel sample was performed combining GC × GC-FID and GRAM method. A standard addition sample of jet fuel was made by spiking the three aromatic compounds to 60 mL of the neat jet fuel such that the added concentration (w/w) of each compound was 0.756% for isopropylbenzene, 0.351% for propylbenzene, and 0.742% for 1,3,5-trimethylbenzene. The total time for each GC × GC separation was only 2.8 min, while 14.4 min was the time spent for the analysis by a GC method reference. Moreover, only two of the three aromatic isomers were adequately resolved for quantification using the reference GC method, while all three isomers in the GC × GC separation could be quantified by chemometric analysis. Standard addition method and retention time alignment were used to correct retention time deviations, since for an accurate GRAM predictive model, both retention time and peak width need to remain constant between the sample data set and calibration standard data set. The standard addition method amends possible significant changes that can occur in retention times and in peak widths when analytes of interest are determined in a chemical matrix different from the sample used as standard in a GRAM model. In order to evaluate the impact of retention time alignment on the predictive results, GRAM models were built for each one of the isomers studied with and without the alignment. GRAM quantification results of each compound were the average of 25 models originated from the combination of 5 neat and 5 spiked jet fuel data sets. The results showed that the retention time alignment algorithm dramatically improved the precision for 1,3,5-trimethylbenzene by a factor of four times. Additionally, the accuracy of GRAM quantification for propylbenzene and 1,3,5-trimethylbenzene when compared to the results of the GC reference method was also improved with the application of the alignment algorithm. Isopropylbenzene could not be determined by the GC reference method because of significant signal overlap [39].

The suitability of PARAFAC for mathematically resolving overlapped peaks in GC × GC-TOFMS chromatograms and then performing qualitative and quantitative determination with the deconvoluted signals was demonstrated by Synovec and coworkers in 2004 [40], which probably represents the first application of PARAFAC to GC × GC data to build multiway calibration models. An environmental sample containing fuel, pesticides, and natural products was chosen for the study, because it represents a complex matrix. This environmental sample was analyzed by GC × GC-TOFMS. One subregion of the GC × GC-TOFMS TIC chromatogram containing two compounds not fully resolved was taken with the aim of rightly deconvoluting these compounds and making the identification using the MS information. A PARAFAC model initialized by trilinear decomposition and imposing nonnegativity constraint on all dimensions was developed with four factors. Two factors are due to the two components present in that subregion, the third due to some background, and the fourth because of baseline offset, which was not subtracted prior to the chemometric analysis. The background factor is composed of additional, unknown interferents. PARAFAC deconvolution results enabled the identification of the two components by the comparison of each resolved mass spectrum to NIST library spectra. The two compounds were identified as being 1-methoxy-2-propyl acetate and chlorobenzene. After that, chlorobenzene was used for a standard addition method in the environmental sample in order to perform a quantitative study in which PARAFAC deconvolution was carried out on a region of the standard addition data set containing the spiked standard. Then, the quantitative determination was performed by reconstructing the chlorobenzene peak and summing all of the mass channels to generate a TIC chromatogram for both the sample and the sample + standard addition sample. Finally, the volumes for the sample and standard addition peaks were calculated and the original concentration of chlorobenzene in the environmental sample was obtained. In the end of this study, a four-step procedure was proposed to deconvolute, identify, and quantify analytes of interest that are not fully resolved nor have a fully selective mass channel through the combination of PARAFAC algorithm and GC × GC-TOFMS analysis. First, two data files are collected: a sample and a sample + standard addition, in which quantitative amounts of all the analytes of interest are spiked into the standard addition. PARAFAC is performed on the region around each analyte of interest, which will give individual peak profiles and mass spectra for both data sets. The analytes in the sample data set are identified by comparing their deconvoluted mass spectra to those of the standard addition sample or to a MS library. Quantification is then achieved by reconstructing the signal of the analytes of interest in both data sets and then applying signal integration and the usual mathematical techniques for quantification via standard addition. The fact that the deconvolution is performed separately on both the sample and the standard addition loosens the retention time alignment requirements, thus simplifying and improving the quantification process significantly.

After the paper described above, the same research group turned their attention to generate multiway calibration models in the metabolomics field using PARAFAC combined to GC × GC analysis. Metabolomics is a relatively new field that aims to extract, identify, and/or quantify metabolites from a biological sample produced in different conditions (e.g., solar and artificial light) to understand which metabolic route is preferred or active under such conditions. Unlike genomics or proteomics, in metabolomics, the amount of different metabolites produced by distinct live organisms is enormously greater than other "omics"; hence, a new protocol must be developed for each sample. For volatiles and semivolatile metabolites, GC × GC-TOFMS is the best choice to analyze such high complex sample and identify the most important metabolites. However, interpretation of such amount of data is a task almost impossible to carry out manually. Therefore, chemometric tools to compare samples, locate and resolve peaks, or extract pure spectra for identification, followed by quantification, are required.

In this context, Synovec and coworkers coupled GC × GC-TOFMS with chemometric analysis to identify chemical differences in metabolite extracts isolated from yeast cells either metabolizing glucose via fermentation or metabolizing ethanol by respiration [41]. Growth on the ethanol carbon source should lead to production of metabolites different from the growth on glucose. Small polar metabolites extracted from 70 samples (respiration and fermentation) were methoxymated and trimethylsilylated and then analyzed by GC × GC-TOFMS. More than 2500 peaks were recognized, but only data from *m/z* 73, 205, and 387 (trimethylsilyl (TMS) group, TMS carbohydrates, and TMS sugar phosphates, respectively) were exported to diminish the complexity of the sample. Afterward, data were normalized and aligned and PCA was performed individually on each of the selected *m/z* channel. PCA loading plots were ascertained to locate metabolites that present the biggest difference between the fermentation and respiration conditions, revealing 26 most important metabolites.

Since the relative amount of each compound on different metabolic conditions (called concentration ratio) is enough to understand the active metabolic routes used by yeast in such situation, the exact concentration of each compound was not determinate. However, overlapping compounds could interfere in the identification and quantitation steps; thus, a PARAFAC graphical user interface developed in-house, which allows the user to quickly choose the subregion of the chromatograms for the analysis, as well as the *m/z* to use for the deconvolution of overlapping mass spectra, was developed to deconvolute interfering and analyte signals. Only fewer than 10 selected *m/z* channels were submitted to PARAFAC; otherwise, the entire *m/z* range would require excessive computation time and memory. Authors highlight information lost in selecting channels, but it had no effect on the final results of this study because it is expressed as the concentration ratio between fermentation and respiration. The concentration ratio was calculated using the peak volume of

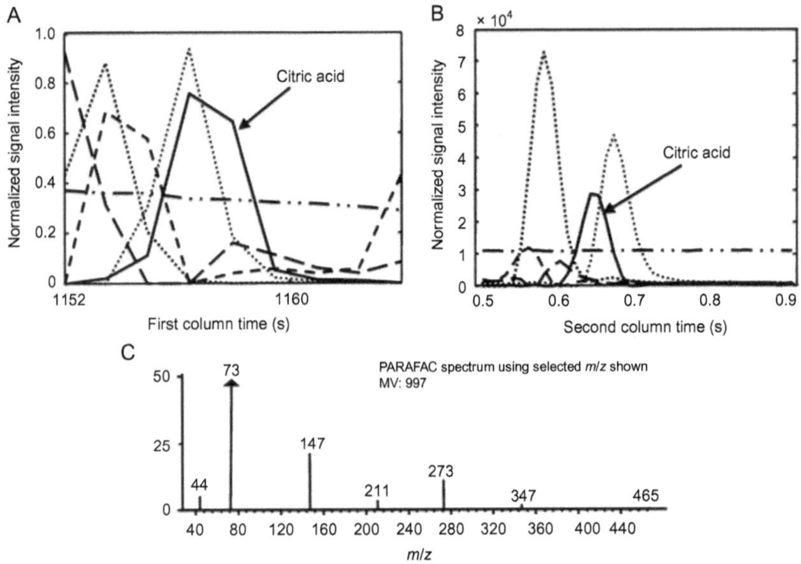


**FIGURE 4** Pure component peak profiles for citric acid 4TMS (solid lines) on column 1 (A) column 2 (B) and the select *m/z* channel mass spectrum (C). *Reprinted from Ref. [41], copyright 2006, with permission from American Chemical Society.*

the compound, which is the outer product of $^1$D and $^2$D peak profiles extracted after deconvolution with PARAFAC. The quantification of the peak volume is not an easy task due to background noise and coeluting compounds. For instance, the pure component profiles for citric acid 4TMS on both columns using 7 *m/z* channels are presented in Figure 4A and B, being separated on both dimensions from five overlapping interferences (represented by dotted and broken lines), as well as the resulting mass spectrum using the selected fragments (C).

Nine of the 26 compounds were not detected in fermentation condition so concentration ratio could not be calculated for these compounds. Glycerol, glucose, and 3 glucose-6-phosphates isomers, all related to the glycolytic pathway, were elevated in respiration condition. The remainders, including trehalose and citric, malic, and glycolic acids, were found to be more abundant in fermentation samples. Concentration ratio ranged from 0.02 for glucose to 67 for trehalose. Although the focus in this study was the demonstration of the feasibility of GC × GC-TOFMS combined with chemometric analysis for the fast separation, identification, and relative quantification of metabolites in extremely complex samples of yeast extracts, the derived chemical information led to results that are consistent with the expected biological ones.

As the entire cube of data was not comprehensively utilized, a similar study was carried out using the Fisher ratio method to locate peaks responsible for differentiation from respiration and fermentation conditions [42] instead of PCA. The Fisher ratio is defined as the class-to-class variation divided by the sum of the within-class variations: it reduces the entire data in an automated fashion with little preprocessing to locate differentiating peak locations. The algorithm calculates a Fisher ratio at every point in the chromatogram for each mass channel (*m*/*z*) of that baseline-corrected, normalized data. A set of third-order GC × GC-TOFMS chromatograms are reduced to a two-dimension *sum of Fisher ratios plot*, similar to a GC × GC chromatogram, where only the peaks that potentially differentiate the sample classes are presented. The data submitted to PARAFAC analysis are dramatically reduced after Fisher ratio: From the average of 590 detected peaks in the chromatograms at *m*/*z* 73 (because of the trimethylsilyl fragments) for respiration and fermentation samples, only 157 peaks were important to distinguish classes and over than 400 peaks are insignificant to this task. The peak volume of each important metabolite was determined using $^{1}$D and $^{2}$D peak profiles extracted by PARAFAC, and the concentration ratio was determined. A total of 26 identified compounds were abundant in only one condition and 28 metabolites were present in both classes and are statistically different, whereas there were several unknown peaks. The number of important metabolites with Fisher ratio is higher than described in the initial study using PCA [41], probably, because PCA does not use the entire GC × GC-TOFMS information as performed by the Fisher ratio algorithm.

Impurity profiling of illicit drugs for forensic purposes has been studied and reported in the scientific literature for over two decades [43]. Researchers have shown that the impurity profiles of illicit drugs are useful for sample matching, source identification, and synthetic route deduction [43]. Precursor dimethyl methylphosphonate (DMMP) is a widely used flame retardant that is listed as a Schedule 2 compound by the Chemical Weapons Conventions (CWC) because it is a well-known chemical weapon precursor and not produced in large quantities. The use of chemometric strategies and GC × GC-TOFMS analysis was described for the determination of the organic impurity profile of commercial DMMP samples in order to characterize each sample according to its origin, which can be useful for crime scene investigations providing valuable forensic information. The sample set consisted of six bottles of DMMP obtained from five chemical suppliers, in which each sample was analyzed in triplicate. Based on the manufacturer's information, 29 impurities were searched for in the GC × GC-TOFMS chromatogram of each replicate of each sample. PARAFAC algorithm was employed to the mathematical resolution of overlapped GC × GC peaks, ensuring clean spectra for the identification of many of the detected analytes by spectral library matching. After that, using a reference mass spectrum for each analyte, target PARAFAC [44] was applied for mean peak volume determination (across the three replicates for each sample). Subsequent data processing focused on using the peak volume information to find significant similarities and differences between the various samples by two different approaches. In the first one, statistical pairwise comparison revealed that the impurity profiles of two samples were identical and the profiles of all other samples were different. In the second approach, nonnegative matrix factorization indicated that there were only five distinct DMMP sample types; consequently, two of them belonged to the same class and these two samples were the same that the pairwise comparison had identified as having identical impurity profiles. The two indistinguishable DMMP samples were confirmed by their chemical supplier to be from the same bulk source. These results demonstrated that the matching of synthesized products from the same source is possible using impurity profile obtained from chemometric analysis of GC × GC-TOFMS data set [45].

Comprehensive separation techniques with three temporal dimensions are not as common as comprehensive 2D chromatographic techniques. In this way, the development and evaluation of a comprehensive, 3D gas chromatographic instrument (GC × GC × GC or $GC^3$) coupled with flame ionization detection (FID) were reported in 2007 [46]. The MPs for sampling column one by column two and column two by column three are set so that a minimum of three slices (more commonly four or five) are acquired by the subsequent dimension resulting in both comprehensive and quantitative data. The complete analyte signal, as viewed in 3D space, is nominally an ellipsoid and, as such, exhibits a trilinear data structure. A 26-component control mixture was prepared in order to evaluate the $GC^3$ proposed. Even with an additional chromatographic column in relation to GC × GC, three of the 26 components were severely overlapped in at least two of the three dimensions. Then, a three-factor PARAFAC model was successfully applied for resolution of these three-overlapped chromatographic signals. After that, PARAFAC were also applied for the quantification of toluene standards. The toluene standards were prepared in acetone and serially diluted, providing a data set composed of five samples. The chemometric quantitative results were compared to the ones obtained by traditional integration method applied to the raw, baseline-corrected data. This comparison demonstrated that PARAFAC may provide a 10-fold improvement in the signal-to-noise ratio.

In GC × GC separation, deviations from trilinearity can be observed under nonideal chromatographic conditions (e.g., analyte overloading and/or severely broadened peaks in the first dimension) or during rapidly changing separation conditions on the first dimension as during a fast temperature program to deal with the general elution problem, which prevents too short or too long retention times on the second column for the target analytes. In most practical implementations of GC × GC-TOFMS, the two GC columns are normally temperature programmed at about the same rate or nearly so; thus, the temperature increase per time for the first column will simultaneously correlate to a temperature increase on the second column. This temperature increasing on the second column throughout a run causes analyte retention times to decrease from slice-to-slice taken from the first column (assuming a constant flow) and thus induces a shift in second-column retention times across a peak. In this way, two different approaches to handle this shift behavior were described. In the first approach, a preprocessing method that corrects the shifts and, thus, restores the trilinearity was applied. This is done using an integer retention time shift correction that utilizes the mass spectral information to guide the shift correction. For the shift-corrected data, PARAFAC was used to find qualitative and quantitative information. The second approach was to apply PARAFAC2, an extended and less-restricted version of PARAFAC designed to handle shift (or other nonlinear phenomena) located within one specific dimension. Although the application of PARAFAC2 in GC × GC-FID chromatograms had already been described in the literature by that time [28,29], this was the first publication proposing the use of PARAFAC2 to build multiway calibration models from GC × GC-TOFMS data set [47]. Seven bromobenzene solutions in hexane were prepared with different concentrations, and each one was injected in the GC × GC-TOFMS instrument in four replicates. A shift of two data points was intentionally caused to evaluate the capabilities of the algorithms in dealing with it. The quantification of bromobenzene was performed by five different multiway calibration models: PARAFAC with raw, shifted, and restored data (after retention time shift correction) and PARAFAC2 with raw and shifted data. The results obtained presented the largest deviation from linearity over the concentration range for PARAFAC modeling of artificially shifted data, which was expected. On the other hand, PARAFAC2 modeled the shifted data almost as well as the raw data, which was also expected and demonstrates the capability of this algorithm in dealing with deviations in one mode. However, PARAFAC performed better on the raw data, with a slightly better linearity over the four largest concentrations and with the ability of modeling the right analyte at lower concentrations; no PARAFAC2 model was able to provide a match value of 750 at the lower concentrations for the comparison of the deconvoluted mass spectrum to a mass spectra library. Furthermore, it was observed that PARAFAC, despite being strictly trilinear, is able to model small systematic shifts in a very efficient way. At last, the shift method correction worked very well for the higher concentrations improving the predictability of PARAFAC in comparison to the model using the raw data, but the shift correction generated a worse PARAFAC model at low concentrations, because the alignment quality starts to drop when the signal-to-noise ratio decreases. In conclusion, PARAFAC was found to be more robust at lower signal-to-noise ratios and was capable in detecting and modeling the target analyte at lower concentration than PARAFAC2, while PARAFAC2 proved to be a very good alternative by removing the requirement of a signal-to-noise-dependent alignment step before the subsequent modeling. However, after that, other applications of PARAFAC2 to build multiway calibration models from GC × GC data set were not found in scientific database.

As previously mentioned, in spite of the great number of research work describing the use of MCR in the analysis of complex mixtures through different analytic techniques, it had not been employed to generate multiway calibration models from GC × GC analysis until 2011, when Augusto and coworkers [48] reported a quantitative determination of rosemary essential oil in samples containing unknown interferents using MCR-ALS to generate the calibration model from GC × GC-FID chromatograms. The quantification of lemongrass essential oil in commercial perfume samples combining MCR-ALS and GC × GC-FID analysis was also performed. In order to build the MCR-ALS model, the GC × GC-FID chromatograms were unfolded from a matrix to a vector; then, the vectors of all samples (calibration and validation ones) were placed in the lines of a 2D matrix (**D**), which was decomposed into the matrix of concentration profiles (**C**) and the matrix with the chromatograms of each pure component (**S**). Finally, after ALS optimization, the vectors obtained in matrix **S** were reshaped into 2D GC × GC-FID chromatograms, and the calibration curve was built using the data contained in matrix **C**. Figure **5** exemplifies this approach but for a two component mixture. The unfolded chromatograms were aligned using peakmatch routine [49] before the model development for each study.

In the first study, the calibration samples were prepared by dilution of the essential oil of rosemary in cereal alcohol at the concentrations of 2.5%, 5.0%, 7.5%, 10.0%, and 15.0% (v/v). For the validation set (10 samples in total), a pineapple essence or a commercial perfume was added as an interference. The first interference was chosen to simulate a complex "perfume"-like sample. This particular mixture is not used in any commercial perfume known to the authors. The second interference was chosen to provide a higher complexity sample, in order to evaluate the chemometric model. The GC × GC-FID obtained for essential oil of rosemary, pineapple essence, and perfume are shown in Figure **6**, where two different areas of the 2D chromatograms that

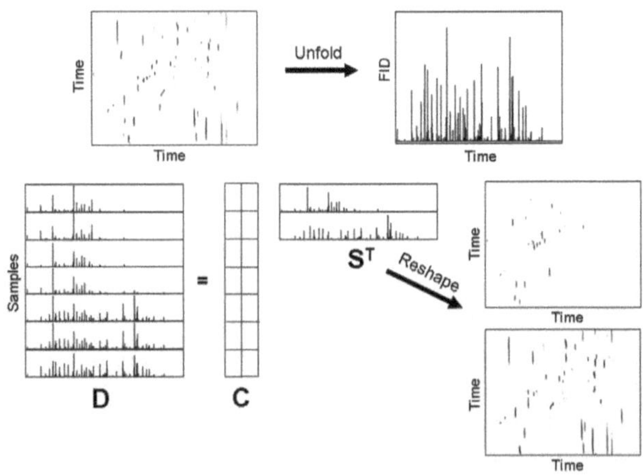


**FIGURE 5** Scheme for MCR-ALS analysis of GC × GC-FID chromatograms. *Reprinted from Ref. [48], copyright 2011, with permission from Elsevier.*

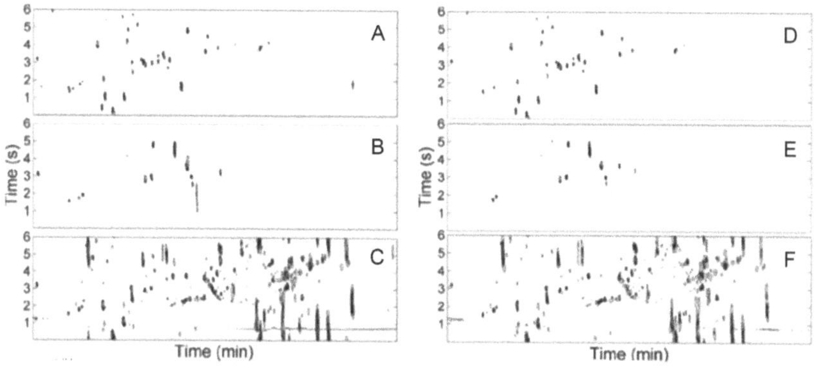


**FIGURE 6** Typical GC × GC chromatograms with the excluded areas suppressed of (A) the essential oil of rosemary, (B) pineapple essence, and (C) commercial perfume. Chromatograms recovered by MCR-ALS for (D) the essential oil of rosemary, (E) pineapple essence, and (F) commercial perfume. *Adapted from Ref. [48], copyright 2011, with permission from Elsevier.*

were excluded for the elaboration of the model were not included. The first excluded area only presented the solvent employed and was not used because it did not present any quantitative information. The second excluded area was a coelution of glycols that, because of their highly polar nature, they presented very strong interactions with the second-dimension column and, consequently, they eluted in $^2$D as extremely large and tailed peaks. The chromatograms obtained for the pure samples were used for the initial estimative of the experimental data. During the ALS optimization of the model, selectivity constraint for concentrations and nonnegativity constraint for concentrations and chromatograms were applied. The chromatograms resolved by the model for the essential oil of rosemary, pineapple essence, and commercial perfume (perfume #1) are showed in Figure **6**, where it is possible to see a high similarity between the chromatograms for the pure samples and the chromatograms resolved by the MCR model. Then, a calibration curve was carried out using the concentration results obtained from the model and the reference concentration of the calibration samples, where a correlation coefficient of 0.996 was obtained. Finally, the concentration of rosemary essential oil in the validation samples was obtained by the interpolation into the calibration curve of the concentration results provided by the MCR model for these samples. The root mean square error of the percentage deviation (RMSPD) and the root mean square error of prediction (RMSEP) values obtained were 7.2% and 0.4% (v/v), respectively.

For evaluation of the accuracy of the proposed MCR-ALS method, a second data set was built to quantify the essential oil of lemongrass in a local commercial perfume (perfume #2), which contains this essential oil. The essential oil of lemongrass, an essence containing the major constituents of this perfume (without the lemongrass essential oil), and two samples of the perfume from different batches were supplied by the perfume manufacturer. Firstly, a calibration model was built by diluting the essential oil at the concentrations of 2.5%, 5.0%, 7.5%, 10.0%, and 15.0% (v/v) in cereal alcohol. The validation set (four samples in total) was prepared by introducing pineapple essence and the essence of the commercial perfume as interferences, to simulate low-complexity and high-complexity samples, respectively. Afterward, the model was used to quantify the amount of essential oil of lemongrass in the perfume #2 samples. The approach used in this second study was the same applied for the first one. Pure sample chromatograms were used as initial experimental estimates, and selectivity constraint for concentrations and nonnegativity constraint for concentrations and chromatograms were applied. The GC × GC-FID chromatograms resolved by the model were very similar to the ones acquired for the pure samples and observed in the first study. A correlation coefficient of 0.983 was obtained in the calibration curve. The RMSPD and RMSEP values for the prediction of the validation samples were 6.9% and 0.5% (v/v), respectively. Finally, the MCR-ALS prediction of lemongrass essential oil in the two batches of the commercial perfume was 8.7% and 8.2%, which is in agreement with the range of 8.0–9.0% indicated by the supplier. Therefore, the results achieved by MCR-ALS in both studies demonstrated that the proposed approach can be used to resolve GC × GC chromatograms even from complex samples.

Alternative and renewable fuels have received increased attention due to the predicted shortage in oil supplies and consequent rise in oil price and to the effects associated with ambient air pollution. In this context, biodiesel represents one of the most significant alternatives to conventional petrodiesel fuel. It is defined as a mixture of fatty acid methyl, or ethyl, esters and is obtained from biological materials such as several vegetable oils, recycled cooking oils, animal fats, and waste products. Biodiesel is frequently mixed with petroleum distillates to attain blends defined by BXX, where XX stands for the percentage of biodiesel (v/v). Therefore, the accurate determination of the percentage of biodiesel content in diesel fuel is really important in order to guarantee that the biodiesel content declared by the BXX denomination of a supplier is the really amount present in the diesel sample. It has been done following the ASTM D7371 procedure, which employs infrared spectroscopic for the analysis. Furthermore, the determination of this percentage can also be carried by using specific fatty acid methyl esters (FAME) present in the biodiesel. Hence, the elucidation of the FAME profile in diesel blends is necessary. According to the UNI EN 14331 procedure, the FAME profile is determined by GC using a capillary column coated with a polar stationary phase, such as poly(ethylene glycol), after an LC preseparation step. In this report, MCR-ALS modeling of biodiesel samples analyzed by GC × GC-FID for the prediction of the BXX mixtures was described [50]. GC × GC-qMS analysis for qualitative purpose was also performed. Five different vegetable oils (soybean, corn, canola, and sunflower) were used for the preparation of biodiesel. Samples of soybean biodiesel were also prepared using unrefined raw oil and recovered oil previously used on conventional cooking. The biodiesels were prepared by ultrasound-accelerated solvolysis. Six separated batches of biodiesel from each different source were prepared and analyzed. Prior to the chemometric analysis, the 2D chromatograms were unfolded from matrix to vectors, and they were aligned using correlation optimized warping method [51].

Before the elaboration of the MCR-ALS method, an initial evaluation of the data set was carried out using multiway principal components analysis, in which chromatographic areas corresponding to each source of the biodiesel and the areas independent of the vegetable oil used were identified. In this way, for the building of the MCR-ALS calibration model, the chromatographic areas dependent on the vegetable oil source were masked (all signal values set to zero). Calibration samples were prepared by mixing soybean biodiesel and mineral diesel, resulting concentrations ranging from 2.0 to 30.0% (v/v) of the former. BXX samples with concentrations ranging from 3.0% to 20.0% (v/v) of assorted biodiesels (soybean, canola, corn, and sunflower) were used to validate and assess the robustness of the calibration model regarding biodiesel source.

The MCR-ALS multiway calibration model was generated using nonnegativity and unimodality constraints, the GC × GC-FID chromatograms of pure soybean oil and pure mineral diesel as initial estimative of the instrumental data, and the chromatograms obtained for the calibration sample set. A calibration curve was plotted using the concentration profiles provided by the chemometric model and the correlation coefficient obtained was 0.997. Then, the prediction of the biodiesel content in the validation samples was performed. The correlation coefficients between the real and predicted biodiesel concentrations were 0.998, for both soybean BXX mixture and nonsoybean BXX mixture. The RMSPD and RMSEP values for the validation sets were 2.7% and 0.4% for soybean BXX mixture and 5.0% and 0.3% for combined BXX mixture, respectively. Therefore, the proposed methodology was able to predict accurately the biodiesel concentration in the BXX samples regardless of its origin.

Tauler and coworkers [52] used MCR-ALS algorithm to GC × GC-TOFMS data in order to resolve and quantify polycyclic aromatic hydrocarbons (PAH) in heavy fuel oil (HFO). The MCR algorithm was operated in a different way from that described previously for GC × GC-FID data [48] because the MS detection gives a four-way data set (${}^{1}t_{R} \times {}^{2}t_{R} \times m/z \times$ sample) for several samples/standards, in which the entire first column chromatogram is sliced into a series of high-speed short secondary chromatograms of a length equal to the MP that are continuously recorded by TOFMS detector. The four-way sample set composed of the GC × GC-TOFMS analysis of the standards and the unknown samples presents dimensions equal to ($I$, $J$, $K$, $L$), where $I$ is the number of collected data points (retention times) in second column, $J$ is the number of $m/z$, $K$ is the number of slices taken from the first column of the total run, and $L$ is the number of analyzed samples. The samples were rearranged by joining the augmented data matrices of each sample with dimensions ($I \times K, J$) so as to give the column-wise superaugmented data matrix with dimensions ($I \times K \times L, J$) for the simultaneous analysis of the different $L$ samples (Figure 7).

Due to the complexity of the HFO, the number of components in each short 2D chromatogram can be different, making the quantification of target compounds very challenging. In contrast to PARAFAC papers, in which the metabolite concentration was measured according to the 3D peak volume, precise quantification was based on the summation of modulated peak areas. In GC × GC, a single compound eluted from 1D is converted in two or more modulate peaks in the short 2D chromatograms. In this MCR study, the summation of the peak areas using the resolved elution profiles in 2D was used. Calibration curves of 10 PAH were built ranging from 0.02 to 5.0 mg $L^{-1}$,

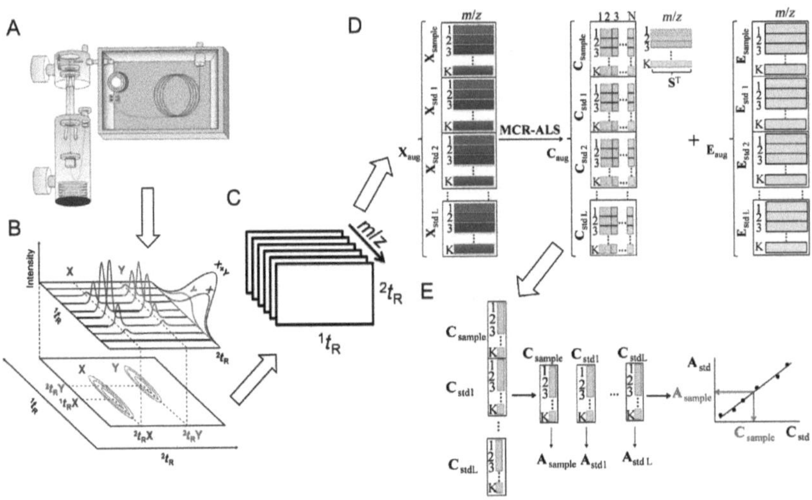


FIGURE 7 (A) GC × GC-TOFMS instrumentation design, (B) signal acquisition of two overlapped components **X** and **Y** eluting from the first column and injected to the second column and corresponding 2D contour plot of the second-column collected chromatograms, (C) three-way data arrangement of the whole GC × GC-TOFMS signal acquired during the analysis of a single sample, (D) $\mathbf{X}_{aug}$ the column-wise superaugmented matrix built using all augmented data matrices corresponding to different samples. $\mathbf{X}_{std}$ are the individual augmented data matrices corresponding to the GC × GC-TOFMS analysis of every standard mixture samples and $\mathbf{X}_{sample}$ is the individual augmented data matrix corresponding to the unknown HFO sample. In all cases, data matrices were arranged with their mass spectral mode in common. $\mathbf{C}_{aug}$ is the matrix of elution profiles of every component in the standard mixture samples, $\mathbf{C}_{std}$, and in the unknown sample, $\mathbf{C}_{sample}$, and $\mathbf{S}^T$ is the pure mass spectra of every component resolved by MCR-ALS. $\mathbf{E}_{aug}$ is the residual matrix and (E) external calibration strategy to obtain quantitative information from the peak areas of the MCR-ALS resolved elution profiles. *Reprinted from Ref. [52], copyright 2011, with permission from American Chemical Society.*

and quantitative results from MCR-ALS method and commercial ChromaTOF software were compared. Relative errors for the phenanthrene and anthracene estimated concentrations, a pair of overlapped compounds, were, respectively, 2.24% and 5.63% using the proposed MCR strategy. In exception of 1-methylnaphthalene, relative errors with MCR were better than ChromaTOF (lower than 6%). PARAFAC algorithm was applied to the same data set, with no previous data alignment, and it resulted in worse resolution and quantification than MCR-ALS, as well as longer time analysis. These results confirmed that deviation from trilinear model affected PARAFAC results, whereas MCR bilinear model was able to deal with such data.

## 2.2 Multiway Calibration II: Direct Approach

In the previous item, algorithms that perform the deconvolution of overlapped signals in GC × GC chromatograms were highlighted. Then, using the area or the volume of the deconvoluted chromatographic peaks, a calibration model is generated employing univariate approaches. In other words, it is necessary two different steps in order to generate a multiway calibration model of GC × GC chromatograms using this kind of algorithm, which demands considerably time and effort of the analyst to accomplish the whole calibration task. On the other hand, a multiway calibration model of GC × GC data can be directly generated using U-PLS or N-PLS algorithms that are extensions of first-order PLS toward higher-order data. The construction of either N-PLS or U-PLS models is straightforward and there is no external univariate regression step involved.

PLS regression is a method for building calibration models between independent (experimental data) and dependent (property of interest) variables, which are organized in two separate matrices. First, each matrix is decomposed into score and loading matrices; then, the score matrices of independent and dependent variables are interactively adjusted to find the maximum covariance between them. Finally, applying the model to unknown samples yields prediction of the property of interest [53]. In the U-PLS algorithm, a three-way GC × GC-FID sample set or a four-way GC × GC-MS sample set is unfolded to a 2D matrix. After that, the algorithm runs in the same way as a regular PLS. Therefore, U-PLS algorithm is just a PLS that unfolds the high-order sample set to a two-way one before starting the development of the calibration model [19].

Differently from U-PLS algorithm, in the N-PLS one, there is no need of unfolding the data before the generation of the calibration model. In this case, a three-way sample set made up of GC × GC-FID chromatograms, for example, is decomposed into a set of triads. A triad consists of a score vector, related to the mode of the data corresponding to samples, and two loading vectors related to the two dimensions of the GC × GC-FID chromatograms. These vectors are calculated to have the maximum covariance with the dependent variable (property of interest) as soon as observed in PLS algorithm. Then, the developed model can be applied to predict the property of interest studied in unknown samples [20].

The main advantages of both U-PLS and N-PLS are their ease of use in the development of a multiway calibration model and the lack of need for trilinear data structure, which means a great plus for the generation of calibration models using GC × GC chromatograms because of retention time deviations usually observed in 2D chromatographic separations. Furthermore, these algorithms do not require peak identification while taking advantage of the redundant measurements of each analyte peak with GC × GC separation. In spite of the U-PLS requirement for unfolding the data, both N-PLS and U-PLS have been widely used to build multiway calibration models using GC × GC-FID or GC × GC-MS data set. However, the second-order advantage is not achieved by both U-PLS and N-PLS, so the composition of the calibration sample set must be carefully designed in order to guarantee that all possible interferences of the prediction sample set (unknown samples) must be present in the calibration step [12,24].

### 2.2.1 Examples of Application

Synovec and coworkers performed the first study using direct multiway calibration to GC × GC separations, and it refers to the quantitative analysis of aromatic and naphthene content in naphtha samples [54]. One of the objectives was to reduce the time analysis from nearly 100 min run using conventional GC to a 6-min GC × GC-FID run and N-PLS analysis. Time reduction was achieved based on the improved resolution and selectivity of GC × GC compared to conventional GC. The proposed method has the potential to be simpler than the conventional GC method. Moreover, PLS calibration models are well suited to quantification of aromatic and naphthene class contents, because unlike GRAM, PLS methods are able to model properties arising from multiple peaks. Three or four repetitions of 19 samples were carried out by GC × GC-FID, and the whole GC × GC data were used to build the N-PLS models using leave-one-out cross validation (LOOCV). The same samples were also analyzed by conventional GC standard method and conventional quantification using peak areas was used to validate and calibrate the GC × GC quantification. The aromatic and naphthene content ranges were 2.25%–14.65% and 5.62%–31.6%, respectively. While the prediction of the highest and lowest contents for both classes presented a relatively poor precision because the calibration set does not cover the full concentration range when those samples were left out in LOOCV calibration, all the other samples were correctly modeled by the N-PLS models.

In a further study, the same research group improved the chemometric method through data alignment and area selection prior to PLS model building [55]. The percent weight of naphthalenes in jet fuel samples was determined by GC × GC-FID and N-PLS. Windowed rank minimization alignment was applied to GC × GC to correct retention time shifts. Five replicates of fast 5-min GC × GC analysis were performed for 14 jet fuel samples, all of them with known weight percentages of naphthalenes (3% on average) previously obtained through ASTMD1840 method. Nine samples were used to build the calibration subset, whereas the remainders were used to evaluate the model predict ability. As naphthalenes elute in a closely space of the entire GC × GC due to chromatographic structure, models using the entire data or selected regions were built and results were compared. The influence of the alignment was also investigated. The number of latent variables was influenced by alignment and lower RMSEP values were obtained for three latent variables, excepting selected region chromatogram unaligned data that reach no clear minimum RMSEP values as additional latent variables were added to the model. The prediction error nearly matches the uncertainty of the reference data percent volume values used to build the calibration models. Therefore, the combined approach of area selection and alignment procedure of GC × GC-FID data followed by N-PLS modeling is able to rapidly and accurately determine the naphthalene content in samples such as jet fuel.

As just highlighted, PLS models are able to quantify complex mixtures as individual components of samples or single compounds. This concept was adopted in a more complex study in order to detect potential adulteration of commercial Brazilian gasoline and quantify the amount of gasoline in adulterated samples [56]. GC × GC-FID data and N-PLS were combined to correlate the concentration of gasoline on samples with chromatographic data. In Brazil, commercial gasoline type C corresponds to a mixture of type A gasoline and ethanol (25%, v/v), and the most usual adulterants are ethanol in excess than the legally prescribed amounts and some petrochemical derivatives. The complexity of gasoline and the fact that several compounds present in type C gasoline can be also found in many possible adulterants (white spirit, kerosene, and paint thinner) lead to an overwhelming analytic problem. The complexity of these samples can be seen in Figure **8**, where the GC × GC chromatograms obtained for a pure sample of type C gasoline and pure white spirit, kerosene, and paint thinner adulterants are presented.

Blends with different concentrations of nonadulterated type C gasoline and type adulterants (plus ethanol) were employed for calibration and prediction set. Twenty-five calibration samples with the concentration for each one of the three adulterants varying from 0% to 30% (v/v) were prepared by mixing adequate volumes of the solvents and gasoline, resulting in gasoline content samples ranging from 30% to 100% (pure type C gasoline). A test sample consisted of 14 samples was prepared similarly to the calibration set samples. In addition, a second test set with 13 samples of type C gasoline collected from local gas stations and analyzed by the official method of the Brazilian National Agency of Petroleum, Natural Gas and Biofuels (Agência Nacional do Petróleo, Gás Natural e Biocombustíveis—ANP) was also evaluated. N-PLS models for gasoline and adulterants were developed without previous

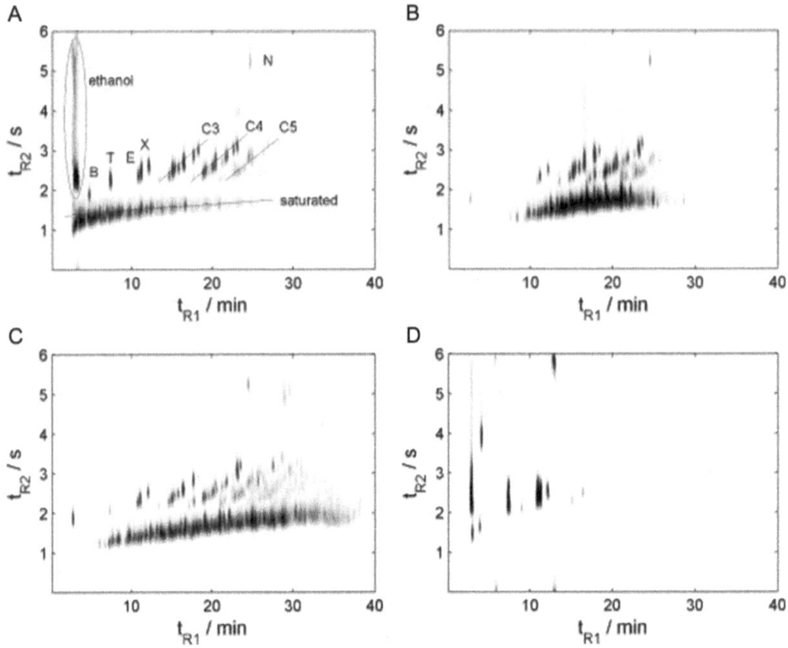


**FIGURE 8** Typical GC × GC-FID chromatograms of pure (A) type C gasoline, (B) white spirit, (C) kerosene, and (D) thinner. Band identification: B, benzene; T, toluene; E, ethylbenzene; X, xylene isomers; C3, benzene C3-substituted; C4, benzene C4-substituted; C5, benzene C5-substituted, and N, naphthalenes. *Reprinted from Ref. [56], copyright 2008, with permission from Elsevier.*

data treatment (peak alignment or noise filtering). In the first moment, the predictive capability of each N-PLS models was evaluated through the LOOCV procedure, and the models were selected having as few latent variables as possible to minimize the RMSECV: three for white spirit, four for thinner, and five for the remainder. In this way, it was avoided an overfitting of the calibration models. The RMSECV values obtained were 5.7% (v/v) (gasoline), 2.7% (thinner), and 2.8% (white spirit, kerosene). The correlation coefficients $r^2$ for LOOCV prediction versus real concentration curves corresponding to the N-PLS model curves were 0.985 (gasoline), 0.991 (kerosene), 0.986 (thinner), and 0.982 (white spirit). The accuracy of the GC × GC-FID and N-PLS method to predict the gasoline C content was evaluated through the root mean square error of prediction (RMSEP) obtained for the validation sample set. The overall RMSEP value was 6.2% (v/v). Samples spiked with ethanol resulted in average RMSEP values greater than samples without ethanol, 8.2% (v/v) and 3.3% (v/v), respectively. The predicted gasoline concentration for the former set was overestimated for all samples, with larger errors for samples containing only ethanol as adulterant. Ethanol was not one of the adulterants added to the calibration samples and N-PLS does not present the second-order advantage, but, even with such error in ethanol spiked gasoline samples, these samples were still detected as adulterated, which was the aim of the study.

Finally, the GC × GC-FID and N-PLS approach was applied to analyze samples of type C gasoline collected from local gas stations and previously examined by using the standard procedures, to check the equivalence of the proposed methodology and the official procedure. For eight samples flagged as adulterated, the content on type C gasoline measured using GC × GC-FID and N-PLS calibration ranged from 48% to 85% (v/v). As % (v/v) gasoline obtained was significantly less than 100%, it is a strong indication of the validity of the proposed approach.

Low gasoline concentration was predicted for two samples (called N1 and N2), considered as nonadulterated by the official analysis method, but this inconsistency does not challenge the reliability of the described approach. The chromatogram of N1 and N2 samples was compared to a pure type C gasoline in Figure **9**. Sample N1 is clearly different from nonadulterated gasoline; thus, problems related to the conservation of the sample are more likely to explain such difference. On the other hand, the GC × GC-FID chromatogram obtained from N2 sample is not so different from the one obtained for the pure gasoline C. However, a closer inspection shows some remarkable differences not present in chromatograms for other samples. The peaks corresponding to saturated and cyclic alkanes in sample N2 chromatogram are less intense as compared to nonadulterated samples (especially for those in the range 10 min-$\leq {}^1t_R \leq$ 20 min). Also, several peaks that are not present in model gasoline C or in any other sample are present, for example, there is an intense peak at ${}^1t_R = 19.57$ min and ${}^2t_R = 2.23$ s that does not appear in model gasoline, as well as several other smaller peaks such as a pair of unresolved species at ${}^1t_R = 18.00$ min and ${}^2t_R = 2.25$ s. Therefore, it is highly possible that sample N2 was adulterated and not detected by the official methods. In conclusion, all samples with % type C gasoline under 80% (v/v), according to GC × GC-FID + N-PLS, were pointed out as adulterated by ANP tests; furthermore, the samples with % type C gasoline of at least 95% (v/v) were classified by ANP as nonadulterated. Therefore, for % type C gasoline less than 80% or over 95%, the chance of a false positive or a false negative can be considered as negligible. Samples in the range 80% $\leq$ % type C gasoline < 95% were detected as either adulterated or nonadulterated; therefore, within this concentration range—and taking into account only this sample set—there would be a considerable possibility of either false-positive or false-negative results.

Whenever huge matrix data are obtained, matrix reduction is one way to reduce data processing time and to select parts of the chromatogram related to significant information. This concept is frequently applied in target analysis with PARAFAC, as described on Section 2.2. For direct calibration, where complex mixtures or chemical properties are quantified as single species, data

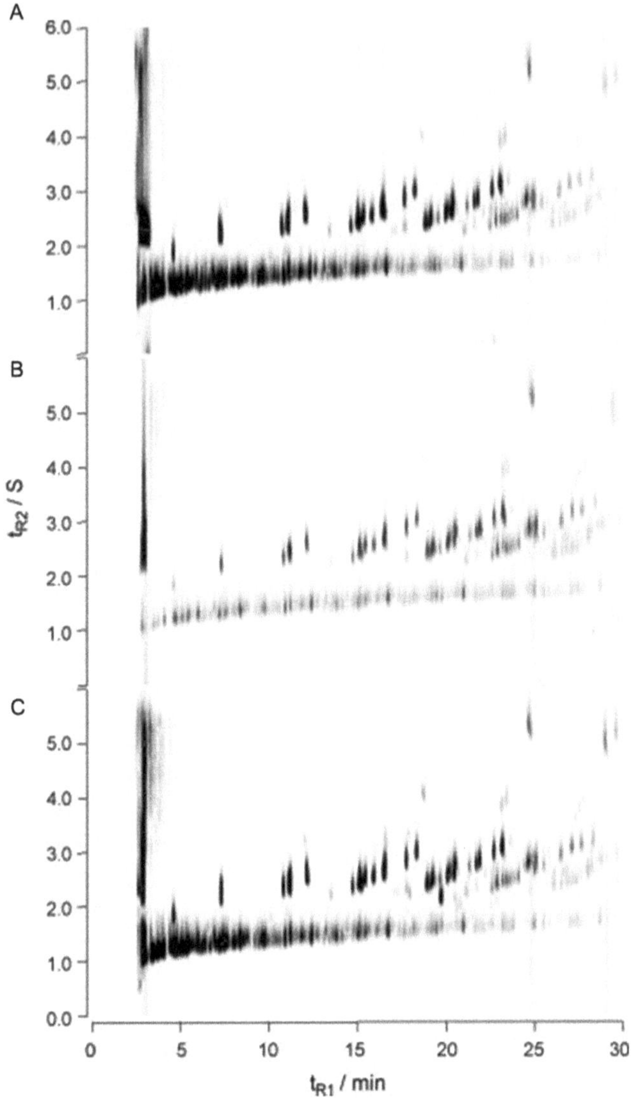


**FIGURE 9** GC × GC-FID chromatograms for typical nonadulterated type C gasoline (A), test sample N1 (B) and test sample N2 (C). Data for ${}^{1}t_{R} > 30.0$ min not shown. *Reprinted from Ref. [56], copyright 2008, with permission from Elsevier.*

reduction may not be a simple task. Sometimes, one or more intervals of the instrumental data set provide more reliable regressions. Discovering which part of the chromatogram is related to the investigated characteristic of the sample may depend on the previous sample knowledge.

De Godoy *et al.* [57] used different chemometric strategies to predict by GC × GC-FID analysis some physicochemical properties of gasoline measured according to ASTM standard procedures performed by ANP. The investigated properties were density and points of the distillation curve, which are officially determined by ASTM D86 and ASTM D4052 tests, respectively. Calibration set consisted of 30 samples and 21 samples was selected as a prediction set. GC × GC chromatograms were unfolded and analyzed without preprocessing.

The boiling point at 10%, 50%, and 90% (v/v) of distillation and the final point of distillation of gasoline are important tests used by ANP to evaluate the quality of gasoline. The boiling point for each percentage of distillation is not related to the whole chromatogram, but to the compounds that boil in the respective temperature windows and present concentration variance among the samples. Therefore, synergy interval PLS (siPLS) was employed for the selection of the intervals presenting the compounds that have impact on the distillation point being analyzed. The siPLS algorithm splits the data set and then calculates all possible PLS model combinations of two, three, or four intervals. Then, the combination of intervals with the lowest RMSECV is selected. Furthermore, genetic algorithm (GA) was used for variable selection inside the intervals selected by siPLS for 10% and 50% (v/v) points of the distillation curve. GA is a method that finds the subset of independent variables most consistent with the dependent variables. The number of latent variables, intervals, combinations, and GA configurations (when employed) was optimized during the development of the calibration models. The intervals selected in each model are shown in Figure **10**. Differently from distillation points, the density is related to the whole chromatogram, because it depends on the entire composition of the sample. Thus, PLS was used to build the calibration model.

The final prediction results were expressed as the RMSEP and RMSPD values. Parameters and results from these models are presented in Table **2**, as well as the reproducibility limits of the official methods. The variation

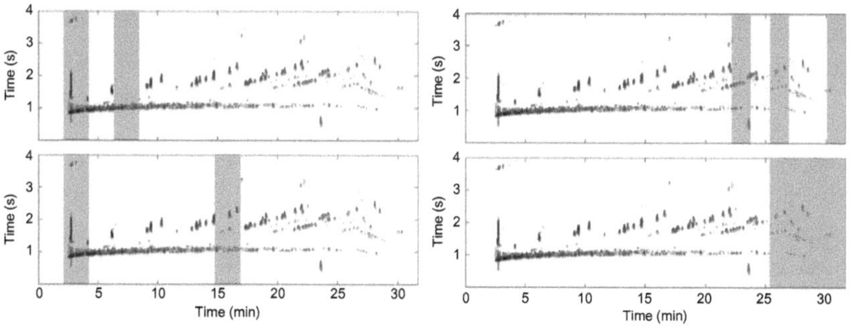


**FIGURE 10** Typical GC × GC chromatogram of gasoline showing the selected intervals (dark areas) by siPLS for 10% (v/v) (left-top), 50% (v/v) (left-bottom), 90% (v/v) (right-top) of distillation curve, and distillation final point (right-bottom). *Reprinted from Ref. [57], copyright 2011, with permission from Elsevier.*

**TABLE 2 Parameters of the siPLS Models and Prediction Errors**

| Parameter | 10% | 50% | 90%[a] | Final point[a] | Density[a] |
|---|---|---|---|---|---|
| Number of intervals | 15 | 15 | 20 | 15 | – |
| Interval combinations | 2 | 2 | 3 | 3 | – |
| RMSEP | 0.4 °C | 0.3 °C | 1.5 °C | 2.1 °C | 1.7 kg $m^{-3}$ |
| RMSPD | 0.8% | 0.5% | 1.1% | 1.7% | 0.2% |
| ASTM limits | 3.2 °C | 1.9 °C | 4.1 °C | 6.8 | 2.0 kg $m^{-3}$ |

[a]*GA not applied.*

obtained with GC × GC-FID and multivariate analysis is comparable to the official tests and presented lower values for RMSEP than a previously reported multivariate analysis of GC-FID (not shown here [58]). Therefore, the distillation points and the relative density of a gasoline sample can be predicted with only one GC × GC-FID run, which require lower sample volumes than the standard procedure and does not involve manual operations.

The purpose of the majority of the studies combining GC × GC data and PLS multiway calibration is to quantify a group of peaks as single species or correlate a single sample property, which depends on the entire sample to the chromatographic data. As presented in the prediction of some points of the gasoline distillation curve, even when multiple peaks are related to the measured property, it is not always desirable to submit the entire data set to chemometric analysis, since only small portions of the chromatogram may be relevant to a specific problem. Another situation in which the analytic information is grouped in a small part of the whole chromatogram is target analysis, and submitting the entire data to multiway calibration is time-consuming; besides, it may lead to poor results.

Based upon arguments just stated, a target analysis using subregions of GC × GC chromatograms was proposed by Augusto and coworkers [59]. They presented a new approach for target quantitative analysis for GC × GC, called interval multiway PLS (iNPLS), in which the 2D GC × GC-FID chromatogram is split in small sections and each of these pieces is treated as an independent new chromatogram. Separated conventional N-PLS calibration models for the concentration of the target analyte were built for each of the pieces of the whole chromatogram, and the best model is selected for quantitative analysis. Both interval partial least squares (iPLS) and siPLS [60] could be used through the unfolding of the second-order data into a first-order data. However, specifically for GC × GC data, some problems can arise in this procedure. As in GC × GC, a single compound is transformed in several small peaks during the modulation process; in iPLS or siPLS methods, peaks from other compounds that had coeluted in the first dimension with the target compound but were separated by the second column will also be included in the selected interval(s) and it may bias the calibration adjustment. In the case of siPLS, considering the conventional time analysis and detector frequency, the number of combinations would be greater than $10^7$, which will take a long time to provide reliable results. Therefore, the main advantage of the iNPLS is to split the data matrix in intervals in both dimensions without losing the second-dimension separation in order to form new reduced matrix in a much faster pace than siPLS.

The iNPLS was preliminarily evaluated using solutions of model compounds with different chemical properties. Synthetic samples containing different concentrations (0.0–3.0%, v/v) of five model compounds (1-octanol, undecane, 2-octanone, cyclohexanone, and toluene) were prepared and selected as a calibration set (17 samples) and a prediction set (8 samples). High correlation coefficients for the relationships between reference concentration and predicted concentration by iNPLS for the five standards (higher than 0.984) and low RMSECV and RMSEP values (all lower than 0.25%) denote that the iNPLS algorithm selected the correct region for each model and the prediction of the concentration for the standards were performed accurately. Subsequently, iNPLS was evaluated by quantifying the allergens geraniol, citronellol, and benzyl alcohol in perfume samples that is a more complex sample in which the coelution may affect the interval selection, as well as the calibration model. Calibration set consisted of 17 samples and the prediction set was composed of 8 samples, with concentration range of the allergens from 0 to 100 ppm. The complexity of the chromatogram of perfume sample spiked with allergens and the size of the selected data is presented in Figure **11**.

As can be seen in this figure, the benzyl alcohol peak (A) is partial overlapped with the constituents of the matrix (perfume). Because of the partial coelution, the 1D interval length was the shortest possible to avoid selection of other peaks. Once again, the iNPLS algorithm selected the correct region for each allergen. The RMSEP values obtained with iNPLS models were compared to those obtained with iPLS and N-PLS in order to verify the advantage of not unfolding the GC × GC matrix data and to use only one part of the chromatograms instead of the entire ones, respectively. RMSEP values for iNPLS were 10.2, 4.5, and 6.5 ppm for benzyl alcohol, citronellol, and geraniol, respectively. Results for iPLS were two to three times bigger than iNPLS, because iPLS models selected parts of the chromatogram that had been separated in the second dimension and were not related to the analyte concentration. The worst RMSEP values were obtained with N-PLS, which were three to eight times higher than iNPLS. As the N-PLS does not have a second-order advantage, the iNPLS algorithm has the same limitation.

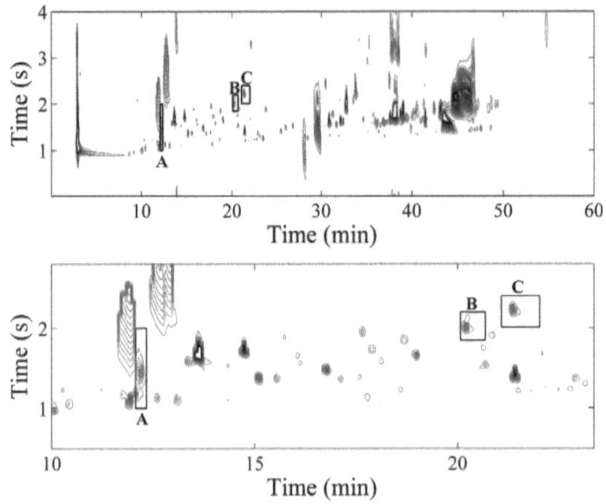


**FIGURE 11** GC × GC chromatogram and the regions selected by the iNPLS algorithm to build the calibrations models of benzyl alcohol (A), citronellol (B), and geraniol (C). *Top* all the chromatogram, *bottom* a zoom in the area of interest. *Reprinted from Ref. [59], copyright 2011, with permission from Elsevier.*

However, in the iNPLS, it is possible to analyze an unknown sample that contains interferents not present in the calibration set since those interferents are not included in the submatrix selected by the iNPLS to build the model.

According to the examples just described, GC × GC technique has shown impressive ability to achieve more resolution of complex mixtures as petrochemical derivatives or fragrances, providing a tremendous amount of data that is not available in conventional GC and, obviously, high-order algorithms must be resorted to interpretation. Hall et al. [10] employed GC × GC-FID and PLS to investigate the weathering processes that acted on the crude oil released in the Gulf of Mexico after the *Deepwater Horizon* rig explosion. The hydrocarbons were exposed to various weathering as dissolution, biodegradation, and photo-oxidation process, leading to a decrease in saturated and aromatic compounds and increase in oxygenated hydrocarbons (OxHC). Comparison between nonweathered and weathered oils can provide an indirect means to determine which compounds were transformed through oxygenation processes. However, thin-layer chromatography–flame ionization detection (TLC-FID) method usually employed to analyze the amount of OxHC only provides total fraction results (saturated, Sat; aromatic, Aro; and two distinct OxHC groups, the less polar, OxHC1, and the more polar, OxHC2). Conventional GC allows better separation than TLC, but severe peak overlapping still takes place due to the sample complexity and, thus, peak identification is problematic. The aim of this study was to determine the compounds whose disappearance in the chromatograms of weathered oil correlates with the formation of OxHCs. For that, 41 samples were analyzed by GC × GC-FID and GC × GC-TOFMS, the last only to help in the identification step. A typical

GC × GC chromatogram of nonweathered oil is presented in Figure **2**. Each chromatogram was unfolded to a single 1,287,000-element row per sample and then a separate PLS was conducted for each TLC-FID fraction (saturated, aromatic, and OxHC fractions). In other words, the multiway chemometric U-PLS algorithm was applied. Initial models were examined for outliers by comparing Hotelling's $T^2$ values to Q residuals and no oil was excluded. Cross validation of results was performed by six random subsets conducted in two iterations. The final model was fit with eight latent variables and captured 85.4% of the variance in the X block and 96.3% of the variance in the Y block. Parameters of the four PLS models are depicted in Table **3**.

Low RMSEC and RMSECV obtained for saturated, aromatic, and the OxHC2 fractions show that a regression of these fractions utilizing the GC × GC data is possible. The OxHC2 fraction did not reach a good fit, probably because they are intermediate between precursor compounds and the most polar oxygenated fraction. Some candidate compounds of precursors for the less polar oxygenated fraction can be extracted from PLS loadings of validated models provided.

Another information that could be extracted from weathered fuels is how long a specific sample had been exposed to weathered process. Such information is important for environmental and forensic investigations. A study to estimate the age of weathered gasoline samples based on GC × GC-FID analysis and PLS, nonlinear PLS (PolyPLS), and locally weighted regression (LWR) was performed by Zorzetti *et al.* [61]. In contrast to PLS, both PolyPLS and LWR are nonlinear regression techniques. Samples of gasoline with varying octane ratings and from several vendors were weathered during 4 days under controlled conditions and their composition monitored over time. In order to reduce the number of variables, peaks were allocated in 36 groups by their proximity in GC × GC chromatogram, taking advantage from chromatographic structure. Training set consisted of two-thirds of the samples randomly chosen and the remainders were selected as test set.

**TABLE 3** Range of Content, RMSEC, RMSECV, and $R^2$ for Best Fit Lines of Calibration and Cross Validation for the Different Fractions

| Parameter | Saturated | Aromatic | OxHC1 | OxHC2 |
|---|---|---|---|---|
| Content fraction range | 0.1–0.5 | 0–0.4 | 0.1–0.4 | 0–0.6 |
| $R^2$ cal. | 0.93 | 0.95 | 0.88 | 0.99 |
| RMSEC | 0.019 | 0.015 | 0.012 | 0.015 |
| $R^2$ cv | 0.77 | 0.72 | 0.15 | 0.78 |
| RMSECV | 0.038 | 0.049 | 0.038 | 0.066 |

Initially, PLS-DA was used to classify the 1152 analyzed samples in "fresh" and "old," resulting in 797 samples that were labeled as fresh (<12 h exposure) samples and 405 as old (>20 h exposure) ones. Separated regression analyses were performed on each class. For fresh samples, the RMSEP value ranged from 24 to 55 min, whereas for old ones, the RMSEP value was roughly one order of magnitude higher, varying from 282 to 370 min. For both regression models, LWR outperforms both PLS and PolyPLS regression methods due to smaller and narrower RMSEP value.

## 2.3 Two-Step Approach × Direct Approach

Two papers described quantitative comparisons between the two types of chemometric algorithms mentioned earlier (the deconvolution ones and the multivariate calibrations ones) in order to generate multiway calibration models from GC × GC data sets.

In the first one, essential oil markers were quantified in full perfumes [28], because for flavor and fragrance industry, quantification of the markers, especially in trace levels, is of high importance since the identification and the quantification of essential oils are usually done through the markers. A set of seven different perfume mixtures for different purposes (detergents and personal care) were selected by an industry. The samples contained 12 target compounds, but this study was limited to the quantification of essential-oil markers, that is, γ-terpinene, citronellyl formate, dimethyl anthranilate, lavandulyl acetate, eucalyptol, and (−) menthone. The other six components were not reported due to confidentiality issues. The samples were diluted 10-fold with 1-propanol containing accurately weighted concentrations of approximately 0.25% *n*-decane as internal standard. Solutions were prepared in triplicate. Calibration mixtures of all 12 components were prepared in the same internal-standard solution with concentrations at five levels ranging from 10 to 1500 mg/kg. All calibration solutions were measured in duplicate. To assess the accuracy of the quantification methods, a second calibration mixture was made containing the same standards, but at concentrations of approximately 200 mg/kg, and it was analyzed under slightly different conditions (lower head pressure) to induce different peak shapes. Prior to the quantification studies, the chromatograms were aligned using a shifting routine described in this paper. PARAFAC, PARAFAC2, and N-PLS were the chemometric algorithms used to perform the quantification, and their results were compared to the ones of conventional integration peaks. A calibration curve for each one of the six essential oil markers studied was plotted and the correlation coefficients obtained showed that all methods resulted in very good linear relationships. Then, using the chromatograms obtained for the second calibration standard, the concentration for all the six markers was predicted (Figure **12**). Integration performed best for (almost) all components. PARAFAC2 and N-PLS tended to overestimate the concentrations. In the

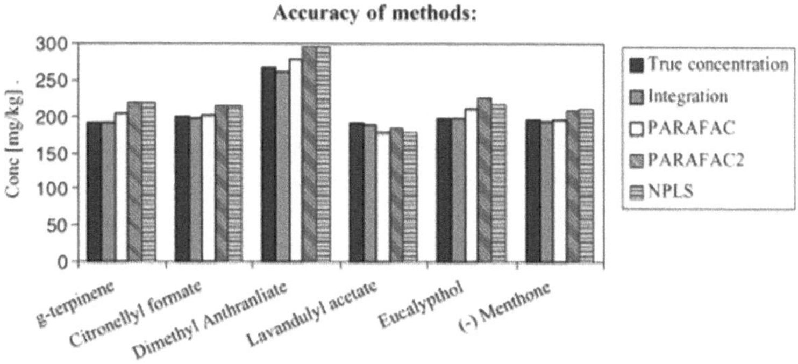


**FIGURE 12** Accuracy of the various quantitation methods based on the analysis of a reference mixture with known analyte concentrations. *Reprinted from Ref. [28], copyright 2003, with permission from Elsevier.*

present case, PARAFAC was the most accurate among the multiway methods used. The influence of the peak shape seemed to be more detrimental for PARAFAC2 than for PARAFAC. This result is surprising, since PARAFAC2 should theoretically be capable of dealing with shifted peaks. Afterward, four real samples were analyzed. Assuming that integration provides the right answer, the multiway methods overestimated the concentrations in many cases, especially at low values (<10 mg/kg). Surprisingly, the highest concentrations were found with PARAFAC2. In conclusion, integration was the most accurate method for the quantitative predictions, but this method is very time-consuming, labor-intensive, and when there are interferents coeluting with the target analyte that is not possible to apply conventional integration because of the impossibility for accurately determining the areas of the chromatographic peaks. Moreover, the integration method required about eight hours more than the multiway algorithms to provide the quantitative information. Among the chemometric methods, PARAFAC provided the most accurate predictions, followed by N-PLS and, then, PARAFAC2.

The second report comparing multiway calibration models elaborated from GC × GC-FID data set with two algorithms that perform deconvolution of signals versus an algorithm that performs the calibration in a direct way using the dependent and independent variables was published by Poppi and coworkers [29]. In this report, kerosene in gasoline was quantified because kerosene is one of the solvents that can be employed to make gasoline adulteration. PARAFAC, PARAFAC2, and N-PLS were the algorithms employed to generate the multiway calibration models from GC × GC-FID chromatograms. Simulated adulterated gasoline was prepared by mixing type C gasoline (supplied and certified as nonadulterated by the Brazilian National Agency of Oil, Natural Gas and Biofuels—ANP) and commercial kerosene to obtain samples with 0%, 1%, 2%, 4%, 6%, 10%, 14%, 18%, 22%, 26%, 30%, 40%, and 50% (m/m) of kerosene in the gasoline.

Each sample was injected in triplicate, and the resulting GC × GC data were employed to generate the calibration model.

The PARAFAC model was built using nonnegativity constraint and two factors, which were chosen based on the model fit and on the Core Consistency Diagnostic (CORCONDIA) test. A cross validation procedure with leave-one-out samples was performed to verify calibration model viability. In this procedure, 39 models were developed and, in all of them, good regression curves were obtained (regression coefficients higher than 0.982). However, some samples presented a deviation from the linearity in the calibration curve, but it did not impede the quantification by the cross validation. The RMSECV obtained was 2.98%. A two-factor nonnegativity PARAFAC2 model was developed. Different from PARAFAC calibration curve, it was not observed linearity deviations in the PARAFAC2 calibration curve. Correlation coefficients higher than 0.982 were obtained in the leave-one-out cross-validation procedure. The RMSECV obtained was 2.65%. The N-PLS model was built using three latent variables, which was chosen based on the model fit and on the RMSECV values. It was not observed linearity deviations and the RMSECV obtained was 2.08%. A comparison of the quantitative results indicated that N-PLS was the best model to fit this data, probably because the N-PLS algorithm uses both the independent (GC × GC-FID chromatograms) and dependent (kerosene concentration) variables to build the model and it can handle trilinearity deviations in the data set. PARAFAC2 was slightly better than PARAFAC, because the PARAFAC2 model allowed small deviations in a mode of data. It is important to highlight that the GC × GC-FID chromatograms were not aligned before the chemometric analysis in order to evaluate the ability of each algorithm in dealing with small retention time deviations.

Therefore, the quantitative results obtained in these two articles showed that both deconvolution and multivariate calibration algorithms are equally capable for the generation of multiway calibration models. The analyst must evaluate the aims of the study and the sample composition in order to correctly plan the whole task for obtaining the calibration model, since the sample preparation and GC × GC configuration until the need of applying preprocessing methods to the data and the choice of the appropriate chemometric algorithm.

## 3 CONCLUSIONS

In this chapter, multiway calibration models built using data set acquired by comprehensive 2D gas chromatography coupled with flame ionization or mass detectors were highlighted. The GC × GC technique was described only in 1991, but robust GC × GC equipments became commercially available roughly in the beginning of the 2000s. Therefore, the association of chemometric algorithms and third- or fourth-order GC × GC data to generate multiway calibration models can be considered new in terms of science, being the first paper dated from 1998. After a discrete start observed until approximately the first half of the 2000s, when practically only one group had published research works describing this association, its use has presented a huge increase since then due to the diffusion of the GC × GC technique. Nowadays, there are some research groups around the world routinely applying chemometric strategies to transform the enormous amount of the information provided by GC × GC analysis into calibration models. Prof. Robert E. Synovec (the University of Washington, the United States) or Prof. Ronei J. Poppi (the University of Campinas, Brazil) headed great part of the applications described in this chapter.

As could be seen in the items discussed before, the algorithms can be classified in two different types. The first one comprises the algorithms that perform deconvolution of chromatographic signals and then a calibration curve employing univariate regression techniques is plotted. The second type corresponds to algorithms that perform a direct calibration using both the independent (experimental) and dependent (property of interest) variables to generate the model. Each algorithm presents several advantages and disadvantages, so it must be chosen based on the aims of the study and on the characteristics of the sample being analyzed.

Both types of algorithms have been applied for building multiway calibration models from GC × GC data sets. It was demonstrated in the examples showed in this chapter that the association of GC × GC separation technique and multivariate analysis has the ability to deal with a broad range of complex samples, for example, petrochemical derivatives and perfumes, in order to provide an accurate calibration model.

In conclusion, the association of GC × GC techniques and multiway chemometric algorithms presents a huge potential and tends to be more and more applied due to the global dissemination of the GC × GC instruments, turning them cheaper and more robust, and to new chemometric interfaces that are making the multiway algorithm friendlier. Moreover, the number of professionals with advanced knowledge in at least one of these areas has been growing recently, and we expect that teamwork with researchers of the two areas will become more common.

## REFERENCES

[1] Deans DR. A new technique for heart cutting in gas chromatography. Chromatographia 1968;1:18–22.

[2] Liu Z, Phillips JB. Comprehensive two-dimensional gas chromatography using an on-column thermal modulator interface. J Chromatogr Sci 1991;29:227–31.

[3] Marriott P, Shellie R. Principles and applications of comprehensive two-dimensional gas chromatography. TrAC, Trends Anal Chem 2002;21:573–83.

[4] Beens J, Brinkman UAT, Dalluge J. Comprehensive two-dimensional gas chromatography: a powerful and versatile analytical tool. J Chromatogr A 2003;1000:69–108.

[5] Górecki T, Harynuk J, Panić O. The evolution of comprehensive two-dimensional gas chromatography (GC × GC). J Sep Sci 2004;27:359–79.

[6] Mondello L, Tranchida PQ, Dugo P, Dugo G. Comprehensive two-dimensional gas chromatography-mass spectrometry: a review. Mass Spectrom Rev 2008;27:101–24.
[7] Adahchour M, Beens J, Vreuls RJJ, Brinkman UAT. Recent developments in comprehensive two-dimensional gas chromatography (GC × GC): I. Introduction and instrumental set-up. TrAC, Trends Anal Chem 2006;25:438–54.
[8] Adahchour M, Beens J, Vreuls RJJ, Brinkman UAT. Recent developments in comprehensive two-dimensional gas chromatography (GC × GC): II. Modulation and detection. TrAC, Trends Anal Chem 2006;25:540–53.
[9] Marriott PJ, Massil T, Hügel H. Molecular structure retention relationships in comprehensive two-dimensional gas chromatography. J Sep Sci 2004;27:1273–84.
[10] Hall GJ, Frysinger GS, Aeppli C, Carmichael CA, Gros J, Lemkau KL, et al. Oxygenated weathering products of Deepwater Horizon oil come from surprising precursors. Mar Pollut Bull 2013;75:140–9.
[11] Escandar GM, Goicoechea HC, Muñoz de la Peña A, Olivieri AC. Second- and higher-order data generation and calibration: a tutorial. Anal Chim Acta 2014;806:8–26.
[12] Escandar GM, Olivieri AC, Faber N(K)M, Goicoechea HC, Muñoz de la Peña A, Poppi RJ. Second- and third-order multivariate calibration: data, algorithms and applications. TrAC, Trends Anal Chem 2007;26:752–65.
[13] Bro R. Multivariate calibration. Anal Chim Acta 2003;500:185–94.
[14] Amador-Muñoz O, Marriott PJ. Quantification in comprehensive two-dimensional gas chromatography and a model of quantification based on selected summed modulated peaks. J Chromatogr A 2008;1184:323–40.
[15] Bro R. PARAFAC. Tutorial and applications. Chemom Intell Lab Syst 1997;38:149–71.
[16] Bro R, Andersson CA, Kiers HAL. PARAFAC2—part ii. Modeling chromatographic data with retention time shifts. J Chemom 1999;13:295–309.
[17] Sanchez E, Kowalski BR. Generalized rank annihilation factor analysis. Anal Chem 1986;58:496–9.
[18] Tauler R, Smilde A, Kowalski B. Selectivity, local rank, three-way data analysis and ambiguity in multivariate curve resolution. J Chemom 1995;9:31–58.
[19] Wold S, Geladi P, Esbensen KIM, Ohman J. Multi-way principal components- and PLS-analysis. J Chemom 1987;1:41–56.
[20] Bro R. Multiway calibration. Multilinear PLS. J Chemom 1996;10:47–61.
[21] Zeng Z, Li J, Hugel HM, Xu G, Marriott PJ. Interpretation of comprehensive two-dimensional gas chromatography data using advanced chemometrics. TrAC, Trends Anal Chem 2014;53:150–66.
[22] Matos JTV, Duarte RMBO, Duarte AC. Trends in data processing of comprehensive two-dimensional chromatography: state of the art. J Chromatogr B Anal Technol Biomed Life Sci 2012;910:31–45.
[23] Murray JA. Qualitative and quantitative approaches in comprehensive two-dimensional gas chromatography. J Chromatogr A 2012;1261:58–68.
[24] Sinha AE, Prazen BJ, Synovec RE. Trends in chemometric analysis of comprehensive two-dimensional separations. Anal Bioanal Chem 2004;378:1948–51.
[25] Synovec RE, Prazen BJ, Johnson KJ, Fraga CG, Bruckner CA. Chemometric analysis of comprehensive two-dimensional separations. Adv Chromatogr 2003;42:1–42.
[26] Carroll JD, Chang J-J. Analysis of individual differences in multidimensional scaling via an n-way generalization of "Eckart-Young" decomposition. Psychometrika 1970;35:283–319.
[27] Harshman RA. Foundations of the PARAFAC procedure: models and conditions for an "explanatory" multimodal factor analysis. UCLA Work Pap Phon 1970;16:1–84.

[28] van Mispelaar VG, Tas AC, Smilde AK, Schoenmakers PJ, van Asten AC. Quantitative analysis of target components by comprehensive two-dimensional gas chromatography. J Chromatogr A 2003;1019:15–29.

[29] de Godoy LAF, Ferreira EC, Pedroso MP, Fidélis CHdV, Augusto F, Poppi RJ. Quantification of kerosene in gasoline by comprehensive two-dimensional gas chromatography and N-way multivariate analysis. Anal Lett 2008;41:1603–14.

[30] Llamas NE, Garrido M, Di Nezio MS, Fernández Band BS. Second order advantage in the determination of amaranth, sunset yellow FCF and tartrazine by UV-vis and multivariate curve resolution-alternating least squares. Anal Chim Acta 2009;655:38–42.

[31] Schiozer AL, Março PH, Barata LES, Poppi RJ. Exploratory analysis of *Arrabidaea chica* deoxyanthocyanidins using chemometric methods. Anal Lett 2008;41:1592–602.

[32] Carneiro RL, Braga JWB, Poppi RJ, Tauler R. Multivariate curve resolution of pH gradient flow injection mixture analysis with correction of the Schlieren effect. Analyst 2008;133:774–83.

[33] Gómez V, Miró M, Callao MP, Cerdà V. Coupling of sequential injection chromatography with multivariate curve resolution-alternating least-squares for enhancement of peak capacity. Anal Chem 2007;79:7767–74.

[34] Terra LA, Poppi RJ. Monitoring the polymorphic transformation on the surface of carbamazepine tablets generated by heating using near-infrared chemical imaging and chemometric methodologies. Chemom Intell Lab Syst 2014;130:91–7.

[35] Jaumot J, Marchán V, Gargallo R, Grandas A, Tauler R. Multivariate curve resolution applied to the analysis and resolution of two-dimensional [1H,15N] NMR reaction spectra. Anal Chem 2004;76:7094–101.

[36] Augusto F, Poppi RJ, Pedroso MP, Antonio L, De Godoy F, Hantao LW. GC × GC-FID for qualitative and quantitative analysis of perfumes. LC GC Eur 2010;23:430–8.

[37] Hantao LW, Aleme HG, Pedroso MP, Sabin GP, Poppi RJ, Augusto F. Multivariate curve resolution combined with gas chromatography to enhance analytical separation in complex samples: a review. Anal Chim Acta 2012;731:11–23.

[38] Bruckner CA, Prazen BJ, Synovec RE. Comprehensive two-dimensional high-speed gas chromatography with chemometric analysis. Anal Chem 1998;2700:2796–804.

[39] Fraga CG, Prazen BJ, Synovec RE. Comprehensive two-dimensional gas chromatography and chemometrics for the high-speed quantitative analysis of aromatic isomers in a jet fuel using the standard addition method and an objective retention time alignment algorithm. Anal Chem 2000;72:4154–62.

[40] Sinha AE, Fraga CG, Prazen BJ, Synovec RE. Trilinear chemometric analysis of two-dimensional comprehensive gas chromatography–time-of-flight mass spectrometry data. J Chromatogr A 2004;1027:269–77.

[41] Mohler RE, Dombek KM, Hoggard JC, Young ET, Synovec RE. Comprehensive two-dimensional gas chromatography time-of-flight mass spectrometry analysis of metabolites in fermenting and respiring yeast cells. Anal Chem 2006;78:2700–9.

[42] Mohler RE, Dombek KM, Hoggard JC, Pierce KM, Young ET, Synovec RE. Comprehensive analysis of yeast metabolite GC x GC-TOFMS data: combining discovery-mode and deconvolution chemometric software. Analyst 2007;132:756–67.

[43] Nic Daéid N, Waddell RJH. The analytical and chemometric procedures used to profile illicit drug seizures. Talanta 2005;67:280–5.

[44] Hoggard JC, Synovec RE. Parallel factor analysis (PARAFAC) of target analytes in GC x GC-TOFMS data: automated selection of a model with an appropriate number of factors. Anal Chem 2007;79:1611–9.

[45] Hoggard JC, Wahl JH, Synovec RE, Mong GM, Fraga CG. Impurity profiling of a chemical weapon precursor for possible forensic signatures by comprehensive two-dimensional gas chromatography/mass spectrometry and chemometrics. Anal Chem 2010;82:689–98.

[46] Watson NE, Siegler WC, Hoggard JC, Synovec RE. Comprehensive three-dimensional gas chromatography with parallel factor analysis. Anal Chem 2007;79:8270–80.

[47] Skov T, Hoggard JC, Bro R, Synovec RE. Handling within run retention time shifts in two-dimensional chromatography data using shift correction and modeling. J Chromatogr A 2009;1216:4020–9.

[48] De Godoy LAF, Hantao LW, Pedroso MP, Poppi RJ, Augusto F. Quantitative analysis of essential oils in perfume using multivariate curve resolution combined with comprehensive two-dimensional gas chromatography. Anal Chim Acta 2011;699:120–5.

[49] Johnson KJ, Wright BW, Jarman KH, Synovec RE. High-speed peak matching algorithm for retention time alignment of gas chromatographic data for chemometric analysis. J Chromatogr A 2003;996:141–55.

[50] Mogollon NGS, Ribeiro FAdL, Lopez MM, Hantao LW, Poppi RJ, Augusto F. Quantitative analysis of biodiesel in blends of biodiesel and conventional diesel by comprehensive two-dimensional gas chromatography and multivariate curve resolution. Anal Chim Acta 2013;796:130–6.

[51] Nielsen N-PV, Carstensen JM, Smedsgaard J. Aligning of single and multiple wavelength chromatographic profiles for chemometric data analysis using correlation optimised warping. J Chromatogr A 1998;805:17–35.

[52] Parastar H, Radovic JR, Jalali-Heravi M, Diez S, Bayona JM, Tauler R. Resolution and quantification of complex mixtures of polycyclic aromatic hydrocarbons in heavy fuel oil sample by means of GC × GC-TOFMS combined to multivariate curve resolution. Anal Chem 2011;83:9289–97.

[53] Brereton RG. Introduction to multivariate calibration in analytical chemistry. Analyst 2000;125:2125–54.

[54] Prazen BJ, Johnson KJ, Weber A, Synovec RE. Two-dimensional gas chromatography and trilinear partial least squares for the quantitative analysis of aromatic and naphthene content in naphtha. Anal Chem 2001;73:5677–82.

[55] Johnson KJ, Prazen BJ, Young DC, Synovec RE. Quantification of naphthalenes in jet fuel with GC × GC/Tri-PLS and windowed rank minimization retention time alignment. J Sep Sci 2004;27:410–6.

[56] Pedroso MP, de Godoy LAF, Ferreira EC, Poppi RJ, Augusto F. Identification of gasoline adulteration using comprehensive two-dimensional gas chromatography combined to multivariate data processing. J Chromatogr A 2008;1201:176–82.

[57] de Godoy LAF, Pedroso MP, Ferreira EC, Augusto F, Poppi RJ. Prediction of the physicochemical properties of gasoline by comprehensive two-dimensional gas chromatography and multivariate data processing. J Chromatogr A 2011;1218:1663–7.

[58] Flumignan DL, de Oliveira Ferreira F, Tininis AG, de Oliveira JE. Multivariate calibrations in gas chromatographic profiles for prediction of several physicochemical parameters of Brazilian commercial gasoline. Chemom Intell Lab Syst 2008;92:53–60.

[59] De Godoy LAF, Pedroso MP, Hantao LW, Poppi RJ, Augusto F. Quantitative analysis by comprehensive two-dimensional gas chromatography using interval multi-way partial least squares calibration. Talanta 2011;83:1302–7.

[60] Norgaard L, Saudland A, Wagner J, Nielsen JP, Munck L, Engelsen SB. Interval partial least-squares regression (iPLS): a comparative chemometric study with an example from near-infrared spectroscopy. Appl Spectrosc 2000;54:413–9.

[61] Zorzetti BM, Harynuk JJ. Using GC × GC-FID profiles to estimate the age of weathered gasoline samples. Anal Bioanal Chem 2011;401:2423–31.

Chapter 12

# Exploring the Multidimensionality of High-Resolution Luminescence Spectroscopy to Generate High-Order Data for Multivariate Calibration Methods

Andres D. Campiglia*,[1], Héctor C. Goicoechea†, Anthony F.T. Moore* and Walter B. Wilson‡

*Department of Chemistry, University of Central Florida, Orlando, Florida, USA

†Laboratorio de Desarrollo Analítico y Quimiometría (LADAQ), Cátedra de Química Analítica I, Facultad de Bioquímica y Ciencias Biológicas, Universidad Nacional del Litoral, Ciudad Universitaria, Santa Fe, Argentina

‡National Institute of Standard and Technology, Gaithersburg, Maryland, USA

[1]Corresponding author: e-mail: andres.campiglia@ucf.edu

## Chapter Outline



Data Handling in Science and Technology, Vol. 29. http://dx.doi.org/10.1016/B978-0-444-63527-3.00012-6

# 1 INTRODUCTION

The multidimensional nature of photoluminescence provides fluorescence and phosphorescence techniques with unique potential for the direct analysis of target compounds in complex matrixes. The combination of excitation and emission spectra to lifetime information within the fluorescence or phosphorescence time domain gives at least three qualitative parameters for compound determination, namely, one excitation and emission wavelength and one lifetime. The simplicity of the experimental procedure makes room-temperature fluorescence the most popular approach. Fluorescence is readily observed from liquid solutions with no need for sample deoxygenation. Room-temperature phosphorescence is rarely observed from liquid solutions, but it can be enhanced with the use of solid substrates [1–4], micelle-stabilized aqueous solutions [5–7], or heavy-atom salts [8,9]. In all three phosphorescence approaches, sample deoxygenation is crucial to avoid quenching of the triplet excited state.

Considerable efforts have been made to develop analytical techniques capable to determine trace concentration levels of organic pollutants in environmental samples. Particular attention has been paid to the 16 polycyclic aromatic hydrocarbons (PAHs) included in the US Environmental Protection Agency (EPA) priority pollutant list [10–19]. One of the main reasons for this motivation is the carcinogenic and toxicological nature of some EPA-PAHs. Their monitoring in air, water, and soil samples is recommended to prevent human exposure to contamination. The time-consuming procedures of traditional methodology make the development of screening techniques particularly attractive for the routine monitoring of numerous samples. Screening techniques capable to provide a "yes or no" answer to PAHs contamination prevent unnecessary scrutiny of uncontaminated samples via conventional methods, reduce analysis cost, and expedite turnaround time for decision making and remediation purposes.

A recent trend for the direct determination—i.e., no chromatographic separation—of targeted species in matrixes of unknown composition refers to processing multidimensional spectroscopic data with second-order multivariate calibration methods [20–29]. The determination of phenanthrene and benzo[*k*]fluoranthene in urban runoff water samples [30]; benzo[*a*]pyrene and dibenz[*a,h*]anthracene in underground, tap, and mineral water samples [31]; and the analysis of chrysene, benzo[*b*]fluoranthene, benzo[*k*]fluoranthene, benz[*a*]anthracene, benzo[*a*]pyrene, and dibenz[*a,h*]anthracene in river water and sludge samples [32] has been demonstrated on the basis of room-temperature fluorescence excitation–emission matrices (EEMs) combined to either parallel factor analysis (PARAFAC) or unfolded partial least-squares/residual bilinearization (U-PLS/RBL).

# 2 HIGH-RESOLUTION LUMINESCENCE SPECTROSCOPY

The main limitation of room-temperature fluorescence and phosphorescence techniques toward the selectivity of analysis is the broad nature of excitation and luminescence (fluorescence and phosphorescence) spectra. The diffuse character of such spectra limits their information content and drastically reduces the selective potential for the analysis of target compounds in complex matrixes. Several strategies exist to improve the specificity of luminescence techniques, including reducing the sample temperature to enhance vibronic spectral resolution [33], collection of multidimensional data formats [34], and time-resolved spectroscopy (lifetime measurements) [35–37].

Reducing the sample temperature offers several advantages. Luminescence quantum yields often increase, and the complications of oxygen quenching and energy transfer are eliminated. But the temperature effects on luminescence are specially pronounced in the so-called high-resolution techniques, namely, Shpol'skii spectroscopy and fluorescence line narrowing spectroscopy. In these techniques, the sharp spectra with vibrational information result from homogeneous and inhomogeneous band-broadening reduction. Homogeneous broadening affects all solute molecules to the same extent. It arises from vibronic coupling to the rapidly fluctuating surrounding matrix and from the limited lifetimes of the states involved in the electronic transition. In an ideal situation, i.e., when all solute molecules experience exactly the same environment, the only remaining source of band broadening (apart from instrumental contributions) is homogeneous. Inhomogeneous broadening results from a distribution of solute molecules with different vibronic transition energies. For amorphous matrixes in the condensed phase, characterized by their lack of long-range order, the local conditions (microenvironments) that influence the transition energies of the solute molecules (predominantly through electron–electron interactions) differ from one solute molecule to another. The resulting differences in electronic transition energies lead to a Gaussian broadening of excitation and emission spectra [33,38,39].

## 3 LASER-EXCITED TIME-RESOLVED SHPOL'SKII SPECTROSCOPY

Shpol'skii spectroscopy has long been recognized for its capability in providing direct determination of individual PAHs in environmental samples without the need of previous chromatographic separation. Its capability results from the resolution observed from excitation and fluorescence spectra in Shpol'skii matrices. The term Shpol'skii matrix refers to a dilute solution of a guest molecule (PAH) in a solvent host (usually an $n$-alkane) where the solvent freezes to 77 K or below into an ordered polycrystalline matrix. If the dimensions of the PAH and solvent match up well enough, PAH molecules occupy a small number of crystallographic sites in the host matrix. Matrix isolation of guest molecules reduces inhomogeneous band broadening. The combination of reduced thermal and inhomogeneous broadening produces vibrationally resolved spectra with sharp line widths with tremendous potential for qualitative and quantitative determination. Higher spectral resolution can be obtained by site-selective excitation, which refers to the excitation of guest molecules occupying one type of crystallographic site. PAHs occupying the same crystallographic site produce identical excitation and emission spectra. PAHs occupying different sites produce identical spectral profiles, which are slightly shifted by small wavelength increases (typically $<1000\ cm^{-1}$). Multisite excitation produces spectra with contributions from PAH molecules in all crystallographic sites. The simplest spectra, and therefore the narrowest full width at half maxima (FWHM), are obtained by site-selective excitation, which is best accomplished with narrow-band laser sources.

Prior to our involvement with Shpol'skii spectroscopy, most of the applications toward the analysis of PAHs were based on the interpretation of fluorescence spectra. Although time-resolved spectroscopy had shown to reduce spectral overlapping, information on fluorescence lifetimes as qualitative parameters for PAHs identification was practically inexistent. The same was true for phosphorescence spectra and lifetimes. One contribution we have made to this area is the introduction of methodology to efficiently collect multidimensional data formats during the lifetime decays of fluorescence (nanoseconds to microseconds) and phosphorescence (milliseconds to seconds) emission [40–42]. Our approach—which we have coined laser-excited time-resolved Shpol'skii spectroscopy (LETRSS)—takes advantage of the full dimensionality of luminescence spectroscopy combining spectral and lifetime information in multidimensional formats known as wavelength time matrices (WTM) and time-resolved excitation–emission matrices (TREEMs).

Adding the temporal dimension to the highly resolved Shpol'skii spectra provides an extremely selective tool for the determination of structural isomers in complex mixtures with numerous PAHs. A single instrument was developed to collect multidimensional data formats in both the fluorescence (microseconds to nanoseconds) and phosphorescence (milliseconds to seconds) time domains. Unambiguous isomer identification is made possible on the bases of spectral and lifetime analysis. In addition to providing a qualitative parameter for PAHs identification, fluorescence decays report on spectral peak purity, an essential condition for the accurate quantitative determination of PAHs without previous chromatographic separation. TREEM gives the analyst the opportunity to select the best time window for minimum (or even none) spectral overlapping in highly complex matrixes. Another significant improvement we made was the introduction of cryogenic fiber optic probes for sample freezing at 4.2 K. It is now possible to easily perform reproducible measurements at liquid nitrogen and liquid helium temperatures in a matter of seconds.

## 4 CRYOGENIC FIBER OPTIC PROBES

The classic sample preparation procedure for 77 K measurements consists of immersing a solution-filled small diameter tube into an optical Dewar filled with liquid nitrogen. The fragility of the Dewar flask, bubbling in the cryogen at irregular intervals from small ice particles that act as nucleation sites, and condensation on the outside of the Dewar are nuisances. But scattering at each of the interfaces encountered by the excitation light on its way to the sample is a much more severe problem, as stray light in the emission monochromator degrades the reproducibility of measurements and the limits of detection. The classic approach typically involves three air/glass interfaces, two liquid nitrogen/glass interfaces, and one matrix/glass interface. The number of scattering interfaces can be reduced with closed-cycle refrigerators or Joule–Thomson miniature refrigerators that employ contact cooling or cold vapor directed onto the sample cell. Trade-offs include higher cost, reduced sample throughput, and possibly less efficient or slower cooling of the sample. Depending on the cooling device and final temperature, and sample size and solvent, freezing times can take between 40 and 100 min per sample. The constancy of the freezing rate is a concern because it can affect the population distribution in the various sites and the corresponding signal intensity for a given set of excitation and emission wavelengths. Several attempts have been made to improve closed-cycle helium refrigerators. Renge and Wild [33] developed a user-friendly closed-cycle refrigerator that enables the simultaneous cooling of four 10 μL samples placed in a gold-plated copper holder surrounded with sapphire windows. The high thermal conductivity of the sample holder material provides fast sample cooling for good-quality Shpol'skii spectra.

We use fiber optic probes for cryogenic measurements at both 77 and 4.2 K. Our probes are similar to the one reported by Hieftje and co-workers [43] for phosphorescence measurements at 77 K. Hieftje's probe consisted of one excitation and six emission fibers bundled in a quartz tube with the opposite end connected to a plastic sample vial. Sample excitation and emission collection was made through a quartz window placed between the quartz tube and the sample vial. Our probes retain the simplicity of "dunking" the sample into the liquid cryogen for fast and reproducible freezing but eliminate all interfaces that could scatter excitation light into the detection system. Figure 1 shows examples of cryogenic probes for measurements at 77 and 4.2 K. At the sample end, the fibers are epoxied in a six-around-one configuration with the delivery fiber in the center. At the collection end, the excitation and collection fibers are separated and vertically aligned with the spectrograph entrance slit. After the sample is introduced into the sample tube, the tip of the probe is positioned above the solution surface as the sample tube is lowered into a container filled with liquid cryogen. The cell is allowed to cool for 90 s prior to luminescence measurements to ensure complete sample freezing. Because there is no physical contact between the fiber and the sample, fiber cleanup between samples is not necessary. Preparing frozen samples for luminescence measurements at 77 and 4.2 K is now a routine technique. Samples are frozen in a matter of seconds.

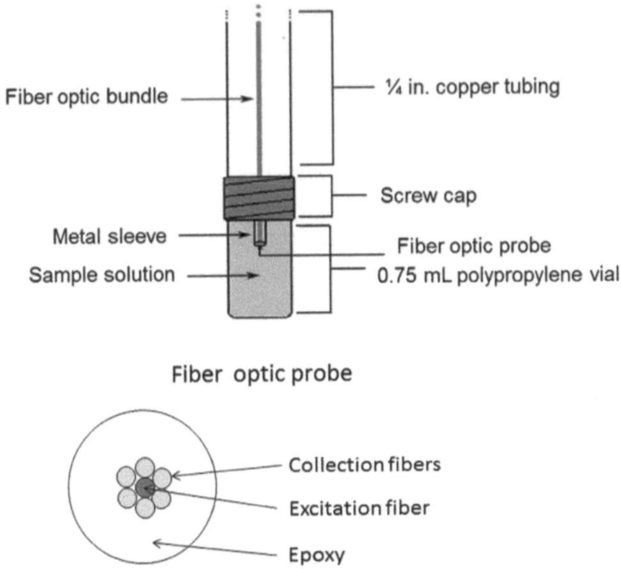


**FIGURE 1** Fiber optic probe (FOP) for 4.2 K laser-induced fluorescence spectrometry. *Adapted with permission from Analytical Chemistry, 73, A. J. Bystol, A. D. Campiglia, G. D. Gillispie, Laser-induced multidimensional fluorescence spectroscopy in Shpol'skii matrixes with a fiber-optic probe at liquid helium temperature, pp. 5762–5770, Figure 1, p. 5763. Copyright 2001 American Chemical Society.*

## 5 INSTRUMENTATION FOR HIGH-RESOLUTION MULTIDIMENSIONAL LUMINESCENCE SPECTROSCOPY

Pulsed excitation sources offer prospects for time-resolving the emission to improve signal-to-background ratios and reduce spectral interference. Early reports on laser-excited Shpol'skii spectrometry implemented time resolution at fixed delay time intervals by using laboratory-constructed gated integrators or commercial boxcar averagers [44–49]. More recently, advantage has been taken of commercially available delay generators and intensified linear photodiode arrays. Multichannel detectors can acquire emission spectra at good signal to noise much faster than is possible with a scanning monochromator. Collecting the entire emission spectrum at once avoids problems associated with pulse-to-pulse fluctuation, laser intensity drift, and photodecomposition that could degrade analytical figures of merit.

Our group has shown that the combination of a pulsed tunable dye laser, a pulsed delay generator, a spectrograph, and an intensifier-charged coupled device (ICCD) is well suited for the rapid collection of WTMs, EEMs, and TREEMs in both the fluorescence and the phosphorescence time domains. Figure 2 shows a schematic diagram of the instrumental setup mounted in

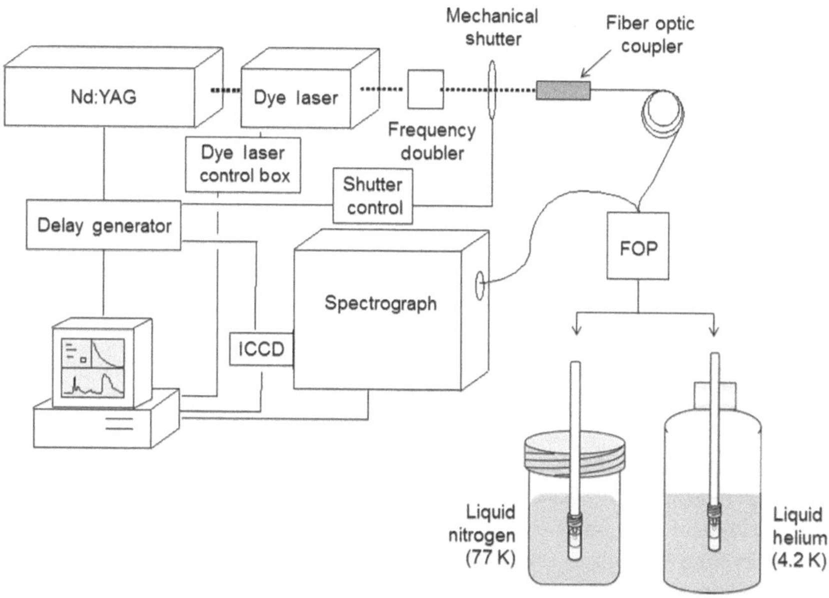


**FIGURE 2** Instrumental system for 4.2 K laser-induced fluorescence spectrometry with single-channel and multichannel fluorescence detection. *Adapted with permission from Analytical Chemistry, 73, A. J. Bystol, A. D. Campiglia, G. D. Gillispie, Laser-induced multidimensional fluorescence spectroscopy in Shpol'skii matrixes with a fiber-optic probe at liquid helium temperature, pp. 5762–5770, Figure 1, p. 5763. Copyright 2001 American Chemical Society.*

our lab. As an excitation source, we have used a compact frequency-doubled pulsed tunable dye laser whose bandwidth (<0.03 nm) is well matched for selective excitation of narrow Shpol'skii excitation spectra. The ICCD is coupled to a spectrograph to rapidly collect a series of emission spectra at different delay times between the laser firing and the opening of the gate on the ICCD. The integration of the mechanical shutter in the path of the dye laser excitation beam facilitates the collection of multidimensional data in the phosphorescence time domain. The shutter has a rise time (open time) of 1.5 ms and a fall time (close time) of 3.0 ms with a maximum pulse width of 6.5 ms. The shutter is controlled either manually (always open or always closed) via a switch on the controller front panel or programmatically through a transistor-transient logic (TTL) pulse. This TTL pulse is programmed through the A+B outputs of the digital delay generator via a general purpose interface bus interface.

Fluorescence measurements are made with the mechanical shutter in the open position. Time resolution is achieved with the intensifier in front of the ICCD, which acts as a superfast shutter with a minimum gate of 2 ns (FWHM). Once triggered by the laser, the pulse delay generator uses this information to determine when the image intensifier in the detector head is gated on (gate delay, *D*) and for how long it is gated on (gate width, *G*). These parameters are entered on the control computer with commercial software. The ICCD acquires data, while the intensifier is gated on. While the intensifier is gated off, the acquired data are transferred from the detector head to the controller card (32-bit Intelligent Bus-Mastering PCI card) in the computer. Typically, fluorescence spectra from PAHs result from the accumulation of emission of 100 laser pulses. This process takes approximately 10 ms/40 nm spectrum. If a wavelength range larger than 40 nm is of interest, the spectrograph is tuned to the new wavelength range and the process is repeated.

Depending on the strength and the duration of the phosphorescence emission, spectra are collected by one of two procedures. For PAHs that generate enough phosphorescence photons per laser pulse and have relatively short lifetimes (less than 40 ms), the use of the mechanical shutter is not necessary. The sequence of events is the same as the one used for collecting fluorescence spectra, but the delay and gate times are set at a much longer time scale. Typical delay and gate times are of the order of 20 and 40 ms, respectively. This delay is long enough to eliminate sample and background fluorescence, and the gate time is the maximum gate allowed for each firing of the laser. Our YAG laser fires every 100 ms (10 Hz), and because approximately 50 ms are needed between shots for data transfer and storage, gate times longer than 40 ms would result in convolution of the collected data with the next laser shot. The use of a faster computer could certainly decrease the data transfer and storage time. The mechanical shutter is useful for PAHs that exhibit weak phosphorescence intensity and relatively long phosphorescence lifetimes. The pulse delay generator controls the timing for both the shutter and the intensifier on the ICCD. During the excitation cycle, the pulse generator is triggered by the laser to set the shutter to the open position for a certain number of pulses (typically 20 pulses). During the emission cycle, the shutter is closed and the phosphorescence decay is recorded. The closing of the shutter sets the "zero reference time" for the delay time and the gate time on the ICCD. This method allows one to build up the triplet-state population to an acceptable signal-to-noise ratio. It also allows for gate times in excess of 40 ms to be used without concern of overlapping laser pulses.

## 6 WAVELENGTH TIME MATRICES

A WTM consists of a series of emission spectra recorded under one excitation wavelength and different time delays after the laser excitation pulse. Recording WTMs during the fluorescence or phosphorescence decay of the sample provides an additional parameter for PAH identification (lifetime).Unambiguous PAH determination is made possible on the basis of spectral and lifetime information. Fluorescence or phosphorescence lifetimes also report on spectral peak purity, i.e., an essential condition for the accurate quantitative determination of PAHs without previous chromatographic separation [50–62]. Previous reports on WTM collection utilized a rather tedious and time-consuming procedure, in which the wavelength of either the laser source or the emission monochromator was incremented, while the luminescence intensity–luminescence decay waveform was collected at each wavelength. By setting the ICCD gate step parameter, our system automatically increments the time interval between successive fluorescence scans so that WTM is easily built up. The duration of the steps by which the gate delay is progressively increased in the course of the sequence of acquisitions is entered on the control computer with commercial software. For collecting phosphorescence WTMs from PAHs with strong phosphorescence and relatively short phosphorescence lifetimes (less than 40 ms), the procedure is the same with the delay and the gate set at much longer time scales (milliseconds).

For collecting WTMs from PAHs with weak phosphorescence and relatively long phosphorescence lifetimes (longer than 40 ms), the series of phosphorescence spectra is taken at incremental delay times from the closing of the shutter. For relatively long phosphorescence lifetimes, this method results in very long acquisition times, typically 1 h or more per WTM. The alternative method involves exciting the PAH for several seconds to build up the triplet-state population, closing the shutter, and then collecting the time-resolved emission over several seconds. The resulting WTM consists of many spectra with a delay step equal to the firing rate of the laser (100 ms). Since each spectrum is an accumulation of only one laser shot, many WTMs (typically 100) need to be accumulated to average out the effects of noise. This averaging is, in effect, similar to accumulating 100 laser shots at each delay increment, as in fluorescence WTM collection. Because the delay is eliminated from each acquisition cycle, the time it takes to collect a full WTM is considerably reduced. Collecting 100 WTMs over a total period of 5 s each takes approximately 9 min.

Based on the reciprocal linear dispersion ($R_L = 0.8$ nm mm$^{-1}$) of the spectrograph, the number of active pixels ($690 \times 256$) in the array, and the dimensions ($18 \times 6.7$ mm) of the active area of the CCD chip, the spectral range of one pixel should give a 0.02-nm limiting resolution. Because of the detector cross talk and the intensifier, the factual limiting resolution corresponds to 4–5 pixels, i.e., 0.08–0.10 nm. This resolution is sufficient to collect accurate spectra from both single-site and multiple-site PAH/*n*-alkane systems. Figure 3 shows a 4.2 K WTM recorded from a multisite PAH/*n*-alkane system. Upon fast cooling in *n*-heptane, benzo[*a*]pyrene occupies four crystallographic sites that yield four distinct fluorescence spectra slightly shifted by small wavelength differences. The maximum wavelengths of the quartet emissions from the 0–0 transitions appear at 402.3, 402.7, 403.1, and 403.8 nm. The maximum wavelengths and the relative intensities of the fluorescence peaks correlate well to those previously reported with conventional methodology [38,39].

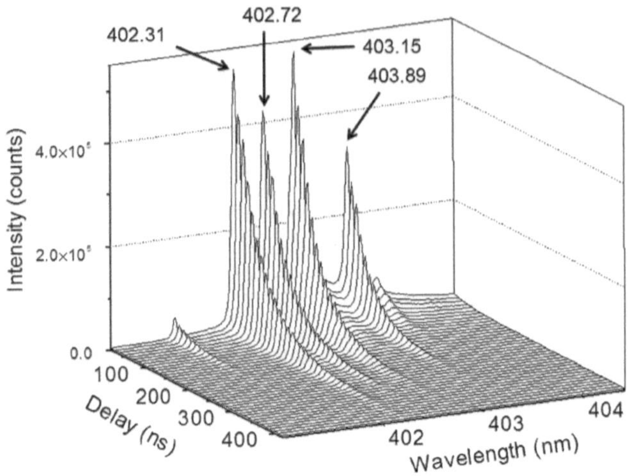


**FIGURE 3** 4.2 K WTM of 10 μg mL$^{-1}$ benzo[*a*]pyrene in *n*-heptane recorded with the fiber-optic probe, spectrograph, and ICCD camera. Excitation wavelength was 366 nm. Delay and gate times were 20 and 200 ns, respectively. Spectrograph entrance slit was 25 μm. Arrow shows characteristic fluorescence peaks (see Table 1).

Fluorescence and phosphorescence lifetimes are obtained via a three-step procedure: (1) full sample and background WTM collection, (2) background decay curve subtraction from the fluorescence or phosphorescence decay curve at a wavelength of maximum emission for each PAH, and (3) fitting of the background corrected data to single exponential decays. Commercial software is used for curve fitting of fluorescence and phosphorescence lifetimes. Fitted decay curves $y = y_0 + A_1 e^{[-(x-x_0)t_1]}$ are obtained by fixing $x_0$ and $y_0$ at a value of zero. Interesting to note are the significant lifetimes differences we have observed from PAH molecules occupying different crystallographic sites in the same matrix. Some examples are summarized in Table 1. No other reports exist on this phenomenon, but our observations are consistent with the well-known sensitivity of fluorescence lifetimes to the microenvironment of the fluorophore. It is generally accepted that the multiplet structure of fluorescence spectra arise from the different orientations of the PAH molecule in the solid matrix. The different orientations lead to nonequivalent crystal field effects and hence wavelength shifts. Most likely, the field effects are also sensed by the PAH fluorescence lifetime. The lifetime differences from site to site are usually larger than their wavelength shifts. This fact helps to distinguish between crystallographic sites with very close maximum wavelengths [41]. Although we have not unfolded the analytical potential of site-selective lifetime analysis yet, the rather large lifetime differences anticipate a promising contribution to enhancing the specificity of site-selective excitation.

**TABLE 1** 4.2 K Fluorescence Lifetimes ($\tau$) of PAH Molecules Occupying Different Crystallographic Sites in the Frozen Matrix

| | *n*-Hexane | | *n*-Heptane | | *n*-Octane | |
|---|---|---|---|---|---|---|
| PAH[a] | $\lambda_{em}$[b] (nm) | $\tau$ (ns) | $\lambda_{em}$[b] (nm) | $\tau$ (ns) | $\lambda_{em}$[b] (nm) | $\tau$ (ns) |
| Benzo[*a*]pyrene | 403.56 | 27.02 ± 0.50 | 402.31 | 52.09 ± 0.93 | 403.10 | 36.04 ± 0.61 |
| | 403.94 | 43.44 ± 1.00 | 402.72 | 46.93 ± 0.49 | 403.20 | 45.68 ± 0.84 |
| | 404.53 | 31.95 ± 0.89 | 403.15 | 42.77 ± 0.89 | | |
| | | | 403.89 | 41.29 ± 1.18 | | |
| Pyrene | 371.63 | 503.98 ± 6.01 | | –[c] | | –[c] |
| | 372.45 | 577.43 ± 3.47 | | | | |
| Benz[*a*]anthracene | 383.34 | 53.63 ± 0.50 | | –[c] | | –[c] |
| | 383.94 | 56.43 ± 0.76 | | | | |

[a]*PAH solutions were prepared at 0.5 μg mL$^{-1}$.*
[b]*Wavelength of maximum fluorescence emission. All data obtained using $\lambda_{exc}$ = 288.50.*
[c]*Spectra and decay data were not measured for this PAH/solvent system.*
Reprinted with permission from Analytical Chemistry, 73, A. J. Bystol, A. D. Campiglia, G. D. Gillispie, Laser-induced multidimensional fluorescence spectroscopy in Shpol'skii matrixes with a fiber-optic probe at liquid helium temperature, pp. 5762–5770, Table 3, p. 5766. 

## 7 EXCITATION–EMISSION MATRICES

Recording two-dimensional fluorescence spectra—i.e., plots correlating fluorescence intensities to emission wavelengths—from a mixture with numerous fluorescence components only provides partial information on the total fluorescence of the sample. The emission profile of a mixture with numerous fluorescence components varies with the relative position of the excitation wavelengths to the excitation maxima of the fluorophores in the sample. Individual fluorophore contributions to the total fluorescence spectrum of the sample also depend on the fluorescence quantum yields of the fluorophores and possible quenching due to synergistic effects. EEMs gather all this information in a single data format to provide a true signature of the total fluorescence of a multifluorophore mixture.

Most of the work on low-temperature EEMs has been performed with high-resolution spectrometers that built up fluorescence EEMs with numerous repetitive scans [33], an approach that is often prone to long acquisition times. Our system is well suited for rapid collection of low-temperature fluorescence and phosphorescence EEMs. Because of the spectrograph and the ICCD, both types of EEMs are collected in short analysis time. The tunable dye laser provides intense excitation energy in a narrow excitation bandwidth. Its scanning capability is essential to take full advantage of the specificity obtainable from the highly resolved PAH excitation spectra in Shpol'skii matrixes. Its intense excitation energy promotes strong PAH fluorescence to recording spectra with optimum spectrograph slits for maximum spectral resolution.

## 8 TIME-RESOLVED EXCITATION-EMISSION MATRICES

In comparison to EEMs, chemical analysis based on TREEMs is potentially more selective because it adds the temporal dimension to the orthogonal spectral dimension of EEMs. The choice of an appropriate time window for TREEMs collection enhances the fluorescence or phosphorescence of targeted compounds over the emission interference of sample concomitants. Two different instrumental procedures have been used in our lab to record TREEMs. In the first approach, TREEMs were recorded as series of WTMs acquired at different excitation wavelengths. The superposition of the WTMs produced a third-order data that we named excitation-modulated WTM (EMWTM) [63]. Figure 4A shows a representation of an EMWTM data format. In the second approach, TREEMs were recorded as series of EEMs acquired at different delay times after the short duration of the excitation pulse. All the intensity values—as a function of excitation and emission wavelengths—for a particular decay time were assembled into an EEM format specific for that decay time. The superposition of EEMs produced a third-order data that we named time-resolved excitation–emission cube (TREEC) [64]. A graphic illustration of a TREEC data format is shown in Figure 4B.

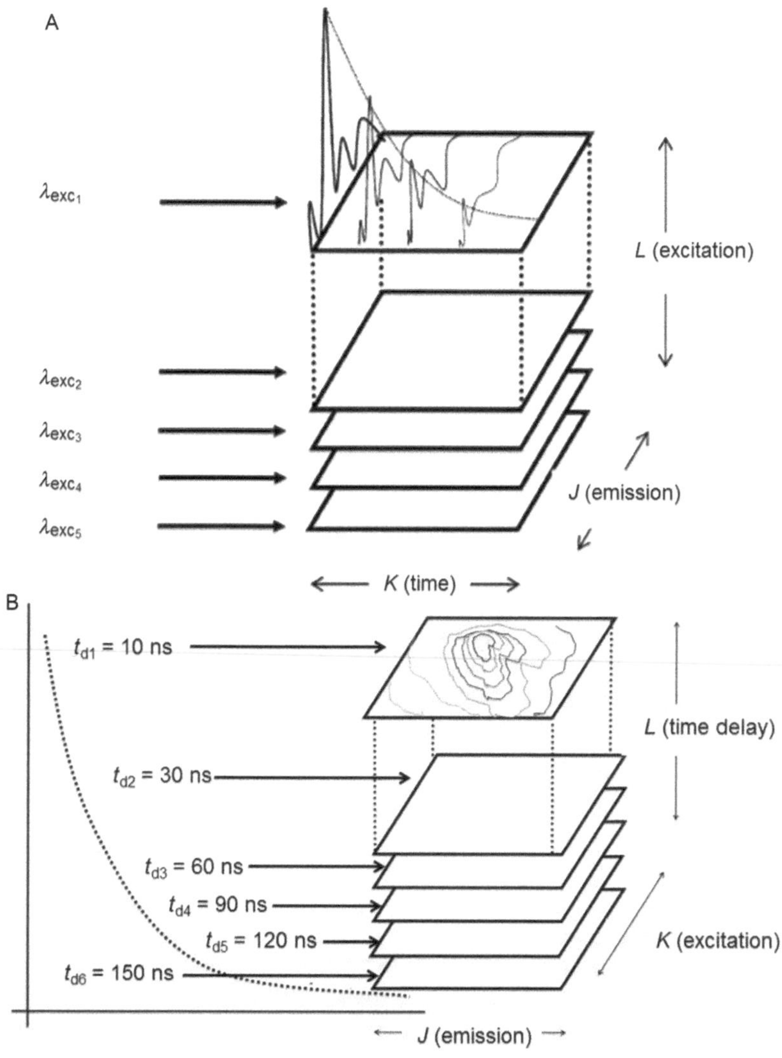


**FIGURE 4** (A) Schematic representation of the third-order tensor (EMWTM) obtained for each analyzed sample. (B) Schematic representation of a TREEC array. *Reprinted with permission from Analytical Chemistry, 77, H. C. Goicoechea, S. Yu, A. C. Olivieri, A. D. Campiglia, Four-way data coupled to parallel factor model applied to environmental analysis: determination of 2,3,7,8-tetrachloro-dibenzo-para-dioxin in highly contaminated waters by solid-liquid extraction laser-excited time-resolved Shpol'skii spectroscopy, pp. 2608–2616, Figure 2, p. 2613. Copyright 2005 American Chemical Society.*

## 9 PARAFAC MODELING OF FOUR-WAY 4.2 K PHOSPHORESCENCE EMWTM ARRAYS FOR THE DIRECT DETERMINATION OF 2,3,7,8-TETRACHLORO-DIBENZO-*PARA*-DIOXIN IN WATER SAMPLES

Collection of EMWTM from any given sample produces a third-order data $\boldsymbol{F}$ of dimensions $J \times K \times L$ ($J$, $K$, and $L$ denote the number of data points in each dimension: $J$ is the number of emission wavelengths, $K$ is the number of time data points, and $L$ is the number of excitation wavelengths). This type of data can be analyzed in two ways: (1) by joining the $I_{\text{cal}}$ training arrays and the unknown sample data $\boldsymbol{F}_{\text{u}}$ into a four-way data array $\boldsymbol{F}$, whose dimensions are $[(I_{\text{cal}}+1) \times J \times K \times L]$ or (2) by combining the test sample data with those for $I_{\text{std}}$ standard additions to the test sample, producing a four-way array of dimensions $[(I_{\text{std}}+1) \times J \times K \times L]$. In any case, if $\boldsymbol{F}$ follows a quadrilinear PARAFAC model, it can be mathematically written in terms of four vectors for each responsive component. These vectors are designated as $\mathbf{a}_n$, $\mathbf{b}_n$, $\mathbf{c}_n$, and $\mathbf{d}_n$ and collect the relative concentrations ($I \times 1$, where $I$ is either $I_{\text{cal}}+1$ or $I_{\text{std}}+1$), as well as the emission ($J \times 1$), time ($K \times 1$), and excitation profiles ($L \times 1$) for component $n$, respectively. The column vectors ($\mathbf{a}_n$, $\mathbf{b}_n$, $\mathbf{c}_n$, and $\mathbf{d}_n$) are usually normalized to unit length and collected into their corresponding loading matrixes, namely, **A**, **B**, **C**, and **D**. If the high-dimensional array $\boldsymbol{F}$ can be uniquely decomposed, this provides access to individual spectral and time profiles, as well as relative concentrations (**A**) of mixture components, whether they are chemically known or not. This constitutes the basis of the so-called second-order advantage, a particularly useful property for the direct analysis of a targeted chemical in complex samples of unknown composition. Identification of the chemical constituent under investigation is done with the aid of the extracted profiles **B**, **C**, and **D**, in comparison with those for analyte standards. Absolute analyte concentrations are obtained after calibration since only relative values (**A**) are provided by decomposing the data array. Experimentally, this can be done by two alternative procedures: (1) external calibration and (2) standard addition. A stepwise description of data treatment processed under the external calibration method and the standard addition method is provided in Table 2.

The first application of PARAFAC modeling to EMWTM arrays was geared toward the analysis of 2,3,7,8-tetrachloro-dibenzo-*para*-dioxin (2,3,7,8-TCDD) in water samples [63]. Environmental monitoring of 2,3,7,8-TCDD is extremely relevant due to its high toxicity to man. 2,3,7,8-TCDD is the most toxic isomer among all the polychlorinated dibenzo-*para*-dioxins and its sole determination is sufficient to establish pollution risk. Unambiguous determination of 2,3,7,8-TCDD calls for high-resolution GC/MS, which requires extensive cleanup procedures. These often include liquid chromatography with combinations of several adsorbents such as modified silica gel, aluminum oxide, and activated carbon. The two main cleanup

**TABLE 2 Stepwise Description of Two Alternative Calibration Modes for Four-Way Data Using PARAFAC**

| Step | External calibration mode | Standard addition mode |
|---|---|---|
| 1 | Build an $(I_{cal}+1)\times J\times K\times L$ array with the first $I$ TREEMs for the training samples and the last one for the unknown | Build an $(1+I_{std})\times J\times K\times L$ array with the first TREEM for the unknown sample and the $N_s$ for the standard additions |
| 2 | Decompose the array and obtain **A**, **B**, **C**, and **D** | Decompose the array and obtain **A**, **B**, **C**, and **D** |
| 3 | Identify the $n$th analyte of interest from **B**, **C**, and **D** profiles | Identify the $n$th analyte of interest from **B**, **C**, and **D** profiles |
| 4 | Regress the first $I_{cal}$ elements of column $\mathbf{a}_n$ against known standard concentrations $\mathbf{c}_{cal}$ of analyte $n$: $[a_{1,n}\ldots a_{I_{cal},n}]=k_1\times\mathbf{c}_{cal}$ (pseudo-univariate calibration) | Regress the first $(1+I_{std})$ elements of column $\mathbf{a}_n$ against known standard concentrations $\mathbf{c}_{cal,s}$ of analyte $n$: $[a_{1,n}\ldots a_{1+I_{std},n}]=k_2\times[0,\mathbf{c}_{cal,s}]+h$ (pseudo-univariate calibration) |
| 5 | Convert relative to absolute concentration of $n$ in the unknown, starting from the last element of column $\mathbf{a}_n$: $c_{unk}=a_{I_{cal}+1,n}/k_1$ | Convert relative to absolute concentration of $n$ in the unknown: $c_{unk}=h/k_2$ |



objectives are to remove coextracted organic matter from the analytical sample and to eliminate the interference of PAHs, polychlorinated biphenyls (PCBs), and polychlorinated dibenzofurans (PDBFs). Our approach was tested with two sample matrixes of unknown composition, namely, Red River water and the output water stream of the American Petroleum Institute (API) separator at a North Dakota refinery. Red River water of the North Basin (Minnesota, North Dakota, and South Dakota) was collected at different locations within the Fargo-Moorhead area. Previous analysis of this water showed a high content of dissolved organic matter [65]. The API water sample was collected from a sampling valve on the output pipe of the API separator. API water is used to desalt and scrub crude oil and oil fractions in the refinery process and shows a visible surface petroleum layer at the separator trough. Previous HPLC–LETRSS analysis of this water had shown heavy contamination with fluorescent pollutants [66,67].

Solid–liquid extraction (SLE) of aqueous solutions was performed with a syringe kit following a previously described procedure that assured extraction efficiencies of ~100% [68,69]. LETRSS measurements from extraction membranes were made as previously described by placing the extraction disk in the sample vial of the low-temperature cell. The membrane was then spiked with 100 μL of *n*-heptane using a microliter syringe. 2,3,7,8-TCDD shows strong phosphorescence in *n*-heptane. The maximum excitation and emission wavelengths appear at 317 and 408 nm, respectively, and the phosphorescence decay follows a single exponential decay with lifetime value of 236 ± 6 ms ($N$ = 3). Figure 5A shows the 4.2 K phosphorescence WTM of 2,3,7,8-TCDD collected from an extraction membrane previously spiked with 100 mL of *n*-heptane. The spiked membrane was excited at 317 nm. Figure 5B shows the first phosphorescence spectrum of the WTM collected after 1 μs of the

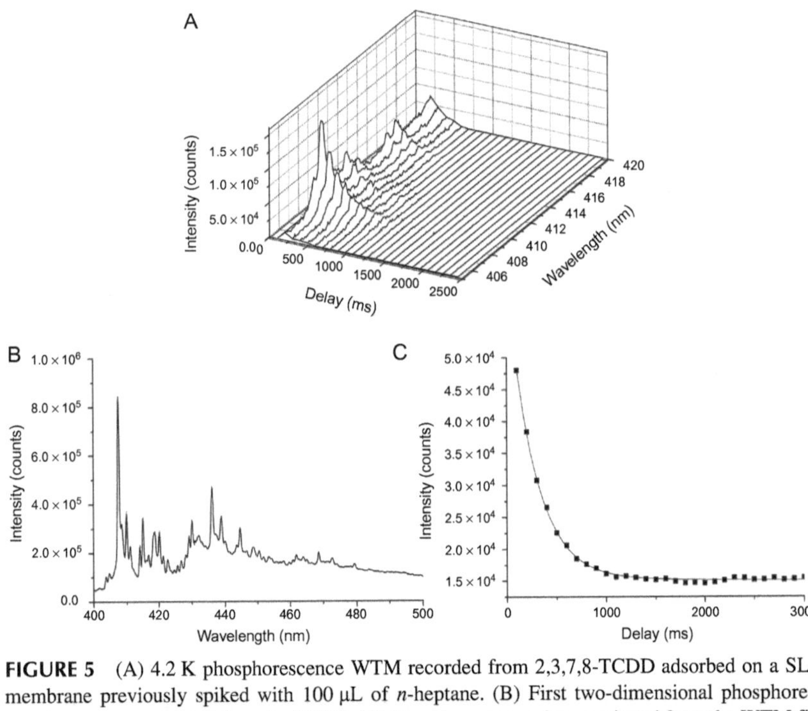


**FIGURE 5** (A) 4.2 K phosphorescence WTM recorded from 2,3,7,8-TCDD adsorbed on a SLE membrane previously spiked with 100 μL of *n*-heptane. (B) First two-dimensional phosphorescence spectrum extracted from the WTM. (C) Phosphorescence decay stripped from the WTM fitting the phosphorescence intensity at 408 nm as a function of delay time. The compound was extracted from a 10 ng $mL^{-1}$ standard aqueous solution. Excitation wavelength was at 317 nm. The delay after the pulse and the gate time were 50 μs and 40 ms, respectively. Each spectrum represents the sum of 100 laser shots, taken at 100 ms intervals. Spectrograph entrance slit was 42 μm. *Reprinted with permission from Analytical Chemistry, 77, H. C. Goicoechea, S. Yu, A. C. Olivieri, A. D. Campiglia, Four-way data coupled to parallel factor model applied to environmental analysis: determination of 2,3,7,8-tetrachloro-dibenzo-para-dioxin in highly contaminated waters by solid-liquid extraction laser-excited time-resolved Shpol'skii spectroscopy, pp. 2608–2616, Figure 1, p. 2613. Copyright 2005 American Chemical Society.*

closing of the shutter. The spectrum shows the vibronic structure characteristic of Shpol'skii spectra with a sharp, maximum intensity peak at 408 nm. The spectral profile is identical to the one observed in *n*-heptane. The plot of phosphorescence intensity versus gate delay stripped from the WTM is shown in Figure 5C. The phosphorescence decay follows a single exponential decay with lifetime value statistically equivalent to the one in *n*-heptane ($p=95\%$, $N_1=N_2=3$). The narrow spectra and the well-behaved single exponential decay on spiked membranes result from analyte partitioning into the organic layer above the surface of the membrane. This phenomenon was first observed with PAHs [55] and then PCBs and PDBFs [54]. The same type of behavior was observed for 2,3,7,8-TCDD.

Water standards and water validation samples were prepared by placing a reduced volume of *n*-heptane standard solution of analyte and interfering agents in a volumetric flask, then evaporating the organic solvent with a nitrogen stream and carrying them up to the mark with Nanopure water. The resultant solution was placed in a sonicator for 15 min. In the cases in which standard addition was performed, the volumetric flask was completed with the water being analyzed. Twelve validation samples were prepared using different waters (Nanopure, API, and Red River waters) and interfering agents. The origin and total concentration of the interfering agent can be seen in Tables 3 and 4.

A set of five excitation wavelengths (317.2, 317.3, 317.4, 317.5, and 317.6 nm) was selected to preferentially enhance the phosphorescence of 2,3,7,8-TCDD in the complex matrixes. Phosphorescence was recorded from 400 to 440 nm, generating a total of 1024 data points. Each sample was analyzed collecting one WTM per excitation wavelength. The WTMs were collected using a gate delay of 50 μs, a gate width of 1100 ms, and a gate step of 100 ms to generate 12 data points per decay curve. The superimposition of the five WTMs produced a third-order EMWTM array of 61,440 data points per sample. Figure 6 shows the profiles **B**, **C**, and **D** (loading for emission, time decay mode, and excitation) provided by PARAFAC when processing sample mix 1—see composition in Table 4—using an external calibration set of five 2,3,7,8-TCDD standard water solutions, namely, 30, 60, 90, 120, and 150 ppb. Three factors were selected using the core consistency test (see Chapter 1). The components were labeled according to their contribution to the overall variance. In this particular case, the interferences appear in the first and second places, indicating that they are the main source of phosphorescence intensity across this particular data array. Table 4 supports this observation because mix 1 contains more than five times of each interfering agent than 2,3,7,8-TCDD.

Comparison of Figures 5 and 6 shows the remarkable matching between the phosphorescence spectrum of 2,3,7,8-TCDD and the third component extracted by PARAFAC. Similar considerations can be made when analyzing the lifetime decays and the excitation spectra. As the complexity of the

**TABLE 3** Compounds Studied as Interferences When Testing the Accuracy of the Proposed Method

| Polychlorinated dioxins | Polychlorinated dibenzofurans | Polycyclic aromatic hydrocarbons |
|---|---|---|
| 1-Chloro-dibenzo-*para*-dioxin | Dibenzofuran | Acenaphthylene |
| 2-Chloro-dibenzo-*para*-dioxin | 2,8-Dichloro-dibenzofuran | Acenaphthene |
| 8-Chloro-dibenzo-*para*-dioxin | 1,2,3,4-Tetrachloro-dibenzofuran | Anthracene |
| 2,7-Dichloro-dibenzo-*para*-dioxin | 2,3,7,8-Tetrachloro-dibenzofuran | Benz[*a*]anthracene |
| 1,6-Dichloro-dibenzo-*para*-dioxin | 1,3,6,8-Tetrachloro-dibenzofuran | Benzo[*b*]fluoranthene |
| 1,3-Dichloro-dibenzo-*para*-dioxin | 1,2,3,4,7,8-Hexachloro-dibenzofuran | Benzo[*k*]fluoranthene |
| 1,2,3-Trichloro-dibenzo-*para*-dioxin | 1,2,3,4,6,7,8,9-Octachloro-dibenzofuran | Benzo[*ghi*]perylene |
| 2,3,7-Trichloro-dibenzo-*para*-dioxin | | Benzo[*a*]pyrene |
| 1,7,8-Trichloro-dibenzo-*para*-dioxin | | Dibenz[*a,h*]anthracene |
| 1,2,3,4-Tetrachloro-dibenzo-*para*-dioxin | | Fluoranthene |
| 1,2,7,8-Tetrachloro-dibenzo-*para*-dioxin | | Fluorene |
| 1,3,7,8-Tetrachloro-dibenzo-*para*-dioxin | | Indeno[1,2,3-*cd*]pyrene |
| 1,2,6,7-Tetrachloro-dibenzo-*para*-dioxin | | 2-Methylnaphthalene |

| | | |
|---|---|---|
| 1,2,8,9-Tetrachloro-dibenzo-*para*-dioxin | | Naphthalene |
| 1,3,7,9-Tetrachloro-dibenzo-*para*-dioxin | | Phenanthrene |
| 1,3,6,8-Tetrachloro-dibenzo-*para*-dioxin | | |
| 1,2,3,7,8-Pentachloro-dibenzo-*para*-dioxin | | |
| 1,2,4,7,8-Pentachloro-dibenzo-*para*-dioxin | | |
| 1,2,3,4,7,8-Hexachloro-dibenzo-*para*-dioxin | | |
| 1,2,3,4,6,8,9-Heptachloro-dibenzo-*para*-dioxin | | |
| 1,2,3,4,6,7,9-Heptachloro-dibenzo-*para*-dioxin | | |
| 1,2,3,4,6,7,8,9-Octachloro-dibenzo-*para*-dioxin | | |

**TABLE 4** Results Obtained by Applying the Present Method to Both Artificial- and Real-Contaminated Samples

| | Interfering agents | | 2,3,7,8-TCDD (ng $mL^{-1}$) | | |
|---|---|---|---|---|---|
| Sample | *Origin* | *Added (ng $mL^{-1}$)* | *Added* | *Found*[a] | *Recovery (%)* |
| Mix 1 | Nanopure water | 450 | 90 | 93 | 103 |
| | 1,3,7,8-TCDD | 450 | | | |
| | 1,2,7,8-TCDD | | | | |
| Mix 2 | Nanopure water | 3570 | 100 | 98 | 98 |
| | 22 Different PCDXs | | | | |
| Mix 3 | Nanopure water | 1780 | 200 | 210 | 105 |
| | 22 Different PCDXs | 1050 | | | |
| | 8 Different PDBFs | | | | |
| Mix 4 | Nanopure water | 1780 | 100 | 109 | 109 |
| | 22 Different PCDXs | 1050 | | | |
| | 8 Different PDBFs | 1500 | | | |
| | 17 PAHs | | | | |
| Mix 5 | Nanopure water | 1500 | 100 | 99 | 99 |
| | 17 PAHs | | | | |
| Mix 6 | Nanopure water | 1500 | 150 | 155 | 103 |
| | 17 PAHs | | | | |
| Mix 7 | Red River water | – | – | ND | – |
| Mix 8 | Red River water | – | 50 | 48 | 96 |
| Mix 9 | Red River water | | 100 | 105 | 105 |
| | 22 Different PCDXs | 1780 | | | |
| | 8 Different PDBFs | 1050 | | | |
| Mix 10 | API water | – | – | ND | – |
| Mix 11 | API water | – | 100 | 104 | 104 |
| Mix 12 | API water | | 100 | 98 | 98 |
| | 22 Different PCDXs | 1780 | | | |
| | 8 Different PDBFs | 1050 | | | |

[a]*ND, not detected.*

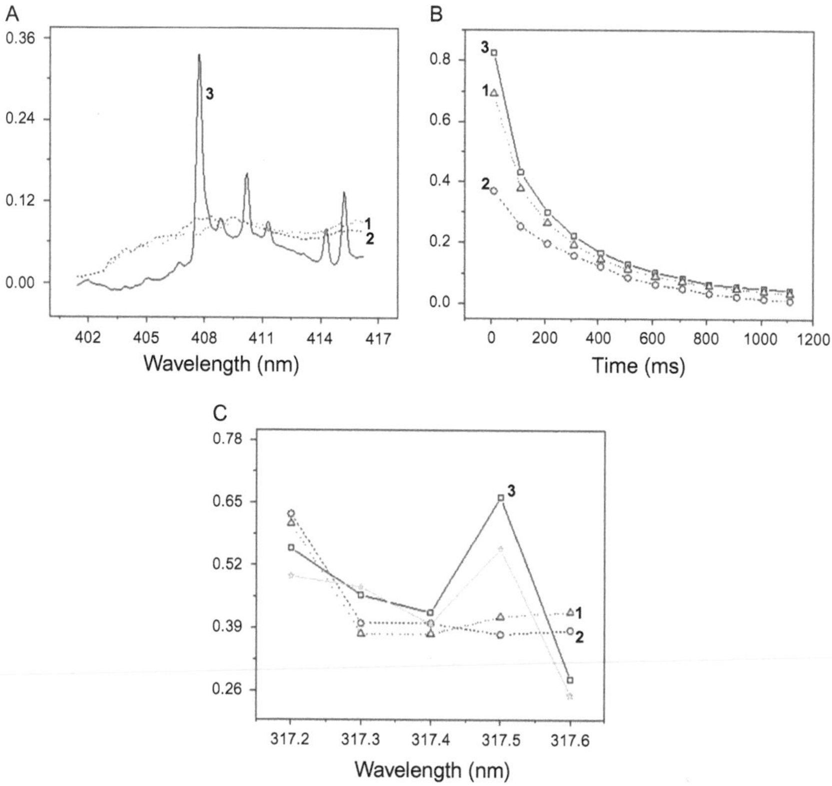


**FIGURE 6** Profiles extracted by the PARAFAC model on sample mix 1 (see Table 4). (A) Emission spectral profiles, (B) time profiles, and (C) excitation spectral profiles. The numbers indicate the order of the factor. In (C), the gray star line corresponds to the normalized excitation spectrum of 2,3,7,8-TCDD. *Reprinted with permission from Analytical Chemistry, 77, H. C. Goicoechea, S. Yu, A. C. Olivieri, A. D. Campiglia, Four-way data coupled to parallel factor model applied to environmental analysis: determination of 2,3,7,8-tetrachloro-dibenzo-para-dioxin in highly contaminated waters by solid-liquid extraction laser-excited time-resolved Shpol'skii spectroscopy, pp. 2608–2616, Figure 3, p. 2614. Copyright 2005 American Chemical Society.*

sample increased, PARAFAC needed standard addition in order to avoid the inner filter effect originated by the high concentration of interfering agents. Figure 7 shows the phosphorescence spectrum of mix 4, which contains 22 PCDXs, 8 PDBFs, and 17 PAHs. Although the main characteristic peak of 2,3,7,8-TCDD is still observed at 408 nm, its direct determination via the external standard calibration method would lead to erroneous concentrations. Reasonable good results—with average recoveries of 102$\pm$4.1%—were obtained for this and the other samples related in Table 4. It should be noted that the number of factors selected in all cases were inferior to 5.

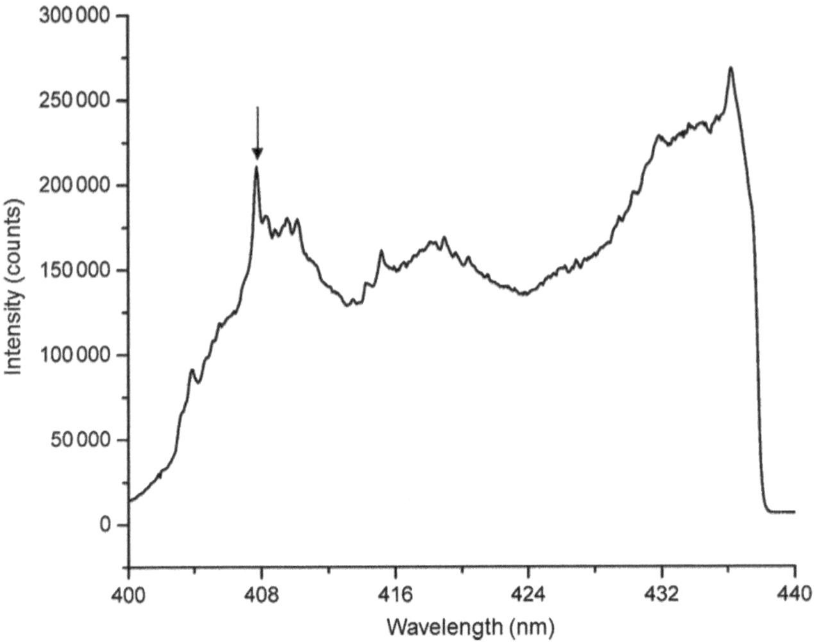


**FIGURE 7** 4.2 K phosphorescence spectrum of sample mix 4 (see Table 4) recorded with the fiber optic probe, spectrograph, and ICCD camera. Excitation wavelength was 317 nm. Delay and gate times were 50 μs and 40 ms, respectively. Spectrograph entrance slit was 42 μm. Spectrum was recorded from extraction membrane previously spiked with 100 μL of *n*-heptane. Sample volume was 100 mL. Arrow shows characteristic phosphorescence peak of 2,3,7,8-TCDD. *Reprinted with permission from Analytical Chemistry, 77, H. C. Goicoechea, S. Yu, A. C. Olivieri, A. D. Campiglia, Four-way data coupled to parallel factor model applied to environmental analysis: determination of 2,3,7,8-tetrachloro-dibenzo-para-dioxin in highly contaminated waters by solid-liquid extraction laser-excited time-resolved Shpol'skii spectroscopy, pp. 2608–2616, Figure 4, p. 2614. Copyright 2005 American Chemical Society.*

## 10 PARAFAC AND U-PLS/RTL MODELING OF 4.2K FLUORESCENCE TREECs FOR THE QUANTITATION OF EPA-PAHs IN SOIL SAMPLES

The first application of PARAFAC and U-PLS/RTL modeling of 4.2 K fluorescence TREECs was geared toward the analysis of EPA-PAHs in soil samples [64]. Soil is one of the most important reservoirs for PAHs, which are deposited in the gaseous state or associated to air-born particles, even at sites far from the petroleum industry. Once present in the soil, PAHs become a long-term source of environmental health risk due to their rather low water solubility, intrinsic chemical stability, and high resistance to biodegradation. Similar to EMWTMs, the decomposition of the four-way TREEC data array with PARAFAC allows the analyst to extract emission, excitation, and time profiles of PAHs along with their relative concentrations. The score of each

PAH is then used to predict its concentration in the unknown sample. The principle of operation of U-PLS/RTL is fundamentally different as the original cube data are transformed into unidimensional arrays (vectors) by concatenating (unfolding) the original three-dimensional information. Concentration information is first used with no data from the unknown sample. A usual U-PLS model is calibrated with the included data and the vector of calibration concentrations. If there are no unsuspected interferences in the test sample, the concentration of each PAH can then be estimated using the same parameters used in the calibration step. The sample scores will be unsuited for PAH prediction when uncalibrated components occur in the test sample. In this case, the residuals of the U-PLS prediction step are abnormally large in comparison to the typical instrumental noise assessed by replicate measurements. The residual trilinearization procedure—which is based on a Tucker3 decomposition—makes possible to model interference effects and accurately predict the concentration of each PAH in the unknown sample. EPA-PAH

EPA-PAHs were extracted from soil samples via a sonication screening procedure developed in our lab [70]. A known amount (0.05 g) of soil sample was mixed with 250 μL of *n*-octane in the vessel of the FOP. The mixture was submitted to 30 min of sonication at room temperature in a sonication bath. After 5 min of settling time, the sample extract was analyzed via LETRSS. The data arrays used for PAHs determination result from the superposition of four TREECs recorded from soil extracts containing different concentrations of 15 EPA-PAHs. Acenaphthylene was the only priority pollutant with no fluorescence emission in frozen octane. Its lack of fluorescence persisted after sample deoxygenation or analyte freezing in *n*-pentane, *n*-hexane, *n*-heptane, and *n*-nonane. The concentrations of the remaining PAHs were adjusted via the multiple standard addition procedure. The first TREEC reflected the original PAH composition of the soil sample, i.e., no standard addition. The other TREEC—one per standard addition—was obtained after adding 5, 10, and 15 μL of a standard EPA-PAH mixture to the 250 μL volume of soil extract in the sample vial of the FOP. All standard additions were made at room temperature.

The 4.2K fluorescence TREEMs used to generate the four TREECs were recorded using an excitation range (280–295 nm) common to the 15 EPA-PAHs. The tunable dye laser was stepped at 0.5 nm increments, generating a total of 22 excitation wavelengths. Fluorescence was recorded within the 300- to 500-nm wavelength range using the following delay ($D$) times: 10, 30, 60, 90, 120, and 150 ns. The gate width for the first TREEM ($D=10$ ns) was 20 ns. The gate widths for the remaining TREEMs were 30 ns. These parameters provided the following time windows ($D-D+G$) for TREEM collection: $\text{TREEM}_1=10$–30 ns, $\text{TREEM}_2=30$–60 ns, $\text{TREEM}_3=60$–90 ns, $\text{TREEM}_4=90$–120 ns, $\text{TREEM}_5=120$–150 ns, and $\text{TREEM}_6=150$–180 ns. Figure 8 depicts $\text{TREEM}_1$, $\text{TREEM}_3$, and $\text{TREEM}_6$ recorded from the soil extract with no standard addition. Visual comparison of the spectral profiles

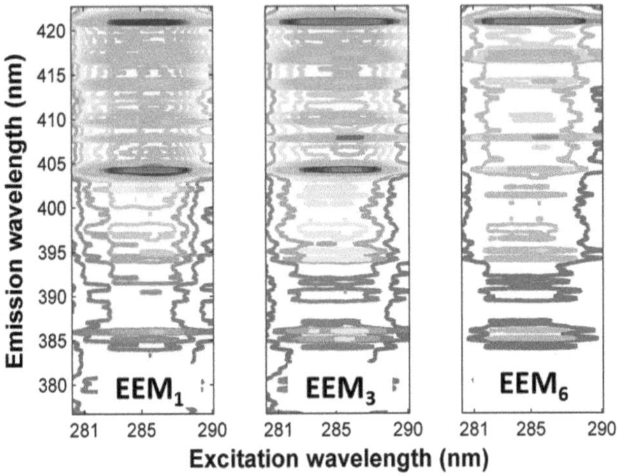


**FIGURE 8** 4.2 K TREEM recorded from soil extract with the following time windows: 10–30 ns ($EEM_1$), 60–90 ns ($EEM_3$), and 150–180 ns ($EEM_6$). Excitation wavelength was stepped at 0.5 nm increments. Each emission spectrum corresponds to the accumulation of 100 laser pulses. *Reprinted from Talanta, 101, H. C. Goicoechea, S. Yu, A. F. T. Moore, A. D. Campiglia, Four-way modeling of 4.2 K time-resolved excitation emission fluorescence data for the quantitation of polycyclic aromatic hydrocarbons in soil samples, pp. 330–336, Figure 1, p. 333. Copyright 2012, with permission from Elsevier.*

of EPA-PAHs within the emission range of the TREEM (375–425 nm) leads to the following PAHs as the main contributors of the total fluorescence of the extract: benz[*a*]anthracene, anthracene, dibenz[*a*,*h*]anthracene, benzo[*b*]fluoranthene, benzo[*a*]pyrene, benzo[*k*]fluoranthene, and fluoranthene. The relative contribution of each PAH to the total fluorescence of the sample varies with the time window of the TREEM. The fluorescence of benzo[*a*]pyrene—which appears in $TREEM_1$ and $TREEM_3$ at 401.2 nm—practically vanishes in $TREEM_6$. Benzo[*a*]pyrene is a medium-lived PAH with $\tau = 42.8 \pm 1.6$ ns. The same is not true for the prominent fluorescence at 422.5 nm. The main reason for its presence in $TREEM_6$ is that belongs to benzo[*ghi*]perylene, a long-lived PAH ($\tau = 123.8 \pm 2.5$ ns). The superposition of $TREEM_{1-6}$ leads to a TREEC with 540,672 data points. The superposition of $TREEM_{1-6}$ corresponds to the TREEC of the soil extract.

The emission and excitation wavelength ranges, the time windows, the number of PARAFAC factors (responsive components), and the number of latent variables for U-PLS/RTL modeling (U-PLS/RTL factors) are listed in Table 5. The number of U-PLS/RTL factors was estimated via the cross-validation method (see Chapter 1). Each PLS model was built by subtracting the TREEC of the sample (zero standard addition) to those recorded after the first, second, and third standard additions. The number of PARAFAC factors was estimated following the internal parameter procedure—also known as the

**TABLE 5** Parameters for PARAFAC and U-PLS/RTL Analysis

| PAH | Excitation[a] (nm) | Emission[a] (nm) | EEM[b] # | Time windows[b] (ns) | PARAFAC factors | U-PLS/RTL factors |
|---|---|---|---|---|---|---|
| Naphthalene | 280.5–285.5 | 319.9–320.7 | 4, 5 | 90/120, 120/150 | 2 | 1 |
| Acenaphthene | 283.0–289.5 | 319.3–319.1 | 1–3 | 10/30, 30/60, 60/90 | 3 | 2 |
| Fluorene | 281.5–287.5 | 303.6–305.1 | 1, 2 | 10/30, 30/60 | 2 | 1 |
| Phenanthrene | 286.0–290.0 | 350.4–351.7 | 1, 2 | 10/30, 30/60 | 2 | 1 |
| Anthracene | 282.5–295.0 | 383.6–384.4 | 1–5 | 10/30, 30/60, 60/90, 90/120, 120/150 | 3 | 2 |
| Fluoranthene | 280.0–290.0 | 402.7–403.8 | 1–5 | 10/30, 30/60, 60/90, 90/120, 120/150 | 3 | 2 |
| Pyrene | 280.5–290.0 | 370.1–371.2 | 4, 5 | 90/120, 120/150 | 3 | 2 |
| Benz[*a*]anthracene | 281.0–288.5 | 381.6–382.4 | 1–3 | 10/30, 30/60, 60/90 | 3 | 2 |
| Chrysene | 280.0–286.5 | 362.1–362.8 | 1–3 | 10/30, 30/60, 60/90 | 3 | 2 |
| Benzo[*b*]fluoranthene | 284.5–290.0 | 398.6–399.5 | 1–4 | 10/30, 30/60, 60/90, 90/120 | 3 | 2 |
| Benzo[*k*]fluoranthene | 280.0–289.5 | 401.5–402.4 | 1, 2 | 10/30, 30/60 | 3 | 2 |
| Benzo[*a*]pyrene | 283.0–287.5 | 400.5–401.5 | 1, 2 | 10/30, 30/60 | 3 | 2 |

*Continued*

**TABLE 5** Parameters for PARAFAC and U-PLS/RTL Analysis—Cont'd

| PAH | Excitation (nm) | Emission (nm) | EEM # | Time windows (ns) | PARAFAC factors | U-PLS/RTL factors |
|---|---|---|---|---|---|---|
| Indeno[1,2,3-*cd*]pyrene | 283.5–290.0 | 401.5–402.4 | 1–3 | 10/30, 30/60, 60/90 | 2 | 1 |
| Dibenz[*a,h*]anthracene | 281.0–295.0 | 390.9–392.1 | 1–3 | 10/30, 30/60, 60/90 | 3 | 2 |
| Benzo[*ghi*]perylene | 283.0–288.5 | 404.6–405.3 | 1–5 | 10/30, 30/60, 60/90, 90/120, 120/150 | 3 | 2 |

[a]*Wavelength ranges are fractions of recorded EEMs. The complete wavelength ranges of the recorded EEMs were 280–295 nm for excitation and 300–500 nm for emission.*

[b]*EEM # refers to the TREEM recorded during the following delay (D)/delay+gate (D+G) time windows: 1=10/30 ns, 2=30/60 ns, 3=60/90 ns, 4=90/120 ns, 5=120/150 ns, 6=150/180 ns.*

Reprinted from Talanta, 101, H. C. Goicoechea, S. Yu, A. F. T. Moore, A. D. Campiglia, Four-way modeling of 4.2 K time-resolved excitation emission fluorescence data for the quantitation of polycyclic aromatic hydrocarbons in soil samples, pp. 330–336, Table 2, p. 334. Copyright 2012, with permission from Elsevier.

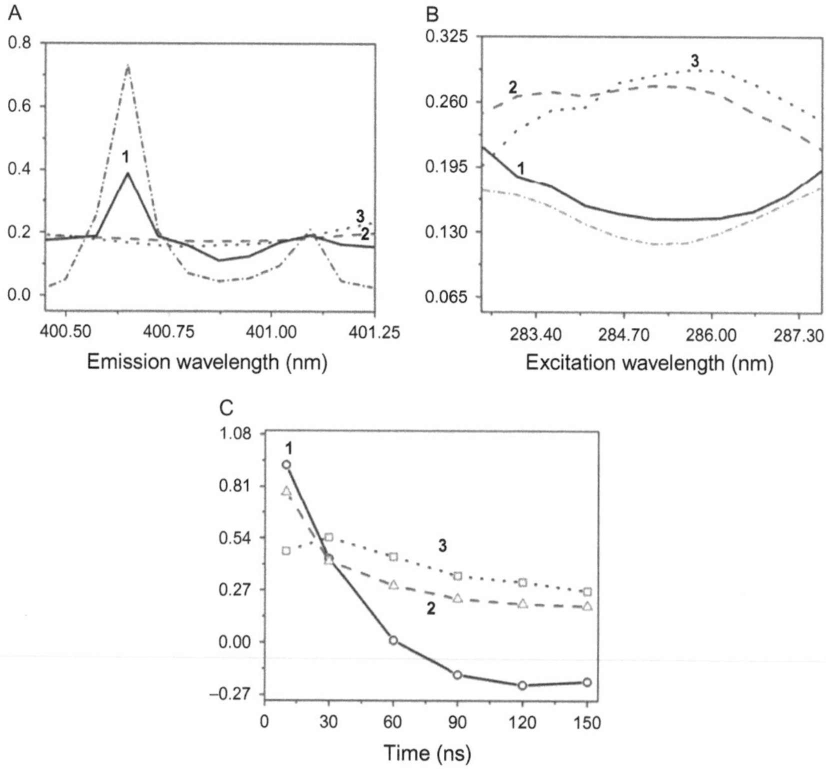


**FIGURE 9** PARAFAC modeling profiles obtained from soil extract. (A) Emission spectral profile, (B) excitation spectral profile, and (C) fluorescence decay profile of benzo[a]pyrene. Number 1, 2, and 3 indicate the factor order. The short dash doted green (gray in the print version) line in (A) corresponds to the normalized emission spectrum of benzo[*a*]pyrene. *Reprinted from Talanta, 101, H. C. Goicoechea, S. Yu, A. F. T. Moore, A. D. Campiglia, Four-way modeling of 4.2 K time-resolved excitation emission fluorescence data for the quantitation of polycyclic aromatic hydrocarbons in soil samples, pp. 330–336, Figure 2, p. 335. Copyright 2012, with permission from Elsevier.*

core consistency procedure—originally introduced by Bro [71]. Figure 9 shows the loading matrices for emission (A), excitation (B), and time decay (C) modes for benzo[*a*]pyrene predicted with the settings in Table 5. The labeling of components **1**, **2**, and **3** follows the order assigned by the model in the four-way array data of the sample. The order reflects their relative contribution to the overall variance. Benzo[*a*]pyrene (**1**) is the main fluorescence contributor in the four-way array data. The fluorescence contribution of interferences **2** and **3** is practically negligible. The similarity among the predicted spectral profiles and the experimental spectra of benzo[*a*]pyrene—denoted as short dash/doted gray lines —is remarkable. The predicted value of the fluorescence lifetime (34.2 ns) obtained with $D$ is close to the experimental fluorescence lifetime of benzo[*a*]pyrene ($38.6 \pm 0.3$ ns, $N=3$). The fact that the decay profile follows a well-behaved, single exponential decay confirms the negligible contributions of **2** and **3** to the total fluorescence of the sample. Similar results were obtained for the remaining EPA-PAHs.

Table 6 summarizes the PAH recoveries obtained via TREEC/chemometric modeling and HPLC. Chromatographic analysis followed reported methodology implemented in our lab [70]. In all the cases, PAH extraction was carried out via the sonication screening method. The statistical comparisons of PAH recoveries at different concentration levels was carried out with the Bonferroni's adjustment test [72]. The alpha value—$\alpha' = 1 - (1 - \alpha')^{1/k}$—was calculated for an overall significance level of 0.05 and 15 PAH comparisons ($k = 15$). Because the calculated value ($\alpha' = 3.5 \times 10^{-3}$) was considerably smaller than the critical $t$ value ($t_{(\alpha',3-1)} = 13.9$), it was concluded that the experimental recoveries of the 15 EPA-PAHs obtained with the three methods—i.e., TREEC/PARAFAC, TREEC/U-PLS-RTL, and HPLC—were statistically equivalent. The main difference is the inability of the HPLC method to detect the presence of anthracene and indeno[1,2,3-*cd*]pyrene. The observed difference can be attributed to the small mass of soil used for screening PAHS. The extraction of 0.05 g of soil probably leads to extract concentrations lower than the limits of detection of HPLC. The same is not true for the two TREEC/chemometric modeling methods. The limits of detection of 4.2 K LETRSS make possible to determine the presence of both anthracene and indeno[1,2,3-*cd*]pyrene in the extracts of 0.05 g of soil.

## 11 CONCLUSIONS

While research efforts employing established methodology continue to advance our knowledge on organic pollutants, the increasing awareness and public concern with environmental and occupational exposure to contaminants calls for analytical approaches capable to handle numerous samples in short analysis time. Data collection from statistically meaningful population sizes should benefit the development of efficient environmental remediation processes, which is extremely relevant for a globally sustainable environment. Although chromatographic techniques have dominated the environmental and toxicological scenarios, alternative solutions to this challenge require a new generation of analysts with knowledge in a plethora of analytical techniques and instrumentation. A recent trend for the direct determination—i.e., no chromatographic separation—of targeted species in matrixes of unknown composition refers to processing multidimensional spectroscopic data with second-order multivariate calibration methods. Under this prospective, particular attention has been paid to photoluminescence techniques. The combination of excitation and emission spectra to lifetime information within the time domains of fluorescence or phosphorescence provides a minimum of three qualitative parameters for compound determination, namely, one excitation and emission wavelength and one lifetime.

**TABLE 6** Comparison of PAH Recoveries Obtained via TREEC/PARAFAC, TREEC/U-PLS/RTL, and HPLC Analyses[a]

| PAH | HPLC (%) | TREEC/PARAFAC (%) | TREEC/U-PLS/RTL (%) | *t*-Student[b] | | |
|---|---|---|---|---|---|---|
| | | | | $t_1$ | $t_2$ | $t_3$ |
| Naphthalene | 44.8±4.0 | 42.2±3.8 | 45.6±3.9 | 0.17 | 0.13 | 0.22 |
| Acenaphthene | 40.9±2.3 | 39.6±1.7 | 39.2±1.6 | 0.52 | 0.70 | 0.14 |
| Fluorene | 43.0±0.3 | 44.3±0.8 | 44.1±0.8 | 2.73 | 2.19 | 0.29 |
| Phenanthrene | 40.5±1.4 | 42.9±1.3 | 41.4±1.4 | 1.31 | 0.52 | 0.82 |
| Anthracene | – | 45.7±2.8 | 45.1±2.8 | – | – | 0.08 |
| Fluoranthene | 24.3±0.8 | 27.0±1.0 | 26.5±0.9 | 3.29 | 3.03 | 0.06 |
| Pyrene | 30.3±2.3 | 31.1±1.8 | 31.3±1.9 | 0.18 | 0.29 | 0.06 |
| Benz[*a*]anthracene | 31.8±1.3 | 27.9±1.8 | 30.9±1.8 | 1.58 | 0.36 | 0.92 |
| Chrysene | 58.5±5.6 | 55.4±4.3 | 57.8±4.1 | 0.12 | 0.02 | 0.13 |
| Benzo[*b*]fluoranthene | 55.4±5.0 | 51.9±3.5 | 53.8±3.4 | 0.18 | 0.10 | 0.16 |
| Benzo[*k*]fluoranthene | 39.3±1.1 | 43.6±1.4 | 41.3±1.4 | 2.71 | 1.26 | 1.17 |
| Benzo[*a*]pyrene | 49.9±2.1 | 40.4±2.1 | 48.9±1.9 | 2.15 | 0.25 | 2.11 |
| Indeno[1,2,3-*cd*]pyrene | – | 46.5±3.6 | 47.1±3.4 | – | – | 0.05 |
| Dibenz[*a,h*]anthracene | 37.5±3.2 | 46.7±3.1 | 43.1±3.4 | 0.95 | 0.88 | 0.34 |
| Benzo[*ghi*]perylene | 38.6±0.6 | 40.5±1.0 | 38.4±0.6 | 2.75 | 0.55 | 3.09 |

[a] *HPLC, TREEC/PARAFAC, and TREEC/U-PLS/RTL analyses were performed in soil extracts obtained via the sonication screening procedure. Each reported value is the average of three independent extractions of soil samples.*

[b] *t-Student, experimental values calculated for* $N=3$ *and* $p<0.05$; $t_1$, *comparison between HPLC and TREEC/PARAFAC;* $t_2$, *comparison between HPLC and TREEC/U-PLS/RTL;* $t_1$ *corresponds to comparison between HPLC and TREEC/PARAFAC;* $t_3$, *comparison between both chemometric models. Reported values should be compared to t critical*$=2.776$.

Reprinted from Talanta, 101, H. C. Goicoechea, S. Yu, A. F. T. Moore, A. D. Campiglia, Four-way modeling of 4.2 K time-resolved excitation emission fluorescence data for the quantitation of polycyclic aromatic hydrocarbons in soil samples, pp. 330–336, Table 3, p. 335. 

Research in our lab has focused on the advancement of line-narrowing photoluminescence techniques. Line-narrowing techniques deal with the measurement of fluorescence and phosphorescence at cryogenic temperatures. Reducing the sample temperature to 20 K or below often improves spectral resolution. The main reason for seeking spectral narrowing is to differentiate the spectral features of chemically related compounds with significantly different carcinogenicity and toxicity properties. We have developed instrumentation to efficiently collect multidimensional data formats in frozen matrixes during the fluorescence and phosphorescence lifetime decays. Fluorescence and phosphorescence decays report on compound identification and spectral peak purity, an essential condition for the accurate quantitation of targeted contaminants without previous chromatographic separation. The inconvenience of sample freezing procedures was eliminated with the aid of cryogenic fiber optic probes. It is now possible to easily perform reproducible and accurate measurements at liquidnitrogen (77 K) and liquid helium (4.2 K) temperatures in a matter of seconds. Our instrumentation takes advantage of the full dimensionality of luminescence spectroscopy combining spectral and lifetime information in multidimensional data formats known as WTMs and TREEMs. Processing WTMs and TREEMs with second-order multivariate calibration algorithms—such as PARAFAC and U-PLS/RTL—provides a general solution to unpredictable spectral interference, a ubiquitous problem in samples of unknown composition.

## ACKNOWLEDGMENTS

H.C.G. acknowledges the Universidad Nacional del Litoral, CONICET (Consejo Nacional de Investigaciones Científicas y Técnicas), ANPCyT (Agencia Nacional de Promoción Científica y Tecnológica, Project No. PICT-2011-0005).

## REFERENCES

[1] Escandar GM. Phosphorescence properties of p-aminobenzoic acid immobilized on a nylon membrane. Appl Spectrosc 2004;58:836–42.

[2] Traviesa-Alvarez JM, Costa-Fernandez JM, Pereiro R, Sanz-Medel A. Flow-through solid phase energy transfer-room temperature phosphorescence for orthophosphate determinations at trace levels. Talanta 2004;62:827–33.

[3] Arruda AF, Goicoechea HC, Santos M, Campiglia AD. Solid-liquid extraction room temperature phosphorimetry and pattern recognition for screening polycyclic aromatic hydrocarbons and polychlorinated biphenyls in water samples. Environ Sci Technol 2003;37:1385–91.

[4] Mendonsa SD, Hurtubise RJ. A comparative study of the solid-matrix phosphorescence of heterocyclic aromatic amines in glucose glasses as a function of temperature. J Lumin 2002;97:19–33.

[5] Kuijt J, Ariese F, Brinkman UAT, Gooijer C. Room temperature phosphorescence in the liquid state as a tool in analytical chemistry. Anal Chim Acta 2003;488:135–71.

[6] Arancibia JA, Escandar GM. Determination of naproxen in pharmaceutical preparations by room-temperature phosphorescence. A comparative study of several organized media. Analyst 2001;126:917–22.

[7] Diaz BC, Terrones SC, Carretero AS, Fernandez JMC, Gutierrez AF. Comparison of three different phosphorescent methodologies in solution for the analysis of naphazoline in pharmaceutical preparations. Anal Bioanal Chem 2004;379:30–4.

[8] Thompson AL, Hurtubise RJ. Solid-matrix fluorescence quenching of benzo[e]pyrene and ()-anti-dibenzo[a, l]pyrene diolepoxide-DNA adducts. Appl Spectrosc 2005;59:126–33.

[9] Castillo AS, Carretero AS, Fernandez JMC, Jin WJ, Gutierrez AF. Heavy atom induced room temperature phosphorescence: a tool for the analytical characterization of polycyclic aromatic hydrocarbons. Anal Chim Acta 2004;516:213–20.

[10] Purcaro G, Moret S, Bucar-Miklavcic M, Conte LS. Ultra-high performance liquid chromatographic method for the determination of polycyclic aromatic hydrocarbons in a passive environmental sampler. J Sep Sci 2012;35:922–8.

[11] Cai SS, Stevens J, Syage JA. Ultra high performance liquid chromatography-atmospheric pressure photoionization-mass spectrometry for high-sensitivity analysis of US Environmental Protection Agency sixteen priority pollutant polynuclear aromatic hydrocarbons in oysters. J Chromatogr A 2012;1227:138–44.

[12] Shitamichi O, Matsui T, Hui YM, Chen WW, Imasaka T. Determination of persistent organic pollutants by gas chromatography/laser multiphoton ionization/time-of-flight mass spectrometry. Front Environ Sci Eng 2012;6:26–31.

[13] Parastar H, Radovic JR, Jalali-Heravi M, Diez S, Bayona JM, Tauler R. Resolution and quantification of complex mixtures of polycyclic aromatic hydrocarbons in heavy fuel oil sample by means of GC × GC-TOFMS combined to multivariate curve resolution. Anal Chem 2011;83:9289–97.

[14] Pandey SK, Kim KH, Brown RJC. A review of techniques for the determination of polycyclic aromatic hydrocarbons in air. Trends Anal Chem 2011;30:1716–39.

[15] Tarpani L, Vocci A, Selvaggi R, Pellegrino R, Ruspolini F, Taglieri L, et al. Solid-phase analysis of polycyclic aromatic hydrocarbons by fluorimetric methods. Appl Spectrosc 2011;65:1342–7.

[16] Hutzler C, Luch A, Filser JG. Analysis of carcinogenic polycyclic aromatic hydrocarbons in complex environmental mixtures by LC-APPI-MS/MS. Anal Chim Acta 2011;702:218–24.

[17] Zhang J, Dong XL, Cheng JS, Li JH, Wang YS. Efficient analysis of non-polar environmental contaminants by MALDI-TOF MS with graphene as matrix. J Am Soc Mass Spectrom 2011;22:1294–8.

[18] Nolte T, Andersson JT. Capillary electrophoretic methods for the separation of polycyclic aromatic compounds. Polycycl Aromat Compd 2011;31:287–338.

[19] Ballesteros-Gomez A, Rubio S. Recent advances in environmental analysis. Anal Chem 2011;83:4579–613.

[20] JiJi RD, Booksh KS. Mitigation of Rayleigh and raman spectral interferences in multiway calibration of excitation-emission matrix fluorescence spectra. Anal Chem 2000;72:718–25.

[21] Cao YZ, Chen ZP, Mo CY, Wu HL, Yu RQ. A PARAFAC algorithm using penalty diagonalizaiton error (PDE) for three-way data array resolution. Analyst 2000;125:2303–10.

[22] Saurina J, Tauler R. Strategies for solving matrix effects in the analysis of triphenyltin in sea-water samples by three-way multivariate curve resolution. Analyst 2000;125:2038–43.

[23] Saurina J, Leal C, Compañó R, Granados M, Tauler R, Prat MD. Determination of triphenyltin in sea-water by excitation-emission matrix fluorescence and multivariate curve resolution. Anal Chim Acta 2000;409:237–45.

[24] Moberg L, Robertson G, Karlberg B. Spectrofluorimetric determination of chlorophylls and pheopigments using parallel factor analysis. Talanta 2001;54:161–70.

[25] Esteves da Silva JCG, Litao JMM, Costa FS, Ribeiro JLA. Detection of verapamil drug by fluorescence and trilinear decomposition techniques. Anal Chim Acta 2002;453:105–15.

[26] Cao YZ, Mo CY, Long JG, Chen H, Wu HL, Yu RQ. Spectrofluorimetric resolution of closely overlapping drug mixtures using chemometrics methods. Anal Sci 2002;18:333–6.

[27] Arancibia JA, Olivieri AC, Escandar GM. First- and second-order multivariate calibration applied to biological samples: determination of anti-inflammatories in serum and urine. Anal Bioanal Chem 2002;374:451–9.

[28] Muñoz de la Peña A, Espinosa-Mansilla A, González-Gómez D, Olivieri AC, Goicoechea HC. Interference-free analysis using three-way fluorescence data and the parallel factor model. Determination of fluoroquinolone antibiotics in human serum. Anal Chem 2003;75:2640–6.

[29] Olivieri AC, Arancibia JA, Muñoz de la Peña AA, Durán-Merás I, Espinosa Mansilla A. Second-order advantage achieved with four-way fluorescence excitation-emission-kinetic data processed by parallel factor analysis and trilinear least-squares. Determination of methotrexate and leucovorin in human urine. Anal Chem 2004;76:5657–66.

[30] Nahorniak ML, Booksh KS. Excitation-emission matrix fluorescence spectroscopy in conjunction with multiway analysis for PAH detection in complex matrices. Analyst 2006;131:1308–15.

[31] Bortolato SA, Arancibia JA, Escandar GM. Chemometrics-assisted excitation-emission fluorescence spectroscopy on nylon membranes. Simultaneous determination of benzo[a]pyrene and dibenz[a, h]anthracene at parts-per-trillion levels in the presence of the remaining EPA PAH priority pollutants as interferences. Anal Chem 2008;80:8276–86.

[32] Bortolato SA, Arancibia JA, Escandar GM. Chemometrics-assisted fluorimetry for the rapid and selective determination of heavy polycyclic aromatic hydrocarbons in contaminated river waters and activated sludges. Environ Sci Technol 2011;45:1513–20.

[33] Renge I, Wild UP. Principles of matrix-induced high-resolution optical spectroscopy and electron-phnon coupling in doped organic crystals. In: Gooijer C, Ariese F, Hofstraat JW, editors. Shpol'skii spectroscopy and other site-selection methods. 1st ed. New York: John Wiley & Sons; 2000. p. 19–72.

[34] Warner IM, Patonay G, Thomas MP. Multidimensional luminescence measurements. Anal Chem 1985;57:463A–483A.

[35] Knorr FJ, Harris JM. Resolution of multicomponent fluorescence spectra by an emission wavelength-decay time data matrix. Anal Chem 1981;53:272–6.

[36] McGown LB. Fluorescence lifetime filtering. Anal Chem 1989;61:839A–845A.

[37] Smalley MB, Shaver JM, McGown LB. On-the-fly fluorescence lifetime detection in HPLC using a multiharmonic Fourier transform phase-modulation spectrofluorimeter. Anal Chem 1993;65:3466–72.

[38] Benzo[a]pyrene. In: Karcher W, Fordham RJ, Dubois JJ, Glaude PGJM, Ligthart JAM, editors. Spectral atlas of polycyclic aromatic compounds. Boston: Kluwer Academic Publishers; 1985. p. 531–551.

[39] Nakhimovsky LA, Lamotte M, Joussot-Dubien J. Handbook of low-temperature electronic spectra of polycyclic aromatic hydrocarbons. Amsterdam: Elsevier; 1989.

[40] Bystol AJ, Campiglia AD, Gillispie GD. Time-resolved laser-excited Shpol'skii spectrometry with a fiber-optic probe and ICCD camera. Appl Spectrosc 2000;54:910–7.

[41] Bystol AJ, Campiglia AD, Gillispie GD. Laser-induced multidimensional fluorescence spectroscopy in Shpol'skii matrixes with a fiber-optic probe at liquid helium temperature. Anal Chem 2001;73:5762–70.

[42] Campiglia AD, Bystol AJ, Yu S. Instrumentation for multidimensional luminescence spectroscopy and its application to low-temperature analysis in Shpol'skii matrixes and optically scattering media. Anal Chem 2006;78:484–92.
[43] Madison RT, Carroll MK, Hieftje GM. Convenient fiber-optic-based sample cell for Shpol'skii and low-temperature phosphorescence spectrometry. Appl Spectrosc 1989;43:422–5.
[44] Yang Y, D'Silva AP, Fassel VA, Illes M. Direct determination of polynuclear aromatic hydrocarbons in coal liquids and shale oil by laser excited Shpol'skii spectrometry. Anal Chem 1980;52:1350–1.
[45] Yang Y, D'Silva AP, Illes M, Fassel VA. Laser excited Shpol'skii spectrometry (LESS) for the selective excitation and determination of polynuclear aromatic hydrocarbons. Proc Soc Photo Opt Instrum Eng 1981;286:126–30.
[46] Renkes GD, Walters SN, Woo CS, Illes MK, D'Silva AP, Fassel VA. Determination of aromatic hydrocarbons in particulate samples by high-temperature extraction and Shpol'skii spectrometry. Anal Chem 1983;55:2229–31.
[47] Elsaiid AE, Walker R, Weeks S, D'Silva AP, Fassel VA. Site-selective Shpol'skii spectrometry of sulfur-, oxygen-, and nitrogen-containing aromatic compounds in complex samples. Appl Spectrosc 1988;42:731–40.
[48] Hofstraat JW, Engelsma M, De Roo JH, Gooijer C, Velthorst NH. Photodiode array detection in analytical high-resolution molecular luminescence spectroscopy. Appl Spectrosc 1987;41:625–31.
[49] Martin TL, Campiglia AD. Fiber-optic system for 77 K phosphorescence lifetime measurements of polycyclic aromatic compounds in Shpol'skii matrices. Appl Spectrosc 2001;55:1266–72.
[50] Bystol AJ, Whitcomb JL, Campiglia AD. Solid-liquid extraction laser excited time-resolved Shpol'skii spectrometry: a facile method for the direct detection of 15 priority pollutants in water samples. Environ Sci Technol 2001;35:2566–71.
[51] Martin TL, Arruda AF, Campiglia AD. Time-resolved laser-excited phosphorimetry at liquid helium temperature for the direct analysis of 2,3,7,8-tetrachloro-dibenzo-para-dioxin in complex mixtures. Appl Spectrosc 2002;56:1354–60.
[52] Bystol AJ, Whitcomb JL, Campiglia AD. A novel approach for solid-liquid extraction laser-excited time-resolved Shpol'skii spectrometry. Talanta 2002;57:1101–11.
[53] Bystol AJ, Thorstenson T, Campiglia AD. Laser-induced multidimensional fluorescence spectroscopy in Shpol'skii matrixes for the analysis of polycyclic aromatic hydrocarbons in HPLC fractions and complex environmental extracts. Environ Sci Technol 2002;36:4424–9.
[54] Arruda AF, Yu S, Campiglia AD. Shpol'skii spectroscopy at the interface of two non-polar microenvironments: a novel approach for the analysis of polycylic aromatic compounds. Talanta 2003;59:1199–211.
[55] Bystol AJ, Yu S, Campiglia AD. Analysis of polycyclic aromatic hydrocarbons in HPLC fractions by laser-excited time-resolved Shpol'skii spectrometry with cryogenic fiber-optic probes. Talanta 2003;60:449–58.
[56] Yu S, Campiglia AD. Laser-excited time-resolved Shpol'skii spectroscopy for the direct analysis of dibenzopyrene isomers in liquid chromatography fractions. Appl Spectrosc 2004;58:1385–93.
[57] Yu S, Campiglia AD. Direct determination of dibenzo[a, l]pyrene and its four dibenzopyrene isomers in water samples by solid-liquid extraction and laser-excited time-resolved Shpol's-kii spectrometry. Anal Chem 2005;77:1440–7.

[58] Wang H, Campiglia AD. Determination of polycyclic aromatic hydrocarbons in drinking water samples by solid-phase nanoextraction and high-performance liquid chromatography. Anal Chem 2008;80:8202–9.

[59] Wang H, Yu S, Campiglia AD. Solid-phase nano-extraction and laser-excited time-resolved Shpol'skii spectroscopy for the analysis of polycyclic aromatic hydrocarbons in drinking water samples. Anal Biochem 2009;385:249–56.

[60] Wang H, Campiglia AD. Direct determination of benzo[a]pyrene in water samples by a gold nanoparticle-based solid phase extraction method and laser-excited time-resolved Shpol'skii spectrometry. Talanta 2010;83:233–40.

[61] Wilson WB, Campiglia AD. Analysis of co-eluted isomers of high-molecular weight polycyclic aromatic hydrocarbons in high performance liquid chromatography fractions via solid-phase nanoextraction and time-resolved Shpol'skii spectroscopy. J Chromatogr A 2011;1218:6922–9.

[62] Wilson WB, Campiglia AD. Determination of polycyclic aromatic hydrocarbons with molecular weight 302 in water samples by solid-phase nano-extraction and laser excited time-resolved Shpol'skii spectroscopy. Analyst 2011;136:3366–74.

[63] Goicoechea HC, Yu S, Olivieri AC, Campiglia AD. Four-way data coupled to parallel factor model applied to environmental analysis: determination of 2,3,7,8-tetrachloro-dibenzo-para-dioxin in highly contaminated waters by solid-liquid extraction laser-excited time-resolved Shpol'skii spectroscopy. Anal Chem 2005;77:2608–16.

[64] Goicoechea HC, Yu S, Moore AFT, Campiglia AD. Four-way modeling of 4.2 K time-resolved excitation emission fluorescence data for the quantitation of polycyclic aromatic hydrocarbons in soil samples. Talanta 2012;101:330–6.

[65] Arruda AF, Campiglia AD. Screening potential of solid-phase extraction room temperature phosphorimetry for the analysis of polychlorinated dibenzofurans in water samples. Environ Sci Technol 2000;34:4982–8.

[66] Hagestuen ED, Campiglia AD. Phosphorimetric detection of polycyclic aromatic hydrocarbons on solid-phase extraction membranes. Appl Spectrosc 1998;52:1096–102.

[67] Arruda AF, Campiglia AD. Determination of trace levels of polychlorinated biphenyls on reversed phase octadecyl bonded silica membranes. Anal Chim Acta 1999;386:271–80.

[68] Hagestuen ED, Campiglia AD. New approach for screening polycyclic aromatic hydrocarbons in water samples. Talanta 1999;49:547–60.

[69] Hagestuen ED, Arruda AF, Campiglia AD. On the improvement of solid-phase extraction room-temperature phosphorimetry for the analysis of polycyclic aromatic hydrocarbons in water samples. Talanta 2000;52:727–37.

[70] Yu S, Goicoechea HC, Campiglia AD. Sonication-laser-excited time-resolved Shpol'skii spectrometry: a facile method for the direct screening of fifteen priority pollutants in soil samples. Appl Spectrosc 2007;61:165–70.

[71] Bro R. A new efficient method for determining the number of components in PARAFAC models. J Chemom 2003;17:274–86.

[72] Some important hypothesis tests. In: Massart DL, Vandeginste BGM, Buydens LMC, De Jong S, Lewi PJ, Smeyers-Verbeke J, editors. Handbook of Chemometrics and Qualimetrics: Part A. New York: Elsevier; 1998. p. 93–120.

Chapter 13

# Figures of Merit in Multiway Calibration

**Alejandro C. Olivieri[1], Santiago Bortolato and Franco Allegrini**
*Instituto de Química Rosario (CONICET-UNR), Facultad de Ciencias Bioquímicas y Farmacéuticas, Universidad Nacional de Rosario, Rosario, Argentina*
*[1]Corresponding author: e-mail: aolivier@fbioyf.unr.edu.ar; olivieri@iquir-conicet.gov.ar*

## Chapter Outline



Data Handling in Science and Technology, Vol. 29. http://dx.doi.org/10.1016/B978-0-444-63527-3.00013-8

# 1 INTRODUCTION

## 1.1 Figures of Merit in Modern Analytical Chemistry: Why Are They Important?

As previously presented in this book, measuring and processing multiway data provides analytical chemists with a number of advantages, such as (1) improved sensitivity, derived from noise averaging multiple measurements of redundant data, (2) increased selectivity, because each new data mode provides an additional degree of partial selectivity, and (3) modeling the analyte contribution and its quantitative determination in the presence of unknown interferences, absent in calibration samples (second-order advantage) [1]. Regarding items 1 and 2, a question which immediately emerges is how figures of merit like sensitivity, selectivity, and even the limit of detection (LOD) should be estimated when dealing with multivariate and multiway data? As analytical chemistry is the science of chemical measurements, finding a reliable way to judge them properly is not a minor issue, and this is the reason why this chapter is focused on trying to give a response to this question.

In modern analytical chemistry research, the search for new estimators to improve analytical figures of merit is an important driving force, with the sensitivity (SEN) and the LOD occupying prominent places among these figures. Finding an expression to calculate consistent estimators for these figures has a relevant influence on different activities, such as (1) comparison of the performance of different experimental procedures, (2) optimization of a given methodology under various experimental conditions, and (3) development of official protocols of validation and analysis, as documented in international standards [2,3].

The LOD has gained significant popularity as a descriptor of the quality of a method in applied analytical chemistry. This is because of two reasons: (1) it is expressed in concentration units, allowing to easily comparing in a direct way different methodologies and (2) it is needed for assessing detection capabilities which are of fundamental importance in certain specific areas: doping control in sports, monitoring traces of contaminants in environmental samples, etc. However, at the core of LOD definition lies the sensitivity, which is also a key parameter in the estimation of other figures of merit: (1) analytical sensitivity, which is important for the comparison of methodologies based on widely different signals, because it is independent of the instrument and technique applied, (2) selectivity, which helps to assess the possibility of analyte quantitation in the presence of interferences, and (3) prediction uncertainty, which gives an idea of the precision of a certain prediction.

## 1.2 From Univariate to Multivariate and Multiway Calibration

### 1.2.1 Sensitivity

According to the International Union of Pure and Applied Chemistry (IUPAC), in the classical single-constituent or univariate calibration, the sensitivity is defined as "the change in the response of the instrument divided by the corresponding stimulus (the concentration of the analyte of interest)," i.e., the slope of the calibration curve [4]. In first-order multivariate calibration, unfortunately, the situation regarding the definition of sensitivity becomes more complex [5]. The difficulty arising in this case is the fact that an intense signal may be useless under severe spectral overlapping with signals from other concomitant constituents. This leads to an important property of multivariate sensitivity: analyte specificity, which means that a certain sensitivity parameter corresponds to each analyte of interest.

As explained in other chapters, multiway calibration involves the measurement of data matrices per sample (or data arrays with three or more modes) for analyte calibration purposes and constitutes a powerful generalization of multivariate calibration [6]. In this field, several different sensitivity expressions have been proposed, some of them based on the extension of the first-order net analyte signal (NAS) concept to further data modes [7–10]. The NAS is the portion of the total signal which can be uniquely ascribed to the analyte of interest, and hence, the slope of the pseudounivariate NAS concentration graph, or the NAS at unit concentration, appears to be a reasonable definition of sensitivity. However, this kind of intuitive extension has caused several difficulties, as there are various competing NAS definitions, with no clear relationships among them [11–13]. Moreover, extrapolation to higher-order calibration led to serious underestimation of sensitivities.

As an alternative to the NAS approach, a general expression emerged in recent years, based on the analysis on how the uncertainty in instrumental signal propagates to the uncertainty in predicted concentrations [14–16]. As will be explained in more detail below, thanks to the developments in this direction, it is now possible to cast all the available sensitivity expressions into a general mathematical equation encompassing all possible degrees of data complexity, from univariate to multiway, and in the latter case for most multiway algorithms [17].

The general expression is in accordance with the fact that the multiway sensitivity displays even more intriguing properties in comparison with the univariate and first-order counterparts: it is not only analyte specific but also strongly dependent on the test sample and on the data processing algorithm. This implies that each test sample can have its own qualitative chemical composition, leading to a specific value of sensitivity. Likewise, the

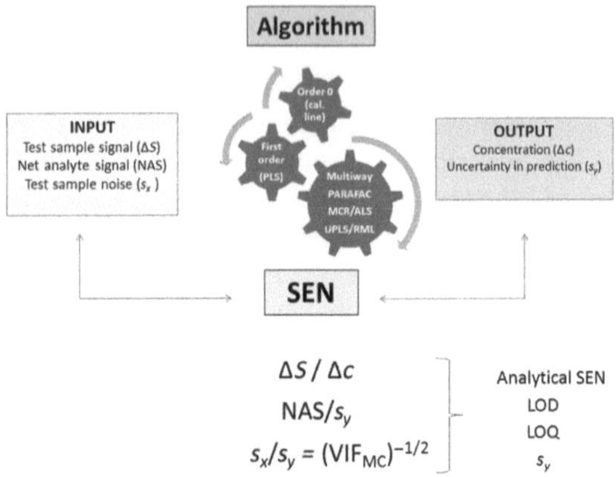


**FIGURE 1** Schematic representation of the different possible ways to define sensitivity according to the data order and the corresponding algorithms.

computational tools employed by each algorithm will affect the analyte sensitivity, and hence, they should be regarded as an integral part of a multiway analytical protocol. Figure 1 summarizes the different ways in which sensitivity, generally understood as the relation between an input and an output, can be defined depending on the order of the data under analysis and the figures of merit resulting from the former.

### *1.2.2 Limit of Detection*

Regarding the LOD, the transition between univariate and multivariate calibration requires a special attention, as in the case of SEN. The terms associated with detection capabilities have been present in the scientific literature for at least the past 100 years. Numerous terms, definition, and calculation approaches have been presented during the latter time period. A recent review with a historical overview about this topic has been recently published [18].

Currently, IUPAC adopts the definition given by the International Standardization Organization (document ISO 11843) [19] for the capability (or limit) of detection as "the lowest quantity of substance that can be distinguished from the absence of that substance (a blank value) within a stated confidence limit" [20–22]. This implies that the LOD is the minimum quantity detectable with a preset probability of false positives (false detects, α- or Type I errors) and false negatives (false nondetects, β- or Type II errors) [20–22]. When the analytical signal is univariate and analyte specific, the estimator is well defined and the LOD can be directly estimated from the univariate calibration line. The recommended detection rule is based on Neyman–Pearson test that considers false detects and false nondetects for the null hypothesis "there is no analyte" and the alternative hypothesis "there is analyte" [20].

However, when dealing with multivariate calibration, as is the case of the well-known partial least-squares (PLS) regression analysis, some aspects which remain outside the field of application of the ISO norm need to be considered [23]. In fact, there is still no generally accepted LOD estimator for PLS studies. Nevertheless, there is a high interest in the topic [17], undoubtedly tied to the inclusion of PLS regression in many commercial instruments, particularly those based on near-infrared spectral measurements [24], in addition to the continuous emergence of new and more sensitive analytical techniques, and the release of regulations on human or environmental exposure to low levels of chemical health hazards.

The main difficulty to estimate a multivariate LOD is that instrumental signals are not specific for a particular analyte. For this reason, the composition of the background of the sample, when extrapolating to zero analyte concentration, plays a fundamental role. As will be developed further in this chapter, a reasonable proposal based on an LOD estimator which adopts the form of a detection interval was recently investigated to try to overcome the previously stated difficulties [25].

### *1.2.3 Other Figures of Merit*

As postulated in Section 1.1, the sensitivity constitutes the core from which other figures of merit can be defined. Among them, the LOD is one of the most well known and widely used. However, in some cases, it might be useful to define other figures in order to emphasize certain features and differences in the data, the samples, or the methodology under analysis. Examples of this kind of figures of merit are the analytical sensitivity, the uncertainty in prediction, and the limit of quantification. A more detailed description of the corresponding estimators will be presented in Section 5.

## 2 A BRIEF INSIGHT INTO DATA PROPERTIES, MODELS, AND ALGORITHMS

Knowledge of the properties and the structure of measured multiway data allow selecting a corresponding model and a data processing tool. As previously stated, this will significantly affect the achieved figures of merit, leading to another peculiar feature of multiway calibration: the algorithm specificity of these figures. This is why a brief reference to the main features of multiway data will be made at this point.

### 2.1 Bilinear and Trilinear Models

As discussed in other chapters, the simplest data array that can be found is a matrix for a single sample, leading to second-order data: if there are $N$

responsive constituents in the sample, a generic element $x_{ij}$ of these data matrices can be written as:

$$x_{ij} = \sum_{n=1}^{N} b_{jn} c_{kn} + e_{ij} \tag{1}$$

where $b_{jn}$ and $c_{kn}$ define the specific properties at instrumental channels $j$ and $k$ for constituent $n$, and $e_{ij}$ is an error term. Any data matrix can be expressed as the product of two matrices; however, if the mathematical rank, i.e., the number of bilinear components required to model the data, is small (ideally equal to the number of chemical constituents), the matrix is considered to be bilinear (strictly speaking, low-rank bilinear). When the matrix cannot be expressed as a sum of a few bilinear terms, it is considered as nonbilinear. In general, bilinearity is lost when the phenomena occurring in the two instrumental modes are mutually dependent [26].

Second-order data are the basic ingredient of a three-way array, which is the simplest multiway data. A three-way data array is low-rank trilinear if it can be expressed as a sum of a few trilinear components when the mixture contains a few constituents. Excitation–emission fluorescence (EEFM) data are typical examples where trilinearity applies. If a number of EEFM ($I$) are stacked in the sample mode, creating a three-way array $\mathbf{X}$, and the samples are mixtures of $N$ fluorescent constituents, a specific signal $x_{ijk}$ at sample $i$, emission wavelength $j$, and excitation wavelength $k$ can be written as:

$$x_{ijk} = \sum_{n=1}^{N} a_{in} b_{jn} c_{kn} + e_{ijk} \tag{2}$$

where $a_{in}$ is proportional to the concentration of constituent $n$ in sample $i$, $b_{jn}$ to the emission quantum yield at wavelength $j$, and $c_{kn}$ to the absorption coefficient at excitation wavelength $k$.

## 2.2 Models to Deal with Trilinearity Deviations

Although it is not directly apparent in Equation (2), for practical purposes it is useful to outline the trilinearity demands: (1) individual data matrices should be bilinear, i.e., **b** and **c** profiles should not depend on each other and (2) **b** and **c** profiles should not depend on the sample, i.e., there should be unique **b** and **c** vectors in both instrumental modes in all samples. In chromatographic-spectral matrix data, elution profiles are not always exactly reproducible from sample to sample. Because of this, a three-way array composed of these latter data matrices will not be, in general, trilinear. Nevertheless, since individual data matrices are bilinear, an augmentation along the elution time mode can be performed. This leads to a chromatographic-spectral matrix augmented in the time direction ($\mathbf{X}_{\text{aug}}$), which is also bilinear and can be formulated as:

$$x_{\mathrm{aug},pk} = \sum_{n=1}^{N} b_{\mathrm{aug},pn} c_{kn} + e_{\mathrm{aug},pk} \tag{3}$$

where $x_{\mathrm{aug},pk}$, $b_{\mathrm{aug},pn}$ and $c_{kn}$ are elements of the $\mathbf{X}_{\mathrm{aug}}$ matrix, and of the $\mathbf{b}_{\mathrm{aug}}$ and **c** profiles respectively, with the index $p$ runs from 1 to $IJ$, because the size of the augmented matrix is $IJ \times K$ ($I$ = number of samples, $J$ = number of elution times, $K$ = number of wavelengths or other spectral sensors). The spectral profile $\mathbf{c}_n$ (in the nonaugmented mode) is unique for each constituent and common to all samples, whereas $\mathbf{b}_{aug,n}$ is the augmented time profile in the augmented elution time mode and is composed of $I$ successive time subprofiles with $J$ times each.

As a consequence of the above discussion, it is possible to classify three-way data as (1) trilinear, (2) nontrilinear with a single trilinearity breaking mode and unfoldable to a bilinear augmented matrix, and (3) other nontrilinear. This latter category refers to data structures in which two trilinearity breaking modes can be identified or cases in which the individual matrices are nonbilinear, as discussed in previous chapters.

Regarding the available algorithms to deal with the different data types previously described, they can be classified into three main groups according to a simple connection between their underlying models and the different data categories discussed above: (1) a multilinear model, (2) a bilinear model for an augmented matrix, and (3) a latent variable model. Group 1 includes parallel factor analysis (PARAFAC) [27], group 2 multivariate curve resolution coupled to alternating least-squares (MCR-ALS) [28] particularly in the extended version [29], and group 3 unfolded and multiway partial least-squares (U-PLS and N-PLS) [30,31].

As already explained in other chapters, both PARAFAC and MCR-ALS achieve the second-order advantage by simultaneously processing the multiple calibration samples and unknowns because their internal algorithmic models are able to decompose the contribution of the potential interfering agents and the analytes to the total signal. However in the case of the PLS-based methodologies, the achievement of the second-order advantage is a postcalibration activity based on a procedure called residual multilinearization (RML) [32–36].

As a final note for this section, it is important to consider that sometimes the border between these algorithms in terms of application fields is not entirely clear, and there may be a considerable overlapping. However, it is likely that future developments will take into account sensitivity and LOD considerations as a helpful decision-making tool in this regard.

### 2.3 From Homoscedastic to Correlated and Heteroscedastic Noise

While many chemometric tools have been designed to extract information from multivariate chemical measurements, one issue that has been somewhat ignored is the role of multivariate measurement errors in this process. However, in recent years, a new branch of chemometrics has gained place: the development of methodologies to try to uncover, understand, and characterize the error structure of a given data set [37–39].

Most of the models and algorithms to deal with multivariate and multiway calibration have been conceived under the simplistic assumption that the error term, as presented in Equations (1–3), is identical and independently distributed (*iid*). In many cases, especially when measurement errors are small or the assumptions are approximately valid, traditional chemometric tools can be applied with excellent results, but in other cases, consideration of the measurement error structure can mean the difference between the success and the failure of the data analysis [40,41]. The same discussion could be extended to the estimation of figures of merit. Although the sensitivity obtained by the uncertainty propagation, the approach (see below) is defined assuming that the input noise is *iid*, when the uncertainty in predictions is to be estimated in the context of correlated and heteroscedastic errors, the scenario is different, and some extra considerations are needed. This is why further research will be required to integrate the information about the error structure to the current available estimators of multivariate uncertainty.

## 3 SENSITIVITY EXPRESSIONS BASED ON NET SIGNAL CHANGES

### 3.1 Univariate Calibration

In this case, prediction of the analyte concentration ($y$) in a test sample from its signal ($x$) proceeds through the known expression [4]:

$$y = (x - n_0)/m_0 \qquad (4)$$

where $m_0$ and $n_0$ are the slope and intercept, respectively, of the zeroth-order linear calibration graph. The slope $m_0$ is the sensitivity since it measures the change in signal for a unit change in concentration.

### 3.2 First-Order Calibration

The concept of NAS, graphically represented in Figure 2, was helpful on the evaluation of the sensitivity in first-order calibration, by extending the univariate definition to the change in NAS for a unit change in analyte concentration [42]. In order to fully understand the NAS concept and its consequences, it is highly useful to consider, as a simple example, a binary mixture where two constituents occur, with the vector signal, e.g., a spectrum for a test sample measured at a number of sensors and given by:

$$\mathbf{x} = y_1\mathbf{s}_1 + y_2\mathbf{s}_2 \qquad (5)$$

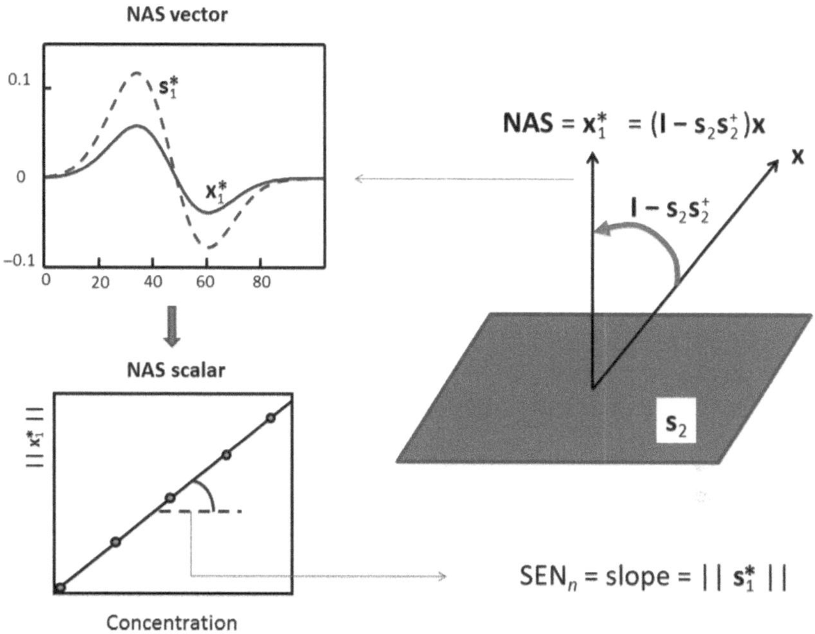


**FIGURE 2** Graphical representation of the NAS concept. *Adapted with permission from Ref. [17]. Copyright 2014 American Chemical Society.*

where $y_1$ and $y_2$ are the constituent concentrations, and $\mathbf{s}_1$ and $\mathbf{s}_2$ are the pure constituent profiles at unit concentration. Equation (5) assumes that: (1) the studied signal is additive (i.e., the total signal is the sum of the individual contributions from both sample constituents) and (2) the constituent signals are proportional to their concentrations, meaning that Beer's law (or similar ones) applies. Focusing on analyte 1 as the constituent of interest, the contribution from constituent 2 can be removed from Equation (5) by left-multiplying both sides by an orthogonal projection matrix $\left[\mathbf{I} - \mathbf{s}_2\left(\mathbf{s}_2^{\mathrm{T}}\mathbf{s}_2\right)^{-1}\mathbf{s}_2^{\mathrm{T}}\right]$, where $\mathbf{I}$ is an appropriately dimensioned unit matrix, because $\left(\mathbf{I} - \mathbf{s}_2\mathbf{s}_2^+\right) \times \mathbf{s}_2 = \mathbf{s}_2 \times \mathbf{s}_2 = 0$, with $\left[\mathbf{s}_2\left(\mathbf{s}_2^{\mathrm{T}}\mathbf{s}_2\right)^{-1}\mathbf{s}_2^{\mathrm{T}}\right]$ designated as $\mathbf{s}_2^+$ (the superscript "+" indicates the generalized inverse). Notice that knowledge of $\mathbf{s}_1$and $\mathbf{s}_2$ is assumed, which is only possible in the context of first-order methodologies such as classical least-squares (CLS) analysis, where the pure spectra are either supplied to the model from separate measurements on pure constituents or adequately retrieved by analysis of mixtures of pure constituents. Thus, Equation (5) leads to Equation (6), and a two-constituent problem has become a virtual single-constituent problem:

$$\left(\mathbf{I} - \mathbf{s}_2\mathbf{s}_2^+\right)\mathbf{x} = y_1\left(\mathbf{I} - \mathbf{s}_2\mathbf{s}_2^+\right)\mathbf{s}_1 \tag{6}$$

Indeed, the left-hand side of the latter equation defines the NAS for constituent 1 in the mixture ($\mathbf{x}_1^*$), as being proportional to its NAS at unit concentration ($\mathbf{s}_1^*$), the proportionality constant being the analyte concentration $y_1$:

$$\mathbf{x}_1^* = y_1 \mathbf{s}_1^* \tag{7}$$

The relevant result to be gathered from Equation (7) is that the NAS vector $\mathbf{x}_1^*$ is parallel to the NAS at unit concentration $\mathbf{s}_1^*$. The real usefulness of the NAS lies in the fact that a plot of the length of the NAS vector ($||\mathbf{x}_1^*||$, also called the scalar NAS, and given as the square root of the sum of the squared elements of the $\mathbf{x}_1^*$ vector) as a function of analyte concentration is linear, the slope being the length of the NAS vector at unit concentration ($||\mathbf{s}_1^*||$). This result immediately leads to an intuitive definition of sensitivity for analyte 1 [43–45]:

$$\mathrm{SEN}_1 = ||\mathbf{x}_1^*||/y_1 = ||\mathbf{s}_1^*|| = \left[\mathbf{s}_1^{\mathrm{T}}\left(\mathbf{I} - \mathbf{s}_2\mathbf{s}_2^{+}\right)\mathbf{s}_1\right]^{1/2} \tag{8}$$

In summary, if both vectorial signals at unit concentration for the pure constituents of a mixture are known or can be estimated from the analysis of mixtures of pure constituents, simple matrix manipulation allows one to define precisely the sensitivity toward a given constituent. A useful relationship between the sensitivity based on this NAS approach and the complete matrix of pure constituent signals can be found by invoking the theory of block pseudoinverse operations [46]: Equation(8) can be generalized to the $n$th constituent of interest in a multiconstituent sample in different forms. One useful form is expressed as a function of the pure profiles for all constituents, ubiquitous in CLS studies:

$$\mathrm{SEN}_n = \left[\delta_n^{\mathrm{T}}\left(\mathbf{S}_{\mathrm{CLS}}^{\mathrm{T}}\mathbf{S}_{\mathrm{CLS}}\right)^{-1}\boldsymbol{\delta}_n\right]^{-1/2} \tag{9}$$

where $\boldsymbol{\delta}_n$ is an $N \times 1$ vector selecting the analyte of interest, and the matrix $\mathbf{S}_{\mathrm{CLS}}$ contains $N$ columns, each with the pure constituent profile $\mathbf{s}_n$ for the $n$th constituent.

A different useful generalization of the multivariate first-order sensitivity can be developed in terms of the vector of regression coefficients, which is specific for a given analyte in a mixture ($\boldsymbol{\beta}_{\mathrm{CLS},n}$). This vector provides the analyte concentration from the below predictive equation

$$y_n = \boldsymbol{\beta}_{\mathrm{CLS},n}^{\mathrm{T}}\mathbf{x} \tag{10}$$

From Equation (10), the sensitivity can be expressed as:

$$\mathrm{SEN}_n = \left(\boldsymbol{\beta}_{\mathrm{CLS},n}^{\mathrm{T}}\boldsymbol{\beta}_{\mathrm{CLS},n}\right)^{-1/2} \tag{11}$$

Interestingly, Equation (11) provides a useful link to estimate the SEN for first-order algorithms based on inverse models, such as inverse least-squares

(ILS), principal component regression (PCR), and PLS [11]. In contrast to the direct approach of the classical Beer's law, inverse calibration models relate concentrations to signals $\left(y_n = \boldsymbol{\beta}_n^{\mathrm{T}}\mathbf{x}\right)$, and these are able to provide a vector of regression coefficients $\boldsymbol{\beta}_n$ from a suitable set of calibration mixtures, and thus, the analogue of Equation (11) is a useful means of estimating the sensitivity for these methodologies. In this way, sensitivity expressions for both direct and inverse calibration first-order methodologies can be brought into a common form.

## 3.3 Multiway (Higher-Order) Calibration

Inspired in the useful first-order NAS philosophy, it is possible to estimate the SEN parameter in three-way (second-order) calibration, calculate the NAS, and remove the contribution of constituents other than the analyte of interest using orthogonal projection matrices. One intriguing aspect of this multiway NAS approach is the fact that, in principle, these projections can be carried out in different ways, leading to competing NAS definitions [9].

A matrix signal $\mathbf{X}$ defined in two different instrumental modes for a simple binary mixture can be written as:

$$\mathbf{X} = y_1\mathbf{M}_1 + y_2\mathbf{M}_2 \tag{12}$$

where $\mathbf{M}_1$ and $\mathbf{M}_2$ are matrix signals at unit concentration for each analyte, and, as before, signal additivity and signal concentration linearity are assumed. If the signals are bilinear and the profiles in both data modes are designated as $\mathbf{b}$ and $\mathbf{c}$, the expression for $\mathbf{X}$ would be:

$$\mathbf{X} = y_1\mathbf{b}_1\mathbf{c}_1^{\mathrm{T}} + y_2\mathbf{b}_2\mathbf{c}_2^{\mathrm{T}} \tag{13}$$

where $\mathbf{b}_1$ and $\mathbf{b}_2$ are the pure constituent profiles in the first data mode, and $\mathbf{c}_1$ and $\mathbf{c}_2$ those in the second data mode.

According to the NAS approach, the contributing matrix signal for constituent 2 may be removed from Equation (13) by these simultaneous operations: left-multiplication with a projection matrix orthogonal to $\mathbf{b}_2$ and right-multiplication with an analogous matrix orthogonal to $\mathbf{c}_2$. This outcome leads to one particular sensitivity expression known as HCD (the acronym follows the authors' final initials) [7], which is valid in a certain calibration scenario (see Table 1).

There is an alternative procedure, which involves first unfolding the matrix $\mathbf{X}$ into a vector and then removing the contribution of constituent 2 with a single removing matrix, orthogonal to the unfolded space spanned by constituent 2. This approach leads to a different second-order sensitivity definition, the MKL sensitivity [8], which is valid in a different calibration situation in comparison with the HCD sensitivity (see Table 1). Although the original works on HCD and MKL sensitivity did not employ NAS

**TABLE 1 Different Three-Way Sensitivity Definitions Based on Extensions of the NAS Concept**

| | |
|---|---|
| **HCD sensitivity** | |
| Authors and ref. | C.N. Ho, G.D. Christian, E.R. Davidson [7] |
| Comments | Valid for one calibrated constituent in the presence of unexpected constituents |
| Expression[a] | $\mathrm{SEN}_n = m_n\{[(\mathbf{B}^\mathrm{T}\mathbf{B})^{-1}]_{nn}\,[(\mathbf{C}^\mathrm{T}\mathbf{C})^{-1}]_{nn}\}^{-1/2}$ |
| **MKL sensitivity** | |
| Authors and ref. | N.J. Messick, J.H. Kalivas, P.M. Lang [8] |
| Comments | Valid in the absence of unexpected constituents |
| Expression[a] | $\mathrm{SEN}_n = m_n\left\{\left[\left(\mathbf{B}^\mathrm{T}\mathbf{B}\right) * \left(\mathbf{C}^\mathrm{T}\mathbf{C}\right)\right]^{-1}\right\}_{nn}^{-1/2}$ |
| **FO sensitivity** | |
| Authors and ref. | A.C. Olivieri, N.M. Faber [9] |
| Comments | Valid for any number of calibrated constituents in the presence of unexpected constituents |
| Expression[a] | $\mathrm{SEN}_n = m_n\left\{\left[\left(\mathbf{B}_\mathrm{exp}^\mathrm{T}\left(\mathbf{I} - \mathbf{B}_\mathrm{unx}\mathbf{B}_\mathrm{unx}^{+}\right)\mathbf{B}_\mathrm{exp}\right) * \left(\mathbf{C}_\mathrm{exp}^\mathrm{T}\left(\mathbf{I} - \mathbf{C}_\mathrm{unx}\mathbf{C}_\mathrm{unx}^{+}\right)\mathbf{C}_\mathrm{exp}\right)\right]^{-1}\right\}m_n^{-1/2}$ |

[a]*The symbol "*" indicates the Hadamard matrix product, and the subscript "nn" denotes the (n,n) diagonal element of the matrix. The parameter m is the total signal of the analyte considered at unit concentration, while the matrices **B** and **C** collect the loadings (profiles for the sample constituents in both data modes, normalized to unit length). The subscripts "exp" and "unx" for the FO expression refer to expected and unexpected, respectively.*

arguments for their derivation, the results are identical to those provided by the above NAS-inspired procedures.

Both HCD and MKL equations were condensed into the more general FO definition [9], conceived to take into account all possible calibration situations, including cases not covered by the former two expressions (see Table 1). The derivation required a complicated series of steps, which combined removal of other sample constituents, partly in matrix form and partly in unfolded form. However, the approach could not be straightforwardly extended to four-way (third-order) calibration, where it is apparent that even more alternative NAS definitions may exist. This situation prompted the finding of an alternative solution to the estimation of the multiway sensitivity.

# 4 SENSITIVITY EXPRESSIONS BASED ON UNCERTAINTY PROPAGATION

## 4.1 The General Sensitivity Expression

A different definition of sensitivity can be given in terms of uncertainty propagation: the sensitivity parameter $SEN_n$ is considered to measure the degree of output noise from a system for a given input noise. More sensitivity is achieved if low output noise is obtained for a given input noise. In this way, the $SEN_n$ parameter can be defined as the ratio of input to output noise:

$$SEN_n = \sigma_x / \sigma_y \tag{14}$$

where $\sigma_x$ and $\sigma_y$ are the uncertainties in signal and concentration, respectively. This uncertainty propagation approach assumes that the input noise is *iid* and employs a small, perturbing noise value to interrogate how the latter is propagated to prediction. However, it does not imply specific assumptions regarding the properties of the real experimental noise. When calibration is precise, the main source of uncertainty in the predicted concentration is the one stemming from the test sample signals, and the ratio of these uncertainties is a good measure of the $SEN_n$.

Equation (14) can be used to apply a Monte Carlo additive noise simulation for estimating sensitivities for any calibration model, whether univariate, multivariate, or multiway, as has been recently done [14–16]. This allowed to obtain operational values for the sensitivity in different calibration scenarios, although it does not provide a closed-form sensitivity equation. Equation (14) was also invoked to derive expression for estimating the sensitivity in most of the relevant multiway calibration models, including PARAFAC, MCR-ALS, and PLS/RML, with results which are (1) compatible with the second-order HCD, MKL, and FO (see Table 1 for application scenarios), (2) in agreement with Monte Carlo additive noise simulations, and (3) extendable to data with increasing number of ways. According to these results, it was possible to write an expression for casting all sensitivity equations into a single unified scheme, covering from zeroth-order (univariate calibration) to calibration models based on data of any order and ways [17]. The main result is appropriately condensed into the following expression:

$$SEN_n = \left\{ \mathbf{g}_n^T \left[ \mathbf{Z}_{exp}^T \left( \mathbf{I} - \mathbf{Z}_{unx} \mathbf{Z}_{unx}^+ \right) \mathbf{Z}_{exp} \right]^{-1} \mathbf{g}_n \right\}^{-1/2} \tag{15}$$

The different factors appearing in Equation (15) are summarized in Table 2 for the most commonly used models in each calibration order. Both the matrix $\mathbf{Z}_{exp}$ (the subscript "exp" stands for expected) and the analyte-specific vector $\mathbf{g}_n$ correspond to the calibration phase. The matrix $\mathbf{Z}_{exp}$ collects profiles (either in pure form or as linear combinations) for the expected constituents present in the calibration set, while $\mathbf{g}_n$ adequately selects or

**TABLE 2** General Sensitivity Formula and Corresponding Parameters for Different Data Orders and Algorithms

$$SEN_n = \left\{ \mathbf{g}_n^{\mathrm{T}} \left[ \mathbf{Z}_{\mathrm{exp}}^{\mathrm{T}} \left( \mathbf{I} - \mathbf{Z}_{\mathrm{unx}} \mathbf{Z}_{\mathrm{unx}}^{+} \right) \mathbf{Z}_{\mathrm{exp}} \right]^{-1} \mathbf{g}_n \right\}^{-1/2}$$

| | | *Analyte selector* | *Calibration components* | *Unexpected components* | | |
|---|---|---|---|---|---|---|
| *Model* | *Order* | $g_n$ | $Z_{exp}$ | $Z_{unx}$ | *Parameters description* | *References* |
| **Univariate calibration** | | | | | | |
| Univariate | 0 | 1 | $m_0$ | – | $m_0$ = slope of univariate graph | [4] |
| **Multivariate calibration** | | | | | | |
| CLS | 1 | $\boldsymbol{\delta}_n$ | $\mathbf{S}_{CLS}$ | – | $\boldsymbol{\delta}_n$: $N \times 1$ vector selecting the analyte of interest (all values of 0, except a 1 at analyte index)<br>$\mathbf{S}_{CLS}$: matrix of pure constituent profiles | [4] |
| ILS | | $\mathbf{y}_{cal,n}$ | $\mathbf{X}_{cal}$ | | $\mathbf{y}_{cal,n}$: vector of calibration analyte concentrations<br>$\mathbf{X}_{cal}$: matrix of calibration signals | |
| PCR | | $\mathbf{v}_{PCR,n}$ | $\mathbf{P}_{PCR}$ | | $\mathbf{v}_{PCR,n}$: vector of latent PLS coefficients<br>$\mathbf{P}_{PCR}$: matrix of PCR calibration loadings | |
| PLS | | $\mathbf{q}_{PLS,n}$ | $\mathbf{W}_{PLS}$ | | $\mathbf{q}_{PLS,n}: \left(\mathbf{P}_{PLS}^{\mathrm{T}} \mathbf{W}_{PLS}\right)^{-1} \mathbf{v}_{PLS,n}^{\mathrm{T}}$<br>$\mathbf{v}_{PLS,n}$: vector of latent PLS coefficients<br>$\mathbf{W}_{PLS}$: matrix of PLS calibration weights<br>$\mathbf{P}_{PLS}$: matrix of PLS calibration loadings | [4,11] |

| | | | | | | |
|---|---|---|---|---|---|---|
| **Multiway calibration** | | | | | | |
| MCR-ALS | 2 | $\boldsymbol{\delta}_n$ | $(m_n/J^{1/2})\mathbf{C}_{exp}$ | $\mathbf{C}_{unx}$ | $J$: no. of sensors of each submatrix in augmented mode<br>$m_n$: slope of pseudounivariate plot<br>$\mathbf{C}_{exp}$: profiles in nonaugmented mode for expected constituents<br>$\mathbf{C}_{unx}$: profiles in nonaugmented mode for unexpected constituents | [15,17] |
| PARAFAC | 2,3,… | $\boldsymbol{\delta}_n$ | See parameters description | Second-order: Equation (16); third-order: Equation (26); fourth-order: same pattern as Equations (16) and (26) adding the extra data mode | Second-order: $\mathbf{Z}_{exp} = m_n \mathbf{C}_{exp} \odot \mathbf{B}_{exp}$[a]<br>Third-order: $\mathbf{Z}_{exp} = m_n\, \mathbf{D}_{exp} \odot \mathbf{C}_{exp} \odot \mathbf{B}_{exp}$<br>Fourth-order: $\mathbf{Z}_{exp} = m_n\, \mathbf{E}_{exp} \odot \mathbf{D}_{exp} \odot \mathbf{C}_{exp} \odot \mathbf{B}_{exp}$ | [14,17] |
| U-PLS/RML | | $\mathbf{v}_{UPLS,n}$ | $\mathbf{P}_{UPLS}$ | | $\mathbf{P}_{UPLS}$: matrix of U-PLS calibration loadings<br>$\mathbf{v}_{UPLS,n}$: vector of latent PLS coefficients | [16,17] |

[a]*The symbol ⊙ refers to the Khatri–Rao product. For two matrices* **A** *and* **B***, the ith. column of* $\mathbf{A} \odot \mathbf{B}$ *follows from the ith. columns of* **A** *and* **B** *as* vec($\mathbf{b}_i^T \mathbf{a}_i$).

combines the latter information, making it specific for the $n$th analyte of interest. The final factor in Equation (15) is the matrix $\left(\mathbf{I}-\mathbf{Z}_{\text{unx}}\mathbf{Z}_{\text{unx}}^{+}\right)$, which is the mathematical manifestation of the second-order advantage, and thus, it only appears in higher-order (three-way and beyond) calibration methodologies (see Table 2). $\mathbf{Z}_{\text{unx}}$ is a block matrix built from the extracted profiles of the unexpected constituents, and for second-order data it can be expressed in a general way as:

$$\mathbf{Z}_{\text{unx}} = \left[\mathbf{c}_1 \otimes \mathbf{I}_{\text{b}} | \mathbf{I}_{\text{c}} \otimes \mathbf{b}_1 | \mathbf{c}_2 \otimes \mathbf{I}_{\text{b}} | \mathbf{I}_{\text{c}} \otimes \mathbf{b}_2 | \cdots\right] \tag{16}$$

The purpose of $\left(\mathbf{I}-\mathbf{Z}_{\text{unx}}\mathbf{Z}_{\text{unx}}^{+}\right)$ is to correct the matrix of profiles for the expected constituents ($\mathbf{Z}_{\text{exp}}$), for the overlapping effect of the profiles for the unexpected constituents (hence the subscript "unx"), or potential interfering agents. Specifically, the matrix $\left(\mathbf{Z}_{\text{unx}}\mathbf{Z}_{\text{unx}}^{+}\right)$ only appears when achieving the second-order advantage because only in this case is such information available. This is why in Table 2, only the entries corresponding to multiway algorithms are filled with respect to this part of Equation (16). The profiles for the unexpected constituents may be: (1) true constituent profiles (or approximations to them) provided, for example, by MCR-ALS, PARAFAC, and other all multilinear decompositions or (2) latent profiles (linear combinations or loadings), as retrieved by RML. What is relevant is that $\left(\mathbf{I}-\mathbf{Z}_{\text{unx}}\mathbf{Z}_{\text{unx}}^{+}\right)$ defines a projection orthogonal to the space spanned by the unexpected constituents, because $\mathbf{Z}_{\text{unx}}$ only contains information relative to the signals for the latter agents.

The fact that closed expressions for $\mathbf{Z}_{\text{exp}}$, $\mathbf{g}_n$, and $\left(\mathbf{I}-\mathbf{Z}_{\text{unx}}\mathbf{Z}_{\text{unx}}^{+}\right)$ can be written for all calibration methodologies (see Table 2) from zeroth- to any order indicates that Equation (15) is a completely general expression for estimating sensitivities. It is also worth noticing the properties of the multiway sensitivity defined by Equation (15): (1) it is analyte specific because the factor $\mathbf{g}_n$ depends on the analyte of interest; (2) it is sample specific because the composition of each test sample is unique as regards the unexpected constituents, generating a unique $\mathbf{Z}_{\text{unx}}$ matrix; and (3) it is algorithm specific because each data processing methodology provides a set of specific $\mathbf{Z}_{\text{exp}}$, $\mathbf{g}_n$, and $\left(\mathbf{I}-\mathbf{Z}_{\text{unx}}\mathbf{Z}_{\text{unx}}^{+}\right)$ factors.

## 4.2 Univariate Calibration

For the case of univariate calibration, the terms of the general Equation (15) are scalars, except $\mathbf{Z}_{\text{unx}}$, since no unexpected constituents are possible in this methodology and, therefore, does not defined: $\mathbf{g}_n=1$, and $\mathbf{Z}_{\text{exp}}=m_0$, leading to $\text{SEN}_n=m_0$, the slope of the calibration graph. This agrees with the IUPAC definition and also with the simple and intuitive uncertainty analysis of Equation (4): if the calibration were precise, uncertainties in $x$ will propagate to $y$ through $\boldsymbol{\sigma}_y=m_0^{-1}\boldsymbol{\sigma}_x$, and thus $\text{SEN}_n$ will be equal to $m_0$ (see Table 2).

## 4.3 First-Order Calibration

The definitions of $\mathbf{Z}_{\text{exp}}$ depend on the specific data processing algorithm used in first-order calibration, but since no unexpected constituents should appear in the test samples, $\mathbf{Z}_{\text{unx}}$ does not exist in none of the cases, and thus $\left(\mathbf{I}-\mathbf{Z}_{\text{unx}}\mathbf{Z}_{\text{unx}}^{+}\right)=\mathbf{I}$. Therefore, the general Equation (15) becomes Equation (9) for CLS and analogous expressions for ILS, PCR, and PLS [17], in full agreement with the NAS-based sensitivity approach.

Uncertainty propagation allows one to achieve the same results, directly from the general predictive equation for analyte $n$:

$$y_n=\boldsymbol{\beta}_n^{\text{T}}\mathbf{x} \tag{17}$$

where $\boldsymbol{\beta}_n$ is the vector of regression coefficient for any first-order methodology. If only $\mathbf{x}$ carries uncertainty, it follows that the uncertainty in concentration is given by:

$$\sigma_y=\left(\boldsymbol{\beta}_n^{\text{T}}\boldsymbol{\beta}_n\right)^{1/2}\sigma_x \tag{18}$$

From Equation (18), $\text{SEN}_n=\left(\boldsymbol{\beta}_n^{\text{T}}\boldsymbol{\beta}_n\right)^{-1/2}$ immediately follows through the uncertainty propagation approach. This sensitivity parameter is analyte specific but does not depend on the composition of the test sample, because the vector of regression coefficients stems from the processing of the calibration data only.

## 4.4 Multiway (Higher-Order) Calibration

### 4.4.1 Multilinear Algorithms

Multilinear algorithms such as PARAFAC [27] and its variants based on the multilinear model [47–49] provide approximations to pure constituent profiles, whether they belong to the category of expected (calibrated) or unexpected (potentially interferent), thus achieving the second-order advantage. Each constituent is characterized by instrumental profiles describing their behavior in the different data modes. In the usual setting, these profile vectors are normalized to unit length, and thus, the scaling factor with respect to analyte concentration is left to the slope ($m_n$) of the pseudounivariate prediction graph (the latter is a plot of the scores or relative concentrations of a given analyte vs. its nominal calibration concentrations). As a function of the relevant parameters for multilinear multiway calibration, the recently derived expression for the sensitivity in multilinear models is [50]

$$\text{SEN}_n=m_n\left\|n\text{th row of }\left[\mathbf{Z}_{\text{exp}}^{\text{T}}\left(\mathbf{I}-\mathbf{Z}_{\text{unx}}\mathbf{Z}_{\text{unx}}^{+}\right)\mathbf{Z}_{\text{exp}}\right]^{+}\right\|^{-1} \tag{19}$$

The matrix $\mathbf{Z}_{\text{exp}}$ is defined as a function of a loading matrices for second- (two data modes), third- (three data modes), and fourth-order (four data modes), while the $\mathbf{Z}_{unx}$ is defined as a function of the loading profiles of the unexpected constituents in the various data modes [17]. It may be noticed that for three-way (second-order) calibration, Equations (15) and (19) appear to be different than the MKL, HCD, and FO expressions (see Table 1); however, the latter numerical results are identical to those provided by Equation (15), indicating that all previous approximations based on the NAS are special cases of the general uncertainty propagation expression [14].

### 4.4.2 Multivariate Curve Resolution-Alternating Least-Squares

The corresponding $\mathrm{SEN}_n$ expression for the MCR-ALS algorithm applied in the extended mode has been recently derived [15].

$$\mathrm{SEN}_n = m_n \left[ J \left( \mathbf{C}^{\mathrm{T}} \mathbf{C} \right)^{-1}_{nn} \right]^{-1/2} \tag{20}$$

In Equation (20), $J$ is the number of data points in each submatrix in the augmented mode, and $m_n$ is the slope of the MCR-ALS pseudounivariate graph (built in a similar manner to PARAFAC, i.e., plotting analyte scores vs. nominal calibration concentrations). Assuming successful decomposition of the augmented matrix $\mathbf{X}_{aug}$ into two matrices ($\mathbf{B}_{aug}$ and $\mathbf{C}$), containing the constituent profiles in the augmented mode and in the nonaugmented mode, respectively, the sensitivity depends on the nonaugmented profiles $\mathbf{C}$, which can be further separated into $\mathbf{C}_{exp}$ and $\mathbf{C}_{unx}$, containing the profiles for the expected (present in calibration) and unexpected constituents, respectively.

The MCR-ALS sensitivity expression can also be shown to be adequately covered by the general Equation (15). In Equation (20), $J$ is the number of data points in each submatrix in the augmented mode. Since each data matrix is assumed to be of size $J \times K$, this also assumes that augmentation has been performed columnwise. In the case of row-wise augmentation, $J$ should be replaced by $K$ in Equation (20). On the other hand, the matrix $\mathbf{C}$ contains the profiles for all sample constituents in the nonaugmented data mode, and the shorthand notation $\left(\mathbf{C}^{\mathrm{T}}\mathbf{C}\right)^{-1}_{nn}$ implies selecting the $(n,n)$ diagonal element of the inverse of matrix ($\mathbf{C}^{\mathrm{T}}\mathbf{C}$). To adapt Equation (20) to the present approach, the matrix $\mathbf{C}$ is divided into two blocks, one for the constituents present in calibration ($\mathbf{C}_{exp}$) and another one for the unexpected constituents ($\mathbf{C}_{unx}$):

$$\mathbf{C} = \left[ \mathbf{C}_{exp} | \mathbf{C}_{unx} \right] \tag{21}$$

It can further be shown that:

$$\begin{aligned} \left(\mathbf{C}^{\mathrm{T}}\mathbf{C}\right)^{-1} &= \left( \left[ \mathbf{C}_{exp} | \mathbf{C}_{unx} \right]^{\mathrm{T}} \left[ \mathbf{C}_{exp} | \mathbf{C}_{unx} \right] \right)^{-1} \\ &= \left[ \mathbf{C}_{exp}^{\mathrm{T}} \left( \mathbf{I} - \mathbf{C}_{unx} \mathbf{C}_{unx}^{+} \right) \mathbf{C}_{exp} \right]^{-1} \end{aligned} \tag{22}$$

Then, the $(n,n)$ diagonal element of the latter matrix can be found using the vector $\boldsymbol{\delta}_n$ as selector:

$$\left(\mathbf{C}^{\mathrm{T}}\mathbf{C}\right)^{-1}_{nn} = \boldsymbol{\delta}_n^{\mathrm{T}}\left[\mathbf{C}_{\mathrm{exp}}^{\mathrm{T}}\left(\mathbf{I} - \mathbf{C}_{\mathrm{unx}}\mathbf{C}_{\mathrm{unx}}^{+}\right)\mathbf{C}_{\mathrm{exp}}\right]^{-1}\boldsymbol{\delta}_n \tag{23}$$

and the sensitivity is given by the equation:

$$\mathrm{SEN}_n = m_n J^{-1/2}\left\{\boldsymbol{\delta}_n^{\mathrm{T}}\left[\mathbf{C}_{\mathrm{exp}}^{\mathrm{T}}\left(\mathbf{I} - \mathbf{C}_{\mathrm{unx}}\mathbf{C}_{\mathrm{unx}}^{+}\right)\mathbf{C}_{\mathrm{exp}}\right]^{-1}\boldsymbol{\delta}_n\right\}^{-1/2} \tag{24}$$

Finally, $\mathbf{Z}_{\mathrm{exp}}$ and $\mathbf{Z}_{\mathrm{unx}}$ from Equation (15) are equal to $\mathbf{C}_{\mathrm{exp}}$ and $\mathbf{C}_{\mathrm{unx}}$, respectively, and the vector $\mathbf{g}_n$ is equal to the multilinear selector $\boldsymbol{\delta}_n$, thereby achieving an expression equivalent to general Equation (15).

### 4.4.3 *Partial Least-Squares/Residual Multilinearization*

For multiway algorithms with a latent-based calibration, such as PLS/RML, the corresponding sensitivity expression has already been developed in the same format as the general equation (15). For U-PLS calibration, for example, $\mathbf{Z}_{\mathrm{exp}}$ is composed of columns which are the calibration loadings, which is understandable, since they represent the behavior of the calibrated constituents in signal space. Here, the vector $\mathbf{g}_n$ does not act as selector of a particular analyte loading, but appropriately combines the loadings in a manner, which specifically reflects the behavior of the analyte of interest. It is equal to the vector of analyte-specific regression coefficients, defined in the space of the latent variables. The above discussion concerning the properties of the $\left(\mathbf{I} - \mathbf{Z}_{\mathrm{unx}}\mathbf{Z}_{\mathrm{unx}}^{+}\right)$ matrix is also pertinent in this case. The uncertainty propagation approach fully agrees with the expression for the U-PLS/residual bilinearization (RBL) sensitivity, which was previously derived from NAS considerations [51]. An analogous expression can be derived for N-PLS/RBL [16].

## 4.5 Other Multiway Algorithms

The general Equation (15) has been applied to assess the sensitivity for several algorithms commonly employed for multiway calibration. However, there are additional methodologies, based on eigenvector–eigenvalue analysis [52,53], which are somewhat less employed. In addition, the latter ones always achieve the lowest HCD sensitivity, even when various constituents are calibrated [51], probably due to the very limited information provided to the model for the single calibration sample, in contrast to methodologies relying on multiple calibration samples.

Other algorithms for which sensitivity studies are lacking are multilinear least-squares/RML (MLLS/RML) [32,34,35,52,53] and PARAFAC2, a variant of PARAFAC conceived to cope with nonmultilinear multiway data with one trilinearity breaking mode, e.g., chromatographic-spectral second-order data, whose sensitivity properties have yet to be explored. Recently, a new version of PARAFAC (Augmented PARAFAC), specifically designed for four-way nonmultilinear data, was used to resolve different third-order chromatographic systems [54]. In the latter work, sensitivities were calculated using expressions based on the multilinear approach, i.e., as if the system were processable by quadrilinear four-way PARAFAC [14]. Thus, the results should in principle be rather overoptimistic. However, it is now possible to postulate an expression for sensitivity that fits better to the three-way augmented PARAFAC model: as for MCR-ALS, Equation (15) becomes similar to Equation (23), but in this case, it is possible to use the information of the two pure profiles **B** and **C** (see FO sensitivity in Table 1):

$$\mathrm{SEN}_n = m_n J^{-1/2}\left\{\boldsymbol{\delta}_n^{\mathrm{T}}\left[\left(\mathbf{B}_{\mathrm{exp}}^{\mathrm{T}}\left(\mathbf{I}-\mathbf{B}_{\mathrm{unx}}\mathbf{B}_{\mathrm{unx}}^{+}\right)\mathbf{B}_{\mathrm{exp}}\right)\left(\mathbf{C}_{\mathrm{exp}}^{\mathrm{T}}\left(\mathbf{I}-\mathbf{C}_{\mathrm{unx}}\mathbf{C}_{\mathrm{unx}}^{+}\right)\mathbf{C}_{\mathrm{exp}}\right)\right]^{-1}\boldsymbol{\delta}_n\right\}^{-1/2} \quad (25)$$

In Equation (25), all symbols have the same meaning that the MCR-ALS context, but the matrices **B** and **C** contain the profiles for all sample constituents in the two nonaugmented data modes, an option which is not possible in the expression of MCR-ALS sensitivity. Figure 3 shows the sensitivities for simulated four-way data ternary systems, consisting in two analytes and one interferent, estimated through Monte Carlo/augmented PARAFAC and Monte Carlo/MCR-ALS calculations, using Equation (24) for MCR-ALS and Equation (25) for Augmented PARAFAC. Interestingly, the sensitivity expression for augmented PARAFAC leads to better results than for MCR-ALS, meaning that Equation (25) is well suited for the proposed objective.

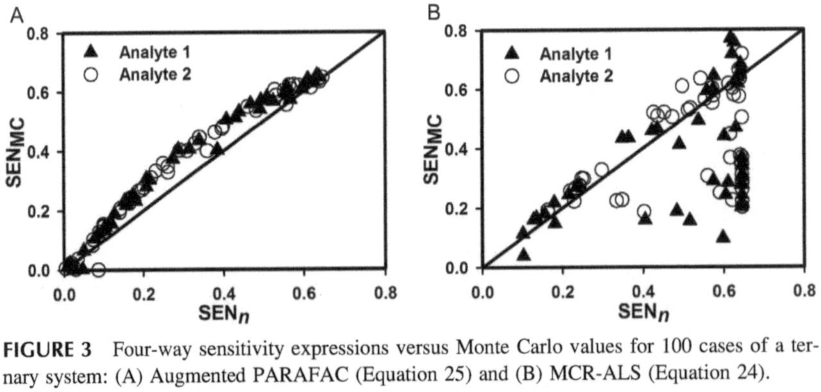


**FIGURE 3** Four-way sensitivity expressions versus Monte Carlo values for 100 cases of a ternary system: (A) Augmented PARAFAC (Equation 25) and (B) MCR-ALS (Equation 24).

### 4.6 Further Multiway Systems

Generalization of $\mathbf{Z}_{\text{unx}}$ in Equation (15) to further data modes is possible by noting, for example, that in four-way (third-order) calibration, the spaces spanned by $\mathbf{Z}_{\text{unx}}$ are the three possible combinations of pairs of modes: for a given interfering constituent, $\mathbf{Z}_{\text{unx}}$ contains blocks of columns for each unexpected constituent, e.g., for the unexpected agent 1, the first block will look as follows [17]:

$$\mathbf{Z}_{\text{unx}} = [\mathbf{d}_1 \otimes \mathbf{c}_1 \otimes \mathbf{I}_{\text{b}} | \mathbf{d}_1 \otimes \mathbf{I}_{\text{c}} \otimes \mathbf{b}_1 | \mathbf{I}_{\text{d}} \otimes \mathbf{c}_1 \otimes \mathbf{b}_1 | \cdots] \tag{26}$$

where $\mathbf{I}_{\text{b}}$, $\mathbf{I}_{\text{c}}$, and $\mathbf{I}_{\text{d}}$ are $J \times J$, $K \times K$, and $L \times L$ identity matrices. This $\mathbf{Z}_{\text{unx}}$ matrix is easily constructed for any number of unexpected constituents.

If a data mode is added (five-way calibration), four different combinations of triads of profiles in each of the possible sets of three modes will provide a $\mathbf{Z}_{\text{unx}}$ matrix for each unexpected agent. For more data modes, specifically, for $(N+1)$-way ($N$th-order) calibration, the blocks of $\mathbf{Z}_{\text{unx}}$ for each unexpected agent will include all possible combination of profiles in $(N-1)$ modes. In sum, $\mathbf{Z}_{\text{unx}}$ conceivably represents the space spanned by the unexpected constituents, but in a nonclassical way, although the systematic block characteristics of this matrix makes it easy to build it for any data order and number of interfering agents.

## 5 OTHER FIGURES OF MERIT

### 5.1 Analytical Sensitivity

One potential problem with the interpretation of the plain sensitivity is that it depends on the specific type of signal employed for developing a calibration methodology. The value of $\text{SEN}_n$ has units of (signal $\times$ concentration$^{-1}$), and therefore, sensitivities derived from spectral and electrochemical measurements cannot be compared on an equal basis. For these reasons, the analytical sensitivity ($\gamma$) has been proposed as a better indicator for comparison purposes, as the ratio between sensitivity and instrumental noise [55]:

$$\gamma_n = \text{SEN}_n / \sigma_x \tag{27}$$

The parameter $\gamma_n$ has units of (concentration)$^{-1}$, is independent of the measured signal, and can be employed to compare different methodologies. Comparison of Equations (14) and (27) implies that $\gamma_n = \sigma_y^{-1}$, and thus the analytical sensitivity has been interpreted as the inverse of the minimum concentration difference which can be appreciated across the linear analytical range, although this appears to be a rather qualitative statement, less rigorous than the detection capabilities to be described below. In any case, having estimated the sensitivity, a measure of the instrumental noise level allows one to compute the analytical sensitivity through Equation (27).

## 5.2 Selectivity

According to IUPAC, selectivity is the extent to which a method can be used to determine particular analytes in mixtures or matrices without interferences from other constituents of similar behavior [56]. However, this qualitative definition does not imply a specific procedure for the estimation of a numerical selectivity parameter [50].

Several requirements have been proposed for a consistent numerical selectivity [50]: (1) a change in the calibration data should be reflected in changes in selectivity, (2) changes in individual analyte selectivities should produce corresponding changes in the selectivity and the amount of these changes should be comparable in size, (3) values such as infinity should not be obtained, (4) a relation between selectivity and prediction uncertainty is desirable, (5) numerical results should be possible for overdetermined systems (having more sensors or wavelengths than components), and (6) generalization to multiway data should be straightforward.

The simplest way in which a selectivity parameter can be defined for most calibration scenarios, complying with the above requirements, is as the dimensionless ratio between two analyte sensitivity values: the sensitivity in a mixture and the sensitivity when all other sample constituents are absent [57].

$$\mathrm{SEL}_n = \mathrm{SEN}_n(\text{in a mixture})/\mathrm{SEN}_n(\text{pure}) \tag{28}$$

In univariate calibration, SEL should be equal to 1 (100%, meaning full selectivity), because no interfering agents are allowed. In first-order CLS calibration, Equation (28) naturally follows as Equation (28) by setting the denominator as a measure of the pure analyte signal [50]. However, for latent variables-based calibration models, no approximations to pure analyte profiles are available, and hence, the selectivity cannot be precisely defined. Although there have been proposals to use the total signal for a given test sample as denominator in Equation (28) in these cases, i.e., this makes the selectivity highly dependent on the unknown samples. Consequently, it may only be sensible to define the selectivity when the pure analyte signal is either adequately retrieved by the processing algorithm or known from separate experiments.

For the multiway analysis world, the defining Equation (28) implies that the multiway selectivity is accessible when the pure analyte signal is adequately retrieved by the processing algorithm [8,11,58], as in the case of multilinear decomposition analysis, for which the selectivity ($\mathrm{SEL}_n$) is directly given by:

$$\mathrm{SEL}_n = \mathrm{SEN}_n/m_n \tag{29}$$

where $m_n$ is the slope of the pseudounivariate calibration graph. The degree by which $\mathrm{SEN}_n$ departs from $m_n$ in Equation (29) is adequately measured by the level of overlapping among the profiles for the various constituents. Since

$\text{SEN}_n < m_n$, the value of $\text{SEL}_n$ continuously varies between 0 (null selectivity) and 1 (100%, full selectivity).

In the case of MCR-ALS, the selectivity is [15] as follows:

$$\text{SEL}_n = \text{SEN}_n J^{1/2}/m_n \tag{30}$$

As was the case with Equation (29), Equation (30) leads to continuous values in the range 0–1, depending on the relative degree of overlapping among the profile for the various sample constituents.

The latter expressions are not the only ones employed for the selectivity calculation in multiway calibration [59–61], which shows that there is some controversy about the numerical concept of selectivity.

## 5.3 Prediction Uncertainty

Prediction uncertainty is based on two proposals for estimating standard errors in multivariate/multiway analysis [62]: (1) resampling techniques such as jack-knife or bootstrap [63] and (2) error propagation, which is preferable because it leads to closed-form expressions and permits better insight into the relative impact of various uncertainty sources on the prediction error [57,64].

The best approximation to concentration variance is the well-known three-term expression (valid for propagation of homoscedastic and uncorrelated noise) [57,64,65]:

$$\sigma_y^2 = \text{SEN}_n^{-2}\sigma_x^2 + h\text{SEN}_n^{-2}\sigma_x^2 + h\sigma_{y_{\text{cal}}}^2 \tag{31}$$

where $\sigma_x^2$ is the variance in instrumental signals, $h$ is the sample leverage, and $\sigma_{y_{\text{cal}}}^2$ is the variance in calibration concentrations. The three terms in the right-hand side of Equation (31) derive, by uncertainty propagation, from the following three sources: (1) instrumental signals for the test sample, (2) instrumental signals for the calibration samples, and (3) calibration concentrations. The first and probably the most relevant of these contributions is directly propagated and is inversely proportional to the squared sensitivity. The second and third terms stem from calibration uncertainties and are scaled by the sample leverage $h$, a dimensionless parameter placing the sample relative to the calibration space. A simple expression exists for $h$ in univariate calibration and in pseudounivariate multiway calibration, whereas in other situations the value of $h$ depends on the presence and levels of additional sample constituents. A general equation is able to appropriately cover all cases, however:

$$h = \mathbf{f}_{\text{test}}^{\text{T}}\left(\mathbf{F}_{\text{cal}}^{\text{T}}\mathbf{F}_{\text{cal}}\right)^{-1}\mathbf{f}_{\text{test}} \tag{32}$$

where $\mathbf{F}_{\text{cal}}$ is a matrix (or vector) corresponding to the calibration set of samples, and $\mathbf{f}_{\text{test}}$ is a vector for the test sample. It is important to notice that $1/I_{\text{cal}}$

($I_{cal}$ is the number calibration samples) should be added to Equation (32) for mean-centering data, and when the model includes an intercept, as in univariate calibration through Equation (4), where $h$ becomes the familiar expression:

$$h = \frac{1}{I_{cal}} + \frac{(y - \bar{y}_{cal})^2}{\sum_{i=1}^{I_{cal}} (y - \bar{y}_{cal})^2} \tag{33}$$

where $y$ is the predicted analyte concentration, $y_i$ is its nominal concentration in the $i$th calibration sample, and $\bar{y}_{cal}$ is the mean calibration concentration. It may be noticed that Equation (31) is accurate for the univariate case [4] and for CLS first-order calibration [66]. For the remaining calibrations, the first term of Equation (31) is accurate [64], while the remaining two terms are appropriate approximations [9,16].

## 5.4 Detection Capabilities

The modern definition of the LOD is due to Currie's pioneering work on hypothesis-based detection limit theory [67] and is an important figure of merit to be reported because it defines the minimum analyte concentration that is detectable with a certain degree of confidence. However, the definition of the LOD officially recommended by IUPAC is somewhat less simpler than the latter idea: it first requires one to define a critical concentration level (CL), which is the level for the detection decision, involving a certain risk of false detects. The LOD is then defined as a CL for which the risk of false nondetects has a probability $\beta$ [68–71]. Following the above cited works, and according to the expressions presented herein, an LOD expression can be achieved, as proportional to the uncertainty in predicted concentration near a blank sample [17,72,73]:

$$\text{LOD}_n = 3.3\left(\text{SEN}_n^{-2}\sigma_x^2 + h_0\,\text{SEN}_n^{-2}\sigma_x^2 + h_0\sigma_{y_{cal}}^2\right) \tag{34}$$

where the subscript $n$ identifies a particular analyte of interest, $h_0$ is the leverage for the blank sample, and the factor 3.3 is the sum of $t$-coefficients accounting for types I and II errors at 95% confidence level [57]. The factor in front of Equation (34) may be corrected for other probabilities and degrees of freedom. Notice the assumptions underlying Equation (34): (1) the $\text{LOD}_n$ is close enough to the blank so that the leverage at the LOD level is equal to the blank leverage $h_0$, otherwise, complex corrections are required [74] and (2) the distance from the blank to the LOD is given as a sum of two confidence intervals; a more rigorous treatment suggests the use of a noncentrality parameter of a noncentral $t$ distribution instead of a sum of classical $t$-coefficients [71]. It is likely, however, that the values provided by Equation (34) and more elaborate statistical approaches do not significantly differ [75,76].

In univariate calibration, the subscript $n$ may be dropped and the LOD characterizes the detection capability toward the analyte under study. However, for first-order methodologies, $\text{SEN}_n$ is analyte specific, as explained above, but the leverage $h_0$ is also sample specific, meaning that different blank samples (samples where the analyte is absent, but contain varying proportions of the remaining constituents) have different associated values of $h_0$. Therefore, the $\text{LOD}_n$ not only becomes analyte specific but also sample specific. Even more, in higher-order calibration, as already discussed, the value of $\text{SEN}_n$ is analyte, sample, and algorithm specific (incidentally, the leverage $h_0$ is not sample specific when pseudounivariate calibration is employed) [10]. For this reasons, the detection capability toward a given analyte depends on various factors beyond the instrumental signals measured for a set of calibration samples. Following this concept, and in order to overcome the sample dependency issue, a recent approach has been proposed based on the interpretation of the LOD as an interval [25]. Although this proposal has been specifically developed for first-order PLS calibration, it could be potentially extended to other algorithms as long as they are consistent with Equation (34). The main ideas underlying this definition are (1) each test sample has a specific LOD value, (2) the universe of test samples is represented by the calibration set of samples, (3) the leverages for the calibration samples can be extrapolated to zero analyte concentration, and (4) a range of LOD values can be estimated for the PLS model as a whole.

As shown in Figure 4, the lower and upper limits of the LOD interval correspond to the calibration samples with the lowest and largest extrapolated leverages to zero analyte concentration and can be calculated using the following expressions [25]:

$$\text{LOD}_{\text{min}} = 3.3\left[\text{SEN}^{-2}\text{var}(x) + h_{0\text{min}}\text{SEN}^{-2}\text{var}(x) + h_{0\text{min}}\text{var}(y_{\text{cal}})\right]^{1/2} \quad (35)$$

$$\text{LOD}_{\text{max}} = 3.3\left[\text{SEN}^{-2}\text{var}(x) + h_{0\text{max}}\text{SEN}^{-2}\text{var}(x) + h_{0\text{max}}\text{var}(y_{\text{cal}})\right]^{1/2} \quad (36)$$

where

$$h_{0\text{min}} = \frac{\bar{y}_{\text{cal}}^2}{\sum_{i=1}^{I} y_i^2} \quad (37)$$

$$h_{0\text{max}} = \max\left(h_{0\text{cal}}\right) \quad (38)$$

It is interesting to note that the leverage in Equation (34) corresponds to the value obtained in univariate calibration with a given calibration set, provided other sample components are absent. In Equation (35), on the other hand, $h_{0\text{cal}}$ refers to the leverages for the projections of all calibration samples onto $H_0$, which is the zero concentration hyperplane in a score space having a

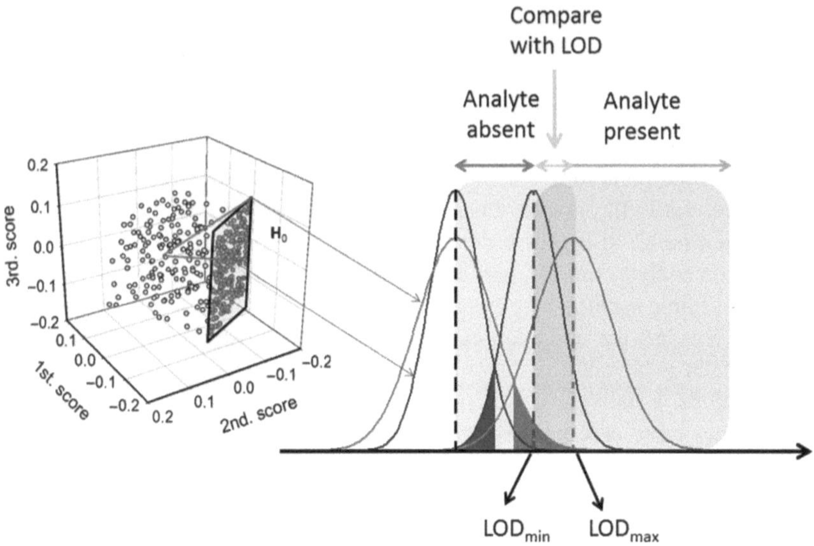


**FIGURE 4** Schematic representation of the limit of detection (LOD) interval approach to calculate the LOD in PLS calibration. Red (gray in the print version) solid lines, leverage and corresponding standard deviation used to calculate $LOD_{min}$; blue (dark gray in the print version) solid lines, leverage and corresponding standard deviation used to calculate $LOD_{max}$. *Adapted with permission from Ref. [25]. Copyright 2014 American Chemical Society.*

number of dimensions equal to the number of calibration latent variables (Figure 4), and $h_{0cal}$ can be calculated as:

$$h_{0cal} = h_{cal} + h_{0min}\left[1 - \left(\frac{y_{cal}}{\bar{y}_{cal}}\right)^2\right] \tag{39}$$

In the concentration range $LOD_{min} < y < LOD_{max}$, a reasonable way to decide if the analyte is detected or not is to estimate a specific LOD value for the corresponding test sample, approximating its real leverage $h$ to the leverage $h_0$, which would correspond to its background components (i.e., in the absence of analyte). This is equivalent to taking the sample as if it were a blank, which is conceivable because its analyte concentration is most probably very low. The obtained LOD value can then be employed to check whether the predicted concentration is below (analyte absent) or above (analyte present) the sample-specific LOD.

Another common approach to calculate limits of detection in multivariate calibration bears some similarities with the univariate approach, and this is why it is sometimes called "pseudounivariate LOD" [77]. In this strategy, the analyte concentrations estimated for the calibration set of samples by the PLS model are plotted against their nominal or reference concentrations. The result is a pseudounivariate calibration graph in which the vertical scale is the estimated analyte concentration instead of either instrumental or latent variables. The graph is processed as in univariate calibration, assuming that the detection limit is insensitive to linear transformations applied to the signal. This leads to an $LOD_{pu}$ value, estimated from the classical univariate Equation (4):

$$LOD_{pu} = 3.3 s_{pu}^{-1} \left[ (1 + h_{0min} + 1/I) \, var_{pu} \right]^{1/2} \tag{40}$$

where $s_{pu}$ is the slope of the pseudounivariate line and $var_{pu}$ is the variance of the regression residuals.

The parameter $LOD_{pu}$ has the advantage of being a single figure of merit characterizing the overall PLS calibration model. However, one critical point about this estimator is the absence of a term accounting for calibration concentration uncertainties. This difference shows up if $LOD_{pu}$ and $LOD_{min}$/$LOD_{max}$ are compared by Monte Carlo noise addition studies, which indicate that the $LOD_{pu}$ distribution is centered at the lower limit $LOD_{min}$ of the presently proposed LOD interval, provided the noise in calibration concentrations is negligible compared to the level of noise in instrumental signals. In contrast, when concentration uncertainties compete with the instrumental noise in relative size, the mutual relationship among $LOD_{pu}$, $LOD_{min}$, and $LOD_{max}$ is less clear.

This can be explained on the basis of how the errors in calibration concentrations var($y_{cal}$) are incorporated into the LOD definitions. In the estimation of both $LOD_{min}$ and $LOD_{max}$, the latter contribution is scaled by the leverage, but in $LOD_{pu}$, it is directly incorporated into the first, test-sample-dependent term of the LOD expression. In the latter case, the "signal" is replaced by the estimated concentration. This means that although the term that takes into account the errors in calibration concentrations is not present in an explicit way, concentrations errors are directly propagated to the standard error in predicted concentrations. In any case, it is important to remark that the conceptual approach to $LOD_{pu}$ is radically different than the presently proposed range of LOD values, which is clearly consistent with the latest advances in error-in-variables theory [11], and leads to a better insight into PLS detection capabilities, because it helps to understand the effect of background and potential interfering agents on the analyte detection for complex samples [25].

The limit of quantitation ($LOQ_n$), in turn, is estimated as the CL for which the relative prediction error is 10% and is easily set at a concentration value which is 10 times the associated prediction uncertainty [57]:

$$LOQ_n = 10 \left( SEN_n^{-2} \sigma_x^2 + h_0 \, SEN_n^{-2} \sigma_x^2 + h_0 \sigma_{y_{cal}}^2 \right) \tag{41}$$

Analogous considerations to those for $LOD_n$ regarding the analyte and sample dependence of the $LOQ_n$ apply.

## 6 COMPARISON OF FIGURES OF MERIT

The measuring and processing of multivariate data leads to a sensitivity increase, derived from multiple redundant measurements and noise averaging. The sensitivity increase can now be precisely computed using the general expression (15). This may help in advanced planning and in anticipating the sensitivity gain for complex multiway experiments. To illustrate the sensitivity gain when increasing the number of instrumental sensors and data orders, a real example will be presented, which shows the usefulness of the general expression to calculate sensitivity and the resulting figures of merit. In this experimental fourth-order data set, excitation–emission fluorescence matrices were measured as a function of time and pH to quantify the amount of the fluorescent pesticide carbaryl, which hydrolyzes in alkaline media to fluorescent 1-naphthol. The calibration samples only contain the analyte carbaryl, but the test samples contain, in addition to the analyte, another fluorescent pesticide as an interfering agent (thiabendazole or fuberidazole). Hence, the second-order advantage is required for successful analyte determination [16].

Given the features of the data, it is possible to explore all the different possibilities provided by second-, third-, and fourth-order calibration. To do this, the corresponding subarrays can be generated by fixing the time, the pH (third-order data), or both pH and time simultaneously (second-order data). When the second-order data were analyzed using U-PLS/RBL, the use of cross-validation rendered an optimum number of calibration latent variables equal to 1. It is important to notice that two chemical components occur in calibration. However, since they are mutually correlated because one component is hydrolyzed to yield the second one, a single U-PLS latent variable is understandable. The same situation stands for higher orders. As can be seen in Figure 5 for two typical test samples with different interferents, the RBL procedure was able to return the corresponding profiles, achieving the second-order advantage. The figures of merit are shown in Table 3 and will be compared below with those corresponding to third- and fourth-order data analysis. When third-order data are generated by fixing the pH value to 10, residual trilinearization (RTL), as in the case of RBL, allowed to model the corresponding interfering agent in each sample. The complete fourth-order data were finally submitted to U-PLS/RQL, with similar qualitative results in comparison with above analysis, but with an additional profile in the pH mode for the interfering agent profiles.

The figures of merit for the analyte carbaryl are reported in Table 3 for comparison with previous methodologies. In comparing the results for second-, third-, and fourth-order data for the studied experimental system, increasing sensitivities and analytical sensitivities are apparent on increasing the data order (Figure 6). A steady improvement in the average concentration error indicators is also observed, as well as in uncertainty in predicted concentrations and detecting capabilities (LOD and LOQ). However, in agreement

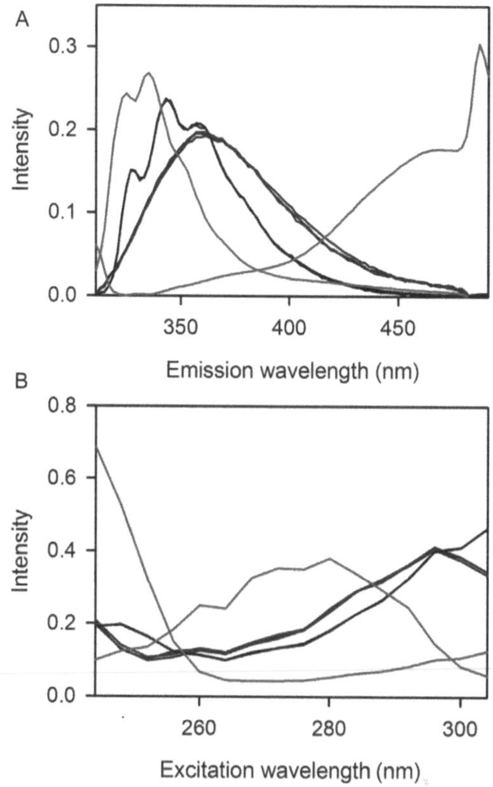


**FIGURE 5** Excitation (A) and emission (B) profiles for the various components of a typical experimental example. Green (gray in the print version) and red (gray in the print version) lines correspond to the experimental spectra for the analyte carbaryl and its hydrolysis product 1-naphthol, respectively. Blue (black in the print version) and black lines (three superimposed similar traces) indicate the profiles for the interfering agents fuberidazole and thiabendazole, as retrieved from test samples containing interferents, by U-PLS/RML analysis of second-, third-, and fourth-order data. *Reprinted with permission from Ref. [16]. Copyright 2012 American Chemical Society.*

with the simulations presented in the previous section, the improvement in concentration uncertainty, LOD, and LOQ is not directly proportional to the gain in sensitivity. Again, this is expected on inspection of Equation (31), where only two of the three terms are directly affected by the sensitivity parameter.

From this example can be clearly seen that, with the expressions now available, it is possible to obtain a reliable quantitative measure of the improvement that could result from the addition of an extra data mode. This gives the analyst a powerful criterion to decide whether it is necessary or not to add extra information to a particular analysis.

**TABLE 3** Figures of Merit for the Experimental Example Using U-PLS/RBL, U-PLS/RTL, and U-PLS/RQL

| Figures of merit[a] | | | |
|---|---|---|---|
| SEN/AFU (L μg$^{-1}$) | 1.3 | 5.5 | 12 |
| γ (L μg$^{-1}$) | 0.7 | 3.1 | 6.7 |
| LOD (μg L$^{-1}$) | 5.3 | 2 | 1.5 |
| LOQ (μg L$^{-1}$) | 16 | 6 | 4.5 |

*[a] $[\mathrm{var}(y_{cal})]^{1/2} = 1$ μg L$^{-1}$; $[\mathrm{var}(x)]^{1/2} = 2$ AFU (arbitrary fluorescence units).*

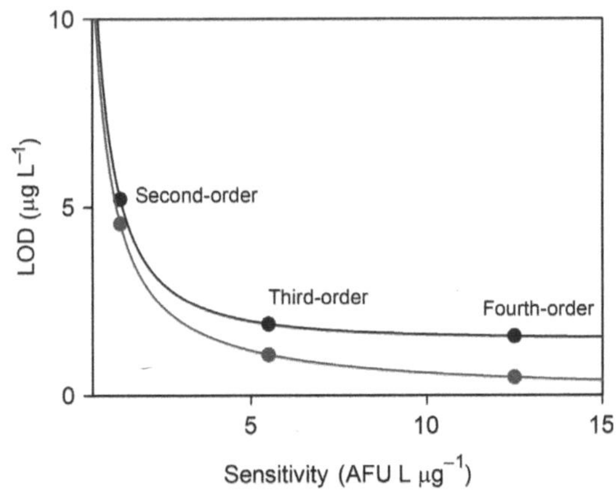


**FIGURE 6** Variation of the limit of detection (LOD) with the sensitivity for the determination of the analyte carbaryl in the experimental example. Red (gray in the print version) circles, LOD values as obtained with U-PLS/RML for the different experimental data orders using as an approximation only the first term of Equation (34). Blue (black in the print version) circles, LOD values from the complete IUPAC's recommended expression (Equation 34), inserting the corresponding values of *h*. *Reprinted with permission from Ref. [16]. Copyright 2012 American Chemical Society.*

## 7 CONCLUSIONS

In this chapter, it has been clearly shown that the task of calculating figures of merit in multivariate and multiway calibration is not as simple as an intuitive and direct extension of univariate expressions and requires special attention. An insight into the data structure clearly shows that it becomes necessary to study and rationalize the definition of estimators, depending not only on the sample being analyzed but also on the structure of the algorithm used to process the corresponding data. In this sense, the derivation of a general formula to calculate sensitivity, applicable to any data order and for the most important processing tools, constitutes an important step to a better understanding of the information needed to develop a reliable validation methodology.

The definition of the LOD is also an important issue, since this figure of merit brings together two important analytical concepts: the sensitivity and the precision in the analytical determination. Regarding this figure of merit, some efforts to define a reliable estimator in first-order calibration have been presented, but further studies should be made to extend it to more complex multiway data.

Finally, one of the main perspectives of this chapter is related to uncertainty estimation. The extension of the proposed expressions to cases in which the error structure is not *iid* is a topic of fundamental importance, since this latter assumption is not always completely adequate. This implies a deep insight into the different sources of instrumental noise that could affect a measurement, and how the noise is propagated through the structure of each calibration algorithm to give a reliable uncertainty value in the estimated concentration.

## ACKNOWLEDGMENTS

Universidad Nacional de Rosario, CONICET (Consejo Nacional de Investigaciones Científicas y Técnicas), and ANPCyT (Agencia Nacional de Promoción Científica y Tecnológica, Project No. PICT-2013-0136) are gratefully acknowledged for financial support. F.A. thanks CONICET for a fellowship.

## REFERENCES

[1] Booksh KS, Kowalski BR. Theory of analytical chemistry. Anal Chem 1994;66:782A–791A.
[2] ISO 5725:1996. Accuracy (trueness and precision) of measurement methods and results. Geneva, Switzerland: International Organization of Standardization; 1996.
[3] Document No. SANCO/12495/2011. Method validation and quality control procedures for pesticide residues analysis in food and feed, Directorate General for Health and Consumer Affairs (SANCO), European Commission; 2012.
[4] Danzer K, Currie LA. Guidelines for calibration in analytical chemistry. Part 1. Fundamentals and single component calibration. Pure Appl Chem 1998;70:993–1014.
[5] Olivieri AC, Faber NM. Validation and error. In: Brown S, Tauler R, Walczak B, editors. Comprehensive chemometrics. vol. 3. Amsterdam: Elsevier; 2009. p. 91.
[6] Sanchez E, Kowalski BR. Tensorial calibration: I. First-order calibration. J Chemometr 1988;2:247–63.
[7] Ho CN, Christian GD, Davidson ER. Application of the method of rank annihilation to fluorescent multicomponent mixtures of polynuclear aromatic hydrocarbons. Anal Chem 1980;52:1071–9.
[8] Messick NJ, Kalivas JH, Lang PM. Selectivity and related measures for *n*th-order data. Anal Chem 1996;68:1572–9.

[9] Olivieri AC, Faber NM. Standard error of prediction in parallel factor (PARAFAC) analysis of three-way data. Chemom Intell Lab Syst 2004;70:75–82.
[10] Olivieri AC, Faber NM. A closed-form expression for computing the sensitivity in second-order bilinear calibration. J Chemometr 2005;19:583–92.
[11] Faber K, Lorber A, Kowalski BR. Analytical figures of merit for tensorial calibration. J Chemometr 1997;11:419–61.
[12] Olivieri AC. Computing sensitivity and selectivity in parallel factor analysis and related multi-way techniques: the need for further developments in net analyte signal theory. Anal Chem 2005;77:4936–46.
[13] Olivieri AC. Analytical advantages of multivariate data processing. One, two, three, infinity? Anal Chem 2008;80:5713–20.
[14] Olivieri AC, Faber K. New developments for the sensitivity estimation in four-way calibration with the quadrilinear parallel factor model. Anal Chem 2012;84:186–93.
[15] Bauza MC, Ibañez GA, Tauler R, Olivieri AC. Sensitivity equation for quantitative analysis with multivariate curve resolution-alternating least-squares: theoretical and experimental approach. Anal Chem 2012;84:8697–706.
[16] Allegrini F, Olivieri AC. Analytical figures of merit for partial least-squares coupled to residual multilinearization. Anal Chem 2012;84:10823–30.
[17] Olivieri AC. Analytical figures of merit: from univariate to multiway calibration. Chem Rev 2014;114:5358–78.
[18] Belter M, Sajnóg A, Barałkiewicz D. Over a century of detection and quantification capabilities in analytical chemistry—historical overview and trends. Talanta 2014;129:606–16.
[19] International Organization for Standardization (ISO). Capability of detectionReport No. ISO 11843–1. Genève, Switzerland: ISO; 1997.
[20] International Organization for Standardization (ISO). Capability of detectionReport No. ISO 11843-2. Genève, Switzerland: ISO; 2000.
[21] McNaught AD, Wilkinson A, IUPAC. Compendium of chemical terminology. 2nd ed. Oxford: Blackwell Scientific Publications; 1997.
[22] Inczédy J, Lengyel T, Ure AM, Gelencsér A, Hulanicki A. IUPAC analytical chemistry division, compendium of analytical nomenclature. 3rd ed. Oxford: Blackwell; 1998.
[23] Ortiz MC, Sarabia LA, Herrero A, Sánchez MS, Sanz MB, Rueda ME, et al. Capability of detection of an analytical method evaluating false positive and false negative (ISO 11843) with partial least squares. Chemom Intell Lab Syst 2003;69:21–33.
[24] Burns DA, Ciurczak EW. Handbook of near-infrared analysis. 3rd ed. Practical spectroscopy series. , vol. 35. Boca Raton, FL: CRC Press; 2008.
[25] Allegrini F, Olivieri AC. IUPAC-consistent approach to the limit of detection in partial least-squares calibration. Anal Chem 2014;86:7858–66.
[26] Escandar GM, Olivieri AC. Practical three-way calibration. 1st ed. Waltham, MA: Elsevier; 2014.
[27] Bro R. PARAFAC: tutorial & applications. Chemom Intell Lab Syst 1997;38:149–71.
[28] Tauler R. Multivariate curve resolution applied to second-order data. Chemom Intell Lab Syst 1995;30:133–46.
[29] Tauler R, Maeder M, de Juan A. Multiset data analysis: extended multivariate curve resolution. In: Brown S, Tauler R, Walczak B, editors. Comprehensive chemometrics. vol. 2. Amsterdam: Elsevier; 2009. p. 473.
[30] Wold S, Esbensen K, Geladi P. Principal component analysis. Chemom Intell Lab Syst 1987;2:37–52.
[31] Bro R. Multi-way calibration. Multi-linear PLS. J Chemometr 1996;10:47–62.

[32] Linder M, Sundberg R. Precision of prediction in second order calibration, with focus on bilinear regression methods. J Chemometr 2002;16:12–27.

[33] Olivieri AC. A combined artificial neural network/residual bilinearization approach for obtaining the second-order advantage from three-way non-linear data. J Chemometr 2005;19:615–24.

[34] Arancibia JA, Olivieri AC, Gil DB, Mansilla AE, Durán-Merás I, de la Peña AM. Trilinear least-squares and unfolded-PLS coupled to residual trilinearization: new chemometric tools for the analysis of four-way instrumental data. Chemom Intell Lab Syst 2006;80:77–86.

[35] Ohman J, Geladi P, Wold S. Residual bilinearization. Part 1: theory and algorithms. J Chemometr 1990;4:79–90.

[36] Maggio RM, de la Peña AM, Olivieri AC. Unfolded partial least-squares with residual quadrilinearization: a new multivariate algorithm for processing five-way data achieving the second-order advantage. Application to fourth-order excitation-emission-kinetic-pH fluorescence analytical data. Chemom Intell Lab Syst 2011;109:178–85.

[37] Leger MN, Vega-Montoto L, Wentzell PD. Methods for systematic investigation of measurement error covariance vatrices. Chemom Intell Lab Syst 2005;77:181–205.

[38] Wentzell PD. Measurement errors in multivariate chemical data. J Braz Chem Soc 2014;25:183–96.

[39] Wentzell PD, Tarasuk AC. Characterization of heteroscedastic measurement noise in the absence of replicates. Anal Chim Acta 2014;847:16–28.

[40] Wentzell PD, Andrews DT, Hamilton DC, Faber K, Kowalski BR. Maximum likelihood principal component analysis. J Chemometr 1997;11:339–66.

[41] Wentzell PD, Andrews DT, Kowalski BR. Maximum likelihood multivariate calibration. Anal Chem 1997;69:2299–311.

[42] Lorber A. Error propagation and figures of merit for quantification by solving matrix equations. Anal Chem 1986;58:1167–72.

[43] Faber NM. Exact presentation of multivariate calibration model as univariate calibration graph. Chemom Intell Lab Syst 2000;50:107–14.

[44] Ferré J, Brown SD, Rius FX. Improved calculation of the net analyte signal in inverse multivariate calibration. J Chemometr 2001;15:537–53.

[45] Ferré J, Faber NM. Net analyte signal calculation for multivariate calibration. Chemom Intell Lab Syst 2003;69:123–36.

[46] Rao CR, Mitra S. Generalized inverse of matrices and its applications. New York: Wiley; 1971.

[47] Chen ZP, Wu HL, Jiang JH, Li Y, Yu RQ. A novel trilinear decomposition algorithm for second-order linear calibration. Chemom Intell Lab Syst 2000;52:75–86.

[48] Xia AL, Wu HL, Fang DM, Ding YJ, Hu LQ, Yu RQ. Alternating penalty trilinear decomposition algorithm for second-order calibration with application to interference-free analysis of excitation–emission matrix fluorescence data. J Chemometr 2005;19:65–76.

[49] Xia AL, Wu HL, Li SF, Zhu SH, Hu LQ, Yu RQ. Alternating penalty quadrilinear decomposition algorithm for an analysis of four-way data arrays. J Chemometr 2007;21:133–44.

[50] Faber NM, Ferré J, Boqué R, Kalivas JH. Quantifying selectivity in spectrophotometric multicomponent analysis. Trends Anal Chem 2003;22:352–61.

[51] Olivieri AC. On a versatile second-order multivariate calibration method based on partial least-squares and residual bilinearization. Second-order advantage and precision properties. J Chemometr 2005;19:253–65.

[52] Sanchez E, Kowalski BR. Generalized rank annihilation factor analysis. Anal Chem 1986;58:496–9.

[53] Sanchez E, Kowalski BR. Tensorial resolution: a direct trilinear decomposition. J Chemometr 1990;4:29–45.

[54] Bortolato SA, Lozano VA, de la Peña AM, Olivieri AC. Novel augmented parallel factor model for four-way calibration of high-performance liquid chromatography-fluorescence excitation-emission data. Chemom Intell Lab Syst 2015;141:1–11.

[55] Cuadros Rodríguez L, García Campaña AM, Jiménez Linares C, Román Ceba M. Estimation of performance characteristics of an analytical method using the data set of the calibration experiment. Anal Lett 1993;26:1243–58.

[56] Vessman J, Stefan RI, van Staden JF, Danzer K, Lindner W, Burns DT, et al. Selectivity in analytical chemistry (IUPAC Recommendations 2001). Pure Appl Chem 2001;73:1381–6.

[57] Olivieri AC, Faber NM, Ferré J, Boqué R, Kalivas JH, Mark H. Uncertainty estimation in spectroscopic multivariate calibration. Pure Appl Chem 2006;78:633–61.

[58] Faber NM, Ferré J, Boqué R, Kalivas JH. Second-order bilinear calibration: the effects of vectorising the data matrices of the calibration set. Chemom Intell Lab Syst 2002;63:107–16.

[59] Arnold MA, Small GW, Xiang G, Qui J, Murhammer DW. Pure component selectivity analysis of multivariate calibration models from near-infrared spectra. Anal Chem 2004;76:2583–90.

[60] Brown CD, Ridder TD. Framework for multivariate selectivity analysis, Part I: theoretical and practical merits. Appl Spectrosc 2005;59:787–803.

[61] Ridder TD, Brown CD, Ver Steeg BJ. Framework for multivariate selectivity analysis, Part II: experimental applications. Appl Spectrosc 2005;59:804–15.

[62] Geladi P. Some recent trends in the calibration literature. Chemom Intell Lab Syst 2002;60:211–24.

[63] Riu J, Bro R. Jack-knife technique for outlier detection and estimation of standard errors in PARAFAC models. Chemom Intell Lab Syst 2003;65:35–49.

[64] Faber K, Kowalski BR. Propagation of measurement errors for the validation of predictions obtained by principal component regression and partial least squares. J Chemometr 1997;11:181–238.

[65] Serneels S, Faber K, Verdonck T, Van Espen PJ. Case specific prediction intervals for tri-PLS1: the full local linearisation. Chemom Intell Lab Syst 2011;108:93–9.

[66] Cabezón M, Olivieri AC. Precision in multi-wavelength spectroscopic analysis using classical least-squares regression. Chem Educ 2006;11:394–401.

[67] Currie LA. Limits for qualitative detection and quantitative determination. Application to radiochemistry. Anal Chem 1968;40:586–93.

[68] Currie LA. Nomenclature in evaluation of analytical methods including detection and quantification capabilities. Pure Appl Chem 1995;67:1699–723.

[69] Currie LA. Detection and quantification limits: origins and historical overview. Anal Chim Acta 1999;391:127–34.

[70] Hubaux A, Vos G. Decision and detection limits for linear calibration curves. Anal Chem 1970;42:849–55.

[71] Clayton CA, Hines JW, Elkins PD. Detection limits with specified assurance probabilities. Anal Chem 1987;59:2506–14.

[72] Boqué R, Larrechi MS, Rius FX. Multivariate detection limits with fixed probabilities of error. Chemom Intell Lab Syst 1999;45:397–408.

[73] Boqué R, Ferré J, Faber NM, Rius FX. Limit of detection estimator for bilinear second-order calibration. Anal Chim Acta 2002;421:313–21.

[74] del Río Bocio FJ, Riu J, Boqué R, Rius FX. Limits of detection in linear regression with errors in the concentration. J Chemometr 2003;17:413–21.
[75] Voigtman E. Limits of detection and decision. Part 1. Spectrochim Acta 2008;63:115–28.
[76] Voigtman E. Limits of detection and decision. Part 2. Spectrochim Acta 2008;63:129–41.
[77] Ortiz MC, Sarabia LA, Herrero A, Sánchez MS, Sanz MB, Rueda ME, et al. Capability of detection of an analytical method evaluating false positive and false negative (ISO 11843) with partial least squares. Chemom Intell Lab Syst 2003;69:21–33.

# Index

Note: Page numbers followed by "*f*" indicate figures, and "*t*" indicate tables.

## A

## B

## D



## E

# Q



# R

## S

# T

## U

## V



## W



## Z

Kolibri 360 GmbH
Ferdinand-Jühlke-Straße 7
99095 Erfurt, Deutschland
produktsicherheit@kolibri360.de